Lecture Notes in Mechanical Engineering

Lecture Notes in Mechanical Engineering (LNME) publishes the latest developments in Mechanical Engineering—quickly, informally and with high quality. Original research reported in proceedings and post-proceedings represents the core of LNME. Volumes published in LNME embrace all aspects, subfields and new challenges of mechanical engineering. Topics in the series include:

- Engineering Design
- Machinery and Machine Elements
- Mechanical Structures and Stress Analysis
- Automotive Engineering
- Engine Technology
- Aerospace Technology and Astronautics
- Nanotechnology and Microengineering
- Control, Robotics, Mechatronics
- MEMS
- Theoretical and Applied Mechanics
- Dynamical Systems, Control
- Fluid Mechanics
- Engineering Thermodynamics, Heat and Mass Transfer
- Manufacturing
- Precision Engineering, Instrumentation, Measurement
- Materials Engineering
- Tribology and Surface Technology

To submit a proposal or request further information, please contact the Springer Editor of your location:

China: Ms. Ella Zhang at ella.zhang@springer.com
India: Priya Vyas at priya.vyas@springer.com
Rest of Asia, Australia, New Zealand: Swati Meherishi
at swati.meherishi@springer.com
All other countries: Dr. Leontina Di Cecco at Leontina.dicecco@springer.com

To submit a proposal for a monograph, please check our Springer Tracts in Mechanical Engineering at https://link.springer.com/bookseries/11693 or contact Leontina.dicecco@springer.com

Indexed by SCOPUS. All books published in the series are submitted for consideration in Web of Science.

Vijay Kumar Gupta · C. Amarnath ·
Puneet Tandon · M. Zahid Ansari
Editors

Recent Advances in Machines and Mechanisms

Select Proceedings of the iNaCoMM 2021

Springer

Editors
Vijay Kumar Gupta
Discipline of Mechanical Engineering
PDPM IIITDM Jabalpur
Jabalpur, Madhya Pradesh, India

C. Amarnath
Department of Mechanical Engineering
Indian Institute of Technology Bombay
Mumbai, Maharashtra, India

Puneet Tandon
Discipline of Mechanical Engineering
PDPM IIITDM Jabalpur
Jabalpur, Madhya Pradesh, India

M. Zahid Ansari
Discipline of Mechanical Engineering
PDPM IIITDM Jabalpur
Jabalpur, Madhya Pradesh, India

ISSN 2195-4356 ISSN 2195-4364 (electronic)
Lecture Notes in Mechanical Engineering
ISBN 978-981-19-3715-6 ISBN 978-981-19-3716-3 (eBook)
https://doi.org/10.1007/978-981-19-3716-3

This Springer imprint is published by the registered company Springer Nature Singapore Pte Ltd.
The registered company address is: 152 Beach Road, #21-01/04 Gateway East, Singapore 189721, Singapore

Preface

This book presents the select peer-reviewed proceeding of the 5th International and 20th National Conference on Machines and Mechanisms (iNaCoMM 2021) organised in the series of biennial conferences under the aegis of the Association for Machines and Mechanism (AMM), which is the Indian national affiliate of the International Federation for the Promotion of Mechanism and Machine Science (IFToMM). The main objective of AMM is to contribute to mechanical design at all levels starting from academic projects to industrial production, thus enhancing the quality and the reliability of indigenous machines. This conference brought together more than hundred researchers, industry experts and students working in various aspects of design and analysis of machines and mechanisms on a single platform for discussion and sharing of the knowledge and expertise.

The chapters included in this book represent broad topics including kinematics and dynamics of machines, compliant mechanisms; gear, cams and power transmission systems; mechanisms and machines for rural, agricultural and industrial applications; mechanisms for space applications; mechanisms for energy harvesting; robotics and automation; human centric robotics; soft robotics; man–machine system, mechatronics and micro–mechanisms; CAD and CAGD; control of machines; vibration of machines and rotor dynamics; acoustic and noise; tribology; condition monitoring and failure analysis; fault diagnosis and health monitoring; biomedical engineering; and composites and advanced materials. Given the broad contents, the book is expected to be a valuable resource for science and engineering students, researchers and industrialists interested in carrying research and development in advanced research areas of mechanical engineering and material science in general and in machines and mechanisms in particular.

Mumbai, India — Prof. C. Amarnath, President, AMM

Jabalpur, India — Prof. Puneet Tandon
Jabalpur, India — Prof. Vijay Kumar Gupta
Jabalpur, India — Dr. M. Zahid Ansari

Contents

About the Editors

Prof. Vijay Kumar Gupta is currently working as Professor in Mechanical Engineering Discipline at PDPM Indian Institute of Information Technology, Design and Manufacturing, Jabalpur. He is having more than 27 years of teaching and research experience. He received his Ph.D. in Mechanical Engineering from the Indian Institute of Technology Bombay, India. His research interests includes smart structure, vibration, design, reliability, finite element analysis, mechatronics and robotics, etc. He has published more than 35 papers in refereed journals and conferences and 14 books/book chapters. He is the recipient of the ISAME K. Suryanarayan Rao Memorial Senior Student Award for R&D in Smart Technology for the year 2003 and JSPS invitation fellowship for the year 2015. He is a member of ASME, IEEE, SRESA, IE, ISTE and other professional bodies.

Prof. C. Amarnath is a Mechanical Engineering Graduate of 1968. He earned his Ph.D. in the area of Mechanism Synthesis and Design in 1976 and joined IIT Bombay as a faculty member soon after. He has held several senior positions like Head of Department and Dean at IIT Bombay. He also headed SINE, the Technology Business Incubator of IIT Bombay. Prof. Amarnath was instrumental in founding and is the current President of the Association for Machines and Mechanisms, the national body affiliated with the International Federation of Mechanism and Machine Theory. He has been associated with various Government of India organizations like DST, DSIR, TIFAC and Chairs some of their committees. Prof. Amarnath is currently working as Emeritus Professor at IIT Bombay, Cummins Chair Professor of Mechanical Engineering at Cummins College of Engineering, Pune and Visiting Professor at IIT Dharwad.

Prof. Puneet Tandon is a joint Professor of Mechanical Engineering and Design Disciplines at PDPM Indian Institute of Information Technology, Design and Manufacturing, Jabalpur, India. His primary research interests include CAx technologies, including CAD/CAM/CAE, BioCAD, and Human Factors in CAD; Innovative Product Design; and Advanced Manufacturing Technologies, including Hybrid,

Dieless, Additive, and Smart Manufacturing. He graduated in Mechanical Engineering from the National Institute of Technology Kurukshetra, India. He received his Master and Doctoral degrees in Mechanical Engineering from the Indian Institute of Technology Kanpur, India. He has more than 33 years of experience in engineering research and education. He has more than 300 publications in peer-reviewed journals/book chapters/international/national conferences, besides being the author of 2 books and 20 patents. He has been editor/guest editor of a few Journals. He has been awarded the 2020 DUO-India Fellowship Award with Padova University, Italy. He has also been awarded prizes in various IMTEX, organized by Indian Machine Tool Manufacturers' Association. Some of his research papers have been among the most downloaded and cited articles and awarded "Certificate of Merit". He was also the organizer of the International Conference on Innovations in Design and Manufacturing (InnDeM) 2012 and Design Workshop (DeW) 2010. He is a fellow of the International Association of Computer Science and Information Technology (IACSIT) and a member of ASME, SME, ACM SIGGRAPH, IAENG, IE and ISTE.

Dr. M. Zahid Ansari is an Associate Professor in Mechanical Engineering Discipline at PDPM IIITDM Jabalpur. He received his B.Tech. degree from the Aligarh Muslim University, India, in 2001, and M.Tech. and Ph.D. degrees in 2006 and 2010, respectively, from the Department of Mechanical Engineering, Inha University, South Korea. He is having more than 15 years of teaching and research experience and published more than 100 journal and conference papers and three patents. His research interests include vibration, design, MEMS and smart materials and structures.

Synthesis of Mechanisms

Studies on Coupler Curves of a 4-Bar Mechanism with One Rolling Pair Adjacent to the Ground

Abhishek Kar and Dibakar Sen

Abstract This paper presents an expository study of the shapes of the coupler curve of a four-bar mechanism with one of its fixed pivots replaced with a rolling pair. Such rolling pairs provide the advantage of being friction and clearance free. In this paper, circle-on-circle and circle-on-line type of rolling pairs have been explored. This arrangement introduces three additional design variables to study the continuous change in the shape of the coupler curve of a conventional 4-bar mechanism. For a given input rotation at the crank, the mechanism does not have a closed-form solution for its configuration. However, providing input rotation at the rolling link allows easy derivation of a closed-form solution for both branches of the configuration. The tracing of the coupler curves is done for arbitrary radius ratios for the rolling link and choice of coupler point on the coupler link. A computer program has been written to study and visualize the coupler curves and its properties; the program not only finds the closure configurations but also identifies the situations of non-closure. Illustrative examples show variety and complexity of coupler curves which are not achieved in linkages. Although these coupler curves of mechanisms with rolling pair are transcendental in nature, the velocity states of the systems are easily derivable. Since the point of contact of the rolling pair is the instantaneous centre for the rolling link with respect to the ground, the velocity of the coupler point is determined using Kennedy's theorem. This helps in characterizing the occurrence of cusps in the coupler curves. The paper presents the geometric conditions for components, crunodes and cusps in the coupler curve with illustrative examples.

Keywords Coupler curves · Planar linkage · Kinematics

A. Kar · D. Sen (✉)
CPDM, Indian Institute of Science, Bangalore, India
e-mail: dibakarsen@iisc.ac.in

A. Kar
e-mail: abhishekkar@alum.iisc.ac.in

V. K. Gupta et al. (eds.), *Recent Advances in Machines and Mechanisms*, Lecture Notes in Mechanical Engineering, https://doi.org/10.1007/978-981-19-3716-3_1

1 Introduction

Variety of mechanisms used in machinery exploits the global and local geometric aspects of the coupler curves. Apparent complexity of form of the coupler curves of a simple 4-bar mechanism has intrigued mathematicians and kinematicians alike since the mid-nineteenth century [1–6]. A comprehensive review of the development of knowledge and application of coupler curves in design is available in [7–9]. Degree of the coupler curve has a strong bearing on the variety and distribution of features of interest in a coupler curve viz. a 4-bar coupler curve cannot have more than 3 cusps or crunodes because it is a 6-degree curve. 6-bar and higher chains offer larger variety [10–12] in terms of their degree, circuits, etc. Geared 5-bar mechanisms, although kinematically equivalent to a 6-bar Stephenson chain, produce many intriguing coupler curves depending upon the gear ratios chosen [13, 14]. Systematic generation of coupler curves with specific geometric properties, as commonly utilized in designs, is presented in [15]. The above studies were presented with an emphasis on their analytical formulation and computer implementation in the earlier days of the use of computer-graphics! Although numerous techniques of optimal synthesis of coupler point path for a given mechanism are available in literature, dedicated studies on the nature of coupler curves are not found in contemporary literature. This paper studies the variation in the *coupler curve of a 4-bar mechanism resulting from replacing one of its joints with a rolling pair*.

Frictionless mechanical devices such as rolamite and roller-band devices as mechanical embodiment of rolling between moving rigid bodies were invented long ago [16–18]. In [19], we can see the emergence of a band-constrained rolling pair and its application for a prosthetic knee. More recent applications of rolling joints are reported in [20–23]. However, a comprehensive kinematic treatment of mechanisms comprising of rolling pairs is available in [24]. The above studies emphasize the engineering advantages of roller joints. However, these studies do not emphasize the implication of roller pairs on the kinematics of the mechanism. The study presented in this paper focuses on the geometric implication on the coupler curve of a 4-bar mechanism with a rolling pair.

In any mechanism, except for the driving link, most other links have a limited kinematic range of motion; joints undergo varying loading due to the inherent nature of motion transmission in a linkage. These are not favourable conditions for conventional bearings. From a kinematics point of view the locus of a point on a link bearing a rolling pair is a transcendental curve viz. trochoid, involute, etc. Hence, the locus of points on other links is also likely to overcome the limitations on shape of algebraic curves produced by linkage mechanisms. The above observations motivated the authors to explore use of rolling pairs on the kinematic behaviour of a mechanism in terms of the coupler curve generated by it.

2 Kinematic analysis

Rolling pairs have been embodied in literature as rolamites (Fig. 1a) or tendons (Fig. 1b), both of which employ a set of presumably flexible and inextensible members for the purpose of articulating the rigid members forming the pair. In the present work, we propose to replace one of the fixed pivots of a 4-bar mechanism with a rolling joint as shown in Fig. 2. The dual tendon arrangement constraining the motion of two convex surfaces as shown in Fig. 1b ensures relative rolling, without slipping over each other. Since gearing ensures rolling motion of the associated pitch circles, *rolling pairs can also be implemented as gears.* The distance between B_0 and O, the centres of curvature of the moving and fixed profiles respectively is a constant when both the contacting profiles are circles. Hence, it is possible to consider B_0 and O as fixed pivots to obtain a direct kinematic inversion of the proposed mechanism; this inversion is nothing but the geared 5-bar mechanism studied in [13, 14]. In this inversion, comparing with Fig. 3, OB_0 and B_0 are fixed pivots and the locus of A_0 and B are circles (algebraic curves); therefore, the coupler curve of a point on link BB_0 is an algebraic curve. However, when link A_0O is fixed, instead of B_0B, although the locus of A is a circle, the locus of B is a trochoid which is a non-algebraic curve. Thus, the locus of C on AB is, theoretically, a non-algebraic curve. Thus, the analytical characteristics of coupler curves studied in the proposed mechanism is different from the ones available in literature, including those of the geared 5-bar mechanisms.

2.1 *Kinematics of the Rolling Pair*

Rolling phenomenon between two rigid bodies in point contact is said to happen when during any finite interval of motion, the length of the trace of the point of contact on both the bodies are equal; when the lengths of the trace are unequal, then the body containing the shorter trace is said to have skidded and the one with longer length slipped! Rolling in plane is characterized by location of the instantaneous centre of velocity being at the point of contact. Two rolling pairs under consideration are shown in Fig. 2 wherein (a) and (b) illustrates change of configuration when a circular profile of a rigid body rolls on a circular profile of another rigid body, and

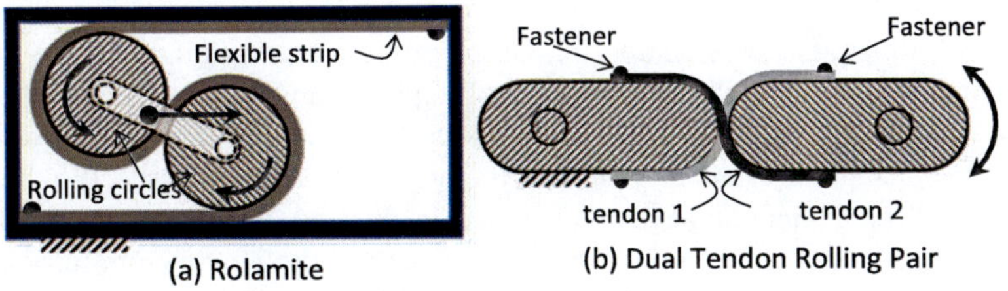

Fig. 1 Embodiments of two rolling pairs

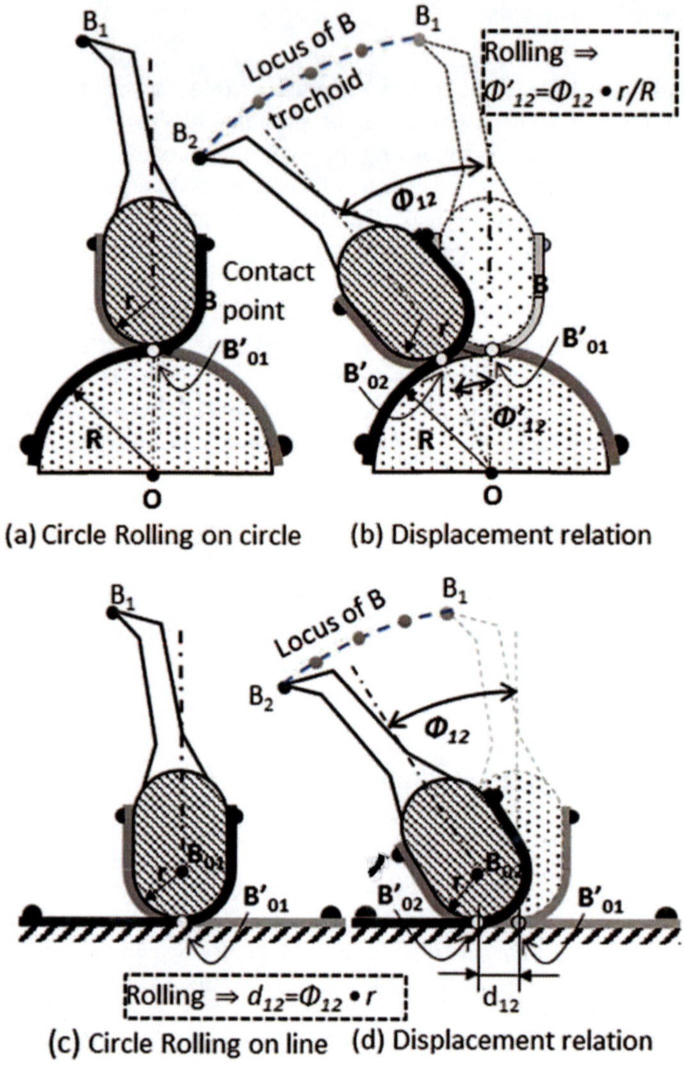

Fig. 2 Configuration of a circle-based rolling pair

(c) and (d) illustrates the situation for a circle rolling over a straight profile. Actual embodiment of the constraints to ensure rolling is unimportant in the analysis. The condition of equal length of the trace between the points of contact B'_{01} and B'_{02} relates the change of orientation, ϕ_{12} of the moving body with respect to the fixed body. The displacement of the point of contact is given by ϕ'_{12}, in case the fixed body is a circle, and d_{12} for the straight one. If the radii of the moving and fixed circles are r and R, respectively, it is easy to see that,

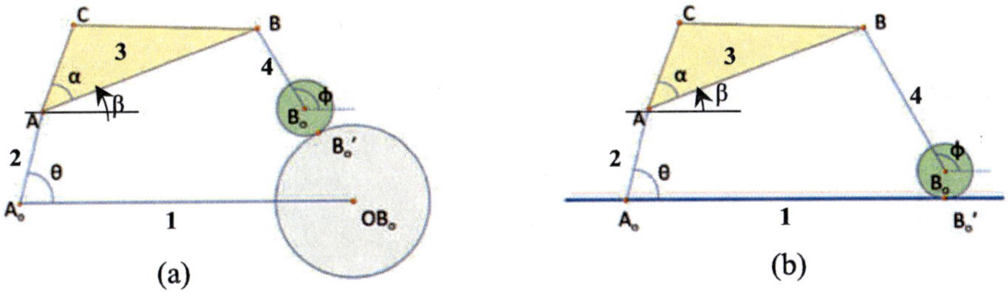

Fig. 3 Coupler point in the 4-bar with a rolling pair

$$\phi_{12}^{'} = \phi_{12} \cdot \frac{r}{R} \tag{1}$$

and,

$$d_{12} = \phi_{12} \cdot r \tag{2}$$

The locus of any point B on the moving body is a trochoid. The instantaneous centre of velocity of the moving body is given by the instantaneous point of contact. These two observations are important for the kinematic analysis of the mechanisms containing these rolling pairs.

2.2 *Position Analysis of the Mechanism*

In this work, we are interested in the two mechanisms shown in Fig. 3a, b which are obtained by adding a dyad, A_0AB, to the rolling pairs shown in Fig. 2a, b, respectively. In conventional 4-bar mechanisms, A_0 and B_0 is fixed; hence, the locus of A and B are circles. Input angle θ gives A directly; B is located at the intersection of two circles centred at A and B_0, and radii AB and B_0B, respectively. The result is obtained easily either geometrically or algebraically by solving a quadratic equation.

In Fig. 3 locus of B is a trochoid therefore, considering input at A_0 requires one to find the intersection point between a circle and a trochoid, as shown in Fig. 4a. This problem cannot be solved in closed form. Hence, we consider an inversion, wherein the output angle, ϕ, is assumed to be known and the corresponding input angle, θ, is determined. Link-4 is rolling over a fixed circle with centre at O and radius R. The change in configuration for link-4 has both rotation (ϕ_{12}) and translation ($B_{01}B_{02}$) components, which determine the location of B on the trochoid. For a known location of B (say B_1), locations of A (A_1 and A'_1) are obtained at the intersections of the two circles as shown in Fig. 4b. This formulation has a closed-form solution, as it involves circles alone.

Let A_0 be the origin and the centre of the fixed profile O be given. Let AA_0 be l_2, AB be l_3, BB_0 be l_4, radius of fixed profile be R and radius of rolling profile be r.

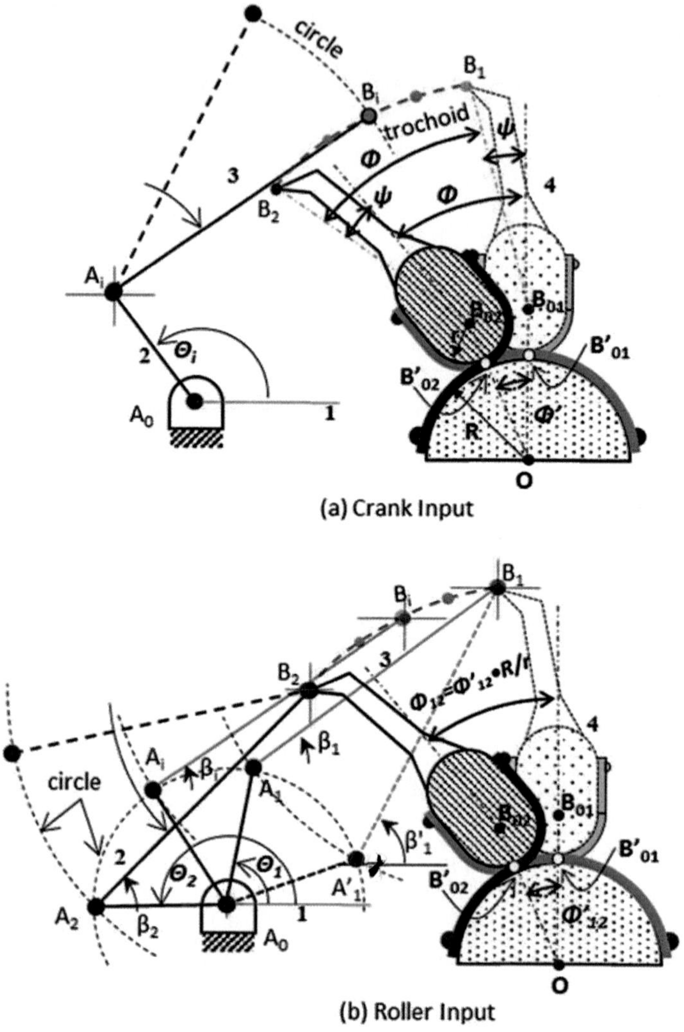

Fig. 4 Displacement analysis of a 4-bar with a rolling pair

Note that the contact point, B'_0, and the centres of the two contacting profiles, O and B_0, are always collinear. A_0, O and the initial location of contact, B'_{01} are given. Let BB_0 make an angle ψ with $B_0B'_0$ and $R = \left\|B'_{01} - \mathrm{O}\right\|$.

The homogeneous transformation matrices $R(\alpha, P)$ and $T(t)$ for rotation, α, about a given point, $P\ (x, y)$, and translation by a vector $t(t_x, t_y)$, respectively, are given as below.

$$\mathbb{R}(\alpha, P) = \begin{bmatrix} \cos\alpha & -\sin\alpha & x\cdot\cos\alpha - y\cdot\sin\alpha - x \\ \sin\alpha & \cos\alpha & x\cdot\sin\alpha + y\cdot\cos\alpha - y \\ 0 & 0 & 1 \end{bmatrix} \tag{3}$$

$$\boldsymbol{T}(t) = \begin{bmatrix} 1 & 0 & t_x \\ 0 & 1 & t_y \\ 0 & 0 & 1 \end{bmatrix} \tag{4}$$

$$\text{Let,}\quad \hat{v} = \frac{1}{R}\left(B_{01}^{'} - O\right) \tag{5}$$

$$\text{Then,}\quad B_{01} = O + (R + r)\hat{v} \tag{6}$$

$$B_1^{'} = O + (R + r + l_4)\hat{v} \tag{7}$$

$$\text{and}\quad B_1 = \mathbb{R}(\psi, B_{01})B_1^{'} \tag{8}$$

Let link-4 be rotated by an angle φ. Let us use then subscript 1 for current locations and 2 for new locations for the points reached because of rotating link-4 by φ.

$$\text{Let,}\quad \hat{k} = \begin{bmatrix} 0 & 0 & 1 \end{bmatrix}^T \tag{9}$$

$$\text{Forthecircle—on—circlecase,}\quad B_{0_2}^{'} = \mathbb{R}\left(\varphi\cdot r/R, O\right)B_{0_1}^{'} \tag{10}$$

$$\text{And,}\quad B_{0_2} = \mathbb{R}\left(\varphi\cdot r/R, O\right)B_{0_1} \tag{11}$$

$$\text{Forthecircle — on — linecase,}\quad B_{0_2}^{'} = \boldsymbol{T}(\varphi\cdot r)B_{0_1}^{'} \tag{12}$$

$$\text{And,}\ B_{0_2} = \boldsymbol{T}(\varphi\cdot r)B_{0_1} \tag{13}$$

$$\text{Then,}\quad B_2 = \boldsymbol{T}\left(B_{0_2} - B_{0_1}\right)\mathbb{R}\left(\varphi, B_{0_1}\right)B_1 \tag{14}$$

$$\text{Let,}\quad D = B_2 - A_0 \tag{15}$$

$$\text{Wherein,}\quad d = \|D\|, \hat{D} = D/d \tag{16}$$

$$\text{Define,}\quad e = \left(l_2^2 - l_3^2 + d^2\right)/(2d) \tag{17}$$

$$\text{Then, usingChace's equation,}\quad A_2 = \pm\sqrt{l_2^2 - e^2}\left(\hat{D}\times\hat{k}\right) + e\hat{D} + A_0 \tag{18}$$

A_1 and A'_1 in Fig. 4b correspond to the two solutions from the '$\pm$' sign in Eq. (10). For the coupler point, C, let, $AC = l_c$, $\angle BAC = \alpha$ and C_2 be the new location of C

$$\text{Let,}\quad \beta_2 = \arg(B_2 - A_2) \tag{19}$$

$$\text{Then,}\quad C_2 = T(A_2)R(\alpha + \beta_2, A_2)\begin{bmatrix} 0 & l_c & 0 \end{bmatrix}^T \tag{20}$$

$$\text{Similarly,}\quad \beta_2' = \arg\left(B_2 - A_2'\right) \tag{21}$$

$$\text{And}\quad C_2' = T(A_2)R\left(\alpha + \beta_2', A_2'\right)\begin{bmatrix} 0 & l_c & 0 \end{bmatrix}^T \tag{22}$$

Thus, the location for the coupler point is determined for both the branches for an incremental input of $d\varphi$ at the roller pair from an arbitrary initial configuration. A representative set of coupler curves generated for the mechanism is shown in Fig. 5

3 Study of Geometry of the Coupler Curves

It can be observed in the representative map of the coupler curves (Fig. 5) that the shape is varied and sometimes complex. Hence, it is important to study the composition in detail (Fig. 6).

3.1 Components or Circuits of Coupler Curves

The coupler curves of 4-bar mechanisms are classified as bicursal and unicursal wherein the coupler curve has two closed components and a single component, respectively. In bicursal coupler curves, a coupler point is topologically constrained to trace one of the components, even if the components have mutual intersections. Hence, understanding the segmentation of components of coupler curves is important.

It may be noted that Eq. (10) has a solution only when the distance $A_0B > (AB + A_0A)$ and $A_0B < (AB - A_0A)$. If we consider an annular region formed by two concentric circles, Ω_{max} and Ω_{min}, with centre at A_0 and radii $r_{max} = (AB + A_0A)$ and $r_{min} = \|AB\text{-}A_0A\|$, respectively, the mechanism will have a closure configuration only when the trochoidal locus of B lies within it. At the intersection of the trochoid with Ω_{max}, the dyad A_0AB is fully stretched out, and at that with Ω_{min}, it is fully folded. Hence, these are ***singular configurations*** of the mechanism. In these two configurations $A_2 = A'_2$, and therefore, $C_2 = C'_2$; i.e. in these two configurations, the two branches of the coupler curve merge to give a ***dichromatic coupler curve*** wherein one segment of the closed curve is traced by one branch of solution while the

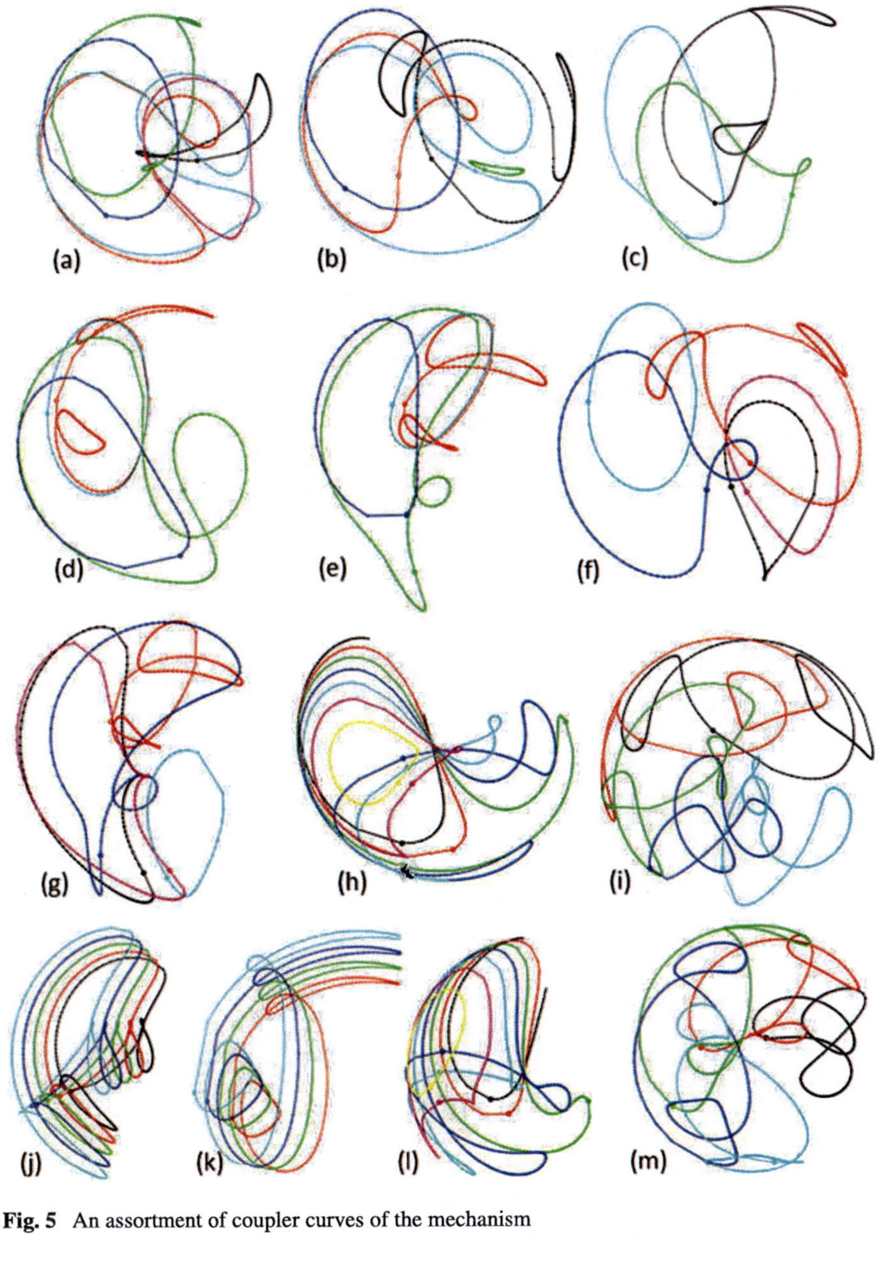

Fig. 5 An assortment of coupler curves of the mechanism

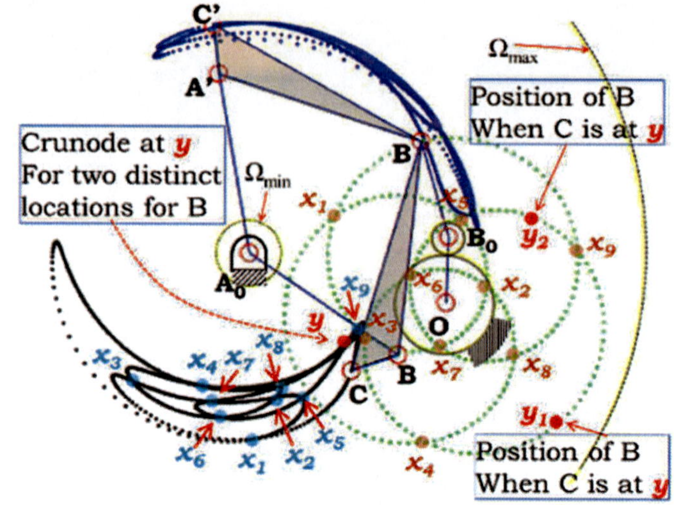

(a) monochromatic bicursal coupler curve with maps of crunodes

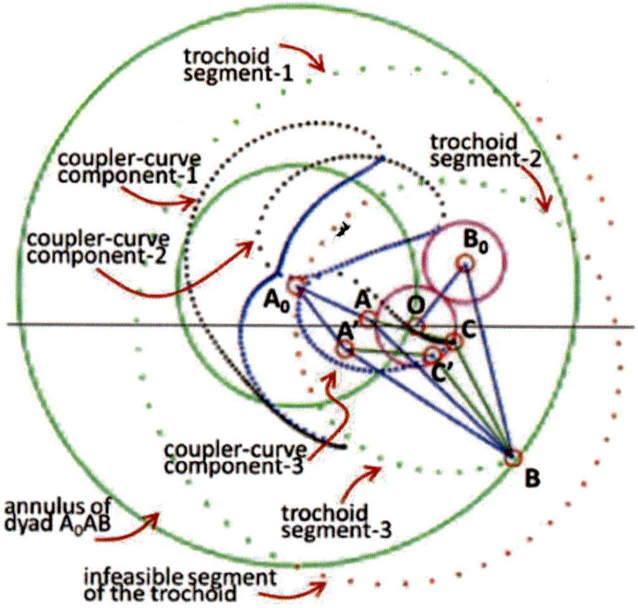

(b) dichromatic multicursal components of coupler curve

Fig. 6 Dyadic annulus and nature of components of coupler curve complementary segment is simultaneously traced by the alternate branch of solution of the coupler point. When all the points on a coupler curve are traced by the same branch of solution, we call it a ***monochromatic coupler curve.***

The trochoidal locus of B as a point on the rolling link is complex although it is symmetric about O; it potentially intersects Ω_{min}, Ω_{max} and fixed circle of the rolling pair multiple times. If the trochoid is such that it does not intersect with Ω_{max} or Ω_{max} over the full range of motion, then the two branches of solution does not merge, and we get *two distinct monochromatic coupler curves* which are traditionally referred to as ***bicursal coupler curves.*** In the circle-on-circle case, the trochoid is bounded by an annulus whose centre is at O and radii are $(OB_0 + BB_0)$ and $||OB_0 - BB_0||$. *We get bicursal coupler curves only when the annulus of the trochoid is completely contained in the annulus of the dyad* A_0AB.

Let us consider a segment of the trochoid intercepted by the circles Ω_{min} and Ω_{max}. Since at these configurations the branches merge, it will generate a dichromatic coupler curve. We get as many closed curves as the number of segments of the trochoid intercepted in the annulus of the dyad A_0AB by the circles Ω_{min} and Ω_{max}. Since the number of component coupler curves is potentially more than two (unlike in the case of a conventional 4-bar coupler curve), we call it ***multi-cursal coupler curves.*** Each component is dichromatic because there is merging of the two branches at singularities.

The number of rotations of the rolling circle that gives a closed trochoid is the lowest common multiple (LCM) of the perimeters of the closed rolling profiles; for circular profiles, it the LCM of the two radii. Thus, B_0 may go around O multiple times before the trochoid closes. This has a bearing on the number of components in the multi-cursal map of the coupler curves. When the number of components is more, naturally the number of intersections a straight line would make with the set of coupler curves is more. Hence, unlike the coupler curves of linkage mechansisms (e.g. 4-bar, 6-bar, etc.) *there is no upper-bound on the number of intersections a straight line can make* with the coupler curve of a 4-bar mechanism with one rolling pair.

When the ratio of the radii of the rolling circle and the fixed circle is an irrational number, then the trochoid is theoretically an open curve of infinite length that would fill the area of the annulus defined by $(OB_0 + BB_0)$ and $||OB_0 - BB_0||$. Correspondingly, the coupler curve is also of infinite length that would fill an area. This filled-area coupler curve will have two components if the condition of bicursal curve is satisfied; otherwise, it will be a single area composed of infinite number of closed dichromatic coupler curves.

3.2 Crunodes in Coupler Curves

A crunode or self-intersection of a curve in plane is easy to perceive visually. From analytical point of view, although a coupler curve is a continuous map of the input

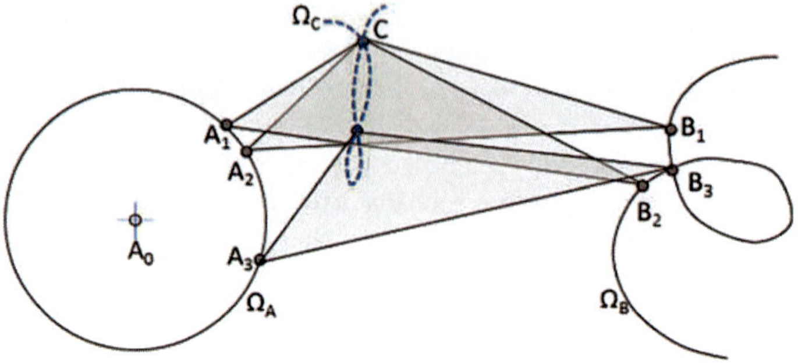

Fig. 7 Conditions of self-intersection

parameter in the Cartesian space through a 1-DoF mechanism, this map is not one-to-one. Each branch of configuration of the mechanism generates distinct curves which we are referring to as components. Since in each embodiment of a mechanism, only one of the branches is operational, only one of the components is generally available for a design. Thus, intersection among the components is not of much significance. There are theories available for self- intersection of a component coupler curve of a 4-bar mechanism. Since no point on link-4 is fixed in the present mechanism, those theories are not directly applicable. Figure 7 shows the situations when the point C will be at a crunode, wherein Ω_A, Ω_B, and Ω_C are the loci of the points A, B and C of the mechanism in Fig. 3, respectively. Ω_C is the coupler curve. Since Ω_B is a trochoid, geometric characterization of the situation is not possible.

In a 4-bar mechanism with a rolling pair as shown in Fig. 3, the location of the coupler point, C, will be same when the location of B is same. B is a point on the trochoid which is a potentially self-intersecting curve. The point of intersection is associated with distinct values of the parameter (ϕ) that generates the trochoid. Thus, ***for every point of self-intersection of the trochoid, there would be a self-intersection of the coupler curve, for each branch Ω_C and $\Omega_{C'}$, simultaneously.*** Figure 6a maps all the crunodes of trochoid and coupler curve. This situation does not occur in conventional linkage mechanisms. It also shows a crunode on the coupler curve without a corresponding crunode on the trochoid.

Since crunodes on a trochoid does not have closed-form expression, a geometric characterization of this situation is also not possible. However, since the locus of B is a trochoid defined by r, R and l_4, it exhibits multiple crunodes. The number of crunodes in a coupler curve, therefore, is not restricted, unlike the algebraic coupler curves of a pure linkage mechanism.

3.3 *Cusps in Coupler Curves*

Although cusps are generally recognized as a sharp point in an otherwise smooth curve, its kinematic characteristic is that at this point on the coupler curve, the velocity

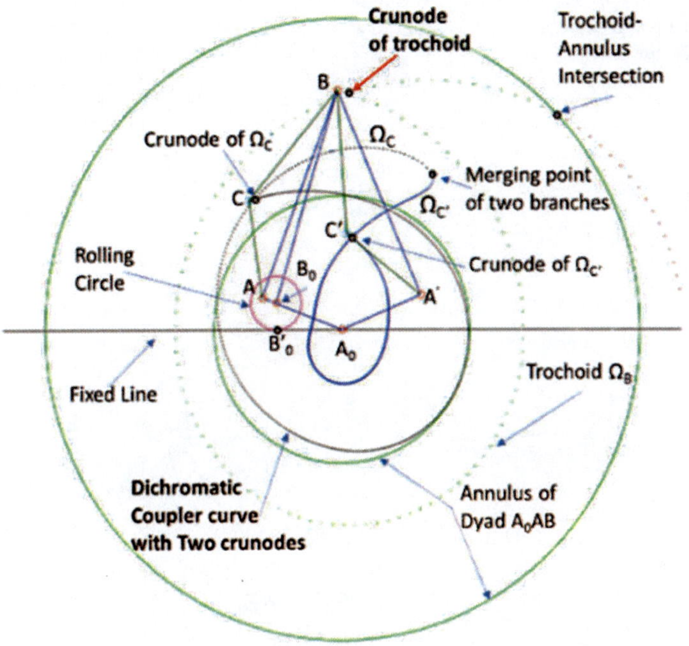

Fig. 8 Crunodes in a coupler curve

of the coupler point is zero for a non-zero input-velocity for the mechanism. Using Arnold-Kennedy's theorem of three instantaneous centres, we get (Fig. 9),

Fig. 9 Velocity of coupler point

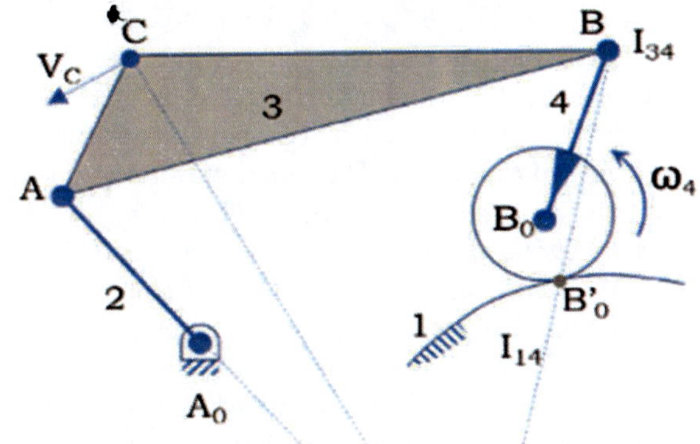

$$\frac{\omega_3}{\omega_4} = \frac{I_{14}I_{34}}{I_{13}I_{34}} = \frac{BB_0'}{BI_{13}}$$

$$V_C = CI_{13} \cdot \omega_3$$

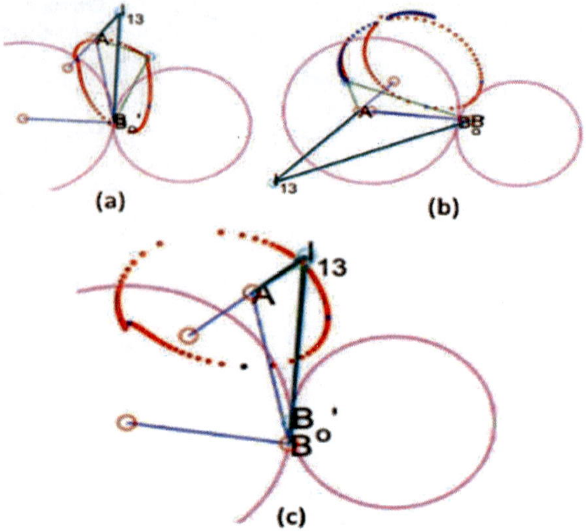

Fig. 10 Sharp and smooth cusps on Ω_C for $l_4 = r$

$$\frac{\omega_3}{\omega_4} = \frac{I_{14}I_{34}}{I_{13}I_{34}} = \frac{BB_0^{'}}{BI_{13}}$$

$$\text{And,} \quad V_C = CI_{13} \cdot \omega_3$$

$$\text{Thus,} \quad V_C = 0 \Rightarrow \text{either } CI_{13} = 0 or, \omega_3 = 0$$

$CI_{13} = 0 \Rightarrow$ Coupler point is coincident with instantaneous centre (Fig. 11),
And, $\omega_3 = 0 \Rightarrow BB_0^{'} = 0$ which is possible only when $l_4 =$ r (Fig. 10).

Note that a cusp occurs at a phase in a mechanism where $l_4 = r$ when the joint *B* coincides with the point of contact as the moving circle rolls over the fixed pivot; then the velocity of the coupler point is zero because the velocity of both ends of the dyad, A_0 and *B*, are zero. The situation is independent of the selection of the coupler point on link-3; hence, the *cusp will occur for the coupler curves of* ***all points in all branches*** *of the mechanism and as many times B touches the fixed profile* (circle or line), ***simultaneously***. This situation is unique for mechanisms with roller pairs and does not occur in conventional 4-bar mechanisms.

In an infinitesimal neighbourhood of a cusp, velocity vector of the coupler point typically has opposite directions. Motion in a mechanical system is continuous. Hence, it is understandable that the magnitudes of velocities of points in the neighbourhood of a cusp will be small. But, there is no causal relation between the directions of velocities of points in the neighbourhood of a cusp; when cusp occurs due to $CI_{13} = 0$, the direction of the tangent to the coupler curve is undefined; but, in case of $l_4 = r$, direction of the tangent is still given by the line perpendicular to CI_{13}. Figure 11c shows a special case when both the conditions are simultaneously satisfied giving a ***smooth cusp*** wherein the coupler point slows down as it approaches

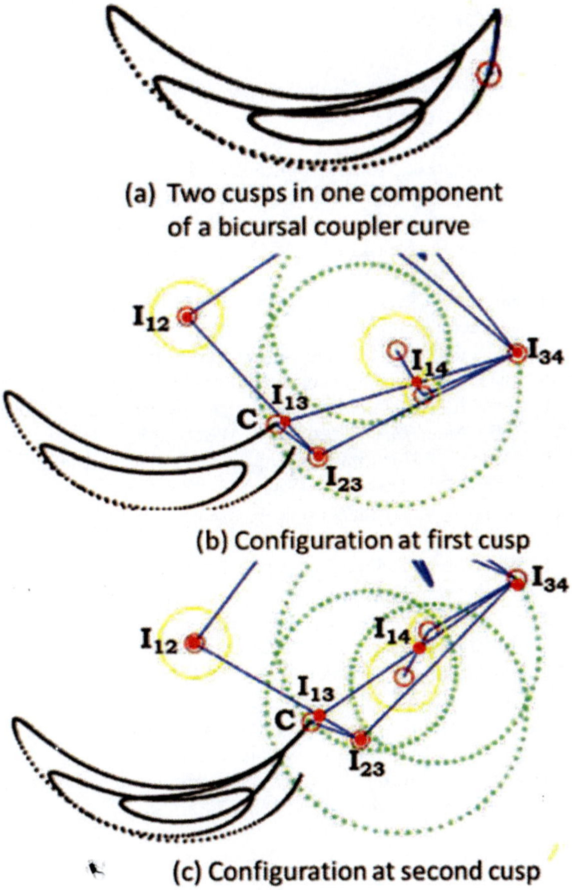

Fig. 11 Sharp cusps (approx) with $CI_{13} \approx 0$

that location. After the stationary phase, it continues to move in the same direction as the velocity with which it entered the stationary phase. Abundance of cusps and crunodes in the coupler curve can be observed in Fig. 5.

4 Conclusion

The geometry of the coupler curve generated by a 4-bar mechanism with one of the fixed pivots replaced with a rolling pair has been studied in this paper. It is shown that the coupler curve of this mechanism has dramatically different geometric attributes in terms of number of circuits, cusps, crunodes, etc. Conditions for cusps and crunodes which are specific to this class of mechanisms have been derived. Consequently, the shapes of the coupler curves are also widely varying which is advantageous for the designers.

References

1. Cayley A (1872) On the mechanical description of certain sextic curves. Proc London Math Soc IV:105–111
2. Roberts S (1875) On three-bar motion in plane space. Proc London Math Soc VII:14–23, https://doi.org/10.1112/plms/s1-7.1.14
3. Cayley A (1876) On three bar motion. Proc London Math Soc VII:136–166
4. Burmester L (1888) Atlas zu Lehrbuch der Kinematik, Leipzig, Fig. 346–353
5. Hrones JA, Nelson GL (1951) Analysis of the four-bar linkage. MIT Press and Wiley. ISBN: 9780262080033
6. Kovacs Z, Recio T, Velez MP (2020) Reasoning about linkages with dynamic geometry. J Symbolic Comput 16–30
7. Nolle H (1875) Linkage coupler curve synthesis: A historical review—I. Developments up to 1875. Mech Mach Theory 9(2):147–168
8. Nolle H (1875) Linkage coupler curve synthesis: A historical review—II. Developments after 1875. Mech Mach Theory 9(3–4):325–348
9. Nolle H (1975) Linkage coupler curve synthesis: a historical review—III. Spat Synth Optim Mech Mach Theory 10(1):41–55
10. Primrose EJF, Freudenstein F, Roth B (1967) Six-bar motion I. The Watt mechanism. Arch Rational Mech Anal 24:22–41
11. Primrose EJF, Freudenstein F, Roth B (1967) Six-bar motion II. The Stephenson-1 and Stephenson-2 mechanisms. Arch Rational Mech Anal 24:42–72
12. Primrose EJF, Freudenstein F, Roth B (1967) Six-bar motion III. Extension of the six-bar techniques to 8-bar and 2n-bar mechanisms. Arch Rational Mech Anal 24:73–77
13. Freudenstein F, Primrose EJF (1963) Geared fife-bar motion part I—Gear ratio minus one. Trans ASME Jl Appl Mech 161–169
14. Primrose EJF, Freudenstein F (1963) Geared fife-bar motion part 2—Arbitrary commensurate gear ratio. Trans ASME Jl Appl Mech 170–175
15. Artobolevskii II, Johnson W (1964) Mechanisms for the generation of plane curves. Pergemon Press, Oxford. ISBN: 9781483152424
16. Wilkies DF (1967) Rolamite: New mechanical design concept, AEC-NASA TECH BRIEF 67-10611
17. Wilkies DF (1969) Roller-band devices, USPTO-3452175
18. Simons LB (1971) Frictionless mechanical motion devices, USPTO-3606795
19. Hillberry BM, Hall AS (1976) Rolling contact prosthetic knee joint, USPTO-3945053
20. Collins CL (2003) Kinematics of robot fingers with circular rolling contact joints. J Robot Syst 20(6):285–296
21. Montierth JR, Todd RH, Howell LL (2009) Analysis of elliptical rolling contact joints in compression. Proc ASME IDETC/CIE, San Diego (USA), Aug 2009
22. Sancisi N, Parenti-Casteli V (2011) Strip-driven devices for the spatial motion guidance of human joints. In: 33rd annual international conference of the IEEE EMBS Boston. Massachusetts USA, Aug 2011
23. Slocum AH (2013) Rolling contact orthopaedic joint design, Thesis, Massachusetts Institute of Technology
24. Kuntz JP (1995) Rolling link mechanisms, Thesis, Delft University of Technology. ISBN-90-370-0126-2

Kinematic Synthesis and Design of a Five Fingered Hand Exoskeleton

Nilavjyoti Sarmah and Shyamanta M. Hazarika

Abstract This paper outlines a method to design an underactuated, linkage-based hand exoskeleton. A planar 8-link mechanism with phalanges and articulations as part of the kinematic chain is used for each finger. The links are attached to the proximal and distal phalanges. For the thumb, a planar 6-link mechanism directly connected at the distal phalanx is used. The kinematic synthesis of the mechanism based on Burmester Theory was performed with two principal considerations: a. To obtain link lengths that ensured no mechanical interference between the mechanisms and the hand and b. The mechanism ensured that each finger follows the natural trajectory. The mechanism is evaluated by comparing the workspace of the anatomical index finger with that of the hand exoskeleton assisted finger.

Keywords Hand exoskeleton · Kinematic synthesis · Pose analysis

1 Introduction

Exoskeletons specifically intended for hand rehabilitation have been reported in the literature. The complexity of designing a device for the human hand increases due to the large number of degrees of freedom (DOF) within a significantly reduced space. The main requirements while designing a hand exoskeleton are: coincidence of the centre of rotation of the hand exoskeleton with the rotational axis of the human joint, the motion should be within the limit of range of motion of the finger and should be light weight so that the patient can wear it without any facing any problem. The hand exoskeleton which is a special kind of robotic hand where forces are applied on the fingers that constrains them to perform a given trajectory and improve the forces that would be naturally exerted. Three aspects of rehabilitation protocols that leads

N. Sarmah (✉) · S. M. Hazarika
Biomimetic Robotics and Artificial Intelligence Lab, Department of Mechanical Engineering, IIT Guwahati, Guwahati 781039, India
e-mail: nilav176103106@iitg.ac.in

S. M. Hazarika
e-mail: s.m.hazarika@iitg.ac.in

V. K. Gupta et al. (eds.), *Recent Advances in Machines and Mechanisms*, Lecture Notes in Mechanical Engineering, https://doi.org/10.1007/978-981-19-3716-3_2

to different solutions are [1]: basic movements of finger joints vs execution of ADLs, gross motion of all fingers versus single finger dexterity and real environment versus virtual environment. The key factor that is to be addressed with respect to kinematics of the hand exoskeleton is the number of DOFs of the mechanism. The second factor is whether to design an underactuated mechanism. Wearability and adaptability are also two major aspects of hand rehabilitation devices. The structure of the exoskeleton needs to be lightweight. The system must have the capacity to adapt to different users.

Hand exoskeletons can be categorized as: 1. multi phalanx hand exoskeleton where forces are exerted differently on each phalanx, 2. single phalanx hand exoskeleton where forces are exerted on the distal phalanx only [2]. Burdea et al. [3] developed a force feedback device which is actuated pnematically to exert force only at the distal phalanx. Sun et al. [4] used a 4-link serial mechanism that is connected to the human finger at the fingertip. In both cases the device is used in haptic application that focused on bidirectional force feedback. However, rehabilitation devices require multi DOF for the training of the human hand and exerting forces at the distal phalanx has limited control over the proximal or middle phalanx [5]. The underactuated mechanism which reduces the requirement of actuator attaches the system either at the proximal or middle segment of the finger. In single phalanx hand exoskeleton, although there is no mechanical coupling among MCP, PIP and DIP joints, the movement of fingertip causes the movement of phalanges and MCP, PIP, DIP joints due to internal DOFs among human joints. The coupling between the joints of a single finger are achieved by various solutions such as cables and pulleys, rigid linkage mechanism, tendon driven mechanism, geared systems etc. The devices that use linkages to transmit the force are stiff enough. The cable driven gloves are more flexible and simpler but require pulleys to achieve high forces and are harder to control at intermediate positions [6]. The series of links can be placed over the human fingers, beside the finger or inside the palm. But placing the linkage mechanism inside the palm has the disadvantage of decreasing the work space of the exoskeleton and also complete grasping of object is not possible. Placing the linkage mechanism beside the finger requires additional space making it difficult for the multi-fingered structure. The mechanical interference between the machine links and the anatomical parts is another critical issue that needs to be addressed while designing a hand exoskeleton. Worsnopp et al [7] solved this issue by adopting the remote centre of motion (RCM) mechanism. RCM is a kind of mechanism which is used to enable the rotation of a body around a fixed axis that is located remotely from the structure of the joint. Different kinds of RCM mechanism are used which can also solve the problem associated with locating the linkage mechanism at the side of the fingers. In this paper, a planar 8-link mechanism for the index finger and 6-link planar mechanism for the thumb of the hand exoskeleton is proposed that is able to move the fingers following the natural trajectory. Linkage based mechanism is preferred as it allows greater transmission of power from the actuator to the fingers.

2 Kinematic Synthesis and Design

Analysis of the hand kinematics is required before proposing any device for the human hand. Figure 1 outlines a method to design a hand exoskeleton starting from the biomechanics of the human hand.

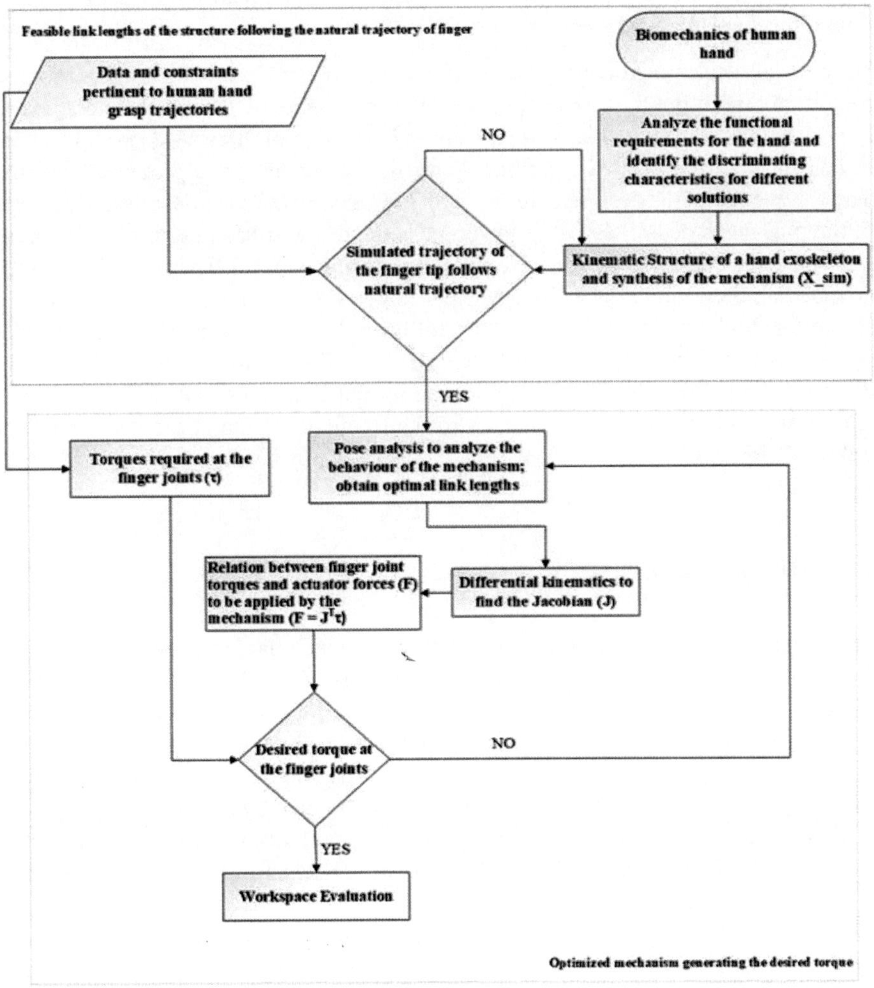

Fig. 1 Biomechanics driven design procedure flowchart

2.1 Analyzing Functional Requirement

Number of Controlled DOF: The DOFs of the hand exoskeleton devices varies based on their design and target application [1]. Most of the hand exoskeleton is made simple with underactuated mechanism [4, 8]. In this paper, the five fingered hand exoskeleton is designed for one active DOF mechanism for each finger. The thumb mechanism also have one active DOF.

Number of mechanical connection (MC): Forces transmitted to the fingers must be minimum. Therefore the connections made to the fingers by the exoskeleton is very important. The hand exoskeletons can have a maximum of three MCs per finger. For our design we have two MCs per finger attached at the distal and proximal phalanx. The complexity of the design increases with higher MCs but the transmission of force to the phalanges decreases. With lower MCs the design becomes simple but stress induced at the contact point increases thereby causing pain [1]. Therefore, our design is traded off between three and one MC per finger. Force component of the total grip strength is highest at the distal phalanx followed by proximal phalanx and middle phalanx respectively [9]. Moreover, the DIP and PIP joints move synchronously during flexion and extension due to the retinacular ligaments that runs from flexor tendon sheath at the proximal phalanx to the terminal tendon at the distal phalanx linking DIP and PIP joints.

2.2 Kinematic Structure

The human finger can be assumed to be a 3R serial chain having three revolute joints that guide the movement through desired trajectory. Table 1 shows the curl angles of joints while grasping a cylindrical object of 4 cm diameter along with the lengths of the phalanges of the index finger of a human hand [1, 10]. Table 2 shows the curl angles and lengths of the thumb phalanges. The lengths of the finger phalanges shown in the tables represent the average length of a healthy person.

Based on the specifications and the above constraints we arrive at the 8-link planar mechanism shown in Fig. 2. The points *D*, *P* and *M* represents the DIP,

Table 1 Dimensions of the phalanges of Index finger and their full flexion joint angles

Distal phalanx	DIP angle	Middle phalanx	PIP angle	Proximal phalanx	MCP angle
2 cm	28°	2.5 cm	62.3°	4 cm	47.4°

Table 2 Dimensions of the phalanges of thumb and their full flexion joint angles

Distal phalanx	IP angle	Proximal phalanx	MCP angle
3.5 cm	49.7°	2 cm	24.8°

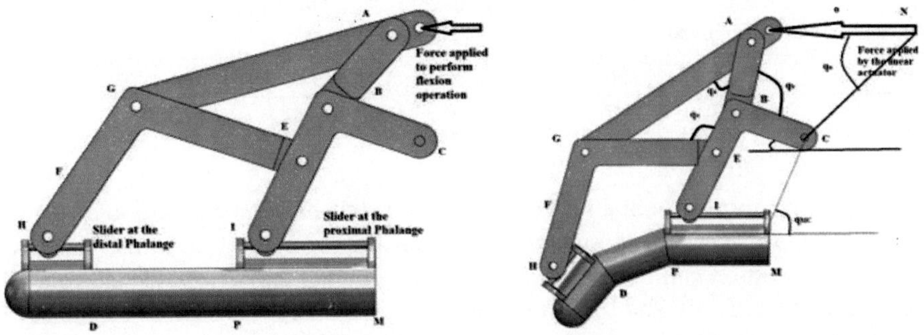

(a) CAD model during extension (b) CAD model during flexion

Fig. 2 Proposed kinematic structure of the 8-link mechanism

PIP and MCP joints respectively. The arrow mark shown in the figure indicates the application of force on the hand exoskeleton by a linear actuator to perform the flexion operation. Mechanical connection between the hand exoskeleton and the finger phalanges is achieved by attaching sliders at the distal phalanx and proximal phalanx. It combines both cylindrical and rotational joint such that the lateral forces produced can be converted into motion in the slider and the perpendicular forces produced are transmitted for movement of phalanges about the finger joints.

2.3 *Kinematic Synthesis*

We synthesize a eight linkage mechanism that is capable of performing flexion and extension of the fingers through the desired trajectory. The problem is to determine the mechanism which placed above the human finger would make the finger follow the desired trajectory and guarantee avoidance of interference between the links and phalanges. The basic module of hand exoskeleton is attached to one human finger and couple the flexion or extension of the three anatomical phalanges. The abduction and adduction of the finger is inhibited. Using Grubler's formula, the relation between the number of links and number of lower pairs is calculated considering 1 DOF planar mechanism.

$$F = 3n - 3 - 2j - h$$

where, F = number of DOF, n = number of links, j = number of lower pairs, h = number of higher pairs. Now, to reduce the complexity the number of higher pairs is considered to be zero ($h = 0$). Using the above relation entries of Table 3 are obtained.

The synthesis follows Burmester Theory: mechanical constraints are introduced to a planar 3R serial chain that guides the end effector movement through a specified

Table 3 Possible combination of the linkage mechanism

n	4	6	8	10	...
j	4	7	10	13	...

set of task positions to obtain a linkage-based mechanism. By specifying the task positions and using inverse kinematics the position of every link in the chain for every task position can be determined. The constraints that can be added to a 3R serial chain are PR, RP and RR. Any of the mentioned three constraints can be used, but for our study focus is on RR constraints. An RR constraint cannot be added to consecutive links, so two links can be connected by first RR constraint. After attaching the first constraint, the second RR constraint can be connected by either of the following two ways: connected to either one of the original links or connected to the link created by the first RR constraint. We initially followed the graphical method for our analysis and obtain a set of feasible link lengths that is able to make the finger with the hand exoskeleton mechanism follow the natural trajectory. From Fig. 3a and b we can observe that the mechanism follows natural trajectory for the index finger. From Fig. 4a and b we can observe that the mechanism follows natural trajectory for the thumb. For the kinematic synthesis, GeoGebra is used and trajectories of DIP, PIP and MCP is considered. The link lengths are obtained by matching the base joints with the trajectory points and iterating the whole process. The synthesis is carried out so that the mechanism is able to follow the desired trajectory and avoid the mechanical interference between the links and the finger joints during flexion. Figure 3a shows the mechanism along with its link lengths that follows the desired trajectory. Figure 3b shows the relative position of the links in the mechanism while performing the flexion operation. The points *N*', *T*' and *V*' represents the proximal

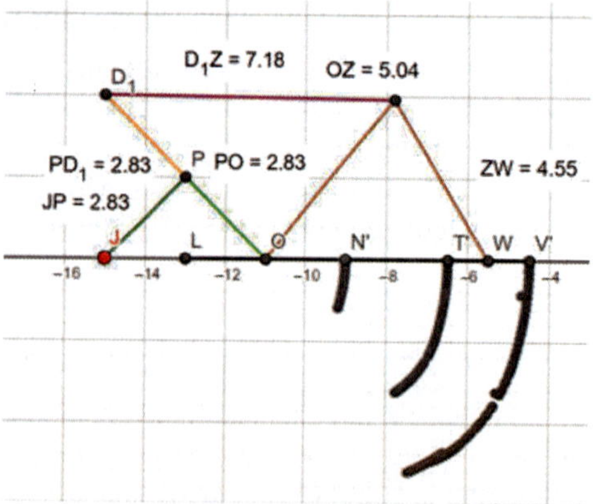

(a) Dimensional Synthesisin GeoGebra

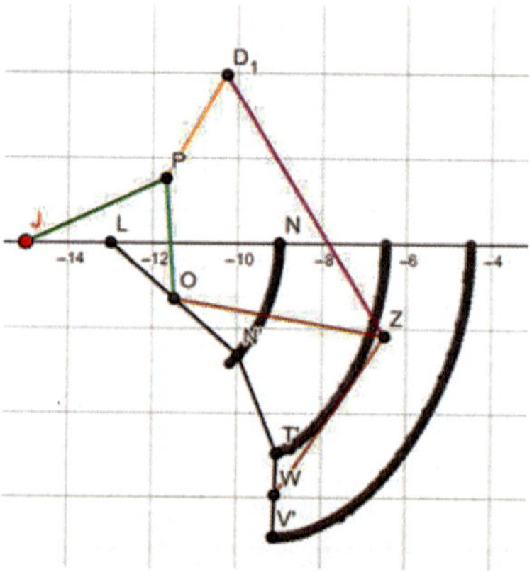

(b) Relative position of the links during flexion

Fig. 3 Kinematic synthesis for index finger

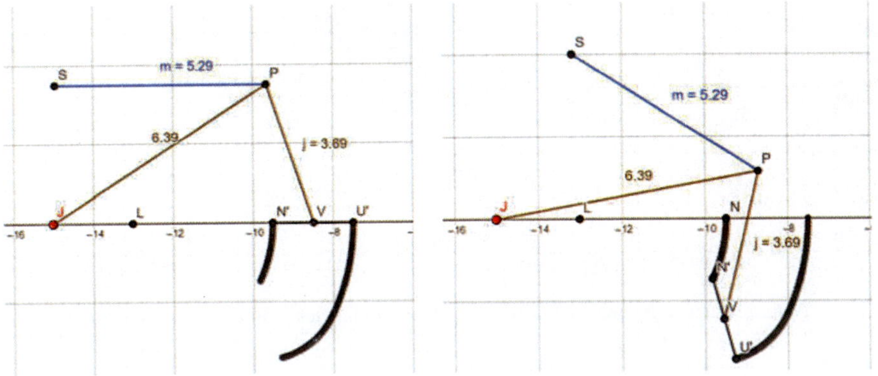

(a) Dimensional Synthesis in GeoGebra (b) Relative position of the links during flexion

Fig. 4 Kinematic synthesis for thumb

interphalangeal joint, distal interphalangeal joint and the fingertip respectively of the index finger. The rest of the fingers i.e. the middle, ring and little fingers follow the same procedure of performing the kinematic synthesis as was followed for the index finger.

The thumb has two phalanges and two joint angles. The mechanism structure is different from rest of the four fingers. This mechanism also satisfies the Grubler criteria as can be seen from Table 3. For the thumb, the link is connected directly to the distal phalanx of the thumb with four links and four joints. For the thumb we have only one mechanical connection (MC) attached.

The first two steps of the synthesis for the thumb mechanism is same as that for the fingers i.e. one DOF (Number synthesis) and planar mechanism (Type synthesis) but a four-bar linkage mechanism. The next step of Dimensional synthesis for the thumb is carried out in GeoGebra based on the proposed mechanism. Figure 4a and b show the mechanism of the thumb along with its link lengths following the desired trajectory from the extension position to full flexion position while grasping a cylindrical object of 4 cm diameter. The interphalangeal (IP) joint and MCP joint flexes by 49.7° and 24.8° respectively during this grasping operation. The proposed mechanism with the mentioned link lengths is also able to avoid mechanical interference between the links and finger joints.

2.4 *Trajectory Comparison*

The trajectory of the actual fingertip movement is obtained from the data and the constraints that are pertinant to grasping. It is then compared with the trajectory of the fingertip for index finger and thumb assisted by the hand exoskeleton as shown in Fig. 5a and b. The blue curve indicates the natural trajectory while the red curve

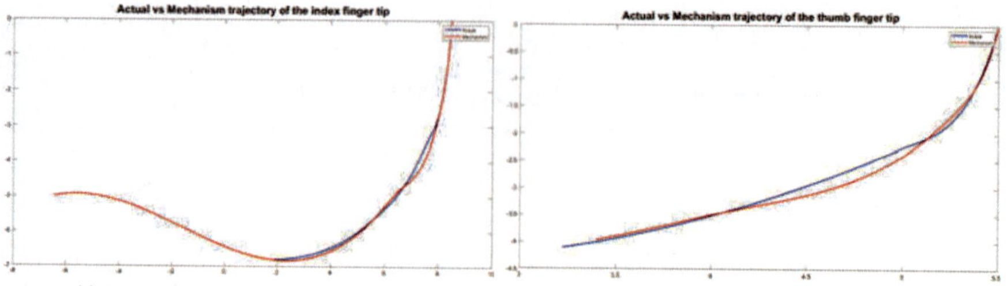

(a) Comparison of trajectories for index finger (b) Comparison of trajectories for thumb

Fig. 5 Natural versus trajectory obtained by synthesized mechanism: **a** Index finger and **b** thumb

indicates the hand exoskeleton assisted trajectory of the index finger and thumb. The trajectory of the fingertip is definite as the movement of the fingers are constrained by coupled joints of the mechanism.

3 Pose Analysis of Mechanism

The behaviour of the proposed device is needed to be analysed and for that pose analysis of the mechanism is performed using the finger joint angles. The finger joint angles are used to define the position of the mechanism. The whole mechanism can be defined using eight unknowns and three known parameters. Eight independent equations are required to obtain unique solution for the eight unknown parameters. Four independent kinematic loops are identified. The loops are chosen so that all the links of the mechanism are covered at least once. The feasible link lengths range are known from the kinematic synthesis. Using those results we can obtain the actuator distance for a particular pose of the finger.

Loop 1 defines the relation between the actuator displacement and the mechanism using a vector loop equation as shown in Fig. 6a. The variables in this loop equation are (l_x, q_b, q_c, q_n).

$$r_A^O + r_B^A + r_C^B + r_N^C + r_O^N = 0$$
$$l_x e^{iq_n} + l_{AB} e^{iq_b} + l_{BC} e^{iq_c} + l_{CN} e^{iq_{cn}} + l_{act} e^{iq_n} = 0$$

Loop 2 defines the motion of the mechanism around MCP using a vector loop equation as shown in Fig. 6b. The variables in this loop equation are (c_1, q_c)

$$r_B^C + r_E^B + r_I^E + r_M^I + r_C^M = 0$$
$$l_{CB} e^{iq_c} + l_{BE} e^{iq_c} + l_{EI} e^{iq_c} + c_1 e^{iq_{MCP}} + l_{MC} e^{iq_{MC}} = 0$$

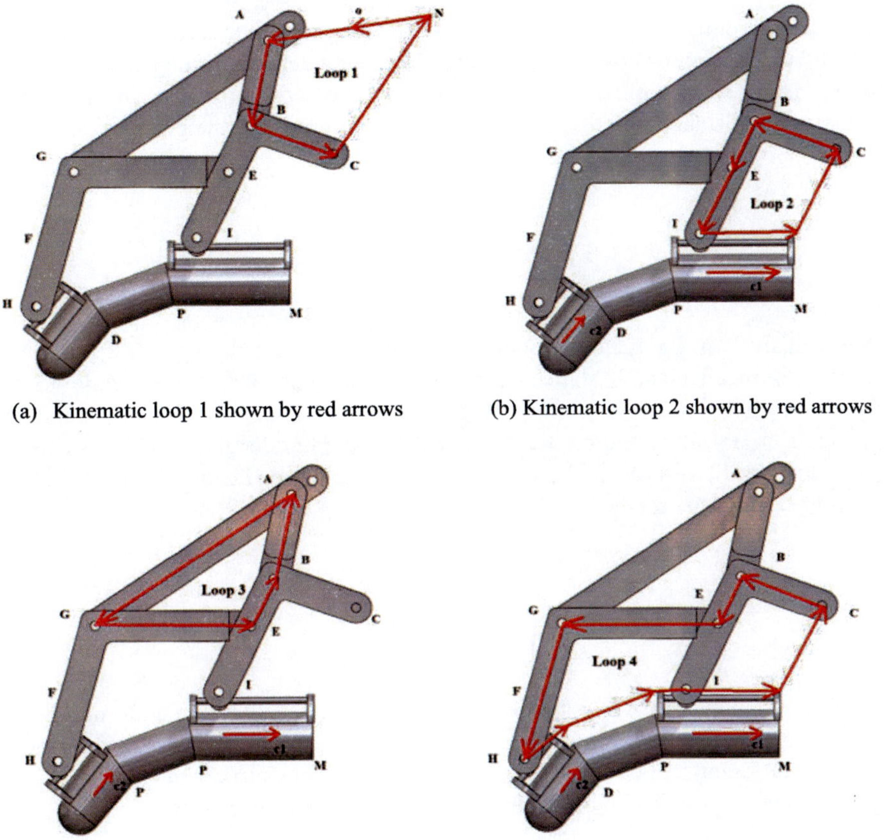

(a) Kinematic loop 1 shown by red arrows

(b) Kinematic loop 2 shown by red arrows

(c) Kinematic loop 3 shown by red arrows

(d) Kinematic loop 4 shown by red arrows

Fig. 6 Kinematic loops 1, 2, 3 and 4

Loop 3 defines the relation between the joints of the mechanism as shown in Fig. 6c. The variables in this loop equation are (q_a, q_b, q_c, q_e)

$$r_A^B + r_G^A + r_E^G + r_B^E = 0$$
$$l_{BA}e^{iq_b} + l_{AG}e^{iq_a} + l_{GE}e^{iq_e} + l_{EB}e^{iq_c} = 0$$

Loop 4 defines the motion of the mechanism around MCP, PIP and DIP as shown in Fig. 6d. The variables in this loop equation (q_c, q_e, c_2)

$$r_B^C + r_E^B + r_G^E + r_F^G + r_D^H + r_P^D + r_M^P + r_C^M = 0$$
$$l_{BC}e^{iq_c} + l_{BE}e^{iq_c} + l_{EG}e^{iq_e} + l_{GF}e^{iq_e} + l_{FH}e^{iq_e}$$
$$+ c_2e^{iq_{DIP}} + l_{DP}e^{iq_{PIP}} + l_{PM}e^{iq_{MCP}} + l_{MC}e^{iq_{MC}} = 0$$

Eight non-linear equations are obtained from the above four vector loop equations. The four loops are so chosen such that each loop has at least one independent variable which is not seen in other loops. After solving the above equations in MATLAB, the actuator distance for the index finger obtained to perform the grasping of a 4 cm diameter cylinder is about 22 mm.

3.1 *Static Analysis*

Static analysis is significant for ensuring stability while grasping an object at any pose of the mechanism. In static analysis, the torque generated at the finger joints are obtained when a required amount of force is transmitted by the linear actuator for grasping an object. Based on the static analysis, the link lengths of the mechanism are optimized. The force applied by the linear actuator and the torques developed at the finger joints are related by the Jacobian matrix.

$$\begin{bmatrix} F_x \\ \tau_n \\ \tau_b \end{bmatrix} = J^T \begin{bmatrix} \tau_3 \\ \tau_4 \\ \tau_5 \end{bmatrix}$$

The Jacobian matrix is obtained by using the derivative of the eight non-linear equations obtained from the kinematic loops of the mechanism. The loop equations are now differentiated w. r. t. time to obtain a relation in matrix form between finger joint velocities ($\dot{q}_{fin}$) and joint velocities of the mechanism ($\dot{q}_m$).

$$[\dot{q}_{fin}] = \begin{bmatrix} \dot{\theta}_3 & \dot{\theta}_4 & \dot{\theta}_5 \end{bmatrix}^T$$

$$[\dot{q}_m] = \begin{bmatrix} \dot{l}_x & \dot{q}_n & \dot{q}_b & \dot{q}_c & \dot{q}_e & \dot{q}_a & \dot{c}_1 & \dot{c}_2 \end{bmatrix}^T$$

The LHS of the equation contains the multiplication of the below two terms of size (8 × 3) and (3 × 1)

$$\begin{bmatrix} 0 & 0 & 0 \\ 0 & 0 & 0 \\ -c_1 \sin\theta_3 & 0 & 0 \\ c_1 \cos\theta_3 & 0 & 0 \\ -l_{PM} \sin\theta_3 & -l_{DP} \sin\theta_4 & -c_2 \sin\theta_5 \\ l_{PM} \cos\theta_3 & l_{DP} \cos\theta_4 & c_2 \cos\theta_5 \\ 0 & 0 & 0 \\ 0 & 0 & 0 \end{bmatrix} \begin{bmatrix} \dot{\theta}_3 \\ \dot{\theta}_4 \\ \dot{\theta}_5 \end{bmatrix}$$

The RHS of the equation contains the multiplication of two terms of size (8 × 8) and (8 × 1)

$$\begin{bmatrix} -\cos q_n & (l_x+5)\sin q_n & l_{AB}\sin q_b & l_{BC} sinq_c & 0 & 0 & 0 & 0 \\ -\sin q_n & -(l_x+5)\cos q_n & -l_{AB}\cos q_b & -l_{BC}\cos q_c & 0 & 0 & 0 & 0 \\ 0 & 0 & 0 & (l_{CB}+l_{BE}+l_{EI})\sin q_c & 0 & 0 & -\cos\theta_3 & 0 \\ 0 & 0 & 0 & -(l_{CB}+l_{BE}+l_{EI}) cosq_c & 0 & 0 & -\sin\theta_3 & 0 \\ 0 & 0 & 0 & (l_{CB}+l_{BE})\sin q_c & (l_{EG}+l_{GF}+l_{FH})\sin q_e & 0 & 0 & -\cos\theta_1 \\ 0 & 0 & 0 & -(l_{CB}+l_{BE}) cosq_c & -(l_{EG}+l_{GF}+l_{FH}) cosq_e & 0 & 0 & -\sin\theta_1 \\ 0 & 0 & l_{AB}\sin q_b & l_{EB} sinq_c & l_{GE} sinq_e & l_{AG} sinq_a & 0 & 0 \\ 0 & 0 & -l_{AB}\cos q_b & -l_{EB}\cos q_c & -l_{GE}\cos q_e & -l_{AG}\cos q_a & 0 & 0 \end{bmatrix}$$

and

$$\begin{bmatrix} \dot{l}_x \ \dot{q}_n \ \dot{q}_b \ \dot{q}_c \ \dot{q}_e \ \dot{q}_a \ \dot{c}_1 \ \dot{c}_2 \end{bmatrix}^T$$

The above matrices are divided into sub-matrix components such that none of the sub-matrix has all the elements as zero. On solving a Jacobian matrix of size (3 × 3) is obtained that relates the finger joint velocities ($\dot{q}_{fin}$) and joint velocities of the mechanism ($\dot{q}_m$). Using the inverse of this Jacobian transpose matrix, the torques at the joints can be calculated by the following equation.

$$\begin{bmatrix} \tau_3 \ \tau_4 \ \tau_5 \end{bmatrix}^T = J^{-1}\begin{bmatrix} F_x \ \tau_n \ \tau_b \end{bmatrix}^T$$

The magnitude and direction of τ_3, τ_4 and τ_5 determines the stability of grasping an object assisted by a linear actuator.

3.2 *Optimization of Link Lengths*

The design of the proposed mechanism is outlined based on the link lengths obtained through the GeoGebra. The link lengths need to optimized to improve the performance of the hand exoskeleton. The optimization procedure to find a set of link lengths is the following:

$$\textbf{Objective function} : \max\sqrt{\tau_3^2+\tau_4^2+\tau_5^2}$$

$$\textbf{such that}\ 0 \le c_1 \le 4$$

$$0 \le c_2 \le 2$$

$$0 \le l_x \le 4$$

τ_3, τ_4 and τ_5 are the joint torques at MCP, PIP and DIP joints respectively. The limitation of c_1 and c_2 are the maximum displacement the slider can move along the proximal and distal phalange respectively. l_x is the stroke distance. The optimised link length that allows the finger to reach the workspace satisfying the constraints are shown in Table 4.

Table 4 Optimized link lengths for the mechanism on index finger

l_{AB}	l_{BC}	l_{BE}	l_{EI}	l_{AG}	l_{GF}	l_{FH}	l_{EG}
3.796336	4.0000	3.50002	1.5002	8.0000	5.004	2.5000	4.0000

Algorithm 1 Optimization of Link Lengths

1: Enter a feasible link length range
2: **for** q_{MCP}= 0:1:47 **do**
3:   **for** q_{PIP}= 0:1:62 **do**
4:   **for** q_{DIP}= 0:1:28 **do**
5:   Solve eight equations obtained from the four kinematic loops and calculate $c_1, c_2, l_x, q_a, q_b, q_c, q_e, q_n$
6:   **if** $0 \leq c_1 \leq 4,\ 0 \leq c_2 \leq 2$ and $0 \leq l_x \leq 4$ **then**
7:   Proceed to next step and calculate torque
8:   **end if**
9:   **else** Go to step 1 and take the next set of link lengths
10:   Calculate the torque at the finger joints using Jacobian for 1 N force applied at the actuator.
11:   Calculate the optimization Objective function $\sqrt{\tau_3^2 + \tau_4^2 + \tau_5^2}$and the set of link lengths for which the optimization function value is maximum will be accepted.
12:   **end for**
13:   **end for**
14: **end for**

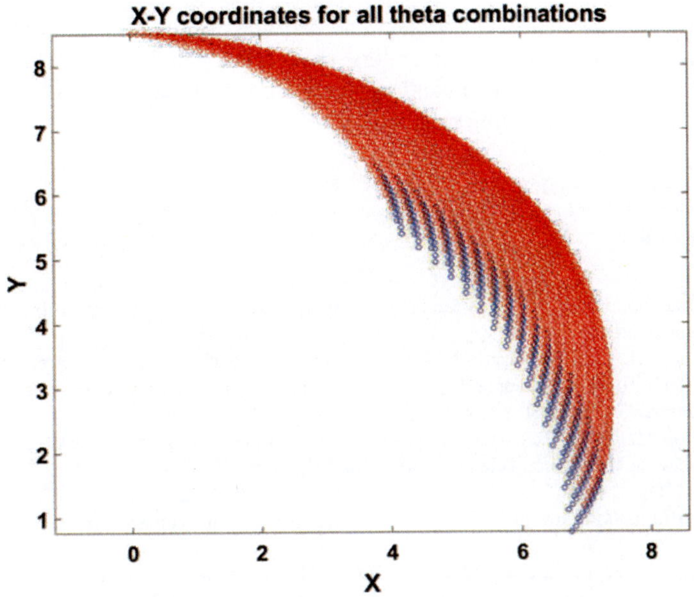

Fig. 7 Comparison of workspace between anatomical and hand exoskeleton assisted finger

3.3 Workspace Evaluation

Fig 7 represent the anatomical workspace of the index finger compared with the hand exoskeleton assisted index finger workspace. The blue marked area indicates the workspace of the index finger without hand exoskeleton and red marked area refers to workspace assisted by hand exoskeleton. It has been observed that the index finger is unable to reach some of the extreme points on wearing the device but it covers almost the whole workspace. This is due to the constraints set by the hand exoskeleton that limits the finger to cover whole of the natural workspace.

4 Conclusion

We have outlined the process to design an underactuated hand exoskeleton to ensure that assisted fingers follow natural trajectory. Kinematic synthesis is carried out to obtain feasible link lengths that ensures no mechanical interference between the mechanism and the hand. Pose analysis is executed to analyse the behaviour of the proposed device and optimize the link lengths. The stability of the mechanism is assured if the forces are transmitted in the right direction while grasping an object. For that static analysis is done. Finger joint torques produced for force generated by the linear actuator for a particular pose was utilized for optimizing the link lengths.

Workspace of the assisted fingers and the anatomical fingers were in close correspondence. The thumb and the index finger synthesized in this paper holds promise for development of a complete hand exoskeleton. Development of a complete hand exoskeleton is part of ongoing research.

References

1. Mozaffari Foumashi M (2013) Synthesis of hand exoskeletons for the rehabilitation of post-stroke patients," alma
2. Fontana M, Dettori A, Salsedo F, Bergamasco M (2009) Mechanical design of a novel hand exoskeleton for accurate force displaying. In: 2009 IEEE international conference on robotics and automation, pp 1704–1709
3. Burdea G, Zhuang J, Roskos E, Silver D, Langrana N (1992) A portable dextrous master with force feedback. Presence Teleoperators Virtual Environ 1(1):18–28
4. Sun Z, Miao X, Li X (2009) Design of a bidirectional force feedback dataglove based on pneumatic artificial muscles. In: 2009 international conference on mechatronics and automation, pp 1767–1771
5. Iqbal J, Khan H, Tsagarakis NG, Caldwell DG (2014) A novel exoskeleton robotic system for hand rehabilitation–conceptualization to prototyping. Biocybern Biomed Eng 34(2):79–89
6. Diez JA, Blanco A, Catalán JM, Badesa FJ, Lledó LD, Garcia-Aracil N (2018) Hand exoskeleton for rehabilitation therapies with integrated optical force sensor. Adv Mech Eng 10(2):1687814017753881
7. Worsnopp TT, Peshkin MA, Colgate JE, Kamper DG (2007) An actuated finger exoskeleton for hand rehabilitation following stroke. In: 2007 IEEE 10th international conference on rehabilitation robotics, pp 896–901
8. Agarwal P, Yun Y, Fox J, Madden K, Deshpande AD (2017) Design, control, and testing of a thumb exoskeleton with series elastic actuation. Int J Rob Res 36(3):355–375
9. Ash HA, Joyce TJ, Unsworth A (1996) Biomechanics of the distal upper limb. Curr Orthop 10(1):25–36
10. Conti R et al (2017) Kinematic synthesis and testing of a new portable hand exoskeleton. Meccanica 52(11–12):2873–2897

Structural Enumeration of Cam-Modulated Linkages

H. Manikandan, Vijayananda Kaup, and Harish Babu

Abstract A kinematic chain is an abstraction of a mechanism constituted by rigid members connected with each other through simple pin joints. They are useful as a means to aid creative thinking in the design of mechanisms. We know that a binary link, along with its pin joints can be directly converted to a higher pair without changing the degrees of freedom of the resulting linkage. This produces new types of linkage in which both lower and higher pairs coexist. There will also be a change in the count of links, joints and types of joints in the resulting linkage. The present work focuses on the synthesis of a family such linkages known as cam-modulated linkages. Kinematic chains, whether it is simple- or multiple-jointed type of chain, can produce many useful cam-modulated linkage by recursively applying the above procedure. The present work deals with the synthesis of cam-modulated linkages from simple jointed kinematic chains only. This paper presents the results of synthesis of cam-modulated linkages obtained from the complete collection of 8-link, 1-freedom simple-jointed kinematic chain. The synthesis is carried out via visual inspection and also by means of a computer program. A total of 131 cam-modulated linkages containing 5, 6 and 7 links and exhibiting one freedom were enumerated. 38, 58, 35 CMLs having 1, 2, 3 higher pairs were synthesized.

Keywords Kinematic chain · Higher pair · Cam-modulated linkage

H. Manikandan (✉)
Mechanical Engineering (VTU-RC), CMR Institute of Technology, Bengaluru 560037, India
e-mail: manikandan.h@cmrit.ac.in

V. Kaup
Department of Mechanical Engineering (VTU-RC), CMR Institute of Technology, Bengaluru, India

H. Babu
Department of Mechanical Engineering, CMR Institute of Technology, Bengaluru, India

V. K. Gupta et al. (eds.), *Recent Advances in Machines and Mechanisms*, Lecture Notes in Mechanical Engineering, https://doi.org/10.1007/978-981-19-3716-3_3

1 Introduction

A successful product design necessitates an effective management of many facets of mechanical design processes. The foremost decision is proper selection of physical configuration of the mechanism under design. This step is termed the structural synthesis of linkages, comprising two discrete phases called the type and number synthesis [1, 2].

Success of this step contributes to wider acceptance of the end product and is popularly referred to as creative synthesis of the machine configuration. Here, the skeletal structure of the mechanism is decided. In order to help machine designers in this stage, many researchers have attempted to provide a catalog of linkage mechanisms. The latter has been recognized as an effective means to meet the manufacturers' demands for novelty or originality in the product [3]. Availability of such a catalog has been found to provide both designers and analysts an opportunity to innovate and optimize their design, in terms of achieving a reduction in product's weight, cost, space, etc., of the designed product.

2 Cam-Modulated Linkages–Systematics

Systematics of linkages is a well-known technique which is found to assist in the creative endeavor of designers in the preliminary stages of the machine design [2]. The technique finds application in the evolution of different types of mechanical artifacts.

Johnson [3] proposed the idea of associated linkages in his book. He demonstrates how simple mechanism consisting of rigid members connected by pin joints can help create a variety of useful mechanical configurations.

For example, a four-bar mechanism can be transformed into slider-crank mechanism by making one of its four links to be infinite length, see Fig. 1a, b. If a binary link in four-bar linkage is made an imaginary link to represent line of force/motion (Fig. 2b), then the result will be cam and follower linkage. A pair of gears in mesh has four-bar linkage as its associated linkage wherein the imaginary link is aligned along the line/path of contact (Fig. 3).

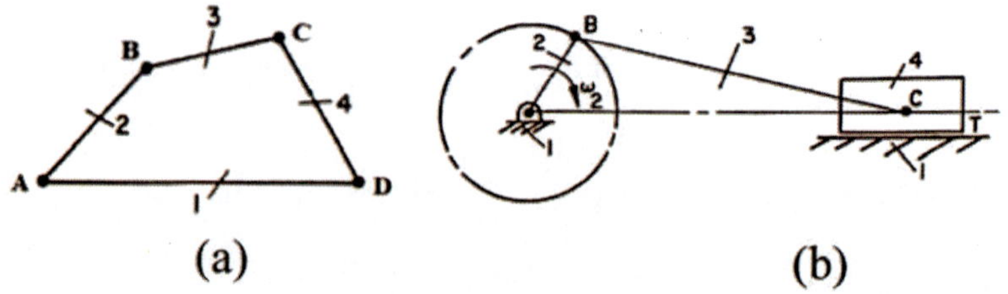

Fig. 1 Associated linkage and its slider-crank inversion

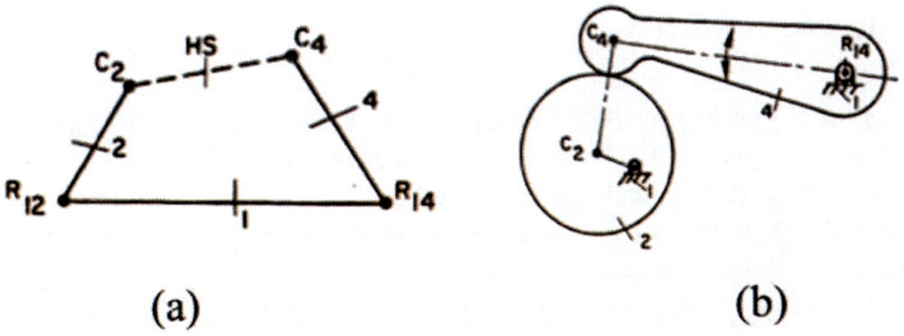

Fig. 2 Associated linkage and its oscillating-follower inversion

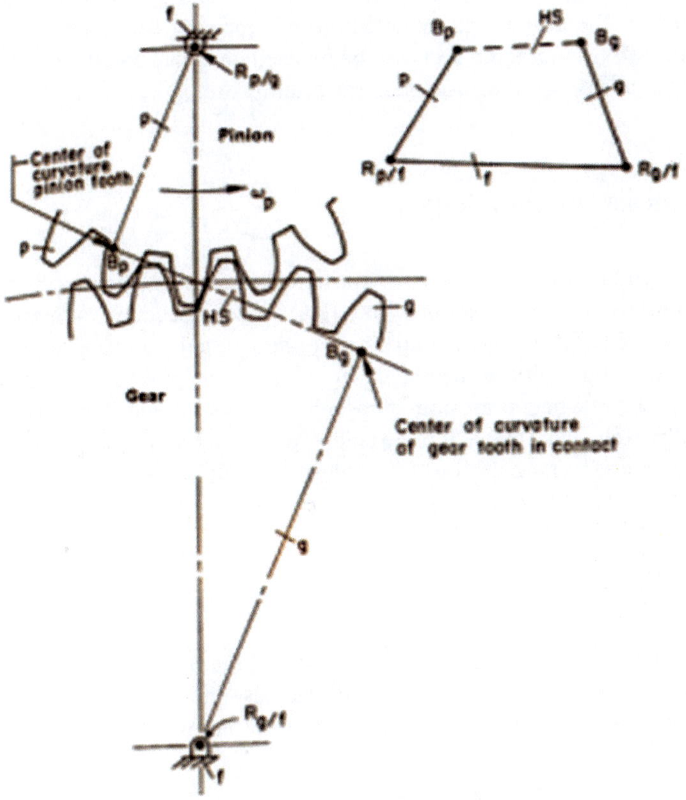

Fig. 3 Four-bar associated linkage and its inversion to gear pair (Johnson [3])

3 Literature Review

There is little literature available on the structural synthesis of CML. In his book, Kurt Hain [4] suggested the methodology of converting binary link into higher pair to obtain CMLs. Using real-life examples, Johnson [3] proposes the concept of associated linkages as a means to synthesize mechanisms having machine elements such as cams, sliders and rack and pinion.

It is known that CMLs exhibit superior kinematic performance in terms of function, path and motion generation when compared with linkage mechanisms with pin joints [5]. This makes them widely accepted and highly adopted in the automation industries and machineries. The tolerance analysis and synthesis are an essential requirement in the design and manufacture of precision cam-modulated linkages. Most of the effort in research seems to be focused on errors and their quantification, sensitivity analysis, optimization, manufacturability, etc.

4 Synthesis Methodology

Structural synthesis of linkages with spring elements, cam pairs, prismatic pairs, belt-and-pulleys and so on starts with SJKCs. The procedure suggested by Hain [4] to convert SJKCs to chains with cams, prismatic pairs and spring elements has been utilized in this work. It may be observed that when the SJKCs from which the synthesis procedure begins are non-isomorphic, the chains that result from each must be mutually exclusive. Hence, it is sufficient to conduct an isomorphism test among only those chains that are derived from the same SJKC.

4.1 Varieties of Conversions

This section tabulates ways in which one or more binary link in a chain is converted to a higher pair. Each such conversion will decrease the link and pin-joint count by one and two, respectively. Figure 4 shows the different ways of converting a binary link into a higher pair.

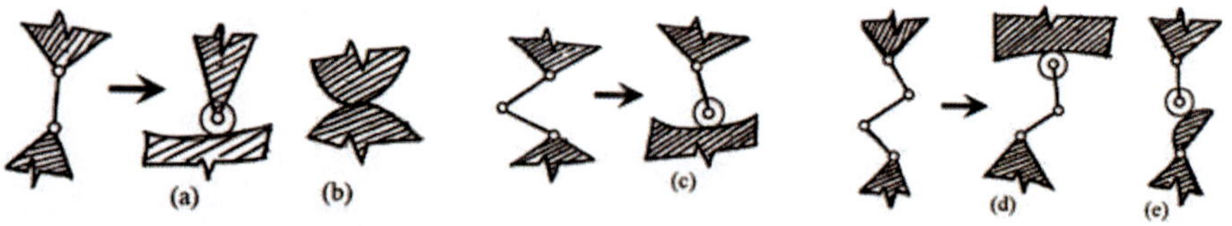

Fig. 4 Converting binary link to higher pair

Fig. 5 Infeasible CML

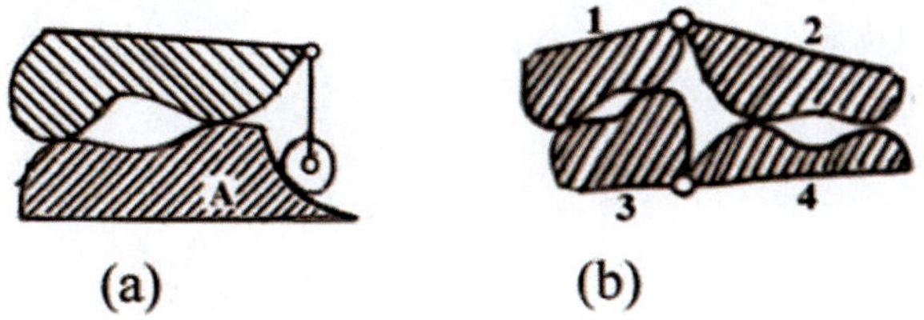

i. Fig. 4a, b. Conversion of a single binary link lying between non-binary links into a higher pair. While a roller is used in Fig. 4a, two profiled surfaces forming a higher pair are shown in direct contact in Fig. 4b. Kinematically, the arrangements shown in Fig. 4a and b are identical.
ii. Fig. 4c. Here, the binary link is part of a dyad. Converting any one of the two binary links will yield an arrangement shown. However, attempt to convert both the binary links into two higher pair leads to physically unrealizable arrangement. Therefore, only one binary link in a dyad is feasible for conversion.
iii. Fig. 4d, e. For reasons explained in item (ii) above, when the three binary are in series (a triad), only one binary link in the series can be converted to a higher pair. Two situations arising from this conversion are shown in Fig. 4d, e.

4.2 *Non-viable CMLs*

Case–1: Each link in a CML must have at least one turning pair. Links joined to other links by higher pairs only lead to an infeasible situation since transmission of motion and force across such a link cannot be accomplished in a constrained manner. One such example is shown in Fig. 5a, although Gruebler's equation evaluates the degree of freedom of the CML to unity. Here, the link-A is characterized by the absence of lower pairs in it.

Case–2: Fig. 5b illustrates an example wherein, although each of the four links has a turning pair, the CML becomes infeasible because the sub-chain made up of links 1 and 2 can be pulled apart to form a disconnected chain. As in case-1, transmission of motion and force across such a link cannot be accomplished in a constrained manner, and hence, such CMLs are to be excluded from the collection.

4.3 *Isomorphism Problem in CML*

Search for presence of isomorphic CMLs is an important step in the enumeration of kinematic structures. The methodology involves systematic search of binary links in simple jointed kinematic chains (SJKC) which can be turned into higher pairs. Due to this, it is understandable that all CMLs generated from one SJKC will be distinct from the ones generated from another non-isomorphic SJKC. Therefore,

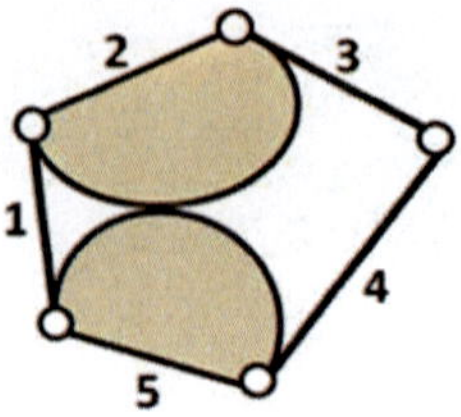

$$\begin{bmatrix} 0 & 1 & 0 & 0 & 1 \\ 1 & 0 & 1 & 0 & 2 \\ 0 & 1 & 0 & 1 & 0 \\ 0 & 0 & 1 & 0 & 1 \\ 1 & 2 & 0 & 1 & 0 \end{bmatrix}$$

Fig. 6 5-link 1-d.o.f. CML with a higher pair; and its matrix representation

check for isomorphism may be limited among the CMLs enumerated from the same SJKC. Since the number CMLs generated from 8-link, 1-freedom SJKC collection is reasonably small, this work uses both visual and computer-based approaches to isomorphism test.

The following section describes computer implementation of the isomorphism test proposed in this work with reference to a 5-link, 1-d.o.f. CML (Fig. 6), along with its matrix representation. It may be noted that the element (2,5) and (5,2) in the matrix is assigned a value of two to distinguish them from the pin joint value of one. These values serve to uniquely represent the structure of the linkage in the matrix notation.

As the first step, for the given CML (containing five links), five rooted trees are generated keeping each of the five links in turn as the reference. Figure 7a shows a rooted tree with reference to link-1 as the root. The rooted trees show two types of arrows: the thin arrow indicates the traversal through a pin joint whereas the thick arrows lying between 2 and 5 and vice-versa in Fig. 7a and b indicate traversal through a higher pair. The reader is referred to [ ] which explains in greater detail the way to obtain the rooted tree from any given chain, and the procedure to convert the rooted tree into an isomorphism index for a generalized simple jointed kinematic chain.

Assignment of the link invariant is an important step in the proposed isomorphism test. All binary links having two pin joints at its either ends will be assigned an invariant 'A'. Thus, links 1, 3 and 4 in Fig. 6a are assigned A as the invariant. Links 2 and 5 are ternary links having three joints by way of a pin joint between links 1 and 2; a pin joint between 2 and 3; and a higher pair between links 2 and 5. Link-2 is assigned and invariant B1 wherein letter B tells that the link is a ternary link and

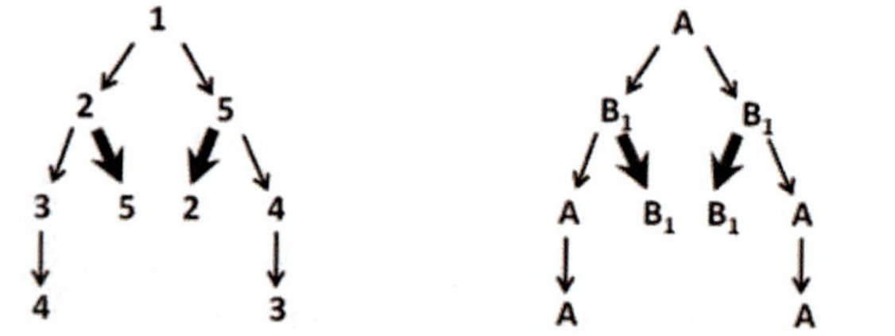

(a) With link labels (b) With link-invariants

Fig. 7 Rooted tree with reference to link-1

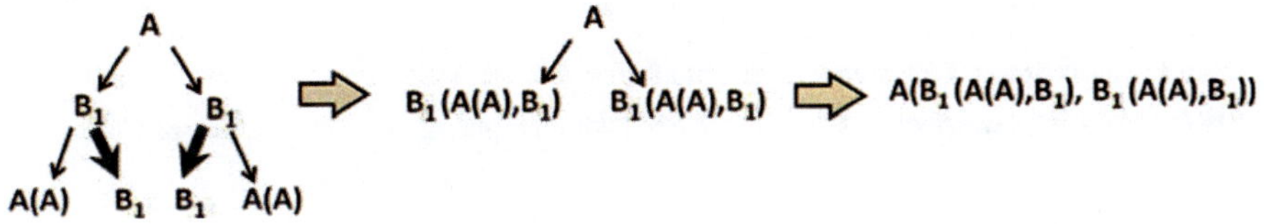

Fig. 8 Reducing rooted tree in Fig. 7b into a string

subscript '1' indicates that one of the three joints on link-2 is a higher pair. For the same reasons, invariant for link-5 will also be B1.

Fig. 7b shows the rooted tree in which link labels have been substituted with link invariants. The next task is to obtain a unique string from the tree in Fig. 7b by reducing the rooted tree levels in a step-wise manner. This is illustrated in Fig. 8 resulting in the string $A(B1(A(A), B1), B1(A(A), B1))$. An ordered set of five such strings, obtained by taking each of the five links as the root link will constitute the isomorphism index of the given CML.

5 Results of Synthesis

16 SJKCs, which is the complete collection of 8-link, 1-degree of freedom chains is considered in this study for enumeration of CMLs. The total number of links in such CMLs depends upon the count of higher pairs present in it. The count of higher pairs in CML equals the number of binary links that is converted into a cam pair in the SJKC. Thus, a CML with n higher pairs comprises of $(N - n)$ number of links and $(J - 2n)$ number of pin joints and will exhibit the same degrees of freedom as the original SJKC with N links.

Table 1 lists the results of synthesis of CMLs generated from 8-link, 1-freedom SJKC. As may be seen, a total of 131 CMLs exhibiting one freedom are enumerated. 38, 58, 35 CMLs having 1, 2, 3 higher pairs were synthesized. Sketches of only some of the CMLs derived from each of 16 SJKCs are listed in Table 2 for lack of space.

Table 1 Collection of CMLs generated from 8-link, 1-DOF SJKCs

SJKC	Link assortments	CLMs with *n* higher pairs			No. of CMLs
		$n = 1$	$n = 2$	$n = 3$	
1	$n_3 = 4; n_2 = 4$	01	02	01	04
2	$n_4 = 1; n_3 = 2; n_2 = 5$	02	03	02	07
3	$n_3 = 4; n_2 = 4$	02	03	02	07
4	$n_4 = 2; n_2 = 6$	01	02	03	06
5	$n_4 = 1; n_3 = 2; n_2 = 5$	03	05	03	11
6	$n_3 = 4; n_2 = 4$	01	02	–	03
7	$n_4 = 1; n_3 = 2; n_2 = 5$	04	06	05	15
8	$n_3 = 4; n_2 = 4$	03	04	02	09
9	$n_4 = 1; n_3 = 2; n_2 = 5$	05	08	04	17
10	$n_3 = 4; n_2 = 4$	01	02	02	05
11	$n_4 = 2; n_2 = 6$	02	04	03	09
12	$n_3 = 4; n_2 = 4$	04	05	02	11
13	$n_3 = 4; n_2 = 4$	02	02	01	05
14	$n_3 = 4; n_2 = 4$	01	01	-	02
15	$n_3 = 4; n_2 = 4$	02	03	01	06
16	$n_4 = 1; n_3 = 2; n_2 = 5$	04	06	04	14
Total CMLs		**38**	**58**	**35**	**131**

Table 2 Sketches of 131 CMLs generated from 8-link, 1-DOF SJKCs

#	SJKC	Derived cam-modulated linkages
1		
2		
3		
4		

6 Conclusions

This research work considers systematic enumeration of CMLs from all the 16 variants of 8-link, 1-freedom SJKCs. The process entails successive identification and conversion of binary links in the SJKC into higher pairs. As the method requires comparison of CMLs generated from the same SJKC for possible duplication, and also since smaller number of CMLs were to be dealt with at any time for the category of CML under synthesis, the task of isomorphism detection is initially accomplished through visual inspection only. 131 CMLs containing 1, 2 and 3 higher pairs were generated and are presented in the form of sketches.

The steps of CML synthesis and the isomorphism test for CMLs were implemented on a computer using Python program. The results generated from the program are found to be in complete agreement with the CMLs derived by the method of visual inspection and comparison. As the next steps, CMLs having even larger number of links may be attempted, and automatic sketching of CMLs may be tried. The sketches will serve as a powerful aid in the conceptual design stage of mechanism design.

References

1. Reuleaux F (1963) The kinematics of machinery. Dover, New York
2. Hartenberg RS, Denavit J (1964) Kinematic synthesis of linkages. Mc-Graw Hill Book Company, Library of Congress Catalog Card Number 64-23251
3. Johnson RC (1971) Mechanical design synthesis with optimization applications. Van Nostrand Reinhold Company, Library of Congress Catalog Card Number 77-131262
4. Hain K (1967) Applied kinematics. McGraw Hill, New York
5. Chang W-T, Wu L-I2013)Tolerance analysis and synthesis of cam-modulated linkagesMath Comput Model57(3–4):641–660

Studying the Effects of Varying Link Lengths in Double Lambda Mechanism and Its Application to Rover Suspension Design

Aniruddha Nayak, Vivek Gupte, and Pravin Singru

Abstract Rocker-bogie suspensions are known to cause the bogie overturn problem, which greatly limits the traversable obstacle height and velocity. To counter this problem, use of linear motion suspension systems has been proposed. The double lambda rocker-bogie suspension is an example of one such system. However, in the double lambda rocker-bogie suspension, the ground clearance is limited due to presence of linkages. We study methods to improve ground clearance of this suspension system. Two different methods and their individual and combined effects on ground clearance as well as on the nature of the coupler curve of the modified mechanisms have been studied.

Keywords Double lambda mechanism · Coupler curves · Rover suspension

1 Introduction

Rovers can have three, four [1], six wheels, or even eight wheels [2]. The suspension redistributes the normal reaction in the different wheels of the rover to prevent it from toppling [3].

The most commonly used suspension system is called the rocker-bogie suspension. It is a simple, yet proven six-wheel suspension system that is currently in use. The rocker-bogie has one significant drawback of bogie overturning [4], especially if it encounters an obstacle at high speeds. In the current exploration rovers, the travel velocity is very low so there the chances of this are less. But if rover is to be designed for high-speed applications, then rocker-bogie might cause problems.

A. Nayak (✉) · V. Gupte · P. Singru
Birla Institute of Technology and Science, Pilani, K.K. Birla Goa Campus, Sancoale, Goa 403726, India
e-mail: f20180850@goa.bits-pilani.ac.in; aniruddhanayak2mar@gmail.com

V. K. Gupta et al. (eds.), *Recent Advances in Machines and Mechanisms*, Lecture Notes in Mechanical Engineering, https://doi.org/10.1007/978-981-19-3716-3_4

2 Literature Review

An in-depth study on lambda mechanism as a rover suspension has not been already done. Barlas et al. have analysed different suspension systems and have pointed out the "bogie overturn problem" that persists in the rocker-bogie system [4]. The author has introduced the use of Chebyshev's lambda mechanism in a suspension. The mechanism converts the circular motion of its crank to a quick return linear motion of the endpoint of its coupler. Two such mechanisms have been coupled together to replace the bogie in the rocker-bogie system. The lambda mechanism follows specific ratios for link lengths. The author has also identified the ground clearance of the system, a parameter which has been tried to be improved in this study.

Miller et al. proposed an improvement on the rocker-bogie system to improve traversal speeds, by using eight wheels instead of six [2]. Barlas et al. have suggested that double lambda suspension can also be used for fast-paced operation in rovers [4]. Hence, the double lambda mechanism could be an improvement on rocker-bogie systems for fast speed operation at similar obstacle capacity.

An example of the use of rocker double lambda bogie suspension is shown in a report by Ozdemir et al. [5]. They have developed a mobile robot for fast-paced rescue operations in rough terrains. The robot employs the use of rocker double lambda bogie suspension with a differential gear mechanism to connect the two rockers. According to the authors, after some design iterations, the suspension system could overcome the environmental and geographical obstacles.

3 Problem Statement

Instead of using a regular bogie, a Chebyshev's lambda mechanism can be used in the bogie. The wheels of the bogie, in this case, can move with respect to each other and transfer normal reactions between each other. In this system, the bogie overturn problem is non-existent.

In the case of Chebyshev's lambda mechanism, the coupler curve is a curve traced by the endpoint of the coupler (end point of E) as the crank A is rotated around the joint (Fig. 1). The curve traced by the endpoint of the coupler consists of a near-vertical line followed by a quick return curve that returns the coupler point to its starting location. In suspension systems, the mechanism is only used in the vertical region as the wheels go over obstacles.

One drawback of double lambda suspension is the lack of ground clearance. The rocker-bogie systems can be designed to have larger ground clearance. However, because of the presence of linkages in the bogie of lambda bogie suspension, there is a limit to which the ground clearance can be increased. There are two main ways to increase ground clearance in the double lambda bogie for a particular set of link lengths:

(a) It can be changed by altering the angle "τ" (Fig. 2).
(b) Ground clearance can be increased by changing link lengths.

Earlier studies [4, 5] have already suggested the first method, however, its implications on factors such as range of motion (ROM) and width of the resulting bogie width have not been studied.

In the second method, the effect of changing certain individual link lengths to get increased ground clearance is investigated. In Chebyshev's lambda mechanism, the linear motion and verticality of the coupler point are lost, if the prescribed ratios are not followed.

The second method might seem detrimental. However, if the increase in ground clearance is important, some compromises in the ideal lambda mechanism might be considered. The aim is to understand which link lengths, if any, are safer to alter than others, and how much of an adverse effect is to be expect so that the system can be designed to handle it.

4 Objectives

- Study the method of changing τ to increase the ground clearance of double lambda mechanism.
- Study effect of changing link lengths on ground clearance.
- Study combined effect of both the methods on ground clearance.
- Study implications changing link lengths on nature of the coupler curve.

5 Methodology

In this study, the uppermost link in the bogie is the crank (*A*). Moving along the clockwise direction, the next link is the coupler (BE), rocker (*C*), and frame (*D*). The length *E* is the coupler extension (Fig. 1). The near-linear quick return coupler curve characteristic of Chebyshev's lambda mechanism is traced by the coupler point at the end of *E*. In the following analysis, first a set of link lengths for Chebyshev's lambda mechanism, assuming different values for crank length was calculated. The following relations were used to obtain the link lengths [4]:

$$\begin{gathered} A = \text{crank length},\ D = f * A \\ \text{where } 1.55 \le f \le 3.0,\ B = 0.5(3 * D - A) \\ (\text{In the lambda mechanism } \mathrm{B} = \mathrm{C} = \mathrm{E}) \end{gathered} \tag{1}$$

For the analysis, four different crank lengths (*A*) were considered, to study the scalability of the results. The remaining link lengths were calculated using four different values of factors (*f*). The modified configurations were obtained by increasing the link lengths of each configuration, one at a time. The lengths were increased by increments of 5% up to 25% total increment. For each configuration, there were 30 variations with 6 cases (0% to 25%) for each link length and a total of five-link lengths (*A*, *B*, *C*, *D*, *E*). Considering all 16 configurations, there were a total of 480 unique cases. Using these 480 cases, coupler curves were plotted for each mechanism using MATLAB [6].

The coordinates of the coupler point for the Lambda mechanism are given by the equations:

$$X = \left(\frac{E}{B}\right) * [C * \cos(\alpha) - A * \cos(\theta)] + C * \cos(\alpha) \tag{2a}$$

$$Y = \left(\frac{E}{B}\right)[C * \sin(\alpha) - A * \sin(\theta) - D] + C * \sin(\alpha) \tag{2b}$$

The crank angle θ is varied and α is obtained from trigonometric relations to get the values of *X* and *Y* to get the coupler curve.

Increasing link lengths led to certain combinations where a closed geometry did not exist for some values of "τ". In some cases, the cranks interfered with each other throughout the working range. Using such link lengths in rover suspension is impractical. Henceforth, these impractical cases are referred to as "infeasibilities".

To calculate ground clearance, the vertical distance between the origins in Fig. 1 to the coupler point, when both wheels of the bogie are horizontal to each other was considered. The variation of ground clearance was checked in the case of 25% increase of link lengths (80 values). Also, τ was varied from 140° to 170° in increments of 5°, to check for effects of increasing τ, as well as combined effects of the variations on ground clearance. The values for ground clearance were calculated graphically, using SOLIDWORKS 2020 [7].

6 Results and Discussion

For the analysis, the ground clearance and horizontal deviation of the coupler point were studied. First, the impact of change of τ and *f* on ground clearance in the stock double lambda mechanism was studied.

To make a comparison, the GC for the configuration with $\tau = 140°$ and $f = 1.55$ was considered as reference. From the Fig. 3, it can be observed that for any given *f* increasing from 140° to 170° ground clearance increases. At a particular, τ increasing the factor value from 1.55 to 3.00 increases the ground clearance. It is seen from Fig. 3 that increasing τ is beneficial. However, increasing has an unwanted effect on a range of motion. Range of motion is not in the scope of this paper, but we have done an extensive study in which the effects were noted.

It is observed that increasing *f* increases the sensitivity of the effects of link length changes on mechanism feasibility. It was observed that link length *C* is the most sensitive to length changes. In the figures, the points where the mechanism becomes infeasible have been neglected.

Figure 4 shows the percentage change in ground clearance with respect to the respective ideal values for each of the four f values. From the graphs, it can be observed that the link length E has the best results showing a maximum increase in ground clearance, and link length C has the worst impact in all cases. Same results in percentage change were obtained for other crank length values for each f.

For any given configuration, the coupler curves formed by one lambda mechanism by varying one particular link were plotted on one graph, and this showed the trend in which the coupler curve deforms. The coupler curves of the default variation have been compared to that of the 25% increased variation to get the deviation.

From the Fig. 5, it can see that link length C has the worst impact on the coupler curve, followed by link length B. The link length E and A have the least impact on the nature of the coupler curve and would almost retain the verticality and linearity of the coupler curve. It was observed that the percentage deflection values were exactly the same for all the crank lengths considered, proving that the results are scalable and uniform.

7 Conclusion

Two methods to improve ground clearance of a double lambda bogie suspension system were studied. Both the methods to increase ground clearance are effective, even more so when used together.

Changing the length of link E is most beneficial as it has the least effect on the linearity and ROM. It provides the greatest increase in ground clearance. Changing link length C is not advisable as it has the largest decrease in linearity, reduction in ROM, leads to the formation of infeasible geometries, and causes a decrease in ground clearance. Changing link lengths increases the width of the bogie and has implications in normal reaction distribution on wheels, and on the climbing and descending capabilities. The effect of changing bogie width needs further research.

The effects of changing multiple links at a time could provide more increase in ground clearance at minimum deviation from the ideal mechanism. This aspect needs to be further studied in depth. Different optimisation algorithms can be designed and implemented to get better results.

Appendix

See Figs. 1, 2, 3, 4, and 5.

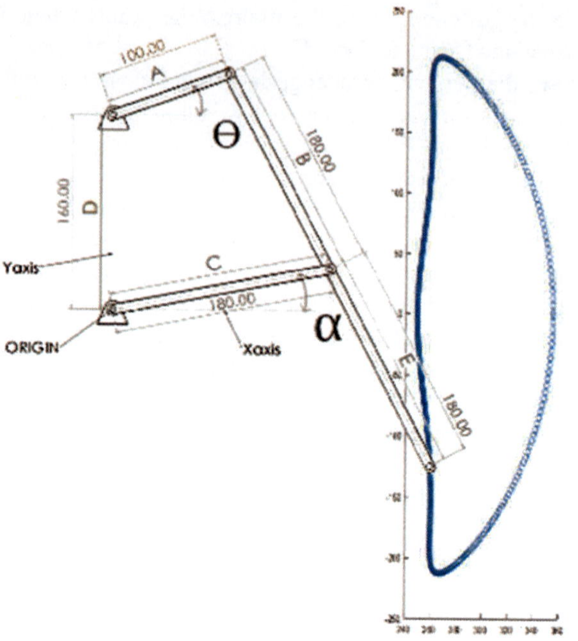

Fig. 1 Chebychev's lambda mechanism with its coupler curve

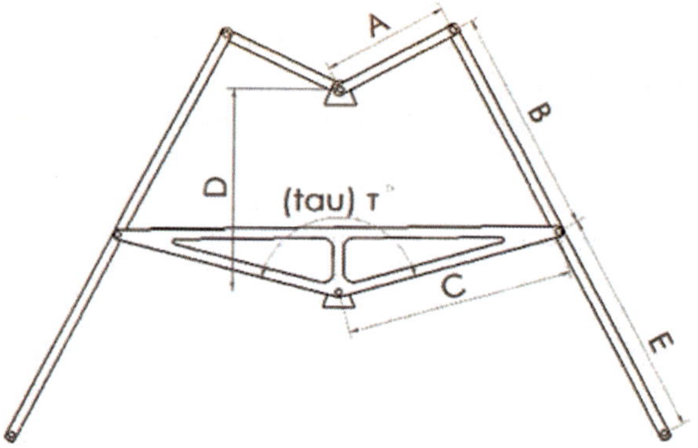

Fig. 2 Double lambda bogie

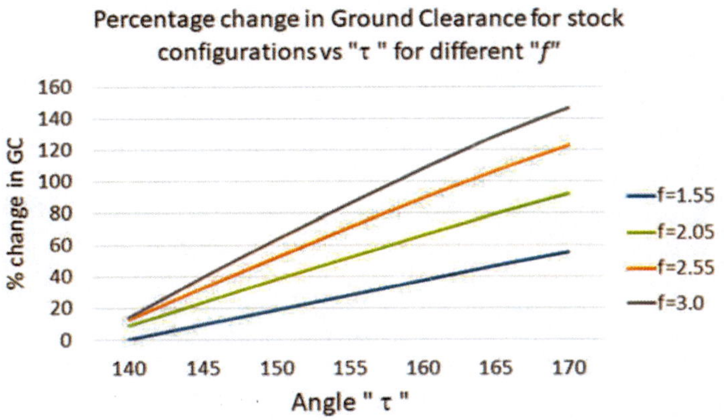

Fig. 3 Percentage change in ground clearance for stock configurations versus τ for different f

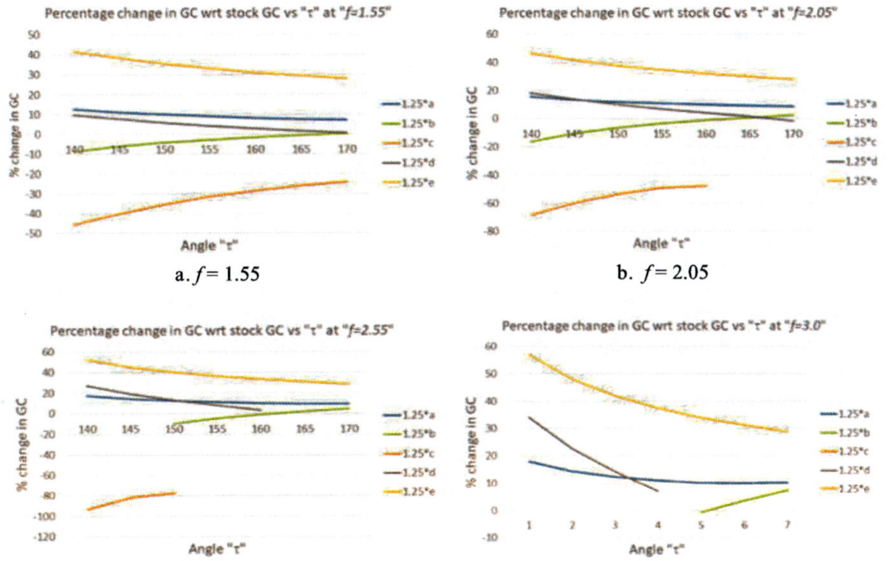

a. $f = 1.55$

b. $f = 2.05$

c. $f = 2.55$

d. $f = 3.0$

Fig. 4 Percentage change in ground clearance for varied link lengths versus τ for different f

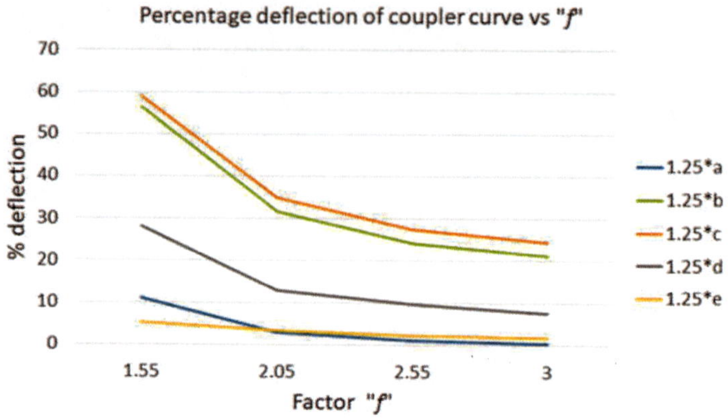

Fig. 5 Horizontal deflection in coupler curve for varied link lengths versus *f*

References

1. Sandin PE (2003) Devices illustrated. In: Vol. Walkers (Issue 4), New York. https://doi.org/10.1036/007142928X
2. Miller DP, Lee TL (2002) High-speed traversal of rough terrain using a Rocker-Bogie mobility system. Space 2002 and Robot 428–434. https://doi.org/10.1061/40625(203)54
3. Harrington BD, Voorhees C (2004) The challenges of designing the Rocker-Bogie suspension for the mars exploration rover. In: Proceedings of the 37th aerospace mechanisms symposium, pp 185–195. http://hdl.handle.net/2060/20040084284
4. Barlas F (2004) Design of a mars rover suspension mechanism 93
5. Ozdemir AC, Akbasak S, Tezel Y, Sert E, Jamal C (2020) Effect of rocker double lambda mechanism on mobile robot. J Young Investigators 37(5). https://doi.org/10.22186/jyi.37.5.51-56
6. MATLAB (2020) Version 9.8.0.1380330 (R2020a). The MathWorks Inc., Natick, Massachusetts
7. SOLIDWORKS (2021) Dassault systèmes, [Dassault systèmes SOLIDWORKS 2020], [2020 SP03], Dassault Systèmes, San Diego

Modelling of Mechanisms

Positional Error Estimation of Five-Bar Mechanism Under the Influence of Tolerances

Darren Alton Dsouza, Ankur Jaiswal, and H. P. Jawale

Abstract This paper presents an analysis of positional error in a five-bar planar mechanism operated with variable link tolerances. As part of this research, a new method is developed to analyze the effect of link tolerance on the end effector's positioning. The proposed method's results are validated by comparing the graphical methods and CAD Model Simulation. The forward and inverse kinematics of the five-bar mechanism are analyzed for ten different positions in the first quadrant. The permutations and combinations of tolerances on the positioning errors are studied, and the maximum error is sought. An effort is made to compensate it.

Keywords Link tolerance · Positional error · Geometrical approach · Graphical approach · Forward and inverse kinematics

Abbreviations

θ_2 and θ_5	Input angles driven by actuators
θ_2 and θ_3	Passive angles
l_1, l_2, l_3, l_4, l_5	Link Lengths
X_P	X Coordinate position of end effector
Y_P	Y Coordinate position of end effector
$l_{BE}, l_{BD}, l_{AC}, l_{OC}$	Construction lines drawn for calculation purposes
φ_i, β_i	Interior angles
$lt_1, lt_2, lt_3, lt_4, lt_5$	Link lengths with tolerance

D. A. Dsouza · A. Jaiswal (✉)
Department of Mechatronics, Manipal Institute of Technology, Manipal Academy of Higher Education (MAHE), Manipal, Karnataka 576104, India
e-mail: ankur.jaiswal@manipal.edu

H. P. Jawale
Department of Mechanical Engineering, Visvesvaraya National Institute of Technology, Nagpur 440010, India

V. K. Gupta et al. (eds.), *Recent Advances in Machines and Mechanisms*, Lecture Notes in Mechanical Engineering, https://doi.org/10.1007/978-981-19-3716-3_5

1 Introduction

Robotic manipulators are commonly used to place objects or tools where needed. As manipulators, several applications use closed-chain mechanisms with planar and spatial configurations. Closed-chain kinematic manipulators are of interest to researchers due to their high pay load, enhanced rigidity, improved precision, accuracy, reduced drive torque, and quicker actuation [1–6]. The closed-chain five-bar mechanism is one such mechanism that may be utilized as a planar manipulator in a specific workspace (Fig. 1). A coupler point *P* on one of the connected links serves as an end effector.

The planar five-bar mechanism is synthesised for function generation, path generation, and rigid body guidance. From simple machine linkages to complex biomedical equipment, assembly, and robot industries, the five-bar mechanism has a wide range of applications. The configuration is primarily employed as a planar mechanism. Many complex mechanisms can be studied using a five-link chain as a primitive configuration.

For the desired task, the performance of the mechanism is crucial. In general, performance is assessed by the deviation of the rigid body being directed from its desired position and orientation [7–10]. A performance deviation is the result of actual configuration compared to ideal. Both cumulative and random factors have an impact on performance of a mechanism. Some of the factors influencing the deviation are joint clearances, finite stiffness for link materials, backlash, assembly defects, link dimension tolerances, and structural constraints [11–18]. The presence of link tolerance or joint clearance affects a mechanisms positional accuracy. Many researchers have considered these issues in the context of performance evaluation [19–25].

Position accuracy, or the ability to reach the desired position, is described in terms of error, which is defined as a deviation from the desired position. It is a critical criterion for assessing the effectiveness of closed-loop mechanisms. The accuracy of a

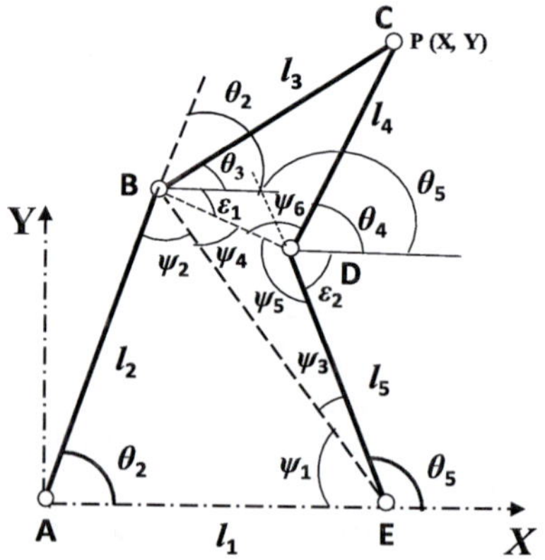

Fig. 1 Forward kinematics of five-bar mechanism

mechanism is determined by a variety of factors related to drives, links, and joints. Drive-related variations can be ascribed to transmission finite stiffness, resolution, and backlash [26]. Drive stiffness and torque, external loads, self-weight, and inertial load on links will all influence the manipulator's positional accuracy as a result of drive stiffness. The resolution of the drive causes a structural error that can be mitigated using the appropriate drive. With the right drive selection and control programming, deviations due to the stiffness and resolution of the drive can be minimized well in advance. Other sources of position error include the properties of the link such as dimensional tolerances, deviations, and extensions. Manufacturing tolerances are an unavoidable part of the process. Link tolerance causes a deviation in position and orientation. Many researchers address the impact of dimensional tolerances on mechanical error in mechanisms when modelling, synthesizing, and analyzing them [27–32]. A configuration's positional accuracy can be a mix of cumulative and random errors. Random error is outside the purview of calibration and compensation, whereas the cumulative contribution estimated during calibration can be corrected. Therefore, it is essential to estimate the maximum possible inaccuracy and comparative performance of a configuration during the design phase under the influence of random error sources [33].

The path generation of four-bar linkage is analyzed using Taguchi method (TM) and the random coordinate search algorithm (RCA). The result obtained is compared to previous works on the same evolutionary algorithm problem [34]. The wavelet analysis approach is used to calculate the actual sizes of a planar four-bar and five-bar mechanism for path synthesis problems [35, 36]. The geometrical approach is used to formuate the forward and inverse kinematics of novel 6-DOF spatial five-bar linkage, namely SUUUS linkage using closed-loop equations is presented. Motion control, workspace, stiffness and dynamics analysis of the mechanism was analyzed [37].

The work presented in this article is organized as follows: the description of the mechanisms, i.e., deduction of forward and inverse kinematics of the mechanism, verification of graphical and geometrical approach in Sect. 2. Effect of link-tolerance on the ideal five-bar mechanism is studied in Sect. 3. Error compensation of angles driven by actuator (input angles) is presented in Sect. 4. Finally, the conclusions are presented in Sect. 5.

2 Kinematic Synthesis

2.1 *Forward Kinematics*

In this approach, simple equations are created using many geometric relationships among the variables in the manner, as shown in Fig 1. Geometric analysis is performed to validate the results of graphical analysis. The relations are used to convert the base joint angles θ_2 and θ_5 into tip coordinates, X and Y.

The interior lengths, l_{BE} and l_{BD}, are computed from the law of cosine. The interior joint angles, φ_{i}, are all determined using the law of cosines or simple geometric relationships. The internal joint angles are then used to determine the angles θ_2 and θ_3. Finally, the end effector's position is obtained by summing the vectors from the origin to the tip.

Applying cosine law to determine θ_3 and θ_4

$$\cos\phi_1 = \left[\frac{l_1^2 + l_{\mathrm{BE}}^2 - l_2^2}{2l_1 \cdot l_{\mathrm{BE}}}\right] \tag{1}$$

$$l_{\mathrm{BE}}^2 = l_1^2 + l_2^2 - 2l_1l_2\cos\theta_2 \tag{2}$$

$$\cos\phi_2 = \left[\frac{l_2^2 + l_{\mathrm{BE}}^2 - l_1^2}{2l_2l_{\mathrm{BE}}}\right] \tag{3}$$

$$\phi_3 = 180 - (\theta_5 + \phi_1) \tag{4}$$

$$\cos\phi_4 = \left[\frac{l_{\mathrm{BD}}^2 + l_{\mathrm{BE}}^2 - l_5^2}{2l_{\mathrm{BD}}l_{\mathrm{BE}}}\right] \tag{5}$$

$$l_{\mathrm{BD}}^2 = L_5^2 + L_{\mathrm{BE}}^2 - 2l_5l_{\mathrm{BE}}\cos\phi_3 \tag{6}$$

$$\cos\phi_5 = \left[\frac{l_5^2 + l_{\mathrm{BD}}^2 - l_{\mathrm{BE}}^2}{2l_{\mathrm{BD}}l_5}\right] \tag{7}$$

$$\cos\phi_6 = \left[\frac{l_4^2 + l_{\mathrm{BD}}^2 - l_3^2}{2l_{\mathrm{BD}}l_5}\right] \tag{8}$$

$$\theta_3 = \cos^{-1}\left[\frac{l_3^2 + l_{\mathrm{BD}}^2 - l_4^2}{2l_{\mathrm{BD}}l_3}\right] + \partial_1 \tag{9}$$

$$\partial_1 = 180 - (\theta_2 + \phi_4 + \phi_2) \tag{10}$$

$$\theta_3 = 360 \pm (\partial_2 + \phi_5 + \phi_6) \tag{11}$$

$$\partial_2 = 180 - \theta_5 \tag{12}$$

$$X_P = l_2C\theta_2 + l_3C\theta_3 = l_4\cos\theta_4 + l_5\cos\theta_5 + l_1 \tag{13}$$

$$Y_P = l_2s\theta_2 + l_3s\theta_3 = l_4\sin\theta_4 + l_5\sin\theta_5 \tag{14}$$

2.2 Inverse Kinematics

The position of the end effector P in the Cartesian coordinated system is given in terms of X and Y in this technique, from which the input angles θ_2 and θ_5 are calculated. Figure 2 depicts the important variables for the inverse kinematic formulation. Internal lengths, l_{AC}, l_{EC}, and l_{OC}, are calculated as the distance between the tip and the equivalent places on the base. The internal angles β_i are calculated using the law of cosines. Finally, the internal angles are used to determine the base joint angles. The algorithm is depicted below.

$$l_{\mathrm{OC}} = Y_P \tag{15}$$

$$l_{0A} = X_C \tag{16}$$

$$l_{\mathrm{AC}} = \sqrt{X_{p^2} + y_{p^2}} \tag{17}$$

$$l_{\mathrm{EC}} = \sqrt{l_{0E}^2 + Y_p^2} \tag{18}$$

$$l_{\mathrm{OE}} = l_{\mathrm{OA}} - l_{\mathrm{AE}} = \left(X_p - l_1\right) \tag{19}$$

$$l_{\mathrm{EC}} = \sqrt{\left(X_p - l_1\right)^2 + Y_p^2} \tag{20}$$

$$\theta_2 = \beta_1 + \beta_2 \tag{21}$$

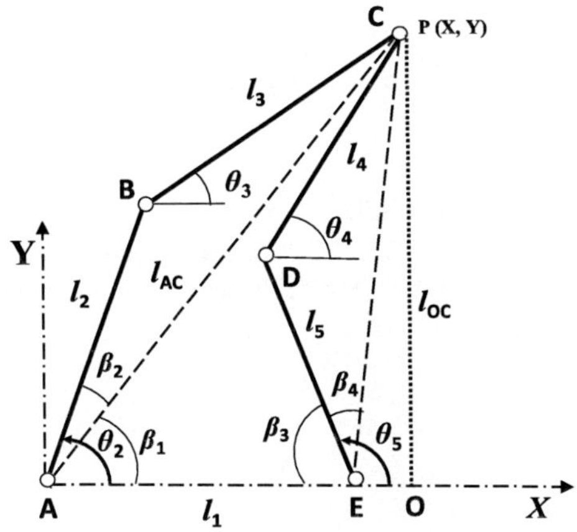

Fig. 2 Inverse kinematics

$$\cos\beta_1 = \left[\frac{X_p^2 + l_{\mathrm{AC}}^2 - Y_p^2}{2l_{\mathrm{AC}}X_p}\right] \tag{22}$$

$$\cos\beta_2 = \left[\frac{l_2^2 + l_{\mathrm{AC}}^2 - l_3^2}{2l_{\mathrm{AC}}l_2}\right] \tag{23}$$

$$\cos\beta_3 = \left[\frac{l_1^2 + l_{\mathrm{EC}}^2 - l_{\mathrm{AC}}^2}{2l_{\mathrm{AC}}l_1}\right] \tag{24}$$

$$\cos\beta_4 = \left[\frac{l_5^2 + l_{\mathrm{EC}}^2 - l_4^2}{2l_{\mathrm{EC}}l_5}\right] \tag{25}$$

$$\theta_5 = 180 - (\beta_3 \pm \beta_4) \tag{26}$$

3 Case Study

The effect of link proportions on the behavior of the 5R mechanism is significant for generality when using the formulation presented in the preceding sections. The initial 5R configuration is designed, with link proportions stated in terms of link length l_1. In this paper, the effect of tolerance on links of a five-bar mechanism was performed and analyzed. Given the constant link lengths ($l_1 = 35$ mm, $l_2 = 100$ mm, $l_3 = 100$ mm, $l_4 = 70$ mm, $l_5 = 70$ mm), experimentation was done over a variety of combination of tolerances on each link length (2430 combinations), and the maximum positional error was obtained.

4 Validation Using CAD Model Approach

The estimations from the geometrical approach are carried out as the equations given in the above sections. The one of the estimations is presented below for 5R mechanisms, for given link proportions at position P5, and tolerance of 0.1 (100 micron) in Fig. 3a, b. The corresponding values are given in Table 1, showing the conformance of the geometrical approach. The same is also found using graphical approach with 2D drafting tool. The graphical estimation plotted with and without tolerances is shown in finger geometrical Fig. 3a and b. Estimates correspond to three decimal places.

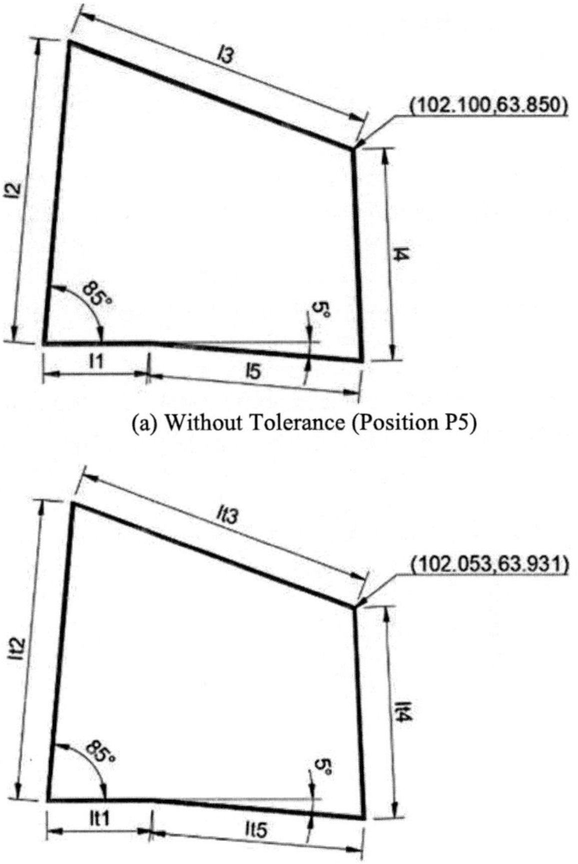

Fig. 3 **a, b** Graphical or CAD model approach

(b) With Tolerance (Position P5)

Table 1 Comparison of results using geometrical and CAD model approach

Input angles			Geometrical approach		CAD model		Graphical error
Position	θ_2	θ_5	$P1_x$	$P1_y$	$P1_x$	$P1_y$	
P1	45	40.52	141.440	0.000	141.420	0.000	0.02
P2	55	−35	133.172	16.707	133.170	16.710	0.004
P3	65	−25	125.377	35.027	125.380	35.030	0.004
P4	75	−15	114.789	50.816	114.790	50.820	0.004
P5	85	−5	102.099	63.850	102.100	63.850	0.001
P6	95	5	87.963	74.062	87.960	74.060	0.004
P7	105	15	72.978	81.534	72.980	81.530	0.004
P8	115	25	57.652	86.471	57.650	86.470	0.002
P9	137.4	101.42	0.000	135.347	0.000	135.350	0.003

5 Results and Discussion

5.1 *Effect of Link Tolerance on Five-Bar Mechanism*

Tolerance ranges taken were from 10 to 100 microns; nine different positions were analyzed graphically as well as geometrically to study the maximum positional error in each case (P1–P9). Table 2 depicts the data of the influence of link tolerance on a five bar mechanism. A noticeable trend was evident; the maximum positional error was the highest for a given position when the tolerance on the given links were maximum as depicted through Fig. 4. Figure 5 shows the increase in positional error with the increase in tolerance for a given position.

The 3D surface plot of position, tolerance, and maximum positional error Fig. 6 was plotted and satisfied the result of the comparison done on maximum positional

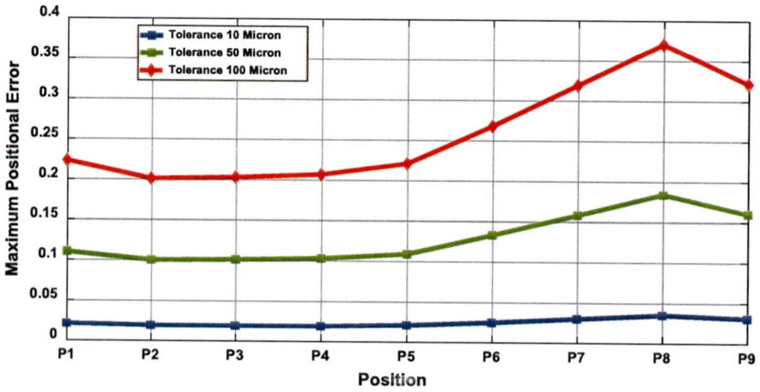

Fig. 4 Position versus maximum positional error

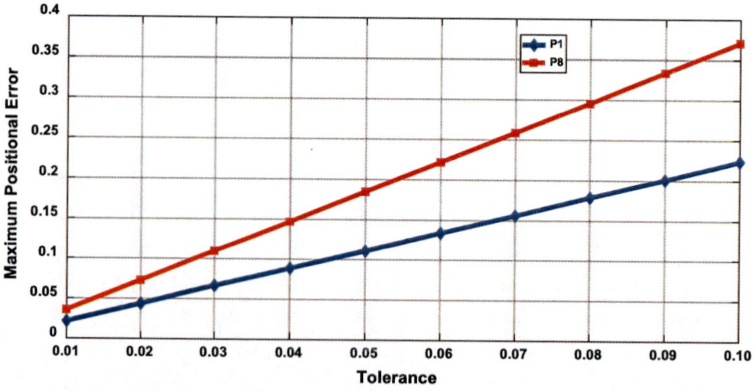

Fig. 5 Tolerance versus maximum positional error

Table 2 Maximum positional error versus tolerances

Maximum positional error									
Tolerance/Position	*P*1	*P*2	*P*3	*P*4	*P*5	*P*6	*P*7	*P*8	*P*9
0.01	0.022	0.02	0.02	0.02	0.022	0.026	0.031	0.036	0.032
0.02	0.044	0.04	0.04	0.041	0.044	0.053	0.063	0.073	0.064
0.03	0.067	0.06	0.061	0.062	0.066	0.08	0.095	0.11	0.096
0.04	0.089	0.08	0.081	0.083	0.088	0.107	0.127	0.147	0.129
0.05	0.111	0.101	0.102	0.104	0.11	0.134	0.159	0.185	0.161
0.06	0.134	0.121	0.122	0.125	0.133	0.161	0.191	0.222	0.193
0.07	0.156	0.141	0.142	0.146	0.155	0.188	0.223	0.259	0.225
0.08	0.179	0.161	0.163	0.167	0.177	0.215	0.256	0.296	0.258
0.09	0.201	0.182	0.183	0.187	0.2	0.242	0.288	0.334	0.29
0.1	0.224	0.202	0.204	0.208	0.222	0.269	0.32	0.371	0.322

error of each position, given the tolerance changes. It was observed through the plot that the positional error increased with the increase in tolerance of link length for each position as well as the dip in positional error for extreme points P1 and P9 was depicted.

Magnification factor is defined as the ratio of the maximum positional error to the tolerance of the link

$$\text{Magnification Factor} = \frac{\text{Maximum Positional Error}}{\text{Tolerance of Link}}$$

Table 3 shows that the magnification factor obtained is a constant value for a given position verified through Fig. 7. The maximum magnification factor was observed in position P8 which has the maximum positional error for given tolerances Fig. 8. Magnification factor versus tolerance was also studied and depicted through Fig. 9.

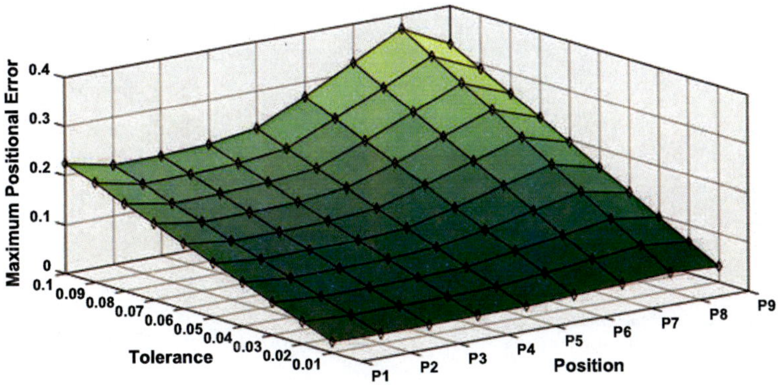

Fig. 6 3D surface plot of position, error, and tolerance

Table 3 Magnification factor for given tolerances

Magnification factor									
Tolerance/Position	*P*1	*P*2	*P*3	*P*4	*P*5	*P*6	*P*7	*P*8	*P*9
0.01	2.200	2.000	2.000	2.000	2.200	2.600	3.100	3.600	3.200
0.02	2.200	2.000	2.000	2.05	2.200	2.650	3.150	3.650	3.200
0.03	2.233	2.000	2.033	2.067	2.200	2.667	3.167	3.667	3.200
0.04	2.225	2.000	2.025	2.075	2.200	2.675	3.175	3.675	3.225
0.05	2.220	2.020	2.040	2.080	2.200	2.680	3.180	3.700	3.220
0.06	2.233	2.017	2.033	2.083	2.217	2.683	3.183	3.700	3.217
0.07	2.229	2.014	2.029	2.086	2.214	2.686	3.186	3.700	3.214
0.08	2.238	2.013	2.038	2.088	2.213	2.688	3.200	3.700	3.225
0.09	2.233	2.022	2.033	2.078	2.222	2.689	3.200	3.711	3.222
0.1	2.240	2.020	2.040	2.080	2.220	2.690	3.200	3.710	3.220

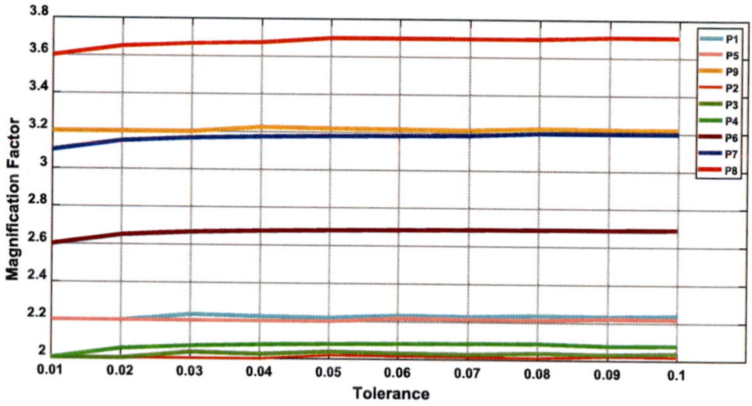

Fig. 7 Magnification factor versus tolerance

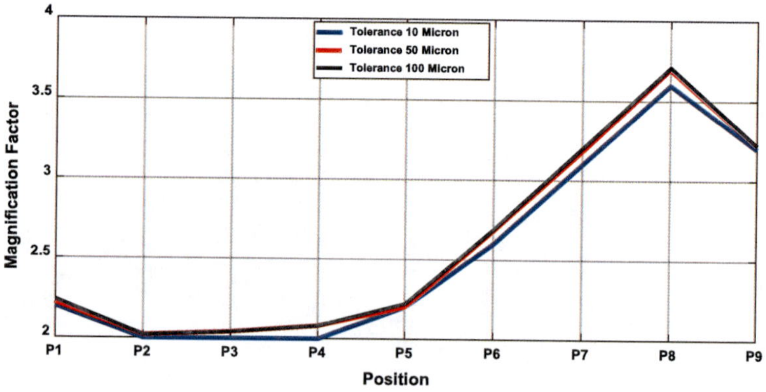

Fig. 8 Magnification factor versus position plot

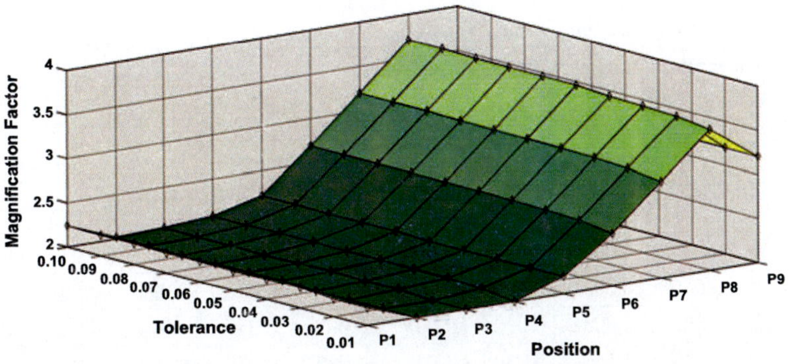

Fig. 9 3D surface plot of position, tolerance, and magnification factor

The maximum positional error of each position grows as the amount of tolerance in the link lengths increases. The greatest positional error obtained was P8, when there was a tolerance of 100 micron given to it. Magnification factor obtained for a given position was a constant.

5.2 *Error Compensation*

For the given mechanism, error compensation is done using inverse kinematics (Fig. 2) by calculating the input angles at maximum error position. Tables 4 and 5 show the error compensation necessary for each position for the given tolerance so that we obtain the original value of tip coordinates X_P and Y_P without any tolerances.

Table 4 Error compensation for input angle θ_1

Error compensation in input angle (θ_1)									
Tol./Pos	*P*1	*P*2	*P*3	*P*4	*P*5	*P*6	*P*7	*P*8	*P*9
0.01	0.008	0.107	0.017	0.033	0.009	0.007	0.002	0.024	0.02
0.02	0.008	0.107	0.017	0.106	0.009	0.025	0.002	0.043	0.02
0.03	0.008	0.107	0.017	0.106	0.009	0.007	0.002	0.024	0.037
0.04	0.008	0.107	0.092	0.033	0.009	0.063	0.071	0.024	0.037
0.05	0.008	0.03	0.092	0.033	0.009	0.007	0.071	0.024	0
0.06	0.008	0.03	0.017	0.106	0.063	0.063	0.071	0.024	0
0.07	0.008	0.185	0.092	0.033	0.063	0.063	0.071	0.091	0
0.08	0.008	0.03	0.058	0.106	0.063	0.063	0.066	0.091	0.094
0.09	0.008	0.03	0.092	0.106	0.063	0.063	0.071	0.091	0.057
0.1	0.008	0.185	0.092	0.106	0.009	0.063	0.071	0.091	0.057

Table 5 Error compensation for input angle θ_4

Error compensation in input angle (θ_4)									
Tol./Pos	*P*1	*P*2	*P*3	*P*4	*P*5	*P*6	*P*7	*P*8	*P*9
0.01	0.035	0.085	0.04	0.073	0.009	0.05	0.055	0.013	0.567
0.02	0.035	0.003	0.12	0.005	0.007	0.025	0.055	0.002	0.554
0.03	0.035	0.003	0.12	0.005	0.016	0.05	0.01	0.013	0.554
0.04	0.035	0.085	0.04	0.034	0.009	0.05	0.01	0.087	0.554
0.05	0.035	0.003	0.04	0.034	0.009	0.02	0.094	0.087	0.567
0.06	0.035	0.003	0.012	0.005	0.009	0.055	0.094	0.087	0.626
0.07	0.035	0.166	0.039	0.044	0.085	0.055	0.094	0.087	1.441
0.08	0.035	0.003	0.199	0.005	0.085	0.055	0.067	0.161	3.175
0.09	0.035	0.079	0.039	0.082	0.085	0.055	0.169	0.161	1.441
0.1	0.035	0.166	0.039	0.082	0.009	0.13	0.169	0.161	1.441

The maximum error compensation required increases as the tolerance limit of the link increases. At a tolerance of 0.08 micron, the greatest compensation angle obtained was 3.175° for P9.

6 Conclusions

This work provided a mechanism for estimating the maximum position error, considering the impact of tolerances on the dimensions of the link. The error is indicated in relation to a set of variations in inputs and tolerances. It was investigated how tolerance affects the individual length of the link. For accurate position error estimation, a simple geometric technique is provided.

Based on this study's findings, the following significant conclusions may be drawn:

1. The mechanical error is proportional to the tolerance size on individual links and the location of the coupling in any specified input position.
2. Changes in tolerance only result in a change in the error sign, not the magnitude of the error.
3. The compensation angles necessary to restore the original position without tolerances may be calculated using the exact tolerance in each individual link as well as the maximum positional error.

References

1. Gergorio Raffale D, Castelli VP (2001) Position analysis in analytical form of 3-PSP mechanism. ASME J Mech Des 123:51–57
2. Cameron S (1990) Stiffness Mapping for Parallel Manipulators. IEEE Trans Robot Autom 6(3):377–382
3. Ecorechard G, Maurine P (2004) Self-calibration of delta robots with elastic deformation compensations. In: IEEE international conference on robotics and automation (ICRA'04)
4. Zhongchao S, Tianxiang Y, Weimin C, Bifeng S (2011) Reliability analysis for close position accuracy of gear door mechanism based on importance sampling. In: International conference on quality, reliability, risk, maintenance, and safety engineering (ICQR2MSE), pp 168–172
5. Wang Z, Zhao Z, Pang Z, Zhang C (2011) Kinematics analysis and simulation of a new 3 degrees of freedom spatial robot mechanism composed by closed chain. In: Second international conference on mechanic automation and control engineering (MACE), pp 1373–1376
6. Olivier C, Pierrot F, Krut S, Baradat C, Nabat V (2011) Par2: a spatial mechanism for fast planar two-degree-of-freedompick-and-place applications. Meccanica 46(1):239–248
7. Hartenberg RS, Denavit J (1964) Kinematic synthesis of linkages. McGraw-Hill, New York
8. Jaiswal A, Jawale HP (2017) Comparative study of four-bar hyperbolic function generation mechanism with four and five accuracy points. Arch Appl Mech 87(12):2037–2054. https://doi.org/10.1007/s00419-017-1310-5
9. Sandor GN, Erdman AG (1984) Advanced mechanism design: analysis and synthesis. Prentice-Hall Inc., Englewood Cliffs
10. Jawale HP, Jaiswal A (2018) Investigation of mechanical error in four-bar mechanism under the effects of link tolerance. J Braz Soc Mech Sci Eng 40(8):1–13
11. Chakraborty J (1975) Synthesis of mechanical error in linkages. Mech Mach Theory 10(2):155–165. https://doi.org/10.1016/0094-114X(75)90016-6
12. Du X, Venigella PK, Liu D (2009) Robust mechanism synthesis with random and interval variables. Mech Mach Theory 44(7):1321–1337. https://doi.org/10.1016/j.mechmachtheory.2008.10.003
13. Chen F-C, Tzeng Y-F, Hsu M-H, Chen W-R (2010) Combining taguchi method, principal component analysis and fuzzy logic to the tolerance design of a dual-purpose six-bar mechanism. TransCan Soc Mech Eng 34(2):277–293. https://doi.org/10.1139/tcsme-2010-0017
14. Lee SJ, Gilmore BJ, Ogot MM (1993) Dimensional tolerance allocation of stochastic dynamic mechanical systems through performance and sensitivity analysis. J Mech Des 115(3):392–402
15. Rhyu JH, Kwak BM (1988) Optimal stochastic design of four-bar mechanisms for tolerance and clearance. J Mech Transm Autom Des 110(3):255–262
16. Sharfi O, Smith M (1983) A simple method for the allocation of appropriate tolerances and clearances in linkage mechanisms. Mech Mach Theory 18(2):123–129. https://doi.org/10.1016/0094-114X(83)90104-0
17. Tsai M-J, Lai T-H (2004) Kinematic sensitivity analysis of linkage with joint clearance based on transmission quality. Mech Mach Theory 39(11):1189–1206. https://doi.org/10.1016/j.mechmachtheory.2004.05.009
18. Wu J, Purwar A, Ge Q (2010) Interactive dimensional synthesis and motion design of planar 6R single-loop closed chains via constraint manifold modification. J Mech Robot 2(3):031012
19. Augustynek K, Adamiec-Wo´jcik I (2012) Analysis of mechanisms with flexible beam-like links, rotary joints and assembly errors. Arch Appl Mech 82(2):283–295
20. Flores P (2015) On the study of the kinematic position errors due to manufacturing and assembly tolerances. In: Kecskeméthy A, Geu Flores F (eds) Interdisciplinary applications of kinematics. Mechanisms and machine science, vol 26. Springer, Cham
21. Li X, Ding X, Chirikjian GS (2015) Analysis of angular-error uncertainty in planar multiple-loop structures with joint clearances. Mech Mach Theory 91:69–85

22. Nzue R-MA, Brethé JF, Vasselin E, Lefebvre D (2010) Comparative analysis of the repeatability performance of a serial and parallel robot. In: 2010 IEEE/RSJ international conference on intelligent robots and systems (IROS). IEEE, pp 63–68
23. Ting K-L, Zhu J, Watkins D (2000) The effects of joint clearance on position and orientation deviation of linkages and manipulators. Mech Mach Theory 35(3):391–401
24. Zhang X, Zhang X (2016) A comparative study of planar 3-RRR and 4-RRR mechanisms with joint clearances. Robot Comput Integr Manuf 40:24–33
25. Zhu J, Ting K-L (2000) Uncertainty analysis of planar and spatial robots with joint clearances. Mech Mach Theory 35(9):1239–1256
26. Jawale HP, Thorat HT (2012) Investigation of positional error in two degree of freedom mechanism with joint clearance. J Mech Robot 4(1)
27. Junfu Z, Xiaoping D (2011) Time-dependent reliability analysis for function generator mechanisms. ASME J Mech Des 133(3):031005
28. Fu-Chen C, Yih-Fong T, Meng-Hui H, Wei-Ren C (2010) Combining Taguchi method, principal component analysis and fuzzy logic to the tolerance design of a dual purpose six bar mechanism. Trans Can Soc Mech Eng 34(2):277–293
29. Xiaoping D, Pavan KV, Deshun L (2009) Robust mechanism synthesis with random and interval variables. Mech Mach Theory 44(7):1321–1337
30. Jun W, Purwar A, Ge QJ (2010) Interactive dimensional synthesis and motion design of planar 6R single-loop closed chains via constraint manifold modification. J Mech Robot 2, 031012; Mach. Theory 18(2):123–129
31. Tsai M-J, Lai TH (2004) Kinematic sensitivity analysis of linkage with joint clearance based on transmission quality. Mech Mach Theory 39:1189–1206
32. Takaaki O (2005) Error compensation system for joints, links and machine frame of parallel kinematics machines. Int J Robot Res 24:1087–1102
33. Jawale HP, Thorat HT (2014) Comparison of open chain and closed chain planar two degree of freedom manipulator for positional error. J Mech Rob 6(2)
34. Kim JW, Jeong S, Kim J, Seo T (2016) Numerical hybrid Taguchi-random coordinate search algorithm for path synthesis. Mech Mach Theory 102:203–216
35. Sun J, Xue N, Liu W, Chu J (2020) A synthesis method for path generation of a planar five-bar mechanism based on dynamic self-adaptive atlas database. Inverse Probl Sci Eng 28(11):1609–1632
36. Sun J, Liu W, Chu J (2016) Synthesis of a non-integer periodic function generator of a four-bar mechanism using a Haar wavelet. Inverse Probl Sci Eng 24(5):763–784
37. Wu X, Wang K, Wang Y, Bai S (2021) Kinematic design and analysis of a 6-DOF spatial five-Bar linkage. Mech Mach Theory 158:104227

Parametric Sensitivity Analysis of Structure Stability: Mathematical Formulation and Analysis

Pranay A. Meka, Harshad Rokhade, Kiran Kumar Mannur, Sangamesh Giramalla Ganiger, and C. V. Chandrashekara

Abstract The present work is on the study of variation in stability region of the beam structure by varying the material, thickness and length. The determined variation in the stability region provides a tool to design engineer in selecting suitable material and size of the beam for a given loading conditions, in establishing stability for the structure. The Bolotin's approach is used to obtain the stability region of the beam. The stability regions are determined analytically using finite element approach. Parametric sensitivity analysis is carried out with different materials, thickness and lengths of a beam. A significant variation in the trend of stability region is observed and reported.

Keywords Beams · Dynamic stability · Parametric load · Stability region · Bolotin's approach

1 Introduction

Dynamic stability of structures is one of the important design criteria in the process of structural design and an indicator of a structural performance. A system with unstable characteristics leads to an increase in response level, resulting in the failure of the system. The structure is designed to exhibit a stable behaviour during working environment and bound by certain limits. The process of designing a structure is to avoid the system unstable behaviour by either varying the loading conditions or by varying the stability region of the system. Varying the loading condition is not feasible in most of the cases, as it is application demanded. Hence, varying the stability region of the system to suit the application plays an important role in designing.

Bolotin [1] authored a textbook summarizing the extensive research work initially carried out by Russian and German researchers, as an English translated version.

P. A. Meka · H. Rokhade · K. K. Mannur · S. G. Ganiger · C. V. Chandrashekara (✉)
Department of Mechanical Engineering, PES University, Bengaluru, India
e-mail: drcvc@pes.edu

P. A. Meka
e-mail: pranayme@udel.edu

V. K. Gupta et al. (eds.), *Recent Advances in Machines and Mechanisms*, Lecture Notes in Mechanical Engineering, https://doi.org/10.1007/978-981-19-3716-3_6

He brought in a step-by-step mathematical procedure to carry out the structural stability analysis. Stability of a structure is determined using the equation of motion expressed in the form of Mathieu-Hill equation in finite element approach for more precise results. Brown et al. [2] carried out a stability analysis of structures using finite element approach. The dynamic characteristics are used to obtain the stability regions of the system. Reddy [3] presented a detailed formulations to obtain dynamic characteristics and developed equation of motion of the beam using finite element method. Majorana [4, 5] developed a finite element mathematical model and obtained stability region of simply supported beam. The Bolotin's formulations are used with finite element method to obtain the boundary frequency equations from equation of motion. Kolukula [6] validated the stability regions by obtaining the response of the system at particular loading conditions, thus validating the formulations used to obtain the stability region of the system.

The present work reports the study of variation in stability region of a beam structure by varying the material, thickness and length. The determined variation in the stability region provides a tool to design engineer in selecting suitable material and size of the beam for a given loading conditions, in establishing stability for the structure. Parametric sensitivity analysis is carried out with different materials, thickness and lengths of a beam. Section 2 details the formulations used to obtain the stability region of the simply supported beam. Section 3 presents the variation in stability region with varying material, thickness and length of the beam. A significant variation in the trend of stability region is observed and are reported.

2 Stability Analysis

A beam of length L, breadth b and thickness t, with density ρ and Young's modulus E, is considered for the mathematical formulations. The beam is pinned at both the ends constraining the transverse degree of freedom at two end nodes of the beam. An external axial dynamic load $P(t)$ is applied along the axis of the column (see Fig. 1).

The axial dynamic load $P(t)$ is given by,

$$P(t) = P_s + P_d \cos \omega t \tag{1}$$

where

P_s Static load component in N

P_d Dynamic load component in N

ω External excitation frequency in rad/s.

The governing equation of motion of the system is given by,

$$M\ddot{q} + Kq - P(t)K_s q = 0 \tag{2}$$

Fig. 1 Beam considered

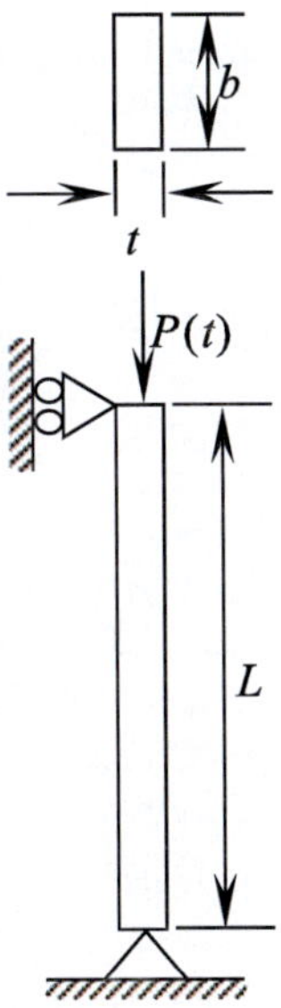

where M, K and K_s are the global mass, stiffness and geometric stiffness matrices of the beam and q is the displacement vector. The boundary frequency equation is obtained using Bolotin's approach [5] and is given by,

$$\omega = 2\sqrt{(M^{-1}K)\left(1 \pm \frac{1}{2}(P_d K^{-1} K_s)\right)} \quad (3)$$

The stability region is obtained using the boundary frequency equation. The region in between the boundary frequency curves gives the various possible combination of dynamic load component P_d and external excitation frequency ω that tends the system to show unstable behaviour.

3 Result and Discussion

Parametric sensitivity analysis for the stability region of the beam is carried out by considering length and thickness of the beam, with two different engineering materials, viz. Aluminium 6082 and structural steel S275N. Table 1 shows the material properties of the materials considered.

Total 16 cases are reported here with first eight cases referring to varying thickness from 0.004 m to 0.007 m with an increment of 0.001 m, while length of the beam being 1 m. Next, eight cases are referring to varying length from 0.7 to 1 m with an increment of 0.1 m, while the thickness of the beam being 0.004 m. For all 16 cases, breadth of the beam is kept constant at 0.03 m. Among 16 cases, first 4 cases and

Table 1 Material properties

Property	Material	
	Aluminium 6082	Structural steel S275N
Density, ρ (kg/m^3)	2700	7850
Young's modulus, E (N/m^2)	70×10^9	2.1×10^{11}
Poisson's ratio	0.3300	0.3046

cases from 9 to 12 refer to beam material Aluminium 6082, and cases 5 to 8 and cases 13–16 are referring to beam material structural steel S275N.

The variation in stability region of the beams for different cases are reported in Sects. 3.1 and 3.2. Section 3.1 focuses on varying thickness cases and Sect. 3.2 reports cases for varying length.

3.1 Varying Thickness

The beam is discretized in to 20 elements, and natural frequencies and critical buckling loads of the beam with Aluminium 6082 material are calculated and are tabulated in Table 2, for cases 1 to 4.

Similarly, the estimated natural frequencies and critical buckling loads for the beam with structural steel S275N material are given in Table 3, for cases 5 to 8.

Using boundary frequency equation, Eq. (3), the stability regions of Aluminium 6082 beam with different thickness values are generated, in MATLAB environment. The excitation frequency ω is considered along x-axis and dynamic load component P_d along y-axis (see Fig. 2).

Table 2 Natural frequencies and critical buckling loads for cases 1 to 4

Mode	Natural frequencies (rad/s)			
	Case 1 ($t = 0.004$ m)	Case 2 ($t = 0.005$ m)	Case 3 ($t = 0.006$ m)	Case 4 ($t = 0.007$ m)
1	58.04	72.55	87.06	101.58
2	233.03	291.28	349.54	407.80
3	531.79	664.74	797.69	930.64
4	1030.50	1288.12	1545.75	1803.37
5	1637.97	2047.46	2456.96	2866.45
6	2590.24	3237.80	3885.36	4532.92
Critical buckling load, P_{cr} (N)				
	110.60	216.01	373.26	592.73

Table 3 Natural frequencies and critical buckling loads for cases 5 to 8

Mode	Natural frequencies (rad/s)			
	Case 5 ($t = 0.004$ m)	Case 6 ($t = 0.005$ m)	Case 7 ($t = 0.006$ m)	Case 8 ($t = 0.007$ m)
1	58.96	73.70	88.44	103.18
2	236.71	295.89	355.06	414.24
3	540.19	675.24	810.29	945.34
4	1046.78	1308.47	1570.17	1831.86
5	1663.85	2079.81	2495.77	2911.74
6	2631.16	3288.95	3946.75	4604.54
Critical buckling load, P_{cr} (N)				
	331.79	648.02	1119.79	1778.18

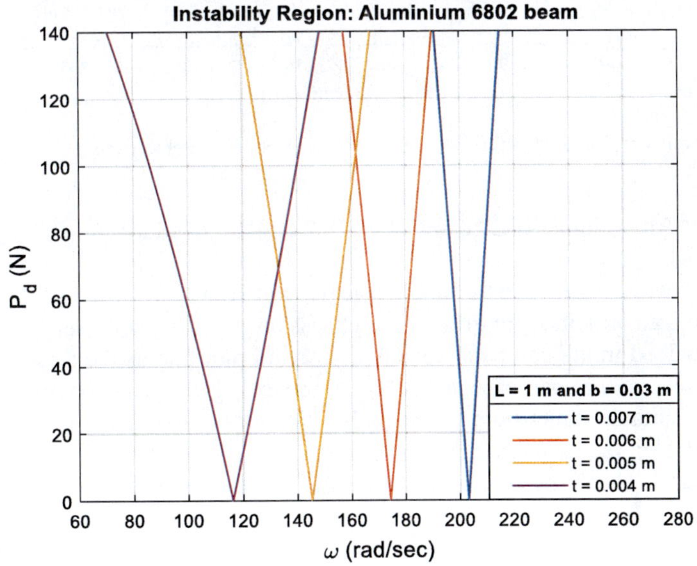

Fig. 2 Stability region of Aluminium 6082 beam with different thickness

Similarly, the stability regions of structural steel S275N beam with different thickness values are generated with excitation frequency ω along x-axis and dynamic load component P_d along y-axis (see Fig. 3).

In both the plots (see Figs. 2 and 3), it noticed that there is a considerable shift in the entire stability region towards right as the thickness of the beam increases. It is obvious due to the reason, as the thickness of the beam increases, the stiffness of the beam increases and is in turn increase in the natural frequency of the system. Further, a comparison of stability regions with varying thickness of beam is generated, considering the ratio of excitation frequency ω and the first natural frequency ω_1 of the beam along x-axis and dynamic load component P_d along y-axis. The stability

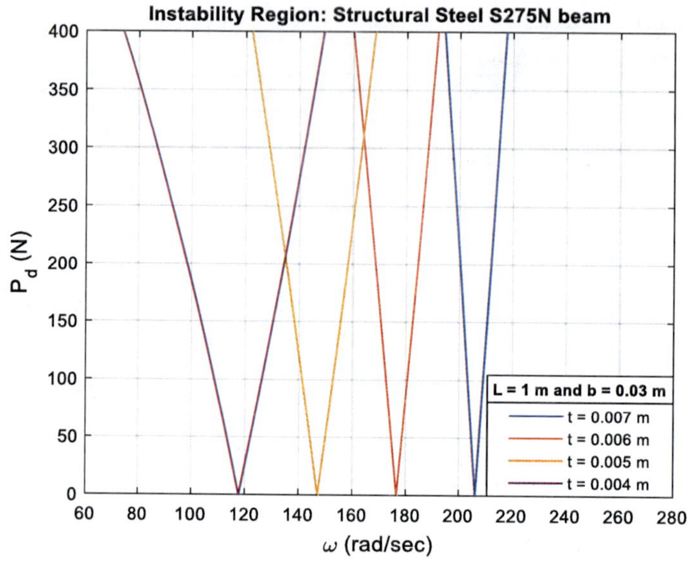

Fig. 3 Stability region of structural steel S275N beam with different thickness

regions of Aluminium 6082 beams of cases 1 to 4 are overlapped in a single graph (see Fig. 4).

Similar way, overlapped the stability regions of structural steel S275N beams of cases 5 to 8 are obtained for different thickness on a single graph (see Fig. 5).

It is noticed in the comparative stability region plot that as the thickness of the beam increases, the span of unstable region reduces, giving a scope for designer to select the suitable dimensions, to meet the stability criteria. The rate at which the unstable region reduction is much faster as the thickness of the beam increases. The variation in the stability region of the beams with different lengths is reported in Sect. 3.2.

3.2 *Varying Length*

The beam is discretized into 20 elements, and the natural frequencies and critical buckling loads of Aluminium 6082 are obtained analytically using MATLAB and are given in Table 4, for cases 9 to 12.

Similar way, the natural frequencies and critical buckling loads of structural steel S275N beams considering *cases* 14 to 16 are given in Table 5. In both the cases, thickness and width of the beam are unchanged.

The stability regions of Aluminium 6082 beam with different lengths as indicated in cases 9 to 12 are plotted with excitation frequency ω along the x-axis and dynamic load component P_d along y-axis (see Fig. 6).

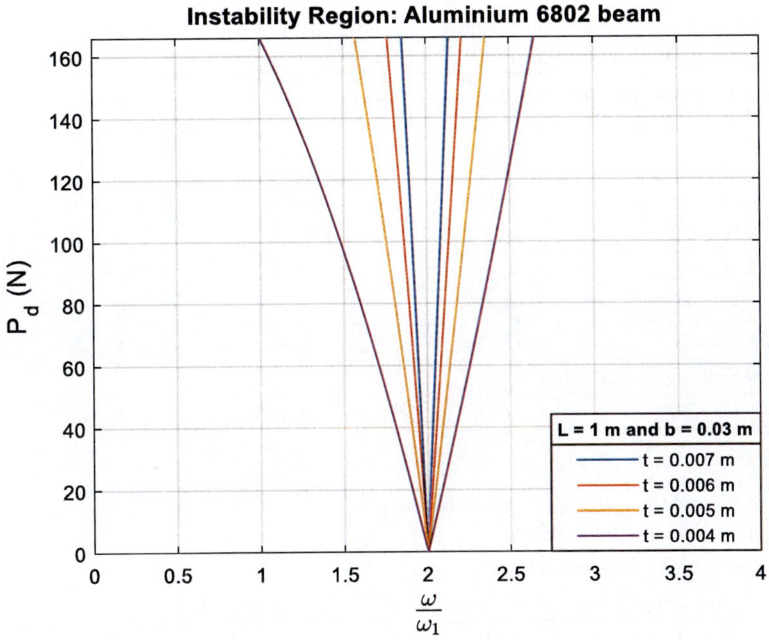

Fig. 4 Comparative stability region of Aluminium 6082 beam of different thickness

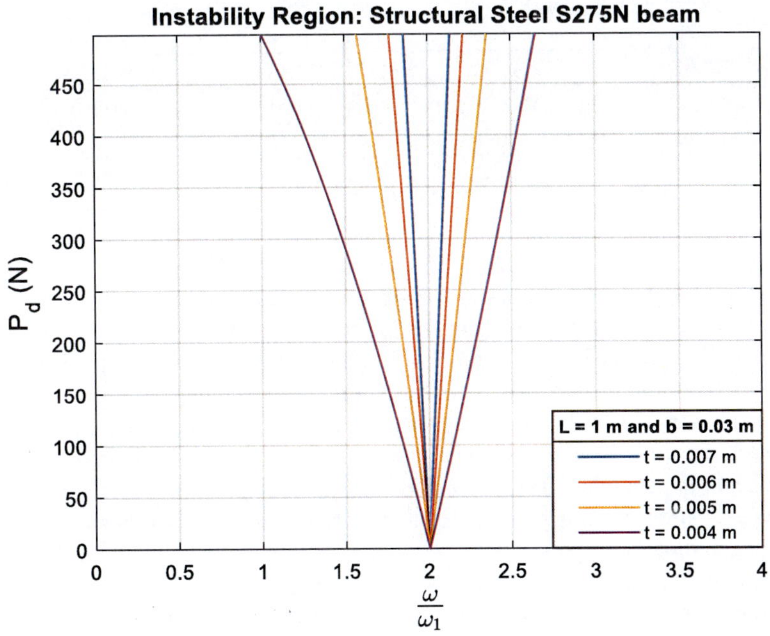

Fig. 5 Comparative stability region of structural steel S275N beam of different thickness

Table 4 Natural frequencies and critical buckling loads of Aluminium 6082 beam of different lengths [$b = 0.03$ m and $t = 0.004$ m]

Mode	Natural frequencies (rad/s)			
	Case 9 ($L = 0.7$ m)	Case 10 ($L = 0.8$ m)	Case 11 ($L = 0.9$ m)	Case 12 ($L = 1.0$ m)
1	118.45	90.69	71.66	58.04
2	475.57	364.11	287.69	233.03
3	1085.29	830.93	656.53	531.79
4	2103.06	1610.15	1272.22	1030.50
5	3342.80	2559.33	2022.19	1637.97
6	5286.21	4047.25	3197.83	2590.24
Critical buckling load, P_{cr} (N)				
	225.71	172.81	136.54	110.60

Table 5 Natural frequencies and critical buckling loads of structural steel S275N beam of different lengths [$b = 0.03$ m and $t = 0.004$ m]

Mode	Natural frequencies (rad/s)			
	Case 13 ($L = 0.7$ m)	Case 14 ($L = 0.8$ m)	Case 15 ($L = 0.9$ m)	Case 16 ($L = 1.0$ m)
1	120.33	92.12	72.79	58.96
2	483.08	369.86	292.23	236.71
3	1102.44	844.05	666.91	540.19
4	2136.28	1635.59	1292.32	1046.78
5	3395.61	2599.77	2054.14	1663.85
6	5369.72	4111.19	3248.35	2631.16
Critical buckling load, P_{cr} (N)				
	677.12	518.42	409.62	331.79

Similarly, the stability regions of structural steel S275N beam with different lengths are plotted with excitation frequency ω along x-axis and dynamic load component P_d along y-axis (see Fig. 7), for cases 13 to 16.

In both the plots (see Figs. 6 and 7), it is noticed that there is a shift in the stability region towards left, i.e. towards lower excitation frequency region as the length of the beam increases. This is due to the reason that decrease in the natural frequency of the system. The comparison of stability regions with varying length for both the material of the beam is generated. The overlapped stability region is plotted considering the ratio of excitation frequency ω and first natural frequency ω_1 of the beam along x-axis and dynamic load component P_d along y-axis. The stability regions of Aluminium 6082 beams of *cases* 9 to 12 are plotted on a single graph (see Fig. 8).

Similarly, overlapped stability regions plot for structural steel S275N beams with dimensions indicated in *cases* 13 to 16 are plotted on the same graph (see Fig. 9).

It is noticed that as the length of the beam increases, the span of unstable region increases. Shorter beam length has more scope for stability region for the structure.

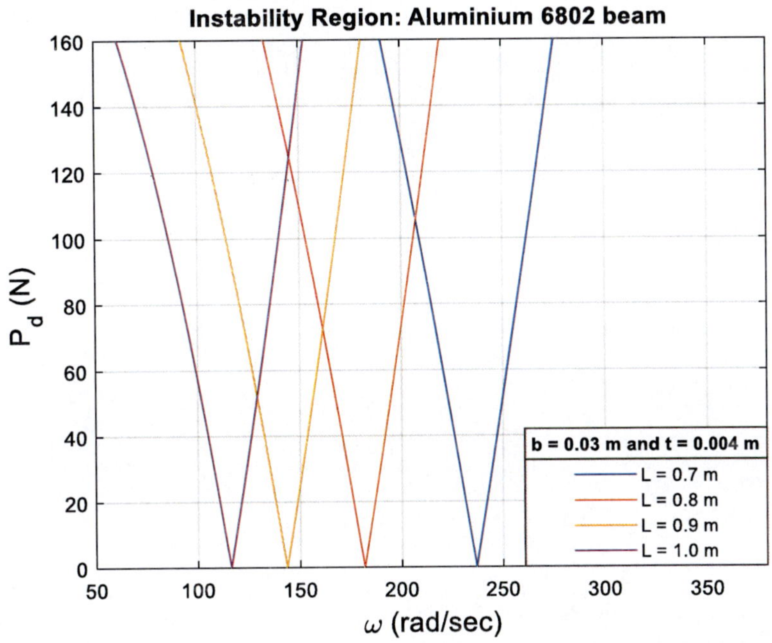

Fig. 6 Stability region of Aluminium 6082 beam of different lengths corresponding to first natural frequency

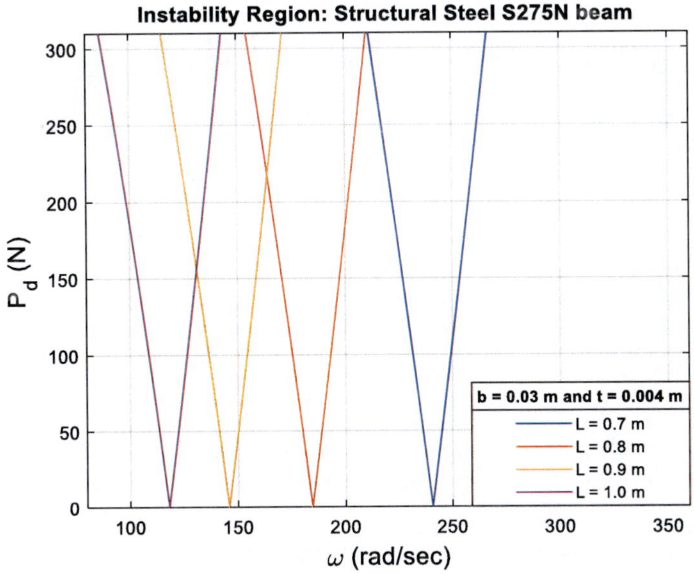

Fig. 7 Stability region of structural steel S275N beam of different lengths corresponding to first natural frequency

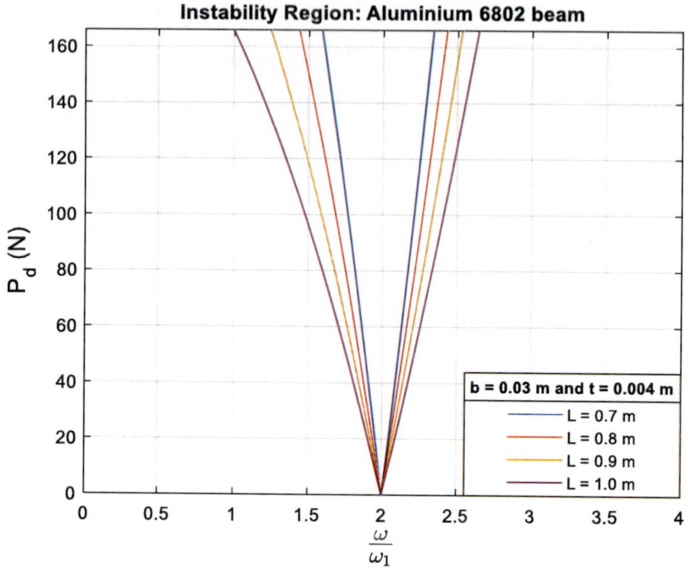

Fig. 8 Stability region of Aluminium 6082 beam of different lengths corresponding to first natural frequency (p_d vs ω/ω_1)

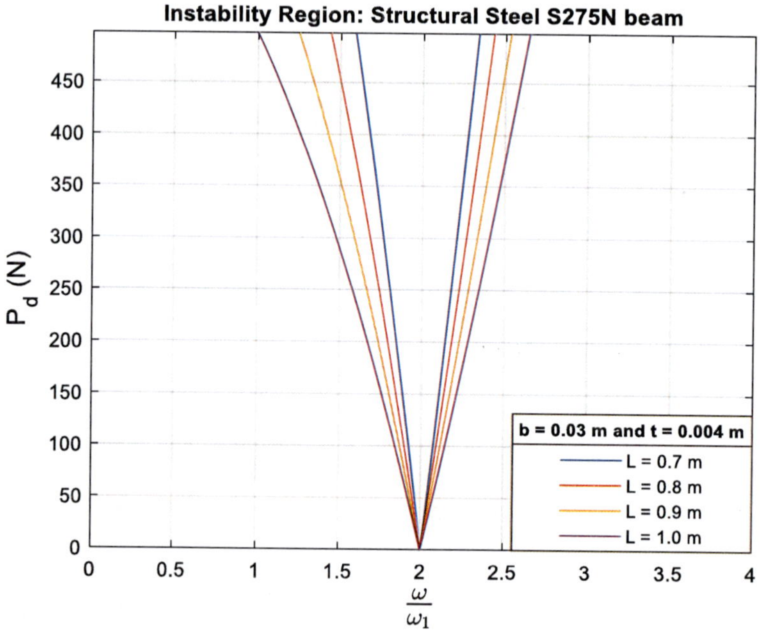

Fig. 9 Stability region of structural steel S275N beam of different lengths corresponding to first natural frequency (p_d vs ω/ω_1)

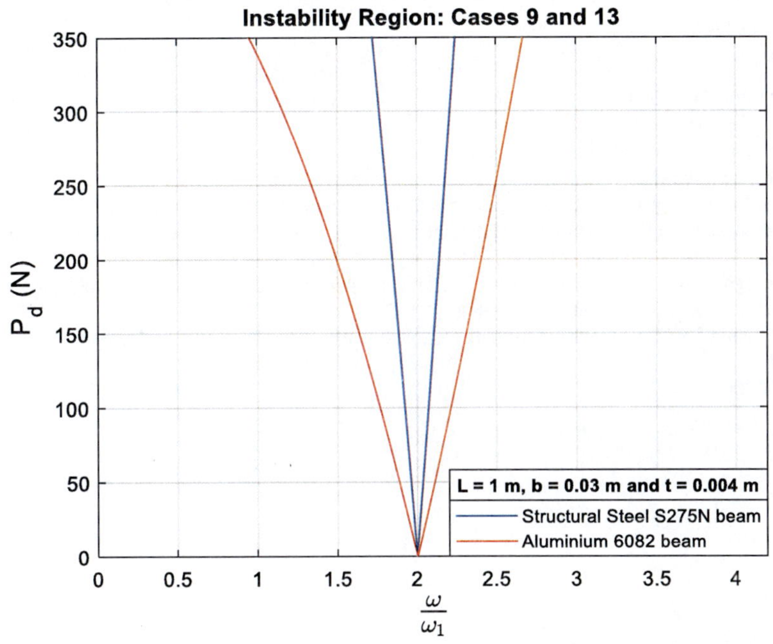

Fig. 10 Stability regions of beams of *cases* 9 and 13 corresponding to first natural frequency

With the available data/plots, the stability region of Aluminium 6082 and structural steel S275N beams with same geometrical parameters as mentioned in cases 9 and 13, respectively, is plotted on a same graph (see Fig. 10).

It is noticed that the structural steel S275N exhibits a lesser unstable region compared to Aluminium 6082, with same dimensions, for same magnitude of dynamic load conditions.

4 Conclusion

Successfully, parametric sensitivity analysis for stability region of two different engineering materials is carried out. A significant shift in the stability region and its span is noticed as the change in the thickness and length of the structure independently changed. Overlapped stability region graph for both varying thickness and length of the beam is reported, and scope for the selection of dimension to meet the stability criteria is indicated. Span of instability region for two engineering materials, viz. Aluminium 6082 and structural steel S275N with same geometrical parameters is obtained and reported.

References

1. Bolotin VV (1964) The dynamic stability of elastic systems, Holden Day. San Francisco, CA, London, Amsterdam
2. Brown JE, Hutt JM, Salama AE (1968) Finite element solution to dynamic stability of bars. AIAA J 1423–1425
3. Reddy JN (2006) An introduction to the finite element method. McGraw-Hill
4. Majorana CE (1997) Pellegrino: dynamic stability of elastically constrained beams: an exact approach. Eng Comput 14(7):792–805
5. Briseghella L, Majorana CE (1998) Pellegrino: dynamic stability of elastic structures: a finite element approach. Comput Struct 69(1):11–25
6. Kolukula SS (2013) Investigation of parametric instability in elastic structures

A Review on Frequency Domain Analysis Approach for Parametric Identification of Nonlinear Joints

Faisal Hussain and Sanjay Ingole

Abstract Structural joints are regarded as a potential source of nonlinear behavior. Parametric identification of joint properties is an important task in predicting the dynamic characteristics of mechanical and structural joints. The objective of this research survey is to present frequency domain dynamic analysis method for nonlinear system parametric identification in structural dynamics and the key developments in this area of research. Finally, identification of findings is done which provides a pathway for the future direction of research work in this area.

Keywords Nonlinear joints · Parametric identification · Dynamic analysis · Frequency domain analysis · Harmonic balance method · Perturbation method · Describing function method

1 Introduction

Most engineering systems are exposed to some degree of nonlinearity, which is induced by a variety of factors including structural joints with looseness or friction characteristics, boundary conditions that impose varying stiffness limits, amplitude-dependent materials, and components such as shock absorbers, vibration isolators, bearings, and linkages, among others. In addition, engineering systems typically show some degree of nonlinearity, and joints are the most common cause of nonlinearity in structures; nonlinearities can modify the structural reaction significantly at times.

In the presence of nonlinearity, nonlinear joint parameters device recognition attempts to create highly accurate mathematical models from input and output measurements taken on the actual structure. Nonlinear system identification is a large field of research that has recently piqued the attention of the structural dynamics community [1, 2]. The exact parameters of a model are required to study the

F. Hussain (✉) · S. Ingole
Department of Mechanical Engineering, Government College of Engineering, Chandrapur, Maharashtra 442403, India
e-mail: fshppn@gmail.com

V. K. Gupta et al. (eds.), *Recent Advances in Machines and Mechanisms*, Lecture Notes in Mechanical Engineering, https://doi.org/10.1007/978-981-19-3716-3_7

behavior of a mechanical device, but determining these parameters using only theoretical methods is extremely difficult [3]. As a result, several experiments have been conducted in this area using experimental approaches to define the parameters [4]. Many mechanical structures, in particular, are made up of several parts joined by different types of nonlinear joints. Although each component's dynamic model is relatively reliable and all subsystems are fairly linear in this situation, the whole system is nonlinear, making the dynamic analysis of the whole system difficult [5].

Most engineering systems are essentially nonlinear rather than linear, such as aerospace devices, rotor bearing components, nonlinear suspension system, and electromechanical systems, and nonlinear dynamic analysis of these systems requires the inherent nonlinearity of the interfaces [6, 7]. Any structural assembly must be connected by certain means, such as bolting, welding, and riveting, or by more complex fastenings, such as smart joints, and the dynamic response of mechanical systems is entirely influenced by joints, which add frictional damping and localized versatility [8, 9]. In certain cases, a linear system can be used to partially analyze a nonlinear system, making nonlinear analysis unavoidable. As a result, nonlinear models are critical in mechanical and structural mechanics for a full understanding and precise prediction of motion [10, 11]. Any linear model would be wrong if there is nonlinearity in the equation, so it is important to identify nonlinearity using experimentally measured responses [12, 13]. Connectors are used in spacecraft to tie and hinge substructures together. Since the connector has nonlinear properties, it has a significant impact on the spacecraft's dynamic characteristics [13]. Nonlinear effects are becoming increasingly significant in modern engineering architectures that are lightweight and scalable [14]. The linear model updating process would not actually converge correctly if nonlinear effects with their parametric detection are not taken into account, and the resulting model predictions will be incorrect [15, 16]. Various criteria, such as the frequency response diagram and the hysteresis loop, are used to calculate the nonlinear influence of the joint contact surface on the overall action of the structure. An optimal equal linear frequency response function can also be used to determine the dynamic parameters of a nonlinear joint [17]. Using force state mapping methods, it is possible to identify a nonlinear joint model [18], Runge–Kutta method, Iwan modeling [19] to represent the nonlinearity of joint, frequency response function (FRF) decoupling method for nonlinear systems (FDMNS) [20] for computing FRFs of a substructure [21] decoupled from a coupled nonlinear structure where it is possible to model nonlinearity as a single nonlinear function, [22]. Also by using the coupling-identification method, the accuracy of joint identification can be improved significantly [23, 24]. Nonlinear structural modification method is always vital in the parametric identification of parameters, especially in large engineering structures [25]. Reduced-order models for nonlinear systems have gained popularity due to their potential to significantly minimize computing costs as compared to full nonlinear finite-element models [26]. Centered on the principle of multi-harmonic equilibrium, a general-purpose methodology for identifying the dynamic characteristics of nonlinear joints has been developed and validated, and the nonlinear characteristics of joints have been successfully identified [27]. Nonlinear parameters can also be found by inverting describing functions, in which the nonlinear variable

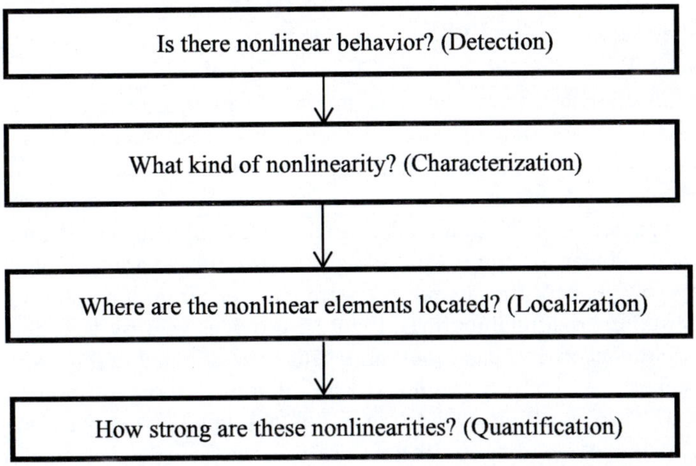

Fig. 1 Identification process [1]

is modeled as a cubic spring with quadratic damping [28]. Substructure synthesis theory was used to derive the frequency equation for nonlinear entity parametric detection, and amplitude response was calculated in the nonlinear area using curve fitting technique, which found a strong agreement between experimental and analytical values [29]. The Iwan model is also used to further explain the principle of nonlinear properties of elements like micro-slip, macro-slip, and pinning, as well as to reliably replicate experimental joint phenomena [30].

As seen in Fig. 1, the recognition process progresses through four phases: diagnosis, characterization, localization, and quantification of parameters [1].

In the above detection method, a complete proof standard operating procedure is established, and it is explicit that if nonlinear behavior is observed, the mechanism is said to be defined in terms of the type of nonlinearity present, its position, and how the functional form of nonlinearity can be extracted. The magnitude of nonlinearities in the device was filtered through the quantification process.

The primary goal of this paper's focused literature review is to assess the current state of parametric recognition of nonlinear joints and to establish a method for identifying research gaps. The aim is to disseminate numerous research areas in the field of nonlinear joints parametric recognition that have been selected by researchers. The proposed study is intended to aid in the identification of specific areas where work or testing is severely lacking. This would be particularly useful for researchers working on parametric recognition of nonlinear joints. The aim of this study is to compare modeling approaches for dynamic analysis of nonlinear structures and to determine the best approach for investigating nonlinear joint parametric detection.

2 Research Methodology

In order to establish a framework for its classification, a review of the literature on parametric recognition of nonlinear joints was undertaken. After consulting studies conducted between 2000 and 2020, research publications in the field of parametric detection of nonlinear joint parameters involving different modeling methods, representative friction models, geometric nonlinear models, and dynamic analysis of nonlinear structures were established through a variety of databases. The related research papers for this project were gathered from a number of reputable peer-reviewed international journals. Figure 2 provides year-by-year figures for the number of articles on nonlinear joint parameters investigated in this article.

Each paper was carefully examined, and the most important observations were recorded. The use of different modeling techniques and their comparisons for investigations of nonlinear joint parameters are presented in detail in the subsequent sections. Finally, the findings of the study have been outlined in order to identify gaps and provide a road map for researchers in this area. Figure 3 depicts the actions taken to carry out the study.

Fig. 2 Number of publication reviewed

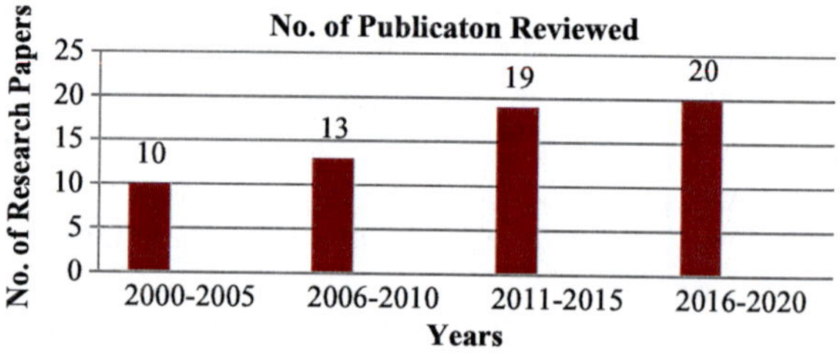

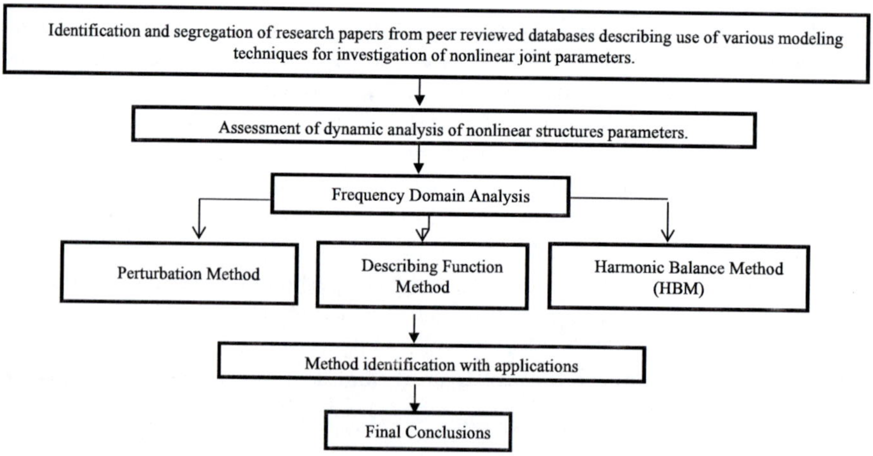

Fig. 3 Footsteps used to conduct the review

3 Dynamic Analysis of Nonlinear Structures

3.1 Frequency Domain Analysis

The concept "frequency domain" refers to the study of mathematical functions or signals in terms of frequency.

A robust finite-element model was set up to provide an understanding of different slip-stick processes in the contact region and the model was used to calculate the dynamic response of assembled assemblies, taking into account the effects of micro- and macro-bolted joint slip. If appropriate information of joint is available, this FE-module can be used to predict the response of assembled structures [31]. Researchers have also changed structural components from ordinary beams to layered beams, which have been studied using ANSYS-Workbench version 17.2, which gives an accurate description of layered structure behavior [32]. The researchers suggested an expanded greenwood model (EGM), which was related to the Iwan model, and the Coulomb friction model and characteristic values were compared [33]. In Ref. [34], the experimental and theoretical research measured the nonlinear effects of a conventional shear lap joint on the dynamics of two systems and it was concluded that the adjusted Iwan beam element (AIBE) has the potential to accurately describe joint effects in the dynamics of real-life structures. The structural model was put through multiple dynamical experiments to determine the nonlinear effects of the joints. The studies reveal some significant effects on the active stiffness and damping of the lap joints [35]. The harmonic balance method which is easy to model was used to investigate nonlinear simulation of bolted flange joints, as well as the mathematics involved in deriving equal stiffness and damping properties and research can be extended by considering these parameters which can be accomplished by probabilistic method [36]. The authors proposed an effective regression-based approach for estimating joint stiffness parameters, in which only natural frequencies are used to extract joint stiffness parameters without considering the rotational nonlinearity which is the area of research [37]. The authors have suggested a recognition approach that uses the calculated frequency response functions (FRFs) specifically to define the joint properties in their research work to minimize measurement noise in frequency response functions [38]. Boundary parametric identification through model updating is tough to parameterize joints and boundary conditions as stiffness parameters have a tendency to make measurements insensitive. Also, geometric parameters have a lot of possibilities in terms of updating. They have a physical meaning, and changes in them affect the measurements [39]; reduced-order characteristic equations on a beam with elastic supports [40] were identified by researchers in their work.

In addition, a two-parameter and three-parameter joint model with cross-coupling for a bolted joint was modeled by assuming a pair of translational and rotational springs, and a frequency equation was derived for the device using the principle of substructure synthesis; further, analytical and experimental findings were compared by case studies [41]. A new approach was adopted using the examples of the beam

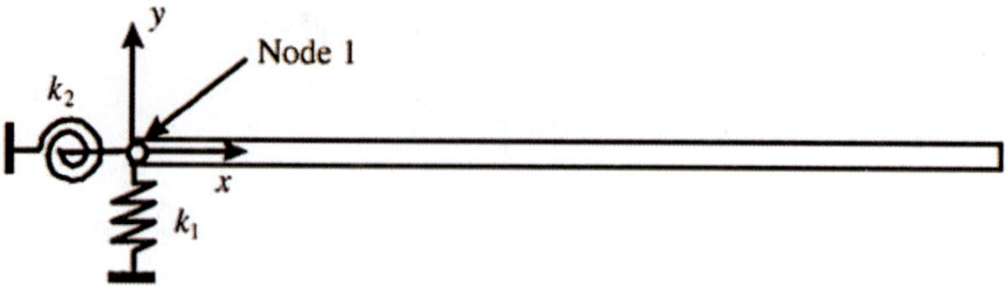

Fig. 4 Beam structure elastically constrained at one end [46]

system and the cylinder plate system, which reduces the influence of flexible construction from the existing paradigm. The modal substructuring is improved with this process [42] as well as the principle of substructure modal synthesis, which was used to reduce the amount of degrees of freedom in a beam configuration [43]. By using the example of a plane frame structure with a reconstruction of interface forces from the complete scheme, a new approach for substructure recognition was validated [44].

The Euler Bernoulli beam was considered in Ref. [45], and an approximate analytical solution was discovered using the perturbation technique. The authors proposed a new model for determining joint stiffness, in which the first reduced-order characteristic polynomial [ROCP] was described in terms of natural frequency and was used to determine elastic constraints imposed on a beam structure, as shown in Fig. 4 [46].

The authors explain the method of nonlinear parametric identification with the application of large transport aircraft by modal testing in their research work, which included single-mode and coupled-mode nonlinear identification, as well as the magnitude of measured and restoring forces [47]. One of the most difficult aspects of recognizing joint properties in depth is that individual errors in the measured data will have a significant effect on the result. The authors have shown a global joint recognition strategy on which full discussions have been performed, and the reaction of the frequency response feature would be exact if the identification is accurate [48]. The authors developed a new method of joint recognition based on rigid body dynamics and frequency response function (FRF) measurements, in which components of stiffness calculation can be performed simultaneously in three translation and three rotational directions, resulting in a 6×6 stiffness matrix for the joint [49]. In Ref. [50], authors considered a bolted lap joint model as shown in Fig. 5 and established a finite-element model by considering micro-slip and micro-slaps subjected to differing loading conditions. A nonlinear thin-layer element was added for modeling touch interfaces, nonlinear parametric identification was completed using the (FRF) sensitivity method, and it was found to have predicted the experimental effects with high precision.

Friction is the primary cause of nonlinearity and energy dissipation in assembled structure interaction interfaces. The primary goal was to create a model that could simulate the dynamics of friction surfaces as well as energy dissipation caused by slipping [51]. The researchers presented a general formulation based on Von Karman's

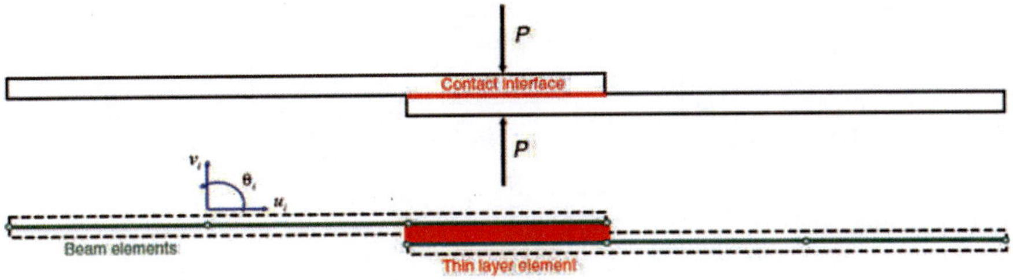

Fig. 5 Structure including a contact interface and its corresponding FE model [50]

equations and suggest a solution approach for forced vibrations based on the Krylov–Bogoliubov linearization method [52]. In addition, using the Valanis elastoplastic principle, the dynamic properties of a beam with frictional contact support were examined, and contact parameters was defined using force state mapping [53] and developed analytical solutions for the steady-state reaction of an infinite beam by using the Euler–Bernoulli beam theory [54].

As a result, it can be concluded that correct simulation of the stiffness and damping characteristics of surface-to-surface contact interfaces is critical for dynamic reaction study of assembled structures. However, owing to the development of nonlinear behaviors such as slip and slap processes, modeling and study of touch interfaces is a challenging task.

The authors used a new method to assess the damping of structural joints, and the FRF was used for joint recognition to solve the flaws involved with slip boundaries identification in the joint interface. This approach was used for bolted structures under both translational and torsional excitations as shown in Fig. 6 under varying loading conditions [55].

The authors define an empirical approach for predicting damping and contact stiffness in bolted joints, in which a torsional contact model is constructed as shown in Fig. 7, which can predict the complex behavior of various types of machine tools subjected to torsional loading conditions, and the proposed method can be used to forecast compressive and shearing vibration modes in bolted connections. The augmented model can be used to anticipate the dynamic behavior of real machine tools, as well as to optimize the design [56].

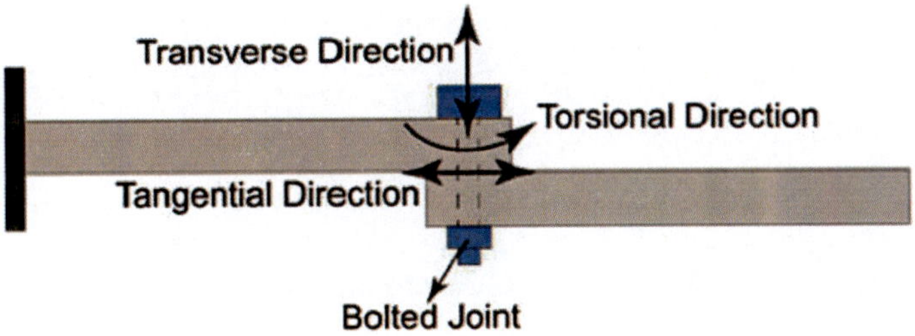

Fig. 6 Schematic of an assembled structure showing the directions of the joint properties [55]

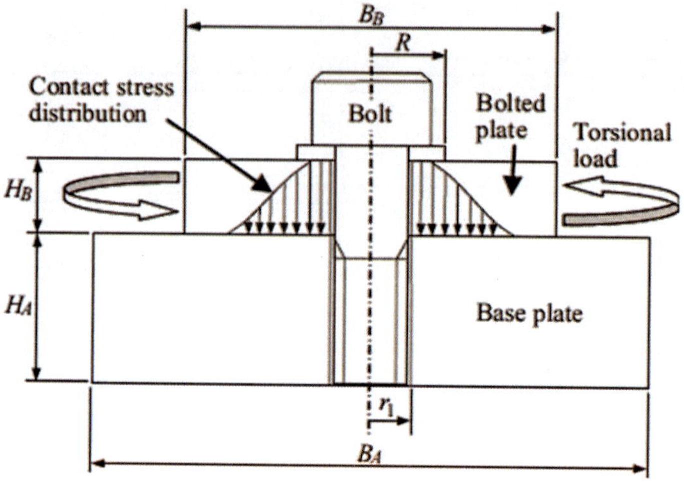

Fig. 7 Bolted plate and base plate with torsional load [56]

To understand the coupling effects under shear and normal directions of contact, the authors have introduced a zero thickness model interface that has six parameters involving stick, micro-slip, slide, and slap actions within the contact surfaces. The model does not require any integration computations, and its implementation is simple [57]; some researchers have modeled the functional behavior of bolted joints by means of joint interface technique as shown in Fig. 8, using experimentally assessed data to define parameters. The suggested method for nonlinear modeling effects in joint interfaces is precise, requires few calculations, and can be implemented with existing FE tools [58].

The researchers [59] have also modeled adhesively bonded joints as shown in Fig. 9 by using two-dimensional finite elements and were concluded that as layer thickness of adhesive increases, the natural frequencies decreases in a slight manner, and the modes obtained were of bending and longitudinal pattern.

Also, a layer-wise plate finite-element method was proposed and extended to model the lap joint in which the joint was treated like a sandwich plate as shown in Fig. 10a and b.

The proposed approach was found to be capable of accurately describing the vibratory existence of lap joints with composite adherents and viscoelastic adhesive, and it was concluded that the adhesive loss factor, overlap area, adhesive thickness,

Fig. 8 FE model of test structure [58]

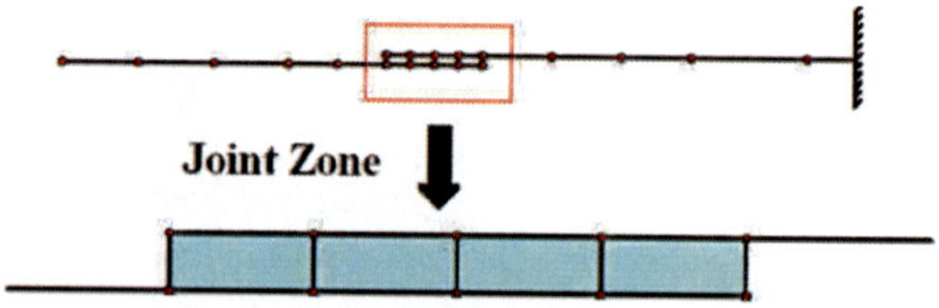

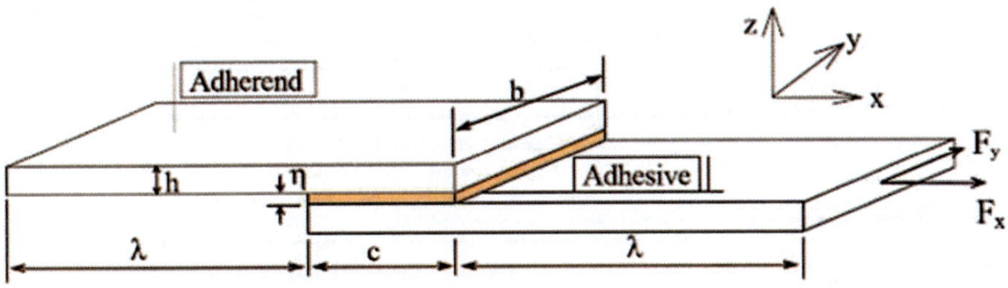

Fig. 9 Adhesive-bonded single-lap joint system [59]

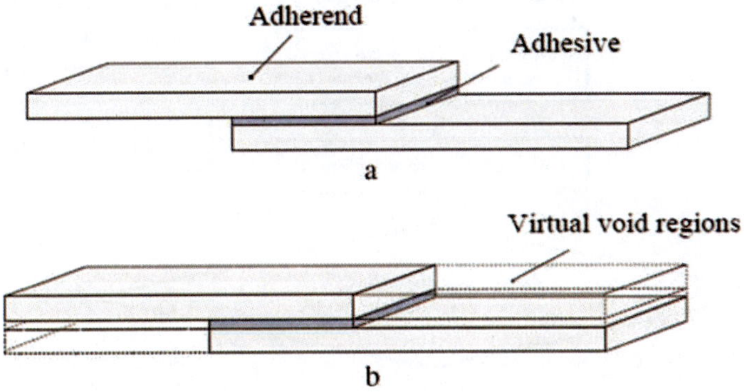

Fig. 10 **a** Single-lap joint [60]. **b**. Equivalent sandwich plate [60]

step height ratio, and step number parameters are all closely related to the joints free vibration behaviors. The sole issue in the current method is that the forecast of modal loss factor for lateral modes of vibration does not always match the corresponding reference result [60].

In Ref. [61], the results of geometrical parameters were calculated using the finite-element method and the back propagation artificial neural network (ANN). The authors demonstrated a collaborative parameter recognition method using a block diagram in which the correct parameters can be identified (Refer Fig. 11). The suggested approach was used to determine the parameters of a longitudinal joint that was used to tie plates together but during damping coefficient determination, the accuracy of identification degrades [62].

To analyze, compute, and model nonlinearities in the structural joint, a hybrid approach is used to perform parametric detection, but it needs the governing equation of dynamic system [63], component mode synthesis (CMS) [64] bispectrum technique to detect the nonlinear behavior of a system which is one of the most powerful tool in frequency domain as it effectively detects the type of nonlinearity in a structural joint [65] Volterra series models for a nonlinear system and interpret the results of the output frequency response [66] and nonlinear output frequency response functions (NOFRFs)-based approach for crack determination. This method

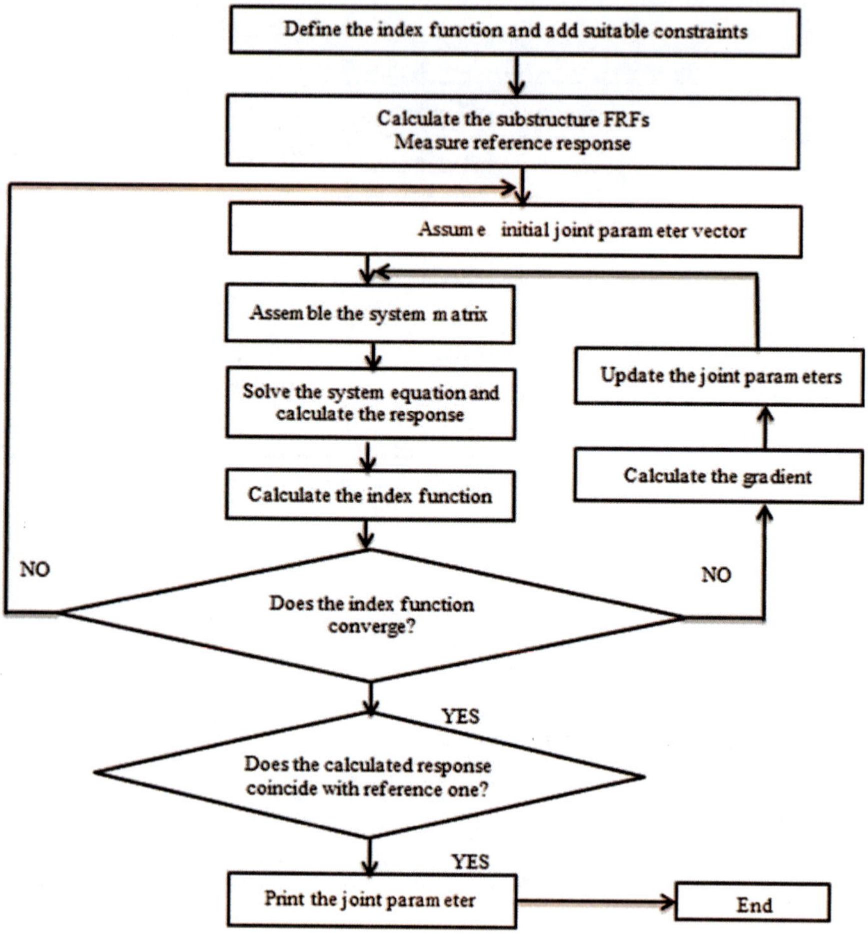

Fig. 11 Joint parameter identification procedure [62]

can be utilized for a variety of nonlinear system frequency investigations, such as vibration fault diagnostics, analysis of nonlinear model and monitoring [67]. In ref [68], authors developed a novel concept by taking example of threaded fasteners, in which resistant to the loosening effect of fasteners under vibration was restricted by modifying the thread geometry of fasteners, and finite-element analysis was done in order to check the deformation is within the elastic zone and to determine the additional torque required to overcome the interference.

Table 1 depicts the various method used in frequency domain analysis for experimental investigation of nonlinear joint parameters.

Table 1 Different methods with their experimental set-up demonstrating process

Method	Set-up	Tools	Remarks
Finite-element method and advanced lumped parameter model [31]	Isolated bolted joint	FE software package	Micro-slip and macro-slip in joints are being investigated
FEM-based ANSYS-workbench version 17.2 [32]	Layered beam	ANSYS model	Bolted joint damping capacity of structures are being investigated
Micro-contacts interaction mechanism [33]	Tribometer set	Expanded green wood model	Experiment with simulation and calculating dissipated energy to determine the damping ratio
Adjusted Iwan beam element [34]	Beam structures with shear lap joint	Modal analysis, energy dissipation analysis	To investigate the effect of a shear lap joint on the dynamic response of a structure
Euler–Bernoulli beam [35]	Single-lap adhesive joint	Euler–Bernoulli beam theory	The effect of the joint was sensed and the device natural frequencies were computed by beam experimentation
Harmonic balance method [36]	Aero engine structure	Finite-element model	Experiments on nonlinear frequency response and nonlinear mutual parameter recognition of a complex structure
Multiple scale method, perturbation technique [45]	Simply supported beam	Perturbation technique	End conditions are described using perturbation theory, and an empirical solution is proposed using the approach of multiple scales
Reduced-order characteristic polynomial [46]	Beam structure	Modeling the structure using reduced-order characteristic polynomial	There is no need for modal matching, and the method mentioned can also be used as an advanced optimization formulation technique

(continued)

Table 1 (continued)

Method	Set-up	Tools	Remarks
Rigid body dynamics theory [49]	Tool holder interface	Rigid body dynamics theory	For six mobility motions, the stiffness of a rigid coupling may be calculated
Finite-element modeling [50]	Bolted lap joint	Finite-element model, frequency response functions (FRF)	Nonlinear parameters of contact surfaces are investigated
Extended Hamilton's principle [51]	Free-frictionally supported beam	Analytical approach	Energy dissipation for bolted beam is evaluated under vibratory motion
Experimental modal analysis (EMA) [55]	Bolted lap joint	Frequency response function and hysteresis loop approach	The detection of damping parameters in structural joints is determined
Analytical model [56]	Structural joint	Efficient calculation algorithm	The forecast of nonlinear contact stiffness and friction damping is described in bolted connections
Analytical contact model [57]	Contact interface model	Reduction technique, L-curve method	Deformation prediction and force restoration in a contact device model
Nonlinear transient module [58]	Bolted lap joint	Eigen-sensitivity method	Nonlinear effects in mechanical joint interfaces are established
Finite-element model [59]	Adhesive-bonded joints	Harmonic analysis, structural damping	Structure dynamic characteristics are calculated
Finite-element method [60]	Adhesive-bonded joints	Finite-element method	Parametric studies to estimate the mechanical properties of an adhesive-bonded lap joint
Finite-element method, artificial neural network (ANN) method [61]	Adhesive joint	Genetic algorithm	A vibration analysis was conducted on the joint
Hybrid method [63]	Single degree of freedom system	Mat lab	Judging dynamic parameters of structural system

(continued)

Table 1 (continued)

Method	Set-up	Tools	Remarks
Component mode synthesis (CMS) method [64]	Beam with damper	Semi-analytical harmonic balance method, model reduction technique	Parametric identification of complex structures
Frequency domain analysis [65]	Mass, spring, damper system (3DOF)	Homogeneity test, coherence function, Hilbert transform of frequency response function (FRF)	Using frequency domain techniques identify and measure nonlinear structural efficiency
Finite-element analysis [68]	Threaded fasteners	Solid modeling and finite-element analysis	Loosening effect of threaded fasteners under vibration is restricted by introducing anti-loosening feature based on threaded bolt geometry

4 Result and Analysis

In the previous part, the practical analysis of nonlinear systems was explored in terms of their methods, which was detailed in numerous figures and table. The definition clearly describes the study type, research area, problem-solving methods, and other factors that the authors used in their respective research projects. The investigation of nonlinear joint parameters in structural joints is comprehensively represented in the literature. It can be shown that each model is equally dependent on their range and is based on operating conditions, the level of precision needed, algorithm shortcomings, realistic use, and so on. Stiffness of the cubic form is the most common. In translational stiffness, nonlinearity is used to illustrate parametric recognition, in which the force–displacement relationship can be written as $F = K_1U + K_3U^3$ known as cubic stiffening effect. The characteristic plot shows that as the forces increase, the displacement decreases while the equivalent stiffness increases as shown in Fig. 12.

Geometric nonlinear models also have polynomial stiffness and piecewise linear stiffness, which can be determined based on their functional use.

5 Conclusions and Future Directions for Work

This survey paper reviews past and recent developments in parametric identification of nonlinear structural joints. This survey paper examines historical and current advances in nonlinear structural joint parametric recognition. In comparison to

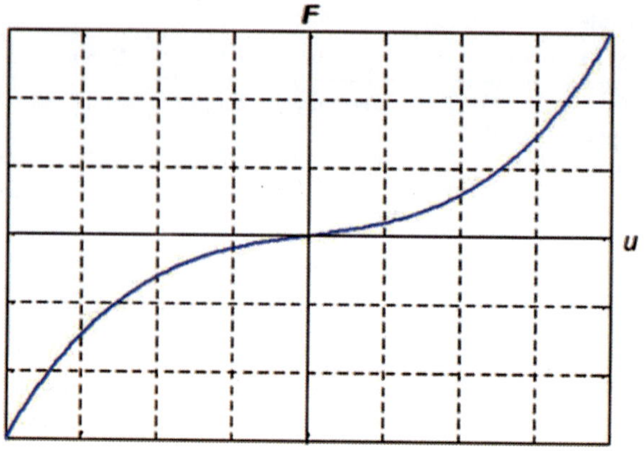

Fig. 12 Characteristic plot of a typical cubic stiffness

the Valanis model, which does not include any physical explanation of the friction phenomenon, more emphasis is placed on Coulomb friction models and Iwan Jerkins beam model. The harmonic balance method has been shown to be effective and precise, and it is capable of dealing of problems with extreme nonlinearity. The Volterra series can be used to parametrically identify structural joints that have nonlinearity characteristics. In addition, comprehensive study on nonlinear system dynamic response characteristics is expected to aid important structural architecture and mechanical applications in future projects. A hybrid approach may also be developed to control the parameters of joint identity if the governing differential equations of a dynamical system are correctly compatible with the depending parameters.

The pathway for researchers is mentioned below:

- The Iwan beam model can be used to investigate bolted joint beam systems in greater depth.
- The damping parameters of the joint can be used in the current model and a satisfactory approach for solving the resulting complex frequency equation can then be proposed.
- For parametric identification of various structural joints, a three-parameter joint stiffness identification model, which is a new approach, can be used.
- Nonlinear joint parameter estimation can also be performed using the substructure synthesis theory as a statistical method, and this process can be used for stiffness detection of different joints.
- In a bolted joint structure, translational nonlinearity is considered, but rotational nonlinearity is not considered. Hence, by considering both the nonlinearities, identification of nonlinear joint parameters for structural joints can be the area of research.

Finally, it can be inferred that parametric identification of nonlinear joints is a very wide area, and various modified methods, methodologies, and algorithms are used for nonlinear system identification in structural joints at the present time.

References

1. Kerschen G, Worden K, Vakakis AF, Golinval JC (2006) Past, present and future of nonlinear system identification in structural dynamics. Mech Syst Signal Process 20(3):505–592. https://doi.org/10.1016/j.ymssp.2005.04.008
2. Noël JP, Kerschen G (2017) Nonlinear system identification in structural dynamics: 10 more years of progress. Mech Syst Signal Process 83:2–35. https://doi.org/10.1016/j.ymssp.2016.07.020
3. Wang ZC, Ding YJ, Ren WX, Wang X, Li D, Li XF (2020) Structural dynamic nonlinear model and parameter identification based on the stiffness and damping marginal curves. Struct Control Heal Monit 27(6):1–24. https://doi.org/10.1002/stc.2540
4. Kim WJ, Lee BY, Park YS (2004) Non-linear joint parameter identification using the frequency response function of the linear substructure. Proc Inst Mech Eng Part C J Mech Eng Sci 218(9):947–955. https://doi.org/10.1243/0954406041991314
5. Chu Y, Wen H, Chen T (2016) Nonlinear modeling and identification of an aluminum honeycomb panel with multiple bolts. Shock Vib 2016. https://doi.org/10.1155/2016/1276753
6. Chintha HP, Chatterjee A (2019) Nonlinear parameter estimation in damping with volterra series through harmonic probing. IOP Conf Ser Mater Sci Eng 624(1). https://doi.org/10.1088/1757-899X/624/1/012003
7. Wang D, Zhang Z (2020) High-efficiency nonlinear dynamic analysis for joint interfaces with Newton-Raphson iteration process. Nonlinear Dyn 100(1):543–559. https://doi.org/10.1007/s11071-020-05522-9
8. Ahmadian H, Jalali H (2004) Identification of nonlinear interface models. Proc 2004 Int Conf Noise Vib Eng ISMA (January):1221–1227
9. Alamdari MM, Li J, Samali B, Ahmadian H, Naghavi A (2014) Nonlinear joint model updating in assembled structures. J Eng Mech 140(7):1–11. https://doi.org/10.1061/(ASCE)EM.1943-7889.0000759
10. Arslan Ö, Aykan M, Nevzat Özgüven H (2011) Parametric identification of structural nonlinearities from measured frequency response data. Mech Syst Signal Process 25(4):1112–1125. https://doi.org/10.1016/j.ymssp.2010.10.010
11. Peng ZK, Lang ZQ, Billings SA, Tomlinson GR (2008) Comparisons between harmonic balance and nonlinear output frequency response function in nonlinear system analysis. J Sound Vib 311(1–2):56–73. https://doi.org/10.1016/j.jsv.2007.08.035
12. Özer MB, Özgüven HN, Royston TJ (2009) Identification of structural non-linearities using describing functions and the Sherman-Morrison method. Mech Syst Signal Process 23(1):30–44. https://doi.org/10.1016/j.ymssp.2007.11.014
13. Wei F, Zheng GT (2010) Nonlinear vibration analysis of spacecraft with local nonlinearity. Mech Syst Signal Process 24(2):481–490. https://doi.org/10.1016/j.ymssp.2009.07.005
14. Da Silva S, Cogan S, Foltête E, Buffe F (2009) Metrics for nonlinear model updating in structural dynamics. J Braz Soc Mech Sci Eng 31(1):27–34. https://doi.org/10.1590/s1678-58782009000100005
15. Lacayo R et al (2019) Nonlinear modeling of structures with bolted joints: a comparison of two approaches based on a time-domain and frequency-domain solver. Mech Syst Signal Process 114:413–438. https://doi.org/10.1016/j.ymssp.2018.05.033
16. Jalali H, Ahmadian H, Mottershead JE (2007) Identification of nonlinear bolted lap-joint parameters by force-state mapping. Int J Solids Struct 44(25–26):8087–8105. https://doi.org/10.1016/j.ijsolstr.2007.06.003
17. Sadati SMS, Nobari AS, Naraghi T (2012) Identification of a nonlinear joint in an elastic structure using optimum equivalent linear frequency response function. Acta Mech 223(7):1507–1516. https://doi.org/10.1007/s00707-012-0656-6
18. Aubert AC (1986) Identification of nonlinear structural elements by force-state mapping. AIAA J 24(1):155–162. https://doi.org/10.2514/3.9236

19. Liao X, Zhang J, Xu X (2016) Analytical model of bolted joint structure and its nonlinear dynamic characteristics in transient excitation. Shock Vib 2016. https://doi.org/10.1155/2016/8387497
20. Kalaycıoğlu T, Özgüven HN (2018) FRF decoupling of nonlinear systems. Mech Syst Signal Process 102:230–244. https://doi.org/10.1016/j.ymssp.2017.09.029
21. Kalaycıoğlu T, Özgüven HN (2017) Dynamic decoupling of nonlinear systems. Conf Proc Soc Exp Mech Ser (January 2017):199–203. https://doi.org/10.1007/978-3-319-54930-9_17
22. Kalaycıoğlu T, Özgüven HN (2016) New FRF based methods for substructure decoupling. Conf Proc Soc Exp Mech Ser 4(January):463–472. https://doi.org/10.1007/978-3-319-29763-7_46
23. Ren Y, Beards CF (1998) Identification of 'effective' linear joints using coupling and joint identification techniques. J Vib Acoust Trans ASME 120(2):331–338. https://doi.org/10.1115/1.2893835
24. Ruzicka JE, Derby TF (1969) Vibration isolation with nonlinear damping. U S Nav Res Lab, Shock Vib Bull 40(5):19–35
25. Kalaycioȷlu T, Özgüven HN (2014) Harmonic response of large engineering structures with nonlinear modifications. Proc 8th Int Conf Struct Dyn EURODYN 2011 (February 2014):3623–3629
26. Tartaruga I, Elliott A, Hill TL, Neild SA, Cammarano A (2019) The effect of nonlinear cross-coupling on reduced-order modelling. Int J Non Linear Mech 116(May):7–17. https://doi.org/10.1016/j.ijnonlinmec.2019.05.006
27. Ren Y, Lim TM, Lim MK (1998) Identification of properties of nonlinear joints using dynamic test data. J Vib Acoust Trans ASME 120(2):324–330. https://doi.org/10.1115/1.2893834
28. Jalali H, Bonab BT, Ahmadian H (2011) Identification of weakly nonlinear systems using describing function inversion. Exp Mech 51(5):739–747. https://doi.org/10.1007/s11340-010-9375-x
29. Ingole SB, Rajurkar S (2019) Non-linear joint stiffness parameter identification. In: 4th International and 19th national conference on machines and mechanisms, pp 556–568
30. Ranjan P, Pandey AK (2019) Iwan model for bolted joint with residual macroslip stiffness and pinning. In: 4th international and 19th national conference on machines and mechanisms. pp 536–542
31. Gaul L, Lenz J (1997) Nonlinear dynamics of structures assembled by bolted joints. Acta Mech 125(1–4):169–181. https://doi.org/10.1007/BF01177306
32. Mina HK, Al-Ashtrai WK (2019) Inducing frictional force to enhance the transient response in beams. Al-Nahrain J Eng Sci 22(2):88–93. https://doi.org/10.29194/njes.22020088
33. Bouchaala N, Dion JL, Peyret N, Haddar M (2013) Micro-slip induced damping in the contact of nominally flat surfaces. Int J Appl Mech 5(1):1–20. https://doi.org/10.1142/S1758825113500051
34. Hartwigsen CJ, Song Y, McFarland DM, Bergman LA, Vakakis AF (2004) Experimental study of non-linear effects in a typical shear lap joint configuration. J Sound Vib 277(1–2):327–351. https://doi.org/10.1016/j.jsv.2003.09.018
35. Ingole SB, Chatterjee A (2011) Vibration analysis of single lap adhesive joint: experimental and analytical investigation. JVC/J Vib Control 17(10):1547–1556. https://doi.org/10.1177/1077546310380429
36. Böswald M, Link M (2005) Identification of non-linear joint parameters by using frequency response residuals. Conf Proc Soc Exp Mech Ser (February)
37. Ingole SB, Chatterjee A (2010) A method for joint stiffness identification. AIP Conf Proc 1298:338–343. https://doi.org/10.1063/1.3516327
38. Wang JH, Liou CM (1991) Experimental identification of mechanical joint parameters. J Vib Acoust Trans ASME 113(1):28–36. https://doi.org/10.1115/1.2930151
39. Mottershead JE, Friswell MI, Ng GHT, Brandon JA (1996) Geometric parameters for finite element model updating of joints and constraints. Mech Syst Signal Process 10(2):171–182. https://doi.org/10.1006/mssp.1996.0012

40. Ahmadian H, Mottershead JE, Friswell MI (2001) Boundary condition identification by solving characteristic equations. J Sound Vib 247(5):755–763. https://doi.org/10.1006/jsvi.2001.3708
41. Ingole SB, Chatterjee A (2017) Joint stiffness identification: a three-parameter joint model of cantilever beam. Int J Acoust Vib 22(1):3–13. https://doi.org/10.20855/ijav.2017.22.1445
42. Allen MS, Mayes RL, Bergman EJ (2010) Experimental modal substructuring to couple and uncouple substructures with flexible fixtures and multi-point connections. J Sound Vib 329(23):4891–4906. https://doi.org/10.1016/j.jsv.2010.06.007
43. Kubomura K (1982) A theory of substructure modal synthesis. J Appl Mech Trans ASME 49(4):903–909. https://doi.org/10.1115/1.3162634
44. Sjövall P, Abrahamsson T (2008) Substructure system identification from coupled system test data. Mech Syst Signal Process 22(1):15–33. https://doi.org/10.1016/j.ymssp.2007.06.003
45. Pakdemirli M, Boyaci H (2003) Non-linear vibrations of a simple-simple beam with a non-ideal support in between. J Sound Vib 268(2):331–341. https://doi.org/10.1016/S0022-460X(03)00363-8
46. Li WL (2002) A new method for structural model updating and joint stiffness identification. Mech Syst Signal Process 16(1):155–167. https://doi.org/10.1006/mssp.2000.1339
47. Fllekrug U (2012) Non-linearity in structural dynamics and experimental modal analysis. Nonlinearity Bifurc Chaos Theory Appl. https://doi.org/10.5772/48812
48. Modak SV, Kundra TK, Nakra BC (2002) Prediction of dynamic characteristics using updated finite element models. J Sound Vib 254(3):447–467. https://doi.org/10.1006/jsvi.2001.4081
49. Wyatt Becker PJ, Wynn RH, Berger EJ, Blough JR (1999) Using rigid-body dynamics to measure joint stiffness. Mech Syst Signal Process 13(5):789–801. https://doi.org/10.1006/mssp.1999.1232
50. Jalali H, Hedayati A, Ahmadian H (2011) Modelling mechanical interfaces experiencing micro-slip/slap. Inverse Probl Sci Eng 19(6):751–764. https://doi.org/10.1080/17415977.2010.531467
51. Asadi K, Ahmadian H, Jalali H (2012) Micro/macro-slip damping in beams with frictional contact interface. J Sound Vib 331(21):4704–4712. https://doi.org/10.1016/j.jsv.2012.05.026
52. Jezequel L (1981) Structural damping by slip in joints. Am Soc Mech Eng 105(81):497–504
53. Ahmadian H, Jalali H, Pourahmadian F (2010) Nonlinear model identification of a frictional contact support. Mech Syst Signal Process 24(8):2844–2854. https://doi.org/10.1016/j.ymssp.2010.06.007
54. Basu D, Kameswara Rao NSV (2013) Analytical solutions for Euler-Bernoulli beam on visco-elastic foundation subjected to moving load. Int J Numer Anal Methods Geomech 37(8):945–960. https://doi.org/10.1002/nag.1135
55. Sanati M, Terashima Y, Shamoto E, Park SS (2018) Development of a new method for joint damping identification in a bolted lap joint. J Mech Sci Technol 32(5):1975–1983. https://doi.org/10.1007/s12206-018-0405-4
56. Shamoto E, Hashimoto Y, Shinagawa M, Sencer B (2014) Analytical prediction of contact stiffness and friction damping in bolted connection. CIRP Ann Manuf Technol 63(1):353–356. https://doi.org/10.1016/j.cirp.2014.03.134
57. Ahmadian H, Mohammadali M (2016) A distributed mechanical joint contact model with slip/slap coupling effects. Mech Syst Signal Process 80:206–223. https://doi.org/10.1016/j.ymssp.2016.04.018
58. Iranzad M, Ahmadian H (2012) Identification of nonlinear bolted lap joint models. Comput Struct 96–97:1–8. https://doi.org/10.1016/j.compstruc.2012.01.011
59. Kaya A, Tekelioğlu MS, Findik F (2004) Effects of various parameters on dynamic characteristics in adhesively bonded joints. Mater Lett 58(27–28):3451–3456. https://doi.org/10.1016/j.matlet.2004.07.001
60. Wang S, Li Y, Xie Z (2019) Free vibration analysis of adhesively bonded lap joints through layerwise finite element. Compos Struct 223(May):110943. https://doi.org/10.1016/j.compstruct.2019.110943
61. Gunes R, Apalak MK, Yildirim M (2007) The free vibration analysis and optimal design of an adhesively bonded functionally graded single lap joint. Int J Mech Sci 49(4):479–499. https://doi.org/10.1016/j.ijmecsci.2006.09.010

62. Lee DH, Hwang WS (2007) An identification method for joint structural parameters using an FRF-based substructuring method and an optimization technique. J Mech Sci Technol 21(12):2011–2022. https://doi.org/10.1007/BF03177459
63. Pakrashi V, Fitzgerald P, O'Leary M, Jaksic V, Ryan K, Basu B (2018) Assessment of structural nonlinearities employing extremes of dynamic responses. JVC/J Vib Control 24(1):137–152. https://doi.org/10.1177/1077546316635935
64. Gaul L, Becker J (2010) Damping prediction of structures with bolted joints. Shock Vib 17(4–5):359–371. https://doi.org/10.3233/SAV-2010-0532
65. Prawin J, Rao ARM (2020) Detection and quantification of non-linear structural behavior using frequency Domain methods. Res Nondestruct Eval 31(2):69–106. https://doi.org/10.1080/09349847.2019.1623958
66. Billings SA (2013) The identification and analysis of nonlinear systems in the frequency Domain. Nonlinear Syst Identif 149–216. https://doi.org/10.1002/9781118535561.ch6
67. Bayma RS, Zhu Y, Lang ZQ (2018) The analysis of nonlinear systems in the frequency domain using nonlinear output frequency response functions. Automatica 94:452–457. https://doi.org/10.1016/j.automatica.2018.04.030
68. Ranjan BSC, Vikranth HN, Ghosal A (2013) A novel prevailing torque threaded fastener and its analysis. J Mech Des Trans ASME 135:1–9. https://doi.org/10.1115/1.4024977

A New Method for Solving Simultaneous Impact Problems in Constrained Multibody Systems

Koushik Kabiraj and Sourav Rakshit

Abstract During simultaneous impacts in linkages connected by joints, there can be several sequences of pair-wise impulse propagation and the choice of the most accurate sequence involves a combinatorial evaluation of all possible impulse sequences. In this paper, a simultaneous impact algorithm for planar frictionless constrained multibody systems is proposed that does not require an impulse propagation sequence to be determined. The formulation is developed by extending the generalized Newtonian restitution model for simultaneous impacts in unconstrained rigid bodies presented in [1] to planar linkages connected by frictionless joints. The algorithm is a computationally efficient alternative to the modeling of collisions in force-based continuous-time domains [2, 3], never results in an increase in kinetic energy (K.E.) [1] and predicts contact separation between bodies having zero pre-impact relative velocity of approach. Results using the proposed algorithm showing various collision scenarios in constrained linkages are included in the paper. Furthermore, the solutions are compared with linear complementarity (LCP) [4]-based approach and simulation results from ADAMS software.

Keywords Simultaneous impact · Constrained multibody systems · Linear complementarity · Newtonian restitution · Coefficient of restitution · ADAMS

1 Introduction

Applications such as machine operator training in a virtual environment or physics-based gaming environments require simulations to be processed in real time. In such cases, computational resources are often a constraint that leads to relaxing accuracy of results. Multibody collisions are frequently encountered in such applications. Collisions are nonlinear short duration phenomenon that can be modeled using a continuous-time force-based approach [3] or discretized time impulse-based approach [5]. Force-based approaches are computationally expensive, and this

K. Kabiraj (✉) · S. Rakshit
Indian Institute of Technology Madras, Chennai, India
e-mail: koushik.atti@gmail.com

V. K. Gupta et al. (eds.), *Recent Advances in Machines and Mechanisms*, Lecture Notes in Mechanical Engineering, https://doi.org/10.1007/978-981-19-3716-3_8

limitation is often overcome by using impulse-based methods which approximates the impact as an instantaneous event. In impulse-based approach, the kinetic energy (K.E.) losses are modeled using coefficient of restitution (COR)—various definitions of COR are summarized in [6]. Application of this approach to constrained multibody systems for a single point impact is described in [7]. Although momentum and energy conservation laws are sufficient to determine the change in velocities of the two participating bodies in single point collisions, additional assumptions are required when more than two bodies collide simultaneously [8]. Linear complementarity (LCP) models [4] address such assumptions by imposing a complementarity condition between contact impulse and velocity for simultaneous impact problems. However, LCP fails to predict contact separation between bodies initially in contact but with zero relative velocity of approach [1, 9]. Pairwise iterative impact models overcome this but frequently produce energetically inconsistent results or break symmetry [9]. Moreover, in these algorithms, the number of collision paths increases exponentially as the number of participating bodies increases.

In this paper, the impact model described in [1] for unconstrained rigid bodies has been extended to planar frictionless linkages connected by joints. Following [1], the impact event has been modeled as a constrained optimization problem, where the constraints describe physical laws of contact mechanics. The solution is the set of impulses that minimize net change in K.E. of the system during the impact. A single restitution inequality using kinematic COR [6] is defined which includes a weighted sum of individual restitution inequalities for all contacts. The proposed method applies to both open-chain and closed-loop linkages approximating each collision as a frictionless point contact. Moreover, the algorithm never results in an increase in K.E., predicts contact separation and does not require an impulse propagation sequence to be determined. In the subsequent sections, first, an equation for impulse-momentum relation in constrained multibody system is derived including equations to calculate change in velocity of contact points; following that, the equations are extended for a collision scenario of multiple sets of constrained multibody systems, and then, the simultaneous impact model along with the restitution inequality is introduced. In the results section, solutions for few impact scenarios in constrained linkage systems using the proposed method are presented and the solutions are compared with results obtained using LCP method and ADAMS software simulation. In the final section, findings using the proposed method are summarized following which scope for further development of the algorithm is discussed.

2 Impulse-Momentum Relations in Constrained Multibody Systems

Impulse-momentum relation for an unconstrained rigid body is given by

$$\mathbf{P}_{\mathrm{ext}} = \mathbf{M}_{\mathrm{rb}} \Delta \mathbf{V}_{\mathrm{cm}} \tag{1}$$

where $\mathbf{P}_{\text{ext}} \in \mathcal{R}^{3\times1}$ is the external impulse and impulsive moment vector, $\mathbf{M}_{\text{rb}} \in \mathcal{R}^{3\times3}$ is the inertia matrix, and $\Delta\mathbf{V}_{\text{cm}} \in \mathcal{R}^{3\times1}$ is the change in linear and angular velocity of center of mass (COM). However, Eq. (1) in this form cannot be used for constrained linkages as those involve constraint impulses and impulsive moments, and the net change in velocity of COM of the link depends on both external and constraint impulses. The constraints for such bodies can be in the form of joints, contacts, etc., and are expressed as unilateral or bilateral equations. In the following section, an equation relating external impulse to change in independent coordinate velocity is derived by eliminating constraint forces and moments from the generalized dynamic equation of a planar frictionless multibody system connected by joints [10]. After that, the equation is extended to calculate the change in velocity of the contact points where the impulses act on the system.

2.1 Response to External Impulse

The generalized dynamic equation of motion for any constrained linkage system depends on the choice of coordinate system. In this section, all equations are derived for a j-link planar system with n degrees of freedom (DOF). If the body coordinate system (BCS) of each link in the j-link system is chosen to be rigidly fixed to the COM, then the equation of motion in global coordinate system (GCS) can be written as shown below [10].

$$\mathbf{M}\ddot{\mathbf{q}} = \mathbf{Q}_e + \mathbf{Q}_c \tag{2}$$

$\mathbf{M} \in \mathcal{R}^{3j\times3j}$ is the inertia matrix, and $\ddot{\mathbf{q}} \in \mathcal{R}^{3j\times1}$ is the matrix of linkage COM accelerations. $\mathbf{Q}_e \in \mathcal{R}^{3j\times1}$ is the equivalent external force and moment vector, and $\mathbf{Q}_c \in \mathcal{R}^{3j\times1}$ is the constraint force and moment vector. In Eq. (2), it is assumed that the displacement coordinates for any link k in GCS are defined by $\mathbf{q}_k = \left[x_k^{\text{cm}}\ y_k^{\text{cm}}\ \theta_k^{\text{cm}}\right]^T$, where x_k^{cm} and y_k^{cm} are the COM coordinates of link k and θ_k^{cm} is the angle between the x-axis of BCS of link k and x-axis of GCS. The system of Eq. (2) has $3j$ equations and $6j - n$ unknowns. The $6j - n$ unknowns are formed by the $3j$ coordinate accelerations and $3j - n$ joint constraint forces and moments. Hence, additional equations are required to find a solution. These equations are obtained from the bilateral displacement constraints that the joints impose on the links. These joint constraints in GCS can be generalized and written as

$$\mathbf{C}(\mathbf{q}, t) = \mathbf{0} \tag{3}$$

where $\mathbf{C} \in \mathcal{R}^{(3j-n)\times1}$ is the vector of linearly independent constraint equations, $\mathbf{q} \in \mathcal{R}^{3j\times1}$ is the vector of linkage displacement coordinates, and t is time. Each equation in $\mathbf{C}$ matrix is a mathematical relation that constraints the relative displacement between two links that are connected by a joint. Equation (2) when combined with Eq. (3) forms a set of differential algebraic equations with $6j - n$ equations and $6j - n$ unknowns. Since the $j - link$ system has n DOF, only n out of $3j$ displacement coordinates are independent and the remaining $3j - n$ are dependent coordinates. In the context of this work, an equation relating external impulse to the change in velocity of independent coordinates $(\mathbf{q}_i \in \mathcal{R}^{n\times 1})$ needs to be derived. For the same, the dependant coordinates $(\mathbf{q}_d \in \mathcal{R}^{(3j-n)\times 1})$ will have to be eliminated from Eq. (2). This is achieved by using the principle of virtual work and coordinate partitioning of $\mathbf{q}$ into $\mathbf{q}_i$ and $\mathbf{q}_d$ [10]. For this, first the kinematic Jacobian form of Eq. (3) is derived as shown below.

$$\mathbf{C}_q \delta\mathbf{q} = \mathbf{0} \tag{4}$$

$\mathbf{C}_q \in \mathcal{R}^{(3j-n)\times 3j}$ is the Jacobian matrix of $\mathbf{C}$, and $\delta\mathbf{q} \in \mathcal{R}^{3j\times 1}$ is the virtual change in $\mathbf{q}$. If rows of $\delta\mathbf{q}$ vector and columns of $\mathbf{C}_q$ matrix in Eq. (4) are rearranged to group all the dependant coordinates $(\mathbf{q}_d)$ and independent coordinates $(\mathbf{q}_i)$ separately, then Eq. (4) can be re-written in the form of Eq. (5).

$$\begin{bmatrix}\mathbf{C}_{qd} & \mathbf{C}_{qi}\end{bmatrix}\begin{bmatrix}\delta\mathbf{q}_d \\ \delta\mathbf{q}_i\end{bmatrix} = \mathbf{0} \tag{5}$$

In Eq. (5), $\delta\mathbf{q}_i \in \mathcal{R}^{n\times 1}$ and $\delta\mathbf{q}_d \in \mathcal{R}^{(3j-n)\times 1}$ are the virtual change in $\mathbf{q}_i$ and $\mathbf{q}_d$, respectively, and $\mathbf{C}_{qi} \in \mathcal{R}^{(3j-n)\times n}$ and $\mathbf{C}_{qd} \in \mathcal{R}^{(3j-n)\times(3j-n)}$ are the columns of $\mathbf{C}_q$ corresponding to $\delta\mathbf{q}_i$ and $\delta\mathbf{q}_d$, respectively. This can further be restructured to obtain an expression for $\mathbf{q}_d$ in terms of $\mathbf{q}_i$ as shown in Eq. (6).

$$\delta\mathbf{q}_d = (-\mathbf{C}_{qd}^{-1}\mathbf{C}_{qi})\delta\mathbf{q}_i \tag{6}$$

By defining a matrix $\mathbf{B}_i \in \mathcal{R}^{3j\times n}$ as

$$\mathbf{B}_i = \begin{bmatrix}-\mathbf{C}_{qd}^{-1}\mathbf{C}_{qi} \\ \mathbf{I}\end{bmatrix} \tag{7}$$

$\delta\mathbf{q}$ can be written in the form of $\delta\mathbf{q}_i$ as shown in Eq. (8).

$$\delta\mathbf{q} = \mathbf{B}_i \delta\mathbf{q}_i \tag{8}$$

The same relation also holds good to calculate the change in generalized coordinate velocities of all the links $(\Delta\dot{\mathbf{q}} \in \mathcal{R}^{3j\times 1})$ from the change in independent coordinate velocities $(\Delta\dot{\mathbf{q}}_i \in \mathcal{R}^{n\times 1})$ as shown in Eq. (9).

$$\Delta\dot{\mathbf{q}} = \mathbf{B}_i \Delta\dot{\mathbf{q}}_i \tag{9}$$

Using the principle of virtual work, since total virtual work done by the frictionless constraint forces and moments $(\mathbf{Q}_c)$ is always zero, Eq. (2) can be reduced to Eq. (10).

However, the $\mathbf{M}\ddot{\mathbf{q}} - \mathbf{Q}_e$ terms in Eq. (10) cannot be equated to zero because $\delta\mathbf{q}^T$ consists of both dependant and independent coordinates.

$$\delta\mathbf{q}^T(\mathbf{M}\ddot{\mathbf{q}} - \mathbf{Q}_e) = 0 \tag{10}$$

Substituting $\delta\mathbf{q}$ from Eqs. (8) to (10) results in Eq. (11).

$$\delta\mathbf{q}_i^T\mathbf{B}_i^T(\mathbf{M}\ddot{\mathbf{q}} - \mathbf{Q}_e) = 0 \tag{11}$$

It should be noted that the rows of $\mathbf{M}$, $\ddot{\mathbf{q}}$ and $\mathbf{Q_e}$ matrices in Eq. (10) must be rearranged to be consistent with the coordinate partitioning done in Eq. (5) before the substitution. Now, $\delta\mathbf{q}_i^T$ can be eliminated from Eq. (11) to form Eq. (12).

$$\mathbf{B}_i^T\mathbf{M}\ddot{\mathbf{q}} - \mathbf{B}_i^T\mathbf{Q}_e = 0 \tag{12}$$

Equation (12) does not have the constraint force and moment vector but still consists of both independent and dependent coordinate accelerations in $\ddot{\mathbf{q}}$. If all external forces on the multibody system are due to collisions, $\mathbf{Q}_e$ can be assumed to be impulsive. Hence, integrating Eq. (12) from t_0 to $t_0 + \Delta t$ while $\Delta t \rightarrow 0$ results in the below equation

$$\mathbf{B}_i^T\mathbf{M}(\Delta\dot{\mathbf{q}}) - \mathbf{B}_i^T\mathbf{P}_e = 0 \tag{13}$$

where $\Delta\dot{\mathbf{q}} \in \mathcal{R}^{3j\times1}$ is the discontinuous change in coordinate velocities and $\mathbf{P}_e \in \mathcal{R}^{3j\times1}$ is the equivalent external impulse and impulsive moment vector. Finally, $\Delta\dot{\mathbf{q}}$ from Eq. (9) can be substituted into Eq. (13) to form Eq. (14), which relates the external equivalent impulse and impulsive moment vector to the change in independent coordinate velocities of the $j - link$ system.

$$\Delta\dot{\mathbf{q}}_i = (\mathbf{B}_i^T\mathbf{M}\mathbf{B}_i)^{-1}\mathbf{B}_i^T\mathbf{P}_e \tag{14}$$

It must be noted that although Eq. (14) does not have explicit terms for reaction impulse and impulsive moments, the reaction impulse and impulsive moments can be calculated after determining $\Delta\dot{\mathbf{q}}_i$. In a planar system, revolute joints only have reaction impulses and prismatic joints have reaction impulse and impulsive moments.

If the j-link systems collide with other bodies at h different points, then an impulse will act at each collision point. Assuming the collisions to be frictionless, a vector $\mathbf{P}_{\text{mbd}} \in \mathcal{R}^{2h\times1}$ can be defined in GCS that it includes the external impulses of each of the h collision points. $\mathbf{P}_e$ can be related to $\mathbf{P}_{\text{mbd}}$ using matrix $\mathbf{J}_{\text{mbd}} \in \mathcal{R}^{3j\times2h}$, as defined in Eq. (15).

$$\mathbf{P}_e = \mathbf{J}_{\text{mbd}}\mathbf{P}_{\text{mbd}} \tag{15}$$

As an example, Fig. 1 shows the j-link system with external impulses acting on k^{th} and $(k+1)$ link. $\mathbf{r}_k^{\text{pt1}}$, $\mathbf{r}_k^{\text{pt2}}$ and $\mathbf{r}_{k+1}^{\text{pt3}}$ are the position vectors of impulse points in GCS; $\mathbf{r}_k^{\text{cm}}$ and $\mathbf{r}_{k+1}^{\text{cm}}$ are the position vector in GCS of COM of link k and $k+1$,

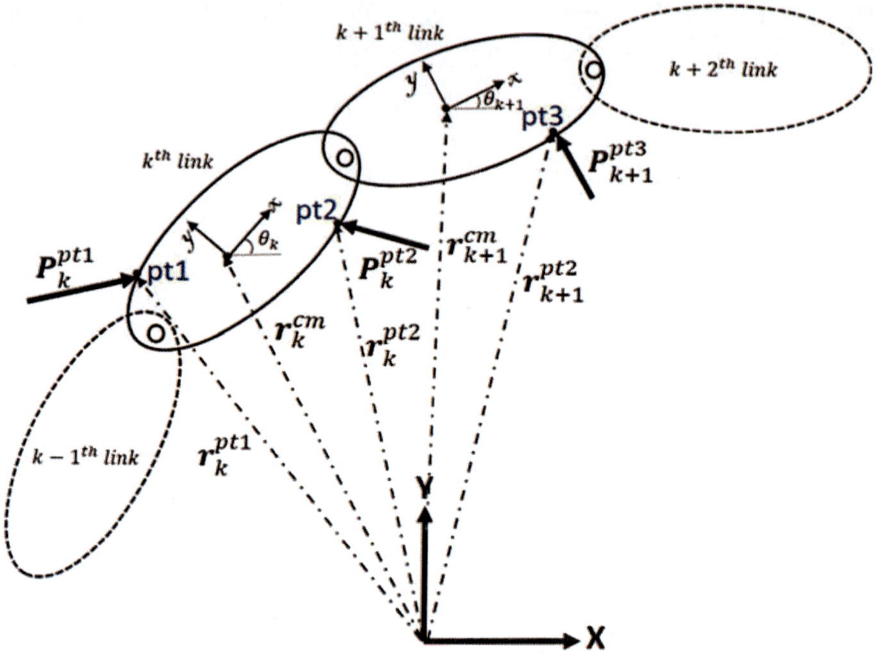

Fig. 1 External impulse on k^{th} and $(k+1)^{\text{th}}$ link of a j link system

respectively. A convention is adopted such that for any position vector $\mathbf{r}_k^{\text{pt}}$, x_k^{pt} and y_k^{pt} are the x and y component, respectively, in GCS. The elements of the column matrix $\mathbf{P}_e$ corresponding to the kth and $(k+1)^{\text{th}}$ link are given by Eq. (16), where $P_k^{x,\text{pt1}}$ and $P_k^{y,\text{pt1}}$ are the x and y component of $\mathbf{P}_k^{\text{pt1}}$ in GCS.

$$\mathbf{P}_e = \begin{bmatrix} \vdots \\ P_k^{x,\text{pt1}} + P_k^{x,\text{pt2}} \\ P_k^{y,\text{pt1}} + P_k^{y,\text{pt2}} \\ \begin{pmatrix} (P_k^{x,\text{pt1}})(y_k^{\text{cm}} - y^{\text{pt1}}) - (P_k^{y,\text{pt1}})(x_k^{\text{cm}} - x^{\text{pt1}}) \\ +(P_k^{x,\text{pt2}})(y_k^{\text{cm}} - y^{\text{pt2}}) - (P_k^{y,\text{pt2}})(x_k^{\text{cm}} - x^{\text{pt2}}) \end{pmatrix} \\ P_{k+1}^{x,\text{pt3}} \\ P_{k+1}^{y,\text{pt3}} \\ (P_{k+1}^{x,\text{pt3}})(y_{k+1}^{\text{cm}} - y^{\text{pt3}}) - (P_{k+1}^{y,\text{pt3}})(x_{k+1}^{\text{cm}} - x^{\text{pt3}}) \\ \vdots \end{bmatrix} \tag{16}$$

Since there are multiple external impulses on the multibody system, the elements of $\mathbf{J}_{\text{mbd}}$ matrix corresponding to the two links are given by Eq. (17) and that of $\mathbf{P}_{\text{mbd}}$ is given by Eq. (18).

$$\mathbf{J}_{\text{mbd}} = \begin{bmatrix} \vdots & \vdots & \vdots & \vdots & \vdots & \vdots & \vdots & \vdots \\ \ldots & 1 & 0 & 1 & 0 & 0 & 0 & \ldots \\ \ldots & 0 & 1 & 0 & 1 & 0 & 0 & \ldots \\ \ldots & \beta_1 & \beta_2 & \beta_3 & \beta_4 & 0 & 0 & \ldots \\ \ldots & 0 & 0 & 0 & 0 & 1 & 0 & \ldots \\ \ldots & 0 & 0 & 0 & 0 & 0 & 1 & \ldots \\ \ldots & 0 & 0 & 0 & 0 & \beta_5 & \beta_6 & \ldots \\ \vdots & \vdots & \vdots & \vdots & \vdots & \vdots & \vdots & \vdots \end{bmatrix}, \tag{17}$$

$$\mathbf{P}_{\text{mbd}} = \begin{bmatrix} \vdots \\ P_k^{x,\text{pt1}} \\ P_k^{y,\text{pt1}} \\ P_k^{x,\text{pt2}} \\ P_k^{y,\text{pt2}} \\ P_{k+1}^{x,\text{pt3}} \\ P_{k+1}^{y,\text{pt3}} \\ \vdots \end{bmatrix} \tag{18}$$

In Eq. (17), $\beta_1 = (y_k^{\text{cm}} - y^{\text{pt1}})$, $\beta_2 = -(x_k^{\text{cm}} - x^{\text{pt1}})$, $\beta_3 = (y_k^{\text{cm}} - y^{\text{pt2}})$, $\beta_4 = -(x_k^{\text{cm}} - x^{\text{pt2}})$, $\beta_5 = (y_{k+1}^{\text{cm}} - y^{\text{pt3}})$ and $\beta_6 = -(x_{k+1}^{\text{cm}} - x^{\text{pt3}})$. Substituting $\mathbf{P}_e$ from Eq. (15) into Eq. (14) gives a relation between the change in velocity of the independent DOF ($\Delta\mathbf{q}_i$) and external impulse vector ($\mathbf{P}_{\text{mbd}}$) as shown in Eq. (19).

$$\Delta\dot{\mathbf{q}}_i = (\mathbf{B}_i^T\mathbf{M}\mathbf{B}_i)^{-1}\mathbf{B}_i^T\mathbf{J}_{\text{mbd}}\mathbf{P}_{\text{mbd}} \tag{19}$$

2.2 Change in Velocity of Collision Point

The simultaneous impact model (described in the next section) requires a relation between change in velocity of collision point and external impulse vector. So Eq. (19) needs to be extended to compute change in velocities of collision points.

In the j-link constrained linkage system with n DOF, the change in velocity in GCS for any point g on link k ($\Delta\mathbf{V}_k^g \in \mathcal{R}^{2\times1}$) can be calculated from the change in coordinate velocities ($\Delta\dot{\mathbf{q}}$). This geometrical relation can be embedded in the form of a matrix ($\mathbf{L}_k^g \in \mathcal{R}^{2\times3j}$) for the gth point on k^{th} linkage as shown in Eq. (20).

$$\Delta \mathbf{V}_k^g = \mathbf{L}_k^g \Delta \dot{\mathbf{q}} \tag{20}$$

To derive a generalized expression for $\mathbf{L}_k^g$, the example in Fig. 1 is used. The position vector of any point g, on link k $(\mathbf{r}_k^g \in \mathcal{R}^{2\times1})$ in GCS, can be written as in Eq. (21), where $\bar{x}_k^g$ and $\bar{y}_k^g$ are the coordinates of g in the BCS of link k and $\mathbf{R}_k \in \mathcal{R}^{2\times2}$ is a rotation matrix form BCS of link k to GCS.

$$\mathbf{r}_k^g = \begin{bmatrix} x_k^{\text{cm}} \\ y_k^{\text{cm}} \end{bmatrix} + \mathbf{R}_k \begin{bmatrix} \bar{x}_k^g \\ \bar{y}_k^g \end{bmatrix} = \begin{bmatrix} x_k^{\text{cm}} \\ y_k^{\text{cm}} \end{bmatrix} + \begin{bmatrix} \cos\theta_k & -\sin\theta_k \\ \sin\theta_k & \cos\theta_k \end{bmatrix} \begin{bmatrix} \bar{x}_k^g \\ \bar{y}_g^k \end{bmatrix} \tag{21}$$

The velocity of point g is obtained by differentiating Eq. (21) w.r.t. time as shown in Eq. (22), where $(\mathbf{R}_k)_\theta \in \mathcal{R}^{2\times2}$ is the Jacobian matrix of $\mathbf{R}_k$ w.r.t. θ.

$$\mathbf{V}_k^g = \begin{bmatrix} \dot{x}_k^{\text{cm}} \\ \dot{y}_k^{\text{cm}} \end{bmatrix} + \dot{\theta}_k (\mathbf{R}_k)_\theta \begin{bmatrix} \bar{x}_k^g \\ \bar{y}_k^j \end{bmatrix} = \begin{bmatrix} \dot{x}_k^{\text{cm}} \\ \dot{y}_k^{\text{cm}} \end{bmatrix} + \dot{\theta}_k \begin{bmatrix} -\sin\theta_k & -\cos\theta_k \\ \cos\theta_k & -\sin\theta_k \end{bmatrix} \begin{bmatrix} \bar{x}_k^g \\ \bar{y}_k^g \end{bmatrix} \tag{22}$$

This can be written in a more concise form as shown in Eq. (23). The equation also holds good to calculate the instantaneous change in $\mathbf{V}_k^g$.

$$\mathbf{V}_k^g = \begin{bmatrix} 1 & 0 & -\bar{x}_k^g \sin\theta_k - \bar{y}_k^g \cos\theta_k \\ 0 & 1 & \bar{x}_k^g \cos\theta_k - \bar{y}_k^g \sin\theta_k \end{bmatrix} \begin{bmatrix} \dot{x}_k^{\text{cm}} \\ \dot{y}_k^{\text{cm}} \\ \dot{\theta}_k \end{bmatrix} \tag{23}$$

For the j-link system as in Fig. 1, $\Delta \mathbf{V}_k^g$ can be calculated from $\Delta \dot{\mathbf{q}}$ as shown in Eq. (24). $\mathbf{L}_k^g$ has all terms zero except for the terms corresponding to k Superscript normal t normal hk^{th} link coordinates.

$$\Delta \mathbf{V}_k^g = \mathbf{L}_k^g \Delta \dot{\mathbf{q}} = \begin{bmatrix} 0 & \ldots & 1 & 0 & -\bar{x}_k^g \sin\theta_k - \bar{y}_k^g \cos\theta_k & \ldots & 0 \\ 0 & \ldots & 0 & 1 & \bar{x}_k^g \cos\theta_k - \bar{y}_k^g \sin\theta_k & \ldots & 0 \end{bmatrix} \begin{bmatrix} \Delta \dot{x}_1^{\text{cm}} \\ \Delta \dot{y}_1^{\text{cm}} \\ \Delta \dot{\theta}_1^{\text{cm}} \\ \vdots \\ \Delta \dot{x}_k^{\text{cm}} \\ \Delta \dot{y}_k^{\text{cm}} \\ \Delta \dot{\theta}_k^{\text{cm}} \\ \vdots \\ \Delta \dot{x}_j^{\text{cm}} \\ \Delta \dot{y}_j^{\text{cm}} \\ \Delta \dot{\theta}_j^{\text{cm}} \end{bmatrix} \tag{24}$$

An expression relating the change in independent coordinate velocities $(\Delta \dot{\mathbf{q}}_i)$ to the change in velocity of point g can be obtained by substituting $\Delta \dot{\mathbf{q}}$ from Eq. (9) into Eq. (24) as shown in Eq. (25).

$$\Delta \mathbf{V}_k^g = \mathbf{L}_k^g \mathbf{B}_i \Delta \dot{\mathbf{q}}_i \tag{25}$$

It must be noted that for the substitution to hold good, the columns of $\mathbf{L}_k^g$ matrix must be rearranged and made consistent with the coordinate partitioning done in Eq. (9). Finally, $\Delta\dot{\mathbf{q}}_i$ from Eq. (19) can be substituted into Eq. (25) to form Eq. (26).

$$\Delta\mathbf{V}_k^g = (\mathbf{L}_k^g\mathbf{B}_i(\mathbf{B}_i^T\mathbf{M}\mathbf{B}_i)^{-1}\mathbf{B}_i^T\mathbf{J}_{\text{mbd}})\mathbf{P}_{\text{mbd}} \tag{26}$$

Equation (26) is a relation between the vector of external impulses $(\mathbf{P}_{\text{mbd}})$ on the j-link system and change in velocity of collision point g on link k $(\Delta\mathbf{V}_k^g)$ in GCS. In a group of bodies, if there are several constrained linkage systems and the j-link system is assigned an index c, Eq. (26) can be written in a concise form as shown in Eq. (27), where $\mathbf{W}_{k,\text{mbd}}^{g,c} \in \mathcal{R}^{2\times 2h}$ is defined in Eq. (28). The superscript c has been added to indicate the equation is written for the multibody system with index c. Equation (27) will be used in the simultaneous impact model.

$$\Delta\mathbf{V}_k^g = (\mathbf{W}_{k,\text{mbd}}^{g,c})(\mathbf{P}_{\text{mbd}}^c) \tag{27}$$

$$\mathbf{W}_{k,\text{mbd}}^{g,c} = (\mathbf{L}_k^g\mathbf{B}_i(\mathbf{B}_i^T\mathbf{M}\mathbf{B}_i)^{-1}\mathbf{B}_i^T\mathbf{J}_{\text{mbd}}) \tag{28}$$

3 The Simultaneous Impact Model

In a restitution-based collision model, COR relates the velocity of approach to the velocity of separation. When body a collides with body b at point s, the change in relative velocity of separation $(\Delta\mathbf{V}_{b-a}^s \in \mathcal{R}^{2\times 1})$ in GCS is given by

$$\Delta\mathbf{V}_{b-a}^s = \Delta\mathbf{V}_b^s - \Delta\mathbf{V}_a^s \tag{29}$$

where $\Delta\mathbf{V}_a^s \in \mathcal{R}^{2\times 1}$ and $\Delta\mathbf{V}_b^s \in \mathcal{R}^{2\times 1}$ are change in collision point velocities of a and b, respectively. If a and b are constrained multibody systems, then it can be assumed that the collision is only at point s between ka^{th} link of a and kb^{th} of b. Therefore, in GCS if impulse $\mathbf{P}^s \in \mathcal{R}^{2\times 1}$ acts on b, $-\mathbf{P}^s \in \mathcal{R}^{2\times 1}$ will act on a. Using Eqs. (27) and (29), an equation relating $\Delta\mathbf{V}_{b-a}^s$ and $\mathbf{P}^s$ can be established as shown in Eq. (30).

$$\Delta\mathbf{V}_{b-a}^s = (\mathbf{W}_{kb,\text{mbd}}^{s,b})\mathbf{P}^s - (\mathbf{W}_{ka,\text{mbd}}^{s,a})(-\mathbf{P}^s) \tag{30}$$

Since this is a single point collision, $\mathbf{P}_{\text{mbd}}^b = \mathbf{P}^s$ and $\mathbf{P}_{\text{mbd}}^a = -\mathbf{P}^s$. Therefore Eq. (30) can also be written as below,

$$\Delta\mathbf{V}_{b-a}^s = (\mathbf{W}_{kb,\text{mbd}}^{s,b})\mathbf{P}_{\text{mbd}}^b - (\mathbf{W}_{ka,\text{mbd}}^{s,a})\mathbf{P}_{\text{mbd}}^a \tag{31}$$

If there are u constrained linkage systems colliding simultaneously at f points, then the column matrix of impulses in GCS for the system can be written as $\mathbf{P} \in \mathcal{R}^{2f\times 1}$. To derive a generalized collision model, a notation has been defined such that each constrained linkage system is assigned a body index. Furthermore, a sign convention has been assumed such that $\mathbf{P}^s$ acts on body a if it is the higher index body among the participating bodies at the sth collision point and $-\mathbf{P}^s$ acts on a if it is the lower index body. Assuming a collides with several other bodies at h_a points in the system, a matrix $\mathbf{A}^a_{\text{mbd}} \in \mathcal{R}^{2h_a \times 2f}$ will need to be defined for a, such that when $\mathbf{A}^a_{\text{mbd}}$ is multiplied with $\mathbf{P}$, it results in $\mathbf{P}^a_{\text{mbd}}$, where $\mathbf{P}^a_{\text{mbd}} \in \mathcal{R}^{2h_a \times 1}$ is the column matrix of impulses acting on a (with proper sign convention). With this convention, change in velocity of separation at point s between a and b ($\Delta \mathbf{V}^s_{b-a} \in \mathcal{R}^{2\times 1}$) can be written as Eq. (32).

$$\Delta \mathbf{V}^s_{b-a} = ((\mathbf{W}^{s,b}_{kb,\text{mbd}})\mathbf{A}^b_{\text{mbd}} - (\mathbf{W}^{s,a}_{ka,\text{mbd}})\mathbf{A}^a_{\text{mbd}})\mathbf{P} = (\mathbf{W}^{s,b-a}_{kb-ka})\mathbf{P} \tag{32}$$

In Eq. (32), $\mathbf{W}^{s,b-a}_{kb-ka} \in \mathcal{R}^{2\times 2f}$ is a matrix which when multiplied with $\mathbf{P}$ vector gives the change in relative velocity of separation at point s which is a collision point between kb^{th} link of b and ka^{th} link of b. Assuming that the first collision in the system of u bodies is between the 1st links of constrained linkage systems 2 and 1, and the fth collision between l^{th} and o^{th} link of constrained linkage system u and $(u-1)$, respectively, the change in velocities of separation for the entire system in GCS ($\Delta \mathbf{V} \in \mathcal{R}^{2f\times 1}$) is given by Eq. (33).

$$\Delta \mathbf{V} = \begin{bmatrix} (\mathbf{W}^{1,2-1}_{1-1}) \\ \vdots \\ (\mathbf{W}^{s,b-a}_{kb-ka}) \\ \vdots \\ (\mathbf{W}^{f,u-(u-1)}_{l-o}) \end{bmatrix} \begin{bmatrix} \mathbf{P}^1 \\ \vdots \\ \mathbf{P}^s \\ \vdots \\ \mathbf{P}^f \end{bmatrix} = \mathbf{W}_{GCS}\mathbf{P} \tag{33}$$

In Eq. (33), $\mathbf{W}_{GCS} \in \mathcal{R}^{2f\times 2f}$ is a matrix that is best viewed as a set of f different $2\times 2f$ matrices, where the $2\times 2f$ matrix corresponding to the sth point is $\mathbf{W}^{s,b-a}_{kb-ka}$. Since COR relates the velocity of approach before and after the collision, a local coordinate system (LCS) for each collision point is defined such that the $\hat{n}_s$ axis points from lower index body to higher index body perpendicular to the sth collision point and $\hat{t}_s$ axis is perpendicular to it. Equation (33) now needs to be changed to LCS of the collision points to define a restitution relation in the collision model. This can be done by using a rotation matrix $\mathbf{R}_{sys} \in \mathcal{R}^{2f\times 2f}$ as shown in Eq. (34).

$$\Delta \mathbf{V}_{\text{LCS}} = (\mathbf{R}^T_{\text{sys}}\mathbf{W}_{\text{GCS}}\mathbf{R}_{\text{sys}})\mathbf{P}_{\text{LCS}} = \mathbf{W}_{\text{LCS}}\mathbf{P}_{\text{LCS}} \tag{34}$$

For the collision scenario described above, $\mathbf{R}_{sys}$ is defined in Eq. (35), where for any point s, θ_s is the angle between the x-axis of GCS and $\hat{n}_s$ axis of LCS of point s. $\mathbf{P}_{\text{LCS}} \in \mathcal{R}^{2f\times 1}$ is the column vector of collision impulses with each $\mathbf{P}^s$ in LCS of collision point s. $\Delta \mathbf{V}_{\text{LCS}} \in \mathcal{R}^{2f\times 1}$ is the column matrix of the discontinuous change in velocity of separation of the system, where for each point s the velocity of separation is in LCS of sth collision point.

$$\mathbf{R}_{sys} = \begin{bmatrix} \cos\theta_1 & -\sin\theta_1 & \cdots & \cdots & & \cdots & \cdots & \cdots & & \cdots \\ \sin\theta_1 & \cos\theta_1 & \cdots & \cdots & & \cdots & \cdots & \cdots & & \cdots \\ \vdots & \vdots & \ddots & & & & & \vdots & & \vdots \\ \cdots & \cdots & \cdots & \cos\theta_s & -\sin\theta_s & \cdots & & \cdots & & \cdots \\ \cdots & \cdots & \cdots & \sin\theta_s & \cos\theta_s & \cdots & & \cdots & & \cdots \\ \vdots & \vdots & & & & \ddots & & \vdots & & \vdots \\ \cdots & \cdots & \cdots & \cdots & & \cdots & \cdots & \cos\theta_f & -\sin\theta_f \\ \cdots & \cdots & \cdots & \cdots & & \cdots & \cdots & \sin\theta_f & \cos\theta_f \end{bmatrix} \tag{35}$$

Following [1], the simultaneous impact event is modeled as a constrained optimization problem where the solution set $\mathbf{P}_{\mathrm{LCS}}$ minimizes the net change in kinetic energy (ΔK.E.) of the system during the impact subject to constraints describing laws of contact mechanics. From [1], the net change in kinetic energy of the system upon collision is given by the below equation

$$\Delta\text{K.E.} = \frac{1}{2}\mathbf{P}_{\mathrm{LCS}}^{T}\mathbf{W}_{\mathrm{LCS}}\mathbf{P}_{\mathrm{LCS}} + \mathbf{P}_{\mathrm{LCS}}^{T}\mathbf{V}_{\mathrm{in}} \tag{36}$$

where $\mathbf{V}_{\mathrm{in}} \in \mathcal{R}^{2f\times 1}$ is the column matrix of relative contact velocities before impact in LCS of collision points and $\mathbf{W}_{\mathrm{LCS}}$ is defined in Eq. (34). To find a solution, Eq. (36) is minimized subject to constraints described below.

Constraint 1—Impulse-momentum relation: Solution set $\mathbf{P}_{\mathrm{LCS}}$ must obey Eq. (34).
Constraint 2—Condition for non-tensile impulse: The normal component of impulse ($\mathbf{P}_{\mathrm{LCS}}^{nr}$) for all collision points in the system is always non-tensile.

$$\mathbf{P}_{\mathrm{LCS}}^{nr} \geq 0 \tag{37}$$

Constraint 3—Velocity constraints: None of the contact points have a positive velocity of separation before the impact and after collision, no contact points have negative velocity of separation (meaning before collision, bodies should be approaching one another and after collision, bodies should be separating from one another). In Eq. (38) and (39), $\mathbf{V}_{\mathrm{in}}^{nr}$ is the normal component of initial velocities and $\mathbf{V}_{fn}^{nr}$ is the normal component of post-impact velocities.

$$\mathbf{V}_{\mathrm{in}}^{nr} \leq 0 \tag{38}$$

$$\mathbf{V}_{fn}^{nr} \geq 0 \tag{39}$$

Constraint 4—Restitution inequality: Following [1], a single Newtonian restitution inequality is used for describing all the f contact points for the system of u bodies as shown below

$$\sum_{k=1}^{f}(V^{nr}_{fn,k})(m_k)^{\alpha} \geq -\sum_{k=1}^{f} e_k(V^{nr}_{\text{in},k})(m_k)^{\alpha} \tag{40}$$

e_k is the kinematic COR at the kth contact; $V^{nr}_{fn,k}$ and $V^{nr}_{\text{in},k}$ are the normal component of post-impact and pre-impact relative velocity, respectively, at the kth contact. m_k [1], called the weightage parameter, is the reciprocal of the terms in $\mathbf{W}_{\text{LCS}}$ matrix which corresponds to the normal component of impact at the kth contact point. α is called allocation parameter and can be assigned any value. It effects the relative weightage given to each term in the restitution inequality (40). $\alpha = 0$ implies that all terms in the restitution inequality have equal weightage, $\alpha = 0.5$ makes each term dimensionally consistent with square root of energy ($\sqrt{m V^2}$), and $\alpha = 1$ changes each term to momentum units. From a physical view point, α can be seen as a tuning parameter that induces a net loss if K.E. post-collision and affects the relative distribution of contact impulses in the system.

4 Results and Discussion

Three cases of simultaneous impacts in multibody systems are presented in this section, and the solutions are compared with results obtained using LCP [4]-based method and ADAMS software simulation.

There are two different algorithms in ADAMS to solve contact problems: the IMPACT function model (contact stiffness-based model) and POISSON restitution model. The IMPACT function algorithm models the contact as a spring-damper system in force-based domain [11], whereas the restitution model uses COR to determine post-impact velocities and uses a penalty parameter to calculate reaction forces. The choice of solution algorithm is an user input while setting up a contact. In the below examples, the solutions are obtained using both restitution and contact stiffness-based algorithm. It was noticed that the solution set in ADAMS is highly sensitive to the chosen time step in both algorithms. For the below simulation experiments, the time steps are so chosen such that the impact events are treated as simultaneous by the software and the results are in close proximity of the solutions obtained using the proposed method. The time step for each scenario has been presented in the following subsections. For the restitution-based solution in ADAMS, the restitution parameter is set to 1 and penalty parameter is kept at the default of 1.0E+05. In the contact stiffness-based solution, stiffness, force exponent and penetration depth parameters are kept to the default of 1.0E+05 N/mm, 2.2 and 0.1 mm, respectively, and damping is changed to 0.0. The solutions obtained using restitution-based method are denoted as ADAMS (COR) and that of contact stiffness method is denoted as ADAMS (CS). Linear complementarity-based results are represented as LCP, and results using the proposed method are denoted by the choice of allocation parameter alpha.

For all the below examples, each constrained linkage system is assigned a body index and each link is numbered. Ground body and link are always indexed as 1.

Hence, k^{th} link in a system of constrained linkage system with index a is denoted by link_k^a. The BCS of any link_k^a is assumed to have its origin rigidly fixed at the COM (x_k^a, y_k^a) and rotated by θ_k^a with respect to the GCS. Furthermore, a convention is adopted such that the x-axis of BCS of link_k^a is aligned along the link and the y-axis is perpendicular to it. It must be noted that this convention is not mandatory to obtain a solution using the proposed method. In the below illustrations, the BCS axis for links having at least one independent coordinate are shown using dotted lines and the GCS axis are shown using solid line. With the above assumptions, the position and orientation of link_k^a are defined by $\left[x_k^a \; y_k^a \; \theta_k^a\right]^T$. All links and unconstrained bodies are assumed to have a mass of 1 unit with the COM at the geometric center. The COR (e) for all contact points is assumed to be 1.

4.1 2-Link Manipulator Hitting a 4-Bar Linkage

Fig. 2 shows a single impact at $cp_1 = \left[5.97 \; 3.48\right]^T$ between a 2-link manipulator and 4-bar linkage. The impact event in this case is inherently simultaneous as the system has three joints with the ground. The 2-link manipulator is designated index a and 4-bar linkage b. The GCS is assigned to have its origin at the joint between link_2^b and ground, as shown in Fig. 2. θ_2^a, θ_3^a and θ_2^b are assumed to be the independent coordinates of the system with initial values of 55.4°, 163.7° and 45°, respectively. Solution is obtained for an initial velocity of $\dot{\theta}_2^a = 1$ rad/s, $\dot{\theta}_3^a = -1$ rad/s and $\dot{\theta}_2^b = 0$

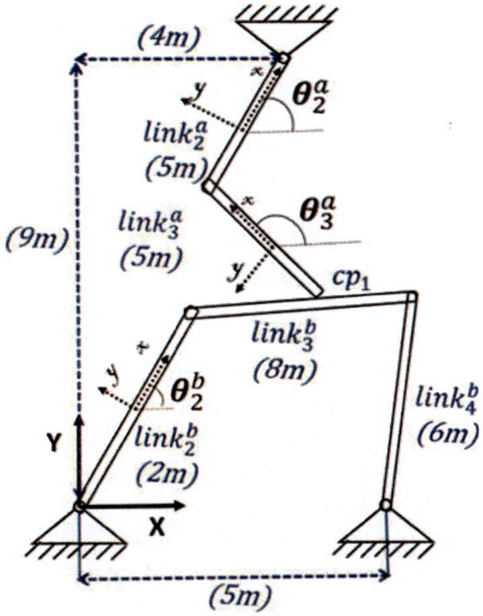

Fig. 2 2-link manipulator colliding with 4-bar linkage

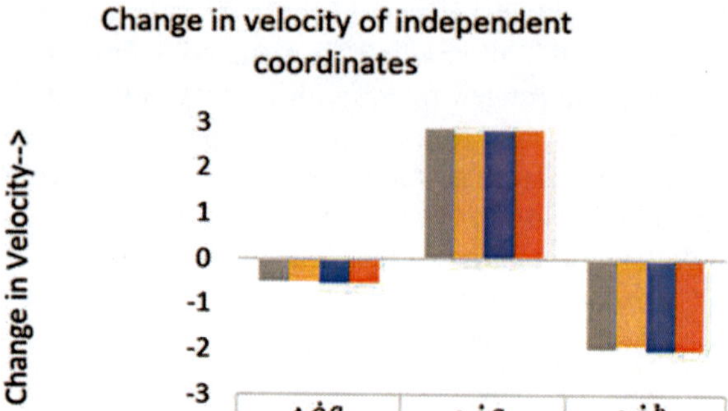

	$\Delta\dot{\theta}_2^a$	$\Delta\dot{\theta}_3^a$	$\Delta\dot{\theta}_2^b$
ADAMS (COR)	-0.514	2.870	-2.002
ADAMS (CS)	-0.495	2.749	-1.910
LCP	-0.538	2.833	-2.009
alpha=1	-0.539	2.832	-2.010

Fig. 3 Results for 2-link manipulator impacting a 4-bar linkage

rad/s. α is chosen to be 1 as it does not influence the results in case of single point collisions. Time step for ADAMS (COR) method is chosen to be 2.00E−03 s and that of ADAMS (CS) method is chosen to be 1.00E−03 s.

The solution set showing change in independent coordinate velocities are presented in Fig. 3. Average deviation of LCP result is 0.1%, ADAMS (COR) is 2.1%, and ADAMS (CR) is 5.4% with respect to the proposed method. Also, using the proposed method ΔK.E. is 0 for this example as there is only 1 collision point.

4.2 2-Link Manipulator Hitting a Slider-Crank Mechanism

Simultaneous impact at $cp_1 = \left[1.061\ 1.061\right]^T$ and $cp_2 = \left[2.407\ 1.297\right]^T$ between a 2-link manipulator and slider-crank mechanism is shown in Fig. 4. The 2-link manipulator is designated index a and slider-crank mechanism b. The system has 5 DOF, where x_2^a, y_2^a, θ_2^a, θ_3^a and θ_2^b are assumed to be the independent coordinates with initial values of 0.969 m, 3.556 m, 92.1°, −72.1° and 45°, respectively. Initial velocities are assigned to be $\dot{x}_2^a = 0$ m/s, $\dot{y}_2^a = 0$ m/s, $\dot{\theta}_2^a = 1$ rad/s, $\dot{\theta}_3^a = -1$ rad/s and $\dot{\theta}2_2^b = 0$ rad/s. The solution is obtained for $\alpha = 0.5$, 1, 1.5 and 2. These values of α are chosen to study the effect of allocation parameter on the solution set. A time step of 4.00E−03 s is chosen for ADAMS (COR) method and 5.00E−06 s for ADAMS (CS) method.

Fig. 5 shows that the change in post-impact independent coordinate velocities depends on α. With respect to the solution using $\alpha = 2$, LCP results on an average deviate by 4%, ADAMS (COR) results by 8.4% and ADAMS (CR) results by 5.8%. Also, the results using $\alpha = 2$ follow the same trend as LCP and ADAMS. For other

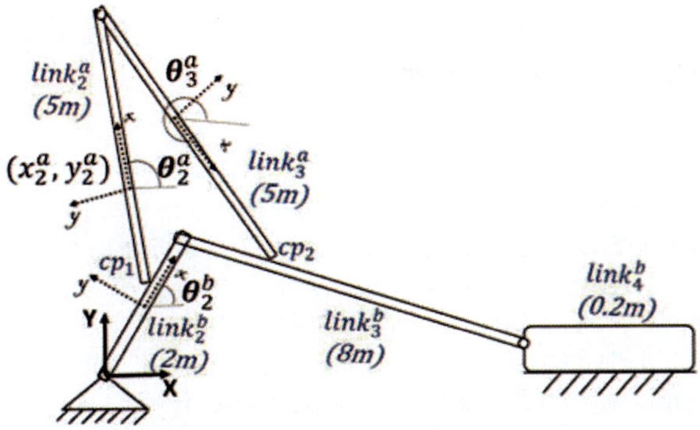

Fig. 4 2-link manipulator hitting a 4-bar linkage

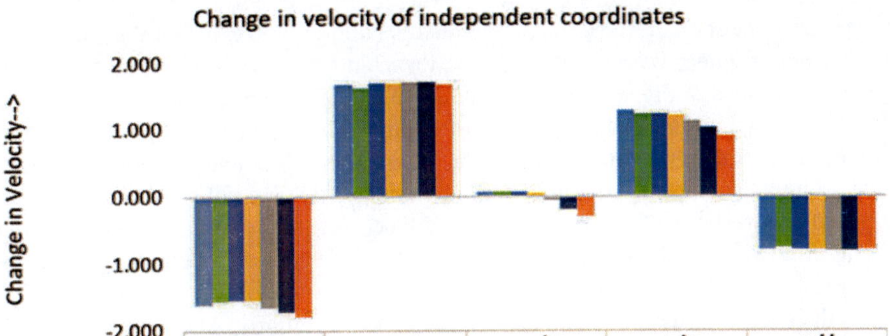

	$\Delta\dot{x}_2^a$	$\Delta\dot{y}_2^a$	$\Delta\dot{\theta}_2^a$	$\Delta\dot{\theta}_3^a$	$\Delta\dot{\theta}_2^b$
ADAMS (COR)	-1.615	1.678	0.066	1.274	-0.803
ADAMS (CS)	-1.565	1.622	0.060	1.227	-0.777
LCP	-1.545	1.700	0.060	1.221	-0.808
alpha=2	-1.555	1.703	0.051	1.215	-0.810
alpha=1.5	-1.650	1.719	-0.064	1.128	-0.821
alpha=1	-1.733	1.715	-0.189	1.016	-0.823
alpha=0.5	-1.788	1.682	-0.310	0.890	-0.812

Fig. 5 Results for 2-link manipulator impacting slider-crank mechanism

values of α, the deviation percentage is higher. α can be best seen as a tuning parameter that induces a net loss of K.E. Though the same value of α is used for both the collision points in this case, different values can be used to tune the system behavior and K.E loss. A mathematical formulation to determine the best value of α is an open problem and subject to further analysis and research.

4.3 Simultaneous Impact of 2 Unconstrained Rigid Bodies on a 2-Link Manipulator

Fig. 8 describes a scenario of a ball hitting a 2-link manipulator at $cp_1 = [8.803\ 8.071]^T$ which is in contact with another ball at $cp_2 = [5.657\ 5.657]^T$. Ball^1 is assigned index a, 2-link manipulator b, and ball^2 is assigned index c. The position and orientation of COM of the 2 balls along with θ_2^b and θ_3^b are assumed to be the independent coordinates. Initial values of x^a, y^a, θ_2^b, θ_3^b, x^c and y^c in GCS are 8.553 m, 8.504 m, 45°, 30°, 6.010 m and 5.303 m, respectively. The solution is obtained for $\alpha = 0, 0.5$ and 1 with ball^1 having an impact velocity of 1 m/s in a direction perpendicular to link$_3^b$. Time step for ADAMS (COR) method is chosen to be 4.00E−04 s and that of ADAMS (CS) method is chosen to be 2.54E−04 s (Fig. 6).

Angular momentum of the system about joint$_{1-2}^b$ is conserved as there is no external moment on the system and the results comply with this. For the balls, only the change in linear velocity of COM is plotted in Fig. 7 as there are no tangential impacts. As seen in Fig. 7, the results for all values of α distinctively differ from the LCP solution, and moreover, $\Delta\dot{\theta}_3^b$ is positive for LCP and negative for all the other solutions. This can be explained by the fact that LCP adds a mathematical constraint that prohibits contact separation between two bodies that are initially in contact with zero relative velocity of approach. Ball^2 and link$_2^b$ form such a pair. However, results using ADAMS simulation are much closer to the results obtained using different values of α. ADAMS (COR) results deviate on an average by 12.5, 14.9 and 17.8% from the results with $\alpha = 0, 0.5$ and 1, respectively. Similarly, ADAMS (CS) results deviate on an average by 16.7, 14.4 and 12.9% from the results with $\alpha = 0, 0.5$ and 1, respectively. As seen, the average deviation of ADAMS results in this example is higher than the previous two examples. This can be explained by the fact that in

Fig. 6 2 balls colliding with 2-link manipulator

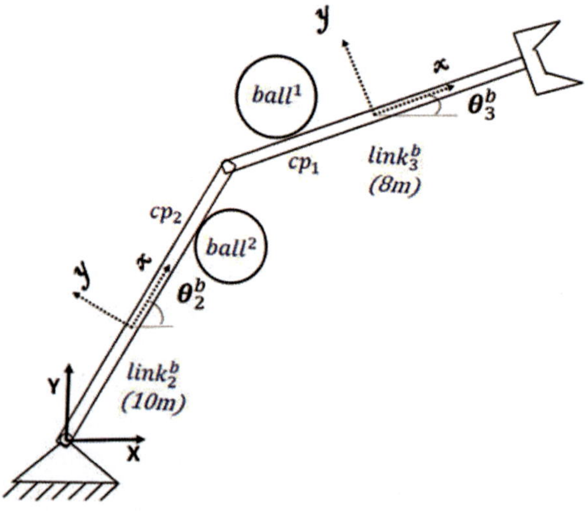

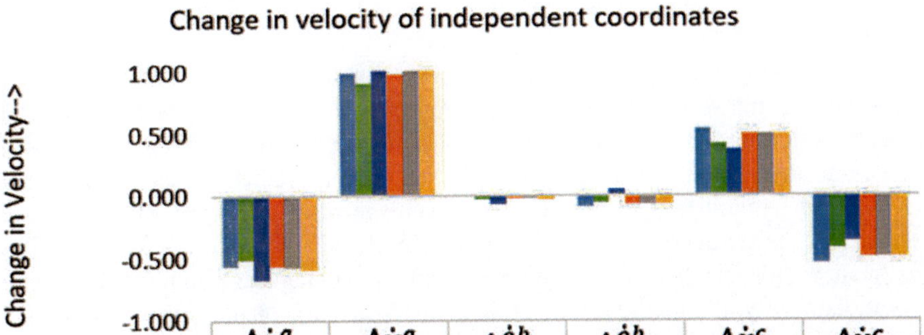

	$\Delta\dot{x}^a$	$\Delta\dot{y}^a$	$\Delta\dot{\theta}_2^b$	$\Delta\dot{\theta}_3^b$	$\Delta\dot{x}^c$	$\Delta\dot{y}^c$
ADAMS (COR)	-0.567	0.982	-0.011	-0.085	0.542	-0.542
ADAMS (CS)	-0.518	0.902	-0.024	-0.054	0.421	-0.421
LCP	-0.679	1.175	-0.064	0.054	0.364	-0.364
alpha=0	-0.562	0.973	-0.018	-0.072	0.496	-0.496
alpha=0.5	-0.578	1.000	-0.021	-0.070	0.498	-0.498
alpha=1	-0.593	1.027	-0.024	-0.067	0.496	-0.496

Fig. 7 Results for 2 balls impacting a 2-link manipulator

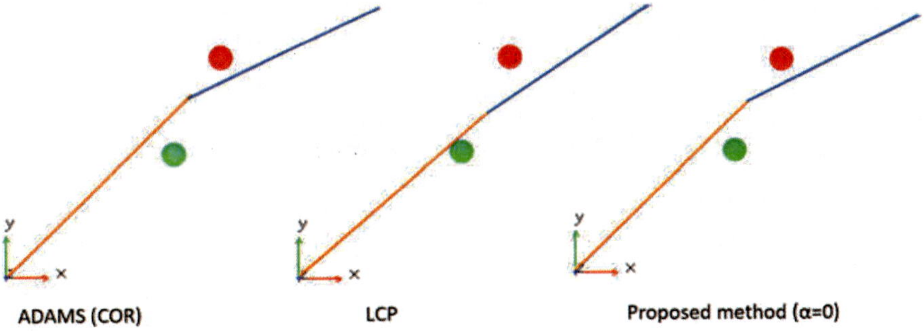

Fig. 8 Position of balls and links after 1 s of impact

ADAMS a contact between two bodies is detected when one body is within a pre-set distance from another body. This pre-set distance is usually small in magnitude but larger than 0. Owing to this, ball^2 in ADAMS receives an impact force from link$_2^b$ only when link$_2^b$ displaces slightly after collision between ball^1 and link$_3^b$ and so the two collisions are not treated as perfectly simultaneous in ADAMS.

However, for all the three values of α, the solution predicts a contact separation between link$_2^b$ and ball^2. This can be seen in Fig. 8 which shows the position and orientation of the system 1 s after the collision using results from ADAMS (COR), LCP and proposed method with $\alpha = 0$. Though α induces a net loss in K.E., the solution is physically intuitive because it predicts contact break and no impact in real-world applications is perfectly elastic.

5 Conclusions

As shown in the results section, the proposed method predicts solutions that are close to solutions obtained using ADAMS software without defying any physical laws. It is also shown that when applicable the method predicts contact break which LCP-based method fails to predict. Moreover, the results do not require a impact sequence to be assumed and solve all impacts simultaneously using a single restitution inequality. This makes it easy to solve problems involving large number of collision point. Currently, the methodology is developed only for planar linkages but it can be extended for spatial linkages using appropriate formulations. As a future scope, an additional constraint to model friction can be embedded into the simultaneous impact model and a method to select the best value of tuning parameter (α) can be developed.

References

1. Rakshit S, Chatterjee A (2015) Scalar generalization of newtonian restitution for simultaneous impact. Int J Mechan Sci 103:141–157
2. Mills JK, Neguyen CV (1992) Robotic manipulator collisions: modeling and simulation. ASME J Dyn Syst Measur Control 114(4):650–659
3. Lankarani HM, Nikravesh P (1994) Continuous contact force models for impact analysis of multibody systems. Nonlinear Dyn 5(2):193–207
4. Anitescu M, Potra FA (1997) Formulating dynamic multi-rigid-body contact problems with friction as solvable linear complementarity problems. Nonlinear Dyn 14(3):231–247
5. Wehage RA, Haug EJ (1982) Dynamic analysis of mechanical systems with intermittent motion. J Mechan Design 104(4):778–784
6. Seifried R, Schiehlen W, Eberhard P (2010) The role of coefficient of restitution on impact problems in multi-body dynamics. In: Proceedings of the institution of mechanical engineers, Part K: Journal of Multi-body Dynamics, vol 224, no 3, pp 279–306
7. Lee SH, Yi BJ, Kwak YK (2000) Modeling and analysis of internal impact for general classes of robotic mechanisms. In: International conference on intelligent robots and systems, pp 1955–1962, IEEE/RSJ
8. Glocker C (2004) Concepts for modeling impacts without friction. Acta Mechanica 168(1):1–19
9. Smith B et al (2012) Reflections on simultaneous impact. ACM Trans Graph 31(4):1–12
10. Shabana AA (2010) Computational dynamics, 3rd edn. Wiley
11. Sapietová A et al (2015) Analysis of the influence of input function contact parameters of the impact force process in the MSC. ADAMS. In: Advanced mechatronics solutions, advances in intelligent systems and computing, vol 393, pp 243–253. Springer, Cham

Design of Mechanisms

Modeling and Optimal Design of Bridge-Type Displacement Amplifier

S. B. Lavanya and G. R. Jayanth

Abstract Bridge-type displacement amplifiers are compliant mechanisms that are widely used in high-precision positioning, where they are typically employed to amplify the input displacement of a piezo-actuator. This paper presents a quasi-static model for the amplifier considering the effect of elastic deformation of the beam and the flexure hinges. Closed-form expressions for the amplification ratio of the amplifier, the angular deformation of the beam, and the necessary force are derived in terms of geometric parameters of the amplifier. The dependence of the amplification ratio on the geometric parameters of the connecting-beam and its tilt angle is studied, and the optimal values of the same are obtained. Additionally, two characteristic length-scales are identified, and guidelines are proposed for choosing the beam-length for a desired amplification in terms of these length-scales. In all cases, the analytical results are validated by comparing them with those of finite element analysis.

Keywords Bridge-type displacement amplifier · Complaint mechanism · Flexure hinge · Mechanical amplifier

1 Introduction

Flexure-based displacement amplifiers are popularly used in precision nano-positioning [1], micro-manipulation applications [2, 3] due to frictionless motion, absence of backlash, and ease of fabrication in micro-scale. These displacement amplifiers are broadly classified as being either lever-type [4] or bridge-type. Among the two, the bridge-type amplifier has negligible cross-coupling motion and possesses a higher natural frequency for the same amplification ratio and similar size. Hence, they are popularly used in piezo-based nano-positioners [5].

S. B. Lavanya (✉) · G. R. Jayanth
Indian Institute of Science, Bangalore 560012, India
e-mail: lavanyas@iisc.ac.in

G. R. Jayanth
e-mail: jayanth@iisc.ac.in

V. K. Gupta et al. (eds.), *Recent Advances in Machines and Mechanisms*, Lecture Notes in Mechanical Engineering, https://doi.org/10.1007/978-981-19-3716-3_9

Given their widespread use, it is important to develop analytical models to assist with designing nano-positioners of the desired amplification ratio. Furukawa's group [6] was one of the first to describe an analytical method for relating the input–output displacements. These, however, were obtained from the geometric analysis without considering the compliance of the flexure hinges and the beam of the amplifier. The reported output displacement was found to be a non-linear function of the input displacement. Over the years, other groups have reported analytical methods for obtaining the amplification ratio using a variety of methods such as Castigliano's theorem [7], lumped parameter model [8], elastic beam theory [9, 10], and matrix method [11, 12]. Among the reported models, the models obtained using the matrix method [11, 12] were the most accurate, but a simple closed-form expression for amplification ratio was not reported. Hence, numerical analysis was necessary to understand the effects of the dimensional parameters. This, however, is relatively computationally intensive and therefore hinders simple, intuitive design. Though closed-form expression for amplification ratio was reported by Ling et al. [10] considering the effects of compliance of connecting-beam, the expression was not valid for an aligned-type bridge amplifier with circular flexural hinges. Furthermore, none of the groups have reported closed-form expression for maximum achievable amplification ratio and the corresponding optimal tilt angle for given amplifier dimensions to aid in the design of the amplifier.

This paper proposes a simple model for an aligned-type bridge amplifier that also considers the compliance of the connecting-beam. The model is used to obtain closed-form expressions for the amplification ratio, the angular deformation of the beam, and the force developed within the flexure-mechanism. Finally, the analytical expression for amplification ratio is employed to determine the optimal tilt angle of the beam to maximize amplification ratio and to determine the minimum length of the beam necessary to achieve a specified amplification. In particular, two characteristic length-scales are identified, which help to generate guidelines for the selection of the length of the beam. All the important results are validated using finite element analysis.

The rest of the paper is divided as follows: Section 2 describes the model of bridge-type amplifier incorporating the compliance of the beam and its verification using finite element (FE)-method. Section 3 reports the optimization for the amplifier and its validation using FE-method. Finally, Sect 4 provides the conclusion.

2 Modeling of Bridge-Type Amplifier

Figure 1a, b shows schematics of two types of bridge amplifiers. The former geometry is termed as aligned-type, while the latter is termed parallel-type amplifier, with the main difference being that the former uses two piezo-actuators while the latter employs just one. Though this leads to a simpler construction for the latter case, it also leads to some disadvantages: It requires the use of a side-beam whose compliance can reduce the amplification ratio. It sets a lower-limit on the size of the flexure based

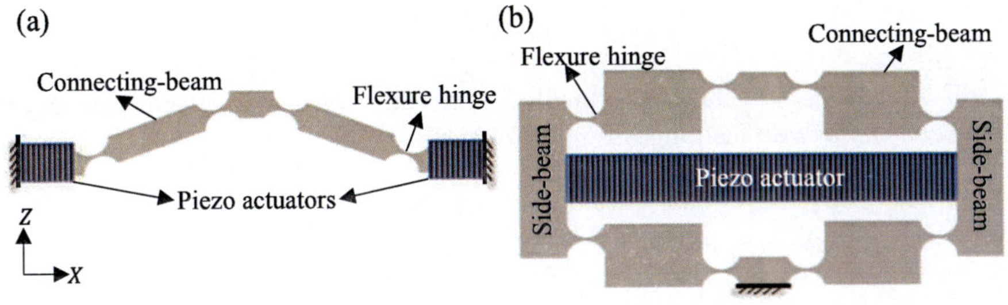

Fig. 1 **a** Aligned-type bridge amplifier. **b** Parallel-type bridge amplifier

on the piezo-actuator that needs to be mounted inside, and thus, potentially leads to lower overall bandwidth of the amplifier. Furthermore, applying the requisite preload is easier in the former case, especially in the event of fabrication errors. Nevertheless, from the point of view of analysis, the quasi-static model for the two would be the same if the compliance of the side-beams is neglected.

Figure 2a shows the model of the bridge-type displacement amplifier. The connecting-beam which is of length l_b is tilted at an angle α and is modeled as an elastic beam, and flexure hinges are modeled as torsional springs of rotational stiffness κ_h. An input force F along X-axis displaces the amplifier by Δx, which results in an amplified output displacement Δz. Due to symmetry, only one side of the amplifier, comprising two hinges and a beam, as shown in Fig. 2b, is considered for analysis purposes. If the two torsional hinges rotate by $\Delta\alpha_1$ and $\Delta\alpha_2$, then by moment balance, it is seen that

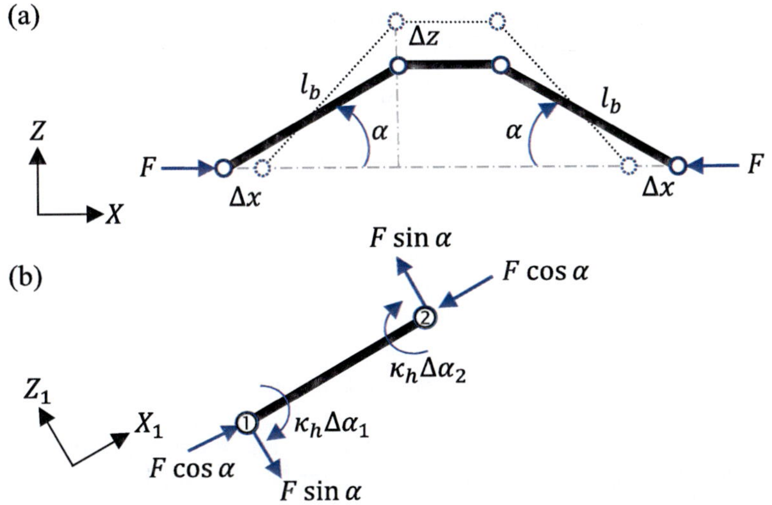

Fig. 2 **a** Lumped parameter model of bridge-type amplifier. **b** Free body diagram of the connecting-beam

$$\kappa_h \Delta\alpha_1 + \kappa_h \Delta\alpha_2 = F l_b \sin\alpha. \tag{1}$$

The force $F\cos\alpha$ acts longitudinally and results in the axial deformation Δl_b of the beam along with the hinges is given by

$$\Delta l_b = -\frac{F\cos\alpha}{k_{lb}}, \tag{2}$$

where k_{lb} is the total axial stiffness of the beam along with hinges which is the series combination of axial stiffness of the beam k_b and the hinges k_h.

Likewise, the force $F\sin\alpha$ along with the moments applied by the torsional spring results in elastic deformation of the beam. The resulting deformation $z_1(x_1)$ can be obtained using the Euler–Bernoulli beam equation, namely

$$\frac{\mathrm{d}^2 z_1}{\mathrm{d}x_1^2} = -\frac{\mathrm{M}(x_1)}{EI}, \tag{3}$$

where $\mathrm{M}(x_1) = -F(l_b - x_1)\sin\alpha + \kappa_h \Delta\alpha_2$ and the boundary conditions are $z_1^{'}(0) = \Delta\alpha_1$, $z_1^{'}(l_b) = \Delta\alpha_2$ and $z_1(0) = 0$. In Eq. (3), E is Young's modulus, and I is the area moment of inertia of the beam. By solving Eqs. (1)–(3), the angles $\Delta\alpha_1$, $\Delta\alpha_2$, and $z_1(x_1)$ can be obtained in terms of the force F and are given by

$$\Delta\alpha_1 = \Delta\alpha_2 = \frac{F l_b \sin\alpha}{2\kappa_h}. \tag{4}$$

$$z_1(x_1) = \frac{F x_1 \sin\alpha}{12\,\kappa_b\,\kappa_h\,l_b}\left(6\kappa_b l_b{}^2 + 3\kappa_h l_b x_1 - 2\kappa_h x_1{}^2\right). \tag{5}$$

In Eq. (5), $\kappa_b = EI/l_b$ is the rotational stiffness of the beam.

By virtue of symmetry, hinge-2 can move only along the Z-axis, i.e., its net X-displacement should be zero. This implies that

$$\Delta x = -\Delta l_b \cos\alpha + z_1(l_b)\sin\alpha. \tag{6}$$

Likewise, the displacement Δz can be obtained in terms of Δl_b and $z_1(l_b)$ and is given by

$$\Delta z = \Delta l_b \sin\alpha + z_1(l_b)\cos\alpha. \tag{7}$$

The amplification ratio or the gain G is defined as the ratio of output displacement Δz to input displacement Δx. The expression for G is obtained from Eqs. (4)–(7) and is given by

$$G = \frac{\Delta z}{\Delta x} = \frac{\cot\alpha(1-\beta)}{1+\beta\cot^2\alpha}, \quad (8)$$

where $\beta = \frac{1}{\left[k_{lb} l_b^2 \left(\frac{1}{2\kappa_h} + \frac{1}{12\kappa_b}\right)\right]}$.

Equation (8) reveals that $G \approx \cot\alpha$ when $\cot^2\alpha \ll 1/\beta$ or equivalently, $\alpha \gg \tan(\sqrt{\beta})$. Likewise, $G \approx (\tan\alpha)(1-\beta)/\beta$ when $\cot^2\alpha \gg 1/\beta$ or equivalently when $\alpha \ll \tan(\sqrt{\beta})$. The former limit represents the "rigid-body" limit of operation where the beam can be assumed to be nearly rigid, while the latter limit represents the "elastic" limit, where G is dominated by elastic effects of the beam.

Since the bridge amplifier is typically actuated by a piezo-actuator, it is convenient to represent the force F and the angles $\Delta\alpha_1$, $\Delta\alpha_2$ in terms of the input displacement Δx and are given by

$$F = \Delta x \left[\frac{k_{lb}}{\cos^2\alpha + k_{lb} l_b^2 \sin^2\alpha\left(\frac{1}{2\kappa_h} + \frac{1}{12\kappa_b}\right)}\right]. \quad (9)$$

$$\Delta\alpha_1 = \Delta\alpha_2 = \Delta x \left[\frac{k_{lb} l_b \sin\alpha}{2\kappa_h \cos^2\alpha + k_{lb} l_b^2 \sin^2\alpha\left(1 + \frac{\kappa_h}{6\kappa_b}\right)}\right]. \quad (10)$$

To validate these results, a bridge amplifier was analyzed using finite element (FE)-analysis, and the results were compared with the analytical predictions. The aligned-type bridge amplifier was assumed to possess a circular flexural hinge of radius $R_c = 50$ μm, and thickness $t_c = 10$ μm. The beam dimensions were the following: length $l_b = 1$ mm, width $w_b = 0.5$ mm, thickness t_b. The elastic modulus of all materials was assumed to be $E = 170$ GPa. The rotational stiffness of the hinge κ_h is determined using the expression by Schotborgh et al. [13], whereas its axial stiffness k_h is obtained using the expression by Yong et al. [13].

Figure 3a shows the deformation profiles for various values of $\alpha = 1°, 5°, 10°$ for $t_c = 20$ μm and $t_b = 30$ μm. For these dimensions, κ_h is comparable to κ_b. The graph shows that the analytical results agree with the computational results with errors of −16.5%, 10.8%, and 11.5% for $\alpha = 1°, 5°, 10°$, respectively. Figure 3b shows the variation of amplification ratio G as a function of α obtained using analytical expression and FE-method for different values of beam thickness $t_b = R_c, 2R_c, 4R_c$. Figure 3b reveals that the analytical results are in good agreement with FE-results. Moreover, there exists an optimum tilt angle α_{opt} at which the amplification ratio is maximum. The results show that increasing t_b, i.e., increasing the rigidity of the beam improves the maximum achievable amplification.

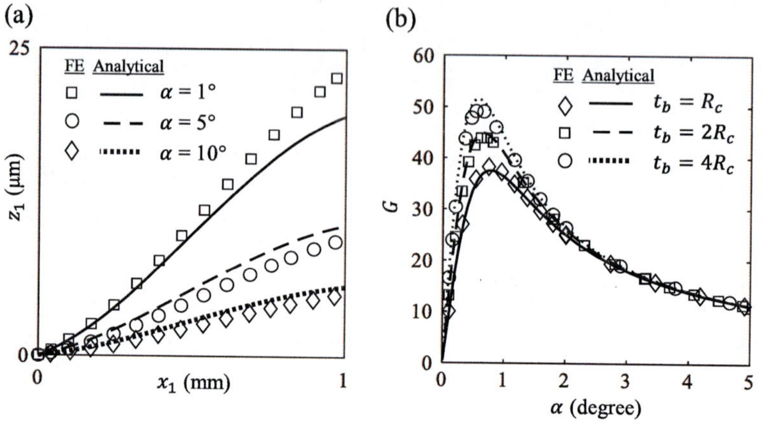

Fig. 3 **a** Plot of linear deformation $z_1(x_1)$ profile of the beam for $\alpha = 1°, 5°, 10°$ for $t_c = 20\ \mu$m and $t_b = 30\ \mu$m. **b** Plot of G versus α for different values of $t_b = R_c, 2R_c, 4R_c$ with for $t_c = 10\ \mu$m

3 Optimization of the Amplifier Geometry

Equation (8) and Fig. 3b reveal that there exists an optimum value of α_{opt} at which the amplification of the amplifier is maximum. α_{opt} is obtained by setting $\frac{dG}{d\alpha} = 0$ and solving for α. Thus, the analytical expression for α_{opt} is given as

$$\alpha_{\mathrm{opt}} = \cot^{-1}\left(\frac{1}{\sqrt{\beta}}\right). \tag{11}$$

Using Eq. (11) in (8), the expression for maximum amplification $G_{\max}$ is given by

$$G_{\max} = \frac{1}{2}\left(\frac{1}{\sqrt{\beta}} - \sqrt{\beta}\right). \tag{12}$$

Since typically $\beta \ll 1$, it is seen that $G_{\max} \approx 1/(2\sqrt{\beta})$. Equation (12) reveals that $G_{\max}$ increases with reduction in β. Since $\beta = 1/\left[k_{lb} l_b^2 \left(\frac{1}{2\kappa_h} + \frac{1}{12\kappa_b}\right)\right]$, it is noted that $G_{\max}$ can be increased with an increase in l_b and by reducing κ_h.

Figure 4 shows the variation of $G_{\max}$ as a function of beam-length l_b obtained using the analytical expression for various beam-thicknesses t_b and shows that $G_{\max}$ is an increasing function of both l_b and t_b. The analytical results were validated with those of FE-analysis for the case $t_b = 2R_c$, and the two are seen to match well.

The figure also reveals that to achieve the desired amplification, the beam should have a certain minimum length $l_{b\min}$ for the given hinge dimension and beam cross-section area. The analytical model can be employed to estimate the minimum length $l_{b\min}$ necessary for the beam to achieve a specified amplification G. In particular, this can be readily done for the case when the rotational stiffness of the beam κ_b is

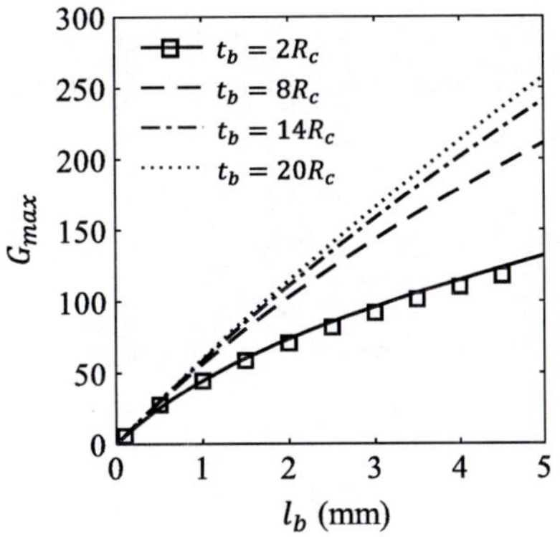

Fig. 4 Plot of maximum amplification ratio $G_{\max}$ for the variation in beam-length l_b for different values of $t_b = 2R_c, 8R_c, 14R_c, 20R_c$ obtained from analytical expression. The square markers correspond to $G_{\max}$ obtained from FE-method for the case $t_b = 2R_c$

much greater than κ_h. In such a case, β can be approximated to be $\beta \approx 2\kappa_h/(k_{lb}l_b^2)$. Furthermore, it is noted that $k_{lb} = k_b||k_h$, where k_b is the axial stiffness of the beam and k_h is the net axial stiffness of the two hinges. Since $k_b = Ew_bt_b/l_b$, and $G_{\max} \approx 1/(2\sqrt{\beta})$, it is possible to analytically relate $G_{\max}$ to l_b. Thus, for a specified amplification ratio G, the minimum length necessary can be derived to be

$$l_{b\min} = 4G^2\left(l_{01} + \sqrt{l_{01}^2 + \frac{l_{02}^2}{G^2}}\right), \tag{13}$$

where $l_{01} = \kappa_h/Ew_bt_b$ and $l_{02} = \sqrt{\kappa_h/k_h}$ are two characteristic lengths of the amplifier.

Though increasing the beam-length improves the amplification ratio, the improvement is not significant for very large values of l_b. Thus, the upper-limit on the beam-length $l_{b\max}$ can be set when the rate of change of amplification ratio with respect to the beam-length, i.e., $\mathrm{d}G_{\max}/\mathrm{d}l_b$ reduces below a specified limit g_o. The approximate expression for $l_{b\max}$ can be obtained for the desired amplification-rate g_o using Eq. (12) under the conditions $\kappa_h \ll \kappa_b$ which is generally valid in practice and is given by

$$l_{b\max} = \frac{1}{64g_o^2l_{01}}\left(1 - 128g_o^2l_{02}^2 + \sqrt{512g_o^2l_{02}^2 + 1}\right). \tag{14}$$

Thus, it is desirable to choose the beam-length l_b such that $l_{b\min} < l_b < l_{b\max}$ when the tilt angle is chosen to be α_{opt}.

As an example, for the amplifier parameters considered in Sect. 2 and the desired amplification ratio of 50, $l_{b\min} = 1.161$ mm as per Eq. (13). This is within an error of 0.5% in comparison with the exact value determined from the numerical solution of Eq. (12). Likewise, if $g_o = 20\ \text{mm}^{-1}$, $l_{b\max}$ is obtained from Eq. (14) to be $l_{b\max}=$

2.867 mm, which is within an error of 8.8% in comparison with the exact value determined from the numerical solution of the equation $\mathrm{d}G_{\max}/\mathrm{d}l_b = g_o$.

4 Conclusion

This paper reported the quasi-static modeling and optimization of the bridge-type displacement amplifier. The reported model takes into account the effect of the compliance of the connecting-beam. Closed-form expressions for the amplification ratio, the necessary force, and the resulting angular deformations of the hinges were derived using the developed model. It was seen that for values of tilt angle below a certain threshold, the elastic effects of the beam become important, whereas for angles larger than this limit, the elastic effects can be neglected, and the beam can be modeled as a rigid element. The model also revealed that the required input force reduced with an increase in the tilt angle. The analytical expressions of amplification ratio and input force were validated by employing the finite element method. Subsequently, an optimization of the amplifier was undertaken wherein analytical expressions for maximum achievable amplification ratio, and the corresponding optimum tilt angle was obtained. The effect of beam-length on the maximum achievable amplification was studied and verified using the finite element method. Two characteristic length-scales were identified, and guidelines for choosing the beam-length in terms of these two length-scales were also described. An analytical expression was provided for the minimum beam-length necessary to achieve a specified amplification ratio. The discussed results are useful for quickly designing the bridge-type amplifier, with minimal iterations, for a specified amplification ratio.

References

1. Clark L, Shirinzadeh B, Bhagat U, Smith J, Zhong Y (2015) Development and control of a two DOF linear–angular precision positioning stage. Mechatronics 32:34–43. https://doi.org/10.1016/j.mechatronics.2015.10.001
2. Wang P, Xu Q (2018) Design and testing of a flexure-based constant-force stage for biological cell micromanipulation. IEEE Trans Autom Sci Eng 15:1114–1126. https://doi.org/10.1109/TASE.2017.2733553
3. Millet O, Bernardoni P, Régnier S, Bidaud P, Tsitsiris E, Collard D, Buchaillot L (2004) Electrostatic actuated micro gripper using an amplification mechanism. Sens Actuat A 114:371–378. https://doi.org/10.1016/j.sna.2003.11.004
4. Iqbal S, Malik A (2019) A review on MEMS based micro displacement amplification mechanisms. Sens Actuat A 300:111666. https://doi.org/10.1016/j.sna.2019.111666
5. Yong YK, Moheimani SOR, Kenton BJ, Leang KK (2012) Invited review article: high-speed flexure-guided nano-positioning: mechanical design and control issues. Rev Sci Instrum 83:121101. https://doi.org/10.1063/1.4765048
6. Furukawa E, Mizuno M, Doi T (1995) Development of a flexure-hinged translation mechanism driven by two piezoelectric stacks. JSME Int J Ser C, Dyn Control Robot Design Manuf 38:743–748. https://doi.org/10.1299/jsmec1993.38.743

7. Lobontiu N, Garcia E (2003) Analytical model of displacement amplification and stiffness optimization for a class of flexure-based compliant mechanisms. Comput Struct 81:2797–2810. https://doi.org/10.1016/j.compstruc.2003.07.003
8. Ma H-W, Yao S-M, Wang L-Q, Zhong Z (2006) Analysis of the displacement amplification ratio of bridge-type flexure hinge. Sens Actuat A 132:730–736. https://doi.org/10.1016/j.sna.2005.12.028
9. Xu Q, Li Y (2011) Analytical modeling, optimization and testing of a compound bridge-type compliant displacement amplifier. Mech & Mach Theor 46:183–200. https://doi.org/10.1016/j.mechmachtheory.2010.09.007
10. Ling M, Cao J, Zeng M, Lin J, Inman DJ (2016) Enhanced mathematical modeling of the displacement amplification ratio for piezoelectric compliant mechanisms. Smart Mater Struct 25:075022. https://doi.org/10.1088/0964-1726/25/7/075022
11. Ling M (2019) A general two-port dynamic stiffness model and static/dynamic comparison for three bridge-type flexure displacement amplifiers. Mech Syst Signal Process 119:486–500. https://doi.org/10.1016/j.ymssp.2018.10.007
12. Li Y, Xu Q (2009) Design and analysis of a totally decoupled flexure-based XY parallel micromanipulator. IEEE Trans Rob 25:645–657. https://doi.org/10.1109/TRO.2009.2014130
13. Yong YK, Lu T-F, Handley DC (2008) Review of circular flexure hinge design equations and derivation of empirical formulations. Precis Eng 32:63–70. https://doi.org/10.1016/j.precisioneng.2007.05.002

Design Exploration of Stewart Platform

Suraj Kumar Mishra and C. S. Kumar

Abstract In this work, kinematic design exploration of a Stewart platform manipulator is carried out. The paper presents the effect of the design parameters of the Stewart platform on its different range of motions. We employ the inverse kinematic model of the manipulator with some design constraints to do this exploration. There are also limits on input actuations. The different kinematic design parameters of the mechanisms are outlined. Out of all the parameters, we select three design parameters for our study. These parameters are then changed one by one to see the effect on the range of motions. Useful trends are generated for each case. These results are then validated by Adams simulations with a very close confirmation. The trends presented in the paper could prove extremely helpful for designers to obtain the configuration with improved range of motion in any of the six degrees-of-freedom motions.

Keywords Stewart platform · Design exploration · Inverse kinematics

1 Introduction

Design process of a compliant mechanism is an iterative process [1]. It typically starts with synthesis of a rigid body mechanism as per the requirement of application and constraints. In the next step, the compliant version of the mechanism is made by using suitable flexure joints. Now, to check whether this compliant version of the mechanism fulfils the design criteria one resorts to analysis of the compliant version of the mechanism. If the compliant model falls short of the objective, initially changes are done on the compliant mechanism. If however these changes does not bring desired results, changes in the rigid body mechanism are carried out.

The micromanipulators are the mechanisms which are used for manipulation at the micro/nano scales with precision. The Stewart platform [2, 3] is one of the most

S. K. Mishra (✉) · C. S. Kumar
Indian Institute of Technology Kharagpur, Kharagpur 721302, India
e-mail: surajkmishra@iitkgp.ac.in; surajkmishra1989@gmail.com

C. S. Kumar
e-mail: kumar@mech.iitkgp.ac.in

V. K. Gupta et al. (eds.), *Recent Advances in Machines and Mechanisms*, Lecture Notes in Mechanical Engineering, https://doi.org/10.1007/978-981-19-3716-3_10

popular parallel manipulators. The design exploration presented in this paper is targeted towards the compliant parallel micromanipulators (CPMs) which are based on Stewart platform. One such CPM was proposed in an earlier work of the authors [4, 5].

Considerable amount of work has been done for range of motion or workspace determination of Stewart platform [6–8]. There are quite a few research papers on finding the optimal geometry to maximize the workspace as well [9–14]. However, the effect of dimensional parameters on range of motion or workspace has not been explored much. Merlet [15] in his paper considered four different designs of Stewart platform: symmetric simplified manipulator (SSM), triangular symmetric simplified manipulator (TSSM), minimal symmetric simplified manipulator (MSSM), and modified Stewart platform (MSP) [16]. The author concluded that in terms of workspace, SSM design gives best performance followed by TSSM and MSSM design. Jiang and Gosselin [8] carried out design exploration for MSSM type Stewart platform. They found out that when the base and platform are similar triangles, the optimum architecture will be obtained when both base and platform are equilateral triangles with the size ratio between the platform and the base is 1/2. Moreover, they discovered that if the base and platform are not similar triangles, then it is difficult to determine the global optimal architecture.

There could be a situation wherein the Stewart platform is required to have better range of motion in certain degree of freedoms (DOFs). In the literature, we did not find any work which investigates the effect of dimensional parameters of the Stewart platform on its range of motion of each DOF. The present work attempts to fill this gap. A general Stewart platform involves numerous design parameters which makes its design and analysis very complex. In their paper on optimal design of Stewart platform, Bhattacharya et al. [17] proposed to reduce the number of parameters to two or three so that the performance can be graphically represented. A similar approach has been followed in this paper.

2 Kinematic Design Parameters of the Stewart Platform

In Fig. 1a, base radius and platform radius are shown as R_b and R_p, respectively. Considering a limit on available space, base radius of the mechanism has been assumed constant in the present exploration. The base angle is one of the important parameters in the design. It is represented by A_b in Fig. 1b. In present study, the base angle is altered while keeping the platform angle A_p (Fig. 1a) constant. The change in base angle changes the platform radius of the mechanism. In Fig. 1b, the parameter A_i depicts vertical inclination of the leg. The change in vertical inclination changes several other parameters as well. The affected design parameters are: platform radius, height of the platform, and platform angle. The legs also have a horizontal inclination represented by A_h in Fig. 1a. Change in A_h alters the platform radius and platform angle. Under the current constraints of the design, the height (H) of platform (see

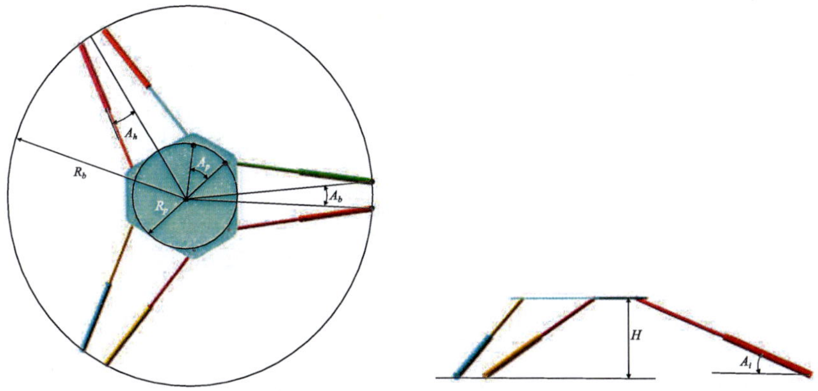

(a) Design parameters of the mechanism (top view)

(b) Design parameters of the mechanism (side view)

Fig. 1 Design parameters of a typical Stewart platform

Fig. 1b) is primarily dependent on the vertical inclination (A_i) of the leg. The effect of change in height can be observed while one changes the inclination of the leg.

2.1 *Selection of Design Parameters for Exploration*

A brief description of all the design parameters of the mechanism was presented above. Since all the design parameters are not independent, a separate study for each parameter would be a redundant exercise. Therefore, in the present work, three parameters have been selected for design exploration; they are: vertical inclination of the leg (A_i), horizontal inclination of the leg (A_h), and base angle (A_b). In what follows the effect of change in these parameters will be explored.

3 Methodology

Figure 2 summarizes the procedure adopted to generate a plot between a design parameter and motion range for a particular DOF. We start by selecting the a value of design parameter d within limit ($d1 < d < d2$). Let us say that we want to check range of motion in X-direction. For a pose in X-direction inside limit ($X1 < X < X2$), inverse kinematics of the updated design gives the input actuations. If these actuations fall within the actuation limit, we call this pose as reachable pose and try the value of X. If any of the actuations are outside the actuation limit, we discard the pose and try the next value of X. Following this process if all poses in the range ($X1 < X < X2$) are covered, then we calculate the difference between the

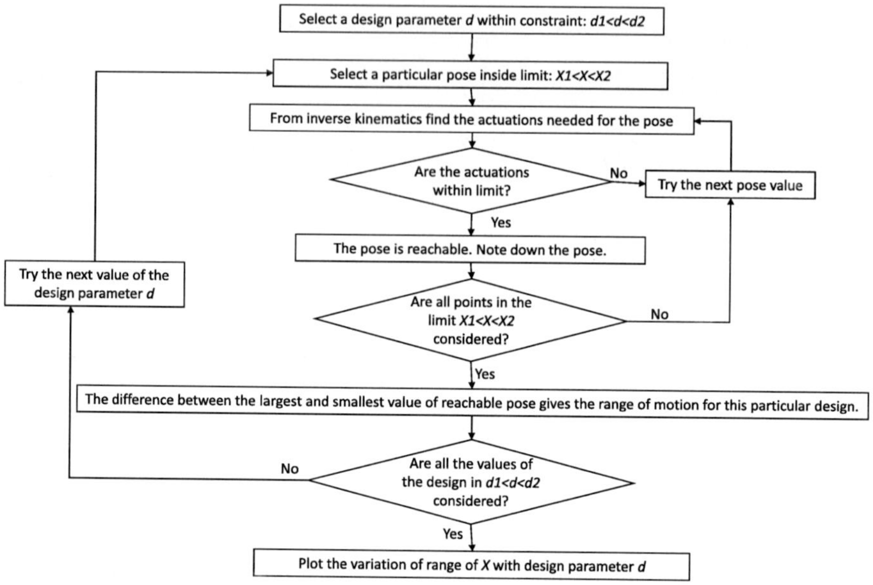

Fig. 2 Flow chart of the methodology followed to obtain the plots

largest and smallest value of reachable pose. This difference will give us the range of X-motion of the platform with d as design parameter.

Following the procedure discussed above, if we cover all the design values in limit ($d1 < d < d2$), then we can generate a plot between values of design parameter (d) and corresponding X-motion ranges. The actuation limit for present exploration is $\pm$ 6.25 μm.

4 Effect of Change in Vertical Inclination (A_i) of the Leg

It was observed earlier that all design parameters are not independent. Table 1 presents the list of dependent and fixed design parameters for the present exploration. Since R_p, A_p and H change with A_i, their expressions need to be derived.

Table 1 Dependent and constant design parameters in case of change in vertical inclination of the leg

Driving design parameter	A_i
Dependent variable parameters	R_p, A_p, H
Fixed parameters	A_b, A_h

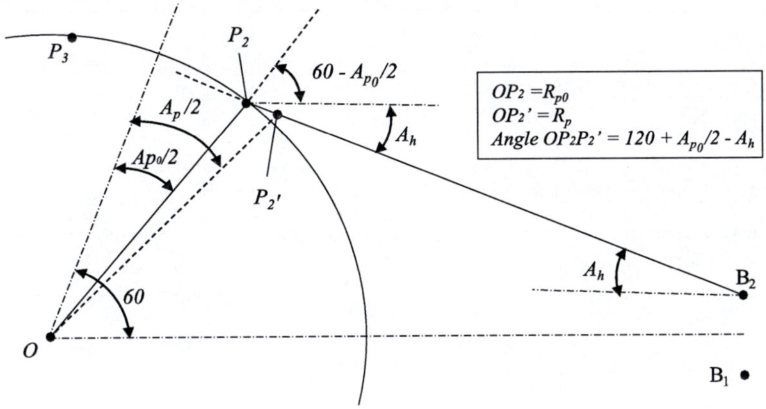

(a) Top view

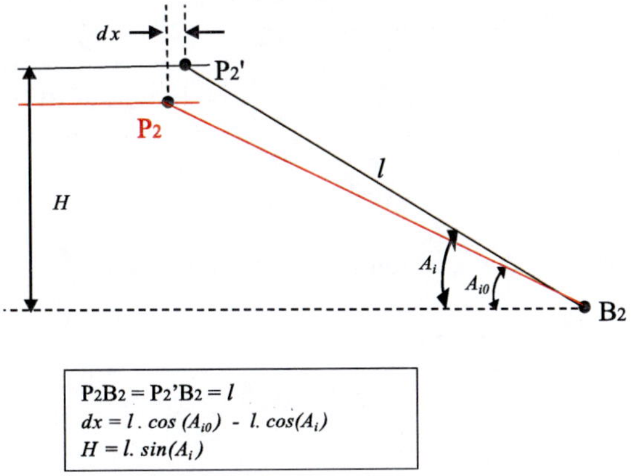

(b) Side view

Fig. 3 Geometrical analysis of change in vertical inclination

$\mathbf{R_p}$: With reference to Fig. 3a, the updated platform radius is OP_2'. To determine it, triangle OP_2P_2' is considered. Here, P_2P_2' is represented in Fig. 3b as dx. From the properties of triangle, R_p can be written as follows:

$$R_p = \sqrt{R_{p0}^2 + \mathrm{dx}^2 - 2\,R_{p0}\,\mathrm{dx}\;\cos(120 - (A_h) + (A_{p0}/2))} \tag{1}$$

$\mathbf{A_p}$: This angle is indicated in Fig. 3a by twice of the angle $(A_{p0}/2 + \angle P_2OP_2')$. The angle P_2OP_2' can be determined by the property of the triangle OP_2P_2'. The final equation for A_p is given below.

$$A_p = A_{p0} + 2\,\sin^{-1}\left(\frac{dx}{R_p}\,\sin\left(120 - A_h + \frac{A_{p0}}{2}\right)\right) \tag{2}$$

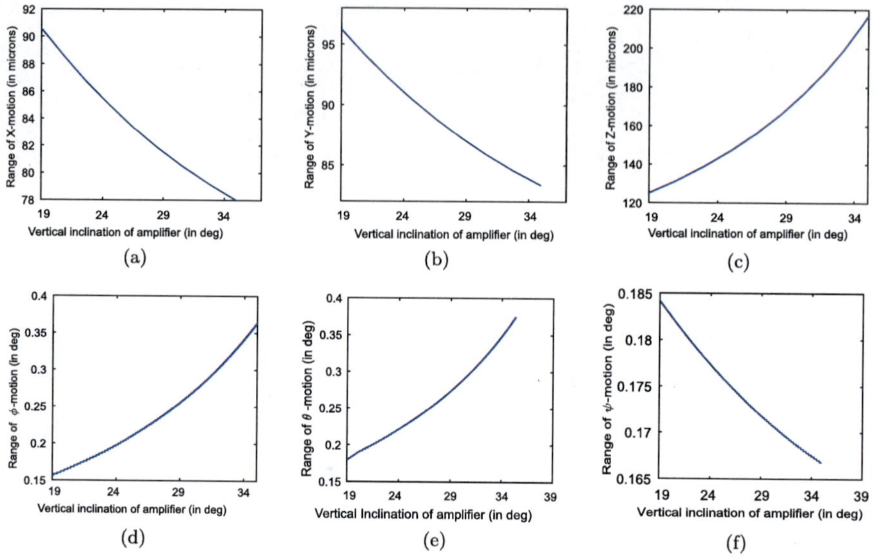

Fig. 4 Effect of change in vertical inclination of leg (A_i) on range of motions

H: With reference to Fig. 3b the equation for H can be written as

$$H = l\ \sin(A_i) \tag{3}$$

4.1 Results

Figure 4 shows the effect of change in vertical inclination (A_i) on the range of motion of each of the 6-DOF motions. It should be noted that original value of A_i (i.e. A_{i0}) was 23°. The range of variation of A_i is: 19°–35°.

5 Effect of Change in Horizontal Inclination (A_h) of the Leg

Table 2 outlines the different design parameters for this operation. The schematic diagram for change in A_h is depicted in Fig. 5. Since R_p and A_p are dependent parameters here, we need to find their expressions. Consider triangles OSP_2' and B_2TP_2' in Fig. 5. For line $P_2'S$, the following equation can be written:

$$l' \sin(A_h) + K_1 = R_p \sin\left(60 - \frac{A_p}{2}\right) \tag{4}$$

Table 2 Dependent and constant design parameters in case of change in horizontal inclination of the leg

Driving design parameter	A_h
Dependent variable parameters	R_p, A_p
Fixed parameters	A_b, A_i, H

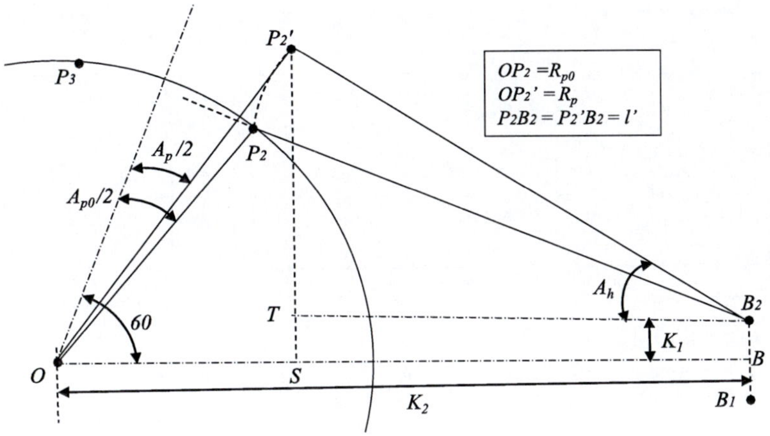

Fig. 5 Geometrical description for horizontal inclination of mechanism (top view)

Similarly for line OB, the following equation is given:

$$l' \cos(A_h) + R_p \cos\left(60 - \frac{A_p}{2}\right) = K_2 \tag{5}$$

Solution of Eqs. 4 and 5 yields the expressions for R_p and A_p.

$$A_p = -60 - 2\tan^{-1}\left(\frac{l' \cos A_h - K_2}{l' \sin A_h + K_1}\right) \tag{6}$$

$$R_p = (l' \sin A_h + K_1)\sqrt{1 + \frac{(l' \cos A_h - K_2)^2}{(l' \sin(A_h) + K_1)^2}} \tag{7}$$

5.1 Results

Figure 6 depicts the effect of change in horizontal inclination (A_h) on the range of motion of each of the 6-DOF motions. The value of A_h for original mechanism was 8°. The range of variation of A_h for present exploration is 0°–12.5°.

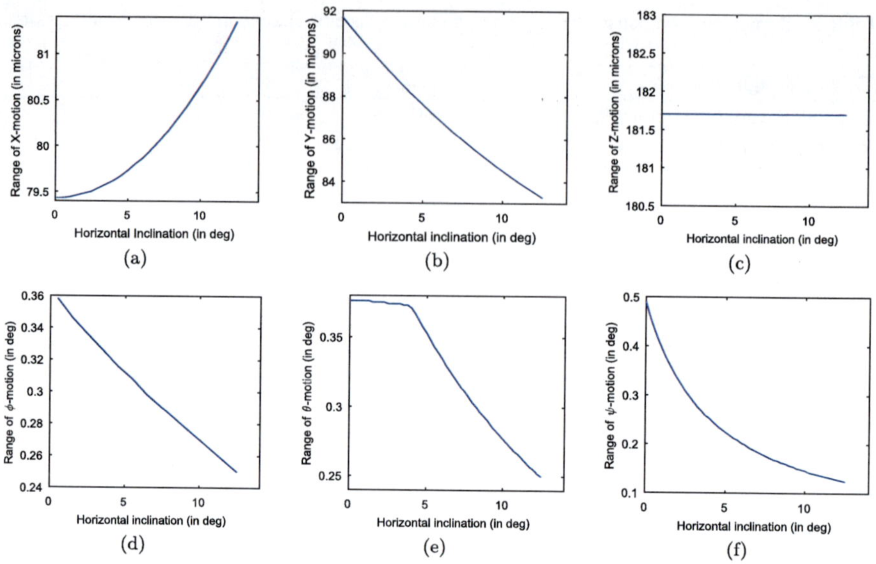

Fig. 6 Effect of change in horizontal inclination of leg (A_h) on range of motions

6 Effect of Change in Base Angle (A_b)

The variation in base angle can be done in many possible ways. In present case, variation platform angle A_p has been assumed to be constant. In such a situation, change in base angle alters R_p and A_h. In what follows, we shall try to find the expression for R_p (Table 3). With reference to Fig. 7, P_2B_2' can be written as

$$P_2B_2' = \sqrt{(l')^2}$$

$$= \sqrt{\left(R_b \cos\left(\frac{A_b}{2}\right) - R_p \cos\left(60 - \frac{A_p}{2}\right)\right)^2 + \left(R_b \sin\left(\frac{A_b}{2}\right) - R_p \sin\left(60 - \frac{A_p}{2}\right)\right)^2} \quad (8)$$

Table 3 Variable and constant design parameters during change in base angle of the mechanism

Driving design parameter	A_b
Dependent variable parameter	R_p, A_h
Fixed parameters	A_p, A_i, H

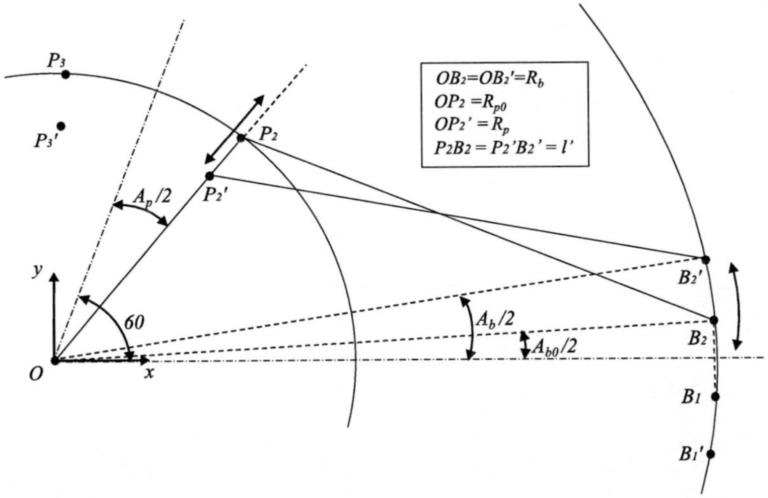

Fig. 7 Geometrical description for change in base angle of the mechanism (top view)

The above equation yields the following quadratic equation in R_p.

$$R_p^2 - 2R_b \cos\left(60 - \frac{A_p}{2} - \frac{A_b}{2}\right) R_p + (R_b^2 - (l')^2) = 0 \tag{9}$$

Solution of this equation is presented below.

$$R_p = \frac{1}{2}\left(2\,R_b \cos\left(60 - \frac{A_p}{2} - \frac{A_b}{2}\right) \pm\sqrt{4\,R_b^2 \cos^2\left(60 - \frac{A_p}{2} - \frac{A_b}{2}\right) - 4(R_b^2 - (l')^2)}\right) \tag{10}$$

In the previous equation, if one takes the positive sign, then R_p corresponds to the condition wherein $R_p > R_b$. As this is physically not possible for the current design of the mechanism. The final acceptable equation for R_p is given as

$$R_p = \frac{1}{2}\left(2\,R_b\, cos\left(60 - \frac{A_p}{2} - \frac{A_b}{2}\right) - \sqrt{\left(4\,R_b^2\, cos^2\left(60 - \frac{A_p}{2} - \frac{A_b}{2}\right) - 4(R_b^2 - (l')^2)\right)}\right) \tag{11}$$

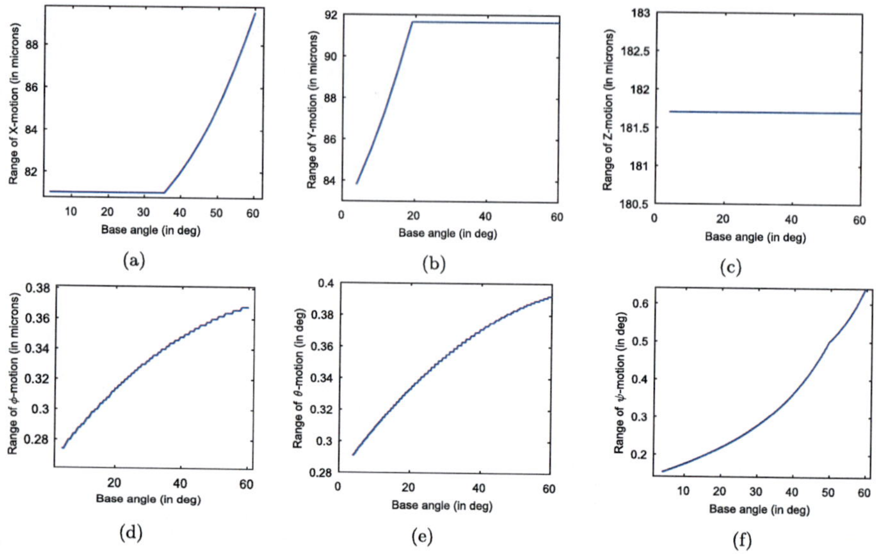

Fig. 8 Effect of change in base angle (A_b) on range of motions

Table 4 Dimensional parameters for the original design (all lengths are in mm)

K_1	l'	l	R_{p0}	R_{b0}	A_{p0}
9.20	99.18	108.24	36.70	127.16	42.40°
A_{b0}	A_{h0}	A_{i0}	H_0	K_2	
8.30°	8°	23°	43.35	126.82	

6.1 Results

The effect of change in base angle (A_b) on the range of motion of each of the 6-DOF motions is summarized in Fig. 8. Base angle for original mechanism was 8.3°. Here, range of variation of A_b is 4°–60°.

7 Validation Through MSC-Adams Simulations

In this section, the correctness of the analytical results will be checked by using MSC-Adams software package. In the previous sections, we obtained the trends for motion range for variation in three design parameters: vertical, and horizontal inclination of the leg, and base angle. In what follows, we shall make three updated designs, each with change in one of the three previously mentioned parameters. Then, with the help of Adams simulation, it will be checked whether they actually produce the outcome as predicted by the analytical model. The values of design parameters for the original design is summarized in Table 4.

(a) Original design: $A_i = 23^o$ (b) Updated design: $A_i = 19^o$

Fig. 9 Comparison of the original and updated design in Adams environment

Table 5 Comparison of analytical and simulation results for change in A_i

Motion type	Range in the original design	Expected range for new design	Range observed from Adams simulation
X-motion	87.22 μm	90.56 μm	90.558 μm
ϕ-motion	0.183°	0.158°	0.1580°

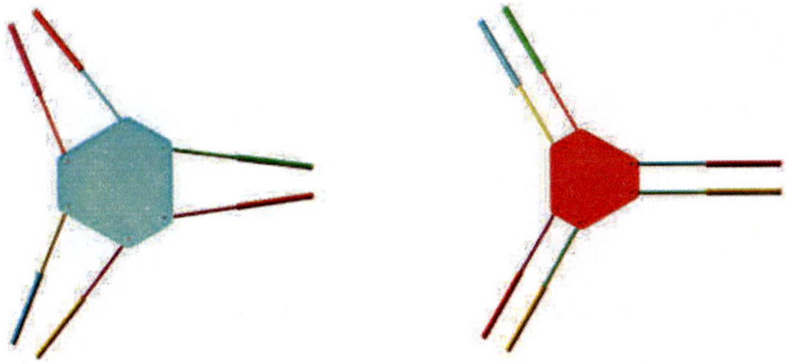

(a) Original design: $A_h = 8^o$ (b) Updated design: $A_h = 0^o$

Fig. 10 Comparison of the original and updated design in Adams environment

7.1 Change in Vertical Inclination

In the original design of the mechanism, vertical inclination of the leg was 23°. Let us consider a design with 19° inclination. The Adams model of the original and the new design is depicted in Fig. 9. The updated design parameters are: $A_i = 19°$, $A_p = 36.14$ mm, $R_p = 34.72$ mm, and $H = 36.33$ mm.

Table 5 presents a comparison of the expected and observed motion ranges for the updated design.

7.2 Change in Horizontal Inclination

The leg in the original design had 8° of horizontal inclination. For the new design, let us assume that there in no horizontal inclination, i.e. $A_h = 0$. Figure 10 compares this new design with the original design. Updated dimensional parameters in this case are: $A_h = 0°$, $R_p = 29.13$ mm, and $A_p = 83.19°$.

Table 6 Comparison of analytical and simulation results for change in A_h

Motion type	Range in the original design	Expected range for new design	Range observed from Adams simulation
Y-motion	85.66 μm	91.73 μm	91.732 μm
θ -motion	0.301°	0.377°	0.3785°

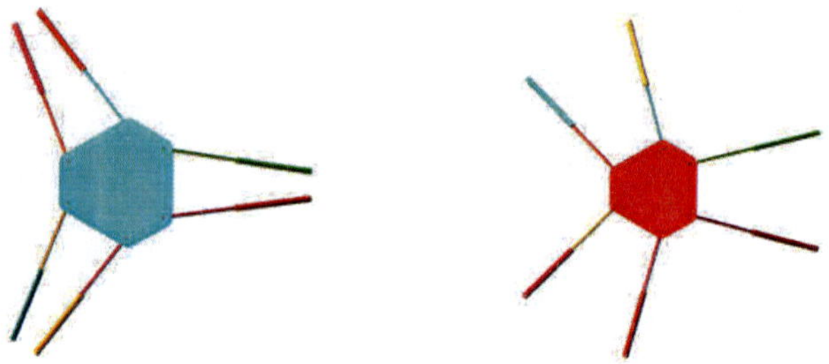

(a) Original design: $A_b = 8.3^o$ (b) Updated design: $A_b = 40^o$

Fig. 11 Comparison of the original and updated design in Adams environment

Table 7 Comparison of analytical and simulation results for change in A_b

Motion type	Range in the original design	Expected range for new design	Range observed from Adams simulation
Z-motion	181.70 μm	181.70 μm	181.701 μm
ψ -motion	0.169°	0.364°	0.3640°

A comparison between expected and observed motion ranges for the updated design is outlined in Table 6.

7.3 *Change in Base Angle*

The value of base angle for the original design was 8.3°. Let us change this to 40°. The resulting new design is illustrated in Fig. 11 along with the original design. For the present case, updated structural parameters are: $A_b = 40°$ and $R_p = 30.05$ mm.

In Table 7, a comparison between the expected motion and observed motion for the new design is summarized.

In this section, we performed the simulations for validation of the analytical results. It was seen that simulation results very closely confirmed to the expected values predicted by analytical model. This proves the reliability of the results presented in Sects. 4.1, 5.1, and 6.1.

8 Conclusions

In present work, kinematic design exploration of a Stewart platform was carried out. We used the inverse kinematics model of the platform with some design constraints to do this exploration. There were also limits on input actuations. The different kinematic design parameters of the mechanisms were presented with the help of rigid-body model of the mechanism. We also discussed about fixed and variable design parameters. Out of all the parameters, we selected three design parameters for our study. These parameters were then changed one by one to see the effect on range of motions. Useful trends were generated for each case. These results were then validated by Adams simulations with very close confirmation.

The trends presented in the work could prove extremely helpful for designers to obtain the configuration with desired range of motion.

References

1. Howell LL, Midha A (1994) A method for the design of compliant mechanisms with small-length flexural pivots. J Mechan Des 116(1):280–290
2. Stewart D (1965) A platform with six degrees of freedom. Proc Inst Mechan Eng 180(1):371–386
3. Gough VE (1962) Universal tyre test machine. Proc FISITA 9th Int Tech Congr Lond 1962:117–137
4. Mishra SK, Kumar CS (2018) Design and kinematics of a compliant Stewart micromanipulator. In: 2018 international conference on manipulation, automation and robotics at small scales (MARSS). IEEE, pp 1–6
5. Mishra SK, Kumar CS (2022) Compliance modeling of a full 6-DOF series–parallel flexure-based Stewart platform-like micromanipulator. Robotica 1–28. https://doi.org/10.1017/S0263574722000327
6. Gosselin C (1990) Determination of the workspace of 6-dof parallel manipulators
7. Pernkopf F, Husty M (2006) Workspace analysis of Stewart-Gough-type parallel manipulators. Proc Inst Mechan Eng Part C: J Mechan Eng Sci 220(7):1019–1032
8. Jiang Q, Gosselin CM (2009) Evaluation and representation of the theoretical orientation workspace of the Gough–Stewart platform. J Mechan Robot 1(2). https://doi.org/10.1115/1.3046137
9. Dwarakanath T, Dasgupta B, Mruthyunjaya T (2001) Design and development of a Stewart platform based force-torque sensor. Mechatronics 11(7):793–809
10. Ceccarelli M, Carbone G, Ottaviano E (2005) Multi criteria optimum design of manipulators. Bull Polish Acad Sci Tech Sci 9–18
11. Jiang Q, Gosselin CM (2009) Geometric optimization of the mssm Gough–Stewart platform. J Mechan Robot 1(3). https://doi.org/10.1115/1.3147202
12. Toz M, Kucuk S (2013) Dexterous workspace optimization of an asymmetric six-degree of freedom Stewart-Gough platform type manipulator. Robot Auton Syst 61(12):1516–1528
13. Xie Z, Li G, Liu G, Zhao J (2017) Optimal design of a Stewart platform using the global transmission index under determinate constraint of workspace. Adv Mechan Eng 9(10):1687814017720880
14. Qiang H, Wang L, Ding J, Zhang L (2019) Multiobjective optimization of 6-dof parallel manipulator for desired total orientation workspace. Math Probl Eng
15. Merlet J-P (1999) Determination of 6d workspaces of Gough-type parallel manipulator and comparison between different geometries. Int J Robot Res 18(9):902–916
16. Stoughton RS, Arai T (1993) A modified Stewart platform manipulator with improved dexterity. IEEE Trans Robot Autom 9(2):166–173

17. Bhattacharya S, Hatwal H, Ghosh A (1995) On the optimum design of Stewart platform type parallel manipulators. Robotica 13(2):133–140

Design of Bistable Arch-Profiles by Using Bilateral Relationship and Shape Optimization

Sabyasachi Dash, Praneet Nallan Chakravarthula, and Safvan Palathingal

Abstract The design of bistable arches is challenging due to their inherent nonlinear behaviour. It is known that the arch profiles of bistable arches in their two force-free equilibrium states are related to each other through a bilateral relationship. The closed-form analytical form of the bilateral relationship helps in designing the as-fabricated profile of the bistable arches when their toggled shape is completely specified. However, in certain cases, the arch profiles cannot be designed solely using the bilateral relationship. In this paper, we address such cases and present a design method to obtain arch-profiles using shape optimization in conjunction with the bilateral relationship. Such cases include: (1) arch-profiles that are partially specified in the first or/and second states; (2) arch-profiles that extremise the travel have a prescribed change in slope and have a specific ratio of travel between stable and unstable states. Towards this, the arch-profiles are represented using a set of finite variables corresponding to the mode weights of the basis set in which their shape is described. We formulate the optimization problem for each design case with the constraint to restrict the solutions in design space where the arches are bistable. After describing the procedure for numerically solving the optimization problem, design of arch-profiles using this method is illustrated through examples.

Keywords Snap-through buckling · Compliant mechanisms · Inverse problem · Mobile-holder

---

S. Dash (✉)
Department of Mechanical Engineering, National Institute of Technology, Tiruchirappalli, Tamil Nadu, India
e-mail: sabyadash21@gmail.com

P. N. Chakravarthula · S. Palathingal (✉)
Department of Mechanical and Aerospace Engineering, Indian Institute of Technology, Hyderabad, Telangana, India
e-mail: safvan@mae.iith.ac.in

V. K. Gupta et al. (eds.), *Recent Advances in Machines and Mechanisms*, Lecture Notes in Mechanical Engineering, https://doi.org/10.1007/978-981-19-3716-3_11

1 Introduction

Stable profiles of bistable arches: initial profile, $H(X)$, and toggled profile, $W(X)$, can be designed analytically by using bilateral relationships [1, 2] when one of them is completely prescribed as summarized in Fig. 1. In this paper, we address three cases in which design using the bilateral relationship can be improved with optimization. One is the case of partially-specified arches as illustrated using an example in Fig. 2. Here, the profile is specified only in a region of interest. In the example considered, specified profile is a flat segment at the mid-span of the initial profile of the arch. This leaves a region of unspecified section in the arch-profile as indicated in Fig. 2. Each valid shape chosen for the unspecified region leads to a different initial profile. By formulating an optimization problem, we seek to identify optimal designs in such scenarios, which is the novelty of this work. Utility of the formulation is illustrated by considering an example, that of a mobile holder, as indicated in Fig. 4. When the phone is pushed against the toggled profile, the arch snaps to the as-fabricated profile that matches with the dimensions of the phone and holds it in place without any external force.

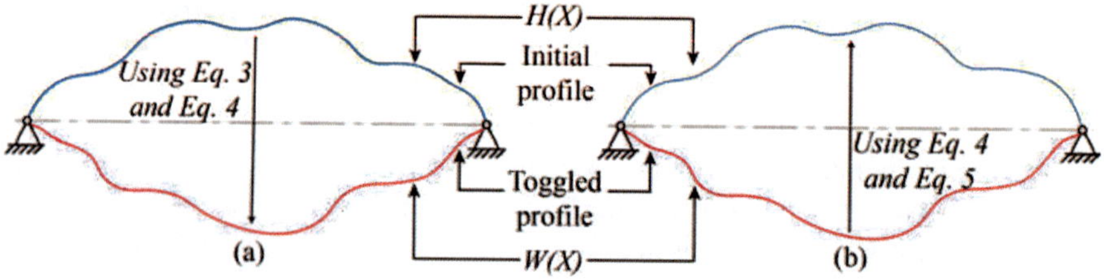

Fig. 1 Designing bistable profiles of an arch with pinned-pinned boundary conditions: **a** toggled profile from prescribed initial profile, **b** initial profile from a specified toggled profile

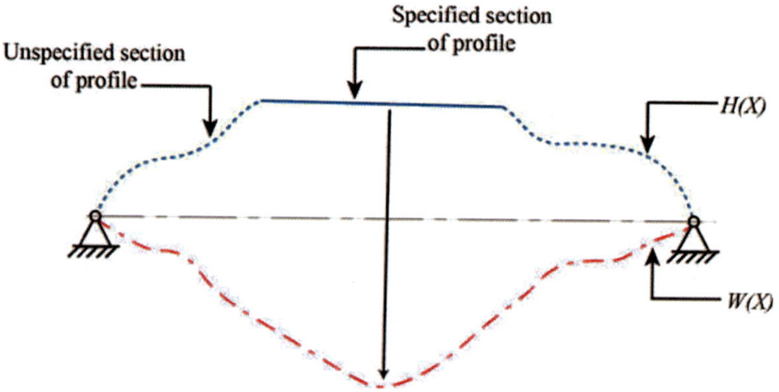

Fig. 2 Bistable arch with a partially-specified initial profile

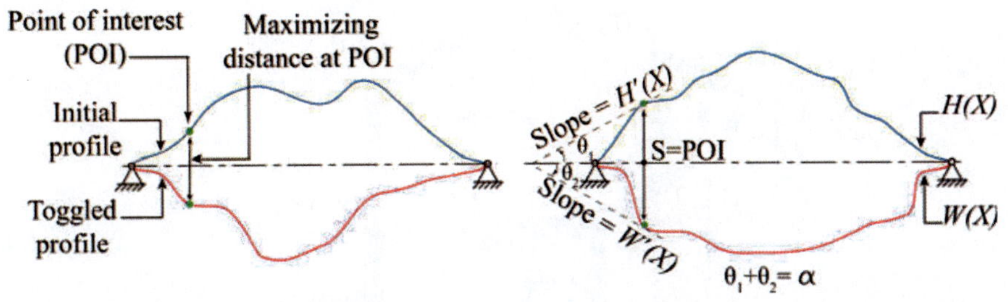

Fig. 3 Optimization of arch-profiles

Second case is concerned with extracting optimal characteristics at a point of interest (POI). We consider three examples here: maximum travel between the two stable states for multiple points along the span of the arch, specified change in slope at a POI when the arch deforms between the two stable states (see Fig. 3), and specific travel between stable and unstable state.

Bistable arches are well studied and analyzed in the literature [3–6]. The optimization-based design presented in this work helps to design optimal arch-profiles for the aforementioned design cases. Such optimal arches will find use in applications for bistable arches [7] such as micro-optical switches and relays [8], grippers [9], chairs for people with arthritis [10], mechanical logic gates [11], and rear trunk lid of cars [12].

The rest of the paper is arranged as follows: in Sect. 2, details on the mathematical preliminaries and the analytical relationships between the as-fabricated and the toggled shapes and the required conditions for bistability are summarized. In Sect. 3, the optimization problem and constraints are defined, key objective functions are identified, and optimal arch-profiles are obtained. These results are discussed in Sect. 4. Finally, the bistability of the optimized profiles is validated using nonlinear finite element analysis.

2 Summary of Bilateral Relationship

The bilateral relationship between the initial and toggled profiles of bistable arches can be described analytically [1]. For pinned-pinned bistable arches, the normalized initial profile, $H(X)$, and the normalized deformed profile, $W(X)$, are taken as a weighted combination of all the buckling mode shapes of a straight pinned-pinned column. We can write the normalized profiles as follows:

$$H\left(X\right)=\sum_{i=1}^{\infty}a_i\sin\left(M_iX\right) \tag{1}$$

$$W\left(X\right)=\sum_{i=1}^{\infty}A_i\sin\left(M_iX\right) \tag{2}$$

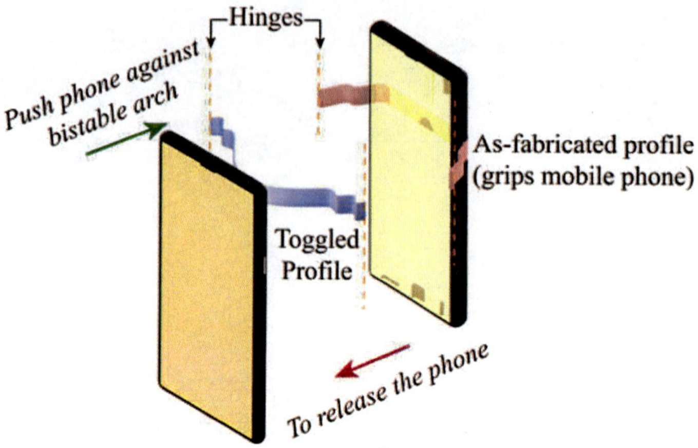

Fig. 4 Using a bistable arch to hold a mobile phone in place without any external force

where $M_i = i\pi$, $X = \frac{x}{L}$, and L is the span of the arch. The equation for the first weight of a force-free deformed profile (A_1) of the arch can be obtained by solving Eq. (3)

$$3Q^2 \sum_{i=1}^{\infty} \left\{ \frac{\frac{1}{i^2}\left(1 - \frac{a_1}{A_1}\right) - 2}{\left[1 - \frac{1}{i^2}\left(1 - \frac{a_1}{A_1}\right)\right]^2} \right\} {a_i}^2 = 1 \tag{3}$$

where $Q = h_{\mathrm{mid}}/t$, h_{mid} is the height at the mid-span of the arch, and t is the in-plane depth of the cross section. Equation (3) can be solved numerically. It is to be noted that the equation yields multiple solutions of A_1, and the solution corresponding to the stable toggled profile needs to differentiated from the unstable arch-profile. The stability of profile can be evaluated by analyzing the positive definiteness of the Hessian matrix, discussed in Sect. 2.1. The mode weights other than A_1 can be found using:

$$A_i = \frac{a_i}{1 - \frac{1}{i^2}\left(1 - \frac{a_1}{A_1}\right)} \tag{4}$$

The initial profile of the arch can also be obtained analytically when a toggled profile is specified with the aid of:

$$a_1 = A_1 \frac{\sum_{i=1}^{\infty} \frac{{A_i}^2}{i^2} - 2\sum_{i=1}^{\infty} {A_i}^2 - \frac{1}{3Q^2}}{\sum_{i=1}^{\infty} \frac{{A_i}^2}{i^2}} \tag{5}$$

The remaining a_i values can be obtained from Eq. (4).

2.1 Stability of the Deformed Profiles

As mentioned in the previous section, to ensure stability of the force-free deformed profile, we obtain the Hessian matrix from the strain energy expression. The strain energy comprises bending and the compressing energies, and we can define the total strain energy as a function of known and unknown mode weights as [1]:

$$\mathrm{SE} = \frac{1}{4}\left(\sum_{i=1}^{\infty} a_i^2 M_i^4 + \sum_{i=1}^{\infty} A_i^2 M_i^4 - 2\sum_{i=1}^{\infty} a_i A_i M_i^4\right) + 3\frac{h_{\mathrm{mid}}^2}{8t^2}\left(\sum_{i=1}^{\infty} a_i^2 M_i^2 - \sum_{i=1}^{\infty} A_i^2 M_i^2\right)^2 \tag{6}$$

Thus, the diagonal terms in the Hessian matrix, **H**, are given by

$$H_{ii} = \frac{M_i^4}{2} - \frac{3Q^2 M_i^2}{2}\left(\sum_{k=1}^{\infty} a_k^2 M_k^2 - \sum_{k=1}^{\infty} A_k^2 M_k^2\right) + 3Q^2 A_i^2 M_i^4 \tag{7}$$

and the other terms are given as

$$H_{ij} = 3Q^2 A_i A_j M_i^2 M_j^2 \tag{8}$$

Positive definiteness of **H** defined in Eqs. (7) and (8) guarantees bistability of the arch.

3 Optimization Problems

In this section, we define the optimization problem used for designing bistable arches with (1) maximum travel at a point, (2) specified change in slope at a point, (3) specific travel between the initial, unstable, and second stable states, and (4) specified region in the arch-profile.

3.1 Travel at a Point Between Initial and Toggled Profiles

The optimization problem defined here maximizes the distance between the arches in the initial and toggled states for a specified value of X. The optimization problem is defined as

$$\max_{a_i} J_1(a_i) = H(S) - W(S)$$

subject to:

$$g_1(A_1) = 3Q^2 \sum_{i=1}^{\infty} \left\{ \frac{\frac{1}{i^2}\left(1 - \frac{a_1}{A_1}\right) - 2}{\left[1 - \frac{1}{i^2}\left(1 - \frac{a_1}{A_1}\right)\right]^2} \right\} a_i{}^2 - 1 = 0$$

$$g_2(A_i) = A_i - \frac{a_i}{1 - \frac{1}{i^2}\left(1 - \frac{a_1}{A_1}\right)} = 0 \tag{9}$$

$$H\,(0.5) = 1$$

$$v^{\mathrm{T}}\mathbf{H}v \geq 0$$

$$l_{\mathrm{b}} \leq a_i \leq u_{\mathrm{b}}$$

data:

$$S, Q, l_{\mathrm{b}}, u_{\mathrm{b}}$$

where the objective function, $J_1(a_i)$, is the distance between the toggled and initial profiles at the point S. $H(X)$ and $W(X)$ are given by Eqs. (1) and (2). The first and second constraints follow from the bilateral Eqs. (3) and (4). The constraint $H\,(0.5) = 1$ ensures that we consider a family of bistable arches with a constant h_{mid}. The upper (u_{b}) and lower (l_{b}) bounds on the design variable, a_i the weights of the fabricated shape, help in the convergence [6]. The inequality constraint ensures that the arch-profiles are bistable.

3.2 *Change in Slope at a Point*

Bistable arches used in applications like a passive gripper require change in slopes at point between the two stable states to be a certain value. Towards this, we pose an optimization problem to design bistable arch-profiles of change in slope α at a point S (see Fig. 3). The objective function can be expressed as

$$J_2(a_i) = \left(H'\,(S) - W'\,(S) - \tan(\alpha)\right)^2 \tag{10}$$

where

$$\alpha = \theta_1 + \theta_2$$

constraints remain as same as in Eq. (9). In the data, we need to specify the required change in slope $\tan(\alpha)$.

3.3 Travel at a Point Between Initial, Unstable, and Toggled Profiles

As mentioned before, bistable arches have a stable initial state, an unstable in-between state, and a stable toggled profile. The displacements between each of these states are of interests to a designer. The distance between the initial and toggled profiles is denoted by Γ_{stable} $(H(S) - W_{\text{stable}}(S))$, and the distance between the initial profile and unstable profile is Γ_{unstable} $(H(S) - W_{\text{unstable}}(S))$. We intend to design the initial profile such that Γ_{unstable} is β times Γ_{stable}. The optimization problem can be written as

$$\min_{a_i} J_3(a_i) = (\beta\,\Gamma_{\text{stable}} - \Gamma_{\text{unstable}})^2$$

subject to:

$$\begin{aligned} &g_1(A_1) = 0 \\ &g_2(A_i) = 0 \\ &g_1(A_1') = 0 \\ &g_2(A_i') = 0 \\ &H\,(0.5) = 1 \\ &v^{\mathrm{T}}\mathbf{H}v \geq 0 \\ &v^{\mathrm{T}}\mathbf{H}'v < 0 \\ &l_{\mathrm{b}} \leq a_i \leq u_{\mathrm{b}} \end{aligned} \tag{11}$$

data:

$$S,\, Q,\, l_{\mathrm{b}},\, u_{\mathrm{b}}$$

where A_i' are the mode weights of the unstable profile and $\mathbf{H}'$ is the corresponding Hessian.

3.4 Partially-Specified Profile

Arches with partially-specified profile in the initial state $(P(S),\ S_a \leq S \leq S_b)$ are relevant in various practical applications as demonstrated by a mobile phone holder example later. The unspecified region of the initial profile is optimized for a shape that snaps into a bistable second state whilst extremise an objective function of choice. Here, we are using travel $(J_1(a_i))$ as the objective function. The optimization problem can be written as

$$\max_{a_i} J_1(a_i) = (H\,(S) - W\,(S))^2$$

subject to:

$$\begin{aligned} &g_1(A_1) = 0 \\ &g_2(A_i) = 0 \\ &H(S) = P(S) \quad \text{for } S_a \leq S \leq S_b \\ &v^{\mathrm{T}}\mathbf{H}v \geq 0 \\ &l_{\mathrm{b}} \leq a_i \leq u_{\mathrm{b}} \end{aligned} \tag{12}$$

data:

$$P(S), S_a, S_b, Q, l_{\mathrm{b}}, u_{\mathrm{b}}$$

4 Results and Discussion

The objective functions are extremised using the Fmincon function in the optimization Toolbox in MATLAB R2019b [13]. The Fmincon function allows to easily incorporate the nonlinear equality and inequality constraints as well as the lower and upper bounds on the design variables defined in the optimization problems in Sect. 3.

Details on the implementation and the optimization results obtained for each of the objective functions are presented next.

4.1 *Travel at a Point Between Initial and Toggled Profiles*

The key step in solving Eq. (9) is to satisfy the nonlinear constraint, $g_1(A_1) = 0$. We use Fsolve function in MATLAB to solve for values of A_1 that satisfy $g_1(A_1) = 0$, which is passed on to the Fmincon routine as a nonlinear constraint. Implementing the linear constraint $g_2(A_i) = 0$ and setting the bounds on the design variables are straight forward in Fmincon. The optimization using Fmincon is carried out by assuming an initial value to the design variable a_i, which is iterated in each step whilst satisfying constraints towards the optimal solution. The optimized profiles are checked for bistability, by analyzing the eigenvalues of the matrix defined in Eqs. (7) and (8) and ensuring they are strictly positive.

For the examples considered here, five mode weights $[a_1, a_2, a_3, a_4, a_5]$ are chosen with the lower and upper bounds $[0.1, -0.2, -0.2, -0.2, -0.2]$ and $[2, 0.2, 0.2, 0.2, 0.2]$, respectively. The initial guess used is $[1, 0, 0, 0, 0]$. The optimal arch-profiles when S is 0.1, 0.9, 0.5, and 0.35 are given in Fig. 5. Note that the shapes of the profiles when S is 0.1 and 0.9 are mirror images of each other. The result obtained for $S = 0.5$ is in agreement with [6]. The optimal a_i values obtained in each of these cases are given in Table 1.

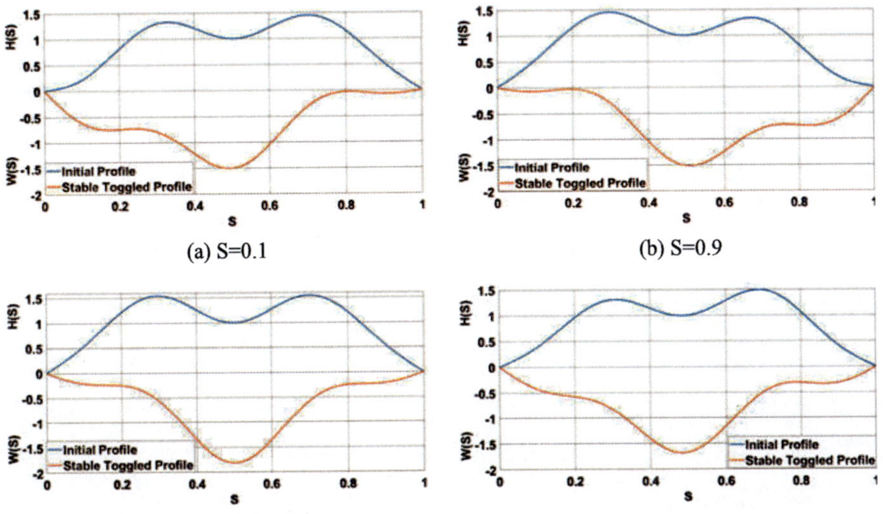

(a) S=0.1

(b) S=0.9

(c) S=0.5

(d) S=0.35

Fig. 5 Arch-profiles with maximum travel at S

Table 1 Optimal a_i for maximum travel at S

Position along span (S)	Optimal a_i
0.1	[1.376, −0.130, 0.177, −0.081, −0.198]
0.35	[1.399, −0.073, 0.199, 0.031, −0.199]

4.2 Change in Slope at a Point

We follow a similar approach to solve Eq. (10), except for the change in the objective function. The arch is optimized to have a difference in slope of $\tan(\alpha)$ at the point S (see Fig. 3). In order to design for cases with change in slope of 3 and -3, we had to use 15 mode weights as design variables. In Fig. 6a, the difference in slopes ($\tan(\alpha)$) between the initial and toggled profiles is 3, corresponding to an angle α of $71.5°$ at $S = 0.25$ along the span. A similar approach was followed for $\tan(\alpha) = -3$ at $S = 0.8$ as shown in Fig. 6b.

4.3 Travel at a Point Between Initial, Unstable, and Toggled Profiles

In the example considered here, we take $S = 0.5$. The algorithm was modified to include the constraints $g_1(A'_1) = 0$ and that the Hessian $\mathbf{H}'$ is negative definite, along

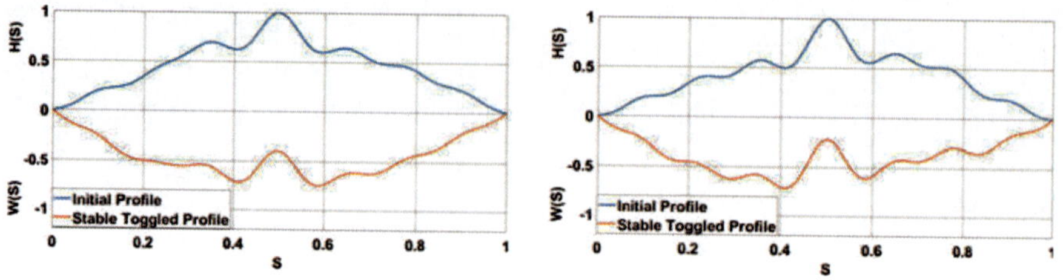

(a) tan(α) = 3 at S = 0.25 (b) tan(α) = -3 at S = 0.8

Fig. 6 Arch-profiles with a specific change in slope at S

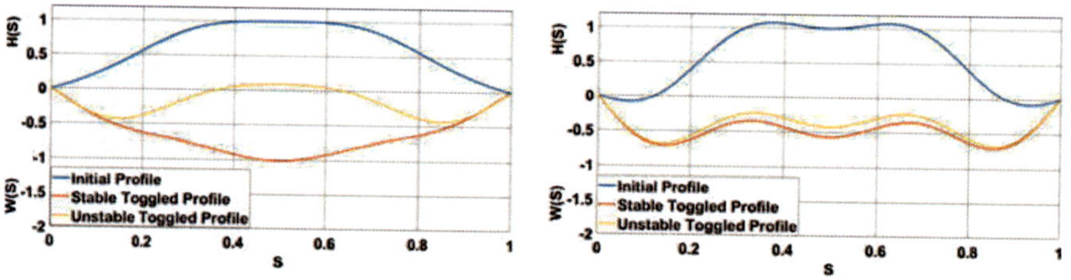

(a) $\beta = 0.45$ (b) $\beta = 0.95$

Fig. 7 Arch-profiles with specific travel between stable and unstable profiles

Table 2 Optimal values of a_i for fixed relationship between stable and unstable profiles

β	Optimized a_i
0.45	[1.018, 0, −0.036, 0, −0.054]
0.95	[1, 0, −0.2, 0, −0.2]

with the stable solution A_1 as in the previous cases. Figure 7a, b shows the plots of the profiles when β is 0.45 and 0.95, respectively. Table 2 summarizes the optimal values for these two cases.

4.4 Partially-Specified Profile

Motivated by the mobile phone holder application (Fig. 4), the initial profile is constrained to have a fixed value for a section at the mid-span of the arch. The equality constraints in the optimization algorithm are therefore modified to have a fixed value throughout the range of values of S given by

$$P(S) = 1 \text{ for } 0.3 \leq S \leq 0.7 \tag{13}$$

The designed bistable arch is shown in Fig. 8.

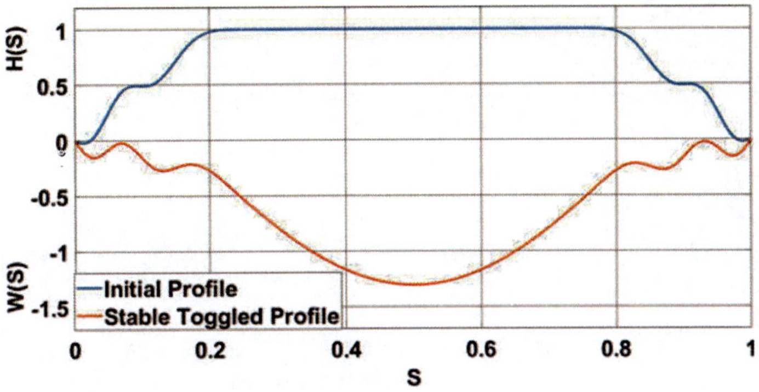

Fig. 8 Partially-specified profile

5 Summary

This work focussed on obtaining optimized designs of bistable arches with (1) maximum travel at a point, (2) specified change in slope at a point, (3) specific travel between the initial, unstable, and second stable states, and (4) specified region in the arch-profile with multiple constraints using the optimization formulation in Sect. 3. Illustrative examples were considered in Sect. 4 to show the utility of these design cases. We verified the optimized bistable profiles obtained in the work by using a finite element model in the commercial software ABAQUS (V2018). Arches were modelled as 3D deformable bodies (Young's modulus, $E = 200$ MPa, and Poisson's ratio, $\nu = 0.3$) using a 20-node quadratic hexahedral brick with reduced integration (C3D20R). The finite element analysis of the mobile holder example (Sect. 4.4) is presented in Fig. 9. Furthermore, bistability of each of the optimized shapes was confirmed from their force-displacement plots. Force displacement curve for the case described in Fig. 5d is presented in Fig. 10. The force changes its direction as the displacement increases and is zero at three points indicating bistability.

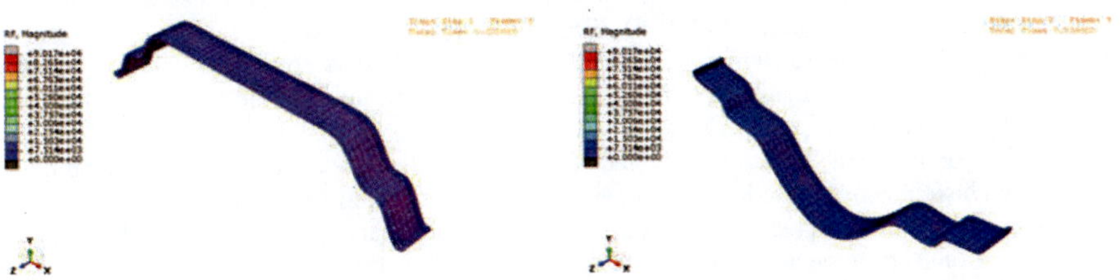

(a) Initial profile (b) Toggled profile

Fig. 9 Finite element model demonstrating the mobile holder

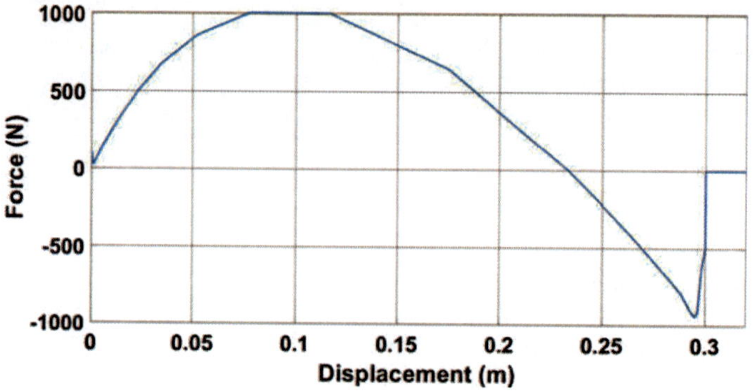

Fig. 10 Force displacement curve for maximum displacement at $S = 0.35$

References

1. Palathingal S, Ananthasuresh G (2018) A bilateral relationship between stable profiles of pinned–pinned bistable shallow arches. Int J Solids Struct 143:183–193
2. Palathingal S, Ananthasuresh G (2019) Analysis and design of fixed–fixed bistable arch-profiles using a bilateral relationship. J Mech Robot 11(3)
3. Fung YC, Kaplan A (1952) Buckling of low arches or curved beams of small curvature. National Advisory Committee for Aeronautics, Washington, DC
4. Hoff NJ, Bruce VG (1953) Dynamic analysis of the buckling of laterally loaded flat arches. Stud Appl Math 32(1–4):276–288
5. Qiu J, Lang JH, Slocum AH (2004) A curved-beam bistable mechanism. J Microelectromech Syst 13(2):137–146
6. Palathingal S, Ananthasuresh G (2017) Design of bistable arches by determining critical points in the force-displacement characteristic. Mech Mach Theory 117:175–188
7. Hu N, Burgueño R (2015) Buckling-induced smart applications: recent advances and trends. Smart Mater Struct 24(6):063001
8. Lee J-H, Lee M-H, Jang W-I, Choi C-A, Joo JW (1999) Bistable planar polysilicon microactuator with shallow arch-shaped leaf springs. In: Micromachined devices and components V, vol 3876. International Society for Optics and Photonics, pp 274–279
9. Nguyen T-A, Wang D-A (2016) A gripper based on a compliant bistable mechanism for gripping and active release of objects. In: 2016 international conference on manipulation, automation and robotics at small scales (MARSS). IEEE, pp 1–4
10. Darshan S, Lassche T, Herder JL, Ananthasuresh G (2015) A compliant two-port bistable mechanism with application to easy-chair design. In: Proceedings of the 14th IFToMM world congress, pp 227–236
11. Banik D, Palathingal S, Ananthasuresh G, Ghosh A (2021) A mechanical or gate using pinned-pinned bistable arches. In: Mechanism and machine science. Springer, pp 861–874
12. Zhang S, Chen G (2011) Design of compliant bistable mechanism for rear trunk lid of cars. In: International conference on intelligent robotics and applications. Springer, pp 291–299
13. MathWorks. Fmincon. https://www.mathworks.com/help/optim/ug/fmincon.html?s_tid=doc_ta

Design and Analysis of a Miniaturized Atomic Force Microscope Scan Head

B. N. Arya and G. R. Jayanth

Abstract The Atomic Force Microscope (AFM) finds widespread applications as a tool for nano-scale characterization studies and atomic manipulation. Here, we propose the design of a miniaturized AFM scan head for 3-axis nano-positioning. The scan head uses parallelogram-based flexures for amplifying displacements in-plane and a bridge-type displacement amplifier for out-of-plane positioning and achieves a displacement range of ± 5 μm along *X*-, *Y*- and *Z*-axes. Subsequently, a lumped parameter model has been obtained for analysing the quasi-static and dynamic characteristics of the different subsystems of the positioner. A comparison of the analytical expressions for the displacement gain and eigen frequencies with Finite Element (FE) analysis revealed match to within 4%. The bandwidth along *Z*-axis is about 5 kHz, which is much larger than that of a conventional AFM scan head. Finally, a feedback control system has been designed to achieve position control using model inversion.

Keywords Compact AFM · Displacement amplifier · Model inversion

1 Introduction

An Atomic Force Microscope (AFM) is a type of scanning probe microscope used for nano-scale characterization, topography imaging and manipulation of conducting and insulating samples, at sub-nanometer resolution. Conventional AFM systems possess certain limitations which affect the quality of their measurement results such as low speed, small scan area and the effect of actuation nonlinearities. Another important limitation arises from its bulky nature. By exploiting the benefits of scaling laws, it has long been known that a compact construction can make the AFM system immune to these limitations [1].

B. N. Arya (✉) · G. R. Jayanth
Indian Institute of Science, Bangalore 560012, India
e-mail: aryanair@iisc.ac.in

G. R. Jayanth
e-mail: jayanth@iisc.ac.in

V. K. Gupta et al. (eds.), *Recent Advances in Machines and Mechanisms*, Lecture Notes in Mechanical Engineering, https://doi.org/10.1007/978-981-19-3716-3_12

The first compact AFM designs featured a single-chip CMOS-based AFM with integrated sensing and actuating mechanisms such as piezoresistive sensing, thermal bimorph actuators [2] and electrothermal actuators [3], while others used a combination of electrostatic and electrothermal actuators, along with an integrated piezoelectric layer for fine positioning along *X*-, *Y*- and *Z*-axes, respectively [4]. Such integrated single-chip systems require complex signal routing circuitry. The use of electrothermal actuators also poses certain limitations such as reduced scan rates due to limited bandwidth, heat dissipation [5] and parasitic resonances. Also, during scanning, AFM tips are found to get damaged frequently, necessitating their replacement. This results in having to discard the entire probe, along with its integrated fine positioning and deflection sensing mechanisms, even when their function remains intact. This contributes to a significantly higher running cost of the instrument. Also, single-chip AFM models impose constraints on the stiffness and geometry of the probe that they can support, which greatly reduces their flexibility.

Therefore, here we propose a compact design for the fine positioning mechanism, on which a conventional AFM probe can be placed, as opposed to integrating them on a consumable such as a cantilever. This enables retaining the high bandwidth and compactness without having to discard these systems when the tip gets blunted. The designs of the fine positioner are based on flexure-based displacement amplifiers actuated by miniature piezo actuators. The use of displacement amplifiers enables a compact construction with large motion ranges along *X*-, *Y*- and *Z*-axes, compared to in-plane positioners that employ compliant beams for just guiding motion and not for amplifying it [6]. Such a compact design also allows the easy integration of any deflection measurement technique such as piezoresistive sensors. The quasi-static and dynamic models of these structures have been developed. Finally, a feedback control system using model inversion technique has been designed for regulating the motion of the positioners by actuating the piezo actuators accordingly.

The rest of the paper is divided as follows. Section 2 discusses the design of the positioning mechanisms. Sections 3 and 4 describe quasi-static modelling using Pseudo Rigid Body Models (PRBM) and eigen-frequency analysis using Rayleigh's technique, respectively. Section 5 discusses feedback control using model inversion. Finally, conclusions are presented in Sect. 6.

2 Design of Fine Positioner of the Miniaturized AFM

Figure 1a shows the geometric model of the fine positioner for a miniaturized AFM system. It consists of flexure-based amplifiers which employs parallel kinematics for achieving nano-scale displacements along *X*-, *Y*- and *Z*-axes. In particular, it consists of three main components, namely the in-plane positioner, which is designed for motion along *X*- and *Y*-axes, the out-of-plane positioner which is designed for motion along *Z*-axis and a decoupling stage in between the two, for decoupling the motion between the in-plane and out-of-plane positioner.

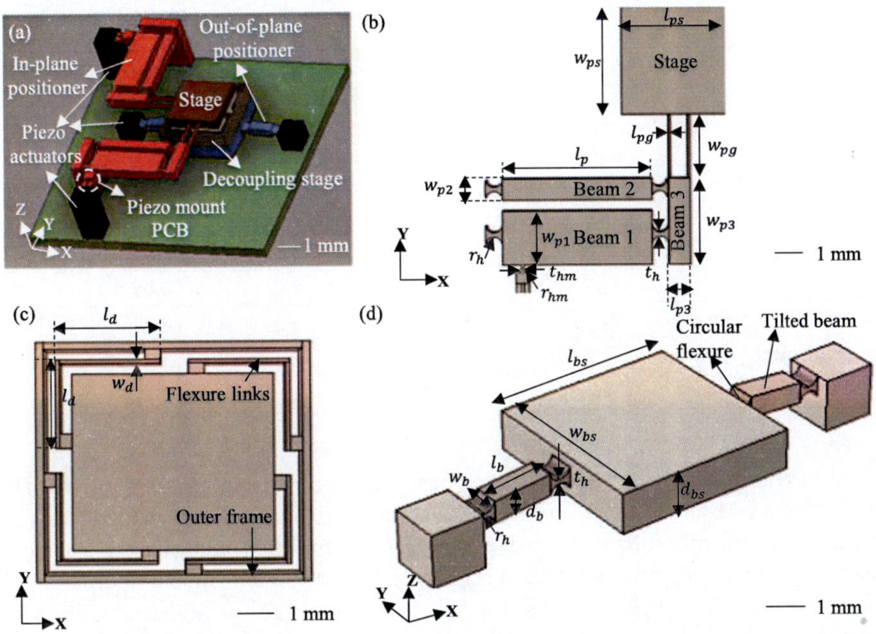

Fig. 1 Computer-Aided Design (CAD) model of **a** fine positioner, **b** in-plane positioner, **c** decoupling stage, **d** out-of-plane positioner

The in-plane positioner consists of parallelogram flexures (Beam 1 and Beam 2 in Fig. 1b) with circular flexure hinges at either ends of the beam, connected to an output beam (Beam 3 in Fig. 1b). They constitute a lever-based amplifier design and achieve amplified motion along *X*- and *Y*-axes. The motion between the two axes is decoupled by using thin flexure guided beams, which connect the output beam of the amplifier to a stage block. Shear chip piezo actuators are chosen for actuating the in-plane positioners. To decouple the angular motion of the parallelogram flexure from the motion of the shear piezo actuator, a circular flexure hinge-based piezo mount with a pair of guided beams is used to connect the two. The guided beams are employed to decouple transverse motion of the parallelogram flexure at its point of connection to the circular hinge from the motion of the piezo actuator (Fig. 1b).

The decoupling stage serves the purpose of decoupling the in-plane motion of the in-plane positioner, from that of the out-of-plane positioner while coupling the out-of-plane motion of the positioner to that of the stage. The decoupling stage is composed of a central platform around which 4 L-shaped flexure links are positioned symmetrically (Fig. 1c). This structure is then connected to an outer frame. The central platform connects to the stage of the in-plane positioner while the outer frame connects to the out-of-plane positioner. The interconnecting flexures are designed to achieve high compliance along *X*- and *Y*- axes and high stiffness along the *Z*-axis.

The out-of-plane positioning mechanism consists of a bridge-type displacement amplifier. The mechanism comprises 3 nearly rigid tilted beams connected by circular

Table 1 Dimensions for the different elements of the fine positioner

	Parameter	Dimension (mm)	Parameter	Dimension (mm)
In-plane positioner	l_p	3.5	t_h	0.1
	w_{p1}	1.25	r_h	0.2
	w_{p2}	0.5	l_{pg}	0.05
	l_{p3}	0.5	w_{pg}	1.5
	w_{p3}	2	$l_{ps} = w_{ps}$	2.5
	t_{hm}	0.05	r_{hm}	0.05
Decoupling stage	l_d	1.3	w_d	0.05
Out-of-plane positioner	l_b	1.1	$l_{bs}=w_{bs}$	3.4
	$w_b=d_b$	0.4	d_{bs}	0.75

flexure hinges (Fig. 1d). The two tilted beams are identical in length and are tilted by the same angle to the horizontal. When an input displacement is provided along *X*-direction by means of shear chip piezo actuators, an amplified output is obtained along *Z*-axis. Here, the displacement gain is only dependent on the angle of inclination of the beam and hence provides greater flexibility in design for higher amplification ratios, without taking up additional space.

The dimensions of the different elements in the positioners are shown in Table 1. The depth of all the elements was taken as 0.5 mm, unless specified otherwise.

3 Quasi-Static Modelling and Analysis of the Fine Positioners

Quasi-static modelling involves developing lumped parameter models for the constituent compliant elements of the positioner. In all cases where bending deformations are involved, Euler–Bernoulli beam theory [7] has been employed to analyse the deformations.

3.1 In-Plane Positioner

The model of the parallelogram flexure consists of two beams (Beam 1 and Beam 2) that form the sides of the parallelogram, connected by circular flexure hinges to an output beam (Beam 3) (Fig. 1b). Beams 1 and 2 are both modelled as elastic beams, and Beam 3 is assumed to be nearly rigid. The length of the beams is denoted by l_p, and their width is denoted by w_{p1} and w_{p2} for Beam 1 and Beam 2, respectively. The circular flexures are modelled as torsional hinges with torsional stiffness k_{θ_p}.

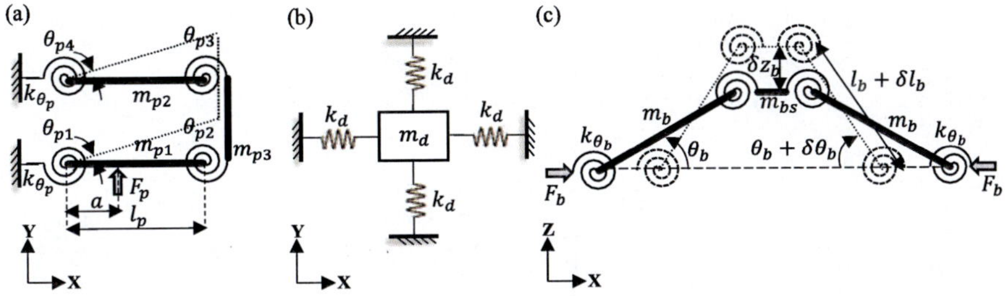

Fig. 2 Lumped parameter model of **a** in-plane positioner, **b** decoupling stage, **c** out-of-plane positioner

The angular displacement of the ith hinges ($i = 1 \ldots 4$) is given by the variables θ_{pi} (Fig. 2a).

The expression for output displacement of the parallelogram flexure was derived by considering a force F_p applied at a distance a from the point of rotation of the fixed hinges to Beam 1. If f_p is the reaction force that Beam 3 applies on Beam 1, and $y_{p1}(x)$ and $y_{p2}(x)$ represent the deformation profiles of the Beams 1 and 2, then the boundary conditions are given by $y_{p1}(l_p) = y_{p2}(l_p)$, $\left(\frac{\mathrm{d}y_{p1}}{\mathrm{d}x}\right)_{x=0} = \theta_{p1}$, $\left(\frac{\mathrm{d}y_{p1}}{\mathrm{d}x}\right)_{x=l_p} = \theta_{p2}$, $\left(\frac{\mathrm{d}y_{p2}}{\mathrm{d}x}\right)_{x=l_p} = \theta_{p3}$, $\left(\frac{\mathrm{d}y_{p2}}{\mathrm{d}x}\right)_{x=0} = \theta_{p4}$. Also, by moment balance it can be seen that $\left(F_p - f_p\right) = \frac{k_{\theta_p}}{a}\left(\theta_{p1} + \theta_{p2}\right)$ and $f_p = \frac{k_{\theta_p}}{a}\left(\theta_{p3} + \theta_{p4}\right)$. The value of k_{θ_p} was estimated from the Paros–Weisbord equations [8] and is dependent on r_h, the radius of curvature, and t_h, the thickness of the hinge (Fig. 1b).

By applying these boundary conditions and using Euler–Bernoulli beam theory, $y_{p1}(x)$ and $y_{p2}(x)$ were obtained to be

$$y_{p1}(x) = \begin{cases} \left(\frac{f_p - F_p}{6EI_{p1}}\right)x^3 + \left(\frac{F_p a - f_p l_p - k_{\theta_p}\theta_{p2}}{2EI_{p1}}\right)x^2 + \theta_{p1}x & \text{for } x \leqslant a \\ \left(\frac{f_p}{6EI_{p1}}\right)x^3 - \left(\frac{f_p l_p + k_{\theta_p}\theta_{p2}}{2EI_{p1}}\right)x^2 + \left(\frac{F_p a^2}{2EI_{p1}} + \theta_{p1}\right)x - \frac{F_p a^3}{6EI_{p1}} & \text{for } x \geqslant a \end{cases} \tag{1}$$

$$y_{p2}(x) = \left(\frac{-f_p}{6EI_{p2}}\right)x^3 - \left(\frac{f_p l_p - k_{\theta_p}\theta_{p3}}{2EI_{p2}}\right)x^2 + \theta_{p4}x \tag{2}$$

where x is the distance from the fixed end of the beam, I_{p1} and I_{p2} are the area moments of inertia of Beams 1 and 2, respectively, and E is the Young's modulus. By applying the boundary conditions, all the variables can be obtained in terms of the input force F_p. Normalizing the reaction force, f_p, and input force, F_p as $\widetilde{f_p} = \frac{f_p}{k_{\theta_p}}a$ and $\widetilde{F_p} = \frac{F_p}{k_{\theta_p}}a$, the angles and the normalized reaction force can be obtained by solving the following equation:

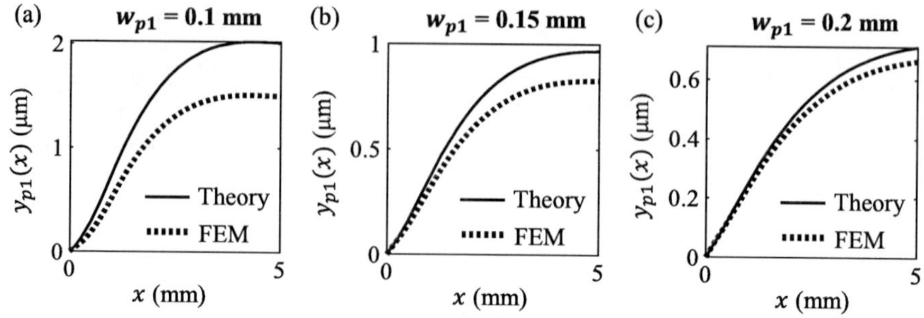

Fig. 3 Comparison of analytical and FEM simulation results for displacement profile of Beam 1, for $F_p = 1$ mN and $l_p = 5$ mm and **a** $w_{p1} = 0.1$ mm, **b** $w_{p1} = 0.15$ mm, **c** $w_{p1} = 0.2$ mm

$$\begin{bmatrix} 0 & 1 & 1 & 1 & 1 \\ 1 & 0 & 0 & -1 & -1 \\ \frac{k_{\theta_p} l_p}{2EI_{p2}} & 0 & 0 & -\left(1+\frac{k_{\theta_p} l_p}{EI_{p2}}\right) & 1 \\ \frac{k_{\theta_p} l_p}{2EI_{p1}} & -1 & \left(1+\frac{k_{\theta_p} l_p}{EI_{p1}}\right) & 0 & 0 \\ \left(\frac{k_{\theta_p} l_p^2}{3EI_{p1}} + \frac{k_{\theta_p} l_p^2}{3EI_{p2}}\right) & -l_p & \frac{k_{\theta_p} l_p^2}{2EI_{p1}} & -\frac{k_{\theta_p} l_p^2}{2EI_{p2}} & l_p \end{bmatrix} \begin{bmatrix} \widetilde{f}_p \\ \theta_{p1} \\ \theta_{p2} \\ \theta_{p3} \\ \theta_{p4} \end{bmatrix} = \begin{bmatrix} 1 \\ 0 \\ 0 \\ \frac{k_{\theta_p} a}{2EI_{p1}} \\ \frac{k_{\theta_p} a(3l_p - a)}{6EI_{p1}} \end{bmatrix} \widetilde{F}_p \tag{3}$$

To validate these results, the deformation profile $y_{p1}(x)$ of Beam 1 has been plotted for three different values of beam width, namely $w_{p1} = 0.1, 0.15, 0.2$ mm (Fig. 3) and the results have been compared with those of Finite Element (FE) analysis.

The percentage error between analytical and FE results is only 6% for $w_{p1} = 0.2$ mm and is less than 30% for slender beams, for an input force of $F_p = 1$ mN. The greater mismatch between the two plots for slender beams suggests incorrect assumption in modelling the flexure hinges as torsional springs for the case of slender beams. However, since wider and stiffer parallelogram beams are preferred for amplified in-plane displacement, the effect of the mismatch is not significant from the point of view of design.

3.2 Decoupling Stage

The decoupling stage consists of 4 L-shaped flexure links connected to a central platform. Each of these flexures is modelled as linear springs of stiffness k_d (Fig. 2b), placed symmetrically around a rigid central platform. During in-plane motion, the longitudinal element of the flexures undergoes axial deformation, and the lateral element undergoes transverse deformation. Thus, the single flexure link can be modelled as a series combination of two linear springs with effective stiffness $k_{d_{\text{eff}}}$ given by

Table 2 Comparison of analytical and FEM results for effective stiffness of decoupling stage and out-of-plane fine positioner

Parameter	Theory	FEM	Error (%)
$k_{d_{\text{eff}}}$ (kN/m)	8.95	9.79	8.58
$k_{b\text{eff}}$ (kN/m)	43.03	42.49	−1.27

$$k_d = \frac{k_{d\,\text{ax}} k_{d\,\text{tr}}}{k_{d\,\text{ax}} + k_{d\,\text{ax}})} \tag{4}$$

where $k_{d\,\text{tr}} = \frac{12EI_d}{l_d^3}$ is the transverse stiffness and $k_{d\,\text{ax}} = \frac{EA_d}{l_d}$ is the axial stiffness of the flexure link. Here, I_d, A_d and l_d are the area moment of inertia, area of cross section and length of the flexure link, respectively. Since all the 4-flexure links have similar dimensions and undergo same amount of deformation, the total effective stiffness is given by $k_{d_{\text{eff}}} = 4k_d$. Due to the symmetric nature of the geometry, estimating the displacement profile along either X- or Y-axis is sufficient. Thus, considering an input force F_{d} applied along X-axis, the output displacement δx_{d}, is given by

$$\delta x_d = \frac{F_d}{k_{d_{\text{eff}}}} = \frac{F_d}{4k_d} \tag{5}$$

For an input force of $F_d = 0.1$ N, the comparison between Finite Element Method (FEM) and analytical results for the effective in-plane stiffness shows a match of 8.6% (Table 2).

3.3 Out-of-Plane Positioner

The bridge displacement amplifier consists of two tilted beams, attached to a cuboidal platform through circular flexure hinges. The tilted beams are modelled as elastic beams of length l_b, and the circular flexures are modelled as torsional hinges of stiffness k_{θ_b}. When an external input force F_b is applied along X-axis at the ends of the tilted beams, they undergo compression by an amount δl_b and the circular hinges undergo rotation by an amount $\delta\theta_b$, resulting in an output displacement, δz_b, of the cuboidal platform along Z-axis (Fig. 2c).

The effective stiffness of the bridge amplifier along Z-axis is then estimated by equating the total potential energy of the amplifier to the potential energy stored in an equivalent lumped model with an effective stiffness $k_{b\text{eff}}$ as,

$$4\left(\frac{1}{2}k_{\theta_b}\delta\theta_b^2\right) + 2\left(\frac{1}{2}k_{b_x}\delta l_b^2\right) = \frac{1}{2}k_{b\text{eff}}\delta z_b^2 \tag{6}$$

where k_{b_x} is the effective longitudinal stiffness of the elastic beam, comprising of the series combination of the longitudinal stiffness of the 2 circular hinges [7] and the tilted flexure beams. Simplifying Eq. (6) and using moment balance equations for a force component acting along the length of the tilted beam, the output displacement, δz_b, can be written as:

$$z_b = F_b \sin 2\theta_b \left(\frac{k_{b_x} l_b^2 - 2k_{\theta_b}}{4k_{\theta_b} k_{b_x}} \right) \tag{7}$$

For an input force of $F_b = 1$ μN, the comparison between FEM and analytical results for the effective Z-axis stiffness shows a match of 1.3% (Table 2).

By incorporating the above design considerations, the individual elements of the fine positioner, namely the in-plane positioner, decoupling stage and the out-of-plane positioner, were assembled along with the piezo actuators and their motion along X-, Y- and Z-axes was studied. The stiffness along the in-plane and out-of-plane directions for each of them was also estimated in FEM by applying a point load of 1 μN. The in-plane positioner offers much lower stiffness along Z-axis (2.21 kN/m) than along X- and Y-axes (46.02 kN/m). The X–Y stiffness is itself much smaller than that of the driving piezo actuators, thereby ensuring that the displacement of the actuators is transmitted almost completely to the stage.

The stiffness of the decoupling stage along X- and Y-axes (9.78 kN/m) is almost 5 times less than that of the in-plane positioner. The decoupling stage has a Z-axis stiffness (94.72 kN/m) which is almost 43 times higher than that of the in-plane positioner, in accordance with the requirement to couple the out-of-plane motion of the Z-positioner placed below it to the sample stage. The Z-axis stiffness of the out-of-plane positioner is 37.70 kN/m, which is much higher than the out-of-plane stiffness of the in-plane positioner but lower than that of the decoupling stage. Therefore, the Z-positioner would couple its displacement to the sample stage, with an attenuation of about 28%.

To verify parasitic motion, the percentage of cross-axis displacements of the fine positioner has been estimated along all the 3 axes and shown in Fig. 4. For X-axis positioning (Fig. 4a), for a maximum output displacement of 23 μm, the percentage

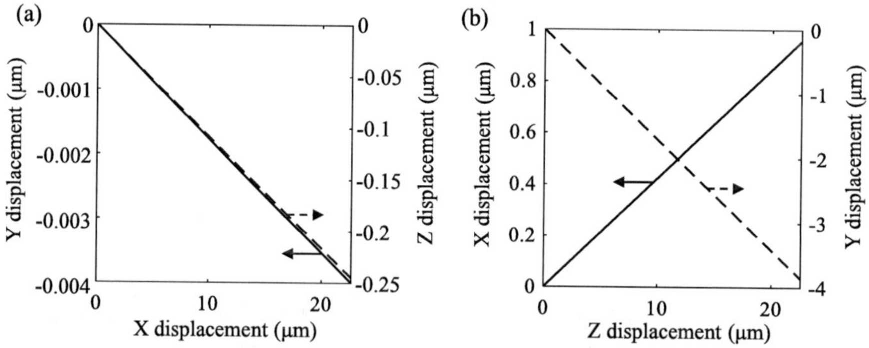

Fig. 4 Plot of **a** Y and Z cross-axis displacement for X-axis positioning and **b** X- and Y-cross-axis displacement for Z-axis positioning, for an input displacement of 5 μm

of Y- and Z-axis cross-axis displacement is 0.02 and 1.08%, respectively. Similar results have also been obtained for Y-axis positioning owing to the symmetry of the in-plane positioner. Similarly, for Z-axis motion, the percentage of X- and Y-axis cross-coupling is 4 and 17%, respectively (Fig. 4b).

4 Eigen-Frequency Analysis of the Fine Positioners

The fundamental eigen frequencies of the different components of the fine positioner were obtained by using the Rayleigh Quotient method [7]. In this method, the first step is to assume an approximate mode shape function, to replicate the first eigen mode, followed by determining the Rayleigh Quotient by equating the maximum kinetic and potential energies of the system. Here, the quasi-static deformation profiles obtained by utilizing the expressions derived in the previous section are employed as the approximate mode shapes; the eigen frequencies of the different positioners are estimated.

4.1 *In-Plane Positioner*

Due to the high lateral stiffness of the parallelogram flexures (for the dimensions mentioned in Table 1), their contribution to the elastic potential energy is negligible. Hence, only the elastic potential energy of the torsional springs is considered. Therefore, the potential energy, $U_{p\max}$, of the in-plane positioner is given by

$$U_{p\max} = 0.5k_{\theta p}\sum_{i=1}^{4}\theta_{pi}^2 \tag{8}$$

The kinetic energy associated with motion of the hinges is negligible in comparison to the beams, and hence, their effects can be ignored. The total kinematic energy would be due to displacements of Beam 1 and Beam 2 which are assumed to be given approximately by Eqs. (1) and (2).

Therefore, the overall maximum kinetic energy $T_{p\max}$ is given by

$$T_{p\max} = \left[\frac{\rho}{2}\left\{\int_0^{l_p}\left(A_{p1}y_{p1}^2(x) + A_{p2}y_{p2}^2(x)\right)\mathrm{d}x + \int_0^{3r_h+l_p+0.5l_{p3}} A_{p3}y_{p3}^2(x)\mathrm{d}x\right\}\right]\omega_p^2 \tag{9}$$

where ω_p is the eigen frequency of the in-plane positioner, ρ is the density of the material used (aluminium), l_{p_3} is the width of Beam 3, A_{p1}, A_{p2} and A_{p3} are the

Table 3 Comparison of analytical and FEM results for eigen frequencies of different elements of the fine positioner

	Theory (kHz)	FEM (kHz)	Error (%)
In-plane positioner	2.63	2.62	−0.38
Decoupling stage	5.18	5.28	1.89
Out-of-plane positioner	6.40	6.71	4.62

area of cross sections, and y_{p1}, y_{p2} and y_{p3} are the maximum modal displacement vectors of Beams 1, 2 and 3, respectively. By writing $T_{p\max} = (t_{p\max}\omega_p^2)$, we can obtain the approximate eigen frequency from Eqs. (8) and (9) as $\omega_p^2 = \frac{U_{p\max}}{t_{p\max}}$.

The comparison of results for FEM and analytical expression shows an error of approximately 0.4% (Table 3).

4.2 Decoupling Stage

The dynamic model of the decoupling stage is obtained by assuming the potential energy of the flexure links alone, owing to their slender geometry as compared to the central platform (Table 1). However, for estimating the kinetic energy, the central stage alone is assumed to contribute, since its mass is significantly higher than that of the flexures. Similar to the method described in Sect. 4.1, Rayleigh technique is used to compute the maximum potential energy of the decoupling stage, $U_{d\max}$, as

$$U_{d\max} = 0.5k_{d_{\text{eff}}}\delta x_d^2 \tag{10}$$

and kinetic energy, $T_{d\max}$,

$$T_{d_{\max}} = 0.5m_d\delta x_d^2\omega_d^2 \tag{11}$$

where m_d and ω_d are the mass and eigen frequency of the decoupling stage, respectively . Here, δx_d and $k_{d_{\text{eff}}}$ are taken from Eq. (5). The comparison of results for FEM and analytical expression shows an error of approximately 2% (Table 3).

4.3 Out-of-Plane Positioner

Here, the potential energy of the model is the combination of the potential energy stored in the 4 circular flexure hinges and the 2 rigid beams. The total kinetic energy is considered to be primarily due to the motion of the central block and the two beams. Thus, the maximum potential energy $U_{b\max}$ is

$$U_{b\max} = 4\left(\frac{1}{2}k_{\theta_b}\delta\theta_b^2\right) + 2\left(\frac{1}{2}k_{b_x}\delta l_b^2\right) \tag{12}$$

and the maximum kinetic energy T_{bmax} is

$$T_{b\max} = 0.5(2m_b + m_{bs})\delta z_b^2\omega_b^2, \tag{13}$$

where m_b and m_{bs} are the masses of the tilted beams and the central stage platform, respectively, and ω_b is the eigen frequency of the bridge amplifier. Here, δz_b and k_{b_x} are taken from Eq. (7). The comparison of results for FEM and analytical expression shows an error of approximately 5% (Table 3).

5 Feedback Control

The 3-axis positioning mechanism needs to be operated in a feedback control loop to compensate for the nonlinearities introduced by the piezoelectric actuators, such as hysteresis and creep. For this purpose, firstly the transfer function of the positioner is obtained from the frequency responses, relating their displacements to the applied voltage inputs along the respective directions, namely *X*-, *Y*- and *Z*-axes. Here, the frequency response plots for motion along *Z*-axis alone are considered, and the exact same steps can be repeated along *X*- and *Y*-axes. The Bode displacement plots are obtained in COMSOL, by providing a frequency sweep of amplitude 2500 V and frequency range of 0–50 kHz, with a step size of 0.5 kHz, to the shear piezo actuators of the bridge amplifier. The resultant transfer function, $P(s)$, obtained by using MATLAB function for model identification, namely Identified Frequency Response Data (idfrd), is given by

$$P(s) = \frac{M_1 s^5 + M_2 s^4 + M_3 s^3 + M_4 s^2 + M_5 s + M_6}{s^6 + N_1 s^5 + N_2 s^4 + N_3 s^3 + N_4 s^2 + N_5 s + N_6} \tag{14}$$

where $M_1 = -2710$, $M_2 = 1.454 \times 10^9$, $M_3 = -5.097 \times 10^{12}$, $M_4 = 4.315 \times 10^{18}$, $M_5 = -1.24 \times 10^{21}$, $M_6 = 2.718 \times 10^{27}$ and $N_1 = 2619$, $N_2 = 4.267 \times 10^9$, $N_3 = 7.221 \times 10^{12}$, $N_4 = 5.699 \times 10^{18}$, $N_5 = 4.71 \times 10^{21}$ and $N_6 = 2.401 \times 10^{27}$.

The Bode plot of the derived transfer function is superimposed with the frequency response obtained from FEM analysis and is seen to match to a good degree, as shown in Fig. 5a. In view of the poorly damped open-loop dynamics of the positioner, model inversion technique is used to cancel the under-damped poles of the plant transfer function and replace them with critically damped or over damped poles. This method helps in achieving a higher closed loop bandwidth, than what is possible by adopting a conventional controller design.

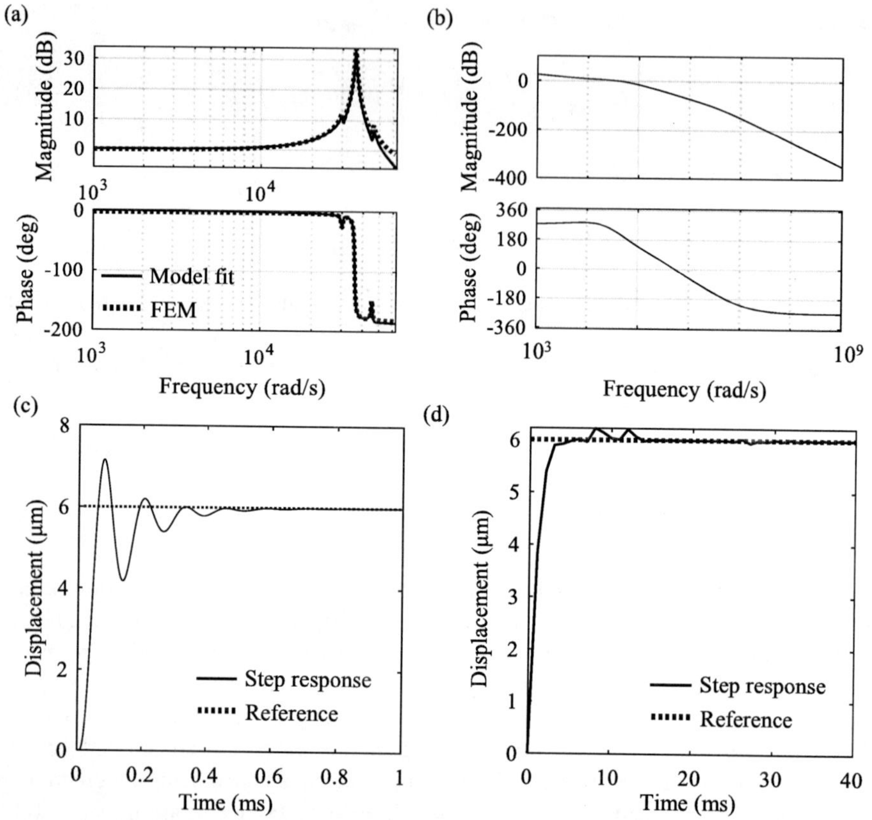

Fig. 5 **a** Model fitting for Z-axis positioning, **b** frequency response plot for Z-axis open-loop gain, **c** step response plot for Z-axis positioning with model inversion, **d** step response plot for Z-axis positioning without model inversion

In this technique, the under-damped minimum-phase poles of the plant, of the form $p_{i\alpha} \pm jp_{i\beta}(i = 1, 2, 3)$, are cancelled by the controller zeros and are replaced by critically damped poles with corner frequencies at $\sqrt{p_{i\alpha}^2 + p_{i\beta}^2}$, $i = 1, 2, 3$. Likewise, the minimum-phase zeros are also cancelled. In other words, for the plant transfer function,

$$P(s) = \frac{\prod_{i=1}^{m}\left(\frac{s}{z_i} + 1\right)}{\prod_{j=1}^{n}\left(\frac{s}{p_j} + 1\right)} \tag{15}$$

the transfer function of the model inversion block, $M(s)$ can be written as

$$M(s) = \frac{\prod_{j=1}^{n}\left(\frac{s}{p_j} + 1\right)}{\prod_{i=1}^{m}\left(\frac{s}{z_i} + 1\right)\left(\frac{s}{\sqrt{p_{i\alpha}^2 + p_{i\beta}^2}} + 1\right)} \tag{16}$$

where z_i and p_j denote the zeros and poles of the plant and m and n are the total number of zeros and poles of the plant, respectively. Subsequently, a PID controller is designed to achieve a phase margin of 40°, with proportional (P), integral (I) and derivative (D) gains being $P = 1.89$, $I = 12{,}600$, $D = 8 \times 10^{-5}$, respectively.

The Bode plot of the overall open-loop system, consisting of the cascade of the plant *P*(*s*), the model inversion block $M(s)$ and the compensator or the controller, shows a closed loop bandwidth of 4×10^4 rad/s (Fig. 5b). The closed loop step response of the model shows a settling time of 0.42 ms and rise time of 32.82 μs (Fig. 5c). In comparison, the settling time for the system without model inversion is about 100 times larger (Fig. 5d). In addition, the response also shows unmodelled oscillations. Therefore, the performance of the system is drastically improved by incorporating a model inversion function into the feedback loop for closed loop position control.

6 Conclusion

This paper presented the design and analysis of a miniaturized AFM scan head, that can achieve fine positioning in three dimensions. For fine positioning along the in-plane directions, a parallelogram flexure-based displacement amplifier with circular flexure hinges was used. A bridge-type displacement amplifier was designed for out-of-plane fine positioning. A decoupling stage with flexure links was then used to decouple the motion of the in-plane positioner from that of the out-of-plane positioner along the in-plane axes and to couple the motion of the latter to the former along the Z-axis. The volume of the proposed compact design of the fine positioner was $1.2 \times 1.2 \times 0.3$ cm^3.

Subsequently, the proposed designs were modelled to study their quasi-static and dynamic behaviour and the theoretically estimated values were found to match with the FEM simulation results with an average error of approximately 4%. Finally, the analytical models of the fine positioners were used to design a feedback control loop using PID control cascaded with model inversion, to enable closed loop position control, with significant improvement over conventional control without model inversion.

References

1. Strathearn D, Sarkar N, Lee G, Olfat M, Mansour RR (2017) The benefits of miniaturization of an atomic force microscope. In: 30th IEEE international conference on micro electro mechanical systems (MEMS), pp 1363–1366. IEEE, Las Vegas, USA
2. Barrettino D, Hafizovic S, Volden T, Sedivy J, Kirstein K, Hierlemann A, Baltes H (2004) CMOS monolithic atomic force microscope. In: Symposium on VLSI circuits. Digest of technical papers (IEEE Cat. No. 04CH37525), pp 306–309. Widerkehr and Associates, Honolulu, HI, USA
3. Sarkar N, Mansour RR, Patange O, Trainor K (2011) CMOS-MEMS atomic force microscope. In: 16th international solid-state sensors, actuators and microsystems conference, pp 2610–2613. IEEE, Beijing, China
4. Ruppert MG, Fowler AG, Maroufi M, Moheimani SOR (2017) On-chip dynamic mode atomic force microscopy: a silicon-on-insulator MEMS approach. J Microelectromech Syst 26:215–225. https://doi.org/10.1109/JMEMS.2016.2628890
5. Bell DJ, Lu TJ, Fleck NA, Spearing SM (2005) MEMS actuators and sensors: observations on their performance and selection for purpose. J Micromech Microeng 15:S153–S164. https://doi.org/10.1088/0960-1317/15/7/022
6. Awtar S (2003) Synthesis and analysis of parallel kinematic XY flexure mechanisms
7. Meirovitch L (2010) Fundamentals of vibrations, 1st edn. McGraw-Hill, New York
8. Yong YK, Lu T-F, Handley DC (2008) Review of circular flexure hinge design equations and derivation of empirical formulations. Precis Eng 32:63–70. https://doi.org/10.1016/j.precisioneng.2007.05.002

Validation of a Steering System Mathematical Model via Test Rig Measurements

Robin Sharma, P. Ganai, V. Pare, H. Kanchwala, and S. J. Srihari

Abstract This paper presents the testing and validation of a steering system model carried out on a steering test rig. The primary parameters considered in this study were the steer torque, steering angle and rack force. These parameters were varied, and the behaviour of the steering mechanism was evaluated. The forces and displacements of steering knuckles were also measured. The rig results were used to validate the multi-body dynamics model of the steering linkage developed using MSC Adams®. The model simulation results correlate well with the test rig measurements, and this ensures the fidelity of the developed model. The parameters under study also give a hint of the human effort required and its resultant effects on the end points of the steering system.

Keywords Steering · Test rig · Rack force · Torque · Steer angle

1 Introduction

Over the years, automobiles have undergone major upgrades specifically in the development of steering systems. With the advancement of steering mechanisms, it is imperative to simultaneously analyse and upgrade their testing techniques as well. There are several ways to test a steering system; most techniques involve development of customised test benches and rigs [1, 2]; and they are then compared with theoretical analysis of mathematical models [3]. Some researchers have focussed on monitoring the response and efficacy of steering assist systems [4, 5] in addition to the development of physical models and simulators based on designs to carry out the tests [6].

R. Sharma (✉) · P. Ganai · V. Pare
Shri G.S. Institute of Technology and Science, Indore, India
e-mail: robinsharma2770@gmail.com

H. Kanchwala
Center for Automotive Research and Tribology, Indian Institute of Technology, Delhi, India

S. J. Srihari
National Automotive Test Tracks (NATRAX), Pithampur, India

V. K. Gupta et al. (eds.), *Recent Advances in Machines and Mechanisms*, Lecture Notes in Mechanical Engineering, https://doi.org/10.1007/978-981-19-3716-3_13

This paper explores testing and validation of steering tests carried on a rig. The rig is capable of simulating real road conditions for any type of steering system with any type of actuator. Tests were performed to obtain the steering kinematic parameters and were used to validate the mathematical model. Variables under study in these experiments are force, torque, angle and displacement. These variables are crucial to define the characteristics and functioning of a steering system; for instance, the torque determines effort required by the driver; the force gives an idea of load on various components; and the displacement determines turning radius and under/oversteer response [7, 8]. The steering geometry can go through various design iterations and redesigned using the results from this rig to meet design requirements.

2 Steering System

2.1 Steering Test Rig

The test rig used for our tests is MTS 335 steering test system [9] (see Fig. 1). The rig has different modes of operation; we used the control mode as steer and set the active mode to angle. Testing was carried out in a stepwise manner; the setup was adjusted according to the test steering linkage, and fixtures were prepared to mount the linkage. The machine is next calibrated with the built-in software. Eventually, the tests were performed and the steering angle, torque, rack force and displacement were measured.

The steering system used for this study was from Maruti-800 vehicle (see Fig. 2). A manual steering was used for conducting the tests for easier calculations. For developing CAD model, measurements were made via tape, protractor and vernier callipers.

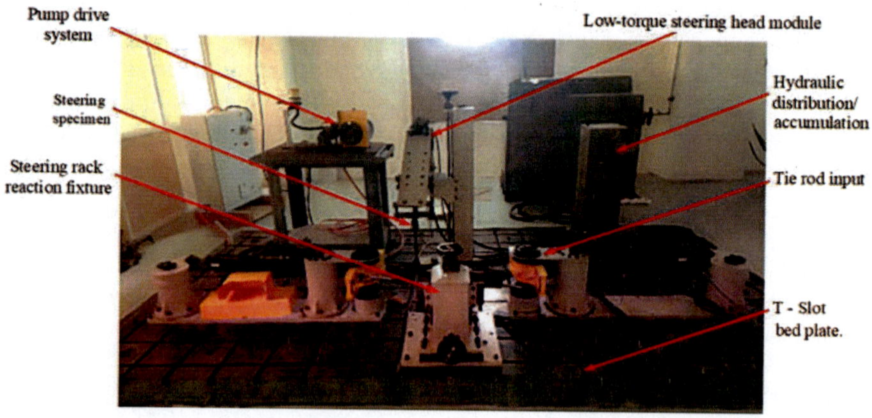

Fig. 1 Steering test rig

Fig. 2 Steering column and rack used for the rig testing

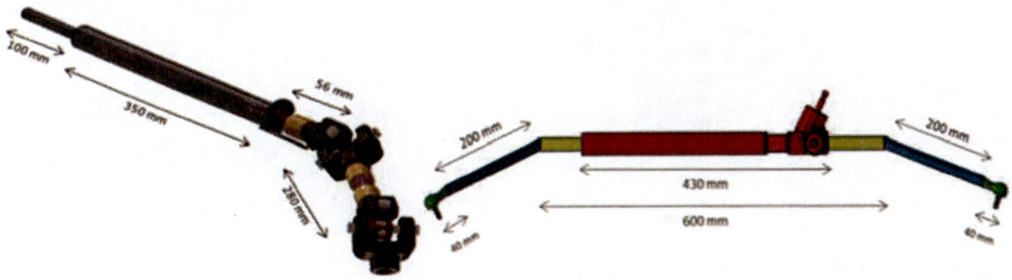

Fig. 3 CAD model of the steering column and rack assembly

Table 1 Components and dimensions for rack and column

S. No.	Component	Dimensions (mm)	S. No.	Component	Dimensions (mm)
1	Steering link	460	3	Rack	600
2	Lever arm	270	4	Push rod	240

2.2 CAD Model

The CAD model of the steering system was prepared in SolidWorks®. The dimensions used for preparing the CAD model is shown in Fig. 3. The steering linkage is connected to the steering wheel by a set of universal joints which are further connected to pinion gear. These joints transmit the angular motion of steering wheel to pinion rotation and eventually the rack translation. The dimensions are given in Table 1.

2.3 Mathematical Model in Adams

The simulation was carried out using MSC Adams to test steering system dynamics [10]. From the CAD model, various steering hardpoints were obtained. The model was prepared according to the physical steering test system; the parameters needed to

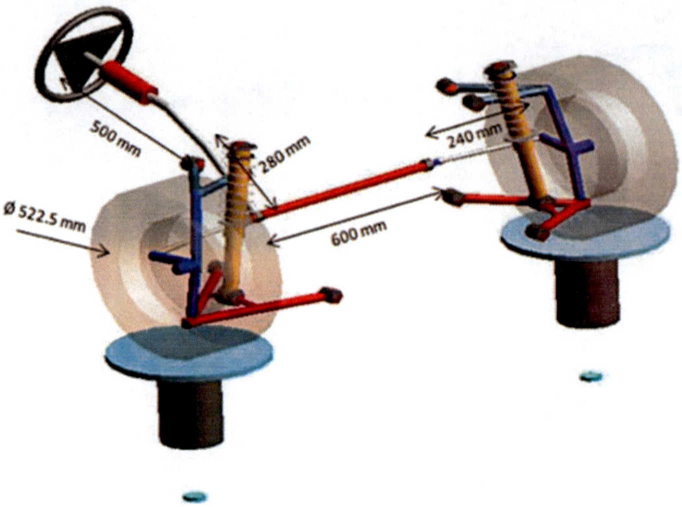

Fig. 4 Adams model of the steering system

Table 2 List of components with their dimensions in the CAE model

S. No.	Component	Dimensions	S. No.	Component	Dimensions
1	Tie rod	42 inches	4	Tyre	145/70 R12
2	Stub	3.03 inches	5	Knuckle Angle	35°
3	Knuckle	3.93 inches			

be studied were selected. In the simulation, steer angle, steer torque and rack forces were primary and the force and displacement on knuckle were secondary parameters. The simulation results are presented and discussed in Sect. 3.2. The steering system modelled in Adams is shown in Fig. 4. The dimensions are given in Table 2.

2.4 Ackermann Principle

The condition where instantaneous centres (ICs) of all the wheels of a four-wheel vehicle meet at one point during a turn is called the Ackermann condition [11]. This point acts as the pivot for the vehicle while completing the turn; at this time, vehicle experiences pure rolling. A steering geometry that complies with this principle is termed as Ackermann geometry. This can be observed through experimental data reported in Table 3. The line diagram of geometry for this steering is illustrated in Fig. 5.

Table 3 Description of geometry

Notation	Description	Value
U	Track width	1.30 m
W	Wheel base	2.36 m
V	Length of rack	1.08 m
s	Inner steer angle	44°
t	Outer steer angle	31.5°
I	I.C. of rotation	

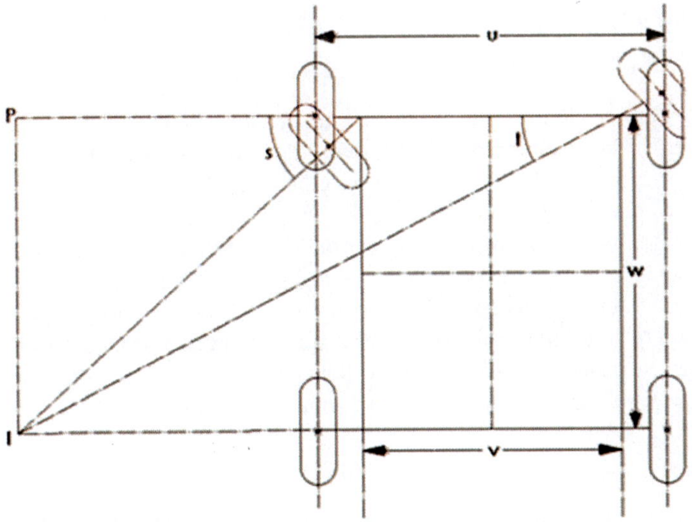

Fig. 5 Steering geometry

2.5 *Formulae Related to the Tests*

The relation between steering torque and rack force can be determined and understood by the formulae mentioned below [12, 13].

$$T_{\text{Steering}} = -\frac{1}{a} F_{\text{Tyre}};\ F_{\text{rack}} = -y.F_{\text{Tyre}}$$

where T_{Steering}: steering torque; F_{Tyre}: tyre force; a: steering ratio; F_{rack}: rack force; and y: pneumatic trail (distance between force due to side slip and centre of contact patch). Expressions for inner and outer wheel angle for Ackermann geometry are as follows [14, 15]:

$$t = \tan^{-1}\frac{W}{R + U/2} \cong \frac{W}{R + U/2}; \quad s = \tan^{-1}\frac{W}{R - U/2} \cong \frac{W}{R - U/2}$$

R is the distance of instantaneous centre (I.C.) from C.G. The approximations are valid for small angles ($\tan^{-1}$ value of the angle ≈ angle itself). This condition can be applied for a wide range of manoeuvres as the road wheel angle is usually small.

3 Results and Discussions

3.1 *Test Rig and Simulation Procedure*

The tests were carried out on the test rig as per the objectives, and their observations were tabulated. Steer torque and steer angle are inputs on the steering wheel, while input force acts on the steering rack laterally. The resultant variables force and displacement were observed on the knuckle in lateral direction. In the first test, the steer angle was set as the control variable and resultant lateral displacement and displacements of both ball joints were measured. The constraint is set at 50° on both negative and positive sides. In second test, the lateral force on rack was the control variable, and the resultant lateral displacement and force on both ball joints of knuckles were measured. Constraint for control variable is 1 kN for force and 20 Nm for torque. In addition to rig tests, model simulation was also performed.

3.2 *Comparison Between Experiment and Simulation Results*

In this section, the experimental and simulation results are compared. For the comparisons, the observations made for dependent variables were plotted against control variables one at a time. In all the figures (6-11), the subplot (a) represents the characteristics for the left flank of the steering system and similarly subplot (b) for the right flank of the steering system. Figure 6a, b is the plots of lateral displacement of the ball joint against the steer angle input. The model simulation results match well with the rig test measurements for the entire range of steering wheel angle input.

Figure 7a, b shows dependence of the lateral force on knuckles on steer angle input. The model simulation results match well with rig measurements for wide angle range.

Figure 8a, b shows plots of lateral displacement of ball joint versus steering rack force. Model simulation matches well with rig measurements for wide range of rack force.

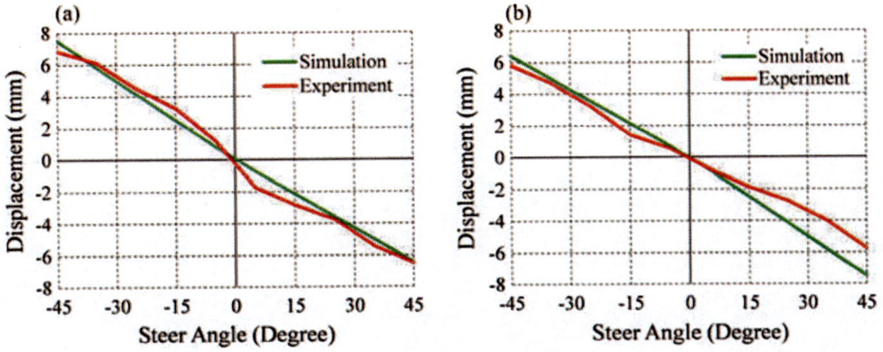

Fig. 6 **a** Left lateral displacement versus steer angle; **b** right lateral displacement versus steer angle

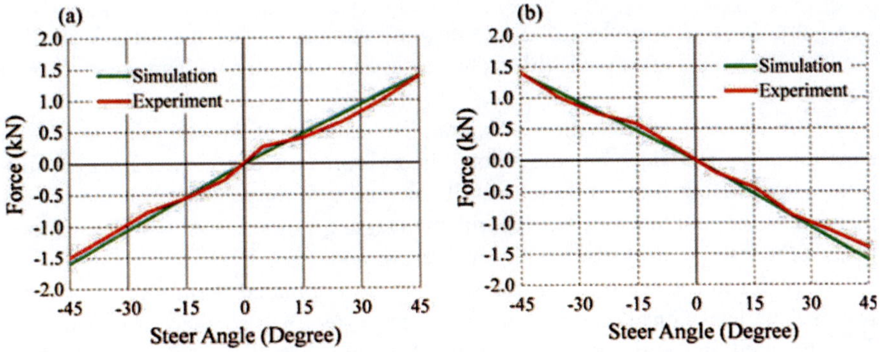

Fig. 7 **a** Left lateral force versus steer angle; **b** right lateral force versus steer angle

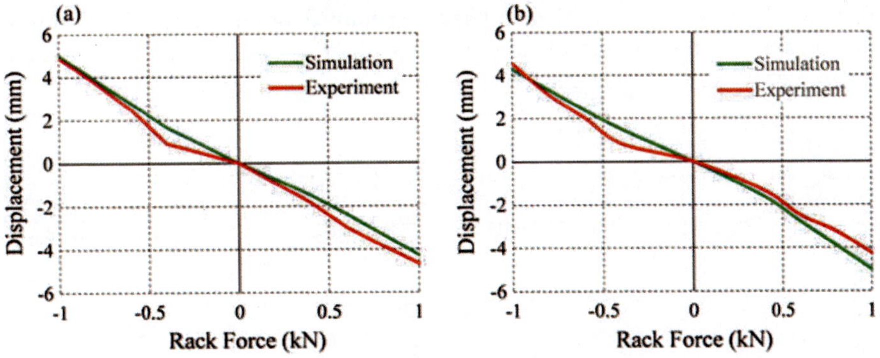

Fig. 8 **a** Left lateral displacement versus rack force; **b** right lateral displacement versus rack force

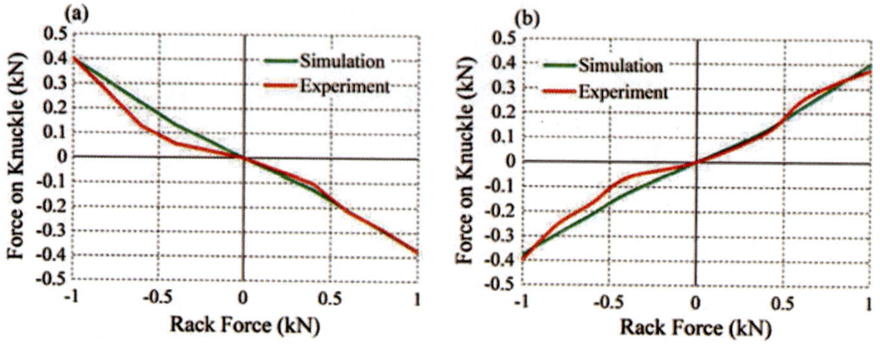

Fig. 9 **a** Left knuckle lateral versus rack force; **b** right knuckle lateral versus rack force

Figure 9a, b shows plots of force on the knuckle against the rack force. The simulation results match well with rig test measurements for a wide range of the rack force input.

Figure 10a, b represents plots of lateral displacement of the ball joint against the steer torque. Figure 11a, b is the plots of lateral force on knuckles against the steer torque input. Both the simulation results are in close agreement with the test rig measurements, and this detailed experimental testing validates our simulation model against the test rig results of the physical steering system.

We performed three tests on the test rig and compared them with simulation results. The tests considered herein involved the steer angle, steer torque and rack force as input and knuckle force and displacement as output. Different tests were performed by keeping one of the variables, viz. angle, torque and force, as constant and varying other two variables. Results obtained from the test rig conformed to the theoretical relation of proportionality between applied torque and force. Apart from that, it was observed during the test that when the force values were very small in magnitude, there is not a significant change in torque and angle.

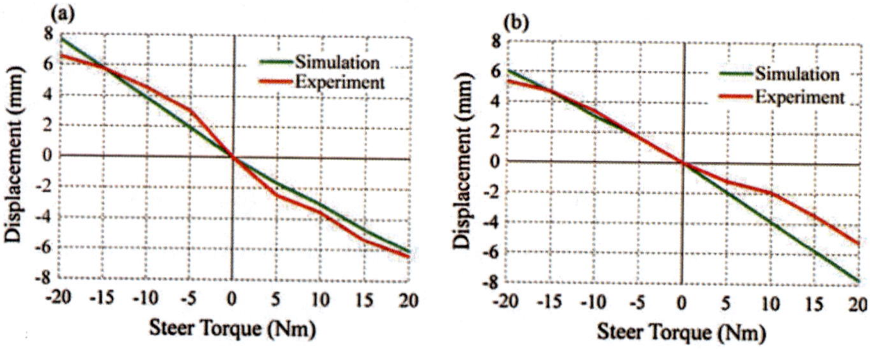

Fig. 10 **a** Left lateral disp. versus steer torque; **b** right lateral disp. versus steer torque

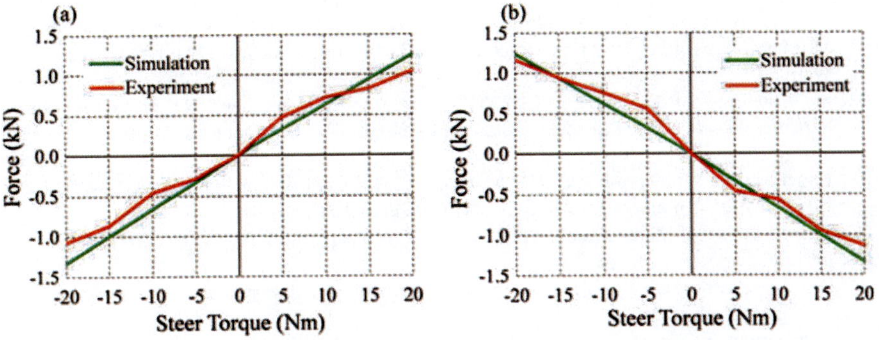

Fig. 11 **a** Left lateral force versus steer torque; **b** right lateral force versus steer torque

4 Conclusion

This study demonstrates the use of a test rig to acquire test data (force, torque, angle and displacement) closest to the real-time performance of the car steering system. The test data acquired was aptly compared with simulation results. The results also show that the given steering confirms with the Ackermann geometry. This methodology can be incorporated in path planning and full vehicle models to input more realistic steering forces in the mathematical model [16, 17]. For future work, a similar approach can be investigated with steering faults. The faults that will be induced in the steering system will be backlash, inaccurate joint phasing of the universal joint, worn out pinion, inaccurate position of steering tie rod, bump and roll steer geometry errors and studying the effect of wheel travel on the steering using the same rig. To observe the effects of all the input parameters (steer torque, steer angle and rack force) simultaneously on the knuckle force, load cells on the knuckle will be attached to obtain exact forces and reactions due to different inputs and this will be eventually compared with the mathematical model simulation results.

References

1. Nippold C, Küçükay F, Henze R (2016) Analysis and application of steering systems on a steering test bench. Autom Engine Technol 1(1):3–13
2. Knapczyk J, Maniowski M (2004) Experimental rig study on resistance forces in car steering system with rack and pinion. Arch Mechan Eng 51(2):259–281
3. Hlaing TZT, Win HH, Thein M (2017) Design and analysis of steering gear and intermediate shaft for manual rack and pinion steering system. Int J Sci Res Publ 7(12):861–881
4. Kuranowski A (2019) Electrical power steering–modelling and bench testing. Czasopismo Techniczne 8:143–158
5. Wang M, Zhang N, Chapman C, Park N, Misra A (2006) Simulation and testingof hydraulic power steering system. In: FISITA world automotive congress. Society of Automotive Engineers of Japan Inc
6. Khan U, Bianchi A (2019) Steering system development using test rig and driving simulators

7. Blundell M, Harty D (2004) Multibody systems approach to vehicle dynamics. Elsevier
8. Milliken WF, Milliken DL (1995) Race car vehicle dynamics. Society of Automotive Engineers, Warrendale, vol 400, p 16
9. MTS Steering tests systems: MTS 335. Accessed: 29th September 2021. Online repository: https://www.mts.com/en/products/automotive/subsystem-component-test-systems/steering-test-systems
10. Kanchwala H, Wideberg J, Alba CB, Marcos D (2015) Control of an independent 4WD electric vehicle by DYC method. Int J Veh Syst Model Test 10(2):168–184
11. Koladia D (2014) Mathematical model to design rack and pinion Ackerman steering geomtery. Int J Sci Eng Res 5(9):716–720
12. Rill G, Castro AA (2020) Road vehicle dynamics: fundamentals and modeling with MATLAB®. CRC Press
13. Ramanujam R, Sainath B, Varunjikar TM, Pramod P (2020) U.S. Patent Application No. 16/033,579
14. Gillespie TD (1992) Fundamentals of vehicle dynamics, vol 400. Society of Automotive Engineers, Warrendale
15. Kanchwala H, Trigell AS (2017) Vehicle handling control of an electric vehicle using active torque distribution and rear wheel steering. Int J Veh Des 74(4):319–345
16. Kanchwala H (2019, July) Path planning and tracking of an autonomous car with high fidelity vehicle dynamics model and human driver trajectories. In 2019 IEEE 10th international conference on mechanical and aerospace engineering (ICMAE), pp 306–313. IEEE
17. Sharma SK, Pare V, Chouksey M, Rawal BR (2016) Numerical studies using full car model for combined primary and cabin suspension. Procedia Technol 23:171–178

Dynamics of Mechanisms

Dynamic Modelling a 6-DOF Compliant Flexure-Based Stewart Micromanipulator

Suraj Kumar Mishra and C. S. Kumar

Abstract In this paper, we present a dynamic analysis of a 6-DOF compliant flexure-based Stewart micromanipulator. First, a planar parallelogram linkage, which is a subsystem of the 6-DOF micromanipulator, is considered for the study. The analytical model for this mechanism is formulated. The mathematical model for the full 6-DOF mechanism is too complex to carry out because of the unavailability of kinematic equations. Therefore, we choose a multibody dynamics simulation software—MSC Adams for modelling. The suitability of the software is first checked by comparing simulations result for parallelogram linkage with those obtained from the analytical model. Close confirmation of the results proves the suitability of the software. We then prepare the PRB model for the full 6-DOF micromanipulator in the Adams environment. The FEA simulations reveal that the outcomes of the Adams PRB model of the 6-DOF micromanipulators are well within acceptable limits. This work proves the effectiveness of multibody dynamics simulation software for dynamic modelling of spatial high DOF flexure-based compliant mechanisms.

Keywords Dynamic modelling · Compliant mechanisms · Flexure joints · Pseudo-rigid-body modelling · MSC Adams

1 Introduction

Many of the applications of a compliant mechanism involve operation at high speed and frequency. Therefore, the dynamic analysis of a compliant parallel micromanipulator (CPM) is really important. The dynamic model helps in finding the natural frequencies of the system and also critical in the design of a suitable controller. The dynamic model can also be used in optimising the geometry of the CPM.

S. K. Mishra (✉) · C. S. Kumar
Indian Institute of Technology Kharagpur, Kharagpur 721302, India
e-mail: surajkmishra@iitkgp.ac.in; surajkmishra1989@gmail.com

C. S. Kumar
e-mail: kumar@mech.iitkgp.ac.in

V. K. Gupta et al. (eds.), *Recent Advances in Machines and Mechanisms*, Lecture Notes in Mechanical Engineering, https://doi.org/10.1007/978-981-19-3716-3_14

In the past few decades, Lagrange-based dynamic modelling of the compliant mechanism has been widely used. The approaches used by researchers can be classified into the following three categories. First one is pseudo-rigid-body (PRB) modelling-based model. In this approach, the compliant mechanism is converted into its rigid-body counterpart. The flexure joints are replaced with their equivalent stiffness. The resulting mechanism can then be handled by exploiting all the procedures available for multibody mechanisms. This method can be found in detail in the following literatures: [1–6]. The next approach is the lumped-parameter dynamic model. This method is a popular method to model small-deflection mechanisms. The stiffness of compliance model of the mechanism is first developed to obtain input and output stiffness. For stiffness modelling, Castigliano's second theorem or compliance matrix method is generally used. Once input or output stiffness is obtained, the expressions for elastic potential energy and kinetic energy can be determined by using kinematic and potential energy equations with input or output motion (degrees-of-freedoms) DOFs as the variable [7–11].

Another method is based on distributed parameter model. In this method, the detailed DOF of each member of the system as variables is considered. This method of modelling can further be classified into the finite element method and dynamically similar rigid-multibody model. First, the discretization of the full model into several subsystems is carried out. Then, a dynamic model of the full system is obtained by formulating the energy equations for each subsystems and including them with Lagrange's equation. This method can be used to handle distributed compliance systems with complex geometry [12–14].

Due to the coupling of kinematic and static behaviour, the analysis of spatial CPMs becomes a challenging task. Moreover, the analytical methods discussed above are more effective only when the system in consideration has a simple, planar design. For more involved parallel and spatial cases, the analytical methods of modelling become extremely cumbersome as they require the availability of the kinematic models. Dynamic modelling techniques are widely explored for compliant mechanisms with three or fewer DOFs. Some of these works can be seen in references: [1, 2, 6, 10, 11, 15]. However, in literature, we could not find much work related to the dynamic modelling of higher 6-DOF flexure-based mechanisms with series-parallel spatial structure.

The present work tries to fulfil this gap and introduce a way to model such spatial designs. We first consider a planar flexure-based parallelogram mechanism, for which analytical modelling is possible. This is a sub-system of a 6-DOF micromanipulator. After the preparation of the analytical model, we will compare it with a model made in a multibody dynamics simulation software—Adams. If the results for both models compare, then it can be established that the Adams modelling can be used for dealing with the more involved designs. We then take up the 6-DOF spatial compliant micromanipulator and show the suitability of the Adams package to model its dynamic behaviour.

The rest of the paper has been organised in the following manner. The next section gives a brief overview of the design of the system. In Sect. 3, we present an analytical dynamic model of a planar parallelogram mechanism using PRB modelling. Section 4 introduces multibody dynamics simulation software—Adams for PRB modelling for the parallelogram. This section also presents the dynamic modelling for the 6-DOF flexure-based Stewart micromanipulator in the Adams environment. Finally, Sect. 5 outlines the validation of the Adams PRB results with FEA simulations. Concluding remarks are discussed in Sect. 6.

2 Design of the 6-DOF Flexure-Based Compliant Stewart Micromanipulator

The CAD model of the system is illustrated in Fig. 1. The micromanipulator consists of six legs arranged in a parallel structure. Each leg includes a compliant universal joint, a parallelogram mechanism, and a spherical flexure. The detailed design and inverse kinematics of the mechanism can be seen in an earlier work of the authors [16, 17].

It can be observed that the mechanism has a spatial series-parallel structure with 6-DOF. As a result, the dynamic modelling of the full configuration would turn out to be an extremely involved task using analytical methods. Therefore, we shall first consider a sub-system of the micromanipulator for which analytical modelling is possible.

3 Analytical Dynamic Analysis of the Parallelogram Mechanism

In this section, we shall explore the analytical method for a sub-system of the original 6-DOF design. The concerned sub-system (parallelogram mechanism) is shown in Fig. 2. The PRB modelling method is adopted for mathematical modelling. The

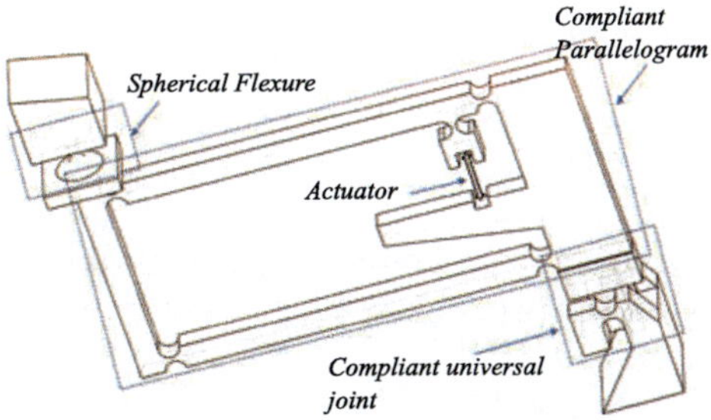

(a) One leg of the micromanipulator

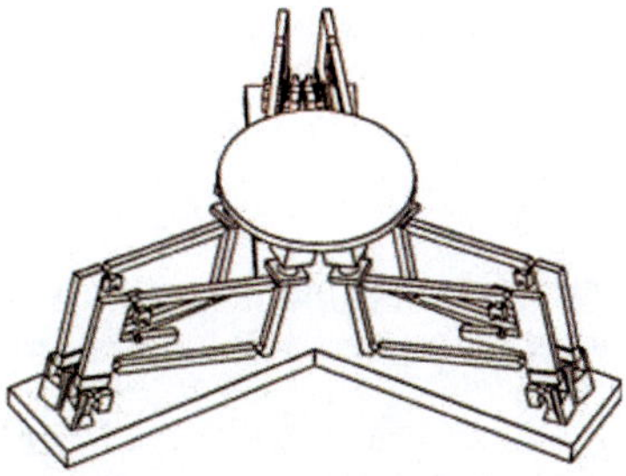

(b) Full model

Fig. 1 A 6-DOF flexure-based compliant Stewart micromanipulator [16]

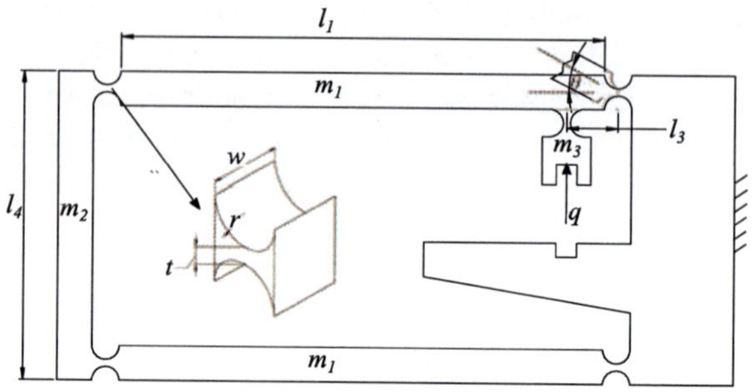

Fig. 2 Parallelogram mechanism

damping has been neglected in the system. Moreover, the potential energies of the members associated with gravity have also not been considered. In what follows, we shall try to write the potential and kinetic energy of the parallelogram for an input motion (q) as shown in Fig. 2.

3.1 *Elastic Strain Energy*

As stated earlier, we shall not consider gravitational potential energy terms. Therefore, the potential energy of the parallelogram involves only the elastic strain energies of the flexures. During the operation, all the flexure hinges rotate about the same angle θ. Additionally, as the motion of the parallelogram is extremely small in comparison to the parallelogram's dimensions, the following assumption has been made to relate the rotation of the flexure with input motion: $\theta = q/l_3$, where θ is the on-axis rotation of the flexure. In general, a flexure has compliance/stiffness in all six axes. In pseudo-rigid-body modelling, a flexure is replaced by only one major stiffness value. This is also known as on-axis stiffness. In the current situation, we have revolute flexures. In the PRB model, this joint will be replaced by a torsional spring which governs its moment-rotation behaviour. Our objective here is to finally find the equivalent stiffness (K_e) of the system.

the elastic strain energy of the parallelogram

$$U = N \times \frac{1}{2} k\theta^2 \tag{1}$$

where N is the number of flexures and k is the flexure's on-axis rotational stiffness of the. k is given by an approximate expression [18] as

$$k = \frac{2Ewt^{2.5}}{9\pi r^{0.5}} \tag{2}$$

$$U = \frac{1}{2}\left(\frac{NEwt^{2.5}}{9\pi r^{0.5} l_3^2}\right) q^2 \tag{3}$$

Hence, the equivalent stiffness of the parallelogram

$$K_e = \frac{NEwt^{2.5}}{9\pi r^{0.5} l_3^2} \tag{4}$$

It should be noted that in the equation above, only, four flexures will contribute to the calculation of the strain energy when we try to determine the natural frequency of the system.

3.2 *Kinetic Energy*

In this section, we shall express the kinetic energy of the members of the parallelogram linkage in terms of input motion, i.e. q. The objective here is to obtain the equivalent mass (M_e) for the system. Referring to Fig. 2, the kinetic energy of the system can be expressed as

$$T = 2\left\{\frac{1}{2}m_1\left(\frac{l_1 + 2r}{2l_3}\dot{q}\right)^2 + \frac{1}{2}I\dot{\theta}^2\right\} + \frac{1}{2}m_2\left(\frac{l_1 + 2r}{l_3}\dot{q}\right)^2 + \frac{1}{2}m_3\dot{q}^2 \tag{5}$$

where $I = \frac{m_1(l_1+r)^2}{12}$.

$$T = \frac{1}{2}\left[2\left\{m_1\left(\frac{l_1 + 2r}{2l_3}\right)^2 + \frac{1}{12}m_1\left(\frac{l_1 + r}{l_3}\right)^2\right\} + m_2\left(\frac{l_1 + 2r}{l_3}\right)^2 + m_3\right]\dot{q}^2 \tag{6}$$

Therefore, the equivalent mass of the system is

$$M_e = 2\left\{m_1\left(\frac{l_1 + 2r}{2l_3}\right)^2 + \frac{1}{12}m_1\left(\frac{l_1 + r}{l_3}\right)^2\right\} + m_2\left(\frac{l_1 + 2r}{l_3}\right)^2 + m_3 \tag{7}$$

3.3 *Results of the Analytical Modelling*

In this section, we shall discuss the results obtained by the analytical model. Table 1 presents the value of the structural parameters of the design defined in Fig. 2. For the selected material—Young's modulus (E), Poisson's ratio, and density are 200 GPa, 0.3, and 8000 kg/m^3, respectively.

The equation of motion for the present system is

Table 1 Dimensional parameters of the parallelogram (in mm)

r	w	t	l_1	l_3	l_4
2.00	4.00	1.00	70	7.4	40

$$M_e\ddot{q} + K_e q = 0 \tag{8}$$

Expressions for K_e and M_e are given in Eqs. 4 and 7, respectively. Finally, we obtain the natural frequency of the parallelogram as

$$f = \frac{1}{2\pi}\sqrt{\frac{K_e}{M_e}} = 221.38\,\text{Hz} \tag{9}$$

4 Dynamic Modelling Using Adams Software

4.1 Motivation Behind Using Adams Software for Dynamic Modelling

We observed in the earlier section that the analytical PRB modelling is an effective way to formulate a dynamic model of the flexure-based system. However, it requires the availability of the kinematic model to write the energy equations [19]. For simple planar mechanisms, such as the current parallelogram, these kinematic equations can easily be determined. But, for our 6-DOF spatial system, the required kinematic equations are challenging to obtain.

As an alternate solution to the issue stated above, one can prepare the PRB model in a multibody dynamics simulation software, for example—MSC Adams. In what follows, we shall explore the suitability of the software for dynamic analysis of the parallelogram linkage. We shall do it by comparing the outcomes of the Adams model to the previously formulated analytical model.

4.2 The Dynamic PRB Model of the Parallelogram in Adams Environment

The Adams model for the parallelogram is shown in Fig. 3. The Adams package allows us to replace flexure joints with conventional revolute joints and includes the stiffness of the flexure joints as torsional spring members. We extracted the natural frequency from the Adams model to check how correctly and closely the model compares with the analytical model. The Adams model predicted the natural

Fig. 3 A PRB model of the parallelogram in Adams environment

frequency as 221.87 Hz, which is very close to the analytical result (221.38 Hz) obtained in Sect. 3.3.

Therefore, we can establish that the Adams model is indeed an accurate method of modelling the PRB model for the flexure-based mechanisms and could be effective in situations, wherein the analytical models are cumbersome to obtain. In the next section, we shall try to model the original 6-DOF compliant Stewart micromanipulator using Adams software.

4.3 *Dynamic Modelling the Full 6-DOF Flexure-Based Stewart Micromanipulator*

For parallelogram linkage, we prepared the complete PRB model in Adams environment itself. This was possible due to its simple geometry. However, a full 6-DOF model, with all its intricate details, is difficult to prepare directly in Adams. Therefore, we shall first take the help of any CAD modelling software, for example—Solidworks. Let us first consider the CAD model of the full mechanism (Fig. 1). All the flexures of this design will be cut-extrude, as shown in Fig. 4, in Solidworks. This converts our monolithic system into a multibody one. This multibody model can now

Fig. 4 Cut-extrude of flexures to convert the monolithic model into a multibody one

Fig. 5 A PRB model of the 6-DOF flexure-based Stewart micromanipulator in Adams environment

Table 2 Comparison of frequencies (in Hz) predicted by FEA model, and Adams PRB model of the micromanipulator

Mode shape	Abaqus	Adams
1	61.73	62.22
2	91.42	93.73
3	91.467	94.56
4	139.06	137.21
5	139.15	141.66
6	146.20	155.56

be exported to the Adams environment. In Adams software, the conventional joints are then added at those previously cut-extruded flexure locations of this multibody model. Moreover, the equivalent stiffness of the joints is also added to joints in the form of torsional springs. Apart from the revolute flexures, this micromanipulator has spherical flexures as well. They connect the legs to the platform or end-effector of the micromanipulator. The stiffness value of these spherical flexures has been taken from the literature [20]. The final Adams model of the micromanipulator is shown in Fig. 5. Dynamic simulation of this Adams model yields its natural frequencies. The natural frequencies extracted from this model are illustrated in Table 2.

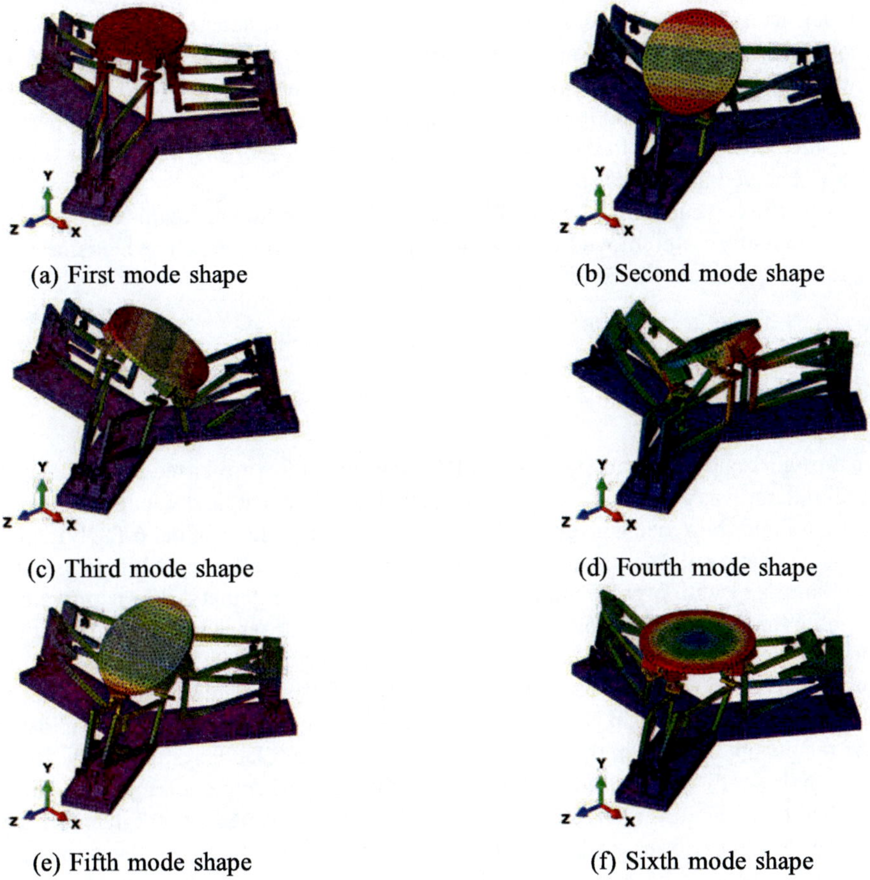

(a) First mode shape (b) Second mode shape (c) Third mode shape (d) Fourth mode shape (e) Fifth mode shape (f) Sixth mode shape

Fig. 6 The initial six mode shapes of the 6-DOF compliant Stewart micromanipulator

5 FEA Validation Using Abaqus Simulations

We shall now resort to FEA simulations to validate the results obtained from the Adams PRB model. The modal FEA analysis of the 6-DOF mechanism is performed with the Abaqus software package. A ten-node tetrahedral element (C3D10) is chosen for meshing. We obtain the natural frequencies and associated modes from simulations. Figure 6 illustrates the first six modes of vibration. A numerical comparison of frequencies predicted by the FEA model and the Adams PRB model has been elaborated in Table 2.

It is observed that there is a slight deviation between Adams PRB result and the result obtained by FEA simulation. The possible sources of the error are as follows. The first source is the inaccuracy of the stiffness equation itself. Secondly, the deformation of the connecting links that were assumed to be rigid in the analytical model. Moreover, in the case of spatial compliant mechanisms, the flexures also undergo a small amount of out-of-axis motion; this adds to the deviation. Finally, the internal material damping of the members also contributes to the error. However, for all practical purposes and for the sake of analysis, errors less than 15% are generally considered acceptable in the analysis of compliant mechanisms, and the maximum error observed here is less than 7%.

Therefore, it can finally be concluded that PRB modelling in Adams environment can efficiently be employed for dynamic analysis of compliant mechanisms with involved spatial configuration.

6 Conclusions

In this work, a dynamic analysis of a 6-DOF compliant flexure-based Stewart micromanipulator was presented. We first formulated an analytical model for a simple planar parallelogram mechanism, which is a sub-system of the original 6-DOF model. We then employed Adams software to exhibit that it also has the capability to produce such a model with high accuracy. This was done to prove that Adams software is reliable for the current problem and can be used in situations, wherein mathematical models are difficult to obtain. The full 6-DOF flexure-based Stewart micromanipulator was then considered for analysis. We successfully prepared the PRB model of this micromanipulator in Adams environment. Finally, FEA simulations validated the results obtained from the Adams PRB model of the 6-DOF system.

The present work showed the limitations of the analytical models and proposed a method to model the high DOF spatial flexure-based micromanipulator in multibody dynamics simulation environment. Models prepared in such a way can also be utilised for control analysis by Adams–Simulink co-simulation. This will be attempted in our future work.

References

1. Boyle C, Howell LL, Magleby SP, Evans MS (2003) Dynamic modeling of compliant constant-force compression mechanisms. Mechan Mach Theor 38(12):1469–1487
2. Yu Y-Q, Howell LL, Lusk C, Yue Y, He M-G (2005) Dynamic modeling of compliant mechanisms based on the pseudo-rigid-body model. J Mechan Des 127(4):760–765
3. Guo Z, Tian Y, Liu C, Wang F, Liu X, Shirinzadeh B, Zhang D (2015) Design and control methodology of a 3-dof flexure-based mechanism for micro/nano-positioning. Robot Comput-Integr Manufact 32:93–105
4. Chen W, Qu J, Chen W, Zhang J (2017) A compliant dual-axis gripper with integrated position and force sensing. Mechatronics 47:105–115
5. Wang F, Liang C, Tian Y, Zhao X, Zhang D (2014) Design of a piezoelectric-actuated microgripper with a three-stage flexure-based amplification. IEEE/ASME Trans Mechatron 20(5):2205–2213

6. Li Y, Wu Z (2016) Design, analysis and simulation of a novel 3-dof translational micromanipulator based on the prb model. Mechan Mach Theor 100:235–258
7. Tang H, Li Y (2013) Development and active disturbance rejection control of a compliant micro-/nanopositioning piezostage with dual mode. IEEE Trans Industr Electron 61(3):1475–1492
8. Zhu X, Xu X, Wen Z, Ren J, Liu P (2015) A novel flexure-based vertical nanopositioning stage with large travel range. Rev Sci Instrum 86(10):105112
9. Zhu W-L, Zhu Z, Guo P, Ju B-F (2018) A novel hybrid actuation mechanism based xy nanopositioning stage with totally decoupled kinematics. Mechan Syst Signal Process 99:747–759
10. Polit S, Dong J (2010) Development of a high-bandwidth xy nanopositioning stage for high-rate micro-/nanomanufacturing. IEEE/ASME Trans Mechatron 16(4):724–733
11. Liu P, Yan P, Zhang Z (2015) Design and analysis of an x-y parallel nanopositioner supporting large-stroke servomechanism. Proc Inst Mechan Eng Part C: J Mechan Eng Sci 229(2):364–376
12. Rösner M, Lammering R, Friedrich R (2015) Dynamic modeling and model order reduction of compliant mechanisms. Precis Eng 42:85–92
13. Shen Y, Chen X, Jiang W, Luo X (2014) Spatial force-based non-prismatic beam element for static and dynamic analyses of circular flexure hinges in compliant mechanisms. Precis Eng 38(2):311–320
14. Ryu JW, Lee S-Q, Gweon D-G, Moon KS (1999) Inverse kinematic modeling of a coupled flexure hinge mechanism. Mechatronics 9(6):657–674
15. Tang X, Pham H-H, Li Q, Chen I-M (2004) Dynamic analysis of a 3-dof flexure parallel micromanipulator. In: IEEE conference on robotics, automation and mechatronics, vol 1, pp 95–100. IEEE, 2004
16. Mishra SK, Kumar CS (2018) Design and kinematics of a compliant stewart micromanipulator. In: 2018 international conference on manipulation, automation and robotics at small scales (MARSS), pp 1–6, IEEE, 2018
17. Mishra SK, Kumar CS (2022) Compliance modeling of a full 6-DOF series-parallel flexure-based Stewart platform-like micromanipulator. Robotica, 1–28
18. Paros JM, Weisbord L (1965) How to design flexure hinges. Machine Design 37:151–156
19. Ling M, Howell LL, Cao J, Chen G (2020) Kinetostatic and dynamic modeling of flexure-based compliant mechanisms: a survey. Appl Mechan Rev 72(3)
20. Lobontiu N (2002) Compliant mechanisms: design of flexure hinges. CRC Press, Boca Raton

Analysis of Connected Shallow Arches Under a Load from a Moving Rigid Wedge

Priyabrata Maharana and G. K. Ananthasuresh

Abstract In this paper, we analyze a pair of shallow arches that are interconnected with each other through their ends by a revolute joint. The other ends of the arches may be pinned or fixed. The two arches have opposite curvature. When one of the arches is subjected to a transverse load, it deforms and eventually flips to its inverted shape. Simultaneously, the deformation of the first arch triggers the switching of the second arch due to a moment transfer between the two arches through the revolute joint, and the second arch flips to its inverted shape too. These arches can be switched back to their as-fabricated shape by applying a transverse load to the second arch. In this way, these connected arches move up and down to take the load at any instant when the switching loads come from a long moving rigid wedge. The motivation to study such connected flipping arches arose from the intent to develop dynamically self-offloading footwear for diabetics suffering from peripheral neuropathy. The purpose of this is to limit the plantar pressure to stay below a prescribed value to prevent ulceration. When a certain region of the sole is offloaded, the array of connected arches helps certain other region to take up the differential load to support the body weight. In this paper, we present a theoretical formulation for two connected arches, solve them semi-analytically, and compare with the results of the geometrically nonlinear finite element analysis.

Keywords Snap-through · Post-buckling · Self-offloading footwear

1 Motivation for Studying Connected Arches

Peripheral neuropathy, i.e., nerve damage that causes loss of sensation, pain, and numbness, is one of the major complications of diabetes. The lack of sensation in the sole of the feet leads to ulceration and recurring wounds which, if left untreated,

P. Maharana (✉) · G. K. Ananthasuresh
Mechanical Engineering, Indian Institute of Science, Bengaluru, India
e-mail: priyabratam@iisc.ac.in

G. K. Ananthasuresh
e-mail: suresh@iisc.ac.in

V. K. Gupta et al. (eds.), *Recent Advances in Machines and Mechanisms*, Lecture Notes in Mechanical Engineering, https://doi.org/10.1007/978-981-19-3716-3_15

necessitate amputation of the feet [1]. This can be easily prevented by using offloading footwear to protect vulnerable regions of the sole. The current clinical practice is to measure the plantar pressure of a patient using pedoscan to identify regions of high pressure and to prescribe customized in-soles. As shown in Fig. 1a, b, either a pocket is carved out in a foam-like material used as the in-sole, or it is contoured to transfer the load to other regions. This is called *static offloading* of the plantar pressure as the high-pressure regions of the sole of the foot are offloaded fully or partially. When a certain region of the sole is thus offloaded, certain other region becomes vulnerable as they bear a larger portion of the body weight. Therefore, ulceration begins elsewhere warranting frequent modification of the in-sole of the footwear. One way to prevent this is to design dynamically offloading footwear that automatically re-distributes the body load evenly on the sole during the entire gait cycle. We had proposed such *self-offloading footwear* by using snapping arches [2] whose concept is depicted in Fig. 2.

Four phases of the gait cycle, namely *heel strike*, *midstance*, *flat foot*, and *heal off*, are shown in Fig. 2a. The manner of offloading is also shown for three types: (i) static

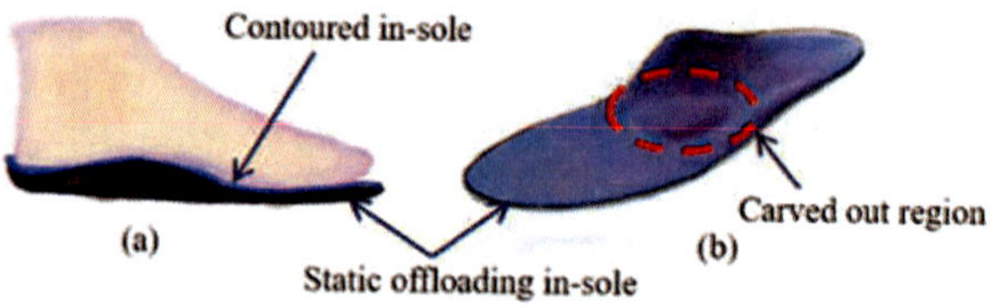

Fig. 1 Static offloading in-sole used in current practice: **a** a contoured in-sole for even distribution of pressure and **b** in-sole with a carved-out region if there is an ulcer

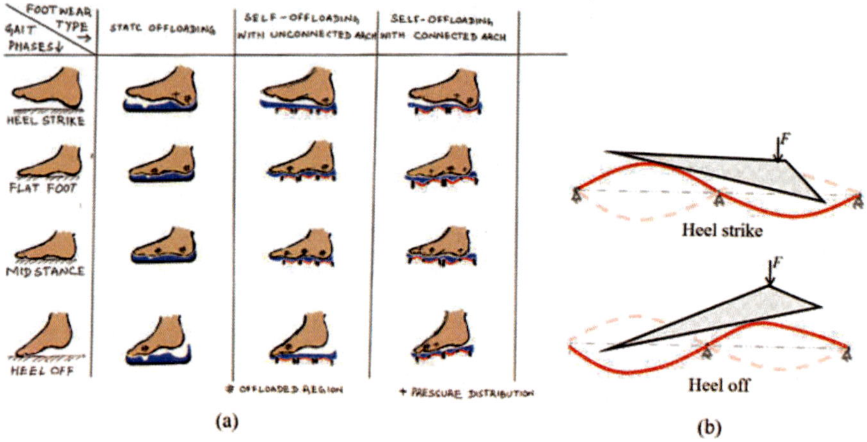

Fig. 2 **a** Comparison of static offloading footwear, self-offloading footwear with unconnected arches, and self-offloading footwear with interconnected arches during four phases of the gait cycle, namely *heal strike*, *flat foot*, *midstance*, and *heal off*. The symbol # shows the offloaded regions and + shows the pressure-redistributed regions during offloading. **b** Flip-flop of the connected arches in heel strike and heel off phase. The dashed line denotes the configurations before switching at the beginning of the particular gait phase

offloading, (ii) self-offloading with unconnected arches, and (iii) self-offloading with interconnected arches. In static offloading, the plantar pressure distribution is preset and remains mostly unchanged. As stated above, this has the drawback of creating high-pressure regions elsewhere. In self-offloading using connected arches [3], the arches (shown in red color in Fig. 2a) snap down along with the deformable in-sole (shown in blue color) such that no region would take up more than the prescribed level of plantar pressure. While this helps to some extent, further improvement is possible if the arches transfer the load to other arches to evenly distribute the body weight. This is possible in interconnected arches as depicted in Fig. 2b, wherein the left and right arches flip-flop up (i.e., they move and down) when the load from the foot (considered as a moving rigid wedge) changes in the walking cycle.

In our earlier work, we have conceptualized and implemented self-offloading with unconnected arches [2, 3]. As shown in Fig. 3a, by placing the in-sole over an array of arches (see Fig. 3b), offloading can be achieved whenever the pressure at a point exceeds the allowed limit (210 kPa in adults). In this arrangement, the arches undergo snap-through by losing contact with the sole of the foot when the load on them is beyond the preset threshold, as shown in Fig. 3d. The arches are designed in such a way that they snap back up during a different phase of the gait cycle. This is illustrated in Fig. 3a wherein the in-sole is supported by four capping plates under which there are arrays of snapping arches and top of which there are two layers of soft-soles to absorb the shock from the in-sole (see Fig. 3c). This prototype was tested at Karnataka Institute of Endocrinology and Research (KIER), Bengaluru. It was found that the self-offloading footwear reduces the peak pressure by about 34% while keeping the plantar pressure during the complete gait cycle below 210 kPa

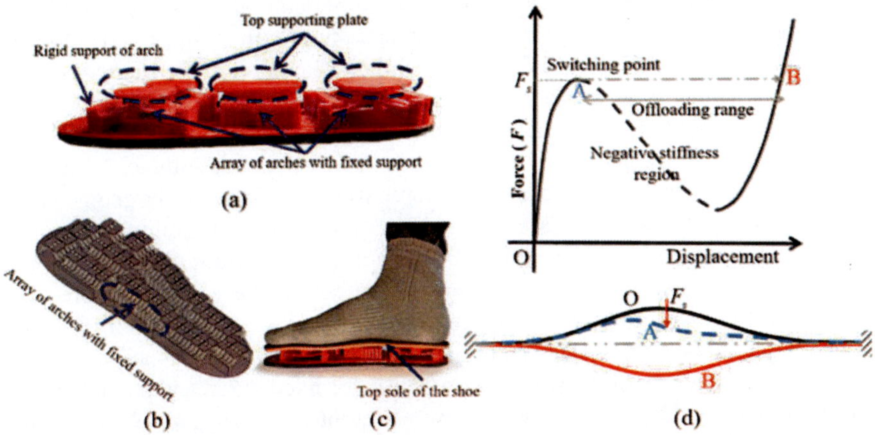

Fig. 3 **a** A self-offloading in-sole using an array of snap-through arches beneath the capping plates, **b** CAD model of the in-sole showing array of fixed–fixed arches, **c** 3D-printed prototype with two layers of top sole (black and orange) to absorb the shock from the in-sole, and **d** force–displacement curve of snap-through arch showing the as-fabricated shape (O), shape at the switching point (A), and the inverted shape (B) after offloading

[3]. It was also observed that further reduction in plantar pressure is possible if the capping plates are interconnected so that they can redistribute the load dynamically among them rather than only offloading individually. This is the motivation for the current study of connected arches.

1.1 Scope of the Paper

As a first study, in this paper, we consider two arches connected with a revolute joint and analyze their snapping behavior. We model the foot as a rigid wedge that comes into contact with the arches at different points during the gait cycle. As a further simplification, we consider the force from the foot only at the midpoints of the two arches.

In this paper, we present a semi-analytical method to investigate the characteristics of planar-connected arches when the two arches have either hinged–hinged or fixed–hinged boundary conditions. This is beneficial in two ways: (i) finite element analysis involves considerable computation because of the inherent geometric nonlinearity, whereas the semi-analytical method will be much faster, which leads to a quicker way to know if the arches undergo flip-flop, and (ii) the analytical method paves the way for closed-form relationships in synthesis of the arch shapes for desired behavior as was done in [4, 5].

We introduce a compatibility condition for the deformed arch profiles of the connected arches such that they maintain the continuity slope at the common revolute joint. We obtain the equilibrium equations of such arches using the potential energy and the derived compatibility condition. We numerically solve these equilibrium equations to get the force–displacement characteristics for the loading and unloading cycle and understand the switching and bistability criteria for different kinds of arch profiles. This analysis helps us in understanding the nature of the connected arches as well as designing them by choosing suitable arch geometry for various applications.

Pre-buckling a straight beam into an arch [6] and arriving at a suitable stress-free arch shape [7–10] are the two ways to get an arch structure. Based on the geometry and boundary conditions, an arch can show either snap-through or bistability phenomena. This nature of the arch is characterized by the physical quantities, like the *switching force*, the *switchback force*, and the *travel* of the arch. These quantities are identified from the nonlinear force–displacement characteristics of the arch. Knowing these quantities helps us in designing the arches for various applications by suitably choosing the arch dimensions. The analysis for fixed–fixed and pinned–pinned arches with general boundary conditions was presented by Qiu et al. [7], and design methods by Palathingal and Ananthasuresh [10], to identify these physical quantities for an arch of arbitrary as-fabricated stress-free shape. The as-fabricated shape of the arch highly influences the bistable nature of the arch. Therefore, Palathingal and Ananthasuresh [4, 5] presented an analytical way of finding bilateral relationships between the two stable configurations of the arch. They also mentioned sufficiency

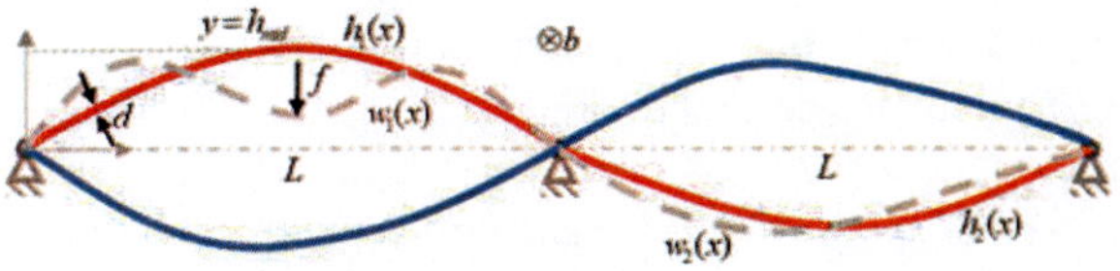

Fig. 4 Connected pinned–pinned arches with all the dimensions showing the as-fabricated shape (red-solid), intermediate deformed shape (gray-dashed), and inverted stable shape (blue-solid)

criteria based on which one can decide the stability of the arches. Though the literature is rich with the analysis of single arches, there is no literature available for an arrangement where two arches are connected through a revolute joint between them, as shown in Fig. 4. But some of the earlier works reported on the analysis of continuous linear beams [9] and periodically connected linear beams [11]. When two arches are connected, one arch acts as a nonlinear spring for the other, and vice versa, due to the coupling of the slope at the common joint.

In Sect. 2, we derive the potential energy, compatibility condition, and the equilibrium equation for the analysis. In this section, we also explain the force–displacement characteristics with a suitable example of pinned–pinned arches and discuss the switching and stability criteria of connected arches. The results of the analytical model are compared with finite element analysis solution done using ABAQUS finite element analysis software (www.simulia.com) and presented in Sect. 3. Summary and concluding remarks of the work are included in the last section.

2 Analysis of Connected Arches

We consider arches with either pinned–pinned or fixed–pinned boundary conditions for the analysis. Let us take two arches of span, L, depth, d, breadth, b that are connected with each other through a common revolute joint in between them as shown in Fig. 4. The mid-span height of the left arch is taken to be h_{mid} where a point load f is applied. The main question we ask is this: Under what conditions, would the second arch flip up when the first arch experiences a load to switch it down?

We begin the analysis by representing the as-fabricated stress-free state, $h_i(x)$, and the deformed shape, $w_i(x)$, of the arches as a linear combination of the buckling mode weights of the corresponding straight beam as follows [8, 9]:

$$h_1(x) = h_{\text{mid}} \sum_{1}^{N} a_i \varphi_i; \;\; h_2(x) = h_{\text{mid}} \sum_{1}^{N} b_i \varphi_i;$$
$$w_1(x) = h_{\text{mid}} \sum_{1}^{N} A_i \varphi_i; \;\; w_2(x) = h_{\text{mid}} \sum_{1}^{N} B_i \varphi_i \tag{1}$$

where φ_i is the mode shape that satisfies the boundary conditions of the arch. For a pinned–pinned arch, the mode shapes are [9]:

$$\varphi_i = \sin\left(i\pi \tfrac{x}{L}\right) \tag{2}$$

and for a fixed–hinged arch, mode shapes are [9]:

$$\varphi_i = \cos\left(M_i \tfrac{x}{L}\right) - \frac{\sin\left(M_i \tfrac{x}{L}\right)}{M_i} + \frac{x}{L} - 1 \tag{3}$$

where $M_i = 1.43\pi, 2.45\pi, 3.47\pi, \ldots$ for $i = 1, 2, 3, \ldots$. For a given as-fabricated mode weight, $a_i s$ and $b_i s$, we determine the deformed profile mode weights, $A_i s$ and $B_i s$, by minimizing the total potential energy of both the arches.

Total potential energy: For arches of Young's modulus, E, and second moment of area, I, the total bending strain energy is given by

$$\mathrm{SE_b} = \frac{1}{2} EI \sum_{i=1,2} \int_0^L \left(\frac{\mathrm{d}^2 h_i}{\mathrm{d}x^2} - \frac{\mathrm{d}^2 w_i}{\mathrm{d}x^2}\right)^2 \mathrm{d}x \tag{4}$$

As we take the arches to be shallow, i.e., $\left(\frac{\mathrm{d}h_i}{\mathrm{d}x}\right)^2 \ll 1$ and $\left(\frac{\mathrm{d}w_i}{\mathrm{d}x}\right)^2 \ll 1$, the nonlinear terms in the bending energy are neglected. When the arches deform in the transverse direction, they develop axial compression, and the net strain energy due to the compression is expressed as

$$\mathrm{SE_c} = \frac{1}{4}\frac{Ebd}{L} \sum_{i=1,2} \left\{ \int_0^{L_1} \left[\left(\frac{\mathrm{d}h_i}{\mathrm{d}x}\right)^2 - \left(\frac{\mathrm{d}w_i}{\mathrm{d}x}\right)^2 \right] \mathrm{d}x \right\}^2 \tag{5}$$

A point load, f, is applied at the mid-span length of the left-side arch. The work potential due to the transverse displacement of the point of application of the load is given by

$$\mathrm{WP} = -f\left[h_1\left(\tfrac{L}{2}\right) - w_1\left(\tfrac{L}{2}\right)\right] \tag{6}$$

The total potential energy of the connected arches includes the strain energy due to bending, compression, and the work potential. It is expressed as

$$\mathrm{PE} = \mathrm{SE_b} + \mathrm{SE_c} + \mathrm{WP} \tag{7}$$

To get the equilibrium equations, we minimize the potential energy w.r.t. the unknown mode weights, $A_i s$ and $B_i s$, for left and right arches, respectively.

2.1 The Compatibility Condition

Due to the interconnection of the arches through the common revolute joint, the deformation of the left arch causes the deformation of the right arch even though it does not experience any external loading and vice versa. As the revolute joint does not resist the rotation of the arch, the deformed shape of both the arches maintains slope continuity at this point. In other words, the rotation of the left arch and the right arch are the same w.r.t. their as-fabricated states at the common revolute joint. Therefore, we impose a rotational compatibility condition at the interconnected revolute joint as

$$\Theta = \frac{\left(\frac{dw_1}{dx}\Big|_L\right) - \left(\frac{dw_2}{dx}\Big|_0\right)}{1 + \left(\frac{dw_1}{dx}\Big|_L\right)\left(\frac{dw_2}{dx}\Big|_0\right)} - \frac{\left(\frac{dh_1}{dx}\Big|_L\right) - \left(\frac{dh_2}{dx}\Big|_0\right)}{1 + \left(\frac{dh_1}{dx}\Big|_L\right)\left(\frac{dh_2}{dx}\Big|_0\right)} = 0 \tag{8}$$

where the term containing $h_i(x)s$ represents the tangent of angle difference, θ_0, between the as-fabricated shape of two the arches at the revolute joint, as shown in Fig. 5. When the arches are identical and have opposite curvature, Eq. (8) reduces to

$$\Theta = \left(\frac{dw_1}{dx}\Big|_L\right) - \left(\frac{dw_2}{dx}\Big|_0\right) = 0 \tag{9}$$

The constraint condition at the common revolute joint eventually leads to concentrated moments, M_{R_1}, on the right end of left arch and M_{L_2}, on the left end of the right arch. Due to the revolute joint, both the moments cancel each other, and the net moment at the point becomes zero, i.e.,

$$M_{R_1} = -M_{L_2} \tag{10}$$

We get the equilibrium configuration of the arch by minimizing the potential energy such that the deformed profile of the arch satisfies the compatibility condition (Eq. 8).

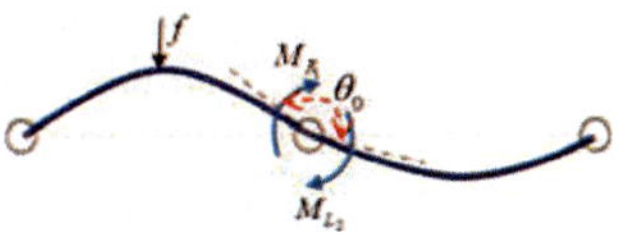

Fig. 5 Connected arch with the reaction moment and the initial angle difference of as-fabricated shape at the revolute joint

2.2 Equilibrium Equations for Connected Pinned–Pinned Arches

In this section, we analyze the pinned–pinned connected arches, obtain the governing equilibrium equations, and study the force–displacement characteristics of the connected arches. We substitute Eqs. (1) and (2) in Eqs. (4) to (6) to get the potential energy expression and minimize the potential energy (Eq. 7) w.r.t. the unknowns $A_i s$ and $B_i s$ along with the compatibility condition (Eq. 8) to get the governing equation of the connected arches as

$$\frac{\partial(\mathrm{PE}+\Lambda\Theta)}{\partial A_i} = \frac{M_i^4}{2}(A_i - a_i) - \frac{3}{2}Q^2 C_1\left(A_i M_i^2\right) + F\left\{-(-1)^{\frac{i+1}{2}}\right\}\Bigg|_{i=1,3,5,\ldots} + \Lambda\left[\left\{(-1)^i M_i\right\} + \tan(\theta_0)\left\{\frac{h_{\mathrm{mid}}}{L}\left[(-1)^i M_i\right]\left[\sum_1^N B_i M_i\right]\right\}\right] = 0 \tag{11}$$

$$\frac{\partial(\mathrm{PE}+\Lambda\Theta)}{\partial B_i} = \frac{M_i^4}{2}(B_i - b_i) - \frac{3}{2}Q^2 C_2\left(B_i M_i^2\right) + \Lambda\left[-M_i + \tan(\theta_0)\left\{\frac{h_{\mathrm{mid}}}{L}M_i\left[\sum_1^N A_i M_i(-1)^i\right]\right\}\right] = 0 \tag{12}$$

$$\frac{h_{\mathrm{mid}}}{L}\left\{\sum_1^N A_i M_i(-1)^i\right\} - \frac{h_{\mathrm{mid}}}{L}\left\{\sum_1^N B_i M_i\right\} + \tan(\theta_0)\left\{1 + \frac{h_{\mathrm{mid}}^2}{L^2}\left\{\sum_1^N A_i M_i(-1)^i\right\}\left\{\sum_1^N B_i M_i\right\}\right\} = 0 \tag{13}$$

where $M_i = i\pi, C_1 = \sum a_i^2 M_i^2 - \sum A_i^2 M_i^2, C_2 = \sum b_i^2 M_i^2 - \sum B_i^2 M_i^2, F = \frac{fL^3}{EIh_{\mathrm{mid}}}$ are the nondimensional force on the arch and Λ is the multiplier corresponding to the continuity equation. Here, the multiplier (Λ) is equivalent to the internal moment at the common revolute joint due to the relative arch deformation. To understand the nature of the force–displacement, i.e., f–δ, characteristic of the connected arch, we numerically solve the preceding set of Eqs. (11) to (13) and trace the curve for a range of displacement values for both loading and unloading cycles.

As an illustrative example, we take the as-fabricated shape of a connected arch consist only the first mode weights such that $a_1 = 1$, and $b_1 = -0.5$. We take $h_{\mathrm{mid}} = 10$ mm, $d = 1$ mm, $b = 4$ mm, $L = 100$ mm, $E = 50$ MPa and substitute them in the equilibrium equations. Using Eqs. (11) to (13), we solve f by varying δ from 0 mm to twice the height of the arch, i.e., 20 mm, to get the f–δ curve for the loading cycle. Similarly, to get the f–δ curve for the unloading cycle, we vary δ in a reverse way. The force–displacement and mode weights-displacement characteristics of these connected arches are shown in Fig. 6.

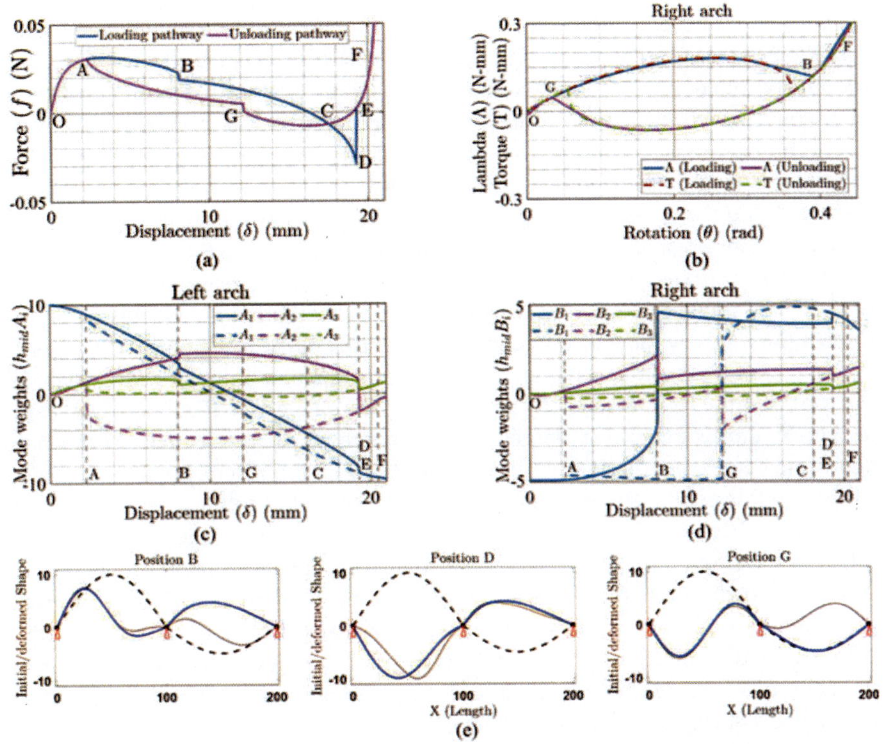

Fig. 6 **a** Force–displacement (f–δ) characteristics of the connected arch showing the loading path and the unloading path, **b** comparison loading and unloading pathways of torque (T)-rotation (θ) characteristics (solid curve) with the multiplier (Λ)-rotation (θ) characteristics (dashed curve) for the right arch. Note that a concentrated torque (T) is applied at the left end of the right arch, **c** mode weights variations of the left arch w.r.t. the displacement, **d** mode weights variations of the right arch w.r.t. the displacement, and **e** as-fabricated and deformed state of the arches at positions B, D, and G. In **c** and **d**, the solid line corresponds to the loading path, and the dotted line represents the unloading path. In **e**, the gray-dashed line shows the as-fabricated shape, gray-solid line shows the deformed shape just before the position B, D, and G, and blue-solid line shows the deformed shape of the arches exactly at positions B, D, and G

In Fig. 6a, O represents the mid-point displacement at as-fabricated stress-free state, and E represents that of the second force-free state. The arch follows the path OABCDEF (blue-solid curve) for the loading cycle and FECGAO (magenta-solid curve) for the unloading cycle. The loading and unloading pathways are different due to the asymmetric nature of the strain energy of the system. The initial phase, O–A, and the post-buckling phase, E–F, are identical for both the loading and unloading cycle, whereas for a displacement between A and E, there are two different arch configurations and force values for the loading and unloading phases, leading to two different pathways.

When we increase the displacement gradually at point O, the magnitude of the first mode weights starts decreasing, and that of the second and third mode starts increasing for both the arches (see Fig. 6c, d). Although these arches and the loading condition are symmetric in nature, the arches deform asymmetrically due to the coupling at the common revolute joint. They act as a nonlinear rotational spring to each other, which leads to asymmetry in the system. As the load increases on the left arch, the slope at either end starts increasing. But due to the stiffness of the rotational spring of the right arch, the motion at the right end of the left arch gets restricted, and the slope at that point starts decreasing. This causes the second mode weight to increase on the positive side. With further increase in the displacement, the force value increases till the switching force value and decreases smoothly (curve O–A–B) till point B. At this point, the force and the stiffness suddenly jump to a lower value. This sudden jump occurs as the rotation of the right arch at the common revolute joint attains the switching rotation value due to a point moment alone (see Fig. 6b), and the right arch switches to its inverted shape (see Fig. 6d, e). As shown in Fig. 6c, the second mode weight of the left arch attains a maximum at this point and starts decreasing gradually till the point D when the displacement is increased further. At this point D, the second mode weight for the left arch suddenly jumps (see Fig. 6e) and becomes negative (see Fig. 6c). This occurs due to the spring stiffening effect of the right arch. This stiffening also leads to a sudden increase in the force value, as shown in Fig. 6a. Further increment in the displacement also increases the force value on the arch. The same argument can also be made for the unloading path FECGAO when we vary the displacement in a reverse direction starting from the point F. We decide the nature of the connected arches based on their switching nature and stability at the inverted shape.

Conditions for switching: The switching point corresponds to the maximum force with respect to the displacement on the force–displacement curve, whereas the switchback point corresponds to the minimum force value. To get these switching points of the arch, we pose an optimization problem as follows.

$$\begin{aligned} &\underset{A_i s,\, B_i s}{\text{Maximize/Minimize:}} \quad F \\ &\text{subject to:} \\ &\qquad \frac{\partial(\text{PE} + \Lambda\Theta)}{\partial A_i} = 0 \\ &\qquad \frac{\partial(\text{PE} + \Lambda\Theta)}{\partial B_i} = 0 \\ &\qquad \Theta = 0 \end{aligned} \tag{14}$$

We numerically solve the preceding optimization problem to get the switching, switchback force, and their corresponding unknown mode weights, $A_i s$ and $B_i s$, so that they satisfy the equilibrium equations which are nothing but the KKT (Karush–Kuhn–Tucker) necessary conditions for the constrained optimization problem. The problem becomes easier when we consider only a few mode shapes to represent the shape of the arches. From our earlier studies [10], we found that the first three mode shapes are adequate to get accurate results. We call our semi-analytical modeling because we numerically solved the KKT conditions. In this short paper, details of the KKT conditions are not included due to paucity of space. We only present the results and their utility in ensuring flip-flopping of the arches.

When the load exceeds these limits, the load-bearing arch (say the left arch) switches to its inverted shape. However, the switching of the left arch does not guarantee the switching of the right arch. The switching of the right arch occurs when either the multiplier, Λ, exceeds the switching moment value of the right arch, or the angular rotation of the left arch at $x = L$, $\frac{dw_1}{dx}\big|_{x=L}$, crosses the critical angular rotation value of the right arch (see Fig. 6b). We get these critical switching moment and rotation value from the torque-rotation characteristics of the right arch alone when it is subjected to a concentrated moment at $x = 0$. For the particular arch considered here, the switching force and switchback force are 0.0315 N and 0.0075 N, respectively. We also found that the right arch switches to an inverted shape when the left arch switches, which is desired.

Conditions for bistability: The second stable state of the connected arch corresponds to a zero-force configuration other than the as-fabricated shape. Also, it represents a minimum on the strain energy landscape. To get the bistable configuration of the connected arches, we solve Eqs. (11) to (13), by substituting $F = 0$. Once we get a solution, we check for the sufficiency condition to confirm the solution indeed has minimum energy. Therefore, we check the determinant of the principal minors of the bordered Hessian matrix, $\mathbf{H_b}$, which is given by [12]

$$\mathbf{H_b} = \begin{bmatrix} 0 & \nabla\Theta \\ \nabla\Theta^T & \mathbf{H}(\mathrm{PE} + \Lambda\Theta) \end{bmatrix} \tag{15}$$

where ∇ represents the gradient and $\mathbf{H}$ represents the Hessian of scalar functions. The value of the determinants of the principal minors becomes negative when the zero-force state corresponds to a stable configuration.

The nature of the connected arches depends on the as-fabricated shape of the individual arches. For certain values of mode weights, $a_i s$ and $b_i s$, both the arches switch to their inverted shape when force and the multiplier exceed the threshold. But these arches may or may not be stable at their inverted shape. We can show from the determinant of the principal minors that it is necessary for the arches to be bistable individually to have an inverted second stable state when they are connected to each other. For the example we are considering, the arches are bistable individually, and they are bistable when they are connected. The studies on these connected arches are verified with the finite element simulation and presented in the next section.

3 Validation Using Finite Element Analysis

The force–displacement characteristics obtained from the analytical modeling show good agreement (within 3% error at the switching and switch back point) with the results of finite element analysis (FEA). The comparison of these results for different as-fabricated connected arches is shown in Fig. 7. The solid curve represents the results obtained from the semi-analytical modeling considering five modes to approximate the deformed shape, and the dashed curve is obtained from FEA. The blue and the magenta curves represent the loading and unloading path, respectively. We have taken $h_{\text{mid}} = 10$ mm, $d = 1$ mm, $b = 4$ mm, $L = 100$ mm, $E = 50$ MPa, and the Poisson's ratio, $\nu = 0.35$ for all the examples. The FEA is performed in ABAQUS using continuum 2D quadrilateral elements, and the curve is obtained from the quasi-static dynamic-implicit analysis. Like the analytical modeling, FEA also shows different pathways for the loading and unloading cycle. Both the analyzes show a kink at the switching and switchback point due to sudden asymmetric mode switching.

For the example shown in Fig. 7a, when the mode weights are of equal magnitude in the as-fabricated shape, the right arch does not switch to its inverted shape and hence is not bistable. As we decrease the magnitude of the right arch, it switches and shows bistability at the inverted shape, as shown in Fig. 7b. Both the analytical and FEA results show a sudden decrease in the slope when the left arch switches to its inverted shape. When we introduce asymmetric modes in the stress-free state, as in example Fig. 7c, the arches invert their shapes but return to their undeformed state as the load is removed, and hence are not bistable. The loss of stability occurs as the left arch does not have its force-free configuration other than the undeformed configuration. A similar analysis can be performed to get the equilibrium equations and stability conditions for the fixed–pinned arches, and the results can be compared with FEA in ABAQUS.

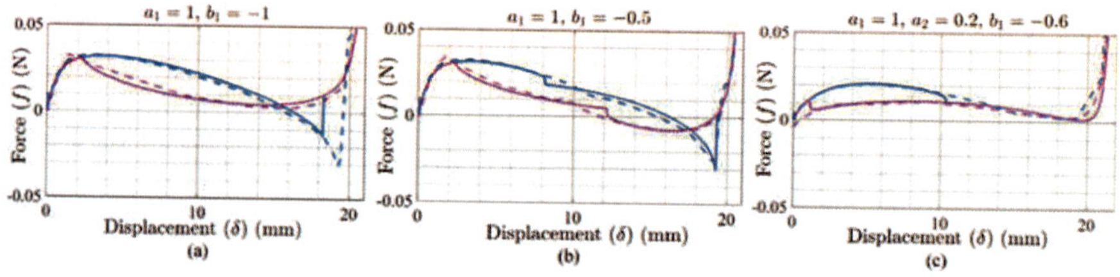

Fig. 7 Comparison of force–displacement characteristics with FEA for different as-fabricated shape, **a** configuration that does not switch to inverted shape and does not show bistability, **b** configuration that switches to inverted shape and shows bistability, and **c** configuration that switches to inverted shape but does not show bistability. The solid lines are the results from analytical modeling, and dashed lines are for FEA. The blue and magenta lines represent the loading path and unloading pathways, respectively.

4 Closure

We presented a semi-analytical method to investigate the characteristics of connected arches. Using the equilibrium equations, we studied the characteristics of interconnected arches and observed different loading and unloading pathways for the connected arches. Based on the switching nature and stability conditions, we noticed three kinds of connected arches based on their as-fabricated shape: (i) When both the arches are identical and do not switch to their inverted shape, they are not bistable in the connected configuration; (ii) when we decrease the magnitude of first mode weight of the right arch to a certain limit, both the arches switch to their inverted shape, and the arrangement shows bistability; and (iii) the arch arrangement may flip but not be stable in their inverted configuration when any of the arches is not bistable individually. This analysis helps us in choosing the as-fabricated shapes of the connected arches for self-offloading footwear and other applications based on the corresponding design requirements.

Acknowledgements The authors gratefully acknowledge the financial support from the Rajiv Gandhi University of Health Science (RGHUS), Karnataka, India. We also thank Jyoti S. Sonawane of Indian Institute of Science for proofreading the paper and giving valuable suggestions. Discussions with Dr. Pavan Belehalli of the Karnataka Institute for Endocrinology Research, Bengaluru, are also gratefully acknowledged in the clinical utility of the connected arches in novel footwear for diabetics.

References

1. Duckworth T, Boulton AJ, Betts RP, Franks CI, Ward JD (1985) Plantar pressure measurements and the prevention of ulceration in the diabetic foot. J Bone Joint Surg Br 67(1):79–85
2. Maharana P, Sonawane J, Belehalli P, Ananthasuresh GK (2019) Analysis of planar bistable and snap-through arches for contact and dynamic loads. In: IFToMM WC 2019. Mechanisms and machine science, vol 73. Springer
3. Maharana P, Sonawane J, Belehalli P, Ananthasuresh G (2022) Self-offloading therapeutic footwear using compliant snap-through arches. Wearable Technologies, 3, E7. https://doi.org/10.1017/wtc.2022.2
4. Palathingal S, Ananthasuresh GK (2019) Analysis and design of fixed-fixed bistable arch-profile using a bilateral relationship. ASME J Mech Robot 11(031002-1):1–18
5. Palathingal S, Ananthasuresh GK (2018) A bilateral relationship between stable profiles of pinned–pinned bistable shallow arches. Int J Solids Struct 143:183–193. ISSN 0020-7683
6. Vangbo M (1998) An analytical analysis of a compressed bistable buckled beam. Sens Actuators A 69(3):212–216
7. Qiu J, Lang JH, Slocum AH (2004) A curved-beam bistable mechanism. J Microelectromech Syst 13(2):137–146
8. Fung YC, Kaplan A (1952) Buckling of low arches or curved beams of small curvature. Technical Note 2840, 78. NACA, Washington, DC
9. Timoshenko SP, Gere JM (1988) Theory of elastic stability, 2nd edn. McGraw-Hill, New York
10. Palathingal S, Ananthasuresh GK (2017) Design of bistable arches by determining critical points in the force-displacement characteristic. J Mech Mach Theory 117:175–188

11. Mead DJ (1970) Free wave propagation in periodically supported infinite beams. J Sound Vib 11(2):181–197
12. Papalambros PY, Wilde DJ (2000) Principle of optimal design: modeling and computation, 2nd edn. Cambridge University Press

Dynamic Analysis of MCF-7 Using Tensegrity Model

B. V. Chandan Bharadwaj, K. Abiram, K. Harish, S. Vivek, and C. V. Chandrashekara

Abstract Cell is a fundamental unit of life. Mechanical stimuli in cells are the driving factor in the many essential life processes of the cell such as growth, proliferation and metastasis. Many dynamic behaviours of cells like natural frequency, response and mode-shape are promising factors giving insight into the wellbeing of a cell. Dynamic behaviour helps in the treatment of many cells-related diseases. Cancer is one such disease leading to abnormal cell growth with the potential to spread to other parts of the body. Despite the rapid medical advancement in developing techniques to cure cancer, these cannot completely cure cancer at all stages and also comes along with side effects. In the present work, a six-strut tensegrity model is developed to mimic the cytoskeleton of MCF-7 cell, considering the effect of prestress. Dynamic analysis is performed, and the first five natural frequencies are in the range of 5–17 kHz. The mode shapes corresponding to the first four natural frequencies are reported.

Keywords Cytoskeleton · MCF-7 · Tensegrity · Cell dynamics · Dynamic analysis

1 Introduction

The fundamental essence of all forms of life is a bio-cell. Human body consists of an average of 30 trillion cells, varying from blood cells to bone cells. Each type of cells is formed to meet a specific functionality through the process of differentiation [1]. The principal functionality for survival of a cell includes providing structural support, facilitating growth through the process of mitosis, transport of nutrients, producing energy, metabolic activities and reproduction. A crucial part in the process of cell division is replication of DNA, any error during this process leads to uncontrolled cell division leading to cancer. The transformation from a normal healthy cell to a malignant cancerous cell affects the various biophysical and biomechanical properties of the cell. The mechanical properties of cellular components are essential in

B. V. Chandan Bharadwaj · K. Abiram · K. Harish · S. Vivek · C. V. Chandrashekara (✉)
Department of Mechanical Engineering, PES University, Bengaluru, India
e-mail: drcvc@pes.edu

V. K. Gupta et al. (eds.), *Recent Advances in Machines and Mechanisms*, Lecture Notes in Mechanical Engineering, https://doi.org/10.1007/978-981-19-3716-3_16

order to obtain the dynamic characteristics of the cell. The dynamic characteristic of cancerous cell helps in the development of a better treatment, to target cancer cells.

Several approaches, such as atomic force microscopy (AFM) and scanning probe microscopy, are utilized by many researchers to obtain the mechanical properties of cells. Yokokawa et al. [2] used scanning probe microscope to obtain the mechanical properties of plasma membrane and nuclear envelope of living HeLa cells in a culture medium. Experimental procedures like in-vivo, in-vitro and in-situ are even though promising methods, ethical restrictions limit the researchers from carrying an in-depth experimental analysis on a living cell. Experimental methodologies are time-consuming, expensive and non-versatile. Researchers have alternatively adopted finite element (FE) methods, which are versatile, cost- and time-effective in simulating the cells to extract their dynamics characteristics. Lim et al. [3] explored various methods to a model a cell and broadly classifies them into two approaches: continuum approach and micro/nanostructural approach. Mubeen et al. [4] presented a continuum approach to obtain the dynamic characteristics of MCF-7 and MCF-10 A cells. The entire cell is considered as a non-homogeneous model, ignoring the role of cytoskeleton. The first four natural frequencies for MCF-7 cell are in the range of 5–9 kHz. Katti et al. [5] presented a computational model to understand the importance of the cytoskeleton in the mechanical response of breast cancer cell. The results indicate, although the cytoskeleton constitutes a small portion of the overall cell volume, its contribution towards the mechanical response of the cell is significant. Ethier and Simon [6] described the cytoskeleton of a cell comprising of discrete elements like actin filaments, intermediate filaments and microtubules. The cytoskeleton helps in maintaining the shape of the cell and in cellular movement. The control of cell functions, including growth, migration and mechano-transduction, depends crucially on stress-induced mechanical changes in cell shape and cytoskeleton.

The tensegrity approach is considered as a prominent alternative modelling method which encapsulate the effect of cytoskeleton. The tensegrity model represents the tension elements as cables and compression elements as struts. The tensegrity model, a type of discrete approach, is very prominent in mimicking the dynamics of the cell by considering the effects of the cytoskeleton. Ingber et al. [7–9] proposed the pivotal idea of modelling a cellular cytoskeleton as a tensegrity model that stabilizes itself by the combinational effect of the discrete elements of the cells. The changes in the orientation and the spacing of the elements represent the mechanism of restoring forces that stabilize the tensegrity. Pugh [10] described the tensegrity as the structure obtained using the discontinuous compression–tension elements, defining a stable volume in space. The tensegrity structure deforms on the application of an external force and returns to its original configuration once the force is removed. Wendling et al. [11] presented a quantitative study on 6-bar tensegrity models to analyse several mechanical parameters involved in the mechanical responses of adherent cells. The tensegrity structures develop a stabilizing tension which is balanced by compression, known as prestress. The general mechanical laws applicable to tensegrity structure are described highlighting the mechanical behaviour of cytoskeleton. Coughlin and Stamenovic [12] modelled the cell as six-strut tensegrity structure to represent the round and spread cell configurations. The struts and the cables of the tensegrity are considered to represent the microtubules and the microfilaments of a cell, respectively. The change in the stiffness of the cell is observed as the configuration change from round to spread cell configuration, because of the changes in geometry and forces. Wee and Voloshin [13] modelled a spread bone cell on a culturing plate using two models, continuum and tensegrity. Dynamic analysis of both these models is carried out on the ANSYS platform. Heyden and Ortiz [14] performed dynamic analysis to verify a spectral gap between the natural frequencies of the healthy and cancerous cells. The existence of spectral gaps assists in selectively targeting the cancer cells and does not disrupt the normal cells. The authors confirm that cell rupture can be selectively induced to cancer cells by the application of tuned ultrasonic harmonic excitation, and the process is termed as oncotripsy. Mittelstein et al. [15] provided experimental confirmation of selective rupture of cancer cells by inducing cyclic low-intensity pulsated ultrasound (LIPUS). The higher death rate of the cancer cells suggests the selective cell rupture by LIPUS can fulfil as a safe non-invasive technique for cancer treatment. The author also highlights that the resonant frequency of the cancer cell is critical to the effectiveness of oncotripsy. The prediction of natural frequency using the FE method calls for an elemental simulation model.

The literature review indicates that there is limited application of tensegrity model to represent breast cancer cell. In the present work, a six-strut tensegrity model is developed to mimic the cytoskeleton of MCF-7 cell, considering the effect of prestress. Dynamic analysis is performed, and the first five natural frequencies and their corresponding mode shapes are reported.

2 Modelling

A discretized six-strut spherical tensegrity model of cytoskeleton of an MCF-7 is developed, using ANSYS APDL 2020 R1 considering six-struts and twenty-four cables. The struts represent the microtubules, and cable represents both intermediate filaments and actin filaments of cytoskeleton. Node points of each strut's end are connected by the cables. Set of two struts is arranged in parallel to each other (see Fig. 1).

The nucleus is considered a solid point mass element, and the nucleus is suspended at the centre of cell using 12 intermediate filaments. With this arrangement, the overall model consists of 42 elements, i.e. 6 struts, 24 cables and 12 intermediate filaments. Struts of cytoskeleton are represented by red in colour, cables are represented by blue in colour and inter-filaments are shown in green in colour.

The distance between the strut is considered half the length of the strut, as reported in many literatures [12, 16]. The relationship between the length of the strut, l_{strut}, and the radius of the MCF-7 cell R is taken as [16],

$$l_{\text{Strut}} = 1.8R \tag{1}$$

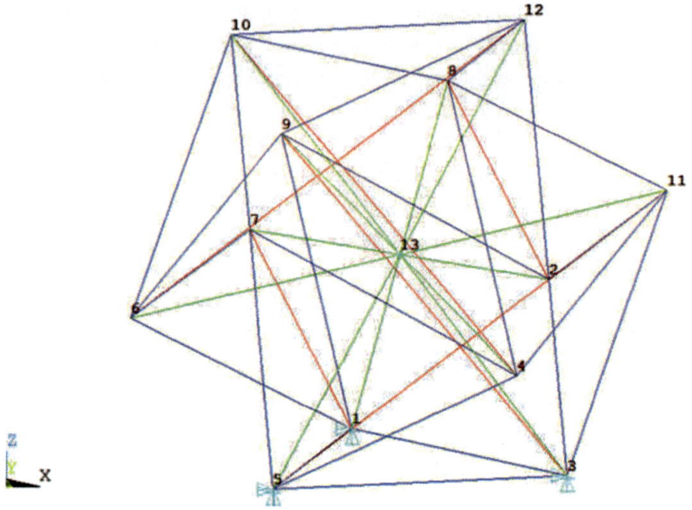

Fig. 1 Six-strut spherical tensegrity model

The volume of the MCF-7 cell is 3375 μm^3 [4], and the radius calculated is 9.31 μm.

The length of the cable is taken as [14],

$$l_{cable} = \sqrt{3/8}\, l_{strut} \quad (2)$$

The modal analysis is performed on the model developed to obtain the natural frequencies and mode shapes. The results are presented and discussed in the following section.

3 Dynamic Analysis

The dynamic characteristics of a cell vary with the age, environment and the ongoing biochemical processes inside a cell. Measuring these properties serve as an indication of the current state of a cell and help in more accurate predictions. The dynamic analysis of a cell is essential for understanding the behaviour of the cell. In this section, the dynamic analysis of the cell model is presented. The material and geometric properties of constituents of the tensegrity structure considered for dynamic analysis are provided in Table 1.

The nucleus is considered as a point mass element, with elastic modulus 150 Pa, density of 1300 kg/m^3 and Poisson's ratio is 0.49. Poisson's ratio for the remaining cellular tensegrity constituents is considered to be 0.3. The mass of the nucleus is

Table 1 Material Properties of tensegrity model [17–19]

Cellular constituent	Elastic modulus (GPa)	Density (kg/m^3)	Diameter (nm)
Microtubule	1.9	1150	25.0
Microfilament	1.3	1150	8.0
Inter-filaments	1.0	1150	10.0

Table 2 Element lengths and mass per unit length of the cytoskeleton elements

Parameter	Value
Strut length (μm)	16.65
Cable length (μm)	10.20
Total length of cytoskeletal elements (μm)	344.69
Total mass of cytoskeleton (ng)	3.88
Mass per unit length (kg/m)	1.126

taken as 1.45 ng. The total mass of the cytoplasm and the membrane of the continuum model are calculated to be 3.88 ng [4].

The small cross sections of the elements of the tensegrity lead to the overall mass of the tensegrity model being minute. To compensate for the difference in mass of the tensegrity model when in comparison with the continuum model, extra masses are added to the struts and cables of the cytoskeleton. The calculation parameters of the mass per unit length are shown in Table 2.

The microtubules, microfilaments and the inter-filaments are considered to be non-bending elements. LINK 180 which supports only axial loading is used to represent all the elements of the developed tensegrity structure, and MASS21 element is used to represent the nucleus. Tension only conditions are applied to the cables and inter-filaments to ensure that the cables are taut at all given time [13]. The added mass per unit length feature is applied to the microtubules and microfilaments. The prestress values of 3.978 and 2.546 kPa are applied to the microfilaments and inter-filaments, respectively. The prestress values are calculated for a pretension value of 1 pN [20].

For the boundary conditions, nodes 1, 3 and 5 are restricted from all degrees of freedom. Modal analysis is performed on the ANSYS APDL 2020 R1 to obtain the dynamic characteristics. For modal analysis, block Lanczos method is employed to obtain the natural frequencies and their corresponding mode shapes are extracted.

The natural frequencies corresponding to the first five modes for a tensegrity model of MCF-7 are presented in Table 3.

The first and second modes (see Fig. 2a, b) correspond to a lateral mode of vibration along z- and x-directions, respectively.

The third and fourth modes (see Fig. 3a, b) correspond to the twisting mode about z- and x-directions, respectively. The results indicated in Table 3 for the first four natural frequencies are in accordance with the first four natural frequencies for a continuum model reported by Mubeen et al. [4], i.e. 5–9 kHz. The difference in the results obtained is a collective effect of the difference in the material properties considered and the boundary conditions provided.

Table 3 Natural frequencies of tensegrity model of MCF-7 cell

Mode	Natural frequency (Hz)
1	5966.49
2	6173.29
3	12,571.17
4	13,517.82
5	17,555.58

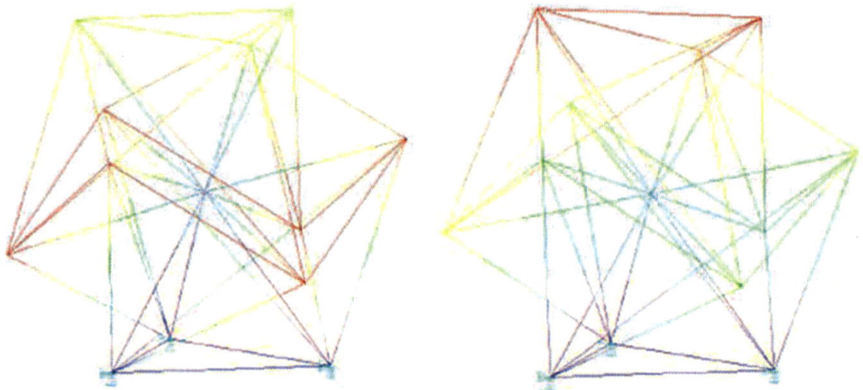

a) Mode 1, 5966.49 Hz   b) Mode 2, 5173.29 Hz

Fig. 2 First and second mode shapes for tensegrity model of MCF-7

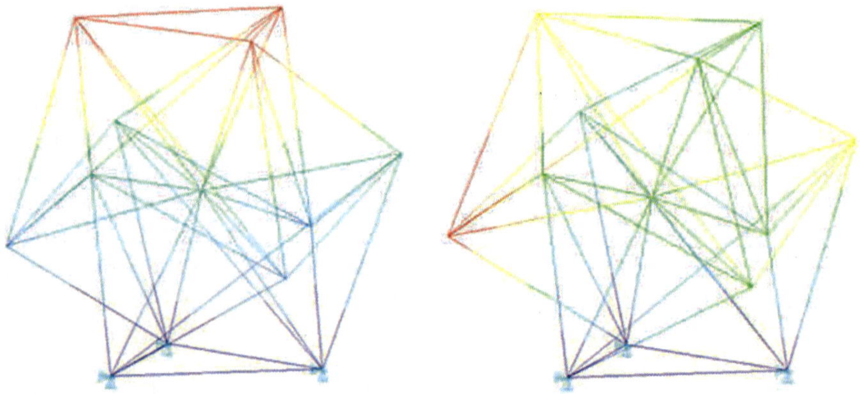

a) Mode 3, 12571.17 Hz   b) Mode 4, 13517.82 Hz

Fig. 3 Third and fourth mode shapes for tensegrity model of MCF-7

4 Conclusion

The current paper establishes a tensegrity model of a breast cancer cell (MCF-7) for modelling and simulation purposes. The nucleus of the cell is modelled as a point mass. The dynamic characteristics are obtained on ANSYS APDL 2020R1 and are reported. The results show that the first five natural frequencies are in the range of 5–17 kHz, which are in the range reported by the continuum modelling approach. The simulation model developed provides promising results that can be utilized in carefully tuning the frequency for the ultrasound technique to selectively disrupt cancer cells. The ultrasound technique of cancer treatment is a promising non-invasive approach to cure cancer with no or very little side effects. The tensegrity modelling provides an alternative approach to the existing modelling techniques being considerate of the importance of the cytoskeleton. Further, considering the viscoelastic properties and the presence of other cell organelles would help in increasing the accuracy of the model.

References

1. Laskey RA et al (2021, March 9) Cell. In Encyclopedia Britannica. Retrieved from https://www.britannica.com/science/cell-biology
2. Yokokawa M, Takeyasu K, Yoshimura SH (2008) Mechanical properties of plasma membrane and nuclear envelope measured by scanning probe microscope. J Microsc 232(1):82–90. https://doi.org/10.1111/j.1365-2818.2008.02071.x
3. Lim CT, Zhou EH, Quek ST (2006) Mechanical models for living cells—a review. J Biomech 39(2):195–216. https://doi.org/10.1016/j.jbiomech.2004.12.008
4. Ahmed Mubeen A, Chaudhary S, Barathwaj R (2020) Dynamic analysis of MCF-10A and MCF-7: a simulation approach. Vibroeng Procedia 30:126–132. https://doi.org/10.21595/vp.2020.21392
5. Katti DR, Katti KS (2017) Cancer cell mechanics with altered cytoskeletal behavior and substrate effects: a 3D finite element modeling study. J Mech Behav Biomed Mater 76:125–134. https://doi.org/10.1016/j.jmbbm.2017.05.030
6. Ethier CR, Simmons CA (n.d.) Introductory biomechanics: from cells to organisms 545
7. Ingber DE (n.d.) Cellular tensegrity: defining new rules of biological design that govern the cytoskeleton 15
8. Ingber DE, Wang N, Stamenović D (2014) Tensegrity, cellular biophysics, and the mechanics of living systems. Rep Prog Phys 77(4):046603. https://doi.org/10.1088/0034-4885/77/4/046603
9. Ingber DE, Heidemann SR, Lamoureux P, Buxbaum RE (2000) Opposing views on tensegrity as a structural framework for understanding cell mechanics. J Appl Physiol 89(4):1663–1678. https://doi.org/10.1152/jappl.2000.89.4.1663
10. Pugh A (1976) An introduction to tensegrity. University of California Press, USA
11. Wendling S, Cañadas P, Chabrand P (2003) Toward a generalised tensegrity model describing the mechanical behaviour of the cytoskeleton structure. Comput Methods Biomech Biomed Engin 6(1):45–52. https://doi.org/10.1080/1025584021000059413
12. Coughlin MF, Stamenovic D (1998) A tensegrity model of the cytoskeleton in spread and round cells. J Biomech Eng 120(6):770–777. https://doi.org/10.1115/1.2834892

13. Wee H, Voloshin A (2014) Dynamic analysis of a spread cell using finite element method. In Barthelat F, Zavattieri P, Korach CS, Prorok BC, Grande-Allen KJ (eds) Mechanics of biological systems and materials, vol 4, pp 135–140. Springer International Publishing, Cham. https://doi.org/10.1007/978-3-319-00777-9_19
14. Heyden S, Ortiz M (2016) Oncotripsy: targeting cancer cells selectively via resonant harmonic excitation. J Mech Phys Solids 92:164–175. https://doi.org/10.1016/j.jmps.2016.04.016
15. Mittelstein DR, Ye J, Schibber EF, Roychoudhury A, Martinez LT, Fekrazad MH et al (2020) Selective ablation of cancer cells with low intensity pulsed ultrasound. Appl Phys Lett 116(1):013701. https://doi.org/10.1063/1.5128627
16. Kenner H (1976) Geodesic math and how to use it. University of California Press, Berkeley
17. Suresh S (2007) Biomechanics and biophysics of cancer cells. Acta Biomater 3(4):413–438. https://doi.org/10.1016/j.actbio.2007.04.002
18. Geltmeier A, Rinner B, Bade D, Meditz K, Witt R, Bicker U et al (2015) Characterization of dynamic behaviour of MCF7 and MCF10A cells in ultrasonic field using modal and harmonic analyses. PLoS ONE 10(8):e0134999. https://doi.org/10.1371/journal.pone.0134999
19. Xue F, Lennon AB, McKayed KK, Campbell VA, Prendergast PJ (2015) Effect of membrane stiffness and cytoskeletal element density on mechanical stimuli within cells: an analysis of the consequences of ageing in cells. Comput Methods Biomech Biomed Engin 18(5):468–476. https://doi.org/10.1080/10255842.2013.811234
20. Chen T-J, Wu C-C, Su F-C (2012) Mechanical models of the cellular cytoskeletal network for the analysis of intracellular mechanical properties and force distributions: a review. Med Eng Phys 34(10):1375–1386. https://doi.org/10.1016/j.medengphy.2012.08.007

Dynamic Modeling of Planar Multi-section Tendon-Driven Continuum Manipulator (TDCM) Using "Euler–Lagrange Formulation"

Vipin Pachouri and Pushparaj Mani Pathak

Abstract The continuum style robots entail a continuous flexible backbone of infinite joints and hypothetically offer limitless degrees of freedom (DOF). Also, the continuum robot structure retains more actuatable DOF than the DOF required for their intended task space. Although this spare freedom enriches maneuverability, flexibility, and dexterity but the modeling and control of these flexible robots turn out to be non-linear and complicated. In this article, the Lagrange construction, which is an energy-based method, is used to devise the dynamic model of a tendon-driven continuum style manipulator. A generalized dynamic model of tendon-driven continuum manipulators (TDCM) is developed based on lumped spring-mass-damper model. The central backbone in-between two adjacent disks is modeled as a flexible arc instead of a rigid link. Also, the stiffness of the central flexible backbone is considered, which performs a considerable role under external intervention and manages the payload. Finally, the proposed model is confirmed experimentally, and a decent match is found between simulation and experimental outcomes.

Keywords Continuum manipulator · Flexible manipulator · Tendon-driven robots · Euler–Lagrange

1 Introduction

The conventional robot manipulators comprise several joints connected by rigid links, and the key benefit of these robotic systems is that their kinematics and dynamics are properly recognized and proven to be effective for a variety of tasks and operations. However, modeling the flexible continuum style robots through the same techniques is difficult to execute. As for these styles of robots, the link length between two joints

V. Pachouri (✉) · P. M. Pathak
Department of Mechanical and Industrial Engineering, Indian Institute of Technology Roorkee, Roorkee, Uttarakhand 247667, India
e-mail: vpachouri@me.iitr.ac.in

P. M. Pathak
e-mail: pushparaj.pathak@me.iitr.ac.in

V. K. Gupta et al. (eds.), *Recent Advances in Machines and Mechanisms*, Lecture Notes in Mechanical Engineering, https://doi.org/10.1007/978-981-19-3716-3_17

is flexible and takes some sort of arc shape. Also, with the integration of flexibility, the preference of mounting actuator directly on robot structure turns out to be infeasible, and hence, its design structure must be improved.

The concept of continuum robot was first introduced by Anderson and his team [1]. They had developed the prototype of a continuum robot named as tensor arm. Later, Mehling et al. [2] developed another continuum robot with a lot more advancement in designs for in-space inspection. Robinson and Davies [3] presented a simplified classification of different possible designs, based on the number of redundancy retained by such robotic systems as discrete, serpentine, and continuum. The robotic systems falling inside the class of continuum do not have any rigid joints and links, instead they have a flexible backbone that bend continuously along its backbone structure. The bending behavior of these robots provides vast advantages with many challenges from its design to control aspects. Some of the features of these types of robotic systems are that they have high dexterity [4] and have better compliance nature, obstacle avoidance capability, and simple design [5, 6]. Rone and Ben-Tzvi [7] had developed the dynamic model that captures curvature variations along a segment using a finite set of kinematic variables of cable-driven continuum robot using Kane's method. Tatlicioglu et al. [8] presented a dynamic model based on the geometric model of extensible continuum robot manipulators with no torsional effects. Falkenhahn et al. [9] developed the dynamic model of a bellow actuated continuum robot, but their model is highly complex and requires a very hefty amount of computation. Two individual researchers Trivedi et al. [10] and Cao and Tucker [11] formulated the dynamics of an elastic rod by using the corset rod theory as a set of PDEs, which are difficult for numerical implementation.

This article presents a generalized dynamic model of tendon-driven continuum manipulators (TDCM) based on lumped spring–mass-damper model utilizing energy centered Lagrangian method. The central backbone between two adjacent disks is modeled as a flexible arc instead of a rigid link as done by the previous researchers. Also, the stiffness of the central flexible backbone is considered in the modeling. Each section is modeled to follow piecewise constant curvature and torsion-free assumption, i.e., only bending motion is permitted throughout the sections.

This paper is presented in four segments. The second section introduces the system's physical and mathematical description, followed by the third section in which the simulation and experimental outcomes are presented. The last section portrays the conclusion and future studies.

2 Physical and Mathematical Description of Tendon-Driven Continuum Manipulator (TDCM)

The tendon-driven continuum manipulator (TDCM) physically consists a central flexible backbone routed through number of disks having central hole. Tendons are routed through the anchor or way points provided on each disk as shown in Fig. 1.

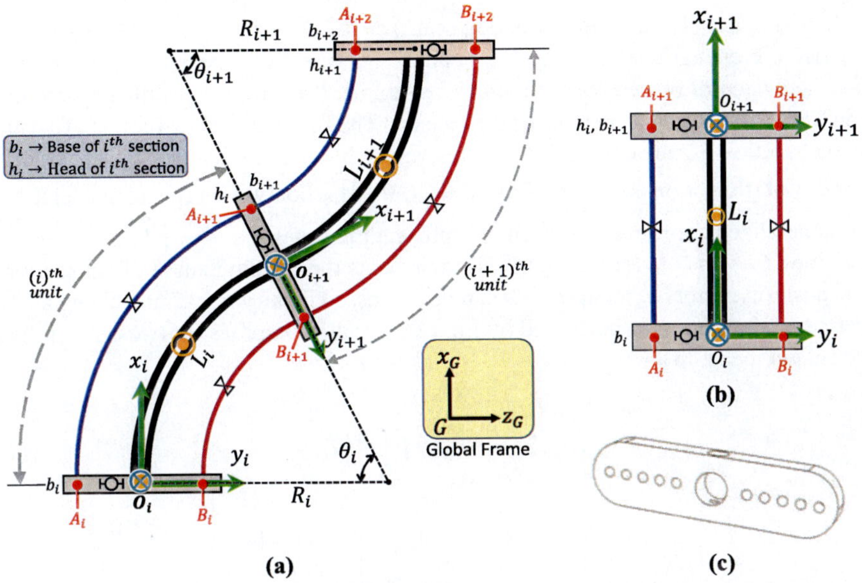

Fig. 1 Schematic of tendon-driven continuum manipulator showing frames⊗, actuating tendons⋈, flexible backbone◉, and disks of ith and $(i+1)$th unit under **a** general scenario, **b** manipulator at singular configuration, and **c** cross-sectional view of the disk

The flexible backbone is actuated by tensioning tendon, which changes the cable length and responsible for forces and moment transmission. For simplification of the mathematical model, the friction acting in-between tendons and disk is neglected, and the flexible backbone is assumed to follow the torsion-free pure-bending assumption. The dynamic equations of TDCM are derived via energy-based Euler–Lagrange formulation.

The local body-fixed reference frame $\{O_i\}$ of ith section is placed at the start of the section as shown in Fig. 1a, and the global frame $\{G\}$ coincides with the local frame of the first section. Mathematically, the position and orientation of $(i+1)$th body-fixed frame relative to ith local frame calculated as shown in Eq. 1 (where s and c correspond to sin and cosines, respectively, and throughout used in this article).

$$
{}^{i}T_{(i+1)} = \begin{bmatrix} c\theta_i & -s\theta_i & 0 & \frac{L_i}{\theta_i} s\theta_i \\ s\theta_i & c\theta_i & 0 & \frac{L_i}{\theta_i}(1-c\theta_i) \\ 0 & 0 & 1 & 0 \\ 0 & 0 & 0 & 1 \end{bmatrix} \tag{1}
$$

Utilizing Eq. (1), the transformation matrixes for two-section TDCM are devised in local reference frames as ${}^{G}T_{1b}$, ${}^{1b}T_{1h}$, and ${}^{1h}T_{2h}$ where *G, h,* and *b* indicate *global, base,* and *head* of respective sections. For example, ${}^{1b}T_{1h}$ represents the transformation from first section head to first section base. The first section head frame $\{1h\}$ and second section base frame $\{2b\}$ coincide over each other at same point. The angular velocity of ith section is expressed as $\omega_i = \begin{bmatrix} 0 & 0 & \dot{\theta}_i \end{bmatrix}^T$. For two-section planar TDCM, the generalized coordinates chosen to represent the states are $q_i = \begin{bmatrix} L_i & \theta_i \end{bmatrix}^T$, where variables $L_i \rightarrow$ is arc length and $\theta_i \rightarrow$ represents the bending angle of ith section. The position vectors of local frames can be expressed in the global frame using arithmetic transformation as portrayed by Eq. (2), and the outcomes for two-section CM are shown in Eq. (3).

$$ {}^{G}T_i = {}^{G}T_{i-1}\prod_{i=1}^{n}{}^{i-1}T_i \tag{2} $$

$$ {}^{G}O_{1b} = \begin{bmatrix} 0 \\ 0 \\ 0 \end{bmatrix};\quad {}^{G}O_{1h} = \begin{bmatrix} \frac{L_1}{\theta_1}s\theta_1 \\ \frac{L_1}{\theta_1}(1 - c\theta_1) \\ 0 \end{bmatrix}; $$

$$ {}^{G}O_{2h} = \begin{bmatrix} \frac{L_1}{\theta_1}s\theta_1 + \frac{L_2}{\theta_2}(s(\theta_1+\theta_2) - s(\theta_1)) \\ \frac{L_1}{\theta_1}(1 - c\theta_1) - \frac{L_2}{\theta_2}(c(\theta_1+\theta_2) - c(\theta_1)) \\ 0 \end{bmatrix} \tag{3} $$

To get the velocities of origin of various frames, Eq. 3 is differentiated with respect to time. The resulting expression is shown by Eq. (4).

$$ {}^{G}\dot{O}_{1b} = \begin{bmatrix} 0 \\ 0 \\ 0 \end{bmatrix};\quad {}^{G}\dot{O}_{1h} = \begin{bmatrix} \frac{\dot{L}_1}{\theta_1}s\theta_1 + L_1\left(\frac{\theta_1 c\theta_1 - s\theta_1}{\theta_1^2}\right)\dot{\theta}_1 \\ \frac{\dot{L}_1}{\theta_1}(1 - c\theta_1) + \frac{L_1 s\theta_1 \cdot \dot{\theta}_1}{\theta_1} - \frac{L_1(1-c\theta_1)\dot{\theta}_1}{\theta_1^2} \\ 0 \end{bmatrix} $$

$$ {}^{G}\dot{O}_{2h} = \begin{bmatrix} {}^{G}\dot{O}_{1b}(1) + \frac{\dot{L}_2}{\theta_2}\{s(\theta_1+\theta_2) - s\theta_1\} + L_2\frac{\mathrm{d}}{\mathrm{d}t}\left(\frac{s(\theta_1+\theta_2)}{\theta_2}\right) - L_2\frac{\mathrm{d}}{\mathrm{d}t}\left(\frac{s\theta_1}{\theta_2}\right) \\ {}^{G}\dot{O}_{1b}(2) - \frac{\dot{L}_2}{\theta_2}\{c(\theta_1+\theta_2) - c\theta_1\} + L_2\frac{\mathrm{d}}{\mathrm{d}t}\left(\frac{c(\theta_1+\theta_2)}{\theta_2}\right) - L_2\frac{\mathrm{d}}{\mathrm{d}t}\left(\frac{c\theta_1}{\theta_2}\right) \\ 0 \end{bmatrix} \tag{4} $$

where

$$ \frac{\mathrm{d}}{\mathrm{d}t}\left(\frac{s(\theta_1+\theta_2)}{\theta_2}\right) = \left(\frac{\theta_2 c(\theta_1+\theta_2)\cdot(\dot{\theta}_1+\dot{\theta}_2) - s(\theta_1+\theta_2)\dot{\theta}_2}{\theta_2^2}\right); $$

$$\frac{\mathrm{d}}{\mathrm{d}t}\left(\frac{s\theta_1}{\theta_2}\right) = \frac{\theta_2 c\theta_1 \cdot \dot{\theta}_2 - s\theta_1 \cdot \dot{\theta}_2}{\theta_2^2}$$

$$\frac{\mathrm{d}}{\mathrm{d}t}\left(\frac{c(\theta_1+\theta_2)}{\theta_2}\right) = \left(\frac{-\theta_2 s(\theta_1+\theta_2)\cdot(\dot{\theta}_1+\dot{\theta}_2) - c(\theta_1+\theta_2)\dot{\theta}_2}{\theta_2^2}\right);$$

$$\frac{\mathrm{d}}{\mathrm{d}t}\left(\frac{c\theta_1}{\theta_2}\right) = \frac{-\theta_2 s\theta_1 \cdot \dot{\theta}_1 - c\theta_1 \cdot \dot{\theta}_2}{\theta_2^2}$$

In a similar way, the velocity vectors of the center of gravity (CG) of each section are computed by differentiating position vectors of each section CG as indicated in Eq. (5).

$$\begin{aligned} {}^{G}O_{\mathrm{CG1}} &= \frac{\mathrm{d}}{\mathrm{d}t}\left(\frac{{}^{G}O_{1h}}{2}\right) = {}^{G}\dot{O}_{1h}/2; \\ {}^{G}O_{\mathrm{CG2}} &= \frac{\mathrm{d}}{\mathrm{d}t}\left(\frac{{}^{G}O_{2h} + {}^{G}O_{1h}}{2}\right) \end{aligned} \tag{5}$$

The kinetic energy ($\mathcal{T}$) for n section TDCM is a combination of linear and rotational energies as generalized by Eq. (6). Let us assume that m_{ti} and m_{di} are the masses of backbone tube and disks, respectively. The flexible tube mass is supposed to act at the CG of each flexible backbone, whereas disk mass is assumed to concentrate at the centroid of each disk.

$$\begin{aligned} \mathcal{T} &= \sum_{i=1}^{n}(\mathcal{T}_{\mathrm{Linear}} + \mathcal{T}_{\mathrm{Rot.}}) \\ \mathcal{T} &= \frac{1}{2}\sum_{i=1}^{n}\Big[\left({}^{G}\dot{O}_{\mathrm{CG}_i}\right)^T m_{\mathrm{ti}}\left({}^{G}\dot{O}_{\mathrm{CG}_i}\right) \\ &\quad + \left({}^{G}\dot{O}_i\right)^T m_{\mathrm{di}}\left({}^{G}\dot{O}_i\right) + \omega_i^T\left({}^{G}R_i\,{}^{i}I_{li}\,{}^{G}R_i\right)\omega_i\Big] \end{aligned} \tag{6}$$

where ω_i is the angular velocity of ith section, ${}^{i}I_{li}$ is the moment of inertia, and ${}^{G}R_i$ is rotation of ith frame expressed in the global frame. The position and inertia tensors expressed in the base frame of ith section are manipulator configuration dependent quantities. So, the moment of inertia of each disk expressed in the global frame attached to each disk, which is a disk orientation-dependent quantity and its local inertia $\left(I_{xx}, I_{yy}\mathrm{and}I_{zz}\right)$ expressed in Eq. (7).

$${}^{G}I_{li} = {}^{G}R_i \begin{bmatrix} I_{xx} & 0 & 0 \\ 0 & I_{yy} & 0 \\ 0 & 0 & I_{zz} \end{bmatrix} {}^{G}R_i^T \tag{7}$$

Based on Eqs. (6) and (7), the expression of kinetic energies for two-section TDCM can be formalized as presented in Eq. (8). The total kinetic energy is formulated as sum to the individual section as symbolized in Eq. (9).

$$\begin{Bmatrix} \mathcal{T}_1 = \frac{1}{2}m_{t1}\left({}^G\dot{O}_{\mathrm{CG}_1}\right)^2 + \frac{1}{2}m_{d1}\left({}^G\dot{O}_{1h}\right)^2 + \frac{1}{2}I_{O_1}\omega_1^2 \\ \mathcal{T}_2 = \frac{1}{2}m_{t2}\left({}^G\dot{O}_{\mathrm{CG}_2}\right)^2 + \frac{1}{2}m_{d2}\left({}^G\dot{O}_{2h}\right)^2 + \frac{1}{2}I_{O_2}\omega_2^2 \end{Bmatrix} \tag{8}$$

where $I_{O_i} = \frac{m_{d_i}L_i^2}{3}$. Here, L_i is arc length if ith section.

$$\mathcal{T} = \sum_{i=1}^{n} \mathcal{T}_i \tag{9}$$

The potential energy ($\mathcal{U}$) retained in the manipulator is framed as combination of gravitation energy, spring energy due to change in tendon lengths, and elastic potential energy of flexible backbone. The potential energy for two-section TDCM can be devised as presented in Eq. (10).

$$\begin{aligned} \mathcal{U} = &-\left(\frac{m_{t1}}{2} + m_{d1}\right)g\left[\frac{L_1}{\theta_1}(1 - c\theta_1)\right] \\ &-\left(\frac{m_{t2}}{2} + m_{d2}\right)g\left[\frac{L_1}{\theta_1}(1 - c\theta_1) - \frac{L_2}{\theta_2}\{c(\theta_1 + \theta_2) - c\theta_1\}\right] \\ &+\frac{1}{2}\mathcal{K}_1^{\mathrm{I}}\left({}^G A_{1_f} - {}^G A_{1_s}\right)^2 + \frac{1}{2}\mathcal{K}_1^{\mathrm{II}}\left({}^G B_{1_f} - {}^G B_{1_s}\right)^2 + \frac{EI_b}{2L_1}\theta_1^2 \\ &+\frac{1}{2}\mathcal{K}_2^{\mathrm{I}}\left({}^G A_{2_f} - {}^G A_{2_s}\right)^2 + \frac{1}{2}\mathcal{K}_2^{\mathrm{II}}\left({}^G B_{2_f} - {}^G B_{2_s}\right)^2 + \frac{EI_b}{2L_2}\theta_2^2 \end{aligned} \tag{10}$$

where $\mathcal{K}_i^{\mathrm{I/II}}$ represents the stiffness of the secondary tendon backbone of ith section for tendon I and II, respectively, ${}^G A_{i_{f/s}}$ and ${}^G B_{i_{f/s}}$ are the final and start (initial) coordinates of tendon anchor or way points expressed in global frame whose final expressions are displayed in Eq. (11). Initial values are retrieved from the inverse kinematics [12] of the manipulator. In Eq. (11), M represents A holding + sign and B holding − sign.

$${}^i M_i = \left[0 \pm \tfrac{d_i}{2}\ 0\right]^T;\quad {}^G M_i = \left({}^G T_{i-1} \prod_{i=1}^{n} {}^{i-1}T_i\right) \times {}^i M_i$$

$${}^G A_{1_f} = \begin{bmatrix} \frac{L_1}{\theta_1}s\theta_1 - \frac{d}{2}s\theta_1 \\ \frac{L_1}{\theta_1}(1 - c\theta_1) + \frac{d}{2}s\theta_1 \\ 0 \end{bmatrix};\quad {}^G B_{1_f} = \begin{bmatrix} \frac{L_1}{\theta_1}s\theta_1 + \frac{d}{2}s\theta_1 \\ \frac{L_1}{\theta_1}(1 - c\theta_1) - \frac{d}{2}c\theta_1 \\ 0 \end{bmatrix}$$

$$ {}^{G}A_{2_f} = \begin{bmatrix} \left(\frac{L_2}{\theta_2} - \frac{d}{2}\right)\mathrm{s}(\theta_1 + \theta_2) + \left(\frac{L_1}{\theta_1} - \frac{L_2}{\theta_2}\right)s\theta_1 \\ -\left(\frac{L_2}{\theta_2} - \frac{d}{2}\right)\mathrm{c}(\theta_1 + \theta_2) - \left(\frac{L_1}{\theta_1} - \frac{L_2}{\theta_2}\right)s\theta_1 + \frac{L_1}{\theta_1} \\ 0 \end{bmatrix}; $$

$$ {}^{G}B_{2_f} = \begin{bmatrix} \left(\frac{L_2}{\theta_2} + \frac{d}{2}\right)\mathrm{s}(\theta_1 + \theta_2) + \left(\frac{L_1}{\theta_1} - \frac{L_2}{\theta_2}\right)s\theta_1 \\ -\left(\frac{L_2}{\theta_2} + \frac{d}{2}\right)\mathrm{c}(\theta_1 + \theta_2) - \left(\frac{L_1}{\theta_1} - \frac{L_2}{\theta_2}\right)s\theta_1 + \frac{L_1}{\theta_1} \\ 0 \end{bmatrix} \tag{11} $$

In this work, TDCM is assumed to follow viscous damping and the dissipation function for two-section CM as symbolized in Eq. (12).

$$ \mathcal{D} = \frac{1}{2}C_1^{\mathrm{I}}\left(\Delta\dot{\ell}_1^{\mathrm{I}}\right)^2 + \frac{1}{2}C_1^{\mathrm{II}}\left(\Delta\dot{\ell}_1^{\mathrm{II}}\right)^2 + \frac{1}{2}C_2^{\mathrm{I}}\left(\Delta\dot{\ell}_2^{\mathrm{I}}\right)^2 + \frac{1}{2}C_2^{\mathrm{II}}\left(\Delta\dot{\ell}_2^{\mathrm{II}}\right)^2 \tag{12} $$

where $C_i^{\mathrm{I/II}}$ represents the damping coefficient of the secondary tendon backbone of ith section for tendon I and II, respectively, and $\Delta\dot{\ell}_i^{\mathrm{I/II}}$ represents the rate of change on tendon length of ith section for tendon I and II. The generalized forces are calculated utilizing expression (13).

$$ Q_i = \sum_{k=1}^{N} F_k\left(\frac{\partial r_k}{\partial q_i}\right) \tag{13} $$

The energy-based Lagrange approach derives the dynamic equations of motion of TDCM systematically independent of the reference frame in terms of generalized coordinates (q_i) using Lagrangian multiplier, $\mathcal{L} = \mathcal{T} - \mathcal{U}$. Finally, dynamic equations are achieved using the expression are shown in Eq. (14).

$$ \frac{\mathrm{d}}{\mathrm{d}t}\frac{\partial\mathcal{L}}{\partial\dot{q}_i} - \frac{\partial\mathcal{L}}{\partial q_i} = Q_i - \frac{\partial\mathcal{D}}{\partial\dot{q}_i} \tag{14} $$

where $i = 1, 2, \ldots, n$ are the number of generalized coordinates. The devised equation of motion based on Lagrangian formulation is in the form as shown in Eq. (15).

$$\begin{bmatrix} M_{11} & M_{12} \\ M_{21} & M_{22} \end{bmatrix} \begin{Bmatrix} \ddot{\theta}_1 \\ \ddot{\theta}_2 \end{Bmatrix} + \begin{bmatrix} C_{11} & C_{12} \\ C_{21} & C_{22} \end{bmatrix} \begin{Bmatrix} \dot{\theta}_1 \\ \dot{\theta}_2 \end{Bmatrix} + \begin{Bmatrix} G_1 \\ G_2 \end{Bmatrix} = \begin{Bmatrix} Q_1 \\ Q_2 \end{Bmatrix} - \begin{Bmatrix} D_1 \\ D_2 \end{Bmatrix} \tag{15}$$

The elements of the above equation are shown in Appendix. The elements of matrix C contain coefficients of the first order derivatives of the generalized coordinates because the system is non-linear, many elements of C contain first order derivatives of the generalized coordinates. G matrix represents elements due to gravitational, spring potential energy of tendons, and elastic potential energy of backbone.

3 Simulation and Experimental Results

The dynamic model proposed in Sect. 2 is simulated for two-sections planar TDCM. The simulation and material parameters are displayed in Table 1. For simplification purpose, only one tendon per section is treated as active, so forces and moments are evaluated based on these active tendons. The developed equations of motion are evaluated for inverse dynamics, and generalized coordinates can be explicitly defined for time ranging from $0 \leq t \leq T_p$ as portrayed in Eq. (16), where $i = 1, 2, \ldots n$ are the number of generalized coordinates and s and f represent the start and final states of the manipulator. The initial conditions, at $t = 0$ s, are $\theta_i(0) = -2°$. The inverse dynamics of TDCM is evaluated considering the initial, and final states are at rest for two cases, (1) $\theta(\mathrm{T_p}) = \left[\frac{\pi}{2}, -\frac{\pi}{9}\right]$ rad and (2) $\theta(\mathrm{T_p}) = \left[\frac{\pi}{2}, -\frac{\pi}{9}\right]$.

$$q_i(t) = q_{i_s}(0) + \frac{q_{i_f}(T_p) - q_{i_s}(0)}{T_p}\left[t - \frac{T_p}{2\pi}\sin\left(\frac{2\pi t}{T_p}\right)\right] \tag{16}$$

An experimental setup is prepared for executing and validates the proposed dynamic model for two-section planar TDCM. The backbone is made up of hollow flexible PU pipe commonly used in pneumatic actuation systems. Several 3D printed disks (material: PLA) are attached along the backbone and way points for cable routing are provided on the disks as shown in Fig. 2. The point where tendons are fixed on the disks is considered as anchor point (A_4, B_4, A_9, and B_9), and rest other points through which tendon passes are called as way points

Table 1 Geometrical and material properties

Parameter	Value
Initial backbone length of each section, (L_i)	242.5 mm
Outer and inner radius of central backbone	(4.5, 2.5) mm
Elastic modulus of flexible backbone, (E)	2e8 N/m^2
Disk and tube masses (m_{di}, m_{ti})	(30, 20) gm
Distance of tendons from central backbone (d)	25 mm

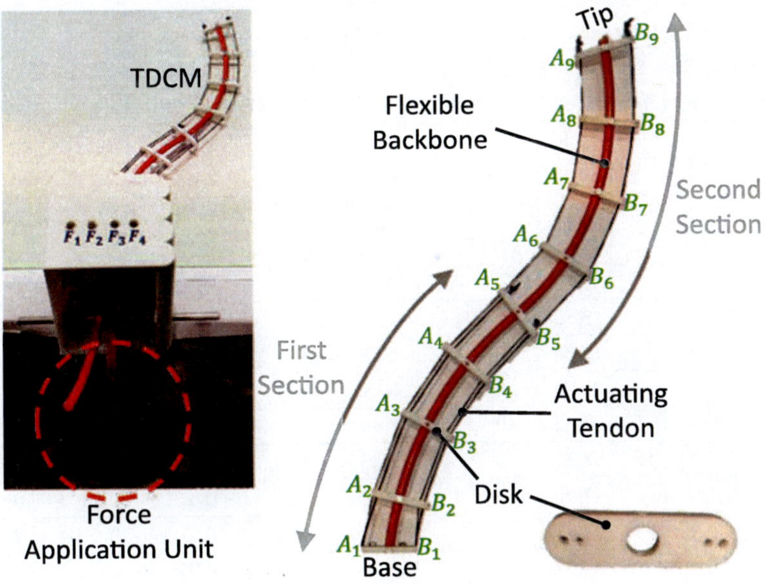

Fig. 2 Schematic exhibiting hardware setup of two-section planar tendon-driven continuum manipulator with actuation unit (left) and continuum manipulator (right)

(elcuding A_4, B_4, A_9, and B_9) illustrated in Fig. 2 (right). The left picture of Fig. 2 shows the TDCM with actuation unit, and the right picture shows various components in TDCM.

On solving the equation of motions for inverse dynamics, i.e., for a given motion calculating the torques or forces responsible for that motion, the torques or tendon forces are evaluated and presented in Table 2. For validating the simulation results, these tendon forces are employed on the TDCM, and their bending profile is compared with the profile obtained from the simulation. As the main contribution of this article is to develop a dynamic model of the TDCM, so this concept is implemented on an open-loop experimental setup. It is obvious that there may be certain positional errors, and they can be reduced with a closed-loop experimental setup and are planned in future.

Figures 3 and 4 demonstrate the results for ***case 1***: $\theta_1 = \pi/2$ and $\theta_2 = -\pi/9$ and ***case 2***: $\theta_1 = -\pi/3$ and $\theta_2 = \pi/3$. Simulation results are shown on the top part, and the lower figure portrays experimental results. The toques, actuating tendon forces, bending angle of each section, rates of bending angle, the motion of two-section TDCM (simulation), and five different bending postures (experimental) of two-section TDCM are shown in Figs. 3 and 4 for two different cases. A good match is obtained between the simulated and experimental bending profiles for two-section TDCM.

Table 2 Tendon forces obtained from simulation for two different cases

Time (s)	Case 1: $\theta = (\pi/2, -\pi/9)$				Case 2: $\theta = (-\pi/3, \pi/3)$			
	F_1 (N)	F_2 (N)	F_3 (N)	F_4 (N)	F_1 (N)	F_2 (N)	F_3 (N)	F_4 (N)
0	0	3.46	0	0.12	0	3.46	0	0.12
1	0	3.55	0	0.16	0	3.44	0	0.12
2	0	4.23	0	0.43	0	3.33	0	0.08
3	0	5.96	0	1.13	0	3.06	0	0.01
4	0	9.05	0	2.36	0	2.60	0.12	0
5	0	13.57	0	4.14	0	2.00	0.26	0
6	0	19.32	0	6.38	0	1.34	0.37	0
7	0	25.84	0	8.89	0	0.74	0.42	0
8	0	32.52	0	11.44	0	0.30	0.38	0
9	0	38.73	0	13.80	0	0.04	0.26	0
10	0	43.96	0	15.78	0.04	0	0.10	0
11	0	47.93	0	17.28	0.01	0	0	0
12	0	50.57	0	18.28	0	0.06	0	0.21
13	0	52.02	0	18.83	0	0.11	0	0.29
14	0	52.58	0	19.04	0	0.14	0	0.32
15	0	52.66	0	19.08	0	0.14	0	0.33

4 Conclusion and Future Studies

This article establishes a generalized dynamic model of tendon-driven continuum manipulators (TDCM) based on lumped spring-mass-damper model employing energy centered Lagrangian method. The central backbone in-between two adjacent disks is modeled as a flexible arc instead of a rigid link, which makes the system highly non-linear and difficult to solve. The systematic approach of deriving the equation of motion through Lagrange method proven to be effective in comparison with other recursive algorithms. Also, the stiffness of the central flexible backbone is considered as it plays a significant role in describing manipulator shape under external intervention, uncertainties, and payload situations. The proposed dynamic model is appropriate for small to larger deflections in continuum manipulators. Finally, the developed theory is validated experimentally, and a decent match is achieved between simulation and experimental outcomes. The same dynamic model is also appropriate to other groups of continuum manipulators [13] such as pneumatic and shape memory actuated manipulators. In future, this model will be stretched for the spatial case, and robust controllers will be analyzed.

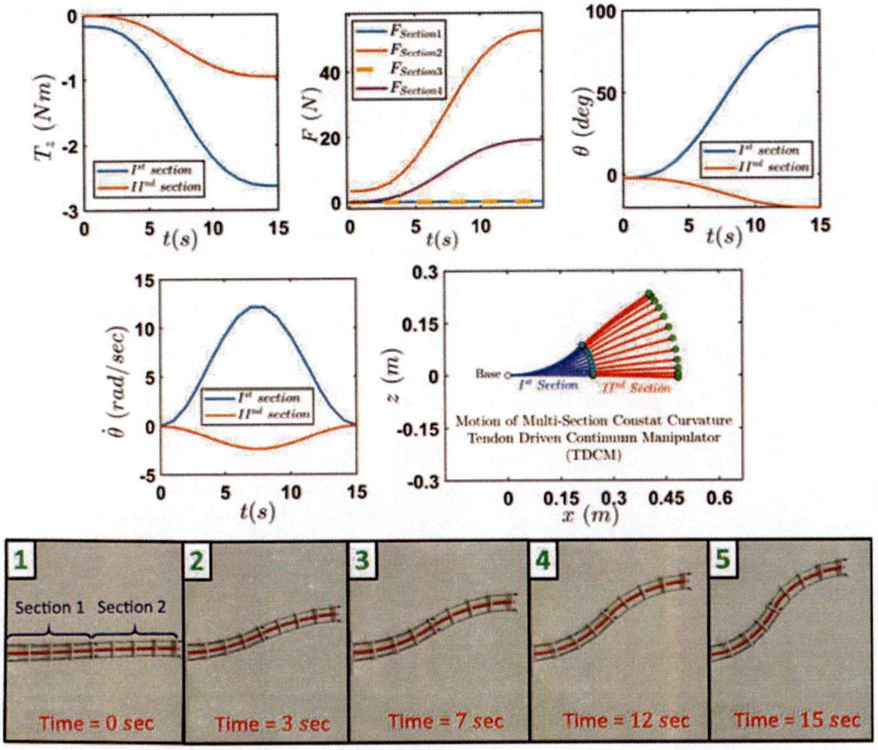

Fig. 3 Simulation and experimental results for case 1

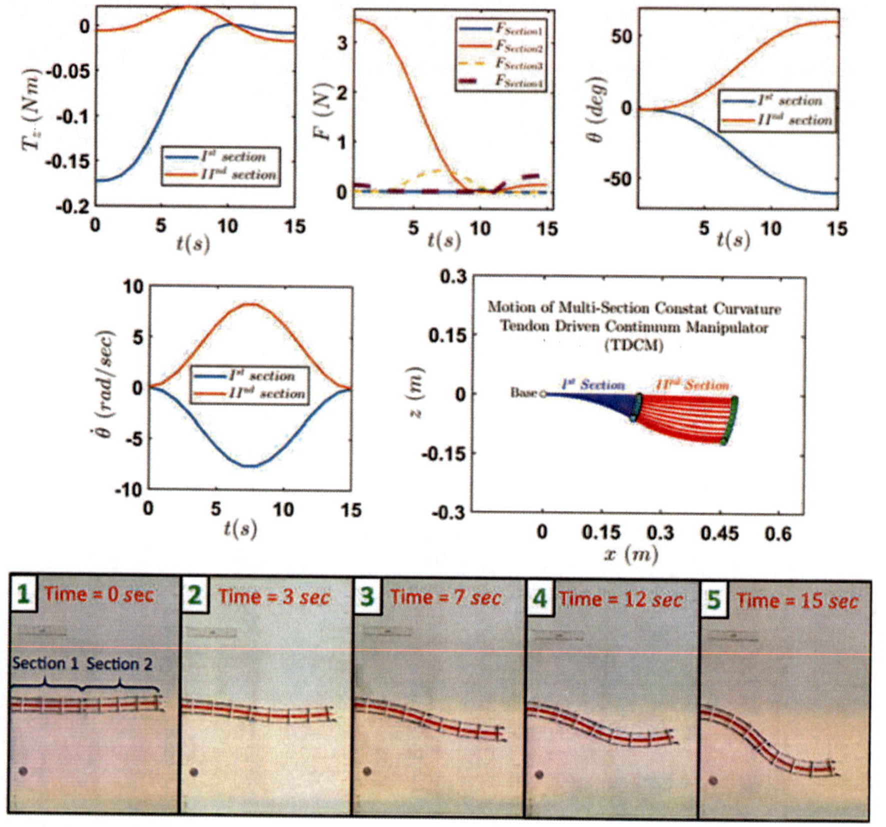

Fig. 4 Simulation and experimental results for case 2

Appendix

For simplifying the EOM's applicable to the robot presented in this article, the following set of rules are used. The simplification rules adopted here are appropriate for the robot presented in this article. $m_{d1} = m_{d2} = m_d$; $m_{t1} = m_{t2} = \left(\frac{1}{2}\right)m_d$; $L_1 = L_2 = L$; $\mathcal{K}_i^{I/II} = K$ and $\mathcal{C}_i^{I/II} = C$.

$c_1 = \cos\theta_1;\ s_1 = \sin\theta_1;\ c_2 = \cos\theta_2;\ s_2 = \sin\theta_2;\ c_{12} = \cos(\theta_1+\theta_2);\ s_{12} = \sin(\theta_1+\theta_2)s_{12} = \sin(\theta_1+\theta_2)$
$M_{11} = \frac{L^2 m_d}{\theta_1^4\theta_2^2}\Big(4\theta_1^4\theta_2^2 + 20\theta_1^3\theta_2 c_2 - 18\theta_1^2 c_2 + 20\theta_1^2\theta_2 s_2 + 20\theta_1^2\theta_2^2 - 41\theta_1\theta_2^2 s_1 - 20\theta_1^2\theta_2 s_{12} - 2\theta_1^3\theta_2 + 20\theta_1^2\theta_2 s_1 + 18\theta_1^4 - 42\theta_2^2 c_1 + 40\theta_2^2\Big)$
$M_{12} = \frac{-5L^2 m_d}{4\theta_1^2\theta_2^2}\big(c_{12} + \theta_2 s_{12} + \big(\tfrac{9}{10}\theta_1^2 - \theta_1\theta_2 - 1\big)c_2 + (\theta_1+\theta_2)s_2 - \tfrac{9}{10}\theta_1^2 - c_1 + 1\big) = M_{21}$
$M_{22} = \frac{9L^2 m_d}{8\theta_2^4}\big(\tfrac{10}{27}\theta_2^4 + \theta_2^2 - 2\theta_2 s_2 - 2c_2 + 2\big)$
$C_{11} = -\frac{5L^2 m_d}{2\theta_1^5\theta_2^3}\Bigg[\frac{\theta_1^3\theta_2^2}{2}(\dot\theta_1+\dot\theta_2)c_{12} - \left(\theta_1^2\theta_2^2\dot\theta_1 + \frac{\dot\theta_1^3\theta_2}{2}\dot\theta_2\right)s_{12} + \frac{\theta_1^3}{2}\Bigg(\Big\{-\frac{9}{5}\theta_1^2 + \theta_1\theta_2 - \theta_2^2\Big\}c_2 + \Big\{-\frac{9}{10}\theta_1^2\theta_2 + \theta_1\theta_2^2 + \theta_2\Big\}s_2 + \theta_2 s_1 + \frac{9}{5}\theta_1^2 - \theta_1\theta_2\Bigg)\dot\theta_2 + \theta_2^2\Bigg(\frac{\theta_1}{2}c_2 + \theta_1^2 s_2 + \Big\{\frac{-\theta_1}{2} + \frac{21}{20}\theta_1^2\theta_2 - \frac{21}{5}\theta_2\Big\}c_1 + \Big\{\theta_1^2 - \frac{21}{5}\theta_1\theta_2\Big\}s_1 - \frac{\theta_1^2}{2} + \frac{21}{20}\theta_1^2\theta_2 + \frac{21}{5}\theta_2\Bigg)\dot\theta_1\Bigg]$

(continued)

(continued)

$$C_{12} = -\frac{5L^2 m_d}{4\theta_1^2\theta_1^3}\left[\left\{\theta_2^2\dot{\theta}_1 + \theta_2^2\dot{\theta}_2 - 2\dot{\theta}_2\right\}c_{12} - \{\theta_2\dot{\theta}_1 + 2\theta_2\dot{\theta}_2\}s_{12} + \left(\left\{-\frac{9}{5}\theta_1^2 + 2\theta_1\theta_2 - \theta_2^2 + 2\right\}c_2\right.\right.$$
$$+\left\{-\frac{9}{10}\theta_1^2\,\theta_2 + \theta_1\theta_2^2 - 2(\theta_1 + \theta_2)\right\}s_2 - 2 + 2c_1 + \frac{9}{5}\theta_2^2\Bigg)\dot{\theta}_2$$
$$\left.+\left(\left\{-\frac{9}{5}\theta_1^2 + \theta_1\theta_2 - \theta_2^2\right\}c_2 + \left\{-\frac{9}{10}\theta_1^2\theta_2 + \theta_1\theta_2^2 + \theta_2\right\}s_2 + \theta_2 s_1 + \frac{9}{5}\theta_1^2 - \theta_1\theta_2\right)\dot{\theta}_1\right]$$

$$C_{21} = \frac{5L^2 m_d}{2\theta_1^3\theta_2^3}\left[\theta_2 c_{12} + \theta_2^2 s_{12} + \left\{-\frac{9}{10}\theta_1^3 + \frac{\theta_1^2\theta_2^2}{2} - \theta_1\theta_2^2 - \theta_2\right\}c_2\right.$$
$$\left.+\left\{\frac{\theta_1^2\theta_2^2}{2} - \theta_2^2 - -\frac{9}{20}\theta_1^3\theta_2 + \theta_1\theta_2\right\}s_2 + \frac{9}{10}\theta_1^3 - \frac{\theta_1^2\theta_2}{2} - \theta_2 c_1 + \theta_2\right]\dot{\theta}_1$$

$$C_{22} = -\tfrac{9L^2 m_d}{8\theta_2^5}\left(\theta_2^2 c_2 - 4\theta_2 s_2 + \theta_2^2 - 4c_2 + 4\right)\dot{\theta}_1$$

$$G_1 = k\left[\frac{ds_{12}}{2} + \boldsymbol{\alpha}_1\right]\left({}^G A_{2_s} - {}^G B_{2_s} - dc_{12}\right) - k\left[\frac{ds_1}{2} + \boldsymbol{\alpha}_3\right]\left({}^G A_{1_s} + {}^G B_{1_s} + \frac{2L(c_1 - 1)}{\theta_1}\right)$$
$$+\frac{EI_b\theta_1}{L} - \frac{5m_d gL}{4\theta_1}\left(s_1 + \frac{c_1 - 1}{\theta_1}\right) + m_d gL\left[\frac{\theta_1(c_1 - c_{12}) - \theta_2(c_1 - 1)}{\theta_1^2\theta_2}\right]$$
$$-m_d gL\left[\frac{\theta_1^2 s_{12} - \theta_1^2 s_1 + 2\theta_2(c_1 - 1) + 2\theta_1\theta_2 s_1}{4\theta_1^2\theta_2}\right]$$
$$-m_d gL\left[\frac{c_1 - c_{12} - \theta_1(s_1 - s_{12}) + \theta_2 s_1}{\theta_1\theta_2}\right]$$

(continued)

(continued)

$G_2 = -\frac{k\boldsymbol{\beta}_1}{2}\left[\frac{1}{\theta_1^2}\left({}^{G}B_{2_s} + \frac{dc_{12}}{2} + \boldsymbol{\beta}_2\right) + \frac{1}{\theta_2^2}\left({}^{G}A_{2_s} - \frac{dc_{12}}{2} + \boldsymbol{\beta}_2\right)\right] + \frac{EI_b\theta_2}{L} - \frac{m_d g L}{4\theta_2^2}(c_{12} - c_1 + s_{12}\theta_2) + \frac{m_d g L}{\theta_1\theta_2^2}(\theta_1(c_1 - c_{12}) - \theta_2(c_1 - 1)) + \frac{m_d g L}{\theta_1\theta_2}(c_1 - 1 - s_{12}\theta_1)$			
$\alpha_1 = \frac{L}{\theta_1^2} - \frac{Lc_1}{\theta_1^2} - \frac{Ls_1}{\theta_1} + \frac{Ls_1}{\theta_2} - \frac{Ls_{12}}{\theta_2};$		$\alpha_2 = -\frac{L}{\theta_1} + \frac{Lc_1}{\theta_1} - \frac{Lc_1}{\theta_2} + \frac{Lc_{12}}{\theta_2};$	$\alpha_3 = \frac{Ls_1}{\theta_1} + \frac{L(c_1-1)}{\theta_1^2};$
$\beta_1 = 2Lc_{12} - 2Lc_1 + 2Ls_{12}\theta_2 + ds_{12}\theta_2^2;$		$\beta_2 = -\frac{L}{\theta_1} + \frac{Lc_1}{\theta_1} - \frac{Lc_1}{\theta_2} + \frac{Lc_{12}}{\theta_2}$	
$Q_1 = \frac{dc_1}{2}(F_1 + F_2)$	$Q_2 \frac{dc_{12}}{2}(F_1 + F_2)$	$D_1 = \frac{CL}{\theta_2\theta_2}\left[2\theta_2(c_1 - 1) + \theta_1 s_1(2\theta_2 - \theta_1) + s_{12}\theta_1^2\right]$	$D_2 = \frac{CL}{\theta_2^2}[c_{12} - c_1 + s_{12}\theta_2]$

References

1. Anderson VC, Horn RC (1970) Tensor arm manipulator
2. Mehling JS, Diftler MA, Chu M, Valvo M (2006) A minimally invasive tendril robot for in-space inspection. https://doi.org/10.1109/BIOROB.2006.1639170
3. Robinson G, Davies JBC (1999) Continuum robots—a state of the art. In: Proceedings of the IEEE international conference on robotics and automation, vol 4, pp 2849–2854. https://doi.org/10.1109/robot.1999.774029
4. Ma RR, Dollar AM (2011) On dexterity and dexterous manipulation. In: IEEE 15th international conference on advanced robotics. New Boundaries Robotics. ICAR 2011, pp 1–7. https://doi.org/10.1109/ICAR.2011.6088576
5. Garriga-Casanovas A, Baena FR (2019) Kinematics of continuum robots with constant curvature bending and extension capabilities. J Mech Robot 11(1):1–12. https://doi.org/10.1115/1.4041739
6. Hannan MW, Walker ID (2003) Kinematics and the implementation of an elephant's trunk manipulator and other continuum style robots. J Robot Syst 20(2):45–63. https://doi.org/10.1002/rob.10070
7. Rone WS, Ben-Tzvi P (2014) Continuum robot dynamics utilizing the principle of virtual power. IEEE Trans Robot. https://doi.org/10.1109/TRO.2013.2281564
8. Tatlicioglu E, Walker ID, Dawson DM (2009) Dynamic modelling for planar extensible continuum robot manipulators. Int J Robot Autom. https://doi.org/10.2316/Journal.206.2009.4.206-3247
9. Falkenhahn V, Mahl T, Hildebrandt A, Neumann R, Sawodny O (2015) Dynamic modeling of bellows-actuated continuum robots using the Euler-Lagrange formalism. IEEE Trans Robot. https://doi.org/10.1109/TRO.2015.2496826
10. Trivedi D, Lotfi A, Rahn CD (2007) Geometrically exact dynamic models for soft robotic manipulators. https://doi.org/10.1109/IROS.2007.4399446
11. Cao DQ, Tucker RW (2008) Nonlinear dynamics of elastic rods using the Cosserat theory: modelling and simulation. Int J Solids Struct. https://doi.org/10.1016/j.ijsolstr.2007.08.016
12. Pachouri V, Pathak PM (2021) Inverse-kinematic model of a cable-driven manipulator. In Advances in industrial machines and mechanisms, pp 553–564
13. Giri N, Walker ID (2011) Three module lumped element model of a continuum arm section. In: IEEE international conference on intelligent robots and systems, pp 4060–4065. https://doi.org/10.1109/IROS.2011.6048577

Dynamic Characteristics of Human Hand-Arm System—Analytical and Simulation Approaches

Srinivaasappa Indira Raj Dhanush, Prajwal Gurunath, Prajwal Kamath, S. Murthy Ninad, and C. V. Chandrashekara

Abstract Investigation on the dynamic characteristics of the human hand-arm system plays a crucial role in assessing the health risks involved due to persistent exposure to hand transmitted vibrations (HTV), during operation of hand-operated tools. Finite element analysis and mathematical modelling are widely used computational tools to simulate dynamics of the hand-arm system and provide insights on the dynamic characteristics such as natural frequencies and mode shapes. For the present work, an anatomically accurate 3D model of the hand-arm system, developed using a CT scan, is simulated in ANSYS to determine the natural frequencies and mode shapes. Mathematical model of a three degree of freedom hand-arm system is developed using lumped mass approach, and dynamic characteristics are determined. Emphasis is given primarily to the torsional aspects of the hand-arm system, to analyze the effects of vibration in the vertical direction. A quantitative comparison of the results obtained through simulation and mathematical approaches is provided. The results obtained show good correlation with existing and literature and show good scope for further research in the area of biomechanical analysis on the hand-arm system.

Keywords Hand-arm system · Mathematical modelling · Finite element simulation · Natural frequencies · Mode shapes

1 Introduction

Workers in various industrial sectors are continuously exposed to unwanted vibrations whilst operating various tools and heavy machinery. This causes immense discomfort during operation and long-term physical pain and psychological distress to the operators, significantly reducing their performance. Exposures to such unwanted vibrations are known to cause neck pain, back pain, vascular, neurological, and musculoskeletal disorders, postural fatigue, respiratory disorders, and possible

S. I. Raj Dhanush · P. Gurunath · P. Kamath · S. M. Ninad · C. V. Chandrashekara (✉)
Department of Mechanical Engineering, PES University, Bengaluru 560085, India
e-mail: drcvc@pes.edu

V. K. Gupta et al. (eds.), *Recent Advances in Machines and Mechanisms*, Lecture Notes in Mechanical Engineering, https://doi.org/10.1007/978-981-19-3716-3_18

cancer. Various industrial sectors require continuous usage of hands whilst handling heavy equipment. The exposure of hands precisely the human full hand-arm system is popularly termed as hand transmitted vibrations (HTV). Prolonged exposure to HTV leads to hand-arm vibration syndrome (HAVS), causing disorders such as vibration white finger, carpal tunnel syndrome, and peripheral neuropathy. The risk of HAVS has attracted researchers over the past eight to nine decades in modelling, analyzing and sought reasonable reasons for cause and thereby suggesting remedies.

Recent epidemiological studies highlight the inaccuracies of the frequency weightings of HTV. Dong et al. [1] reported on the importance of biodynamics for evaluation, measurement, and risk assessment of HTV and noticed that frequency weighting in ISO 5349-1:2001 standard gives importance to low-frequency effect but underestimates the high-frequency effects on human hand and fingers. To help revise the standards, research is being carried out extensively by analyzing the hand-arm system through analytical, simulation, and experimental approaches. Experimentation on human test subjects is limited due to ethical concerns and comparatively expensive. On the other hand, finite element simulation and mathematical modelling are extensively used computational tools in establishing the biomechanical behaviour of the hand-arm system to larger extent and significantly reducing expenses. Development of an accurate 3D model of the hand-arm system for simulation is a challenge faced by researchers. The present work involves performing modal analysis on an anatomically accurate 3D model of the hand-arm skeletal system to extract natural frequencies and mode shapes, primarily in the vertical direction. Mathematical model of the hand-arm system is developed using lumped mass approach based on Newton's Second Law using MATLAB to simulate a three degrees of freedom system of the hand-arm. The mathematical formulation emphasizes on the transverse vibration of the hand-arm system. The obtained results through finite element simulation are validated with those derived from mathematical formulation which shows good correlation considering biological inaccuracies. Developed model is very much relevant to the industrial application in terms of helping the process of creating comfortable and safe work environment for workers who are continuously exposed to high-level vibrational hand-operated equipment.

2 Finite Element Modelling and Simulation

An anatomically accurate 3D model of the hand-arm system is developed through an acquired CT scan using medical imaging software package Materialize Mimics Innovation Suite (consists of Mimics Research 21.0 and 3-Matic Research 13.0). A healthy male member of the team who was aware of the risks involved in the CT scan imaging volunteered to lie down in superman position to obtain clear images of the hand-arm system. 1405 images with slice thickness 1.25 mm for bone and 1405 images with slice thickness 2.5 mm are obtained from the tip of the fingers to the upper shoulder to include the shoulder joint (axial plane). The slice increment for 1.25 mm being 0.625 mm and 2.5 mm for the 2.5 mm slices. 109 and 142 images in coronal and sagittal planes are also similarly obtained. The images are saved in (Digital Imaging and Communication in Medicine) DICOM format. The CT images in DICOM format are shown in Fig. 1a to d.

The obtained CT scan is geometrically reconstructed to a 3D model using Mimics Research 21.0 using functions *segmentation* and *mask creation.* Segmentation process consists of separating various anatomical entities such as cortical bone, trabecular bone, soft tissue, muscle tissue, and skin based on the image pixel's Hounsfield unit (HU). HU is a relative quantitative measurement of radio density. Different masks are created representing the various anatomical entities as mentioned above. The masks are further refined iteratively to be suitable for 3D model reconstruction. An example of the segmented three-dimensional model of the humerus bone is shown in Fig. 2.

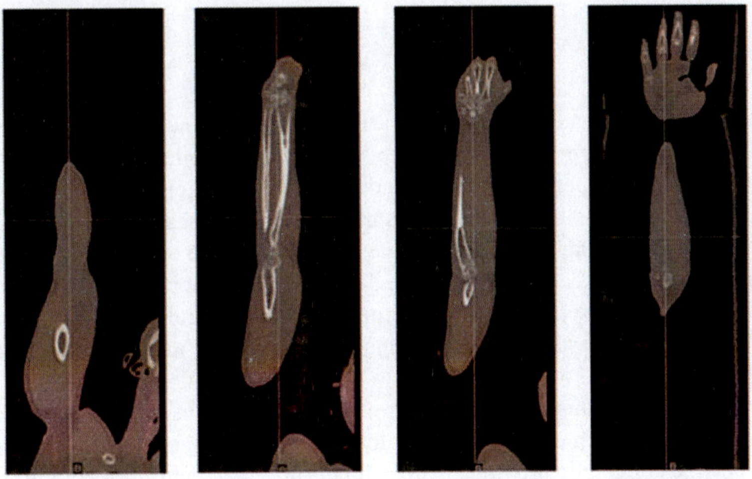

(a) 117.6757 *mm* (b) 141.6015 *mm* (c) 151.3671 *mm* (d) 173.3398 *mm*

Fig. 1 CT images of the hand-arm (coronal view)

Fig. 2 Segmented humerus bone

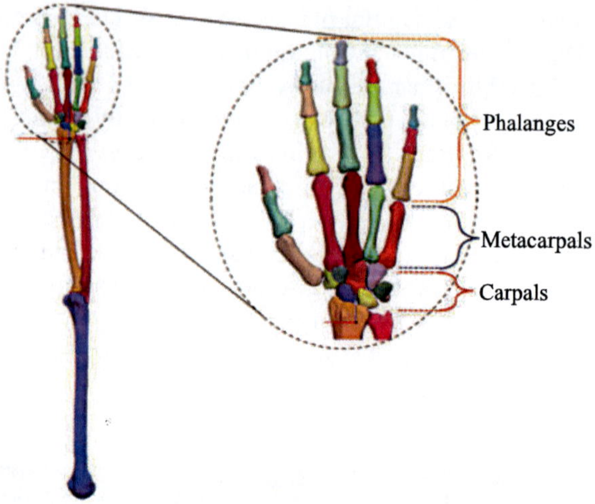

Fig. 3 3D model of the hand-arm system

The developed 3D model of the hand-arm system is borrowed to 3-Matic Research 13.0 for further refinement with accurate surfaces. The model consists of 30 bones, namely humerus (upper arm), radius and ulna (forearm), eight carpal bones, viz., lunate, triquetrum, pisiform, hamate, capitate, trapezoid, scaphoid, trapezium, five metacarpals, and phalanges of each of the five fingers. The completed 3D model of the hand-arm system is shown in Fig. 3. The model is imported to ANSYS 2020 to perform modal analysis to determine natural frequencies and mode shapes.

The material properties of bone are incorporated to the model and are shown in Table 1.

Limitation in establishing the joints between the bones in Mimics Research 21.0 and 3-Matic Research 13.0 gives way to provide the same in ANSYS platform. Joints between bones are provided using a function called connections. Joints play an important role in mimicking the dynamics of the hand-arm system and provided based on the actual working condition of the hand-arm system. Current model involves multiples degrees of freedom at each node point (joints) based on the prominent direction of motion and will have different nature and magnitude of stiffness and damping. The set of relevant stiffnesses and damping are borrowed from various sources of published literature and incorporated to mimic the actual dynamics of the hand-arm.

A body-ground spherical joint is provided at the proximal end of the humerus bone by choosing appropriate surfaces on the bone. The spherical joint provides

Table 1 Material properties of bone [2]

Parameter	Value
Density	2000 kgm^{-3}
Young's modulus	17.5 GPa
Poisson's ratio	0.3

a boundary condition very much near to the actual working condition and mimics the dynamics of the shoulder joint. The body-body spherical joint at the shoulder is shown in Fig. 4.

A body–body revolute joint is provided between humerus and radius-ulna to mimic the elbow joint. Elbow joint's torsional stiffness value of 30.5 Nm/rad [3] is considered and incorporated in the ANSYS model. The revolute joint provided at the elbow is shown in Fig. 5.

A body–body revolute joint is provided between appropriate surface portions of the scaphoid-lunate (lower carpals) carpals and radius-ulna bones to mimic revolute

Fig. 4 Body-ground spherical joint at the shoulder

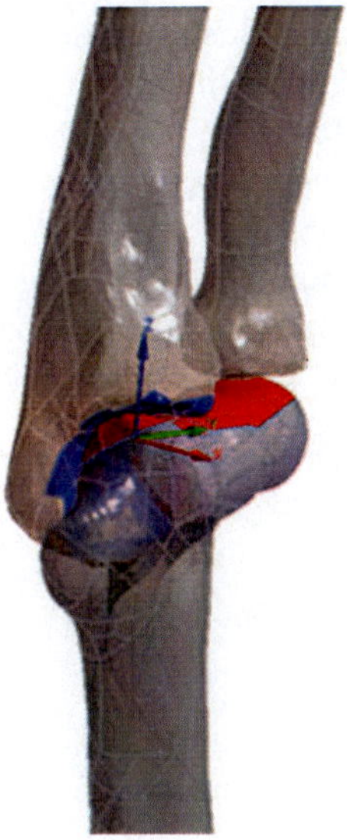

Fig. 5 Body-body revolute joint at the elbow

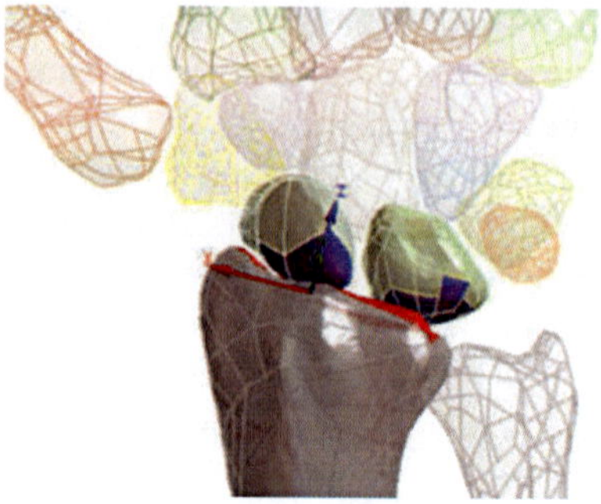

Fig. 6 Revolute joint provided at the wrist

motion of the wrist joint. Torsional stiffness value of 746.3 Nm/rad [3] is provided at the wrist joint. The revolute joint provided at the wrist is shown in Fig. 6.

The nature of the joint between carpal bones are relatively rigid for an adult subject. Similarly, the joints between metacarpals and carpals are relatively rigid. Therefore, for the present work, body–body fixed joints are provided amongst carpals and between carpals and metacarpals. However, these joints to some extent have elastic flexibility. This can be witnessed when only forearm dynamics are simulated. Hence, all the joints between the carpals and between the metacarpals and carpals are deformable. Body–body revolute joints are provided for the interphalangeal joints (joints connecting successive phalanges) and the metacarpophalangeal joints (joints connecting proximal phalanges to the respective metacarpals) for each of the five fingers of the hand. The body-body fixed joints provided between the carpals are shown in Fig. 7. The body–body fixed joint between carpals and metacarpals is shown in Fig. 8.

Values of stiffness of all the distal interphalangeal (DIP) joints are assumed to be same. Similarly, the stiffness of the proximal interphalangeal (PIP) joints is assumed to be the same. Stiffness values are provided for the interphalangeal joints and the metacarpophalangeal joints and are shown in Table 2.

Mesh size for the analysis of final 3D model is finalized after many iterations of running with various mesh size. Various factors such as complexity of model, joints, size, available computer capacity, and so on play an important role in finalizing the

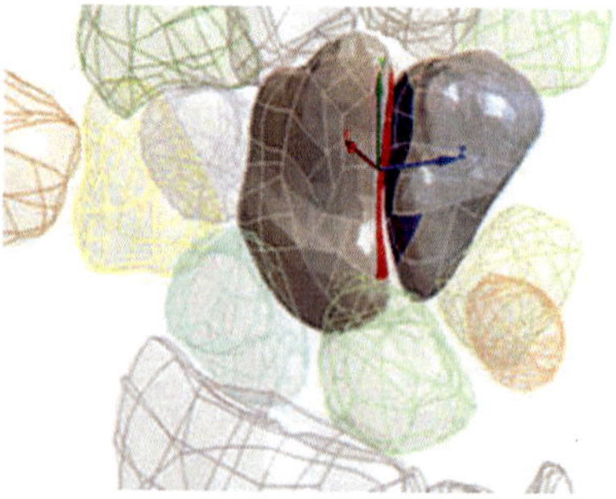

Fig. 7 Fixed joint provided between carpals

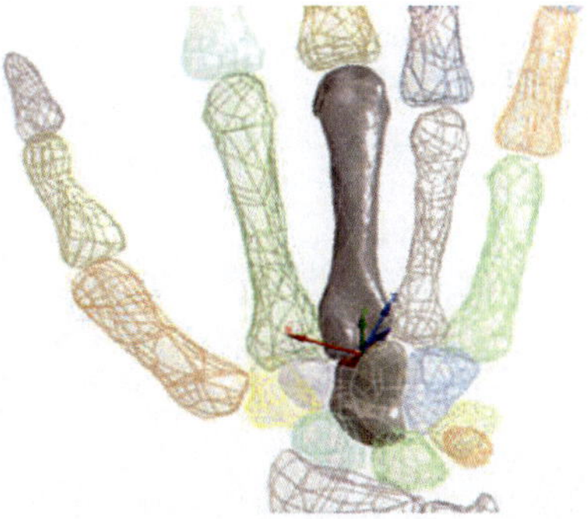

Fig. 8 Fixed joint provided between carpals and metacarpals

Table 2 Stiffness values of interphalangeal and metacarpophalangeal joints [4]

Joint	Torsional stiffness (Nm/rad)
DIP	1437.15
PIP	2813.91
MCP	6037.17

Table 3 Mesh properties

Element size	5 mm
Minimum edge length	3.2017 mm
Growth rate	1.2
Nodes	85,798
Elements	47,728

mesh size. The trade-off between the computer analysis time and convergence of the results are taken into consideration. Convergence of mesh size is a part of the work and is not reported considering the limitation of the number of pages. A mesh size of 5 mm is considered for the final 3D model.

The mesh statistics are shown in Table 3.

The meshed model of the full hand-arm system is shown in Fig. 9.

Considering no pre-stress conditions and upper end of the humerus bone with spherical joint fixed to the ground as the boundary condition, modal analysis is carried out, and natural frequencies and mode shapes are reported.

3 Mathematical Modelling

Mathematical modelling is used to describe the important features of a physical system through mathematical equations. The mathematical model simplifies the system and predicts the system's behaviour. The model gives insight into the dynamic characteristics of the system, such as natural frequency, mode shapes, response, and

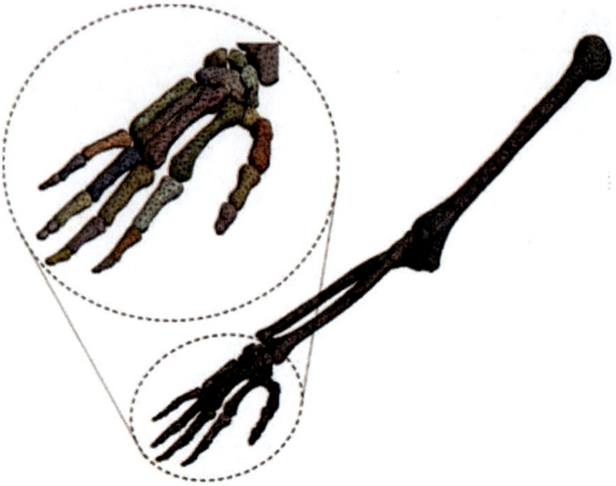

Fig. 9 Meshed model of the hand-arm system

FRFs. The natural frequencies of a system are referred to avoid resonance condition. Resonance of a system causes disruptions to the normal response levels, significantly hindering the performance of the system. The present work focuses on an effort to investigate the effects of vibrations on a complex system, and to study the dynamic characteristics of the hand-arm system.

Discrete model approach is used to build mathematical models of the hand-arm system. Continuous systems are modelled as discrete models to obtain solutions in a simpler manner. Though results obtained through continuous systems are more accurate, analytical models, and solutions available are limited to a precise set of problems and more complex. The hand-arm system is modelled as a three degree of freedom lumped mass system, by considering three masses (hand, forearm, and upper arm). Emphasis is given primarily to the transverse vibration of the hand-arm system, to analyze the effects of vibration in the vertical direction. The three degree of freedom lumped mass model of the hand-arm system is shown in Fig. 10. Torsional springs k_{t1}, k_{t2}, and k_{t3} are shown at the shoulder joint, elbow joint, and wrist joint, respectively. Concentrated mass moment of inertia at the free end of the upper arm, forearm, and hand, respectively, is considered.

The mass moment of inertias of each subsystem of the hand-arm is extracted from ANSYS, of the model used for modal analysis. After a vigorous literature review is conducted, the stiffness properties of the wrist, elbow, and shoulder joints are obtained from Adewusi et al. [3]. The mass moment of inertia of the subsections of the hand-arm is shown in Table 4.

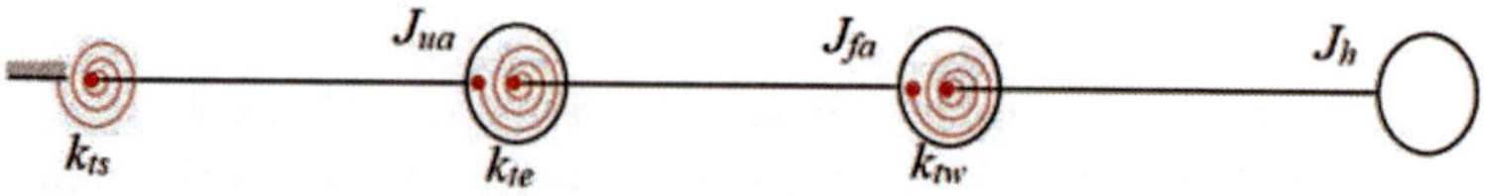

Fig. 10 Three degree of freedom lumped mass model of hand-arm system

Table 4 Mass moment of inertia of hand-arm sections

Hand-arm section	Notation	Mass moment of inertia (kgm^2)
Hand	J_h	0.0012845
Forearm	J_{fa}	0.0062329
Upper arm	J_{ua}	0.075312

Table 5 Rotational stiffnesses of joints

Joint	Notation	Torsional stiffness (Nm/rad)
Wrist	k_{tw}	746.3
Elbow	k_{te}	30.5
Shoulder	k_{ts}	1210.5

The torsional stiffnesses of the various joints are shown in Table 5.

Using the reported values of masses and stiffnesses, the equation of motion is derived using Newton's Second Law of Motion and depicted in matrix form as

$$\begin{bmatrix} J_h & 0 & 0 \\ 0 & J_{fa} & 0 \\ 0 & 0 & J_{ua} \end{bmatrix} \begin{bmatrix} \ddot{\theta}_h \\ \ddot{\theta}_{fa} \\ \ddot{\theta}_{ua} \end{bmatrix} + \begin{bmatrix} k_{tw} + k_{te} & -k_{te} & 0 \\ -k_{te} & k_{te} + k_{ts} & -k_{ts} \\ 0 & -k_{ts} & k_{ts} \end{bmatrix} \begin{bmatrix} \theta_h \\ \theta_{fa} \\ \theta_{ua} \end{bmatrix} = \begin{bmatrix} 0 \\ 0 \\ 0 \end{bmatrix} \quad (1)$$

The eigen values of the system matrices are obtained in MATLAB to obtain the natural frequencies of the system.

4 Results and Discussion

The developed 3D model of the full hand-arm is in good shape in all respect to perform any biomedical engineering analysis, such as finite element analysis, fracture toughness, modal analysis, 3D printing, and impact analysis. The model is borrowed to ANSYS to perform modal analysis to determine natural frequencies and mode shapes. The appropriate material properties of bone are taken into consideration, and analysis is performed. Mathematical model of the hand-arm system is developed using lumped mass approach to determine the natural frequencies. A three degree of freedom hand-arm system is considered for the present work. Emphasis is given primarily to the torsional aspects of the hand-arm system, to analyze the effects of vibration in the vertical direction, and the results obtained through simulation are compared with those obtained through mathematical modelling. The natural frequencies and the mode shape of the hand-arm system in the vertical direction obtained in ANSYS are shown in Fig. 11a to c.

The natural frequencies obtained through simulation and mathematical modelling are shown and compared in Table 6.

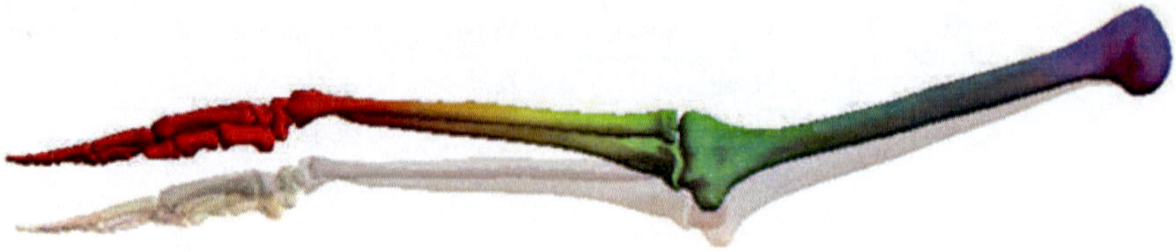

(a) First Natural Frequency = 5.7082 Hz

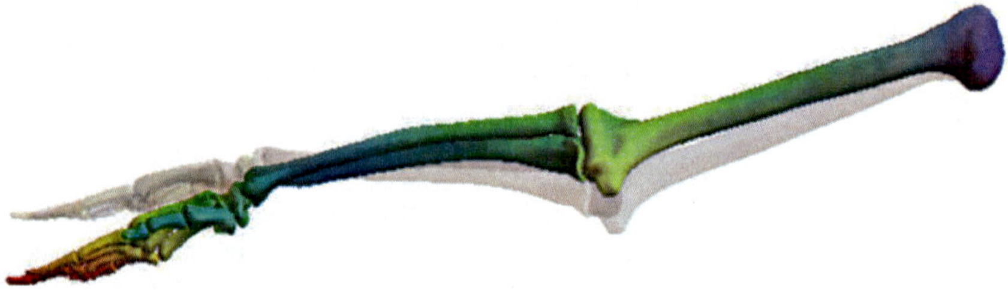

(b) Second Natural Frequency = 62.48 Hz

(c) Third Natural Frequency = 165.58 Hz

Fig. 11 **a** First natural frequency = 5.7082 Hz, **b** second natural frequency = 62.48 Hz, **c** third natural frequency = 165.58 Hz

Table 6 Natural frequencies of hand-arm system obtained through simulation and analytical approaches

Mode	Natural frequency (Hz)	
	Simulation	Mathematical modelling
1	5.7082	5.5019
2	62.48	75.4292
3	165.58	131.2274

The generated mode shapes also correspond to the actual modes of deformation, as indicated in the previous literature.

5 Conclusion

An analytical and simulation approaches are being considered to establish the dynamic characteristics of the human hand-arm system. A CT scan of a live human test subject is taken to develop the three-dimensional model using Mimics and 3-Matic of Materialize Innovation Suite software package. The developed model is borrowed to ANSYS to perform modal analysis and natural frequencies and mode shapes are extracted. Emphasis is given to the mode shapes in the vertical direction. Work carries a scope of involving the muscles around the bone structures, for various application-based analysis. To name few such applications, fatigue analysis, fracture strength analysis, impact analysis, and compatible for 3D printing model for experimentation.

References

1. Dong RG, Wu JZ, Welcome DE (2005) Recent advances in biodynamics of human hand-arm system. Ind Health 43:449–471
2. Liu S, Xu H, Shang Y, Jiang W, Dong J (2020) Finite element modal analysis and harmonic response analysis of human arm grasping model. Comput Methods Biomech Biomed Eng 1–12
3. Adewusi S, Rakheja S, Marcotte P (2012) Biomechanical models of the human hand-arm to simulate distributed biodynamic responses for different postures. Int J Ind Ergonomics 42:249–260
4. Knez L, Slavič J, Boltežar M (2017) A sequential approach to the biodynamic modelling of a human finger. Shock Vibr 1–12

Vibration and Condition Monitoring

Vibration Isolation Characteristics of a Modified Gough-Stewart Platform with the Top Platform Filled with Damping Particles

Nazeer Ahmad, R. Ranganath, D. Poomani, and Ashitava Ghosal

Abstract In this paper, we consider a modified Gough-Stewart platform (MGSP) where two groups of three legs meet at two concentric circles on both top and bottom platforms. The geometry of the MGSP is chosen such that all the first six natural frequencies are equal for a typical payload mounted on the top platform. Additionally, in the top platform, made up of an aluminum honeycomb sandwich, the empty cells of the honeycomb core are filled with damping particles (DPs) to introduce passive damping in the system and to limit the resonance responses. A finite element model (FEM) of the MGSP is developed to quantify the performance in terms of frequency response functions (FRF), resonance peaks, and the damping introduced by the damping particles. The FEM model of the MGSP is combined with the discrete element model (DEM) of the damping particles to compute the effect of the particles on the overall dynamics and damping behavior of the platform. The effect of DPs on the transfer function is evaluated by solving the equations of motion of the DPs and the FEM model of the MGSP simultaneously. Finally, the FRF between the bottom platform and the mass center of the payload is computed for assessing the effectiveness of DPs, the transfer functions between the base excitation and mass center of the payload with respect to four inputs—sine swept inputs X, Z, θ_x, and θ_Z applied separately at the base of the platform—were computed for 25%, 50%, 75%, and 93% fill fractions. The peaks at resonances progressively decrease as the fill fraction was increased. For all the modes, it was seen that the damping introduced

N. Ahmad (✉) · D. Poomani
Structures Group, U R Rao Satellite Centre, Indian Space Research Organization (ISRO), Bangalore 560017, India
e-mail: nazeer@ursc.gov.in

D. Poomani
e-mail: poomani@ursc.gov.in

R. Ranganath
PD SPADEX, U R Rao Satellite Centre, Indian Space Research Organization (ISRO), Bangalore 560017, India
e-mail: rrrr@ursc.gov.in

A. Ghosal
Department of Mechanical Engineering, Indian Institute of Science, Bangalore 560012, India
e-mail: asitava@iisc.ac.in

V. K. Gupta et al. (eds.), *Recent Advances in Machines and Mechanisms*, Lecture Notes in Mechanical Engineering, https://doi.org/10.1007/978-981-19-3716-3_19

by damping particles results in the splitting of the modes and the formation of anti-resonance at resonance peaks. The damping introduced by the particles is more effective in longitudinal direction $\|Z/Z\|$ where there is a reduction from 25 to 8 with increasing fill fractions from 25 to 93%. The cross-axis transfer functions also seen to come down from 98 and 10 to 75 and 5 along the X-axis and Y-axis, respectively.

Keywords Modified Stewart platform · Multi-axis vibration control · Particle impact damping

1 Introduction

The Gough-Stewart platform (GSP) is commonly used in a wide range of applications such as flight simulators, pointing mechanisms, machine tools, force-torque sensors, precision surgery [1, 2], and multi-axis vibration isolation [3, 4]. The Gough-Stewart platform consists of a movable top platform connected to a base platform by six legs. The length of the legs can be changed to provide a desired position and orientation of the top platform. The design of GSP has been based on various performance criteria such as load-carrying capacity, workspace requirements, range of motion dexterity, and isotropy. In the context of kinematics, at an isotropic configuration, the velocity (linear and angular) distribution is a sphere, and the GSP can move with equal "ease" in all directions at these configurations [5], and due to this reason, isotropy is a desired feature in a design. Kinematic isotropy is related to the condition number of a manipulator Jacobian matrix, and several researchers have attempted to design a GSP to obtain kinematic isotropy [6]. In the context of static forces, at an isotropic configuration, the GSP can resist forces and moments equally well in all spatial directions. In the study of statics of the GSP, the so-called force transformation matrix [7, 8] is used instead of the manipulator Jacobian matrix, and again, the goal is to find conditions for which the force transformation matrix has identical singular values. In most of the above work, the researchers discuss various approaches to avoid a basic problem in manipulator Jacobian and force transformation matrices, namely the dimensions of the linear and angular velocity or that of the forces and moments are not the same. In reference [8], the authors derive algebraic conditions for separate force and moment isotropy and claim that it is not possible to obtain identical singular values for both force and moment parts of the force transformation matrices. The dynamic isotropy (DI), the condition of the platform where all the eigenvalues are equal [9], is an important criterion in design of vibration isolator. The dynamic isotropy index is defined as DII $= \omega_{\max}/\omega_{\min}$ is the ratio, as indicated, of the largest to the smallest natural frequency of the GSP. The geometry and inertia conditions for complete dynamic isotropy (i.e., DII $= 1$) of a GSP are practically not realizable.

In a modified version of the GSP (MGSP), the anchorage points on each platform lie on two circles in contrast to a single circle in the GSP. A set of three legs, which are

120° apart, are connected to each circle—this is based on a concept first presented in [10]. An MGSP with ideal joints and rigid links is known to give the configurations that have complete dynamic isotropy, and such an MGSP can be fabricated for real applications. An MGSP with flexural hinges, usually used to avoid friction at the joints, increases the DII or the modal spread. However, the new configurations of the MGSP with flexural joints with DII = 1 also exist [11].

In this paper, we assess the vibration isolation performance of a complete dynamic isotropic MGSP with flexural joints, metallic bellows in the leg for stiffness and a top platform made up of an aluminum honeycomb sandwich (see Fig. 1). The empty cells of the core of the honeycomb sandwich are filled with damping particles to introduce damping in the system and thereby limiting the resonance responses. The performance of the MGSP has been quantified in terms of frequency response functions (FRF), resonance peaks, and the damping introduced by the damping particles.

A finite element model (FEM) of the MGSP with the bellows and flexures, top and bottom platforms, and connectors was developed in ABAQUS® software. The

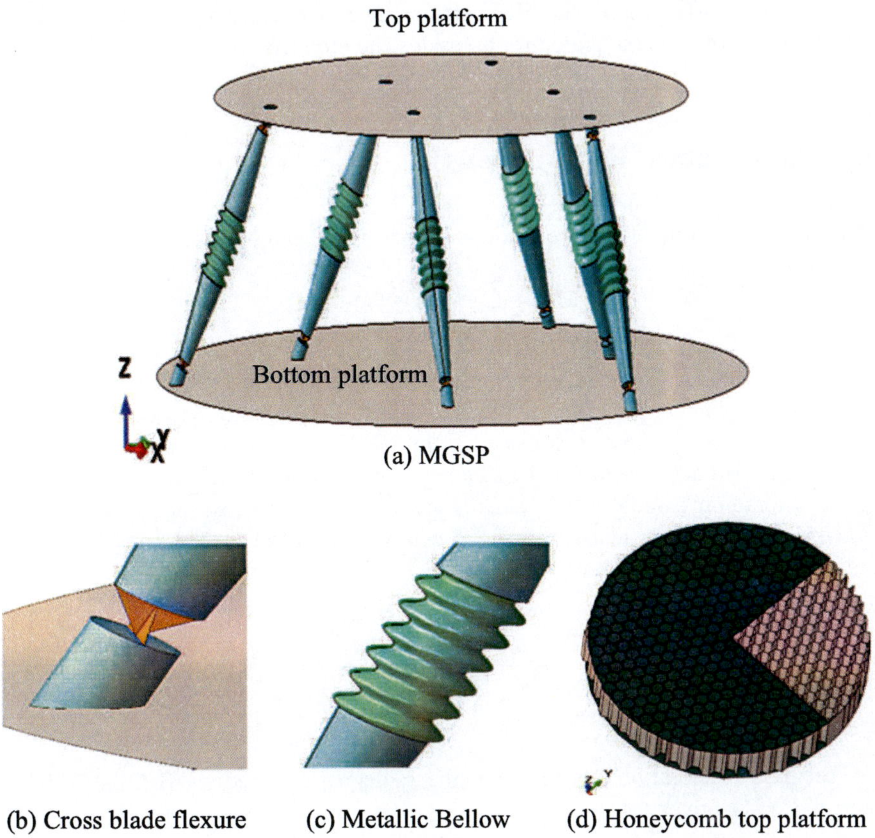

Fig. 1 Modified Gough-Stewart platform (MGSP) with the main components

shell elements S3R were used to mesh the bellows, flexures, and top and bottom platforms. The in-between connectors were model using the solid tetrahedral C3D10 elements [12]. An axis-symmetric payload was modeled using a lumped mass of $m_{\mathrm{pld}} = 10$ kg, with in plane inertias $I_{xx}^{\mathrm{pld}} = I_{yy}^{\mathrm{pld}} = 0.066$ kg m^2, and out of plane inertia of $I_{zz}^{\mathrm{pld}} = 0.1230$ kg m^2. The inertia values are given with respect to a coordinate system with the origin at the center of mass of the payload and axis are parallel to the global axis in its natural pose. The center of mass (CM) of the payload is assumed to be located at 50 mm above the center of the payload platform. The lumped payload mass was connected to the top platform with beam-type multi-point constraint (MPC). The damping particles were modeled using the discrete element method (DEM) wherein the equations of motion of the particles are obtained using Newton's second law. Whenever the damping particles come into contact with the top platform, the contact forces evolving in the process are modeled using dissipative Hertz contact theory. The FEM model of the MGSP and DEM model of the damping particle are coupled through contact forces and were solved simultaneously using the Runge–Kutta method in the MATLAB® (see details of the FEM-DEM coupled model in [13, 14]). Finally, the FRF between the mass center of the payload and the bottom platform was computed for assessing the effectiveness of DPs.

2 Mathematical Formulations

The governing finite element equations of the MGSP can be written as

$$\mathbf{M}\ddot{\chi} + \mathbf{C}\dot{\chi} + \mathbf{K}\chi = \mathbf{f}^e + \mathbf{f}^d \tag{1}$$

where **M**, **C**, and **K** are mass, damping, and stiffness matrices, respectively. $\mathbf{f}^d$ represents the particle damping forces, and $\mathbf{f}^e$ is the external excitation forces. The assembled displacement vector χ consists of the nodal displacement vectors. The χ is arranged such that all the nodal displacements of the bottom platform are in the vector χ_b, and remaining nodal displacements are in the vector χ_s so that $\chi = [\chi_s, \chi_b]$. Accordingly, the other matrices in Eq. 1 can be partitioned as [15].

$$\begin{bmatrix} \mathbf{M}_{ss} & \mathbf{M}_{sb} \\ \mathbf{M}_{bs} & \mathbf{M}_{bb} \end{bmatrix} \begin{Bmatrix} \ddot{\chi}_s \\ \ddot{\chi}_b \end{Bmatrix} + \begin{bmatrix} \mathbf{C}_{ss} & \mathbf{C}_{sb} \\ \mathbf{C}_{bs} & \mathbf{C}_{bb} \end{bmatrix} \begin{Bmatrix} \dot{\chi}_s \\ \dot{\chi}_b \end{Bmatrix} + \begin{bmatrix} \mathbf{K}_{ss} & \mathbf{K}_{sb} \\ \mathbf{K}_{bs} & \mathbf{K}_{bb} \end{bmatrix} \begin{Bmatrix} \chi_s \\ \chi_b \end{Bmatrix} = \begin{Bmatrix} \mathbf{f}_s^e \\ \mathbf{f}_b^e \end{Bmatrix} + \begin{Bmatrix} \mathbf{f}_s^d \\ \mathbf{f}_b^d \end{Bmatrix} \tag{2}$$

The base acceleration is enforced on the χ_b degree-of-freedoms (DOFs). The first part of Eq. 2 is given as

$$\mathbf{M}_{ss}\ddot{\chi}_s + \mathbf{C}_{ss}\dot{\chi}_s + \mathbf{K}_{ss}\chi_s = \mathbf{f}_s^d + \mathbf{f}_s^e - \left(\mathbf{M}_{sb}\ddot{\chi}_b + \mathbf{C}_{sb}\dot{\chi}_b + \mathbf{K}_{sb}\chi_b\right) \tag{3}$$

Using the transformation $\chi_s = \Phi_s q_s$, where the modal matrix Φ_s is obtained from the eigenvalue problem: $(\mathbf{K}_{ss} + \Omega_s^2 \mathbf{M}_{ss})\Phi_s = 0$ and invoking the orthogonality relation with respect to mass and stiffness matrix and assuming viscous damping, i.e., $\mathbf{C}_{ib}\dot{\chi}_b = 0$, Eq. 3 can be written as

$$\ddot{\mathbf{q}} + 2\zeta\Omega\dot{\mathbf{q}} + \Omega^2\mathbf{q} = \Phi^T\left(\mathbf{f}_s^d + \mathbf{f}_s^e\right) - \Phi^T\left(\mathbf{M}_{sb}\ddot{\chi}_b + \mathbf{K}_{sb}\chi_b\right) \tag{4}$$

In Eq. 3, the damping particle force $\mathbf{f}_s^d$ is related to the motions of damping particles impacting against the walls of the honeycomb cell. Let the damping particle i in cells of honeycomb be in contact with n_1 number of neighboring particles and n_2 points with cell walls, then the equations of motion can be written as

$$m_i\ddot{\mathbf{P}}_i = -m_i\mathbf{g} + \sum_{j=1}^{n_1}\mathbf{f}_{ij} + \sum_{w=1}^{n_2}\mathbf{f}_{iw} \tag{5}$$

$$\mathbf{I}\ddot{\Theta}_i = \sum_{j=1}^{n_1}\left(r_i - \frac{\delta_{ij}}{2}\right)\mathbf{n}_{ij} \times \mathbf{f}_{ij} + \sum_{w=1}^{n_2}(r_i - \delta_{iw})\mathbf{n}_{iw} \times \mathbf{f}_{iw} \tag{6}$$

where m_i and $\mathbf{I}$ are the mass and inertia matrix, respectively, of the particle i while $\mathbf{P}_i$ is the position vector and Θ_i is the angular displacement. $\mathbf{g}$, r_i, δ_{ij}, and $\mathbf{f}_{ij}$ represent acceleration due to gravity, radius of the damping particle i, local indention, and contact force between particles i and j, respectively. The Eqs. 4–6 are coupled and must be solved together. For a base excitation problem, it is assumed that the base is excited with a known signal, and thus, $\ddot{\chi}_b$ and χ_b are a known *priori*. The modal vector and natural frequencies along with the sub-matrices appearing in Eq. 4 were obtained from the FEM model. The integration of Eqs. 4–6 was carried out in MATLAB® using a Runge–Kutta method (ODE-45).

3 Numerical Simulations

The MGSP is excited at the base platform by a sine sweep input signal of constant magnitude and frequency ranging from 5 to 300 Hz. The input can be written mathematically as

$$u = \sin\left(2\pi\left(f_s + \frac{(f_{\text{end}} - f_s)t}{2t_{\text{end}}}\right)t\right) \tag{7}$$

where f_s, f_{end}, and t_{end} are, respectively, the start frequency, the end frequency, and the sweep duration. The input signal u can be a linear or angular displacement. A slice of the normalized input signal is given in Fig. 2.

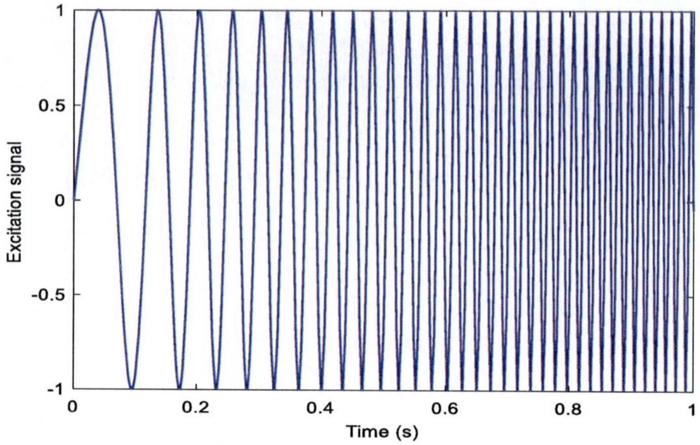

Fig. 2 Input swept sine signal

The input displacement swept sine signal in X, Z, θ_X, and θ_Z directions was applied at the center of the base platform separately, and the output response at the mass center of the payload was computed using the coupled equations of motion. Transfer functions described in the following sections were obtained from the time domain signals using the *tfestimate* function from the signal processing toolbox of MATLAB® for the various amounts of damping particles filled in the cells of the top platform.

3.1 Effect of Fill Fraction on FRFs with Longitudinal Axis (Z-axis) Input

Figures 3, 4, and 5 show the transfer functions between the displacement sine swept input applied in the longitudinal direction (*Z*-axis) and the displacement responses in *X*-axis, *Y*-axis, and *Z*-axis, respectively. The four cases considered are 25, 50, 75, and 93% fill fractions—the fill fraction is defined as the ratio of the filled volume of the honeycomb cell to the total volume of the cell. The damping particles are uniformly filled in the empty cells of the honeycomb core of the top platform. A 25% fill fraction implies that all the cells of the top platform are filled to 25% of the thickness. The amplification at resonance in the longitudinal direction is progressively decreasing—as seen in Fig. 3 that the amplification reduces from 25 to 8 as the fill fraction is increased from 25 to 93%. The maximum reduction in the response at resonance happens when the fill fraction is between 85 to 93% (when the fill fraction is near 100%, the damping particles do not have space for motion, and hence, a lesser number of collisions and subsequently less dissipation of energy take place). It can be seen that the longitudinal mode at 30.43 Hz has split into two modes, and there

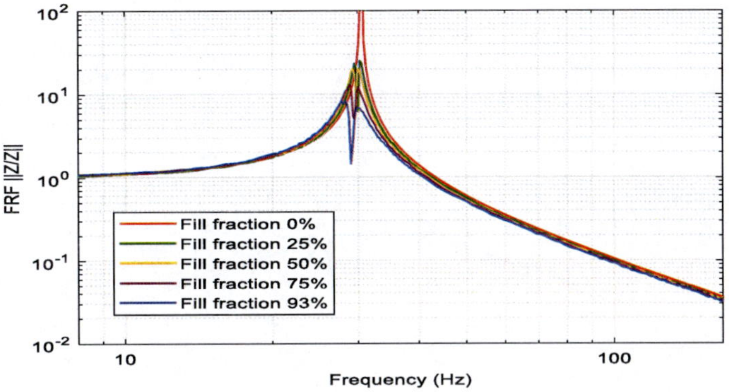

Fig. 3 Transfer function between *Z*-axis input and *Z*-axis output

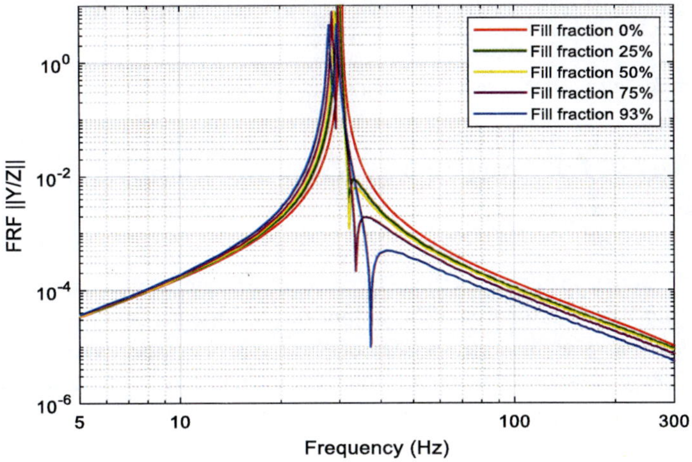

Fig. 4 Transfer function between *Z*-axis input and *Y*-axis output

is an anti-resonance at that frequency after the introduction of damping particles. The cross-axis amplifications at resonance have reduced, but the reduction is not as much as along the *Z*-axis. The $\|X/Z\|$ amplifications have come down from 98 to 75; and $\|Y/Z\|$ amplifications have come down from 10 to 5 when the fill fractions were varied from the 25 to 93%. It can be seen there is also splitting of modes and introduction of two anti-resonances in the isolation region, which is advantageous.

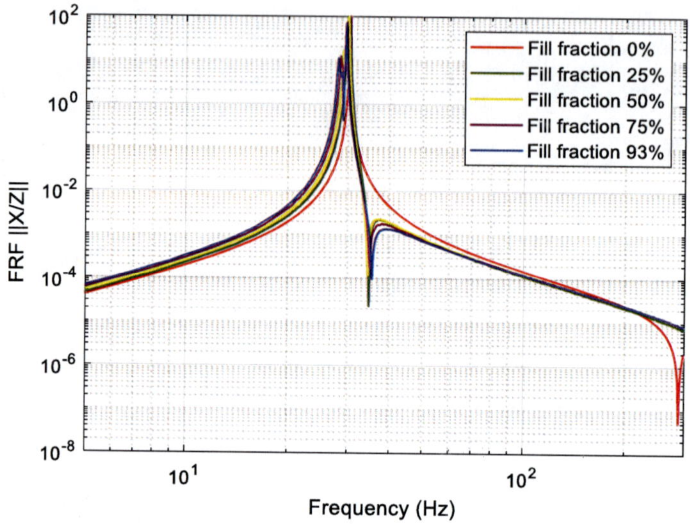

Fig. 5 Transfer function between *Z*-axis input and *X*-axis output

3.2 *Effect of Fill Fraction on FRFs with Lateral Axis (X-axis) Input*

The transfer functions between the displacement sine swept input applied along the *X*-axis at the base platform and the output displacement responses in the *X*-axis, *Y*-axis, and *Z*-axis at the mass center of the payload, respectively, are shown in Figs. 6, 7, and 8. The four cases considered for the damping particles are 25, 50, 75,

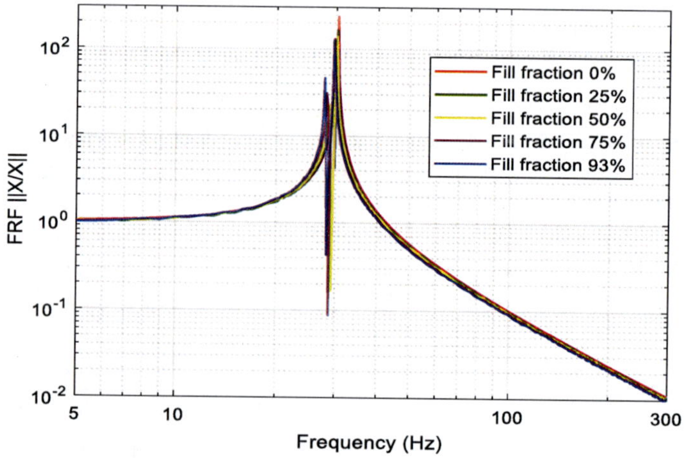

Fig. 6 Transfer function between *X*-axis input and *X*-axis output

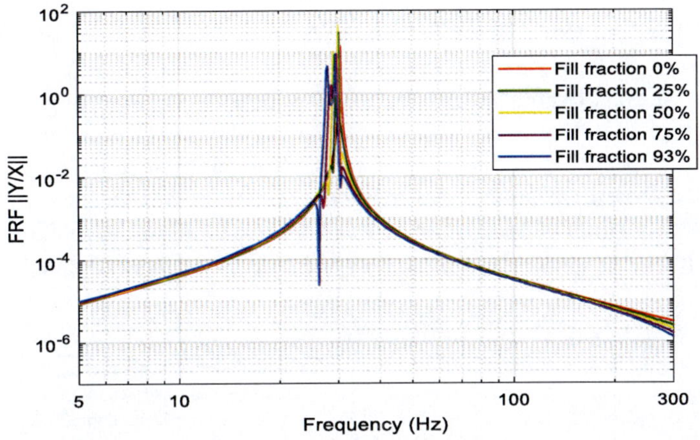

Fig. 7 Transfer function between *X*-axis input and *Y*-axis output

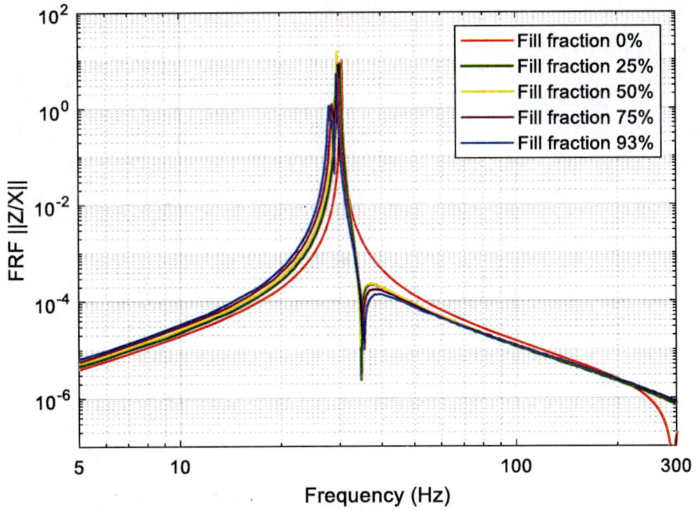

Fig. 8 Transfer function between *X*-axis input and *Z*-axis output

and 93% fill fractions. The resonance responses for in *X*-axis and *Y*-axis are mildly reduced, split, and shifted for *X*-axis excitation. The $\|X/X\|$ FRFs reduced from an amplification of 164 at 25% fill fraction to 130 at 93%. The cross-axis FRFs ||*Y*/*X*|| and ||*Z*/*X*|| show a marginal improvement in the damping at 93% fill fraction. The un-damped to the damped reduction in amplitudes is from 14 to 8.8 and from 9.9 to 7.5 for the ||*Y*/*X*|| and ||*Z*/*X*|| transfer function, respectively. The resonance peak amplifications are seen to increase as the fill fraction is increased from 25 to 50%. The likely reason for this could be the in-phase momentum transfer between the top platform and DPs.

3.3 Effect of Fill Fraction on FRFs with θ_X Input Excitation

Figures 9, 10, and 11 show the transfer functions between the input angular displacement θ_X applied to the base platform and the angular displacement responses θ_X, θ_Y, and θ_Z, respectively, at the top platform. Unlike the translational cases where the input to the cross translational and rotational directions was set zero, in this case, the Z-axis translation of the base plate was kept free to allow the rotation about the X-axis. Again, the transfer functions for the four fill fractions of 25, 50, 75, and 93%

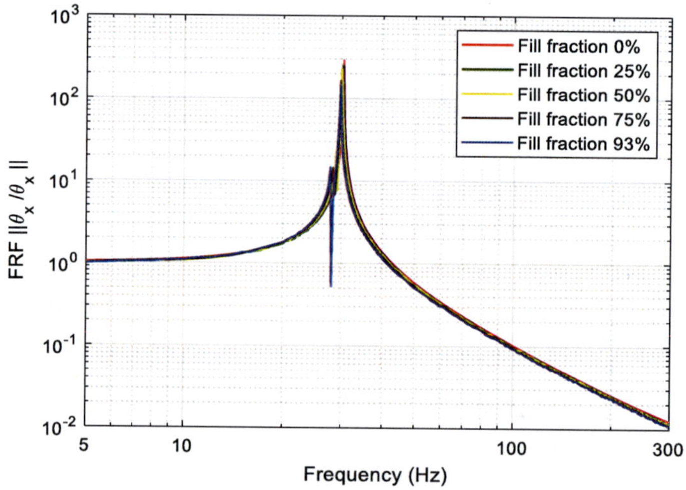

Fig. 9 Transfer function between θ_X input and θ_X output

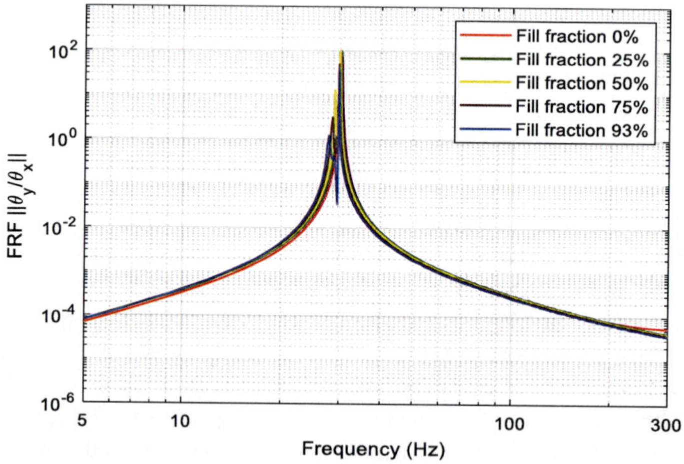

Fig. 10 Transfer function between θ_x input and θ_Y output

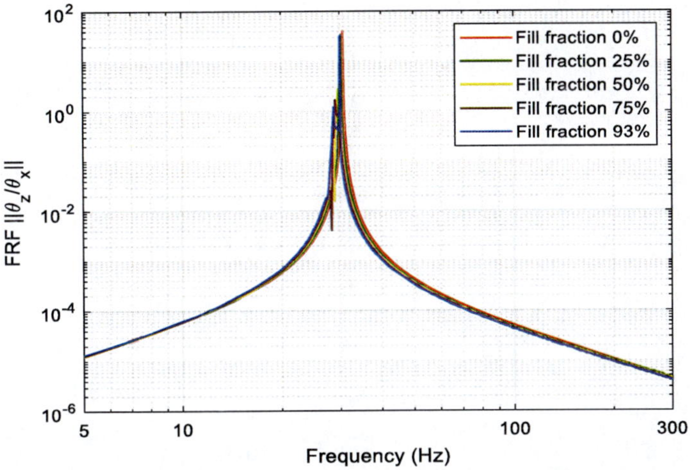

Fig. 11 Transfer function between θ_x input and θ_z output

were considered. As in the previous results, splitting of mode with a mild reduction in magnitude was observed. The reduction in responses θ_X and θ_Y is more in comparison with the θ_Z. The FRFs $\|\theta_X/\theta_X\|$ and $\|\theta_Y/\theta_X\|$ come down from 287 and 73 to 150 and 42, respectively, while the reduction in $\|\theta_Z/\theta_X\|$ is from 40.6 to 32. The reason for the 1.9 times and 1.7 times reduction in $\|\theta_X/\theta_X\|$ and $\|\theta_Y/\theta_X\|$, respectively, is due to the movement of particles at the edge of the top platform due to rotation of the platform about θ_X and θ_Y axis which gives motion in Z-axis. The θ_Y input behavior is expected to be similar to θ_X.

3.4 *Effect of Fill Fraction on FRFs with θ_Z Input Excitation*

The transfer functions between the torsional input θ_Z and responses θ_X, θ_Y, and θ_Z are shown in Figs. 12, 13, and 14, respectively. As expected, there is a significant reduction in resonance amplitude in cross-axis FRFs while it is negligible in the θ_Z direction. The FRF $\|\theta_X/\theta_Z\|$ reduces from 201 to 109, while $\|\theta_Y/\theta_Z\|$ has come down to 140 from 251. The reason for the reduction is that the rotation of the top plate about θ_X and θ_Y axis involves significant motion of the cell at the outer boundary in Z-direction. Since the particle can move in Z-directions, collisions would lead to increased dissipation. In the torsional rotation of the top deck, there is not much scope for the movement of damping particles, and hence, the impact is negligible. The θ_X and θ_Y rotations result in motion in Z-direction at the location away from the axis of rotations, and because of that the DPs move in the cells colliding and rubbing resulting in the improvement of the damping behavior and thus the reduction in the cross-axis resonance amplitudes.

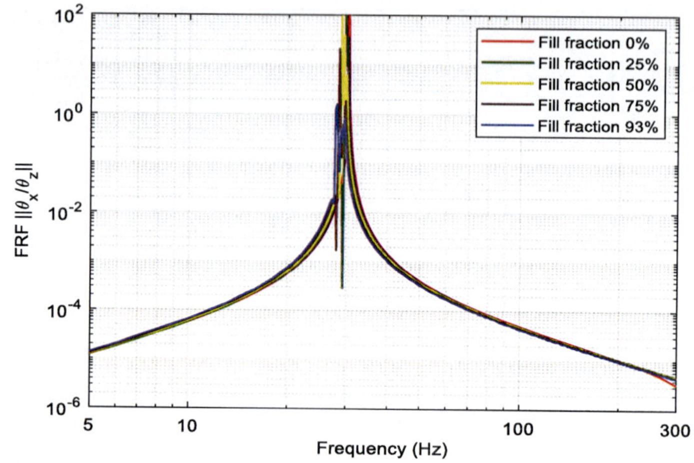

Fig. 12 Transfer function between θ_Z input and θ_X output

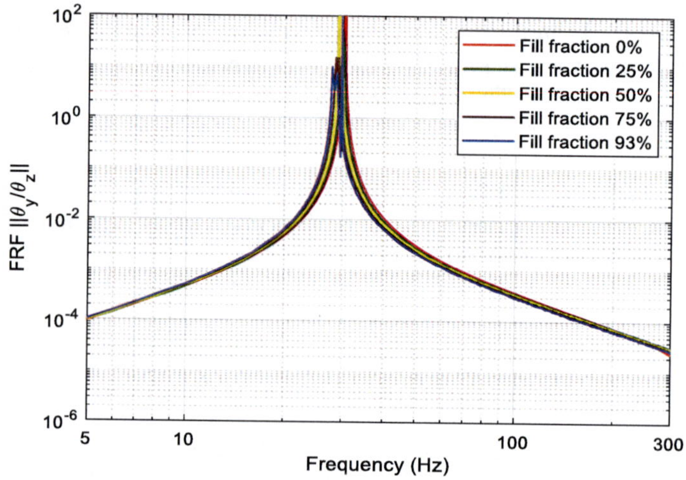

Fig. 13 Transfer function between θ_Z input and θ_Y output

4 Conclusions

The transfer functions between the base excitation and mass center of the payload mounted on the top platform of the MGSP, using the couple DEM-FEM equations of motion of the MGSP, were computed. The cells of the honeycomb core of the top platform were filled with damping particles. The transfer functions FRFs with respect to four inputs, swept sine inputs X, Z, θ_x, and θ_Z applied separately at the base of the platform, were computed for 25, 50, 75, and 93% fill fractions. The behavior with respect to inputs Y and θ_Y is nearly similar to X and θ_x. The peaks at resonances

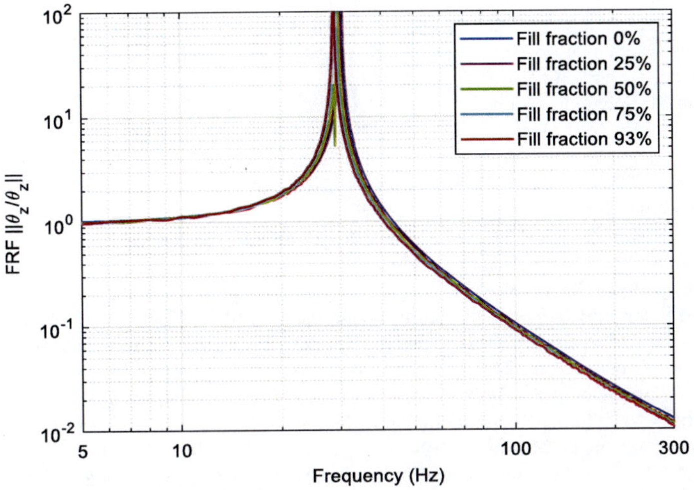

Fig. 14 Transfer function between θ_Z input and θ_Z output

progressively decreased as the fill fraction was increased. For all the modes, it was seen that the damping introduced by damping particles results in the splitting of the modes and the formation of anti-resonance at the frequency of un-damped resonance. As the MGSP without damping particles has all the modes at the same frequency, the splitting at resonance happens due to the presence of close by modes resulting from the addition of mass of the damping particle in the system. The splitting of modes and formation of anti-resonance are advantageous as it reduces the amplification. The damping introduced by the particles is more effective in longitudinal direction $\|Z/Z\|$ where there is a reduction from 25 to 8 with increasing fill fractions from 25 to 93%. The cross-axis transfer functions also come down from 98 and 10 to 75 and 5 along the X-axis and Y-axis, respectively. The reduction in most of the cross-axis transfer functions was seen. The ratio of the damped and un-damped magnitude of the FRFs for θ_X input is [1.9, 1.7, 1.2], and θ_Z input is [1.8, 1.7, 1.03] for θ_X, θ_Y, and θ_Z axis. However, it is not effective in limiting the resonance at torsional mode $||\theta_Z/\theta_Z||$. The magnitude of the FRF $||\theta_Z/\theta_Z||$ at resonance with damping particle and without damping particle is 268.5 and 298.5, respectively.

References

1. Merlet JP (2012) Parallel robots, 2nd edn. Springer Netherlands, Netherlands
2. Mukherjee P, Dasgupta B, Mallik AK (2007) Dynamic stability index and vibration analysis of a flexible Stewart platform. J Sound Vib 307(3–5):495–512. https://doi.org/10.1016/j.jsv.2007.05.036

3. Shyam RBA, Ahmad N, Ranganath R, Ghosal A (2019) Design of a dynamically isotropic Stewart-Gough platform for passive micro-vibration isolation in spacecraft using optimization. J Spacecr Technol 30(2):1–8
4. Furqan M, Suhaib M, Ahmad N (2018) Dynamic analysis of six-axis Stewart platform using flexible joints. Int J Mech Rob Syst 4(3):214–233. https://doi.org/10.1504/ijmrs.2018.095965
5. Klein CA, Miklos TA (1991) Spatial robotic isotropy. Int J Rob Res 10(4):426–437
6. Zanganeh KE, Angeles J (1997) Kinematic isotropy and the optimum design of parallel manipulators. Int J Rob Res 16(2):185–197
7. Tsai K, Huang K (2003) The design of isotropic 6-DOF parallel manipulators using isotropy generators. Mech Mach Theory 38(11):1199–1214
8. Bandyopadhyay S, Ghosal A (2009) An algebraic formulation of static isotropy and design of statically isotropic 6–6 Stewart platform manipulators. Mech Mach Theory 44(7):1360–1370
9. Afzali-Far B, Lidström P (2018) Analytical index of dynamic isotropy and its application to hexapods. Precis Eng 52:242–248. https://doi.org/10.1016/j.precisioneng.2018.01.001
10. Tong Z, He J, Jiang H, Duan G (2011) Locally dynamic isotropy of modified symmetric Gough-Stewart parallel micromanipulators. In: 13th World congress in mechanism and machine science. Guanajuato, México, pp 19–25
11. Ahmad N (2021) Vibration mitigation in spacecraft components using Stewart platform and particle impact damping. PhD Thesis, Department of Mechanical Engineering, Indian Institute of Science, Bangalore, https://etd.iisc.ac.in/handle/2005/4856
12. Abaqus-6.13, Abaqus/CAE user's manual. Available: www.simulia.com
13. Ahmad N, Ranganath R, Ghosal A (2019) Modeling of the coupled dynamics of damping particles filled in the cells of a honeycomb sandwich plate and experimental validation. J Vib Control 25(11):1706–1719. https://doi.org/10.1177/1077546319837584
14. Ahmad N, Ranganath R, Ghosal A (2017) Modeling and experimental study of a honeycomb beam filled with damping particles. J Sound Vib 391:20–34. https://doi.org/10.1016/j.jsv.2016.11.011
15. Wijker J (2004) Mechanical vibrations in spacecraft design. Springer-Verlag, Berlin Heidelberg, New York

Nonlinear Distribution of the Gearbox Dynamic Model Including Tooth Cracks

Vikash Kumar, Subrata Mukherjee, and Somnath Sarangi

Abstract Gearboxes are a widely used component in the industry due to their motion and power transmission flexibilities in compact structures. Lots of work have been already presented regarding the dynamic modeling of the gearbox including defects. But, very few are reported on the nonlinear analysis of the simulated signal. This paper presents the nonlinear analysis of the spur gear dynamic model at different tooth's root crack depths to understand the chaotic complexity present in the signal and its variation with the crack depth. A dynamic model with 8 degrees of freedom (DoF) is developed by incorporating time-varying mesh stiffness (TVMS) model at different crack depths. In addition, 10 dB white Gaussian noise is added to make it the real-world signal. Two widely used data preprocessing techniques such as time-synchronous averaging (TSA) and variable mode decomposition (VMD) are also applied to minimize the effect of noise and preserve the characteristic component in the signal. After that, the chaotic signal complexity is measured by calculating the correlation dimension (CD) on the obtained preprocessed simulated signal. The obtained results show a significant understanding of the chaotic behavior present in the simulated signal at different crack depths in terms of the CD values.

Keywords Dynamic modeling · TVMS · TSA · VMD · Correlation dimension · Chaotic complexity

V. Kumar · S. Mukherjee (✉) · S. Sarangi
Department of Mechanical Engineering, Indian Institute of Technology Patna, Patna, India
e-mail: subrata.re.me@gmail.com

V. Kumar
e-mail: 1821me18@iitp.ac.in

S. Sarangi
e-mail: somsara@iitp.ac.in

V. K. Gupta et al. (eds.), *Recent Advances in Machines and Mechanisms*, Lecture Notes in Mechanical Engineering, https://doi.org/10.1007/978-981-19-3716-3_20

1 Introduction

Dynamic modeling is an effective approach for determining the dynamic properties of gear systems, and numerous studies on the subject have been conducted [1–6]. Researchers either extend the DoF from single to many degrees or account for different practical phenomena like tooth friction, backlash, manufacturing eccentric errors, and so on to efficiently get the dynamic properties of gear systems. Theoretical understanding of the dynamic response obtained from the gearbox dynamic model, specifically the shafts containing the gears, is beneficial for fault diagnosis of the gearbox [1–6]. Different SNR values are incorporated in the model to make the model more realistic in terms of real-life machine applications. The data obtained from the real machine have a lot of noise and inferred signals of other machine components. That is why various data preprocessing techniques are used in literature to preprocess the data [7–12]. Mostly, the severity of faults in the system and their diagnosis is shown through the time domain analysis, frequency domain analysis, and time–frequency domain analysis. But, in recent times, the nonlinear analysis of the system is widely used not only in terms of measuring chaotic system complexity but also use as a feature for fault diagnosis problems. The nonlinear analysis offers much potential for diagnosing any industrial complex failure from vibration signal analysis [13–20]. In references [13–20], authors have studied the chaotic complexity of different systems through CD and used this as a nonlinear feature for fault diagnostic. They also concluded that data preprocessing techniques help in improving the effectiveness of CD in practical cases [13–20].

Therefore, this article presents an improved simulation analysis of calculating the CD of the simulated signal from the 8 DoF spur gear dynamic model at different tooth root crack depths. Appropriate noise of 10 dB is added to make the simulated signal real and denoised by the application of TSA and VMD data preprocessing techniques. After that, the chaotic complexity of the system is presented at different crack depths in terms of CD values, and an effective comparison is made between both preprocessed techniques. Figure 1 is illustrating the detailed flowchart of the proposed method.

The following is a summary of the rest of the paper: Sect. 2 discusses the theoretical background about the dynamic model, TVMS model, TSA, VMD, and CD; Sect. 3 describes the numerical simulation of the proposed model; Sect. 4 explains the chaotic signal complexity of the simulated signal by calculating the CD, and Sect. 5 gives concluding remarks about the paper.

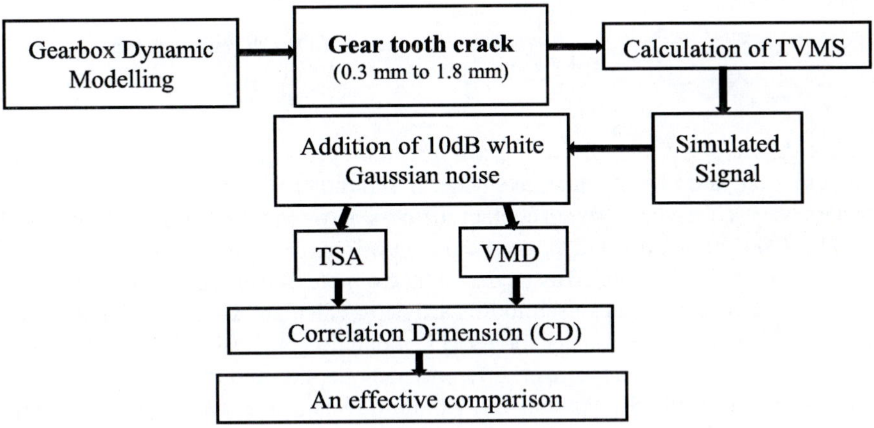

Fig. 1 Proposed methodology for the nonlinear distribution of the gearbox

2 Theoretical Background

2.1 Dynamic Modeling of the Gearbox

The status of tooth deterioration in the gearbox can be evaluated primarily for gear fault detection purposes by introducing a reduction in the TVMS incorporated with the '*N*' DoF dynamic model of the gearbox. This stiffness reduction in dynamic modeling has been used in several research studies to detect faults [1–6]. The TVMS calculation model and tooth pair engagement during meshing are represented in Figs. 2a and b, respectively. The total mesh stiffness $K(t)$ maybe written as:

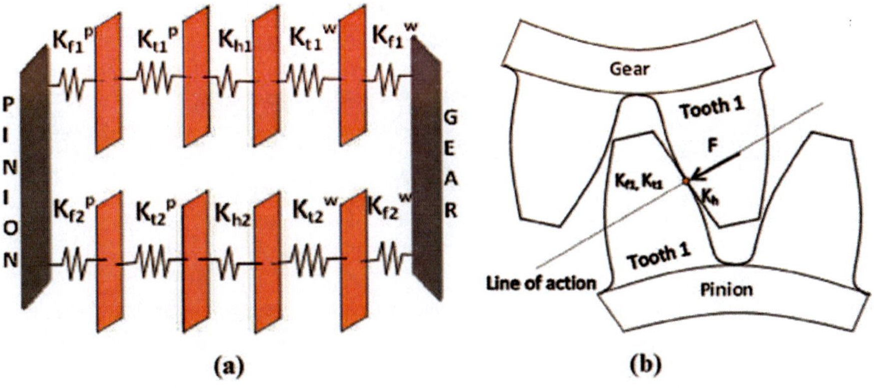

Fig. 2 **a** TVMS calculation model and **b** tooth engagement between gear pair

$$K(t) = \sum_{i=1}^{n} \frac{1}{\frac{1}{K_i^p} + \frac{1}{K_i^w} + \frac{1}{K_{\mathrm{hi}}}}, \quad n = \text{no. of tooth pair in mesh} \tag{1}$$

where K_i^p and K_i^w are the tooth stiffness (comprising of bending, shear, axial compression, and fillet foundation stiffness) of pinion and gear of ith tooth in mesh, respectively. K_{hi} is the Hertzian contact stiffness between the mating gear pair [1–6].

The $K(t)$ for healthy and cracked tooth cases (crack depth 0.3 mm to 1.8 mm at 0.3 mm interval) is calculated using the method of potential energy. All equations and other mathematical studies used in this article to calculate TVMS have been taken from [1, 2]. The gear parameter for TVMS calculation is given in Table 1. Figure 3a shows the TVMS of gear pairs at different crack depths, and it is varying inversely with the crack depth. To generate the vibration response, a dynamic model with 8 DoF [3], including motor and load, is developed. To describe the rotation, each gear disk has 3 DoF (1—rotation, and 2—translation), and each motor disk and load disk has 1 DoF [3]. Figure 3b illustrates the schematic representation of the 8 DoF model, and equations of motion (EQMs) of the model may be given as follows [3, 5]:

The EQMs for the pinion and gear in the 'x' axis are as follows:

$$m_p \ddot{x}_p = -K_{xp} x_p - C_{xp} \dot{x}_p + F_p \tag{2}$$

$$m_g \ddot{x}_g = -K_{xg} x_g - C_{xg} \dot{x}_g + F_g \tag{3}$$

Table 1 Dynamic model input parameters [2]

Parameters	Value	Parameters	Value
Teeth on gear (Z_g) Teeth on pinion (Z_p)	25 30	Mass of gear (m_g) Mass of pinion (m_p)	0.3083 kg 0.4439 kg
Module (m) Tooth width (L)	2 mm 20 mm	Moment of inertia of pinion (I_p) Moment of inertia of gear (I_g)	2×10^{-4} Kgm^2 9.634×10^{-5} Kgm^2
Pressure angle Crack inclination angle	20° 70°	Moment of inertia of motor (I_m) Moment of inertia of load (I_b)	0.023976 Kgm^2 0.01015 Kgm^2
Constant damping ratio (ζ) Poisson's ratio	0.07 0.3	Applied torque on motor (T_m) Load torque on output shaft (T_b)	28.13 Nm 10 Nm
Young's modulus of elasticity	200 GPa	Radial stiffness of bearing (K_x) Radial damping of bearing (C_x)	6.56×10^8 N/m 1.8×10^3 Ns/m
Frequency of pinion (f_p) Frequency of gear (f_g)	33.33 Hz 40 Hz	Viscous friction coefficient (B_v) Gear mesh frequency (GMF)	0.026575 $f_p \times Z_p = 1000$ Hz

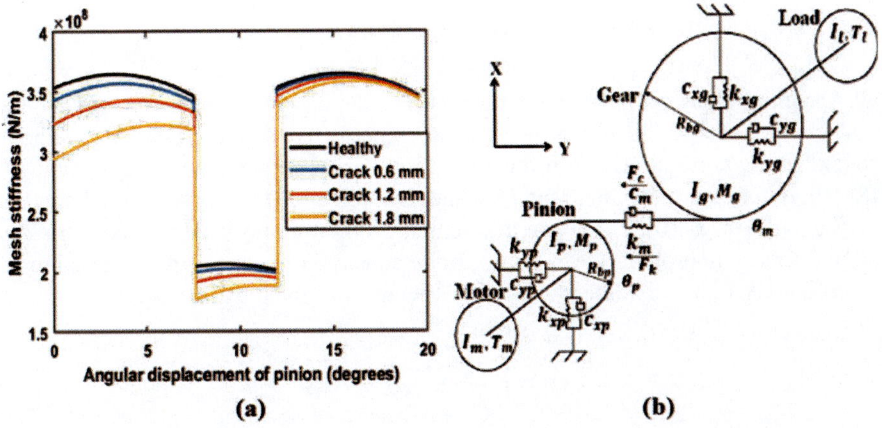

Fig. 3 **a** TVMS plot at varying crack depth and **b** dynamic model with 8 DoF [3]

The EQMs for the pinion and gear in the 'y' axis are

$$m_p\ddot{y}_p = -K_{yp}y_p - C_{yp}\dot{y}_p - N \tag{4}$$

$$m_g\ddot{y}_g = -K_{yg}y_g - C_{yg}\dot{y}_g + N \tag{5}$$

The EQMs for the pinion and gear along 'θ' direction are

$$I_p\ddot{\theta}_p = r_pN + M_p - k_t(\theta_p - \theta_m) - c_t(\dot{\theta}_p - \dot{\theta}_m) \tag{6}$$

$$I_g\ddot{\theta}_g = -r_gN + M_g - k_t(\theta_g - \theta_l) - c_t(\dot{\theta}_g - \dot{\theta}_l) \tag{7}$$

The EQMs for the motor and load along 'θ' direction are

$$I_m\ddot{\theta}_m = -k_t(\theta_m - \theta_p) - c_t(\dot{\theta}_m - \dot{\theta}_p) - B_v\dot{\theta}_m + T_m \tag{8}$$

$$I_l\ddot{\theta}_l = -k_t(\theta_l - \theta_g) - c_t(\dot{\theta}_l - \dot{\theta}_g) - B_v\dot{\theta}_l - T_l \tag{9}$$

where $N = K(t)[(y_p - y_g) - (r_p\theta_p - r_g\theta_g)] + C(t)[(\dot{y}_p - \dot{y}_g) - (r_p\dot{\theta}_p - r_g\dot{\theta}_g)]$, $K(t)$ is the total mesh stiffness evaluated by Eq. (1), and $C(t)$ is the mesh damping coefficient evaluated by $C(t) = 2\zeta\sqrt{K(t)m_pm_g/(m_p + m_g)}$. Figure 3a is clearly showing the sudden jump and fall in the TVMS which itself is a kind of nonlinearity adding toward the stiffness. Though the entire set of EQMs is non-autonomous. But, its effect is further visualized in the vibration signal as it is combined with the machine/state variables and directly coupled with dynamic EQMs (Eqs. 2–9).

2.2 Time Synchronous Averaging (TSA)

TSA, commonly implemented as a time domain averaging technique to extract periodic components, is one of the most potent and effective signal processing methods for extracting periodic signals from a composite signal applied to rotating machinery [8–10]. A detailed study about the TSA and its application on gearbox fault diagnosis can be found in [8–10]. The averaging technique is commonly used to make vibration signals time synchronous (revolution). If we consider the raw vibration signal to be a continuous signal, the linear averaging operation is given by [8–10]:

$$y(t) = \frac{1}{N}\sum_{n=0}^{N-1} y(t - nP) \tag{10}$$

where P is the rotational period and N is the number of averages.

2.3 Variable Mode Decomposition (VMD)

VMD is a non-recursive method for concurrently extracting mode names variable mode functions (VMFs) [11, 12]. In [11], the authors presented a detailed mathematical analysis of the VMD algorithm. A brief mathematical theory [12] of the VMD method's decomposition procedure was described as follows:

Step 1: Initialize $\{\hat{u}_k^1\}, \{\hat{\omega}_k^1\}, \hat{\gamma}^n, n \leftarrow 0$.

Step 2: The value of u_k, ω_k and γ is updated according to the following mathematical formula:

$$\hat{u}_k^{n+1} \leftarrow \frac{\hat{f}(\omega) - \sum_{i<k} \hat{u}_i^{n+1}(\omega) - \sum_{i>k} \hat{u}_i^{n+1}(\omega) + \frac{\hat{\gamma}^n(\omega)}{2}}{1 + 2\alpha\left(\omega - \omega_k^n\right)^2} \tag{11}$$

$$\omega_k^{n+1} \leftarrow \frac{\int_0^{+\infty} \omega\left|\hat{u}_i^{n+1}(\omega)\right|^2 \mathrm{d}\omega}{\int_0^{+\infty} \left|\hat{u}_i^{n+1}(\omega)\right|^2 \mathrm{d}\omega} \tag{12}$$

$$\hat{\gamma}^{n+1}(\omega) \leftarrow \hat{\gamma}^n(\omega) + \tau[\hat{f}(\omega) - \sum_K \hat{u}_i^{n+1}(\omega) \tag{13}$$

Step 3. Repeat step 2 until the function satisfies the condition, $\sum_K \frac{\|\hat{u}_k^{n+1} - \hat{u}_k^n\|_2^2}{\|\hat{u}_k^n\|_2^2} < \epsilon$, where ϵ is the specified precision claim.

2.4 Correlation Dimension (CD)

CD is a tool to quantify the self-similarity of signals [13–20]. Higher CD means more complexity and less similarity. Logan and Matthew [13] described the previous use of correlation measurement in diagnosing bearing failures. The widely used procedure for estimating the CD was developed by Grassberger and Procaccia [14]. The value of the CD is directly proportional to the amount of chaos in the system; therefore, the higher the CD, the higher the chaotic complexity of the system. To calculate CD, the reconstruction matrix proposed in [13–20] is used as the input of the CD algorithm. The detailed steps for calculating the relevant dimensions in [13–20] are as follows:

Step 1: Firstly, a delayed reconstruction $X_{1:N}$ is generated according to embedded dimension and lag in the signal [13–20].

Step 2: The number of points within in range (at the point j) was calculated by Eq. 14,

$$N_j(r) = \sum_{j=1, j \neq k}^{N} 1\left(\left\|X_j - X_k\right\| < r\right) \tag{14}$$

where r is the radius of similarity, 1 is the indicator function, and N is the number of points between the radius range.

Step 3: CD was then estimated as the slope of $C(r)$ versus r, where $C(r)$ is the correlation integral and is given as follows:

$$C(r) = \frac{2}{N(N-1)} \sum_{i=1}^{N} N_i(r) \tag{15}$$

3 Numerical Simulation of Proposed Model

For the healthy case, a MATLAB computer simulation based on ODE45 with a sampling rate of 100 kHz was carried out for 8 s, and then, the simulation for the faulty case (crack depth 0.3 mm to 1.8 mm at 0.3 mm interval) was repeated. The gearbox's vibration signal obtained from the gear dynamic model fundamentally has two parts shown in Fig. 4a, i.e., transition state and the steady state. Figure 4a observed that the vibration signal is coming to its steady-state condition after 3 s. This is mostly because of transient responses and encountered after a certain time or revolution. In this work, the steady-state signal is taken into consideration for further analysis because this signal provides a stable frequency for pinion, which contributes significantly to finding the presence of different characteristic frequencies in the system's response.

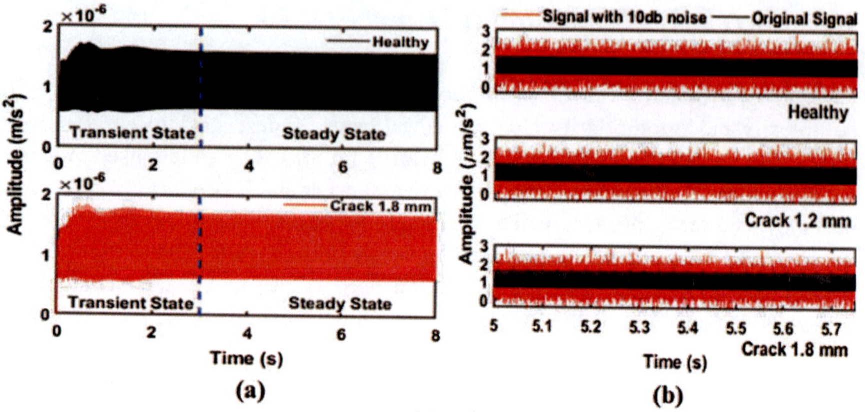

Fig. 4 Dynamic response of the gear dynamic model **a** transient and steady-state vibration response **b** steady-state dynamic response at various crack depths

Further, 10 dB white Gaussian noise [21, 22] is added to the simulated signal to simulate the dynamic response in real-life environments. It is visible from Fig. 4b that the signal without noise has an evident impulse at the cracked tooth position, and its amplitude increases as the crack depth increase. But, with the addition of 10 dB SNR in the simulated signal, these impulses indulge with the noise and affect the quality of faulty signature inside the signal. To overcome this challenge, data prepreprocessing techniques are widely used.

In this work, TSA and VMD both signal processing methods are applied to the noisy simulated signal. Each algorithm provides insightful information on the gearbox vibration, and both algorithms work well with noisy datasets.

Figure 5a shows the TSA of the steady-state simulated noisy signal where the blue color plot was the averaged signal of the black color plot for one cycle of shaft

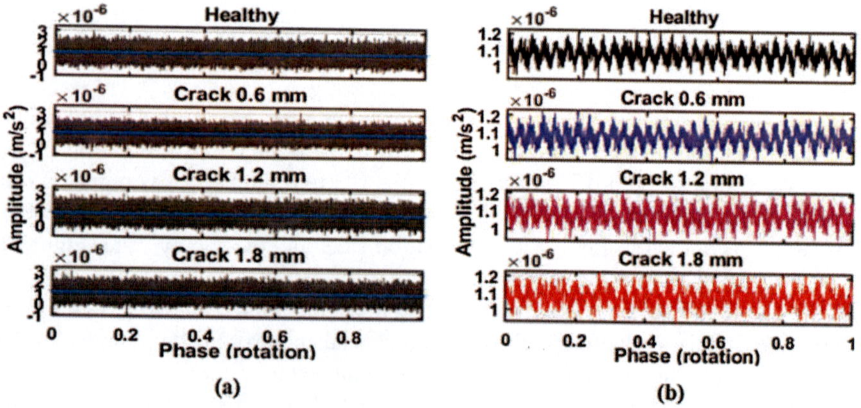

Fig. 5 **a** TSA of the noisy response and **b** zoomed view of the averaged signal

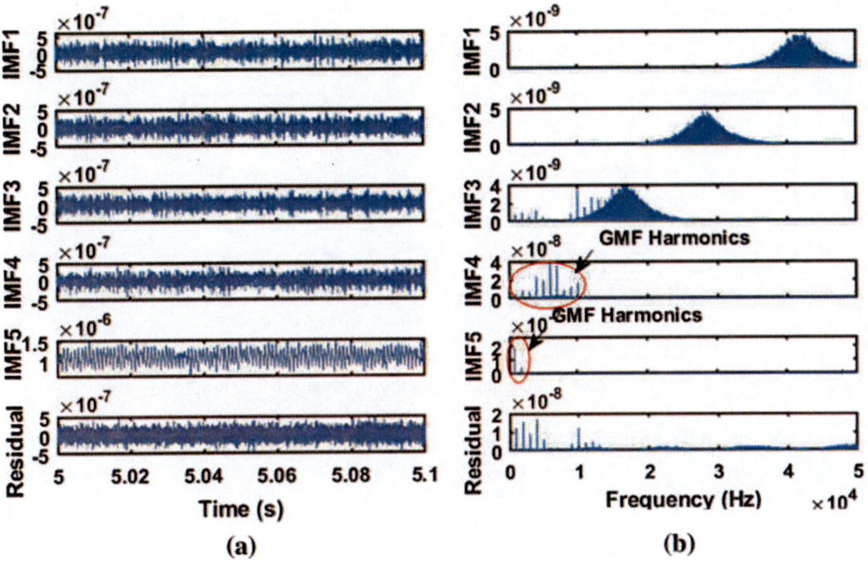

Fig. 6 VMD for the noisy dynamic signal at crack depth 1.8 **a** IMFs and **b** FFT of IMFs

rotation, and its zoom view was presented in Fig. 5b. TSA averages the signal for one cycle of rotation of pinion shafts (time-period 0.03 s) over the entire length of the signal (5 s) to minimize the noise present in the signal. Figure 6 shows the VMD of the noisy steady-state signal with the cracked tooth of 1.8 mm. Figure 6a shows the decomposition of the noisy steady-state signal into IMFS and residual. VMD helps to separate the frequency component in each IMF in decreasing order like IMF1 contains the high-frequency content present in the input signal. Figure 6b demonstrated that the IMF 4 and IMF 5 contain the essential characteristics frequencies, i.e., GMF frequency and harmonics (see Table 1 for value). But, in other IMFs apart from this, different frequencies range is also present. Therefore, for better interpretation, only, IMF 4 and IMF 5 were taken into consideration for further analysis.

4 Nonlinear Distribution of Gear Tooth Crack

Figures 7 and 8 show graphical representations of the CD (for TSA and VMD signals), which makes it possible to extract more sensitive nonlinear characteristics from the dynamic simulation of transmission. The characteristic curve can also reflect the dynamic characteristics of a complex transmission. The calculation results of the correlation measure are arranged in rows and columns, as shown in Table 2. The CD value is small in a healthy state, and the CD value is slightly more in the faulty state. In some cases, the CD of fractals used to analyze nonlinear and transient gear dynamics shows good adaptability. The correlation dimension is used to compare the dynamic

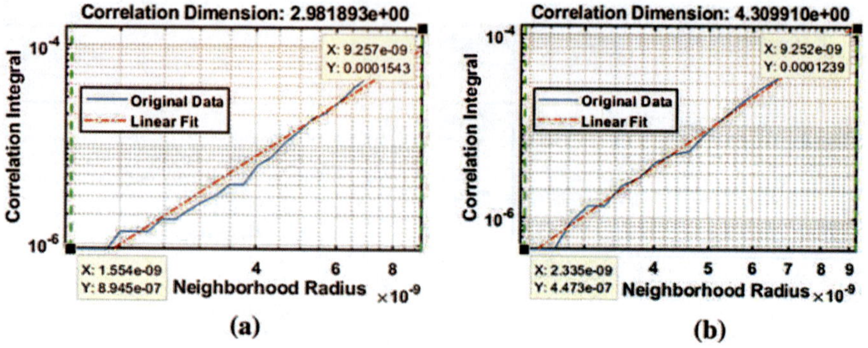

Fig. 7 Correlation dimension of TSA signal **a** healthy tooth and **b** 1.8-mm tooth crack

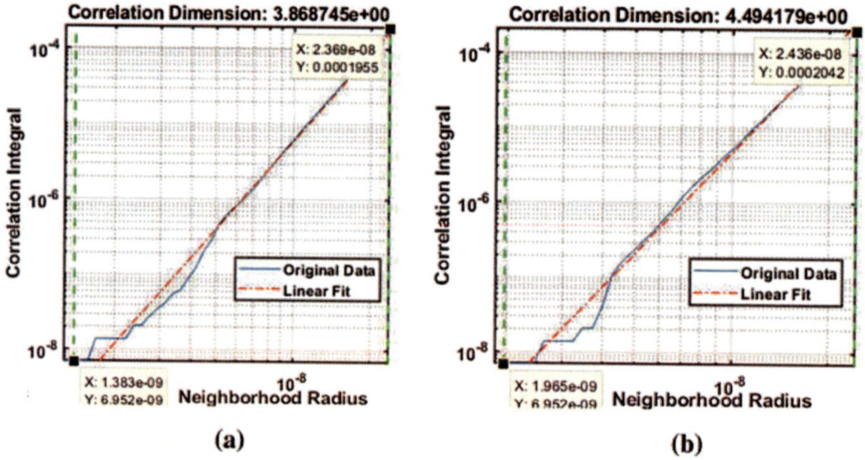

Fig. 8 Correlation dimension of VMD-IMF 4 **a** healthy tooth and **b** 1.8-mm tooth crack

Table 2 Correlation dimension values for healthy tooth and faulty tooth

Crack depth	TSA signal	VMD signal	
		IMF 4	IMF 5
0 (Healthy)	2.952407	3.777088	3.760985
0.3 mm	3.450717	4.115672	3.923388
0.6 mm	3.584351	4.228601	3.810718
0.9 mm	3.663853	4.401500	3.833007
1.2 mm	3.782944	3.785298	3.914113
1.5 mm	4.334656	4.489639	3.734141
1.8 mm	4.237229	4.417905	3.783505

characteristics of the vibration system damage between the structural parameters. For the TSA signal, the CD of a healthy tooth is less than the cracked tooth and almost increasing behavior with increment in the crack depth. This phenomenon shows that the cracked gear tooth has more chaotic signal complexity as compared to the healthy tooth.

The nonlinearity in the system increases as the crack depth increases at the tooth root of the pinion. For the VMD signal, the CD of IMF 4 and IMF 5 is calculated where IMF 4 almost follows a similar trend like TSA except for 1.2-mm crack depth, while IMF 5 shows mixed behavior of nonlinearity present in the system. The small deviation in the trending nature of CD value may occur due to the dependency of CD value on the embedding dimension and the number of points required to obtain a suitable resolution for the neighborhood radius. A detailed analysis of this is a matter of optimizing these parameters, and the same will be reported elsewhere in the separate work.

5 Conclusion

This paper presents an improved study of gearbox fault analysis based on signal complexity analysis. In this work, the nonlinear distribution of the 8 DoF spur gear dynamic model with a crack at the tooth's root is developed and discussed by calculating the CD of the simulated signal. Appropriate noise of 10 dB is added to the simulated signal to experience the real-life environment. From a nonlinear analysis point of view, it is quite difficult to distinguish between random noise and deterministic chaos present in the signal. That is why two data preprocessing techniques (TSA and VMD) are applied to reduce the effect of noise from the signal. After that, the value of CD is calculated for both preprocessed data. The obtained results clearly show how CD value changes with the health condition of the gearbox. For both TSA and VMD techniques, the CD value for a healthy state is low and has less chaotic complexity than the faulty state. Also, in a comparative point of view between TSA and VMD, the TSA shows an approximately increasing pattern with the increment in crack depth than VMD. This work improved the understanding of signal complexity analysis for gearbox fault analysis, but some small deviations in the CD values are still there. Although this work can be treated as an improved version of the Refs. [16–22]. For further study, the limitation of this work will be improved by modifying the TVMS calculation of the gear tooth as well as dynamic study.

References

1. Chaari F, Fakhfakh T, Haddar M (2009) Analytical modeling of spur gear tooth crack and influence on gearmesh stiffness. Eur J Mech A Solids 28(3):461–468
2. Chen Z, Shao Y (2011) Dynamic simulation of spur gear with tooth root crack propagating along tooth width and crack depth. Eng Fail Anal 18(8):2149–2164
3. Mohammed O, Rantatalo M, Aidanpää J (2015) Dynamic modelling of a one-stage spur gear system and vibration-based tooth crack detection analysis. Mech Syst Signal Process 54–55:293–305
4. Mohammed O, Rantatalo M (2016) Dynamic response and time-frequency analysis for gear tooth crack detection. Mech Syst Signal Process 66–67:612–624
5. Kumar V, Rai A, Mukherjee S, Sarangi S (2021) A lagrangian approach for the electromechanical model of single-stage spur gear with tooth root cracks. Eng Fail Anal 129:105662
6. Liang X, Zuo M, Feng Z (2018) Dynamic modelling of gearbox faults: a review. Mech Syst Signal Process 98:852–876
7. Rai A, Upadhyay S (2016) A review on signal processing techniques utilized in the fault diagnosis of rolling element bearings. Tribol Int 96:289–306
8. Mukherjee S, Kaushal R, Kumar V, Sarangi S (2020) A novel approach of gearbox fault diagnosis by using time synchronous averaging and J48 algorithm. Lecture notes in mechanical engineering, pp 927–935
9. Mukherjee S, Kumar V, Sarangi S, Bera T (2020) Gearbox fault diagnosis using advanced computational intelligence. Proc Comput Sci 167:1594–1603
10. Bechhoefer E, Kingsley M (2021) A review of time synchronous average algorithms. In: Annual conference of the PHM Society, vol 1(1). Retrieved from https://papers.phmsociety.org/index.php/phmconf/article/view/1666
11. Dragomiretskiy K, Zosso D (2014) Variational mode decomposition. IEEE Trans Signal Process 62(3):531–544
12. Isham M, Leong M, Lim M, Ahmad Z (2018) Variational mode decomposition: mode determination method for rotating machinery diagnosis. J Vibroeng 20(7):2604–2621
13. Logan D, Mathew J (1996) Using the correlation dimension for vibration fault diagnosis of rolling element bearings—I. Basic concepts. Mech Syst Signal Process 10(3):241–250
14. Grassberger P, Procaccia I (1983) Estimation of the Kolmogorov entropy from a chaotic signal. Phys Rev A 28(4):2591–2593
15. Liu M, Li B, Zhang J, Wang K (2019) An application of ensemble empirical mode decomposition and correlation dimension for the HV circuit breaker diagnosis. Automatika 60(1):105–112
16. Jiang J, Chen J, Qu L (1999) The application of correlation dimension in gearbox condition monitoring. J Sound Vib 223(4):529–541
17. Jing W, Jun D, Bin W (2011) The application of correlation dimensions of vibration signals in analysis of the technological state of diesel engines. In: 2011 third international conference on measuring technology and mechatronics automation
18. Zhao W, Huang D (2011) Using correlation dimension to analyze the signal of microscale crack effect induced self-excited torsional vibration on the experimental platform. In: 2011 third international conference on measuring technology and mechatronics automation
19. Caesarendra W, Kosasih B, Tieu K, Moodie C (2013) An application of nonlinear feature extraction—a case study for low speed slewing bearing condition monitoring and prognosis. In: 2013 IEEE/ASME international conference on advanced intelligent mechatronics

20. Theiler J (1987) Efficient algorithm for estimating the correlation dimension from a set of discrete points. Phys Rev A 36(9):4456–4462
21. Zhao M, Lin J, Miao Y, Xu X (2016) Detection and recovery of fault impulses via improved harmonic product spectrum and its application in defect size estimation of train bearings. Measurement 91:421–439
22. Zhang Y, Tang B, Liu Z, Chen R (2015) An adaptive demodulation approach for bearing fault detection based on adaptive wavelet filtering and spectral subtraction. Meas Sci Technol 27(2):025001

Classification of Worm Gearbox Fault Using Dendrogram Support Vector Machine

Surinder Kumar and Rajesh Kumar

Abstract Worm gearbox has wide range of applications such as in agitators, crushers, mixers, elevators, extruders, and cranes. The fault in gearbox leads to breakdown of the machinery. Early detection of fault in worm gearbox can prompt for economical preventive maintenance which ultimately prevents the breakdown and the production losses. Classification of the fault of worm gear is the first step of ensuring the gearbox protection. In this paper, a robust classification scheme based on autoregression minimum entropy deconvolution (AR-MED) and dendrogram support vector machine (DSVM) has been proposed to classify the faults of worm gearbox. AR-MED filter is used to remove the regular pattern of the gear and enhance the periodic impulsiveness in the signal. Features have been extracted from the filtered signal and used as input in DSVM model for multiclass classification of the fault conditions of worm gearbox. Results reveal that effectiveness of classification of DSVM has been enhanced using AR-MED filter with 100% decoding accuracy.

Keywords Worm gearbox · Vibration signal · Classification · Dendrogram support vector machine · Fault features

1 Introduction

Failure of machine components severely affects the performance, product quality, and risk of the health to operators. Condition monitoring keeps the track of the health of the machineries and accordingly recommends the predictive maintenance to reduce the chance of complete breakdown of the machines. Many condition monitoring techniques such as vibration analysis, acoustic emission analysis, wear debris analysis, and temperature analysis have been developed for fault detection in the gearboxes [1–6]. In general, vibration analysis technique is most popular and effective among the available techniques. Vibration signals carry the signature of the fault, which

S. Kumar · R. Kumar (✉)
Precision Metrology Laboratory, Mechanical Engineering Department, Sant Longowal Institute of Engineering and Technology, Longowal 148106, India
e-mail: rajesh_krs@sliet.ac.in

V. K. Gupta et al. (eds.), *Recent Advances in Machines and Mechanisms*, Lecture Notes in Mechanical Engineering, https://doi.org/10.1007/978-981-19-3716-3_21

can spell out the information about the type of defect and its severity [7, 8]. The vibration analysis has been carried out mainly in three different domains viz: time domain, frequency domain, and time–frequency domain [9–13]. While studying on gear defect, researchers have focused mainly on the spur and helical gearboxes for vibration analysis. Worm and wheel gearbox has advantage of high reduction ratio at single stage over other gearboxes. But only a few literatures are available on its defect analysis through vibration. Rare application of artificial intelligence approaches has been found on defect identification on worm gearboxes. Ümütlü et al., have applied artificial neural network (ANN) to identify the severity of tooth pitting in worm gearbox [14]. Kumar and Kumar have contributed to this field and diagnosed the fault in worm gearbox at its early stage using minimum entropy deconvolution and local cepstrum [15].

Authors have put their effort through this paper to advance the field of condition monitoring and fault diagnosis of worm gearbox. Various seeded defects under different operating conditions have been identified using dendrogram support vector machine (DSVM).

2 Methodology

A raw vibration signal of worm gearbox is preprocessed using AR-MED filter, and then, statistical features have been extracted from filtered signals. Suitable features have been sorted using T-test to avoid additional computational burden. Set of these sorted features has been used as input for DSVM classifier to classify the different fault conditions of the worm gearbox. Flow chart of the methodology is illustrated in Fig. 1.

2.1 *AR-MED Filter*

Autoregression (AR) filter segregates the deterministic pattern from the gear mesh signal. It gives the output as residual: which is difference of the original signal and AR predicted deterministic pattern. But the AR filter does not recognize the abrupt impulses aroused due to localized tooth fault in gear. To improve the effectiveness of the filter, autoregression (AR) and minimum entropy deconvolution (MED) are combined [10]. The output of AR is used as input of MED filter which optimizes the set of coefficients of the filter based on higher kurtosis value. Equations (1 and 2) represent the outcomes of the filter.

$$x_n = (d + w + e_0)*h \tag{1}$$

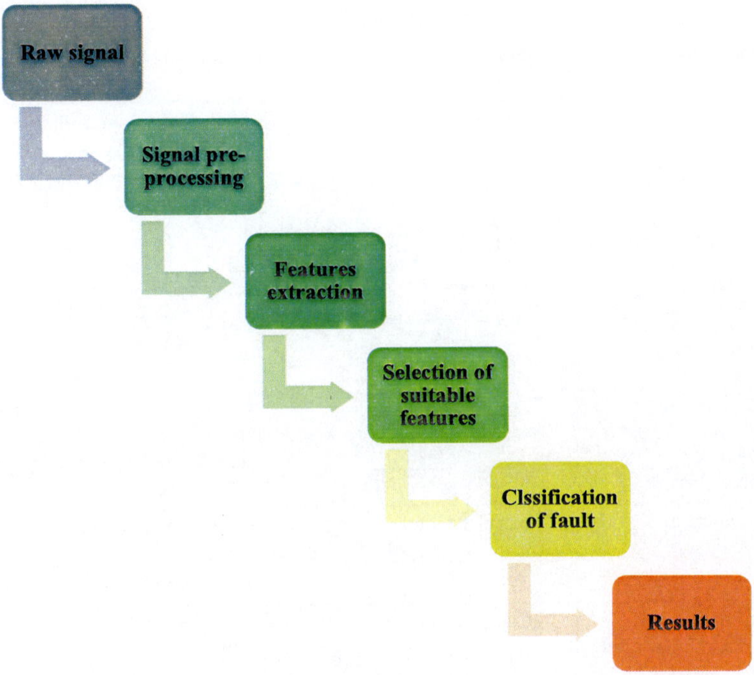

Fig. 1 Flow chart of the methodology

where x_n is the signal of the gearbox. d, w, and e_0 are deterministic pattern, fault signature, and noise, respectively. AR filter segregates the deterministic pattern of the gear and gives output as shown in Eq. (2).

$$e = (w + e_0) * h \tag{2}$$

e is used as input to the MED filter which enhances the periodic impulses in the input signal.

2.2 Statistical Features

Different features have been extracted from filtered signals. Features extracted are listed in Table 1.

Table 1 Various statistical features, their notations, and formulae

Features notation and formula	Features notation and formula	Features notation and formula
Mean $A_1 = \frac{1}{N}\sum_{i=1}^{N} x_i$	Kurtosis $A_{11} = \frac{1}{\sigma^4}\left(\frac{1}{N-1}\sum_{i=1}^{N}(x_i - \mu)^4\right)$	Median absolute deviation $A_{21} =$ median(|x_i − median(x_i)|)
Root mean square $A_2 = \sqrt{\frac{1}{N}\sum_{i=1}^{N} x_i^2}$	Crest factor $A_{12} = \frac{\max(\|x\|)}{\sqrt{\frac{1}{N}\sum_{i=1}^{N} x_i^2}}$	Rate zero crossing $A_{22} = \frac{\text{number of zero crossings}}{\text{total number of points}}$
Root $A_3 = \left(\frac{1}{N}\sum_{i=1}^{N}\sqrt{\|x_i\|}\right)^2$	Shape factor $A_{13} = \frac{\sqrt{\frac{1}{N}\sum_{i=1}^{N} x_i^2}}{\frac{1}{N}\sum_{i=1}^{N}\|x_i\|}$	Entropy $A_{23} = -\sum_{1=1}^{N} h(x_i)\log_2 h(x_i)$
Maximum value $A_4 = \max(\|x\|)$	Impulse factor $A_{14} = \frac{\max(\|x\|)}{\frac{1}{N}\sum_{i=1}^{N}\|x_i\|}$	Histogram upper bound $A_{24} = \max(x_i) + \frac{\max(x_i) - \min\left(\frac{x_i}{N-1}\right)}{2}$
Peak-to-peak $A_5 = \max(\|x\|) - \min(\|x\|)$	Clearance factor $A_{15} = \frac{\max(\|x\|)}{\left(\frac{1}{N}\sum_{i=1}^{N}\sqrt{\|x_i\|}\right)^2}$	Histogram lower bound $A_{25} = \max(x_i) - \frac{\max(x_i) - \min\left(\frac{x_i}{N-1}\right)}{2}$
Standard deviation $A_6 = \sqrt{\frac{1}{N-1}\sum_{i=1}^{N}(x_i - \mu)^2}$	Skewness factor $A_{16} = \frac{1}{\sigma^3}\frac{\left(\frac{1}{N-1}\sum_{i=1}^{N}(x_i-\mu)^3\right)}{\left(\sqrt{\frac{1}{N}\sum_{i=1}^{N} x_i^2}\right)^2}$	Activity $A_{26} = \frac{1}{N-1}\sum_{i=1}^{N}(x_i - \mu)^2$
Median $A_7 = \frac{50(N+1)}{100}$th observation	Kurtosis factor $A_{17} = \frac{1}{\sigma^4}\frac{\left(\frac{1}{N-1}\sum_{i=1}^{N}(x_i-\mu)^4\right)}{\left(\sqrt{\frac{1}{N}\sum_{i=1}^{N} x_i^2}\right)^4}$	Variance $A_{27} = \frac{\sum_{i=1}^{N}(x(i)-x_{\text{avg}})^2}{N-1}$
25th percentiles $A_8 = \frac{25(N+1)}{100}$th observation	Geometric mean $A_{18} = \left(\prod_{i=N}^{N} x_i\right)^{1/N}$	Wavelet energy decomposition $A_{28} = \sum_{n=1}^{N}\|x_i(n)\|^2 / \sum_{k=0}^{2^j-1}\sum_{n=1}^{N}\|x_k(n)\|^2$
75th percentiles $A_9 = \frac{75(N+1)}{100}$th observation	Root sum of squares $A_{19} = \sqrt{\sum_{i=1}^{N}\|x_i\|^2}$	Normal negative log likelihood for single Gaussian $A_{29} = -\sum_{i=1}^{N}\log\left[\frac{1}{\sigma\sqrt{2\pi}}\exp\frac{-(x_i-\mu)^2}{2\sigma^2}\right]$

(continued)

Table 1 (continued)

Features notation and formula	Features notation and formula	Features notation and formula
Skewness $A_{10} =$ $\frac{1}{\sigma^3}\left(\frac{1}{N-1}\sum_{i=1}^{N}(x_i-\mu)^3\right)$	Mean absolute deviation $A_{20} = \text{mean}(\lvert x_i - \mu \rvert)$	

2.3 *Selection of Suitable Features*

To reduce the computational burden and time in processing, suitable features have been sorted based on their score in T-test. A T-test is a statistical test that is used to compare the means of two groups. It is often used in hypothesis testing to determine whether a process or treatment actually has an effect on the population of interest, or whether two groups are different from one another. The T-test score is calculated using Eq. 3.

$$t\text{ score} = \frac{\lvert \overline{x}_1 - \overline{x}_2 \rvert}{\sqrt{\left(\frac{s_1^2}{n_1} + \frac{s_2^2}{n_2}\right)}} \tag{3}$$

where $\overline{x}_1, n_1,$ and s_1 are mean, number of observations and variance, respectively, for group 1. Similarly, $\overline{x}_2$, n_2, and s_2 are mean, number of observations, and variance of group 2, respectively.

2.4 *Dendrogram Support Vector Machine*

A dendrogram support vector machines (DSVM) approach has been utilized for multiclass classification of the faults of worm gearbox in this communication. This method utilizes a taxonomy of classes and decomposes a multiclass problem to a descendant set of binary-class problems. Ascendant hierarchical clustering (AHT) method is used to group all classes in an ascendant hierarchy. This clustering allows to separate the classes and to build different subsets from database of subproblems. Then SVM classifier is applied at each internal node to construct the best discriminant function of a binary-class problem as shown in Fig. 2.

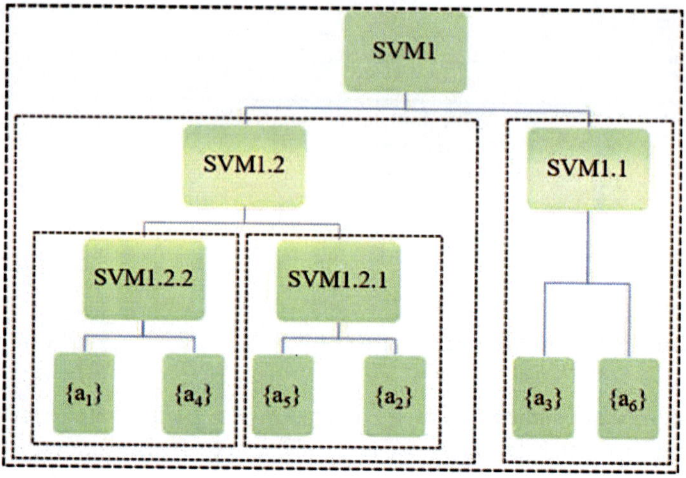

Fig. 2 Structure of classification of DSVM

Table 2 Number of data acquired under different operating and health conditions

Condition	No-load		10 kg-load	
	1800 rpm	2200 rpm	1800 rpm	2200 rpm
Healthy tooth (A)	25	25	25	25
Tooth pitting (B)	25	25	25	25
Tooth missing (C)	25	25	25	25

3 Analysis and Results

3.1 Data Collection

Vibration signals under various operating and health conditions of the gearbox have been acquired using accelerometer and data acquisition system from worm gear test rig. A total number of samples under different operating and health conditions have been presented in Table 2.

3.2 Signal Processing

Raw vibration signals acquired from test rig have been passed through autoregression minimum entropy deconvolution (AR-MED) filter to eliminate the deterministic/regular pattern of the gears and exposing the periodic impulses in the signals as illustrated in Fig. 3.

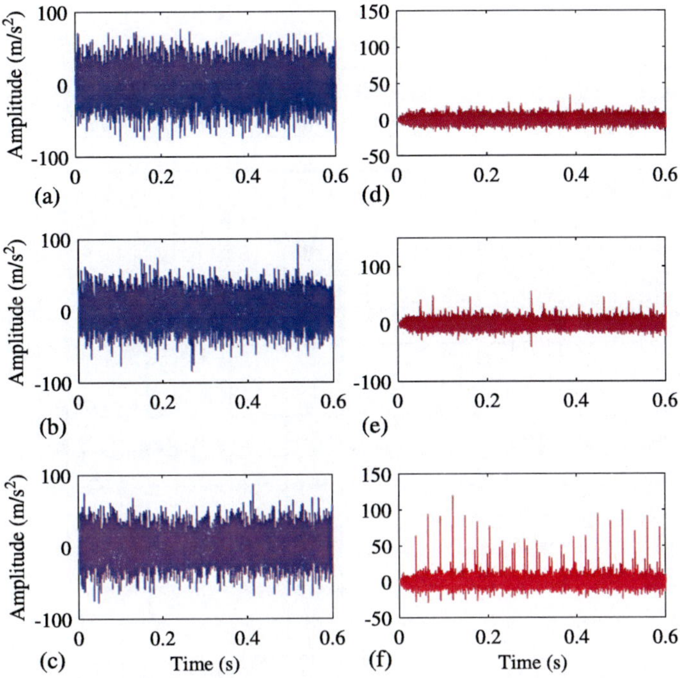

Fig. 3 Typical vibration signals of different health conditions of gearbox: **a** raw signal of healthy tooth condition, **b** raw signal of pitted tooth condition, **c** raw signal of missing tooth condition, **d** filtered signal of healthy tooth condition, **e** filtered signal of pitted tooth condition, and **f** filtered signal of missing tooth condition

3.3 Feature Selection

A total of twenty-nine statistical features listed in Table 1 have been extracted from raw and filtered signals. All the features extracted from filtered signals under three different conditions viz: healthy tooth designated as A, tooth pitting as B, and tooth missing as C are plotted in Fig. 4. Further T-test has been conducted on these features to extract the best features to reduce the computational burden and arranged the features according to their rank as given in Table 3. First five selected features are shown in Fig. 5 and has been observed that these features can classify different health condition of the gearbox.

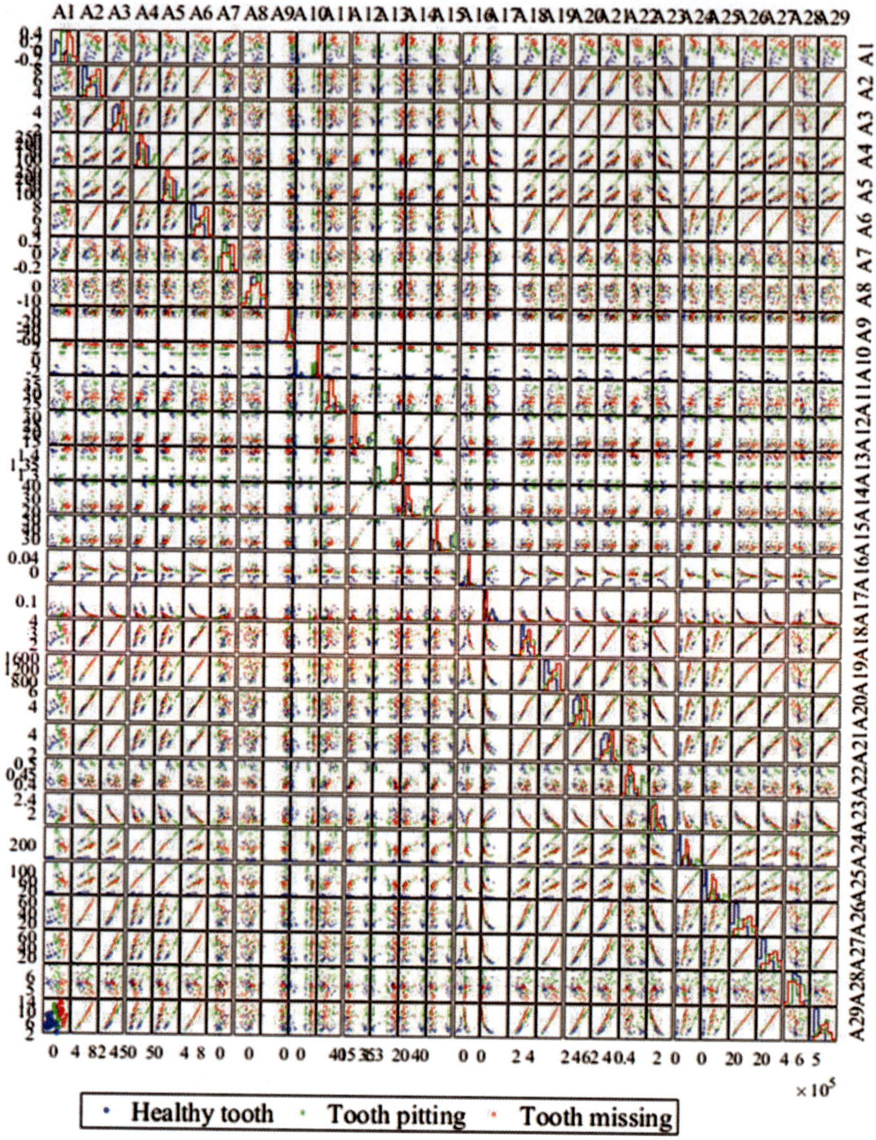

Fig. 4 Interaction of different statistical features

3.4 *Identification of Fault*

Features extracted from raw signals have been used in DSVM to segregate the different health conditions of the worm gearbox. Various parameters required for DSVM are listed in Table 4.

Table 3 Ranking of features using T-test

Conditions / Features	No-load @1800 rpm	10 kg-load @1800 rpm	No-load @2200 rpm	10 kg-load @2200 rpm
A_1	1	1	1	1
A_2	2	2	2	4
A_3	3	3	3	5
A_4	4	4	4	7
A_5	5	5	5	10
A_6	6	6	6	12
A_7	7	7	7	13
A_8	8	10	8	14
A_9	10	12	9	15
A_{10}	12	13	10	22
A_{11}	13	14	11	24
A_{12}	14	15	12	25
A_{13}	15	16	13	28
A_{14}	16	17	14	2
A_{15}	17	18	15	3
A_{16}	18	19	16	6
A_{17}	19	20	17	8
A_{18}	20	21	18	9
A_{19}	21	22	19	11
A_{20}	22	23	20	16
A_{21}	23	24	21	17
A_{22}	24	25	22	18
A_{23}	25	26	23	19
A_{24}	26	27	24	20
A_{25}	27	29	25	21
A_{26}	29	8	26	23
A_{27}	9	9	27	26
A_{28}	11	11	28	27
A_{29}	28	28	29	29

Confusion matrices of features of raw signals at a speed of 1800 rpm with no-load and with load are shown in Fig. 6a and b, respectively. The accuracy $\leq 80\%$ has been observed with features extracted from the raw signals.

Five selected features have been used as input for DSVM to classify various health conditions of the gearbox. To train the system, 80% samples have been used. Further, 20% samples have been used for testing. The test results are presented in terms of confusion matrix and decoding accuracy. From Fig. 7a, b, and d, it has been observed

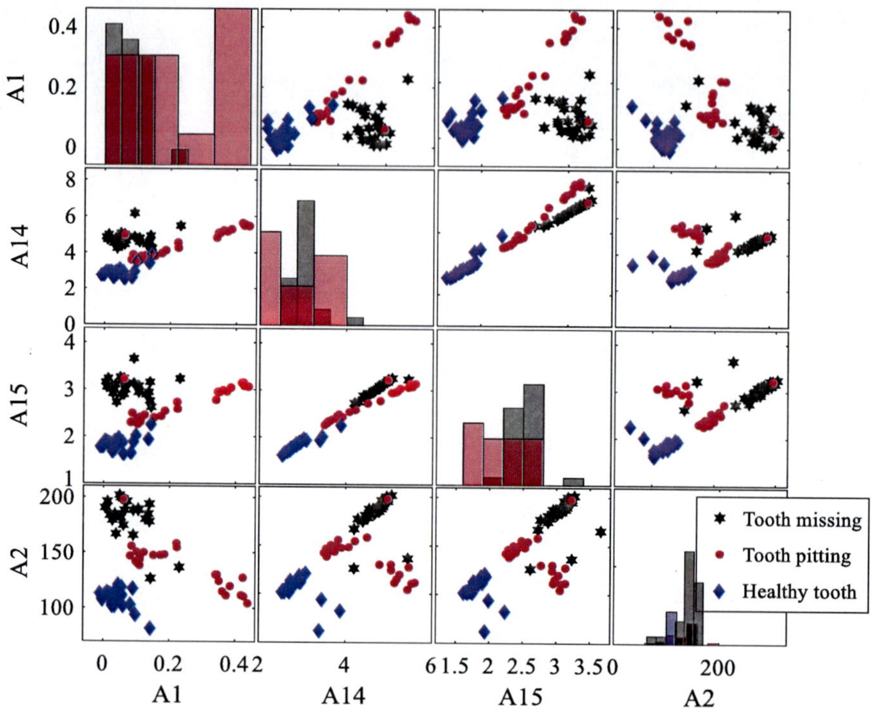

Fig. 5 Interaction of prominent statistical features

Table 4 Parameter of DSVM

Phase	Input parameters	Output parameters
Training	• Train-label (1 × *n* cell vector of training label corresponding to each class, positive numeric) • Train-cell (1 × *n* cell vector of training data corresponding to each class. 80% of total data)	• SVM-structure • Level
Classification	• Label (*n* × 1 vector containing label of n classes) • Test-matrix (test samples 20%) • SVM-structure • Level	• Test-class
Confusion matrix	• Test-class • Labels (*n* × 1 vector of known labels, should be numeric) • Class-name ('HT,' 'PT,' 'MT')	• C-matrix • Decoding accuracy

that DSVM is able to identify all the three different conditions of the gearbox with 100% accuracy, whereas in Fig. 7c, i.e., in no-load at speed of 2200 rpm, some of data have been predicted wrongly, and accuracy of the classifier is reduced to 86%. This weird prediction of the classifier is due to the sudden drop in the amplitude of

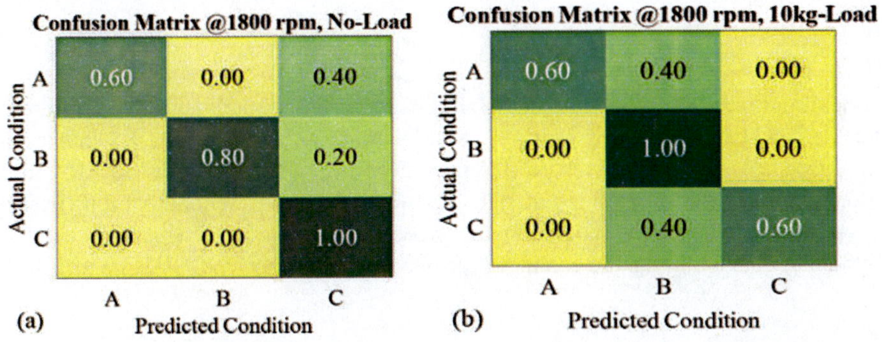

Fig. 6 Confusion matrices of DSVM using features of raw signals at: **a** speed 1800 rpm with no-load, **b** speed 1800 rpm with 10 kg-load

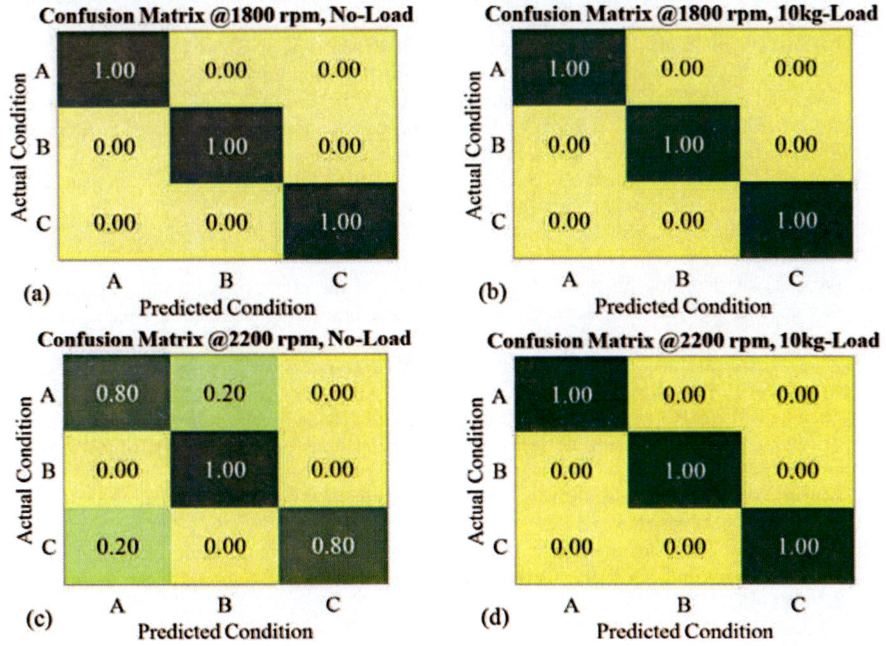

Fig. 7 Confusion matrices of DSVM using features of filtered signals at: **a** speed 1800 rpm with no-load, **b** speed 1800 rpm with 10 kg-load, **c** speed 2200 rpm with no-load, **d** speed 2200 rpm with 10 kg-load

vibration signal for fraction of time in some of the samples acquired under missing tooth condition with no-load at a speed of 2200 rpm.

4 Conclusion

A dendrogram support vector machine has been proposed to classify three different health conditions of worm gearbox (healthy tooth-A, tooth pitting-B, and tooth missing-C). It has been observed that preprocessing of the signal using AR-MED filter improves the capabilities of statistical features to differentiate various health conditions. By T-test, the most influencing features are extracted which reduced the burden of data handling and enhanced the decoding accuracy of DSVM up to 100%.

References

1. Lebold M, McClintic K, Campbell R, Byington C, Maynard K (2000) Review of vibration analysis methods for gearbox diagnostics and prognostics. In: 54th meeting of the society for machinery failure prevention technology. Virginia Beach, VA, pp 623–634
2. Sait AS, Sharaf-Eldeen YI, (2011) A review of gearbox condition monitoring based on vibration analysis techniques diagnostics and prognostics. In: Proulx T (ed) Rotating machinery, structural health monitoring, shock and vibration, vol 5. Conference proceedings of the society for experimental mechanics series. Springer, New York, pp 307–324
3. Kumar A, Gandhi CP, Zhou Y, Kumar R, Xiang J (2020) Latest developments in gear defect diagnosis and prognosis: a review. Measurement 158:107735
4. Elasha F, Cárcel CR, Mba D, Nze GKI, Yebra G (2014) Pitting detection in worm gearboxes with vibration analysis. Eng Fail Anal 42:366–376
5. Peng Z, Kessissoglou NJ, Cox M (2005) A study of the effect of contaminant particles in lubricants using wear debris and vibration condition monitoring techniques. Wear 258:1651–1662
6. Waqar T, Demetgul M (2016) Thermal analysis MLP neural network-based fault diagnosis on worm gears. Measurement 86:56–66
7. Zhang CL, Li B, Chen BQ, Cao HR, Zi YY, He ZJ (2015) Weak fault signature extraction of rotating machinery using flexible analytic wavelet transform. Mech Syst Signal Process 64–65:162–187
8. Lamani GS, Pawar PS, Ranalkar NG, Pawar OP, Patil SV (2018) Vibration analysis of worm and worm wheel gear box. IOSR J Mech Civ Eng 81–84
9. Pawankumar SP, Chandrasekaran M, Arulmozhi P (2020) Gear fault types and condition monitoring indicators for gearbox—a combined assessment. Int J Recent Technol Eng 8:3448–3453
10. Kumar S, Kumar R (2019) Worm and wheel gears fault frequency extraction using minimum entropy deconvolution-based envelope of the vibration signal. J Phys Conf Ser 1240:012073
11. Halim EB, Shah SL, Zuo MJ, Choudhury MAAS (2006) Fault detection of gearbox from vibration signals using time-frequency domain averaging. In: American control conference, vol 6. Minneapolis, MN, USA
12. Vernekar K, Kumar H, Gangadharan KV (2014) Gear fault detection using vibration analysis and continuous wavelet transform. Proc Mater Sci 5:1846–1852
13. Aherwar A (2012) An investigation on gearbox fault detection using vibration analysis techniques: a review. Aust J Mech Eng 10:169–183
14. Ümütlü RC, Hizarci B, Ozturk H (2020) Classification of pitting fault levels in a worm gearbox using vibration visualization and ANN. Sādhanā 45:1–13
15. Kumar S, Kumar R (2021) Diagnosis of an incipient defect in a worm gearbox using minimum entropy deconvolution and local cepstrum. Meas Sci Technol 32:054002

Mechanisms Responsible for Performance Improvements of Pocketed and Textured Lubricated Interfaces

J. C. Atwal, M. R. Pattnayak, R. K. Pandey, P. Ganai, A. Atulkar, V. Bhardwaj, and Niharika Gupta

Abstract Surface textures improve the performance of fluid film bearings, piston rings, ball bearings, gears, etc. In the presence of nano/micro-structures on the lubricated mating surfaces, the researchers have identified different mechanisms which lead to improvement in the performance of the textured lubricated interfaces. Thus, the objective of the paper is to present and discuss the mechanisms for various textured lubricated contacts at one place which would be beneficial for the researchers working in this area.

Keywords Textured surfaces · Micro-pockets · Micro-dimples · Lubricated interfaces · Mechanisms · Performance improvement · Thrust bearing · Fish scale texture

1 Introduction

Surface textures involving the nano- and micro-structures (different sizes and shapes of dimples and pockets) have emerged as a green technology for improving the performance behaviors of lubricated interfaces at the mating surfaces found in the fluid film bearings, piston rings, ball bearings, gears, etc. [1–3]. Authors [4, 5] have performed experiments on the thrust ball bearing having textured race. It has been reported that the bearing with textured race yields low temperature rise, decrease in vibration amplitude and frictional torque as compared to conventional one. In mating gear pair, surface textures have been found to be effective in reducing the vibration amplitude, reduction in surface damage of textured teeth in comparison to conventional case as reported in experimental work of researchers [6, 7].

Researchers [8] have employed combination of pocket and bionic texture on surface of pads of thrust pad bearing. It has been found that together pocket and bionic texture perform better compared to pocketed and plain pad case. The same group of authors employed new pocket design to enhance the performance of water lubricated

J. C. Atwal (✉) · M. R. Pattnayak · R. K. Pandey · P. Ganai · A. Atulkar · V. Bhardwaj · N. Gupta
Department of Mechanical Engineering, I.I.T. Delhi, New Delhi 110016, India
e-mail: Jeewan.Chandra.Atwal@mech.iitd.ac.in

V. K. Gupta et al. (eds.), *Recent Advances in Machines and Mechanisms*, Lecture Notes in Mechanical Engineering, https://doi.org/10.1007/978-981-19-3716-3_22

pad thrust bearings and their study reveal that new pocket design performed better in comparison to conventional rectangular pocket [9]. The performance of powder lubricated journal bearings has been improved by different surface pocket designs, and it has been found that rectangular pocket performance was best among all [10].

The rigid air lubricated bearing performance has been investigated by the researchers employing new surface topographies and their study revealed that surface pattern on the bearing bore improve the performance significantly [11]. Performance of piston ring cylinder liner conjunction has been improved using dimples on the piston, and it has been concluded that textured piston ring reduces the friction by 30% in comparison to conventional one [12].

Through the progress of the research, investigators have identified numerous mechanisms accountable for performance improvement in the presence of tiny structures at the mating surfaces. Thus, the objective of this paper is to explain the various mechanisms through schematic sketches with physical interpretation responsible for performance improvements of fluid film bearings (journal and thrust bearings), spur gearset, piston rings and ball bearings in the presence of surface textures.

2 Mechanisms of Performance Improvements of Textured Pad Thrust and Journal Bearings

This section presents the mechanisms of performance improvements of pad thrust and journal bearings using surface textures. First, various mechanisms for textured thrust pad/slider bearings are discussed, and then it is followed by textured journal bearings.

2.1 Textured Thrust Pad/Slider Bearings

The following four mechanisms are presented herein.

(i) **Inlet suction**

This new mechanism of "inlet suction" for micro-pocketed bearings is presented by the authors [13]. The mechanism can be understood with the help of schematic diagrams Fig. 1a, c, respectively. In Fig. 1a, a micro-pocket can be seen close to the bearing inlet. When the lubricant flows in the direction of sliding, it first encounters a divergent region (due to presence of a micro-pocket), and this induces cavitation in the lubricant. After the divergent region, lubricant passes through a convergent region, and pressure builds up in the lubricant film which generates load capacity in the bearing. These phenomena of cavitation followed by pressure buildup can be seen in Fig. 1c.

Since the cavitation pressure is less than the atmospheric pressure, this causes a pressure difference in inlet region of bearing, and because of this,

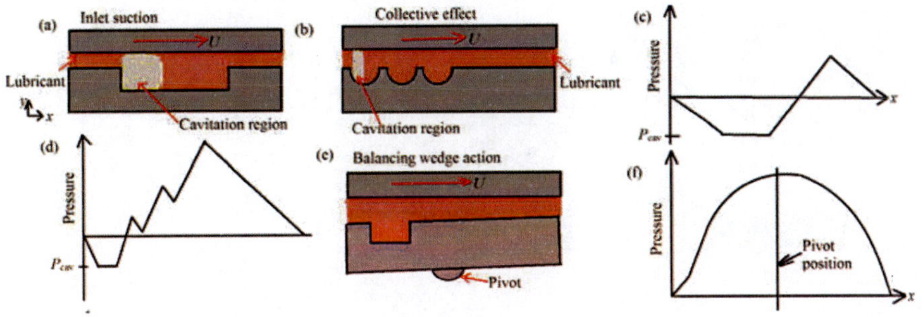

Fig. 1 **a** Micro-pocketed bearing, **b** micro-dimpled bearing, **c** pressure profile for micro-pocketed bearing, **d** pressure profile of micro-dimpled bearing, **e** textured tiling pad with central pivot, **f** pressure profile for textured tiling pad

fluid/lubricant is sucked into the bearing. This inlet suction is believed to be the mechanism for improving the performance of micro-pocketed bearing.

(ii) **Collective effect**

The mechanism of "collective effect" is first presented and discussed by the research group of authors [14]. To understand this mechanism, Fig. 1b, d is presented. In Fig. 1b, an array of three micro-dimples is shown in bearing pad. When the lubricant passes through the first dimple, at first, it cavitates (because of presence of divergent region), and then pressure is generated in convergent zone of first dimple. Now this generated pressure in lubricating film does not allow film pressure to drop down to cavitation pressure in the divergent zone of second dimple as can be seen in Fig. 1d. Thus, pressure generated from previous dimple helps to build more pressure in lubricating film when lubricant passes through the second dimple. This phenomenon continues as fluid passes through the third dimple and over pressure buildup improves as can be observed from Fig. 1d. This mechanism of action for improving the load capacity of bearing is known as "collective effect."

(iii) **Balancing wedge action**

This mechanism of action is first proposed by the researchers [15] for the textured tilting pad bearing with central pivot. Theoretically central pivoted pad has zero load capacity because runner and pad are parallel. In Fig. 1e, a centrally pivoted pad with a micro-dimple at the inlet region is shown. The presence of micro-dimple in the inlet region of bearing changes the pressure profile as can been seen in the Fig. 1f as compared to case of plain pad where the pressure profile remains symmetric about pivot location. In Fig. 1f, pressure profile is not symmetric about the pivot location, and this produces a moment of force about the pivot axis. This generated moment tilts the pad, and a convergent region is formed between moving part and stationary pad as can be seen in Fig. 1e. The convergent region provides a physical wedge, and load capacity in the centrally pivot bearing is generated. This mechanism is known as balancing wedge action.

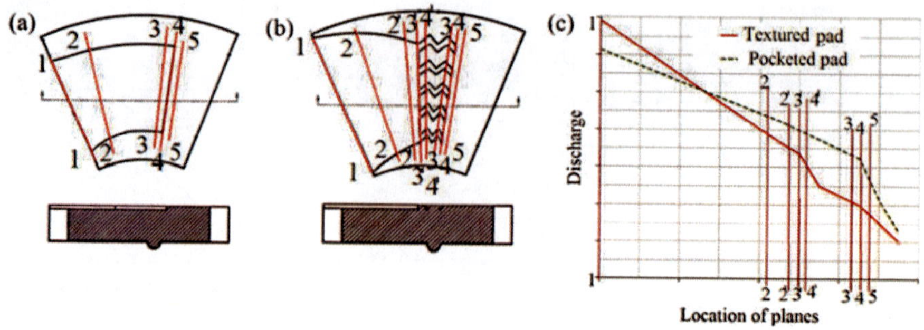

Fig. 2 **a** Micro-pocketed pad, **b** textured pad, **c** variation of discharge with location of planes

(iv) **Flow restrictions due to formation of micro-dams**
This mechanism of performance improvements for micro-pocketed and textured pad thrust bearing is illustrated through Fig. 2a–c. Figure 2a, b shows the two different micro-pocketed pads named as pocketed and textured pads, respectively. To understand the flow restrictions due to formation of micro-dams, five planes are drawn for micro-pocketed pad as shown in Fig. 2a; however, eight planes are drawn for textured pad as depicted in Fig. 2b. In Fig. 2c, flow rate is plotted with different plane location for the cases of micro-pocketed and textured pads. It can be seen from this figure that around 4-4 planes, there is sudden change in flow rate of micro-pocketed pad. This happens because the pocket edge in the location of 4-4 plane acts as a micro-dam which provide the restriction the flow of lubricant and flow rate changes suddenly near this plane location.

Now consider the case of textures pad of Fig. 2b in which there are two sudden changes in flow rate around 3′-3′ and 4-4 planes, respectively. Therefore, in this case, there are formation of two micro-damps which provide flow restrictions about two plane locations. When the fluid flow is obstructed by the formation of micro-dams, there is generation of pressure in lubricating film. Thus, flow restrictions enhance the pressure buildup which improves the performance of bearings. It is also interesting to note here that textured pad performs better than the micro-pocketed pad. The reason is clearly understood from flow rate of Fig. 2c. Textured pad provides flow restrictions at around two planes; however, micro-pocketed pad gives flow restriction at around one plane. For more information on this mechanism of action, readers can view the work of authors [16–19].

2.2 *Textured Journal Bearings*

This section presents the mechanisms for performance improvement of textured journal bearing.

Figure 3a, b depicts the CAD model of a dimpled bearing, various components of journal bearing, respectively. The ongoing worldwide research on textured journal bearing focuses on optimizing texture geometries to improve their tribo-dynamic behaviors. The texture location, density, and shapes of micro-surface topographies in texture significantly influence the performance behaviors of journal bearings. The vital design principles and mechanisms behind the improvement in tribological performance parameters have been summarized herein for ready reference to the readers.

Texturing the entire bearing surface has resulted in decrease in load-sustaining capacity and enhancement in friction. Moreover, shallow dimples placed toward the lubricant inlet region have shown augmented hydrodynamic lift in the bearing.

It has also been established that textures having deeper and denser dimples in the diverging cavitation regime have helped in reducing the coefficient of friction. The primary reason behind the enhancement of minimum film thickness in textured bearings is the lubricant flow constrictions in the circumferential direction in the presence of micro-topographies. The textures behave as micro-dams causing hindrance to the lubricant flow. Due to this, pressure builds up in the lubricating oil, resulting in increased minimum film thickness.

Another theory for the rise in the minimum film thickness is the lubricant sucking action near the cavitation region leading to the rushing of more lubricant into the convergent zone. This builds up pressure in the convergent region and consequently increase the minimum film thickness. The lower temperature rise in the optimized textured journal bearings is due to the enhanced side leakage, which carries away the heat generated through convection. Moreover, the reduction behind viscous friction force in the case of journal bearings with micro-topographies is understood to be the smearing of less lubricant in the divergent region and availability of more lubricant in the minimum film thickness region, which acts as a lubricant reservoir [11, 20].

Figure 3c shows the CAD model of pocketed air foil bearing and naming of various components of this bearing is shown in Fig. 3d. In Fig. 3e, an unwrapped of pocketed top foil is depicted. It has been found that the pocketed foil performs better than the conventional plain top foil [21]. The mechanism responsible is explained

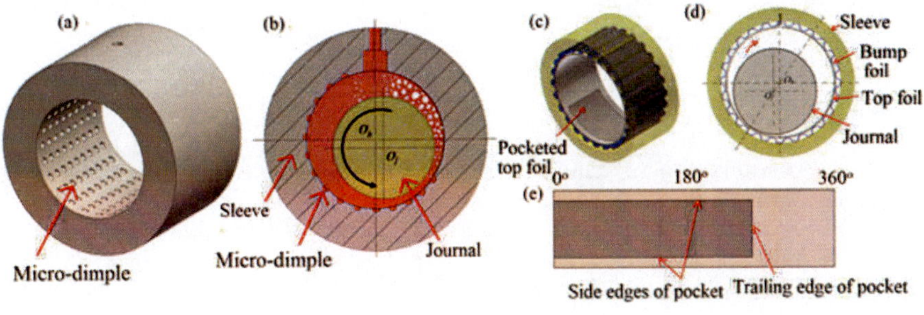

Fig. 3 Textured journal bearing; **a** Isometric view; **b** front view with journal and cavitation region **c** CAD model of a pocketed air foil bearing, **d** schematic view of air foil bearing with different components, **e** unwrapped view of pocketed top foil

with the help of Fig. 3d, e. In an air foil bearing, there are convergent and divergent zones as can be seen in Fig. 3d. In the divergent zone, air pressure remains below atmospheric which helps in suction of air in the clearance space of bearing. The functions of a pocket are twofold. First its side edges reduce the leakage of air, and secondly, the trail edge of pocket acts as a micro-dam which restricts the air flow leading to generation of higher pressure in comparison to conventional case. Thus, it is clear that suction of air in divergent zone followed by reduction of side leakage and formation of micro-dams work synergistically to improve the performance of pocketed foil bearing. More information on foil bearing can be found in Ref. [22].

It is worth mentioning here that the above-mentioned mechanisms are well established and verified using numerical modeling/experimental work. The modeling details in brief are presented below.

2.3 The Governing Lubrication Equation

The above-mentioned mechanisms have been explored by the researchers with the help of the Reynolds equation which is as follows [18]:

$$\frac{\partial}{\partial \overline{x}}\left[\frac{\overline{h}^3}{6\overline{\eta}}\frac{\partial \overline{p}}{\partial \overline{x}}\right]+\frac{L^2}{B^2}\frac{\partial}{\partial \overline{y}}\left[\frac{\overline{h}^3}{6\overline{\eta}}\frac{\partial \overline{p}}{\partial \overline{y}}\right]=\frac{\partial \overline{h}}{\partial \overline{x}} \tag{1}$$

where L, B and h are length, width of bearings and film thickness, respectively. Equation (1) on solution provides film pressure distribution. The above-mentioned equation does not consider the mass-conservation issue due to cavitation. Thus, the mass-conserved form of the equation is written as follows [18, 20]:

$$\frac{\partial}{\partial \overline{x}}\left[\frac{\overline{h}^3}{6\overline{\eta}}\frac{\partial \overline{p}}{\partial \overline{x}}\right]+\frac{L^2}{B^2}\frac{\partial}{\partial \overline{y}}\left[\frac{\overline{h}^3}{6\overline{\eta}}\frac{\partial \overline{p}}{\partial \overline{y}}\right]=\frac{\partial (1-\phi)\overline{h}}{\partial \overline{x}} \tag{2}$$

Equation (2) is subjected to the following constraint condition due to cavitation:

$$\overline{p} \geq 0,\ \phi \geq 0,\ \overline{p}\,\phi = 0 \tag{3}$$

where $\phi = 1 - (\rho/\rho_0)$.

The expression in Eq. (3) can be replaced with an algebraic equation of following from [20]:

$$\overline{p} + \phi - \sqrt{\overline{p}^2 + \phi^2} = 0 \tag{4}$$

Now the task is to solve Eqs. (2) and (3) together using numerical methods/analytical methods for getting the solution. In the past, the researchers have solved the governing lubrication Reynolds equation (either Eqs. 1 or 2) along with film thickness relations of textured bearings for exploring and establishing the inlet suction mechanism (which later verified in the experimental work of authors [23]), staircase effect (i.e., collective effect) mechanism, balancing wedge action mechanism and reductions of flows in circumferential or flow direction. The comprehensive details on these can also be seen in the work of authors [13–16].

3 Mechanisms of Performance Improvements of Textured Piston Ring, Textured Gear, and Textured Race of Ball Bearing

In Fig. 4a–c, CAD models of a textured gear, piston ring and race of ball bearing are depicted. When these textured components form contact with their counterpart in presence of lubricant, performance of contact improves. In the presence of nano/micro-structures on the lubricated mating surfaces, the researchers mainly identified three mechanisms which improve the performance. During the lubrication, the pair of dimpled surfaces form numerous micro-hydrodynamic bearings, as shown in Fig. 5a. The pressure generated at innumerable places in tiny hydrodynamic bearings keeps the loaded solid in the floating condition over the thin film. In this manner, the physical contacts between the solids can be avoided. This leads to the enhancement of tribological and dynamic behaviors of piston rings, ball bearings and gears.

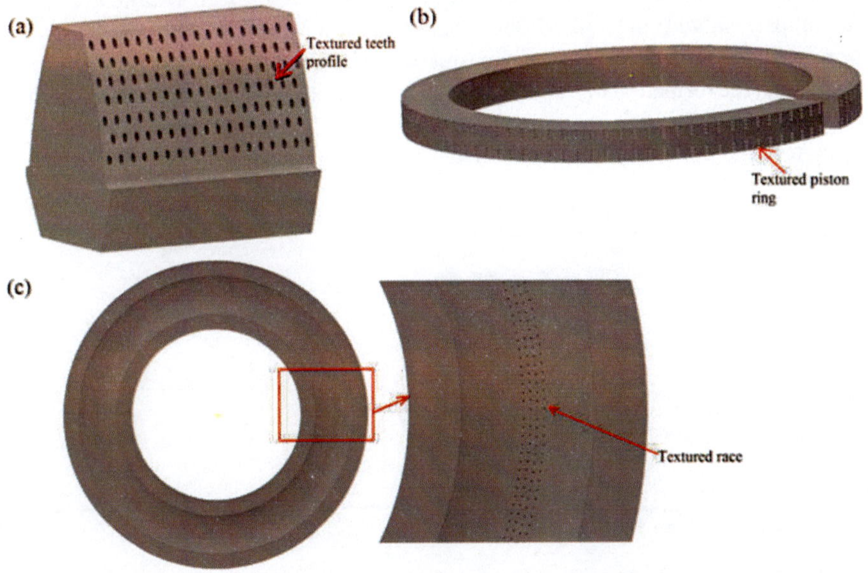

Fig. 4 CAD model of a textured: **a** Gear, **b** piston ring, **c** race of ball bearing

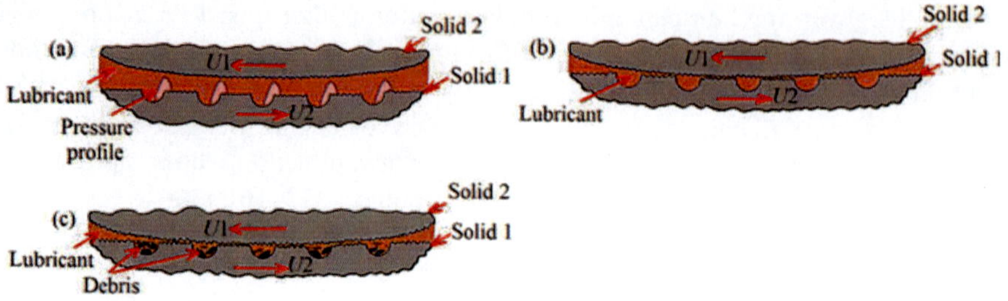

Fig. 5 **a** Development of numerous micro-hydrodynamic bearings, **b** micro-lubricant reservoirs, **c** micro-debris and contaminates trappers

In case of poor film formation because of the lack of lubricant or operating parameters (i.e., at heavy load and low speed), the physical interactions between the mating conventional surfaces happen. In this condition, the friction and vibration enhance at the interface. But in the presence of texture, friction and vibration decrease due to smearing of lubricant available within the innumerable dimples, as illustrated in Fig. 5b. In addition, the debris formed due to the asperity-to-asperity contacts of mating solids get entrapped in the micro-reservoirs (i.e., dimples and pockets), as schematically shown in Fig. 5c. This also benefits in lowering the wear at the solids' interfaces.

In the recent past, authors' group has done experiments on textured/untextured gear pairs to validate the above-mentioned mechanisms [6, 24]. Figure 6a shows the electrical resistance between meshed gear pairs lubricated with grease. It can be seen from Fig. 6a that the electrical resistance has increased with textured gear in comparison with untextured one. This shows the formation of larger film thickness between meshed textured gears in comparison to untextured one. As already discussed, the formation of innumerable micro-hydrodynamic bearings happens with textured surfaces which increases the hydrodynamic pressure and consequently film thickness and therefore electrical resistance [24].

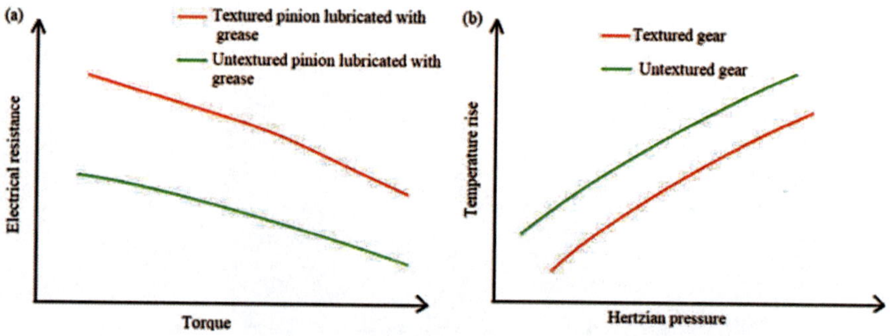

Fig. 6 **a** Comparison of electrical resistance between meshed gears, **b** comparison of temperature rise between untextured and textured gears

In Fig. 6b, bulk temperature rise of lubricated textured gears has been compared with untextured lubricated gears. It is noticed that with textured gear pair, temperature rise has been reduced as compared to untextured one. This indicates that textures present on the surface of gear work as lubricant reservoir and help in reducing the temperature rise; however, the untextured gears lack this feature and contact happens due to scarcity of lubricant in contact zone and temperature increases [6].

The third mechanism of wear debris trapper is also confirmed in experimental work of Ref. [24]. The SEM images (space limitation prohibit to present images here but can be seen in Refs. [6, 24]) of textured gear surface and untextured gear surface were taken, and it was revealed that the textured surface has less damage compared to untextured gear surface which confirm the better lubrication and trap of wear particles inside the textures, and hence, abrasion due to third body was reduced. Complete experimental details of the textured/untextured gear pair are given in refs. [6, 24]. Interested readers are also referred to the work of authors [25, 26] for more information on these mechanisms.

4 Conclusions

In textured slider/thrust/journal bearings, four mechanisms namely inlet suction, collective effect, balancing wedge action and flow reduction due to formation of micro-damps explain the improvement in performance behaviors of these bearings. The performance improvement in textured piston, textured gear and textured race of ball bearings has happened because of formation of innumerable micro-hydrodynamic bearing, the role of texture as lubricant reservoir and debris trapper which have been experimentally validated.

Declaration of Conflict of Interests The authors declare that there are no conflicts of interest in publishing this paper.

References

1. Sudeep U, Tandon N, Pandey RK (2015) Performance of lubricated rolling/sliding concentrated contacts with surface textures: a review. J Tribol 137:031501
2. Gropper D, Wang L, Harvey TJ (2016) Hydrodynamic lubrication of textured surfaces: a review of modelling techniques and key findings. Tribol Int 94:509–529
3. Gachot C, Rosenkranz A, Hsu SM, Costa HL (2017) A critical assessment of surface texturing for friction and wear improvement. Wear 372:21–41
4. Bhardwaj V, Pandey RK, Agarwal VK (2017) Experimental investigations for tribo-dynamic behaviours of conventional and textured races ball bearings using fresh and MoS_2 blended greases. Tribol Int 113:149–168
5. Bhardwaj V, Pandey RK, Agarwal VK (2019) Performance studies of textured race ball bearing. Ind Lubr Tribol 71:1116–1123

6. Gupta N, Tandon N, Pandey RK (2018) An exploration of the performance behaviors of lubricated textured and conventional spur gearsets. Tribol Int 128:376–385
7. Gupta N, Tandon N, Pandey RK, Vidyasagar KC, Kalyanasundaram D (2020) Tribological and vibration studies of textured spur gear pairs under fully flooded and starved lubrication conditions. Tribol Trans 63:1103–1120
8. Atwal JC, Pandey RK (2021) Performance analysis of thrust pad bearing using micro-rectangular pocket and bionic texture. Proc Inst Mech Eng Part J: J Eng Tribol 235:1232–1250
9. Atwal JC, Pandey RK (2021) Performance improvement of water-lubricated thrust pad bearing operating with the turbulent flow using a new micro-pocket design. Tribol Int 154:106738
10. Rahmani F, Pandey RK, Dutt JK (2018) Performance studies of powder lubricated journal bearing having different pocket shapes at cylindrical bore surface. J Tribol 140:031704
11. Pattnayak MR, Pandey RK, Dutt JK (2021) Effects of new micro-pocketed bore surface topographies on the performance behaviours of aerodynamic journal bearing. Surf Topogr Metrol Prop 9:025001
12. Atulkar A, Pandey RK, Subbarao PM (2021) Performance analysis of two-dimensional section of partially textured piston ring with cavitation boundary conditions. Surf Topogr Metrol Prop 9:025025
13. Fowell M, Olver AV, Gosman AD, Spikes HA, Pegg I (2007) Entrainment, and inlet suction: two mechanisms of hydrodynamic lubrication in textured bearings. J Tribol 129:336–347
14. Brizmer V, Kligerman Y, Etsion I (2003) A laser surface textured parallel thrust bearing. Tribol Trans 46:397–403
15. Yagi K, Sugimura J (2013) Balancing wedge action: a contribution of textured surface to hydrodynamic pressure generation. Tribol Lett 50:349–364
16. Aggarwal S, Pandey RK (2017) Frictional and load-carrying behaviours of micro-textured sector shape pad thrust bearing incorporating the cavitation and thermal effects. Lubr Sci 29:255–277
17. Aggarwal S, Pandey RK (2018) Performance investigation of micro-pocketed textured pad thrust bearing. Ind Lubr Tribol 70:1388–1395
18. Atwal JC, Pandey RK (2021) Film thickness and friction investigations in a fluid film thrust bearing employing a new conceived micro-texture on pads. J Tribol 143:061801
19. Atwal JC, Pandey RK (2021) Influence of new surface micro-structures on the performance behaviours of fluid film tilting pad thrust bearings. Proc Inst Mech Eng Part C: J Mech Eng Sci. https://doi.org/10.1177/09544062211030972
20. Pattnayak MR, Pandey RK, Dutt JK (2022) Performance improvement of an oil-lubricated journal bearing using bionic-textures fused micro-pockets. J Tribol 144:041804
21. Ganai P, Pandey RK, Dutt JK (2021) Performance improvement of foil air journal bearing employing micro-pocket and textures on top compliant surface. Surf Topogr Metrol Prop 9:025045
22. Pattnayak MR, Pandey RK, Dutt JK (2020) Performance behaviours of a self-acting gas journal bearing with a new bore design. Tribol Int 151:106418
23. Zhang J, Meng Y (2012) Direct observation of cavitation phenomenon and hydrodynamic lubrication analysis of textured surfaces. Tribol Lett 46:147–158
24. Gupta N, Tandon N, Pandey RK, Vidyasagar KC, Kalyanasundaram D (2022) Tribodynamic studies of textured gearsets lubricated with fresh and MoS_2 blended greases. Tribol Int 165:107247

25. Vidyasagar KC, Pandey RK, Kalyanasundaram D (2021) An exploration of frictional and vibrational behaviors of textured deep groove ball bearing in the vicinity of requisite minimum load. Friction. https://doi.org/10.1007/s40544-021-0495-3
26. Atulkar A, Pandey RK, Subbarao PM (2021) Role of textured piston rings/liners in improving the performance behaviours of IC engines: a review with vital findings. Surf Topogr Metrol Prop 9:023002

Feature Selection Based on Gaussian Ant Lion Optimizer for Fault Identification in Centrifugal Pump

Govind Vashishtha and Rajesh Kumar

Abstract Fault diagnosis of the rotating machinery using vibration signal is largely carried out by experience with some prior knowledge of the signal. The diagnosis process is simplified by using machine learning algorithms. The learning capabilities and classification performance of such machine learning models are mostly influenced by the quantity and quality of the input features. Thus, the appropriate selection of a subset of the most prominent features for machine learning becomes essential to eliminate redundancy of high dimension or irrelevant measurements of the features. In this paper, a filter-based feature selection technique is introduced to select the optimal feature space. A Gaussian ant lion optimization (GALO) is put in with a filter-based selection technique to select the feature subset from a high dimension feature dataset obtained from the vibration signals of centrifugal pump under different health conditions (normal, clogging, wheel cut and blade cut). The K-nearest neighbour (KNN) classifier is applied to the selected feature subset to find the classification accuracy. In addition, the proposed method has been compared with other art of work. The results reveal that the proposed GALO-KNN with filter-based feature selection technique outperforms both in feature reduction and classification accuracy and secures the best feature subset with less computational effort. Thus, the proposed method is capable enough to perform the selection task and shows excellent potential in fault diagnosis.

Keywords Centrifugal pump · k-nearest neighbour · Feature selection · Ant lion optimization · Gaussian mutation strategy

1 Introduction

Centrifugal pump is used to carry water by converting the rotational kinetic energy of the impeller into hydrodynamic energy. The bearings support the shaft on which impeller is straddling. An impeller is housed in an involute casing which forces the

G. Vashishtha · R. Kumar (✉)
Precision Metrology Laboratory, Department of Mechanical Engineering, Sant Longowal Institute of Engineering and Technology, Longowal 148106, India
e-mail: rajesh_krs@sliet.ac.in

V. K. Gupta et al. (eds.), *Recent Advances in Machines and Mechanisms*, Lecture Notes in Mechanical Engineering, https://doi.org/10.1007/978-981-19-3716-3_23

moving water to impart velocity [1–3]. Defects in the centrifugal pump may occur due to corrosion, erosion, metallurgical defects, cavitation and improper lubrication [1, 2, 4–6]. Vibration signal helps in determining the health status of the centrifugal pump and is least affected by environmental noise [7, 8]. Thus, vibration-based signal has been preferred for feature selection to identify the pump defects in the proposed work.

Cao et al. [9] introduced the hybrid technique which is a combination of the principal component analysis (PCA) with Gaussian mixed model (GMM) to diagnose the crack in pump. Jiang et al. [10] developed a pump diagnostic technique using empirical mode decomposition (EWT) and deep learning on a cloud platform. Kumar et al. [11] modified the cost function of convolution neural network (CNN) by adding the divergence-based function to eliminate the redundancy of CNN's hidden layer for classifying the different defects of the pump using acoustic images. Ahmad et al. [12] proposed the discriminant feature extraction technique to analyse defect in pump. Al Tobi et al. [13] extracted the fault features based on discrete wavelet transform and further categorized them by different classifiers.

Pre-processing of raw signal is a crucial stage in feature selection as the results of the classification task are more vulnerable to the dimension of the features [14–20]. The process of feature selection aims at choosing the most prominent features by removing the most irrelevant and unnecessary features which render the classification model. Selection of appropriate features not only reduces the complexity in the learning process but also enhances the accuracy of the classification model. Wang et al. [21] modified ant lion optimization (ALO) to select the features for the hyperspectral image. Emary et al. [16] introduced Levy ALO for selecting appropriate features. Azizi et al. [22] proposed the hybrid feature selection technique that enhances the identification accuracy of the severity of the cavitation of the centrifugal pump. Dave et al. [23] applied ensemble empirical mode decomposition (EEMD), Walsh–Hadamard transform (WHT) and discrete wavelet transform (DWT) on bearing test rig under different health conditions and further utilize common information feature ranking algorithm to rank the features extracted from processed signals which then used in different classifier to identify the defects in bearing. Vashishtha and Kumar [24] proposed the ameliorated salp swarm optimization to choose the optimal parameters of the VMD. At optimal parameters, the VMD decomposes the raw signal under different health condition of the centrifugal pump. Sensitive features have extracted from the processed signals whose ranking is done through Pearson's correlation coefficient (PCC) which then fed into extreme learning machine classifier. Vashishtha and Kumar [25] used F-score technique to rank the features obtained from different health condition of the Pelton wheel and the expectation maximization-based principal component analysis to eliminate the redundancy and ambiguity of the data.

In this paper, a Gaussian ant lion optimization technique has been proposed to select the prominent features to obtain the best classification accuracy in order to identify impeller defect in the pump. The remaining paper has the following order: the preliminary of K-nearest neighbour (KNN) and ant lion optimizer (ALO) has been explained in Sect. 2. The proposed fault feature selection technique along with modifications done in ALO has been summarized in Sect. 3. The experimentation part is explained in Sect. 4. The results and discussion have been explained in Sect. 5. Finally, the conclusion of the work is stated in Sect. 6.

2 Background

2.1 *K-Nearest Neighbour*

The distance of each sample in the new sample set is calculated using KNN classifier. For classification, the sample from the set is assigned to a class that contains most sample of this set closet to K instances.

For calculating the distance, the Euclidean distance approach is applied in the classifier which is calculated between two samples $M_1 = (m_{11}, m_{12}, \ldots, m_{1L})$ and $M_2 = (m_{21}, m_{22}, \ldots, m_{2L})$ using Eq. (1) as given below.

$$\text{Dist}(M_1, M_2) = \sqrt{\sum_{l=1}^{L} (m_{1l} - m_{2l})^2} \tag{1}$$

The KNN is a multiclassification method that differentiate different health states of the centrifugal pump.

2.2 *Ant Lion Optimization (ALO)*

Mirjalili designed the novel bio-inspired optimization algorithm called ant lion optimization (ALO) which imitates the hunting behaviour of antlions [25]. Antlions are classified in Myrmeleontidae family [25]. Initially, the antlion digs a hole in the sand via its lava to catch its prey, whereas prey tries to rush from the trap. Then antlion throws sand on the outer periphery of the dig to slide the victim into the dig. Antlion consumes that prey and makes a new hole to hunt another prey. Generally, the size of the hole depends on the hunger level of the antlion and shape of the moon. When the antlion is hungrier and the moon is full, he digs out a larger trap to increase the chance of his survival [26].

ALO is composed of antlions and ants. Antlions are the fittest ones and thus known as hunters. The antlions only change their location when they replace any ant. On the other hand, the ant searches space through the random walk. While searching space, if it got entrapped in the dig made by antlion, then it can be consumed by the antlion. The ant updates its position as per Eq. (2).

$$\text{Ant}_i^t = \frac{L_A^t + L_E^t}{2} \tag{2}$$

where L_A^t and L_E^t indicates random walk about roulette wheel selected antlion and elite antlion, respectively.

The procedure of random walking of Ant_i^t around the Antlion_j^t is illustrated in the following Eq. (3).

$$L_j^t = \frac{(Y_i - a_i) \times \left(d_i - c_i^t\right)}{\left(b_i^t - a_i\right)} + c_i \tag{3}$$

where L_j^t represents the position of Ant_i after carrying out the random walk around Antlion_j for iteration t. a_i and b_i are minimum and maximum steps of random walk, respectively. c and d are lower and upper bounds of the random walk. Y_i in the above equation is defined in Eq. (4).

$$Y(t) = [0,\ \text{cumsum}(2r(t_1) - 1);\ \text{cumsum}(2r(t_2) - 1);\ \ldots;\ \text{cumsum}(2r(t_T) - 1)] \tag{4}$$

where cumsum is used to calculate the cumulative sum which gathers the successive random steps to form the random walk till time t; $x(t)$ is termed as stochastic function formulated as per Eq. (5)

$$x(t) = \begin{cases} 1, & \text{if} \quad \text{rand} > 0.5 \\ 0, & \text{if} \quad \text{rand} \le 0.5 \end{cases} \tag{5}$$

where rand is the random number obtained from a uniform distribution. c and d are calculated from Eqs. (6) and (7), respectively.

$$c_i^t = \begin{cases} \frac{\text{lb}}{I} + Y_{\text{Antlion}_j}^t, & \text{if} \quad \text{rand} < 0.5 \\ \frac{-\text{lb}}{I} + Y_{\text{Antlion}_j}^t, & \text{otherwise} \end{cases} \tag{6}$$

$$d_i^t = \begin{cases} \frac{\text{ub}}{I} + Y_{\text{Antlion}_j}^t, & \text{if} \quad \text{rand} > 0.5 \\ \frac{-\text{ub}}{I} + Y_{\text{Antlion}_j}^t, & \text{otherwise} \end{cases} \tag{7}$$

where lb and ub represent lower and upper limits for dimension *i*, respectively. The parameter I monitors the exploitation rate which is calculated as per Eq. (8).

$$I = 10^v \frac{t}{T} \tag{8}$$

where the maximum iteration is represented by the T. v depicted in Eq. (8) represents the number of iterations which adjusts its value to update the level of accuracy. In the end, the process of selection is followed when the fitter ant replaces the antlion [14].

3 The Proposed Gaussian Mutation Strategy in Ant Lion Optimization

The modifications done in ALO is presented in this section. By introducing the Gaussian mutation approach that utilizes the idea of population division and rebuilding, the searchability of the ALO algorithm can be improved.

3.1 *Modification in Ant Lion Optimization (ALO)*

The Gaussian mutation technique was suggested by Back and Schwefel [27] to boost the searchability of metaheuristic algorithms [28–31]. New offspring have been developed near the parent candidate solution in this technique. The Gaussian mutation technique makes tiny steps to examine each corner of the solution space and avoid diversity loss during the search. The Gaussian density function is defined as follows:

$$f_{0,\sigma^2}(\theta) = \frac{1}{\sigma\sqrt{(2\pi)}} e^{\frac{-\theta^2}{2\sigma^2}} \tag{9}$$

where σ^2 indicates the variance of the candidate solution. The Gaussian density function is reduced at mean 0 and standard deviation 1. The random vector Gaussian (θ) is generated using the density function as given in Eq. (10).

$$G_i^{'} = G_i(1 + \text{Gaussian } (\theta)) \tag{10}$$

where G_i is ith mutated candidate solution and Gaussian (θ) is a random vector.

3.2 *Fault Feature Selection Technique*

From the literature survey, it has been noticed that with smaller number of features the classification accuracy achieved from different classifiers is almost same. But, if the number of features is further minimized, there may be a chance of reduction in classification accuracy. In order to maintain the same classification accuracy, a fitness function is incorporated having output of classification with number of features as represented in Eq. (11).

$$\text{fitness} = \gamma \text{Err}_{\text{Rate(Tf)}} + \delta \frac{|\text{Nf}|}{|\text{Tf}|} \tag{11}$$

where γ and δ represent the weight parameters whose values $\gamma = [0,1]$ and $\delta = [1 - \gamma]$ are selected on the basis of the assessment function. $\text{Err}_{\text{Rate(Tf)}}$ indicates the error of the KNN classifier. Nf and Tf represent the number of features selected at a given time and the total number of features, respectively. Figure 1 shows the overall model for the proposed GLAO-KNN.

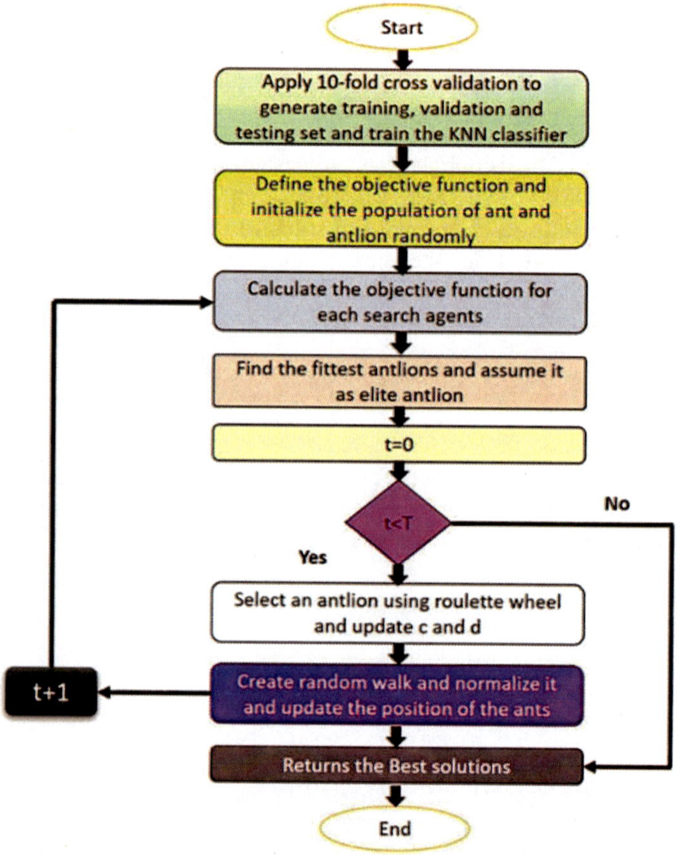

Fig. 1 Model of GALO-KNN

4 Experimentation

4.1 *Fault Feature Selection Technique*

In this segment, the scheme of feature selection has been used over dataset obtained from the centrifugal pump test rig whose pictorial view is shown in Fig. 2. The operating frequency of the pump shaft is 46.67 Hz (i.e. 2800 rpm). In this study, different health conditions of impeller are considered. The impeller has vanes. These vanes suck water axially and give kinetic energy to the water by striking it. Using this kinetic energy, water flows in a radial outward direction. When water strikes the vanes, a chaotic phenomenon generates in the water.

A PCB make uniaxial accelerometer having sensitivity of 100 mV/g is utilized to acquire vibration signals. It has been mounted on the impeller casing. National Instrument make 24-bit 4 channel DAQ is used to acquire the vibration signal, which is set at 70 kHz sampling frequency. Data for 0.1 s having 7000 data points are taken to analyse the signal. The research is done under different conditions of the impeller as shown in Fig. 3. The different health conditions of the impeller are healthy, clogging, blade cut and wheel cut. The raw vibration signals of the corresponding health conditions are shown in Fig. 4.

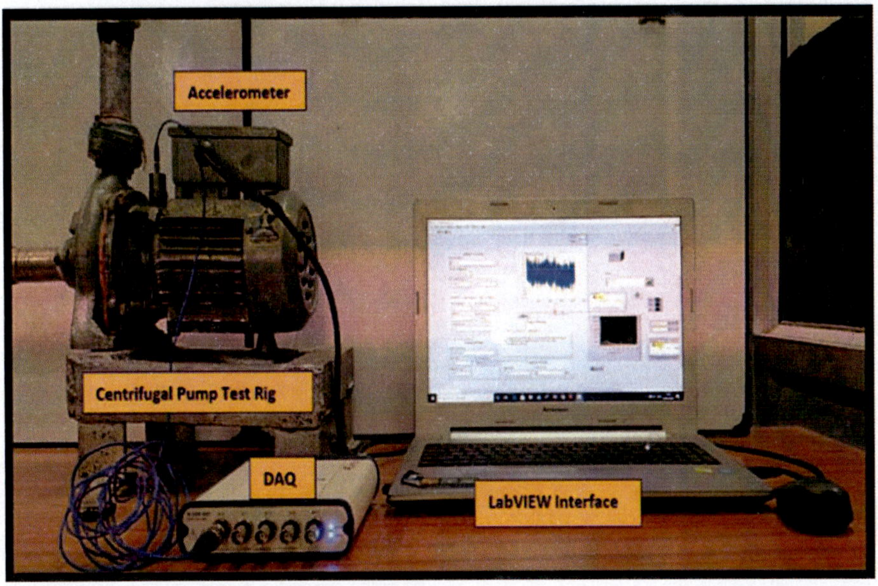

Fig. 2 A pictorial view of centrifugal pump test rig

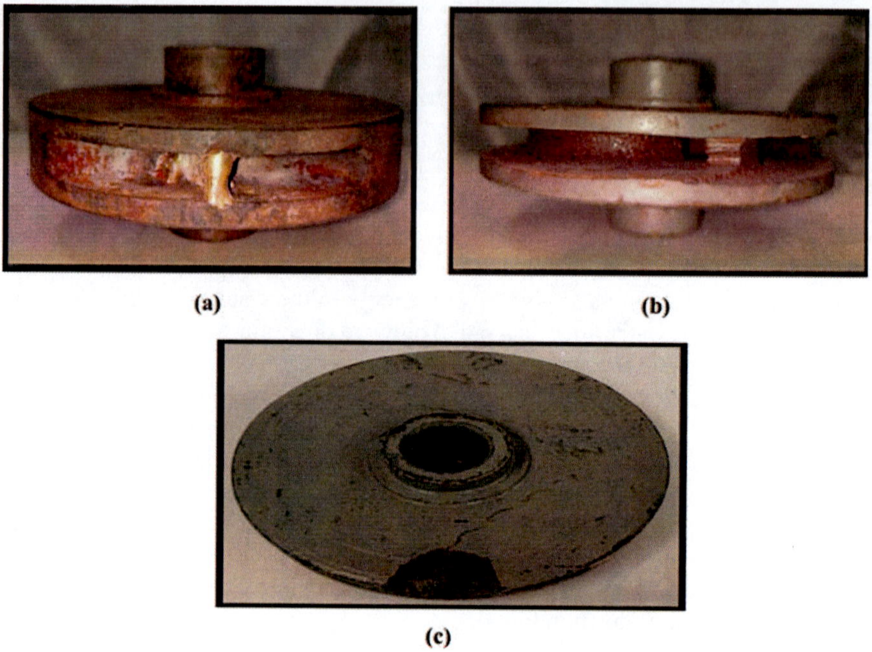

Fig. 3 Different health conditions of impeller **a** clogging, **b** blade cut and **c** wheel cut

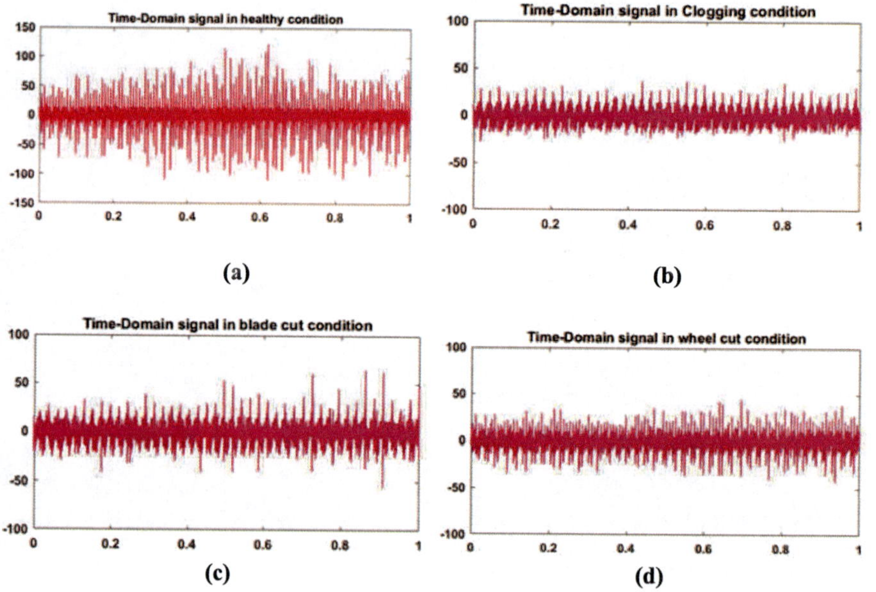

Fig. 4 Raw vibration signal under **a** healthy condition, **b** clogged condition, **c** blade cut condition and **d** wheel cut condition of impeller

4.2 Feature Extraction

A total of 40 vibration signals (10 each from each health condition) are taken into consideration for analysis. Thereafter, twenty-nine features are extracted from these signals. The different features taken into consideration with their definition are tabulated in Table 1. The normalization of features in the range of [0 1] is carried out as per expression given in Eq. (12).

$$\text{Nor\_feat} = \frac{\text{Feat} - \min(\text{Feat})}{\max(\text{Feat}) - \min(\text{Feat})} \quad (12)$$

A KNN with a filter-based feature selection technique is executed to test the accuracy of the chosen features. The dataset obtained from the vibration signals in terms of features is divided into training, test and validation set using tenfold cross-validation.

4.3 Evaluating Measurement Parameters

The following statistical measures are calculated to verify each run of the proposed work and also used for comparison with other algorithms. Each statistical parameter is averaged over 10 runs.

1. Average classification accuracy is formulated as follows:

$$\text{Avg}_{\text{acc}} = \frac{1}{N}\sum_{k=1}^{N} {\text{Avg}_{\text{acc}}}^{k} \quad (13)$$

   ${\text{Avg}_{\text{acc}}}^{k}$ indicates the average accuracy.
2. Mean fitness function is formulated as follows:

$$\text{mean} = \frac{1}{N}\sum_{k=1}^{N} g_k^{*} \quad (14)$$

   It represents average of the mean value.

3. Best fitness

$$\text{Best} = \min_{k} g_k^{*} \quad (15)$$

   It indicates the average minimum value of fitness function.

Table 1 Definitions of functions

Feature's notation and formula	Features notation and formula	Features notation and formula
Mean $A1 = \frac{1}{N}\sum_{i=1}^{N} x_i$	Kurtosis $A11 = \frac{1}{\sigma^4}\left(\frac{1}{N-1}\sum_{i=1}^{N}(x_i-\mu)^4\right)$	Median absolute deviation $A21 = \text{median}(\lvert x\_i - \text{median}(x\_i)\rvert)$
Root mean square $A2 = \sqrt{\frac{1}{N}\sum_{i=1}^{N} x_i^2}$	Crest factor $A12 = \frac{\max(\lvert x\rvert)}{\sqrt{\frac{1}{N}\sum_{i=1}^{N} x_i^2}}$	Rate zero crossing $A22 = \frac{\text{number of zero crossings}}{\text{total number of points}}$
Root $A3 = \left(\frac{1}{N}\sum_{i=1}^{N}\sqrt{\lvert x_i\rvert}\right)^2$	Shape factor $A13 = \frac{\sqrt{\frac{1}{N}\sum_{i=1}^{N} x_i^2}}{\frac{1}{N}\sum_{i=1}^{N}\lvert x_i\rvert}$	Entropy $A23 = -\sum_{1=1}^{N} h(x_i)\log_2 h(x_i)$
Maximum value $A4 = \max(\lvert x\rvert)$	Impulse factor $A14 = \frac{\max(\lvert x\rvert)}{\frac{1}{N}\sum_{i=1}^{N}\lvert x_i\rvert}$	Histogram upper bound $A24 = \max(x_i) + \frac{\max(x_i) - \min\left(\frac{x_i}{N-1}\right)}{2}$
Peak to peak $A5 = \max(\lvert x\rvert) - \min(\lvert x\rvert)$	Clearance factor $A15 = \frac{\max(\lvert x\rvert)}{\left(\frac{1}{N}\sum_{i=1}^{N}\sqrt{\lvert x_i\rvert}\right)^2}$	Histogram lower bound $A25 = \max(x_i) - \frac{\max(x_i) - \min\left(\frac{x_i}{N-1}\right)}{2}$
Standard deviation $A6 = \sqrt{\frac{1}{N-1}\sum_{i=1}^{N}(x_i-\mu)^2}$	Skewness factor $A16 = \frac{1}{\sigma^3}\frac{\left(\frac{1}{N-1}\sum_{i=1}^{N}(x_i-\mu)^3\right)}{\left(\sqrt{\frac{1}{N}\sum_{i=1}^{N} x_i^2}\right)^2}$	Activity $A26 = \frac{1}{N-1}\sum_{i=1}^{N}(x_i-\mu)^2$
Median $A7 = \frac{50(N+1)}{100}$ th observation	Kurtosis factor $A17 = \frac{1}{\sigma^4}\frac{\left(\frac{1}{N-1}\sum_{i=1}^{N}(x_i-\mu)^4\right)}{\left(\sqrt{\frac{1}{N}\sum_{i=1}^{N} x_i^2}\right)^4}$	Variance $A27 = \frac{\sum_{i=1}^{N}\left(x(i) - x_{\text{avg}}\right)^2}{N-1}$
25th percentiles $A8 = \frac{25(N+1)}{100}$ th observation	Geometric mean $A18 = \left(\prod_{i=N}^{N} x_i\right)^{1/N}$	Wavelet energy decomposition $A28 = \frac{\sum_{n=1}^{N}\lvert x_i(n)\rvert^2}{\sum_{k=0}^{2j-1}\sum_{n=1}^{N}\lvert x_k(n)\rvert^2}$
75th percentiles $A9 = \frac{75(N+1)}{100}$ th observation	Root sum of squares $A19 = \sqrt{\sum_{i=1}^{N}\lvert x_i\rvert^2}$	Normal negative log likelihood for single Gaussian $A29 = -\sum_{i=1}^{N}\log\left[\frac{1}{\sigma\sqrt{2\pi}}\exp\frac{-(x_i-\mu)^2}{2\sigma^2}\right]$
Skewness $A10 = \frac{1}{\sigma^3}\left(\frac{1}{N-1}\sum_{i=1}^{N}(x_i-\mu)^3\right)$	Mean absolute deviation $A20 = \text{mean}(\lvert x_i - \mu\rvert)$	

4. Worst fitness function

$$\text{Worst} = \max_{k} g_k^* \tag{16}$$

It represents the average maximum value of fitness function.

5. Average computational time

$$\text{Avg}_{\text{CT}} = \frac{1}{N}\sum_{k=1}^{N} \text{Avg}_{\text{CT}}{}^{k} \tag{17}$$

It computes the average of computational time.

6. Average selected features

$$\text{Avg}_{\text{select}} = \frac{1}{N}\sum_{k=1}^{N} \text{Avg}_{\text{select}}{}^{k} \tag{18}$$

It indicates the average number of features selected at a time.

5 Result and Discussion

Results obtained through the proposed work have been discussed here. The results are presented in Table 2 in terms of accuracy, fitness, computational time and number of features selected at a time for analysis. The proposed method has also been

Table 2 Results of GALO-KNN

Run	Accuracy	Fitness	Number of features	Time (s)
1	**0.88067**	0.1368	15	6.1184
2	0.84925	0.1441	16	6.3400
3	0.80647	0.2470	18	6.1633
4	0.81549	0.1561	22	6.4004
5	0.82791	0.1599	20	6.2147
6	0.80951	0.2452	16	6.5528
7	0.81825	0.2136	26	6.2189
8	0.86439	0.1124	12	6.4680
9	0.81467	0.1880	21	6.2524
10	0.82591	0.1769	20	6.3239

Table 3 Results of PSO-KNN

Run	Accuracy	Fitness	Number of features	Time (s)
1	0.78439	0.2469	26	8.2549
2	0.80794	0.1980	18	7.3685
3	0.81594	0.2029	21	7.5271
4	0.79527	0.2368	22	7.5242
5	**0.82494**	0.1820	20	7.3224
6	0.79527	0.2199	21	7.4009
7	0.77247	0.2471	27	8.1389
8	0.79541	0.2082	19	7.4806
9	0.80462	0.2048	21	7.4712
10	0.78696	0.2257	23	7.4003

compared with PSO-KNN and GA-KNN, and their results are given in Tables 3 and 4. From Table 2, it is observed that the high classification accuracy of 88.07% is achieved while considering 15 features in analysis. The 15 most critical features which contribute to achieving the high accuracy are entropy, geometric mean, median absolute deviation, 75th percentile, shape factor, mean, root, impulse factor, clearance factor, rate of zero crossing, kurtosis, 25th percentile, maximum value, peak to peak and kurtosis factor. Whereas in the case of PSO-KNN and GA-KNN, the highest accuracy achieved is 82.49% and 81.14%, respectively as given in Tables 3 and 4. In both the cases, corresponding number of features are 20.

Table 4 Results of GA-KNN

Run	Accuracy	Fitness	Number of features	Time (s)
1	0.77591	0.2380	24	7.5295
2	0.76948	0.2591	27	802,591
3	0.79275	0.2067	22	7.0043
4	0.78934	0.2479	25	7.5612
5	0.79865	0.2261	23	7.1192
6	0.75321	0.2695	29	8.6542
7	0.80624	0.1973	21	7.1051
8	0.79843	0.2415	23	8.1569
9	**0.81145**	0.1862	20	7.0152
10	0.79725	0.2486	26	8.1135

Table 5 Comparison of measures between different algorithms

Measures	Algorithms		
	GALO-KNN	PSO-KNN	GA-KNN
Average accuracy	0.831252	0.798294	0.789271
Average selected features	18.6	21.8	24
Average computational time	6.30528	7.5889	7.65182
Average fitness function	0.17807	0.21793	0.23209
Best fitness function	0.1124	0.1890	0.1862
Worst fitness function	0.2470	0.2471	0.2591

Table 6 Comparison of measures between different classifiers

Measures	Algorithms		
	GALO-KNN	GALO-random forest	GALO-decision tree
Average accuracy	0.831252	0.75291	0.74126
Average selected features	18.6	23	27
Average computational time	6.30528	8.51923	9.49556
Average fitness function	0.17807	0.23876	0.29292
Best fitness function	0.1124	0.2037	0.2534
Worst fitness function	0.2470	0.2697	0.2997

5.1 Comparison with Other Art of Work

Considering results in Tables 1, 2 and 3, the overall efficacy of the proposed GALO-KNN against other methods such as PSO-KNN and GA-KNN is summarized in Table 5. From Table 5, it is clear that the proposed GALO-KNN outperforms the other art of works in terms of average classification accuracy and computational time. Apart from this, the proposed GALO-KNN attains the finest average value of fitness function. The worst case of the average fitness function of the proposed method is also less than the other art of work. The reason is the great ability of the ALO to avoid strucking in local optima. Also, the proposed algorithm has been compared with different classifiers whose results are tabulated in Table 6. The results indicate the superiority of the proposed method.

5.2 Significance Level of the Proposed Method

The results obtained by GALO-KNN in terms of accuracy have undergone the One-Way-ANOVA test to show whether the results are statistically significant or not. The stated hypotheses of the One-Way-ANOVA test are as follows:

Table 7 Results of One-Way-ANOVA test

	Algorithms			F value	p value	Hypothesis
	GALO	PSO	GA	6.86	0.0039	H1
N	10	10	10			
$\sum X$	186	218	240			
Mean	18.6	21.8	24			
$\sum X^2$	3606	4826	5830			
Std. Dev	4.0332	2.8597	2.7889			

H0 (null hypothesis): It represents that there is no significant difference between the accuracy of the GALO-KNN and other art of work.

H1 (alternative hypothesis): It represents that there is a significant difference between the accuracy of GALO-KNN and other art of work.

In order to show the significance level of the obtained results using the One-Way-ANOVA test, the p value has been compared with the stated value ($\alpha = 0.01$ or $\alpha = 0.05$).

- If the obtained p value is greater than 0.01, then H0 (null hypothesis) is accepted and H1 (alternative hypothesis) is rejected which meant no significant difference exists between GALO-KNN and other art of work.
- If the obtained p value is less than 0.05, then H1 (alternative hypothesis) is accepted and H0 (null hypothesis) is rejected which meant a significant difference exists between GALO-KNN and other art of work.

Table 7 shows the details of the results of the One-Way-ANOVA test. It has been observed from the Table 7 that the attained p value (0.0039) is less than that of α (0.01); that is why H1 hypothesis has been accepted, implying that the obtained results are of significant.

6 Conclusion

A bio-inspired ant lion optimization with a Gaussian mutation strategy is introduced for filter-based feature selection. To determine classification accuracy, the KNN classifier was utilized to classify feature subset. The proposed work is compared to PSO-KNN and GA-KNN in terms of efficacy. The proposed work has attained the highest classification accuracy with less computational time as compared to other art of work. The outcomes of the present study are given below:

- GALO-KNN successfully handles the complex data of different health conditions of centrifugal pump and efficiently classifies the type of impeller defects with an accuracy of 88.06%.

- Fifteen prominent features have been identified through proposed technique that contributes in detecting the impeller defects of centrifugal pump.
- GALO-KNN demonstrated to be a reliable and consistent strategy for finding the most important features among a large number of them.

The outcomes of the study suggested that the GALO has the potential to diagnose impeller defects in centrifugal pump and other rotating components. The filter-based selection technique efficiently handles the high dimensionality data and reduces the redundancy to achieve the highest performance of the classification models.

Acknowledgements Authors are grateful to All India Council for Technical Education (AICTE), New Delhi, India, for financial support (No. 8-29/RIFD/RPS-NDF/Policy-1/2018-19) to carry out this research work.

References

1. Berndt F, Van Bennekom A (2001) Pump shaft failures—a compendium of case studies. Eng Fail Anal 8:135–144
2. Muralidharan V, Sugumaran V, Indira V (2014) Fault diagnosis of monoblock centrifugal pump using SVM. Eng Sci Technol Int J 17:152–157
3. Kumar A, Gandhi CP, Zhou Y, Kumar R, Xiang J (2020) Improved deep convolution neural network (CNN) for the identification of defects in the centrifugal pump using acoustic images. Appl Acoust 167:107399
4. Muralidharan V, Sugumaran V (2012) A comparative study of Naïve Bayes classifier and Bayes net classifier for fault diagnosis of monoblock centrifugal pump using wavelet analysis. Appl Soft Comput J 12:2023–2029
5. Zhang J et al (2018) Experimental investigation on the sharpness reduction of an axial piston pump with reinforced shell. Appl Acoust 142:36–43
6. Pan Y, Li Y, Huang M, Liao Y, Liang D (2018) Noise source identification and transmission path optimisation for noise reduction of an axial piston pump. Appl Acoust 130:283–292
7. Kumar A, Kumar R (2018) Oscillatory behavior-based wavelet decomposition for the monitoring of bearing condition in centrifugal pumps. Proc Inst Mech Eng Part J J Eng Tribol 232:757–772
8. Tan CK, Mba D (2005) Limitation of acoustic emission for identifying seeded defects in gearboxes. J Nondestruct Eval 24:11–28
9. Cao S, Hu Z, Luo X, Wang H (2021) Research on fault diagnosis technology of centrifugal pump blade crack based on PCA and GMM. Meas J Int Meas Confed 173:108558
10. Jiang W, Li Z, Zhang S, Wang T, Zhang S (2021) Hydraulic pump fault diagnosis method based on EWT decomposition denoising and deep learning on cloud platform. Shock Vib 2021:1–18
11. Kumar A, Gandhi CP, Zhou Y, Kumar R, Xiang J (2020) Variational mode decomposition based symmetric single valued neutrosophic cross entropy measure for the identification of bearing defects in a centrifugal pump. Appl Acoust 165:107294
12. Ahmad Z, Rai A, Maliuk AS, Kim J-M (2020) Discriminant feature extraction for centrifugal pump fault diagnosis. IEEE Access 8:165512–165528
13. Al Tobi M, Bevan G, Wallace P, Harrison D, Okedu KE (2020) Faults diagnosis of a centrifugal pump using multilayer perceptron genetic algorithm back propagation and support vector machine with discrete wavelet transform-based feature extraction. Comput Intell, pp 1–26. https://doi.org/10.1111/coin.12390

14. Zawbaa HM, Emary E, Grosan C (2016) Feature selection via chaotic antlion optimization. PLoS ONE 11:1–21
15. Zawbaa HM, Emary E, Parv B (2016) Feature selection based on antlion optimization algorithm. In: Proceedings of the 2015 IEEE world conference complex systems WCCS 2015. https://doi.org/10.1109/ICoCS.2015.7483317
16. Emary E, Zawbaa HM (2019) Feature selection via Lèvy Antlion optimization. Pattern Anal Appl 22:857–876
17. Vashishtha G, Kumar R (2021) An effective health indicator for Pelton wheel using Levy flight mutated genetic algorithm. Meas Sci Technol. https://doi.org/10.1088/1361-6501/abeea7
18. Vashishtha G, Chauhan S, Singh M, Kumar R (2021) Bearing defect identification by swarm decomposition considering permutation entropy measure and opposition-based slime mould algorithm. Measurement 178:109389
19. Chauhan S, Singh M, Kumar Aggarwal A (2020) An effective health indicator for bearing using corrected conditional entropy through diversity-driven multi-parent evolutionary algorithm. Struct Heal Monit. https://doi.org/10.1177/1475921720962419
20. Chauhan S, Singh M, Aggarwal AK (2021) Bearing defect identification via evolutionary algorithm with adaptive wavelet mutation strategy. Measurement 179:109445
21. Wang M, Wu C, Wang L, Xiang D, Huang X (2019) A feature selection approach for hyperspectral image based on modified ant lion optimizer. Knowledge-Based Syst 168:39–48
22. Azizi R, Attaran B, Hajnayeb A, Ghanbarzadeh A, Changizian M (2017) Improving accuracy of cavitation severity detection in centrifugal pumps using a hybrid feature selection technique. Measurement 108:9–17
23. Dave V, Singh S, Vakharia V (2020) Diagnosis of bearing faults using multi fusion signal processing techniques and mutual information. Indian J Eng Mater Sci 27:878–888
24. Vashishtha G, Kumar R (2021) Centrifugal pump impeller defect identification by the improved adaptive variational mode decomposition through vibration signals centrifugal pump impeller defect identification by the improved adaptive variational mode decomposition through vibration sig. Eng Res express 3:035041
25. Vashishtha G, Kumar R (2021) Pelton wheel bucket fault diagnosis using improved Shannon entropy and expectation maximization principal component analysis. J Vib Eng Technol. https://doi.org/10.1007/s42417-021-00379-7
26. Mirjalili S (2015) The ant lion optimizer. Adv Eng Softw 83:80–98
27. Emary E, Zawbaa HM, Hassanien AE (2016) Binary ant lion approaches for feature selection. Neurocomputing 213:54–65
28. Bäck T, Schwefel H-P (1993) An overview of evolutionary algorithms for parameter optimization. Evol Comput 1:1–23
29. Luo J et al (2018) An improved grasshopper optimization algorithm with application to financial stress prediction. Appl Math Model 64:654–668
30. Salgotra R, Singh U (2017) Application of mutation operators to flower pollination algorithm. Expert Syst Appl 79:112–129
31. Yao X, Liu Y, Lin G (1999) Evolutionary programming made faster **3**:82–102

Vibration Monitoring of Defective Shaft Bearing System

Neel Satwara and V. N. Patel

Abstract The most challenging task in a condition monitoring of a shaft bearing system is to identify a bearing fault and a shaft fault in presence of compound faults. In the present experimental study, the vibrations generated by the shaft rotor bearing system in presence of shaft crack(s) and bearing races fault(s) have been analyzed. Vibration signals of healthy and defective shaft bearing system have been captured using of National Instrumentation (NI) data acquisition module and LabVIEW software. The captured vibration signals were analyzed in time domain and frequency domain for the defect detection using MATLAB software. The critical speed of the shaft rotor system has been computed using Dunkerley's equation and also verified experimentally. The presence of peaks at critical frequency (F_{cr}) and its harmonics, while shaft rotates in subcritical speed (1/3, 1/2) range, admit the presence of shaft crack. Moreover, the change in regular shape of orbit plots at the subcritical speed also identifies the cracked shafts. The wavelet transformation has been implemented to reduce the noise from the raw signal used for orbit plots. Six different shafts of 16 mm diameter having 0.2 mm wide and 4 mm or 6 mm depth open crack near to rotor or bearing end or cracks at both locations were used to study the effect of crack location and size on vibration amplitude and frequency. The maximum vibration amplitude has been noticed for the shaft having crack near to bearing end in presence of bearing defect.

Keywords Vibration analysis · Frequency analysis · Fault identification · Wavelet transformation

1 Introduction

The condition monitoring plays vital role in rotating machinery to avoid serious damages or even catastrophic failures. The fault detection in a shaft bearing system before it becomes critical is the greatest challenge during condition monitoring of

N. Satwara · V. N. Patel (✉)
Mechatronics Department, G H Patel College of Engineering & Technology, Vallabh Vidyanagar, Gujarat, India
e-mail: vinodgcet1@gmail.com

V. K. Gupta et al. (eds.), *Recent Advances in Machines and Mechanisms*, Lecture Notes in Mechanical Engineering, https://doi.org/10.1007/978-981-19-3716-3_24

rotating machines. The failure in shaft bearing system may occurs due to the overloading or cyclic loading. The failure of the shaft under cyclic loading condition even though the acting load is below the load causes yielding of the material is known as fatigue failure. The material failure occurs as a result of excessive cyclic loading is known as fatigue fracture. Prior to final fatigue fracture, many different micro fractures are created and due to exertion of cyclic loading the cracks propogates. The fatigue cracks normally initiate at stress concentration and structural discontinuity.

Vibration-based condition monitoring (VCM) is a well-known and proven technique to detect the defect in their early stage. In the rotating machinery, vibrations can be generated due to unbalanced rotating or reciprocating elements, misalignment, wear, and looseness. Moreover, the presence of defect changes the dynamic characteristics of machine elements. Due to the changes in dynamic characteristics, the statistical parameters like RMS value, crest factor, skewness, and kurtosis value also change, which can be noticed from time domain analysis of vibration signal, while the changes in frequency can be noticed from vibration spectra. Moreover, the orbit plots created from the vibration data from two orthogonal measurements taken simultaneously show the path a rotor takes as at vibrates during operation.

Many researchers have performed theoretical and experimental vibration analysis of shaft bearing system in presence of defect. The different approaches/methods like finite element analysis, wavelet transform, artificial intelligence (AI) techniques such as artificial neural networks (ANNs), genetic algorithms (GAs), fuzzy logic (FL), and hybrid techniques adopted by various researchers have been reviewed by Kushwaha and Patel [1]. Three different crack models, like a breathing crack model, a switching crack model, and an open crack model, have been developed, and transient responses have been compared in reference [2]. It has been concluded that breathing model is the most efficient to predict the vibration behavior of cracked rotor. In the experimental vibration study, researchers have observed that the presence of fatigue crack on shaft reduces its first-order critical speed, while high-order harmonic components become dominant in vibration spectra [3]. The authors have also noticed two-circle topological orbit in subcritical speed range of shaft rotation. The unique dynamic behavior of the orbits and frequency spectra is conclusive criteria for practical crack diagnosis [4]. Harmonic analysis has been performed for the shaft bearing modeled in ANSYS, with consideration of shaft crack of different depths and locations by Kumar et al. [5]. They have also observed the variation in vibration displacement amplitude with increase of crack depth. The frequency-time domain (short time Fourier transform) approach has been implemented by Silani et al. to evaluate the dynamic response of a rotor with a breathing crack and detected small crack in shaft [6]. The behavior of whirl orbits in subcritical range has been found useful tool to distinguish breathing crack and open crack model [7]. The envelope analysis has been used to detect transient components of the vibration signals produced by crack [8]. The rotor fault has been detected through auto-correlation and power spectral density functions of the rotor vibration response measured at the bearings [9].

The fatigue cracks may also generate on the elements of rolling bearings. The wavelet analysis can be implemented for vibration signal captured from defective rolling bearings to predict the fatigue crack size of bearing races. The dynamic model to study the vibration responses of defective bearings has been suggested by Shah and Patel [10–12]. The frequency peaks at defect characteristic defect frequencies have been noticed in presence of local and distributed defects on bearing races.

It reveals that the crack fault diagnostics though vibration monitoring play an important role for rotating machinery in the traditional and Industry 4.0 factory. Authors of the present paper could find few literature for experimental study as compared to theoretical study on dynamic behavior of cracked rotor. It has also been observed that the researchers have detected shaft cracks in absence of bearing defect or detected the bearing defect in absence of shaft cracks. Hence, in the present experimental study, the vibration signals generated by cracked shaft in presence and absence of defective bearing races have been captured. The fatigue cracks on shaft and bearings have been identified from vibration spectra and orbit plots.

2 The Experimental Test Rig

The experimental setup used to conduct experiments is shown in Fig. 1. As shown in Fig. 1a, 16-mm shaft of 800 mm length made of EN8 is supported on two SKF UCP202 ball bearings. This shaft receives power from 1HP DC electric motor through belt and pulley arrangement. The DC motor drive was used to change the rotational speed of motor. A 0.2 kg mass rotor (spur gear) is mounted at center of the shaft. Two accelerometers with 104 mV/g sensitivity have been mounted on bearing housing to capture vibration signals (acceleration) from X direction and Y direction. Four channel data acquisition card (9234 NI) receives signals from two accelerometers and transfer to computer for processing through LabVIEW software. It is worth to mention here that 4096 data samples for 2.5 s were captured during each experiment. The captured signals were analyzed in time domain and frequency domain using MATLAB software.

Fig. 1 Experimental setup **a** Shaft bearing system **b** Data acquisition in X and Y direction

3 Preparing for the Experiment

Total six shafts have been used for experimentation. Artificial cracks of 4 mm or 6 mm depth and 0.2 mm width at different locations were created on shafts using EDM wire cutting machine. Table 1 provides the detail of shaft defect. The critical speed of all shafts has been estimated using Dunkerley's equation (Eq. 1),

$$\text{Critical Frequency, } F_{cr} = \frac{0.4985}{\left(\Delta_1 + \Delta_2 + \Delta_{3+\cdots} \frac{\Delta_s}{1.27}\right)^{\frac{1}{2}}} \tag{1}$$

where $\Delta_1 = \frac{mga^2b^2}{3EIL}$, $\Delta_s = \frac{5Mgl^4}{384EI}$.

Table 1 also shows the natural frequency/critical frequency of all shafts estimated by Eq. 1 and tap test. Shaft 1 was supported on bearings and excited by mallet; the vibration signal generated was captured by accelerometer. The frequency peak at 41.56 Hz, natural frequency for transverse vibration can be observed in Fig. 2. The tap test was repeated for all shafts, and results are shown in Table 1. The variation in

Table 1 Description of different crack location on shaft

Shaft No.	Crack depth (mm)	Crack location on shaft	Experimental critical frequency F_{cr}, Hz	Theoretical value of critical frequency F_{cr}, Hz
1	4	Near to bearing end	41.56	33.37 Hz
2	4	Near to rotor	42.37	
3	6	Near to bearing end	39.14	
4	6	Near to rotor	39.14	
5	4	Near to bearing end and near to rotor	43.99	
6	6	Near to bearing end and near to rotor	40.35	

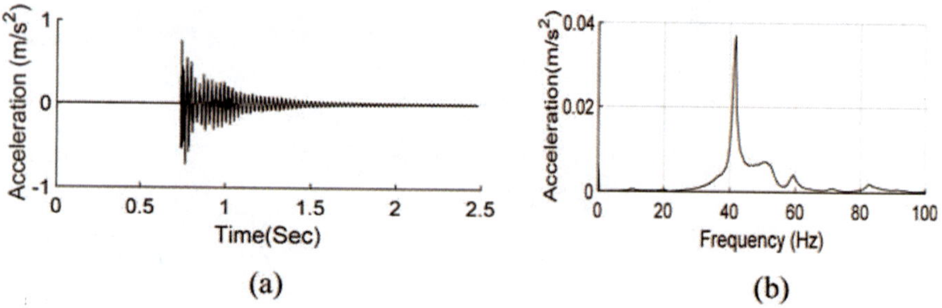

Fig. 2 Vibration signal during tap test of shaft 1 **a** Time domain **b** Frequency domain

computed frequency and measured frequency is due to variation in mass distribution and material removal for mounting the rotor on shaft.

4 Vibration Monitoring of Shaft Bearing System

Vibration acceleration amplitude of each healthy shaft (without fatigue crack) and cracked shaft (after creating artificial crack) at different rotating speeds has been captured. Vibration amplitude versus time data in *X* direction and *Y* direction have been collected and saved for further analysis.

4.1 Vibration Analysis of Health Shaft and Bearing System

The vibrations generated by healthy shafts have been captured and analyzed in time domain and frequency domain. The waterfall diagrams for healthy shafts (before creating crack) are shown in Fig. 3. The shaft rotational speed for each shaft varied from 100 to 2000 rpm. It has been noticed that the frequency peaks at shaft rotation frequency (F_s = rotational speed/60) are clearly visible for all shafts Fig. 3a–f.

The orbit plot (*Y* amplitude vs. *X* amplitude) has been plotted for healthy shafts. The closed loop orbit (elliptical shape) at higher speed is observed for healthy shafts as

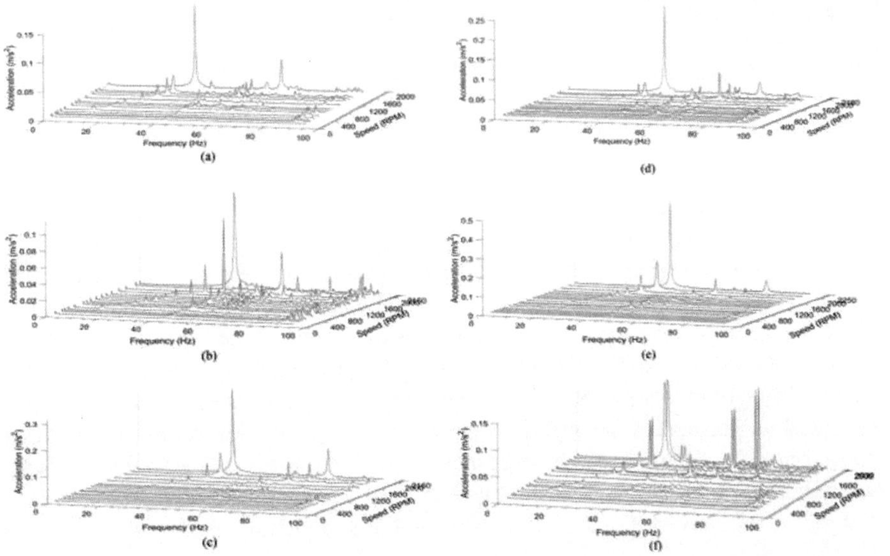

Fig. 3 Vibration responses of healthy shafts at different rotational speeds **a** Shaft 1 **b** Shaft 2 **c** Shaft 3 **d** Shaft 4 **e** Shaft 5 **f** Shaft 6

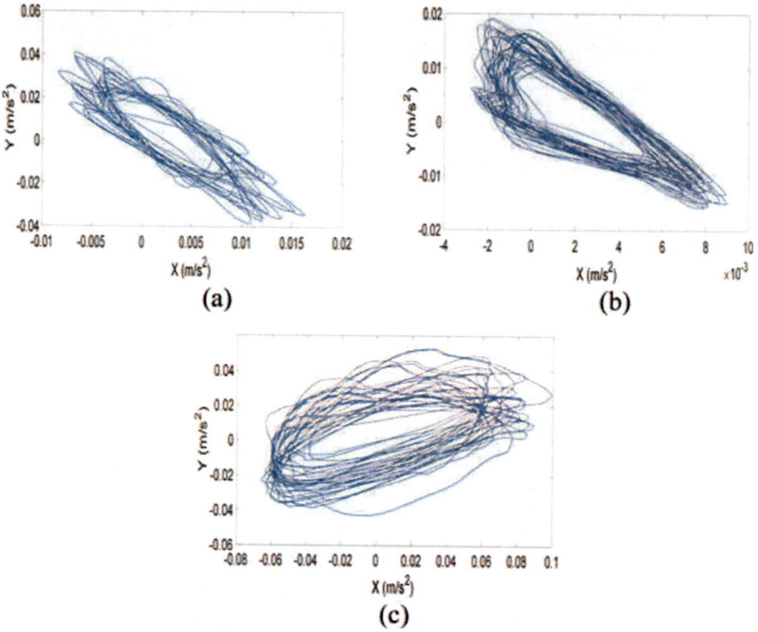

Fig. 4 Orbit plot of healthy shaft **a** Shaft 1 **b** Shaft 2 **c** Shaft 5

shown in Fig. 4. It is worth to mention here that the raw signals have been preprocessed using 'db4' wavelet. The signals were decomposed to level 8 to improve signal-to-noise ratio.

4.2 *Vibration Analysis of Cracked Shaft and Healthy Bearing*

Figure 5 shows the pictorial view of the artificial cracks generated on shaft near to bearing end and near to rotor.

The vibration responses at different shaft rotational speed for the shaft 1 after artificial creation of 4-mm-depth crack near to bearing end have been captured. The vibration spectra at different rotational speeds have been plotted in Fig. 6a. At lower speed, frequency peaks only at system natural frequency 100 Hz are visible. In addition to that at higher shaft rotational speed, frequency peaks at shaft rotational speed (F_s) and its harmonics are clearly visible. The frequency responses at subcritical speed range ($F_{cr}/3$ and $F_{cr}/2$) have been presented in Fig. 6b, c. The frequency peaks at shaft rotational speed 'F_s' and shaft critical speed 'F_{cr}' are clearly visible in Fig. 6b, c. Moreover, the change in orbit plot at subcritical speed is also visible in Fig. 6d.

The vibration responses at different shaft rotational speed for the shaft 2 after artificial creation of 4-mm-depth crack near to rotor have been captured. The vibration

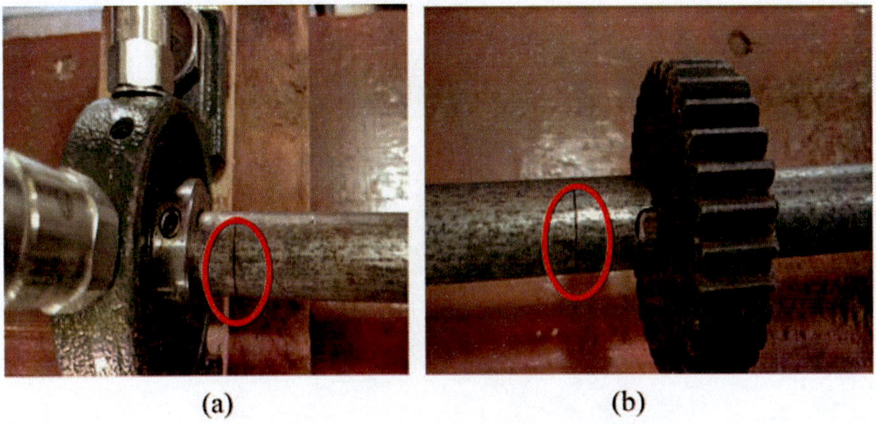

Fig. 5 Artificial fatigue cracks on shaft **a** Near to bearing end **b** Near to rotor

spectra at different rotational speeds have been plotted in Fig. 7a. The shaft rotational speed frequency peaks at shaft rotational speed (F_s) and its harmonics are clearly visible. The frequency responses at subcritical speed range ($F_{cr}/3$ and $F_{cr}/2$) have been presented in Fig. 7b, c. The frequency peaks at shaft rotational speed 'F_s' and shaft critical speed 'F_{cr}' along with their harmonics are clearly visible in Fig. 7b, c. Moreover, the change in orbit plot at subcritical speed is also visible in Fig. 7d. This is the evidence of fatigue crack in shaft.

The vibration responses at different shaft rotational speed for the shaft 3 after artificial creation of 6-mm-depth crack near to bearing end have been captured. The vibration spectra at different rotational speeds have been plotted in Fig. 8a. The shaft rotational speed frequency peaks at shaft rotational speed (F_s) and its harmonics are clearly visible. The frequency responses at subcritical speed range ($F_{cr}/3$ and $F_{cr}/2$) have been presented in Fig. 8b, c. The frequency peaks at shaft rotational speed 'F_s' and shaft critical speed 'F_{cr}' are clearly visible in Fig. 8b, c. Moreover, the change in orbit plot at subcritical speed is also visible in Fig. 8d. The amplitude of vibration increased as compared to shaft 1 due to increase of crack depth.

The vibration responses at different shaft rotational speed for the shaft 4 after artificial creation of 6-mm-depth crack near to rotor end have been captured. The vibration spectra at different rotational speeds have been plotted in Fig. 9a. The frequency responses at subcritical speed range ($F_{cr}/3$ and $F_{cr}/2$) have been presented in Fig. 9b, c. The frequency peaks at shaft rotational speed 'F_s' and shaft critical speed 'F_{cr}' are clearly visible in Fig. 9c. Moreover, the change in orbit plot at subcritical speed is also visible in Fig. 9d. The amplitude of vibration increased as compared to shaft 2 due to increase of crack depth.

The vibration responses at different shaft rotational speed for the shaft 5 after artificial creation of 4-mm-depth cracks near to bearing end and near to rotor have been captured. The vibration spectra at different rotational speeds have been plotted in Fig. 10a. The shaft rotational speed frequency peaks at shaft rotational speed (F_s)

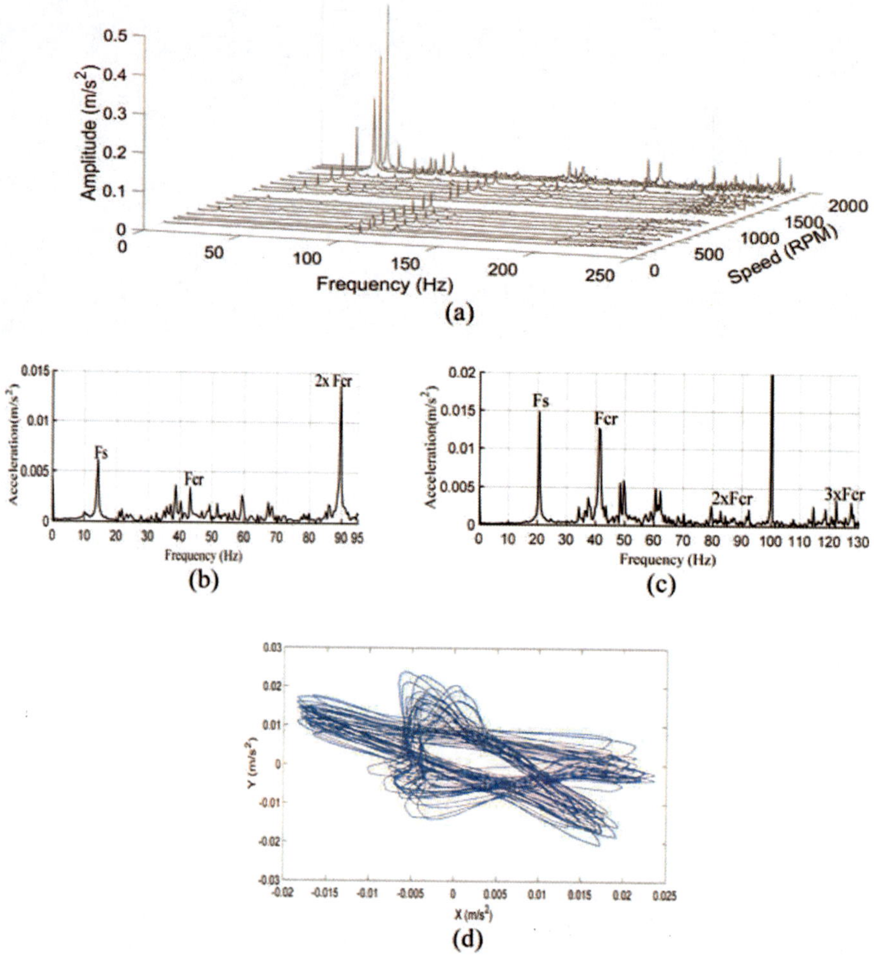

Fig. 6 Vibration response of cracked shaft 1 **a** Frequency response at different shaft rotational speed **b** Frequency response at 830 rpm ($F_{cr}/3$ range) **c** Frequency response at 1240 rpm ($F_{cr}/2$ range) **d** Orbit plot at subcritical speed

and its harmonics are clearly visible. The frequency responses at subcritical speed range ($F_{cr}/3$ and $F_{cr}/2$) have been presented in Fig. 10b, c. The frequency peaks at shaft rotational speed 'F_s' and shaft critical speed 'F_{cr}' are clearly visible in Fig. 10b, c. Moreover, the change in orbit plot at subcritical speed is also visible in Fig. 10d. The amplitude of vibration increased as compared to shaft 1 and shaft 2 due to presence of multiple fatigue cracks.

The vibration responses at different shaft rotational speed for the shaft 6 after artificial creation of 6-mm-depth cracks near to bearing end and near to rotor have been captured. The vibration spectra at different rotational speeds have been plotted

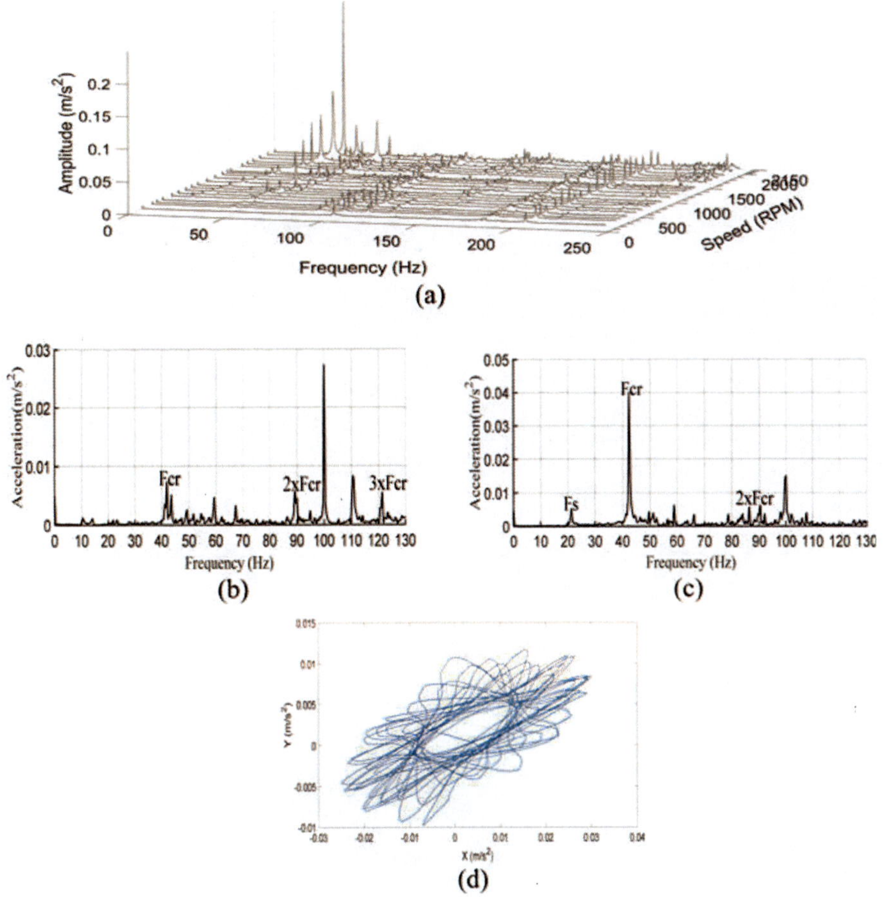

Fig. 7 Vibration response of cracked shaft 2 **a** Frequency response at different shaft rotational speed **b** Frequency response at 845 rpm ($F_{cr}/3$ range) **c** Frequency response at 1270 rpm ($F_{cr}/2$ range) **d** Orbit plot at subcritical speed

in Fig. 11a. The shaft rotational speed frequency peaks at shaft rotational speed (F_s) and its harmonics are clearly visible. The frequency responses at subcritical speed range ($F_{cr}/3$ and $F_{cr}/2$) have been presented in Fig. 11b, c. The frequency peaks at shaft rotational speed 'F_s' and shaft critical speed 'F_{cr}' are clearly visible in Fig. 11b, c. Moreover, the change in orbit plot at subcritical speed is also visible in Fig. 11d. The amplitude of vibration increased as compared to shaft 3 and shaft 4 due to presence of multiple fatigue cracks.

It is worth to mention here that all these observations are in-line with the theoretical and experimental results available in literature [1–8].

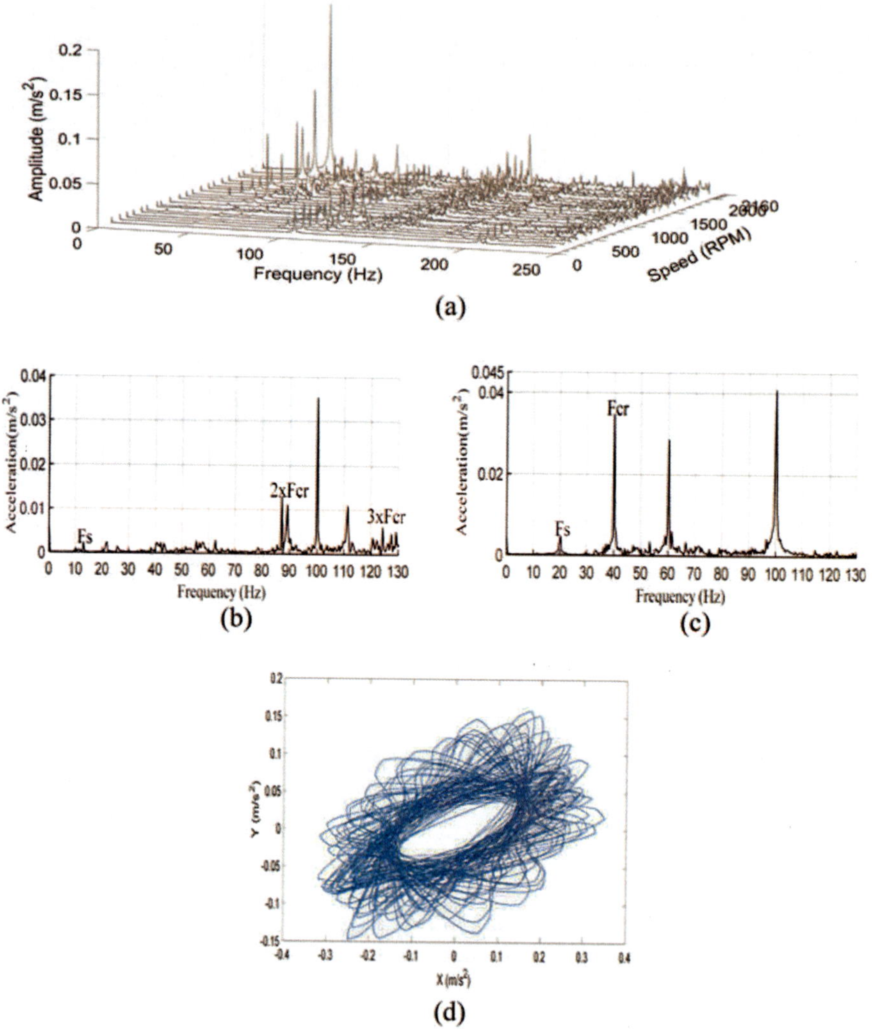

Fig. 8 Vibration response of cracked shaft 3 **a** Frequency response at different shaft rotational speed **b** Frequency response at 780 rpm ($F_{cr}/3$ range) **c** Frequency response at 1200 rpm ($F_{cr}/2$ range) **d** Orbit plot at subcritical speed

4.3 *Vibration Analysis of Cracked Shaft and Defective Bearing*

To study the dynamics of cracked rotor in presence of bearing defect, 0.7-mm-diameter hole was drilled through EDM on outer race of the supporting bearing.

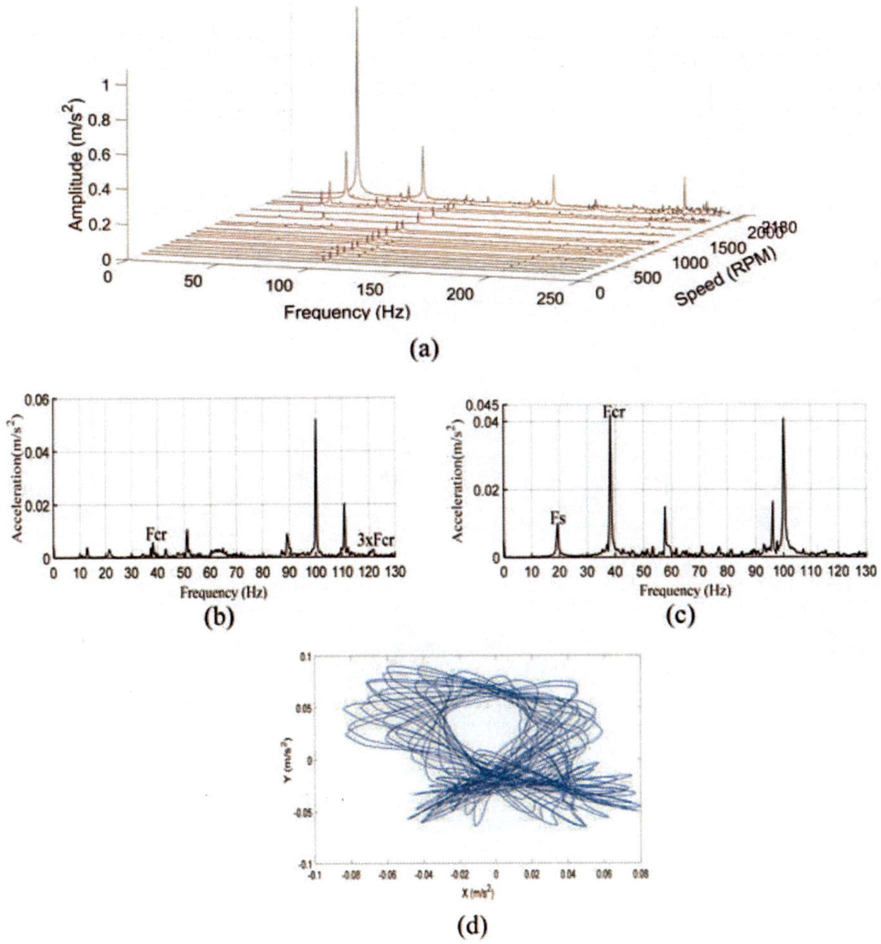

Fig. 9 Vibration response of cracked shaft 4 **a** Frequency response at different shaft rotational speed **b** Frequency response at 780 rpm ($F_{cr}/3$ range) **c** Frequency response at 1150 rpm ($F_{cr}/2$ range) **d** Orbit plot at subcritical speed

The characteristic defect frequency in presence of local defects depends on bearing geometry (Number of ball N_b, ball diameter d, bearing mean diameter D), and shaft rotational speed N_s can be computed using following equation [10–12].

$$\text{BPF0} = \left(\frac{N_b}{2}\right) \times \left(\frac{N_s}{60}\right) \times \left(1 - \left(\frac{d}{D}\right)\right) \tag{2}$$

In present case, the computed ball pass frequency of outer race (BPFO) at shaft speed 1200 rpm is 54.8 Hz, and while at shaft speed 1300 rpm, it is 59 Hz. The vibration responses of the cracked shaft having 4-mm-fatigue crack and 6-mm-fatigue

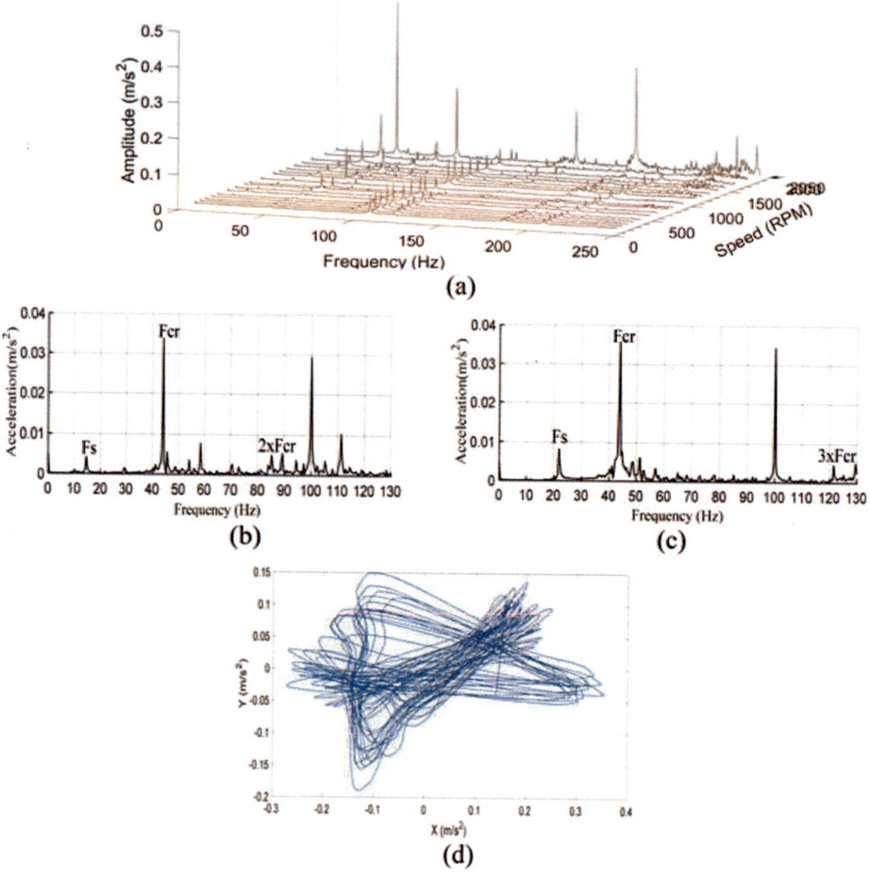

Fig. 10 Vibration response of cracked shaft 5 **a** Frequency response at different shaft rotational speed **b** Frequency response at 850 rpm ($F_{cr}/3$ range) **c** Frequency response at 1300 rpm ($F_{cr}/2$ range) **d** Orbit plot at subcritical speed

crack in presence of bearing defect have been presented in Fig. 12. From Fig. 12, it can be observed that the frequency peaks at shaft rotational speed F_s, shaft critical rotation speed F_{cr}, and ball pass frequency BPFO are clearly visible in presence of outer race bearing defect and shaft cracks.

5 Result Discussion

The vibration spectra of all shafts at subcritical rotational speed have been compared in Fig. 13. The vibration amplitude (RMS values) at subcritical speed has been shown

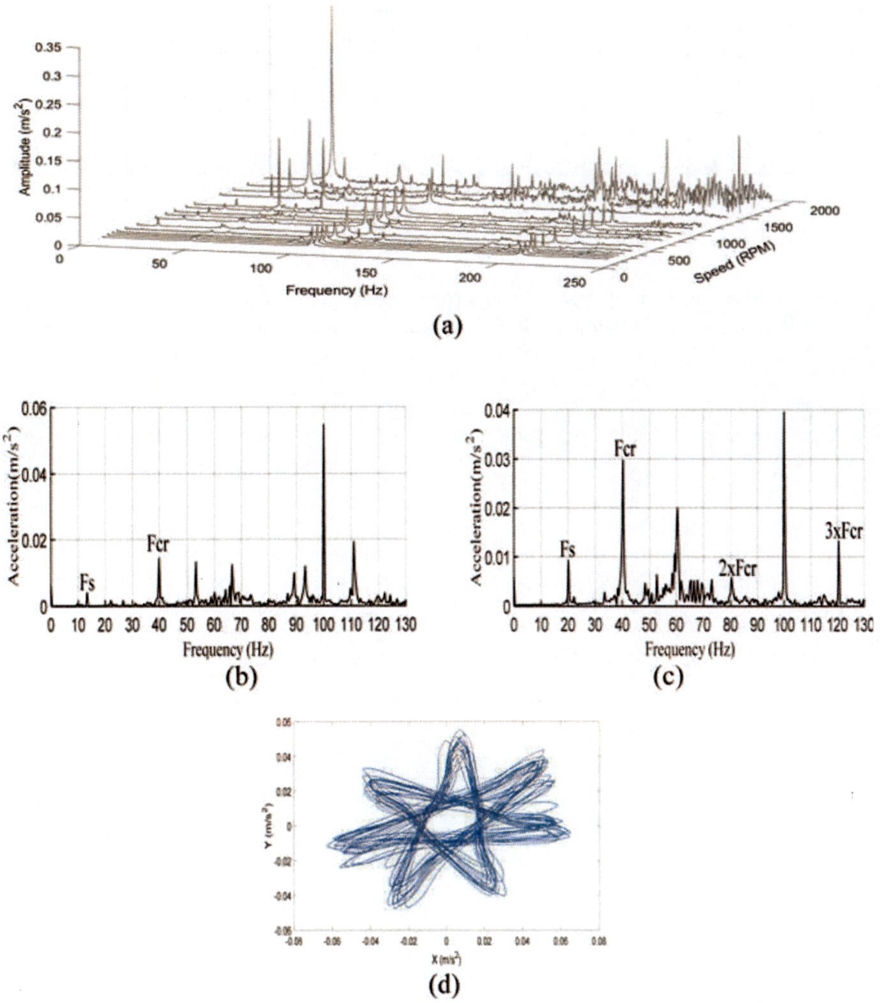

Fig. 11 Vibration response of cracked shaft 6 **a** Frequency response at different shaft rotational speed **b** Frequency response at 805 rpm ($F_{cr}/3$ range) **c** Frequency response at 1210 rpm ($F_{cr}/2$ range) **d** Orbit plot at subcritical speed

in Table 2. It is noticed that the frequency amplitude increased in presence of two 6-mm-depth shaft cracks. The overall vibration amplitude (RMS) increased in presence of shaft cracks and bearing defect. The vibration amplitude is found in case of crack near to bearing end as compared to crack near to rotor.

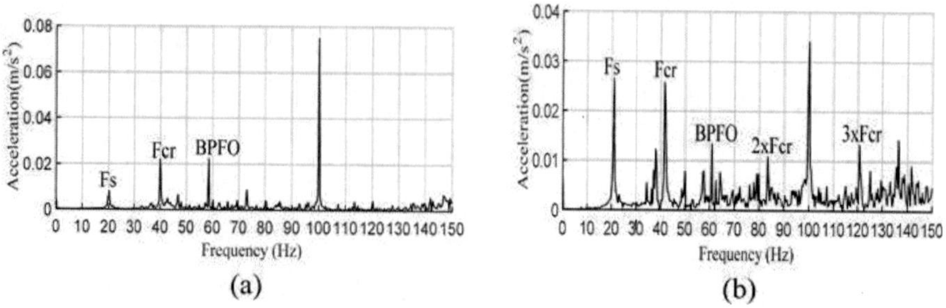

Fig. 12 Vibration response of cracked shaft in presence of bearing defect, shaft rotates at subcritical speed **a** Fatigue crack of 4 mm depth **b** Fatigue crack of 6 mm depth

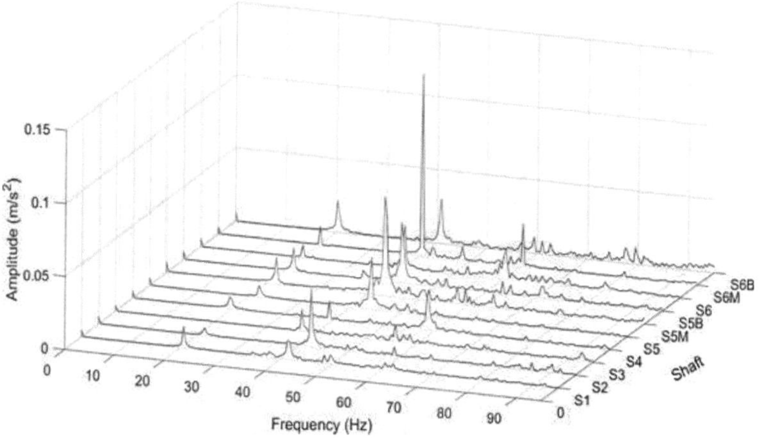

Fig. 13 Waterfall diagram for vibration spectra of all shafts rotating at subcritical speed ($F_{cr}/2$)

Table 2 Vibration amplitude comparison

Shaft	$F_{cr}/2$ rpm	RMS (m/s²)	Shaft	$F_{cr}/2$ rpm	RMS (m/s²)
1	1240	0.1221	5	1300	0.1718
2	1270	0.1153	6	1210	0.1901
3	1200	0.1261	4 mm crack and bearing defect	1200	0.3066
4	1150	0.1537	6 mm crack and bearing defect	1250	0.3066

6 Conclusion

The vibration monitoring of shaft bearing system in presence and absence of shaft fatigue crack(s) has been carried out. The following specific conclusions have been drawn from this experimental study. The overall vibrations amplitude of healthy shaft bearing system increases with increase of shaft rotational speed due to improper foundation of experimental setup. The approximate value of critical speed of shaft has been computed using Dunkerley's equation and verified experimentally. The frequency peaks at shaft rotational frequency have been observed in the vibration spectra of healthy system. The orbit plots in case of healthy systems are found elliptical in shape. The frequency peaks at shaft rotational speed and shaft critical speed have been observed, while shaft rotates at subcritical speed in presence of crack. The orbit plot of petal shapes is observed in presence of crack at subcritical shaft rotation speed. The vibration amplitude increases as the shaft crack depth increases. In addition to frequency peaks at shaft critical speed, frequency peaks at BPFO have also been observed in presence of bearing defect. The authors of this paper believe that this experimental study will provide guidance to practicing engineers and researchers to validate theoretical results

References

1. Kushwaha N, Patel VN (2020) Modelling and analysis of a cracked rotor: a review of the literature and its implications. Arch Appl Mech 90:1215–1245
2. Darpe AK, Gupta K, Chawla A (2004) Transient response and breathing behavior of a cracked Jeffcott rotor. J Sound Vib 272:207–243
3. Zhou T, Sun Z, Xu J, Han W (2005) Experimental analysis of cracked rotor. J Dyn Syst Meas Contr 127:313–320
4. Guo CZ, Yan JH, Bergman LA (2017) Experimental dynamic analysis of a breathing cracked rotor. Chin J Mech Eng 30:1177–1183
5. Sudheer Kumara V, Harikrishnaa Ch, Karthik Saia Ch, Nagaraju Ch (2015) Dynamic analysis of a cracked rotor—an experimental and finite element investigation. In: 4th International conference on materials processing and characterization, materials today: proceedings, vol 2, pp 2131–2136
6. Silani M, Ziaei-Rad S, Talebi H (2013) Vibration analysis of rotating systems with open and breathing cracks. Appl Math Model 37:9907–9921
7. AL-Shudeifat MA (2013) On the finite element modeling of the asymmetric cracked rotor. J Sound Vib 332:2795–2807
8. Zheng GT (1998) Vibration of a rotor system with a switching crack and detection of the crack. J Eng Gas Turbines Power 120:149–154
9. Gradzki R, Kulesza Z, Bartoszewicz B (2019) Method of shaft crack detection based on squared gain of vibration amplitude. Nonlinear Dyn 98:671–690
10. Shah DS, Patel VN (2019) A dynamic model for vibration studies of dry and lubricated deep groove ball bearings considering local defects on races. Measurement 137:535–555
11. Shah DS, Patel VN (2018) Theoretical and experimental vibration studies of lubricated deep groove ball bearings having surface waviness on its races. Measurement 129:405–423
12. Shah DS, Patel VN, Darji PH (2019) Experimental vibration studies of deep groove ball bearings having damaged surfaces. J Inst Eng India Ser C. https://doi.org/10.1007/s40032-018-0497-8

Formulation of Approximate Generalized Generated Database Model for Low-Speed Gear Box Developed Based on Double-Crank Inversion for a Four-Bar Chain

Akshay Anant Pachpor, Jayant Pandurang Modak, and Prashant Brajmohan Maheshwary

Abstract An analysis of double-crank inversion of a four-bar chain shows feasibility for adopting it as a gear box for low-speed applications. It is observed that any point on the coupler of a double-crank inversion of a four-bar chain describes very closely a complete circular path. This circular motion of some appropriately selected points on the coupler can be tapped out using link arrangement which will give output speed variation in the range of 8–12%. In the present paper, an attempt is made to consider many double-crank inversions of a class I four-bar crank with *(a-b)/(c-d)* ratio in the range of 0.1 to 0.6. Here, *a*, *b*, *c* and *d* are the link lengths of class I four-bar chain as per Harding's notation. For each of these inversions, minimum 5–6 points on the coupler would be chosen which will generate circular motion with almost constant angular velocity but average value of angular velocity would be different for a different point on the coupler. Each of these inversions will precipitate a gear box with different range of very low angular velocity variation. There are some industrial operations which do need very low range of angular speed variation depending on properties of raw material being processed with that machine. A very popular example is of motorized shredder of plastics. For all such gear boxes for its output shaft, the expressions are derived for angular acceleration, jerk, pop and crackle. Each one of these expressions would be function of ratio of link lengths, location of point of reference on the coupler and angular velocity of driving crank. As these expressions are established based on generated data regarding kinematics of several inversions of double-crank of four-bar chain, the model formed is nomenclated as GENERALIZED GENERATED DATA-BASED MODEL.

Keywords Mechanisms · Four-bar chain · Double-crank inversion · Gear box · Coupler curves

---

A. A. Pachpor (✉) · J. P. Modak · P. B. Maheshwary
J D College of Engineering & Management, Nagpur, Maharashtra 441501, India
e-mail: pachpor.a@gmail.com

V. K. Gupta et al. (eds.), *Recent Advances in Machines and Mechanisms*, Lecture Notes in Mechanical Engineering, https://doi.org/10.1007/978-981-19-3716-3_25

1 Introduction

Four-bar chains are the foundation of most mechanisms; they provide the basis for synthesis and development [1, 2]. One of the inversions of the class I four-bar chain is the double-crank mechanism. By fixing the shortest link of the four-bar chain of class I, a double-crank inversion can be obtained [1, 2].

Based on a literature review, one can state that significant research has been conducted on the kinematics and dynamics of the crank-rocker inversion of a four-bar chain. Due to the fact that many of the six link chains have two four-bar chains in series, crank-rocker and rocker-rocker, the third inversion double rocker is quite common [3]. These two links have been used to develop many process units [4–7]. Comparatively, applications of the double-crank mechanism are few compared to those of the crank-rocker and double-rocker mechanisms [1]. Some applications of double-crank mechanism are selective bell pepper harvester [8], double-crank external Geneva mechanism [9], a simple-type potted rice seedling transplanting mechanism [10], clamping unit with double-crank [11], etc. Double-crank linkage is typically used to provide non-uniform motion of a driven shaft by coupling two non-inline shafts [12]. With the proper arrangement of links, this assembly might be used as a transmission system for transmission of different speeds with minimal variations in output speeds. By utilizing this concept, the requirement of toothed gears and synchromesh itself for a conventional gearbox could be eliminated [13].

In the present paper, an attempt is made to consider many double-crank inversions of a class I four-bar chain with *(a-b)/(c-d)* ratio in the range of 0.1–0.6. For each of these inversions, appropriate points on the coupler would be chosen which will generate circular motion with almost constant angular velocity but average value of angular velocity would be different for a different point on the coupler. Each of these inversions will precipitate a gear box with different range of very low angular velocity variation.

2 Double-Crank Inversions of Different Configurations

In the Fig. 1, O_1ABO_2 represents a double-crank inversion of a class I four-bar chain where O_1A is input link, AB is coupler link, BO_2 is output link, and O_1O_2 is fixed link. Here, the link lengths are represented as *a*, *b*, *c* and *d* according to Harding's notation, i.e., for $O_1A = 90$ cm, AB = 80 cm, $BO_2 = 100$ cm and $O_1O_2 = 10$ cm, the links are shown as '*b*', '*d*', '*a*' and '*c*,' respectively [1]. The abovementioned link lengths are so chosen that it will result in *(a-b)/(c-d)* ratio as 0.1. In the similar way, six different configurations of double-crank inversion corresponding to *(a-b)/(c-d)* ratios 0.1, 0.2, 0.3, 0.4, 0.5 and 0.6 are generated and the dimensions of the links are tabulated in the Table 1.

For each of the six generated double-crank inversions, 9 coupler points, viz. P_1, P_2, P_3, P_4, P_5, P_6, P_7, P_8 and P_9, are chosen at a distance of 8 cm, 16 cm, 24 cm,

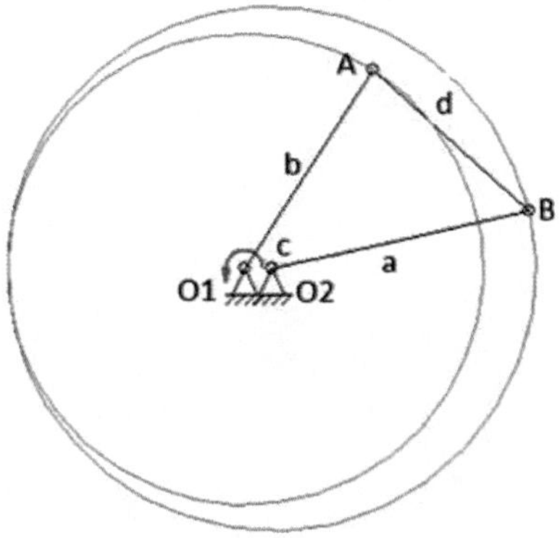

Fig. 1 Double-crank inversion

Table 1 Dimension of links for different configurations

S. No.	*(a-b)/(c-d)* ratio	O_1A	AB	BO_2	O_1O_2
1	0.1	90	80	100	10
2	0.2	90	80	100	30
3	0.3	90	80	100	46.6
4	0.4	90	80	100	55
5	0.5	90	80	100	60
6	0.6	90	80	100	63.3

32 cm, 40 cm, 48 cm, 56 cm, 64 cm and 72 cm, respectively measured from end *A* on the coupler *AB*. These points are then considered for kinematic analysis which is discussed in next section.

3 Kinematic Analysis of Coupler Points on Coupler of Double-Crank Inversion

3.1 Position Analysis of Coupler Points

In this section, a kinematic analysis of different configurations of double-crank inversion for *(a-b)/(c-d)* ratios 0.1, 0.2, 0.3, 0.4, 0.5 and 0.6 is presented. The analysis is carried out at 12 different positions of input crank O_1A where positions of input crank are defined in terms of input crank angle, such as 1st position at $\theta = 30°$, 2nd position $\theta = 60°$ up to 12th position at $\theta = 360°$. Graphically constructing the configuration of double-crank four-bar mechanism O_1ABO_2 corresponding to these 12 positions of input crank O_1A shows that each point on the coupler traces nearly a circular path [13].

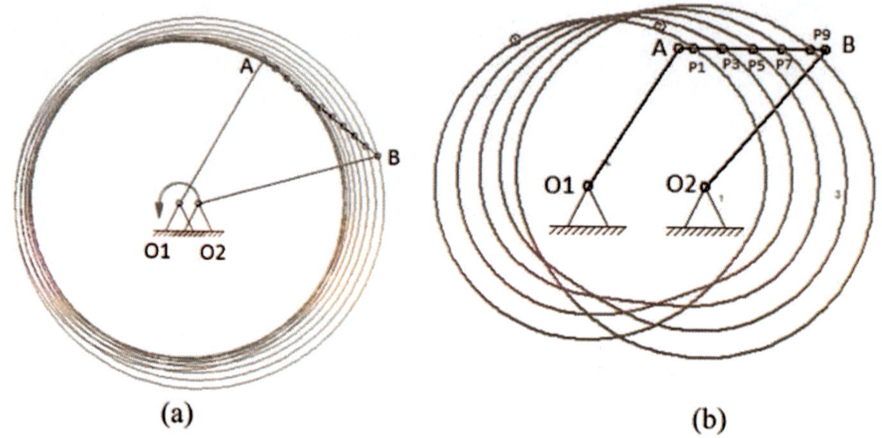

Fig. 2 Position analysis of double-crank inversion

It is observed that for *(a-b)/(c-d)* ratios 0.1, 0.2 and 0.3 the path traced by coupler points P_1, P_2, P_3, P_4, P_5, P_6, P_7, P_8 and P_9 is almost circular [refer Fig. 3]. But in case of double-crank inversions having *(a-b)/(c-d)* ratios as 0.4, 0.5 and 0.6, the path traced by the coupler points is not circular. In fact, the point P_1 follows a path which is nearly circular and other coupler points P_2, P_3, P_4, P_5, P_6, P_7, P_8 and P_9 are not tracing an exact circular path. The path traced by coupler points on double-crank inversion of *(a-b)/(c-d)* ratios 0.6 can be more clearly seen in Fig. 2b. For the sake of clear understanding of the path traced by coupler points, path traced by points P_1 and P_3 is only shown in Fig. 3.

3.2 Velocity Analysis of Coupler Points

In this section, velocity analysis of coupler points P_1, P_2, P_3, P_4, P_5, P_6, P_7, P_8 and P_9 of coupler AB is presented. The velocity analysis is carried out with relative velocity method [13]. Instantaneous angular velocity of coupler points $P_1, P_2, P_3, P_4, P_5, P_6, P_7, P_8$ and P_9 is represented by $\omega_a, \omega_1, \omega_2, \omega_3, \omega_4, \omega_5, \omega_6, \omega_7, \omega_8, \omega_9$ and ω_b, respectively. For each *(a-b)/(c-d)* ratios, viz. 0.1, 0.2, 0.3, 0.4, 0.5 and 0.6, the values of instantaneous angular velocities are calculated at 12 different angular positions of input crank O_1A [13]. The computed values of instantaneous angular velocities are represented graphically in Fig. 4. The percentage angular velocity variations for *(a-b)/(c-d)* ratios 0.1, 0.2, 0.3, 0.4, 0.5 and 0.6 are given in Table 2.

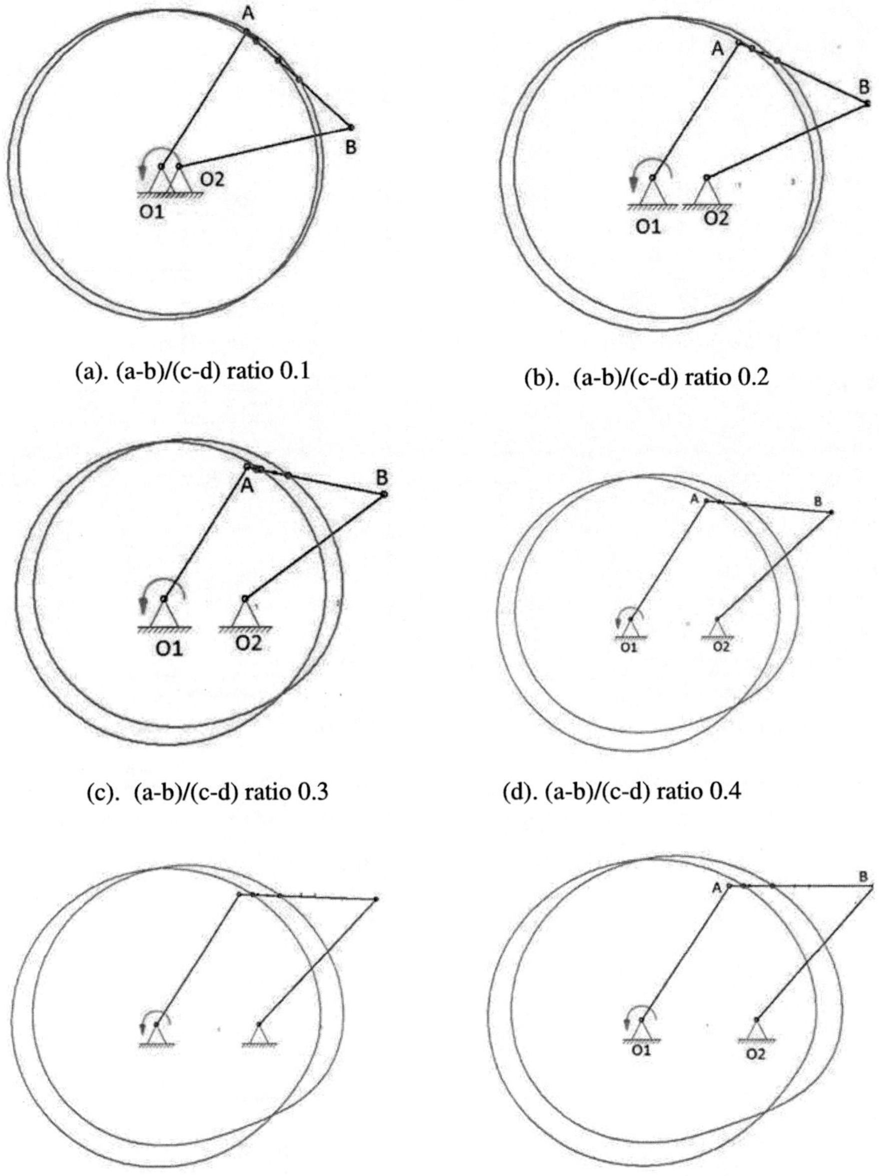

(a). (a-b)/(c-d) ratio 0.1

(b). (a-b)/(c-d) ratio 0.2

(c). (a-b)/(c-d) ratio 0.3

(d). (a-b)/(c-d) ratio 0.4

(e). (a-b)/(c-d) ratio 0.5

(f). (a-b)/(c-d) ratio 0.6

Fig. 3 Schematic views showing path generated by coupler points P_1 and P_3

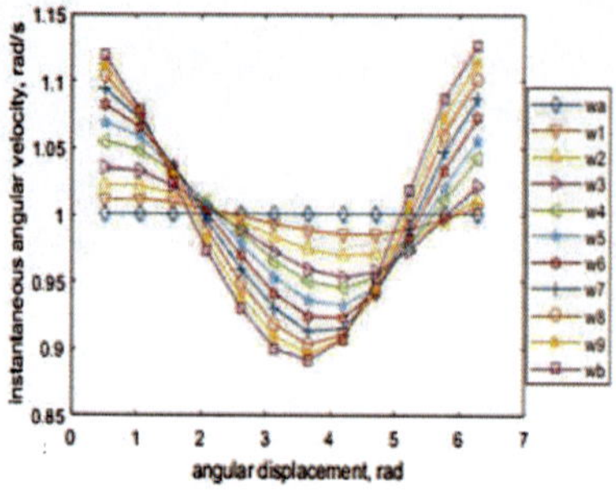

(a). (a-b)/(c-d) ratio 0.1

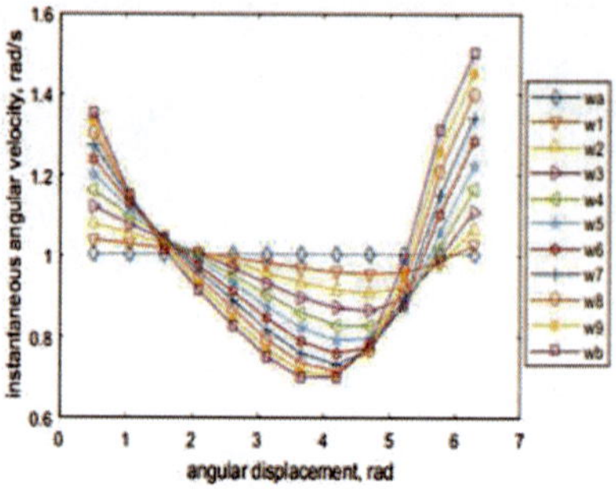

(b). (a-b)/(c-d) ratio 0.2

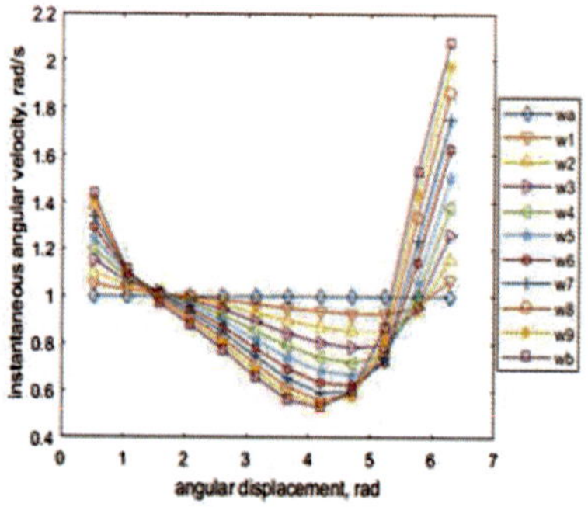

(c). (a-b)/(c-d) ratio 0.3

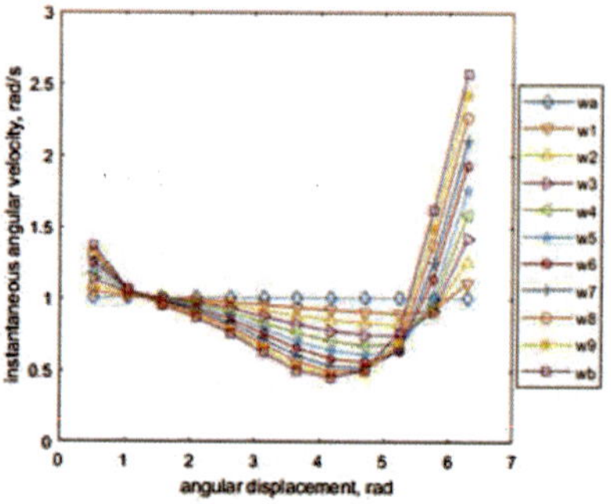

(d). (a-b)/(c-d) ratio 0.4

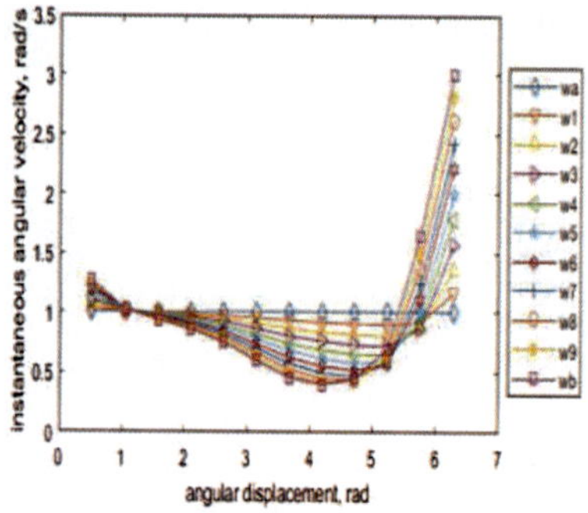

(e). (a-b)/(c-d) ratio 0.5

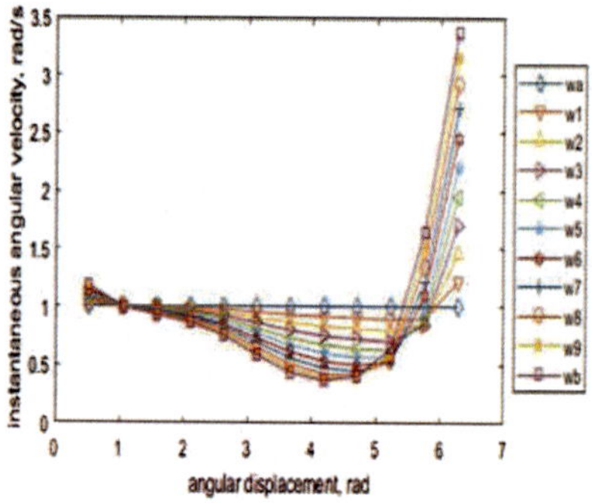

(f). (a-b)/(c-d) ratio 0.6

Fig. 4 Graphical representation of instantaneous velocity variation of coupler points

Table 2 Maximum percentage variation in angular velocity

Ratio	ω_a	ω_1	ω_2	ω_3	ω_4	ω_5	ω_6	ω_7	ω_8	ω_9	ω_b
0.1	0.00	1.34	2.70	4.16	5.68	7.13	8.45	9.60	10.55	11.46	12.50
0.2	0.00	4.5	9.3	14.1	19.1	25.28	31.28	36.89	41.92	46.3	50.0
0.3	0.0	8.7	19.7	32.1	45.2	58.3	70.7	82.0	92.1	100.6	107.8
0.4	0.0	14.1	31.4	50.0	69.1	87.7	105.2	121.0	134.9	146.7	156.4
0.5	0.0	19.7	42.3	66.2	90.3	113.4	135.0	154.4	171.4	185.9	197.7
0.6	0.0	24.7	52.1	80.5	108.6	135.4	160.2	182.6	202.1	218.6	232.1

4 Formulation of Generalized Generated Data-Based Mathematical Models

4.1 Formulation of Mathematical Models for Angular Velocity

In this section, an approach toward forming an exponential type of mathematical models using mathematical modeling technique is discussed. Using this technique, the output angular velocity can be expressed in terms of independent variables such as input link angular velocity, angular displacement of input link and ratios of link lengths [14, 15]. Let us consider the variables output link angular velocity, input link angular velocity, angular displacement of input link, link lengths of link O_2B, link O_1O_2, link O_1A and link AB be represented by $\omega_0, \omega_i, \theta, l_1, l_2, l_3$ and l_4, respectively. Now, output angular velocity can be written as:

$$\omega_o = f[\theta, l_1, l_2, l_3, l_4, \omega_i] \tag{1}$$

Rearranging the terms in Eq. (1) to get dimensionless forms on both sides, one gets

$$\frac{\omega_o}{\omega_i} = k.\theta^a.\left(\frac{l2}{l1}\right)^b.\left(\frac{l3}{l1}\right)^c.\left(\frac{l4}{l1}\right)^d \tag{2}$$

Again, Eq. (2) can be written as

$$\pi_d = k.\pi_1^a.\pi_2^b.\pi_3^c.\pi_4^d \tag{3}$$

where $\pi_d = \omega_o/\omega_i, \pi_1 = \theta, \pi_2 = l_2/l_1, \pi_3 = l_3/l_1, \pi_4 = l_4/l_1.$

Using the values generated from kinematic analysis of the coupler points P_1, P_2, P_3, P_4, P_5, P_6, P_7, P_8 and P_9 of coupler *AB*, the values of the indices *k*, *a*, *b*, *c* and *d* are estimated (Tables 3, 4, 5, 6, 7 and 8).

Table 3 Values of indices of model for (*a-b*)/(*c-d*) ratio 0.1

	P_1	P_2	P_3	P_4	P_5	P_6	P_7	P_8	P_9	B
k	0.894	1.012	0.934	1.094	1.113	1.039	1.190	0.910	1.028	0.980
a	−0.009	−0.018	−0.026	−0.031	−0.035	−0.038	−0.039	−0.038	−0.036	−0.033
b	0.000	0.000	0.000	0.000	0.000	0.000	0.000	0.000	0.000	0.000
c	−0.567	−0.264	0.647	−1.288	0.375	1.378	−1.636	0.617	−1.623	0.788
d	−0.269	0.119	−0.698	0.880	0.156	−0.636	1.391	−0.870	0.743	−0.601

Table 4 Values of indices of model for(*a-b*)/(*c-d*) ratio 0.2

	P_1	P_2	P_3	P_4	P_5	P_6	P_7	P_8	P_9	B
k	1.089	1.004	0.825	1.142	0.802	1.252	0.969	1.336	0.862	1.003
a	−0.027	−0.051	−0.071	−0.086	−0.096	−0.101	−0.101	−0.097	−0.089	−0.080
b	0.000	0.000	0.000	0.000	0.000	0.000	0.000	0.000	0.000	0.000
c	−1.360	−1.033	−0.097	−0.646	−0.632	1.234	0.468	1.278	−0.683	0.248
d	0.934	0.344	−1.035	0.642	−0.973	0.134	−0.644	0.425	−0.589	-0.323

Table 5 Values of indices of model for (*a-b*)/(*c-d*) ratio 0.3

	P_1	P_2	P_3	P_4	P_5	P_6	P_7	P_8	P_9	B
k	1.150	0.995	0.736	0.877	1.119	0.831	0.766	1.069	1.078	1.182
a	−0.036	−0.066	−0.090	−0.108	−0.119	−0.124	−0.123	−0.116	−0.105	−0.089
b	0.000	0.000	0.000	0.000	0.000	0.000	0.000	0.000	0.000	0.000
c	0.065	0.821	−1.086	0.046	−0.029	0.453	−0.641	−1.363	−1.723	0.443
d	0.519	−0.540	−1.029	−0.792	0.336	−1.205	−1.028	0.844	1.089	0.527

Table 6 Values of indices of model for (*a-b*)/(*c-d*) ratio 0.4

	P_1	P_2	P_3	P_4	P_5	P_6	P_7	P_8	P_9	B
k	0.845	1.054	1.198	1.200	1.001	0.932	1.031	1.003	1.087	0.847
a	−0.039	−0.062	−0.084	−0.101	−0.112	−0.117	−0.116	−0.108	−0.095	−0.076
b	0.000	0.000	0.000	0.000	0.000	0.000	0.000	0.000	0.000	0.000
c	−0.624	1.483	−0.235	−0.954	0.525	1.602	−1.358	0.933	0.610	1.759
d	−0.495	−0.524	0.858	1.222	−0.262	−1.050	0.847	−0.300	0.272	−1.329

Table 7 Values of indices of model for (*a-b*)/(*c-d*) ratio 0.5

	P_1	P_2	P_3	P_4	P_5	P_6	P_7	P_8	P_9	B
k	1.314	0.786	1.190	0.993	1.266	0.741	1.228	0.801	0.852	0.830
a	−0.028	−0.053	−0.073	−0.088	−0.098	−0.103	−0.102	−0.095	−0.081	−0.060
b	0.000	0.000	0.000	0.000	0.000	0.000	0.000	0.000	0.000	0.000
c	0.726	−0.567	1.109	−0.275	0.991	−1.139	1.071	−0.351	1.520	0.635
d	0.879	−0.802	0.291	0.170	0.713	−0.623	0.667	−0.500	−1.036	−0.657

Table 8 Values of indices of model for (*a*-*b*)/(*c*-*d*) ratio 0.6

	P_1	P_2	P_3	P_4	P_5	P_6	P_7	P_8	P_9	B
k	0.790	1.069	1.270	0.950	0.899	0.820	1.181	0.826	0.957	0.725
a	−0.022	−0.043	−0.061	−0.075	−0.084	−0.089	−0.089	−0.082	−0.067	−0.044
b	0.000	0.000	0.000	0.000	0.000	0.000	0.000	0.000	0.000	0.000
c	0.198	0.623	1.238	−0.516	−1.275	0.519	1.058	0.744	0.240	0.196
d	−1.124	0.070	0.598	0.186	0.364	−0.819	0.629	−0.719	0.266	−0.864

4.2 *Formulation of Mathematical Models for Angular Acceleration, Jerk and Pop*

Using Eq. (2), expressions for angular velocity, angular acceleration, jerk and pop can be formed. Rearranging Eq. (2), we get an expression for output angular velocity,

$$\omega_o = k \times \theta^a \times \left(\frac{l2}{l1}\right)^b \times \left(\frac{l3}{l1}\right)^c \times \left(\frac{l4}{l1}\right)^d \times \omega_i \tag{4}$$

Differentiating Eq. (4) subsequently, with respect to time, expressions for angular acceleration, jerk and pop can be derived as under:

$$\frac{\mathrm{d}\omega_o}{\mathrm{d}t} = \left[k \times \left(\frac{l2}{l1}\right)^b \times \left(\frac{l3}{l1}\right)^c \times \left(\frac{l4}{l1}\right)^d \times \omega_i\right] \frac{\mathrm{d}(\theta^a)}{\mathrm{d}t} \tag{5}$$

$$\alpha = \frac{\mathrm{d}\omega_o}{\mathrm{d}t} = \left[k \times \left(\frac{l2}{l1}\right)^b \times \left(\frac{l3}{l1}\right)^c \times \left(\frac{l4}{l1}\right)^d \times \omega_i\right] a.\theta^{a-1}\omega_i \tag{6}$$

$$\mathrm{jerk}(j) = \frac{\mathrm{d}^2\omega_o}{\mathrm{d}t^2} = \left[k \times \left(\frac{l2}{l1}\right)^b \times \left(\frac{l3}{l1}\right)^c \times \left(\frac{l4}{l1}\right)^d \times \omega_i\right] a.(a-1).\theta^{a-2}\alpha \tag{7}$$

$$\mathrm{pop} = \frac{\mathrm{d}^3\omega_o}{\mathrm{d}t^3} = \left[k \times \left(\frac{l2}{l1}\right)^b \times \left(\frac{l3}{l1}\right)^c \times \left(\frac{l4}{l1}\right)^d \times \omega_i\right] a.(a-1).(a-2).\theta^{a-3} j \tag{8}$$

Equations (6), (7) and (8) are the equations of angular acceleration, jerk and pop, respectively.

5 Conclusion

In the present paper, an attempt is made to study kinematic properties of 6 double-crank inversions of a class I four-bar crank with (*a*-*b*)/(*c*-*d*) ratio in the range of 0.1 to 0.6. The dimensions of link lengths are selected in such a manner that will result in (*a*-*b*)/(*c*-*d*) ratios as 0.1, 0.2, 0.3, 0.4, 0.5 and 0.6.

For each of these inversions, 9 points are chosen on the coupler AB. It is observed that coupler points for (*a*-*b*)/(*c*-*d*) ratios 0.1, 0.2 and 0.3 on the coupler AB of a double-crank inversion describe very closely a complete circular path. But for (*a*-*b*)/(*c*-*d*) ratios 0.4, 0.5 and 0.6, coupler points except for point P_1 do not seem to trace a complete circular path.

The velocity analysis of coupler points shows that percentage variation in angular velocity increases from point P_1 to point P_9. The least percentage variation in angular velocity is observed in case of double-crank inversion of ratio 0.1. This mechanism is showing a percentage variation in the range of 1.34–12.5%.

The circular motion of some appropriately selected points on the coupler can be tapped out using links arrangement, and this will precipitate into a gear box arrangement [13]. For double-crank inversion of (*a*-*b*)/(*c*-*d*) ratios 0.1, 0.2, 0.3, 0.4, 0.5 and 0.6, there will be 10 coupler points through which we can tap out the circular motion and one can have 10 output shafts for each ratio and average angular velocity for each output shaft will be different.

Based on the concept of GENERALIZED GENERATED DATA-BASED MODEL, mathematical models for angular velocity of each coupler point have been formulated [14, 15]. With the help of these mathematical models, expressions for kinematic properties, i.e., angular velocity, angular acceleration, jerk and pop, have been derived. These kinematic properties are the functions of input angular velocity, angular displacement of input link and ratios of link length. Hence, kinematic properties related to different types of double-crank configurations can be predicted by substituting the link lengths in the mathematical models.

References

1. Hain K (1967) Applied kinematics, 2nd edn. McGraw-Hill Inc., New York
2. Hall A (1961, January) Kinematics and linkage design, 1st edn. Prentice Hall
3. Ju-jiang C, Yu-qi W, Ji-wei S (2012) Preliminary research of crank-group driving mechanism. Appl Mech Mater 229–231:555–558
4. Johns F (1930) Ingenious mechanisms for designers and inventors, vol 1, 1st edn. Industrial Press Inc., Norwalk
5. Johns F (1930) Ingenious mechanisms for designers and inventors, vol 2, 1st edn. Industrial Press Inc., Norwalk
6. Johns F (1930) Ingenious mechanisms for designers and inventors, vol 3, 1st edn. Industrial Press Inc., Norwalk
7. Johns F (1930) Ingenious mechanisms for designers and inventors, vol 4, 1st edn. Industrial Press Inc., Norwalk

8. Shaw LN, The application of an offset double-crank mechanism in a selective bell pepper harvester. Transactions of the ASAE—1975, Paper No. 73-1511
9. Al-Sabeeh AK (1993, September) Double-crank external Geneva mechanism. Transactions of the ASME, 666/vol 115
10. Xin L, Lv Z, Wang W, Zhou M, Zhao Y (2017) Optimal design and development of a double-crank potted rice seedling transplanting mechanism. Trans ASABE 60(1):31–40
11. Wohlrab W (2008, November) Clamping unit with double-crank drive. United States Patent, Patent No.: US 7.445,439 B2
12. Bagci C (1977) Synthesis of double-crank (drag-link) driven mechanisms with adjustable motion and dwell time ratios. Mech Mach Theory 12:619–638
13. Pachpor AA, Modak JP, Maheshwary PB (2021) Feasibility of adoption of double-crank inversion of four bar chain as a substitute for a gear box. In: Rao YVD, Amarnath C, Regalla SP, Javed A, Singh KKF (eds) Proceedings of IPROMM Conference 2020, LMME, ISBN 978-981-16-1769-0. Springer, Singapore
14. Hasammis S (2018) Modelling the design of an industrial enterprise for manufacturing of a new Product. PhD Thesis submitted to R.T.M.N.U., under the supervision of Dr. J. P. Modak and Dr. (Mrs.) U. S. Dange 2017. Degree awarding 2018
15. Chatapalliwar A (2014) Formulation of approximate generalised generated design data based model of a plant for manufacturing biodiesel. PhD Thesis Submitted to R.T.M.N.U. 2013 under the supervision of Dr. J. P. Modak & Dr. V.S. Deshpande. Degree Awarded in 2014

Effect of Balanced Sliding Velocity and Slide-to-Roll Ratio on Surface Wear in Symmetric Profile Modified High Contact Ratio Spur Gears

R. Prabhu Sekar, R. Ravivarman, and Gadi Anil

Abstract Surface damage in gear pairs is primarily due to the influence of sliding surface wear. This paper explores the effect of sliding velocity and slide-to-roll ratio factors on surface wear in symmetric high contact ratio spur gears (contact ratio > 2) with geometric changes induced by profile corrections. In this work, the sum of profile correction factor is maintained constant at 0.2 by distributing the profile shift (x_1) of the pinion from −0.3 to 0.6. The optimal value of x_1 is obtained by balancing the sliding velocity and slide-to-roll ratio at the beginning and end of the contact path. Based on the generalized Archard's wear equation, the respective sliding wear is obtained for the optimal balanced point. The simple formulation of the model allows in obtaining the required profile modification to ensure a pre-established function of minimizing surface wear. The influence of balanced sliding velocity and slide-to-roll ratio on tooth wear is studied and compared. The input for the problem is attained from the 2D elastic model using ANSYS. From this analysis, the surface wear for minimum peak-to-peak in both the face and flank regions has been obtained. The study concludes that modification of profile shift coefficients will provide the gear designers the flexibility to optimize surface wear and its respective efficiency.

Keywords Profile shift coefficient · High contact ratio · Slide-to-roll ratio · Sliding velocity · Surface wear

1 Introduction

Gears are the frequently used transmission component in the mechanical machines. Gears have wide uses going from modest to complex machinery due to their firmness, consistency, peak transmission efficiency and competence to sustain. The spur

R. Prabhu Sekar (✉) · G. Anil
Mechanical Engineering Department, Motilal Nehru National Institute of Technology Allahabad, Prayagraj, India
e-mail: rprabhusekar@mnnit.ac.in

R. Ravivarman
Department of Mechanical Engineering, Pondicherry Engineering College, Pillaichavadi, India

V. K. Gupta et al. (eds.), *Recent Advances in Machines and Mechanisms*, Lecture Notes in Mechanical Engineering, https://doi.org/10.1007/978-981-19-3716-3_26

gear system is the most commonly used gears having an involute profile which needs simple design and ease to manufacture [1]. The gear ratio obtained in spur gear is constant and has an operating efficiency around 98%. Computation of the gears functioning features [2] like deflection, stress and rigidity requires accurate information of the force sustained by each tooth in the gear drive at the contact period. In normal contact ratio gearing system the load taken into a single tooth is seems to be high to overcome this addendum modified [3, 4] high contact ratio gear drive is deployed [5, 6]. In the early findings, numerous investigators have foretold the gear wear [7] using analytical and numerical tools [8] in spur gears. Many investigators including the authors [9–11] have justified the importance of high contact ratio gears and profile modifications. The estimation of wear for a profile modified high contact ratio spur gears is not carried out yet. Given that the main aim of this work was to estimate impending effects of the sliding velocity and slide–roll ratio on the impact of wear depth in the high contact ratio gear drive.

2 Numerical Finite Element Model

A 2D model of the HCR gear drive (Table 1 and Fig. 1) is generated using the FE software ANSYS. Plane strain condition is considered and only five teeth are modelled for analysis to minimize the computational time. The model is meshed using eight nodes and two degrees of freedom with 2D-PLANE 82 plane strain quadrilateral element. The five-teeth full-rim model is constrained completely along the rim of the gear. The point of application of the load and its respective deflections is obtained for the gear and pinion. From the obtained deflections, the corresponding tooth stiffness and its equivalent stiffness are calculated. The individual load-share ratio (LSR) is obtained along the contact path, respectively.

Table 1 Basic dimensions for gearing system

Description	Values
Number of teeth of pinion (z_p)	50
Number of teeth of gear (z_g)	100
Pressure angle (α_o) Degree	20°
Gear ratio (i)	2
Addendum height (h_a) mm	1.25 m
Material elastic modulus (E) GPa	210
Module (m) mm	1
Load (F_N) N	10
Sum of profile correction factor (x)	0.2

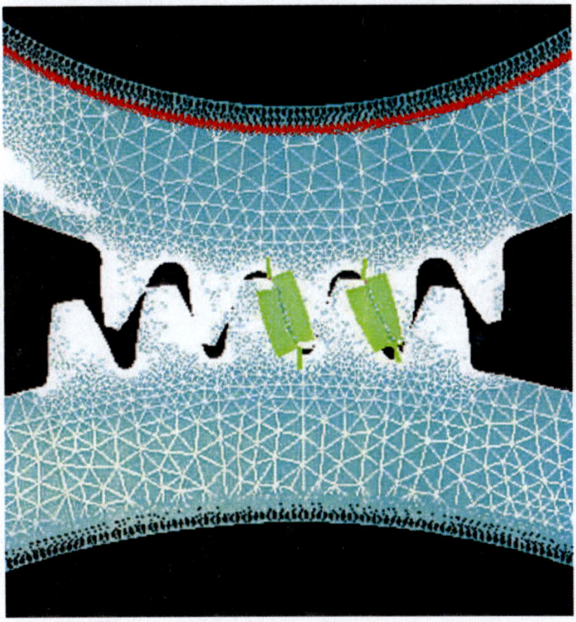

Fig. 1 Five-teeth full-rim FEA model

3 Wear Simulation Model

In this work, Anderson and Erikson [12] suggested single point equation which is used for computing the wear depth (h) along the contact points,

$$h_{i,n} = h_{i,(n-1)} + J_w(\sigma_H)_i 2a_i \left(\frac{(v_p)_i - (v_g)_i}{(v_p)_i} \right) \tag{1}$$

where J_w wear coefficient 5×10^{-16} m^2/N [7], the contact pressure $(\sigma_H)_i$, the half contact width (a_i) and the sliding distances are estimated for the detected contact points.

In the equation prescribed above, the sliding velocity (s) plays an influencing role in determining the wear depth along the tooth profile. Here, the sliding velocity and the slide-to-roll ratio parameters are balanced between the gear and pinion at the beginning and end of the contact and its respective wears are estimated.

$$(v_s)_i = \left| (v_p)_i - (v_g)_i \right| \tag{2}$$

$$(v_e)_i = \frac{1}{2}\left| (v_p)_i + (v_g)_i \right| \tag{3}$$

The slide–roll ratio (SR) is defined at this point as

$$(\mathrm{SR})_i = \frac{(v_s)_i}{(v_e)_i} \tag{4}$$

where

SR—Slide-to-roll ratio

v_s—Sliding velocity

v_e—Entraining velocity.

The peripheral velocity and radius of curvature at any instant point [7] are given as

$$\left(v_p\right)_i = \omega_p \left(R_p\right)_i \tag{5}$$

$$\left(v_g\right)_i = \omega_g \left(R_g\right)_i \tag{6}$$

$$\left(R_p\right)_i = r_{op} \sin \alpha_0 - X_i \tag{7}$$

$$\left(R_g\right)_i = r_{og} \sin \alpha_0 + X_i \tag{8}$$

In Fig. 2, the typical contact points are shown as follows,

A. Lowest point of tooth contact (LPTC)
B. First lowest point of double tooth contact (FLPDTC)
C. First highest point of double tooth contact (FHPDTC)
D. Second lowest point of double tooth contact (SLPDTC)
E. Second highest point of double tooth contact (SHPDTC)
F. Highest point of tooth contact (HPTC).

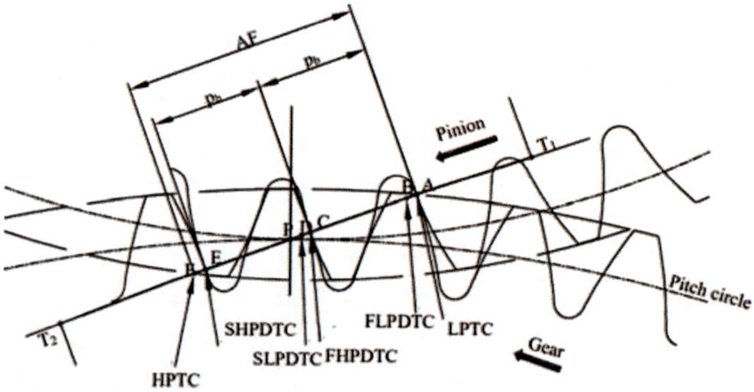

Fig. 2 Typical contact positions for the HCR spur gear drive

3.1 *Balancing of Sliding Velocity at the Beginning and End of the Contact*

The length of approach (LOA-AP) is to be equalized with the length of recess (LOR-PF) for balancing the sliding velocity. Hence, for balancing the sliding velocity, the equations are given as:

$$\sqrt{r_{awg}^2 - r_{bg}^2} - r_{wg}\sin\alpha_w = \sqrt{r_{awp}^2 - r_{bp}^2} - r_{wp}\sin\alpha_w \tag{9}$$

3.2 *Balancing of Slide-to-Roll Ratio at the Beginning and End of the Contact*

The balancing of slide-to-roll ratio is carried out at the beginning and end of the contact which is specified by the equation given below:

$$i^2 = \left(\frac{T_1T_2 - T_1F}{T_1F}\right)\left(\frac{T_1T_2 - T_1A}{T_1A}\right) \tag{10}$$

where

$$T_1T_2 = \left(r_{wp} + r_{wg}\right)\sin\alpha_w \tag{11}$$

$$T_1F = \sqrt{r_{awp}^2 - r_{bp}^2} \tag{12}$$

$$T_1A = T_1T_2 - \sqrt{r_{awg}^2 - r_{bg}^2} \tag{13}$$

where r_b—base circle, r_w—working circle and r_{aw}—addendum circle radius of the pinion and gear.

4 Discussions

This work analyses the gear ratio (i) versus profile correction factor of the drive for different values of the x_1 as shown in Table 2, but the sum of addendum modification coefficient in the drive is maintained at $x_1 + x_2 = 0.2$. The gearing system considered for study was made up by a pinion of 50 teeth and a gear of 100 teeth in order to have the gear ratio (i) equal to 2 for the first case as shown in Fig. 3 and the remaining parameters of these gears are shown in Table 1. In the plot shown above, the values

Table 2 Balanced sliding velocity for *S* gearing under transmission ratio variation

Gear ratio (*i*)	2		3			4		
x_1	LOA (AP)	LOR (PF)	x_1	LOA (AP)	LOR (PF)	x_1	LOA (AP)	LOR (PF)
−0.3	4.00	2.63	−0.3	4.14	2.57	−0.3	2.59	3.88
−0.2	3.75	2.87	−0.2	3.88	2.81	−0.2	2.87	3.67
−0.1	3.49	3.10	−0.1	3.62	3.04	−0.1	3.14	3.45
−0.018	**3.29**	**3.29**	0	3.35	3.26	0	3.42	3.22
0	3.24	3.33	**0.02**	**3.30**	**3.30**	**0.039**	**3.31**	**3.31**
0.1	2.98	3.55	0.1	3.09	3.49	0.1	3.14	3.45
0.2	2.73	3.77	0.2	2.82	3.71	0.2	2.87	3.67
0.3	2.46	3.99	0.3	2.54	3.93	0.3	2.59	3.88
0.4	2.19	4.20	0.4	2.27	4.14	0.4	2.31	4.10
0.5	1.92	4.42	0.5	1.99	4.35	0.5	2.03	4.31
0.6	1.65	4.62	0.6	1.71	4.56	0.6	1.75	4.52

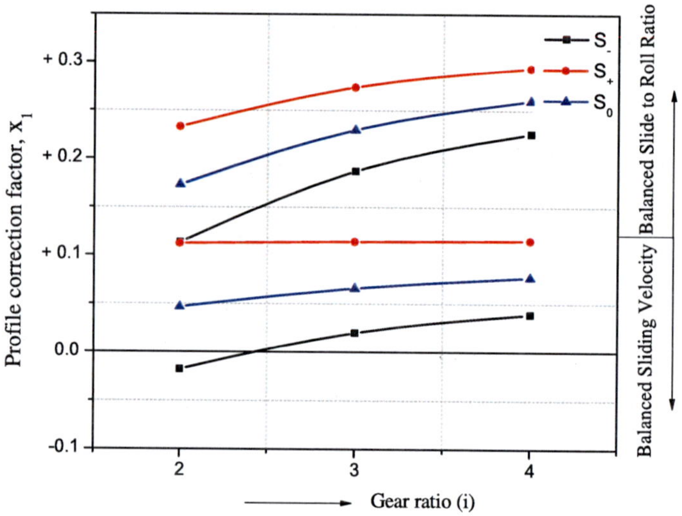

Fig. 3 Influence of profile correction factor on gear ratio (*i*)

of profile correction factor (x_1) for the balanced sliding velocity and balanced slide-to-roll ratio are evaluated and plotted against the increasing gear ratio (i). It can be noticed that the value of profile correction (x_1) for the drive tend to increase for higher gear ratio (i) which is due to the increase in the root size of the tooth which provide resistance to the material removal in terms of balancing the sliding velocity (Table 2) and slide-to-roll ratio.

The wear along the face and flank region of the gear drive is predicted based on Eq. (1). The plot shows the contact points and the wear depth along the involute profile region. From Fig. 4, it can be noticed that the wear along the tooth profile is high for the balanced sliding velocity compared to the slide-to-roll ratio, and also the variation in the maximum wear along the face and the flank regions seems to be low for a balanced slide-to-roll ratio compared to a balanced sliding velocity case. This is primarily influenced by the shift in the sliding distance towards the pitch circle more for a slide-to-roll ratio case compared to the one for the sliding velocity. The maximum wear depth $(h)_{max}$ is computed for the increasing value of gear ratio (i) and represented in Fig. 5. The $(h)_{max}$ keeps on declining for the increasing i value which is mainly due to the reduced sliding distances for the balanced slide-to-roll ratio when compared with the sliding velocity.

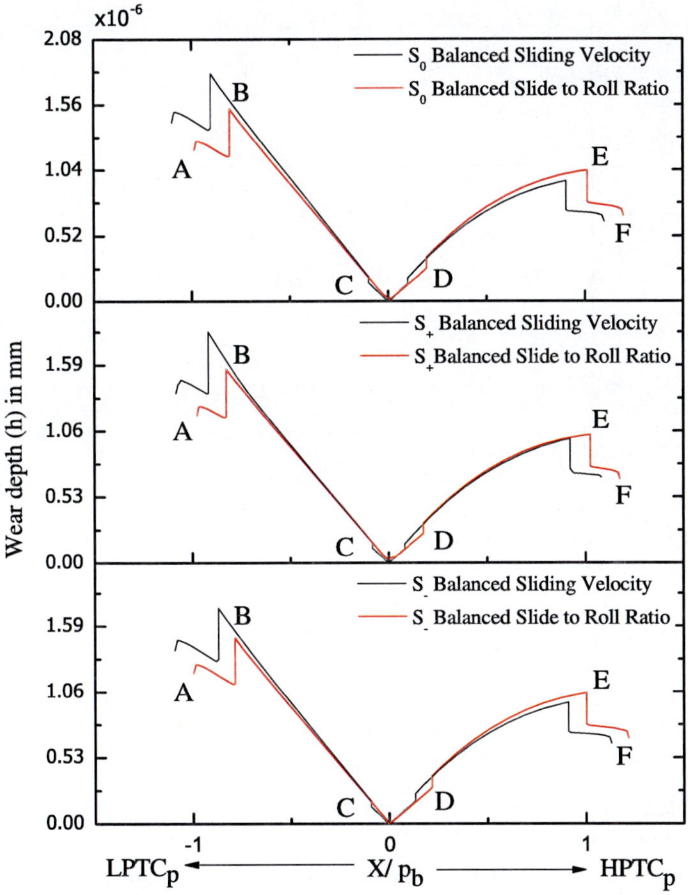

Fig. 4 Influence of profile correction on wear depth (h) for 1000 cycles ($m = 1$, $z_p = 50$, $i = 2$, $\alpha_o = 20°$, $h_a = 1.25$ m and $h_f = 1.5$ m)

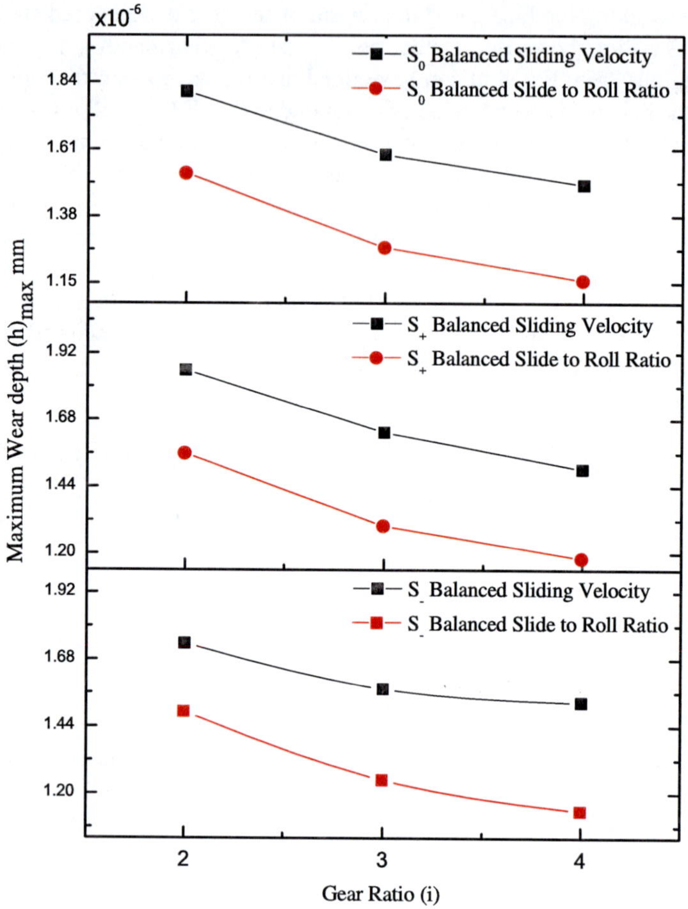

Fig. 5 Effect of gear ratio (i) on maximum wear depth $(h)_{\max}$ ($m = 1$, $z_p = 50$, $\alpha_o = 20°$, $h_a =$ 1.25 m and $h_f = 1.5$ m)

5 Conclusions

This study investigated the development of material loss along the face and flank regions of the gear tooth, with detailed focus on the impact of sliding velocity and slide–roll ratio (SR). The key conclusions can be summarized as follows:

- To reduce the maximum wear depth, it is suggested to vary the sum of addendum modification coefficient.
- The increase of the gear ratio needs an increase of the addendum modification coefficient of the pinion, in order to reach the balanced sliding velocity and slide-to-roll ratio.
- Balancing of slide-to-roll ratio provides more resistance to the tooth compared to the balancing of sliding velocity.

References

1. Dudley DW (1962) Gear handbook. McGraw-Hill, New York
2. Ristivojevic M, Lazovic T, Vencl A (2013) Studying the load carrying capacity of spur gear tooth flanks. Mech Mach Theory 59:125–137
3. Masjedi M, Khonsari MM (2015) On the prediction of steady-state wear rate in spur gears. Wear 343:234–243
4. Imrek H, Duzcukoglu H (2007) Relation between wear and tooth width modification in spur gears. Wear 262(3–4):390–394
5. Sivakumar P, Gopinath K, Sundaresh S (2009) Performance evaluation of high-contact-ratio gearing for combat tracked vehicles—a case study. Proc Inst Mech Eng Part D: J Automob Eng 224:631–643
6. Rameshkumar M, Sivakumar P, Sundaresh S, Gopinath K (2010) Load sharing analysis of high-contact-ratio spur gears in military tracked vehicle applications. Gear Technol 1(3):43–50
7. Flodin A, Andersson S (1997) Simulation of mild wear in spur gears. Wear 207:16–23
8. Siddesh Kumar NM, Shashank TN, Khan N et al (2021) Modal and harmonic analysis of spur gear using FEA. J Fail Anal Preven 21:1855–1865
9. Ravivarman R, Palaniradja K, Sekar RP (2018) Influence of gear ratio on wear depth of nonstandard HCR spur gear drive with balanced fillet stress. Mater Today Proc 5:17350–17359
10. Sekar RP, Ravivarman R (2019) Influence of addendum modification factor on root stresses in normal contact ratio asymmetric spur gears. J Solid Mech 11:210–221
11. Ravivarman R, Palaniradja K, Sekar RP (2019) Performance enhancement of normal contact ratio gearing system through correction factor. J Mech Eng Sci 13:5242–5258
12. Andersson S, Eriksson B (1990) Prediction of the sliding wear of spur gears. In: Proceedings of NORDTRIB 10-13

Application

Design of a Novel Mechanism for Actuation of a Bistable Buckled Beam

Saurav Kumar Dutta, B. Sandeep Reddy, and Santosha Kumar Dwivedy

Abstract The paper presents the design of a novel Scotch Yoke cum beam engine inspired mechanism to alternately actuate a bistable buckled beam between its two stable equilibrium states. The bistable buckled beam is used in a gripper of a pipe climbing robot with an inchworm motion. In the proposed mechanism, a beam engine is modified by replacing its crank with the crank of a Scotch Yoke and by putting additional links. The proposed mechanism gives a completely constrained one degree of freedom motion which can be driven by a single DC motor. The conceptual designs and models of the gripper and the pipe climbing robot are presented. The working of the proposed mechanism is checked and verified in Solidworks 2019.

Keywords Scotch Yoke · Beam engine · Pipe climbing robot · Pole climbing robot · Inchworm motion · Bistable compliant mechanism

1 Introduction

Bistable buckled beam is a type of bistable compliant mechanism. Complaint mechanisms are mechanisms which have flexible bars and flexible joints. These mechanisms are generally one-piece mechanisms and transfer energy through deformation and motion [1, 2]. A bistable compliant mechanism is one which has two stable equilibrium states in its range of motion [3, 4]. Generally, bistable compliant mechanisms are of two types. One is the bistable buckled beam [5, 6], and the other is bistable curved beam [7, 8]. A bistable buckled beam is realized by pre-buckling a straight column and then putting transverse loads on the buckled column so as to toggle it between the two stable equilibrium states. A bistable curved beam is one in which the beam is fabricated as the fundamental mode shape of the straight column buckling problem. Bistable compliant mechanisms are generally used in MEMS as micro-actuators, switches, micro-relays, micro-grippers and micro-valves

Saurav Kumar Dutta (✉) · B. Sandeep Reddy · Santosha Kumar Dwivedy
Department of Mechanical Engineering, Indian Institute of Technology Guwahati, Guwahati 781039, India
e-mail: d.saurav@iitg.ac.in

V. K. Gupta et al. (eds.), *Recent Advances in Machines and Mechanisms*, Lecture Notes in Mechanical Engineering, https://doi.org/10.1007/978-981-19-3716-3_27

[6–9]. They are also used for micromechanical non-volatile memory [10, 11], for morphing and aerospace applications [12–14], for energy harvesting [15–17] and in many consumer products [18]. The bistable compliant mechanisms are actuated between their two stable equilibrium states using shape memory alloys [14, 19–21], dielectric elastomer actuators [22], through pneumatic means [23, 24], piezoelectric actuators [25–27] and electro-thermal actuators [28, 29]. There are some bistable compliant mechanisms which are driven by the force of the user, like in the gripper of Balakuntala et al. [30]. The gripper has four arms which are attached on the bottom of a bistable shell. A switching mechanism, driven by the user opens the gripper. As the gripper makes contact with the object to be grasped, the gripper closes and comes back to its previous equilibrium state. Further, Sarojini et al. [31] developed a compliant sit-to-stand easy-chair. The proposed bistable mechanism is a two-port bistable mechanism. Port 1 force, used to switch the seat from one stable equilibrium state to the other stable equilibrium state, is obtained from the weight of the person, and port 2 force, used to move the seat back to the first stable equilibrium state, is applied using a lever attached to the armrests.

The last paragraph discusses the different applications of bistable compliant mechanisms and the different ways of actuating them. The actuation depends on the type of bistable compliant mechanism and its intended application. Hence, this paper proposes a Scotch Yoke cum beam engine inspired mechanism to alternately actuate a bistable buckled beam, which is used in a gripper for pipe or pole climbing applications. The paper has five sections. Section 2 introduces the model of the gripper for pipe or pole climbing. Section 3 presents the design of the proposed mechanism. Section 4 discusses the model of a pipe climbing robot along with the gripper and its actuating mechanism, and Sect. 5 concludes the paper.

2 Model of the Gripper

Figure 1 shows the three-dimensional (3D) computer aided design (CAD) model of the gripper. The gripper has been modelled in Solidworks 2019 by assuming that the gripper grips a pipe of 105–115 mm diameter. The gripper can grip pipes of other diameters also, but then, the gap between the plates clamping the beam and the plate for affixing the rubber pad has to be changed accordingly, as is evident from Fig. 1. This gap can be changed with the help of four nuts in each bolt. The beam shown in Fig. 1 is in unstressed state. It is stressed longitudinally only once in the beginning till it buckles. During this process, the nuts on one side of the beam are loosened and the beam can be stressed manually. Once the beam buckles, the loosened nuts are tightened and the beam remains in buckled state. Figure 2 shows the trimetric view of the gripper in ungripped and gripped positions. The bistable buckled beam can now be toggled between its two stable equilibrium states or between the gripped and ungripped positions by putting a transverse load on it. The next section discusses about a mechanical drive system which automates the toggling of the bistable buckled beam between its two stable equilibrium states.

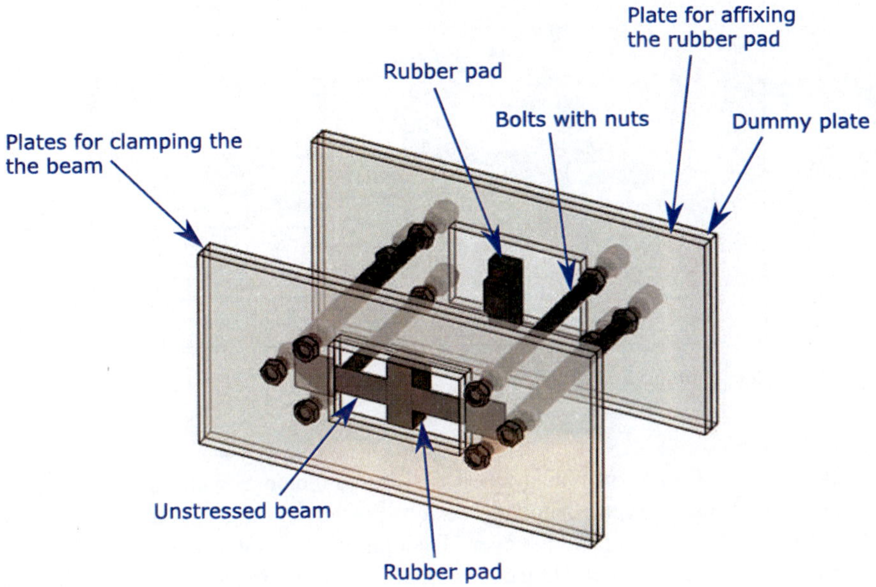

Fig. 1 Trimetric view of the 3D CAD model of the gripper

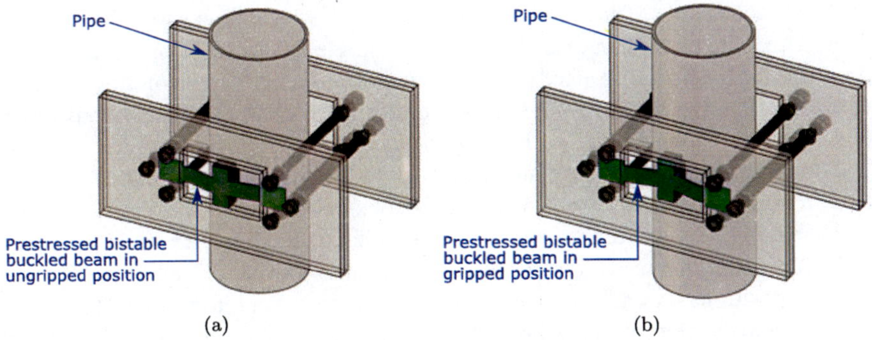

Fig. 2 Trimetric view of the gripper with pipe in **a** ungripped and **b** gripped positions

3 Design of the Actuating Mechanism

The toggling of the two stable equilibrium states of the bistable buckled beam can be automated by using two DC motors. The positions of actuation of the bistable buckled beam are shown in Fig. 3. In Fig. 3, point A corresponds to the central actuation of the bistable buckled beam and point B corresponds to the offset actuation of the bistable buckled beam. The actuator in point A takes the bistable buckled beam from its ungripped position to the gripped position, and the actuator in point B takes the

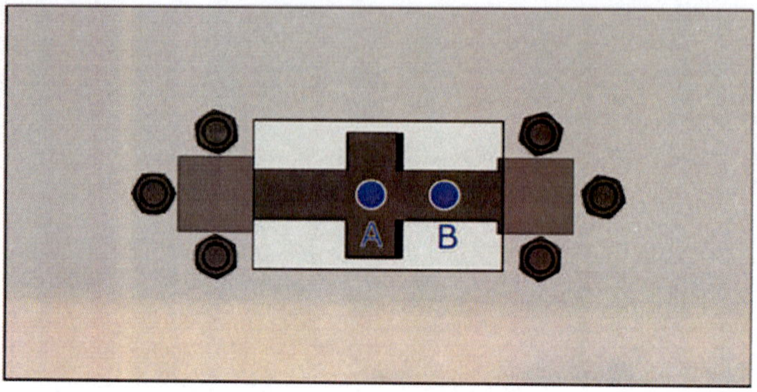

Fig. 3 Front view of the gripper with the actuation points

bistable buckled beam from its gripped position to the ungripped position. Two DC motors with translatory mechanisms could be placed in points A and B. However, to keep the gripper compact and the overall weight of the robot low, it is decided to use one motor for each gripper. Hence, a mechanism is designed by which the bistable buckled beam could be actuated alternately (point A for taking the buckled beam from ungripped to gripped position and point B for taking the buckled beam from gripped to ungripped position), by using a single DC motor.

The mechanism is modelled in Solidworks 2019, which ensures the alternate actuation of the bistable buckled beam using a single DC motor. The working of the proposed mechanism is checked using "animation, basic motion and motion analysis" options in Solidworks 2019. The proposed mechanism has completely constrained one degree freedom of motion. The mechanism shown in Fig. 4 has been modelled by combining the Scotch Yoke mechanism and a beam engine inspired mechanism. Scotch Yoke mechanism is one of the inversions of double slider crank chain [32], and beam engine is one of the inversions of four bar chain [33]. The proposed Scotch Yoke cum beam engine inspired mechanism involves only one DC motor and actuates the two positions (A and B) of the bistable buckled beam alternately. This is also evident from the plots of Fig. 5. From Fig. 5, it can be observed that the sliders of both the Scotch Yoke and the beam engine inspired mechanism operate alternately. The slider of the beam engine inspired mechanism actuates the point A, and the slider of the Scotch Yoke mechanism actuates the point B in the buckled beam through the lever, as can be seen in Fig. 4. Figures 5, 6 and 7 show the displacement versus time, velocity versus time and acceleration versus time plots, respectively, of the sliders of Scotch Yoke cum beam engine inspired mechanism. These plots have been obtained through motion analysis option in Solidworks. The dimensions of the proposed mechanism can be found in Fig. 8. The distance between the crank centre and the pin is 7 mm.

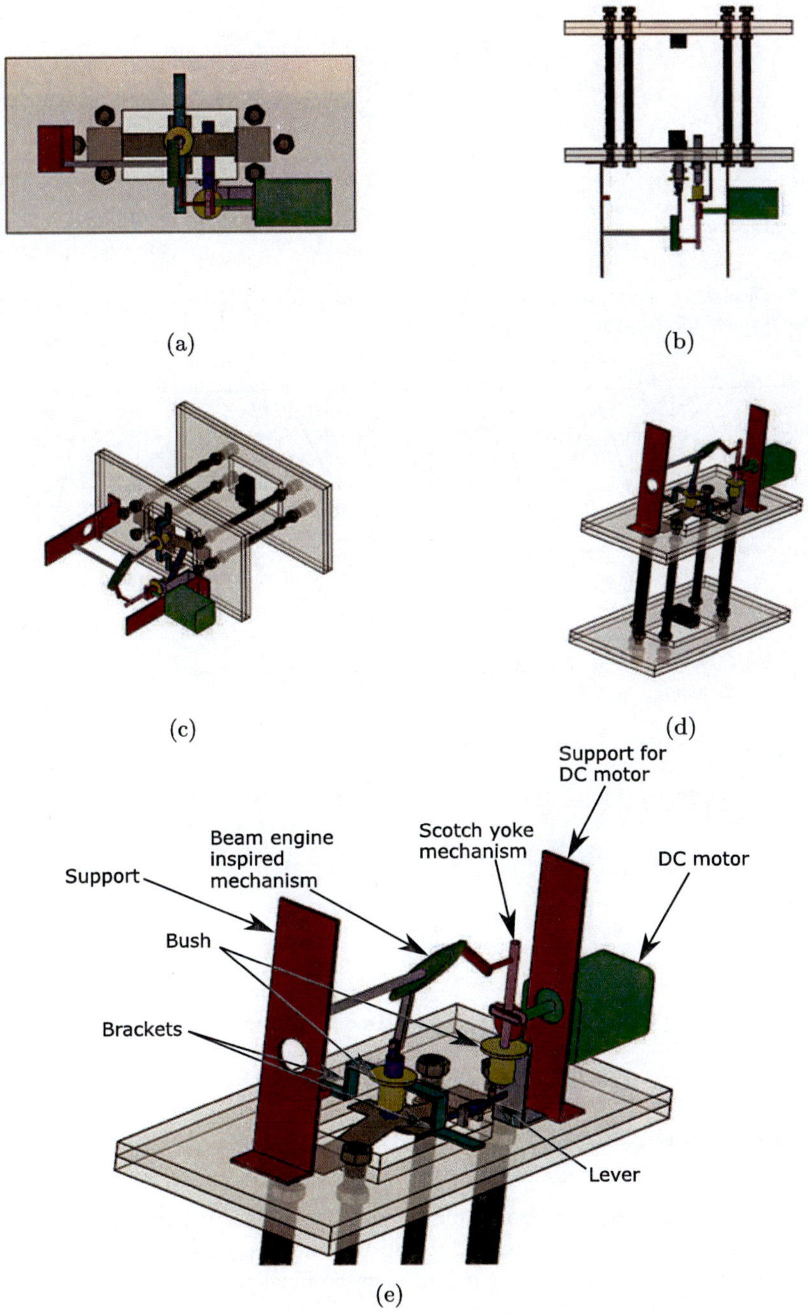

Fig. 4 **a** Front view, **b** top view, **c** isometric view and **d** a different view, **e** close up view of the gripper with the Scotch Yoke cum beam engine inspired actuating mechanism

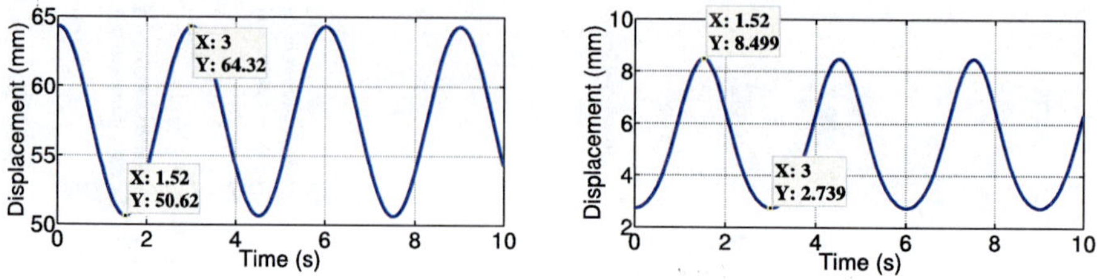

(a) (b)

Fig. 5 Displacement versus time plots of the slider of **a** Scotch Yoke mechanism and **b** beam engine inspired mechanism

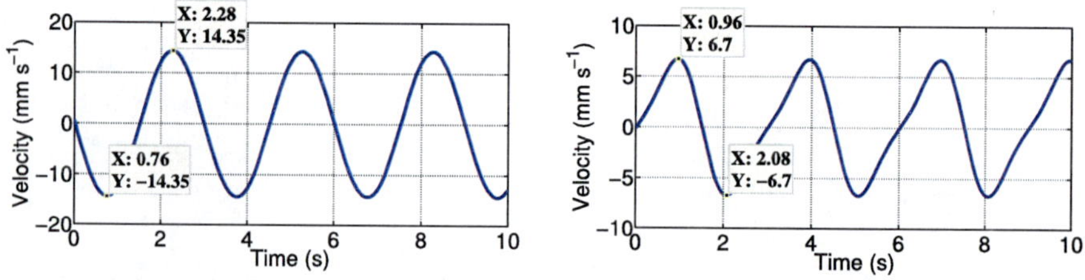

(a) (b)

Fig. 6 Velocity versus time plots of the slider of **a** Scotch Yoke mechanism and **b** beam engine inspired mechanism

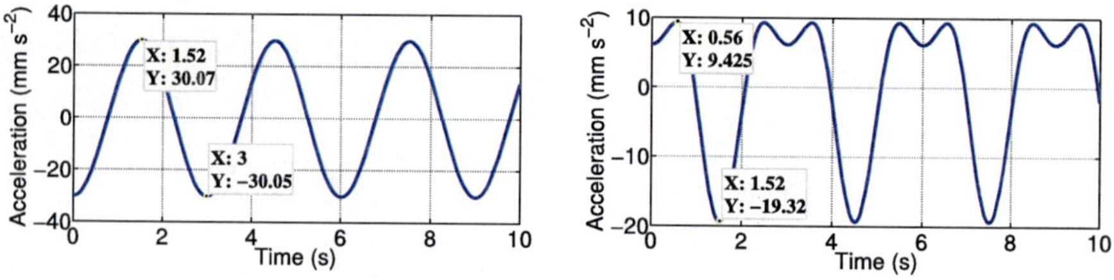

(a) (b)

Fig. 7 Acceleration versus time plots of the slider of **a** Scotch Yoke mechanism and **b** beam engine inspired mechanism

4 Model of the Pipe Climbing Robot

The gripper along with the proposed mechanism can be used in a pipe climbing robot with inchworm motion to automatically grip and ungrip pipes. Pipe climbing robots with inchworm motion can also be called as grasp-push-grasp motion [34] or clamp and push motion [35] or step-by-step motion [36]. The locomotion of a pipe climbing robot with inchworm locomotion is shown in Fig. 9.

The robot in Fig. 9 has two grippers, A and B. After assembling the robot around the pipe, the grippers A and B are in the clamped position at points 2 and 1, respec-

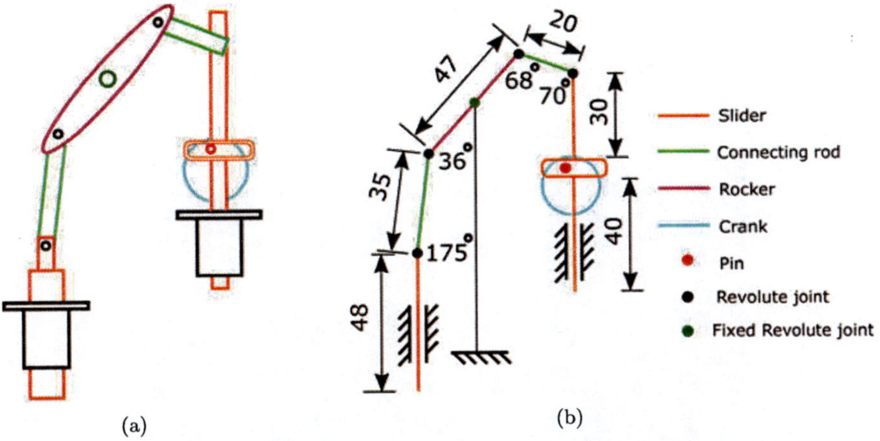

Fig. 8 **a** 2D schematic and **b** kinematic diagram of the proposed Scotch Yoke cum beam engine inspired mechanism

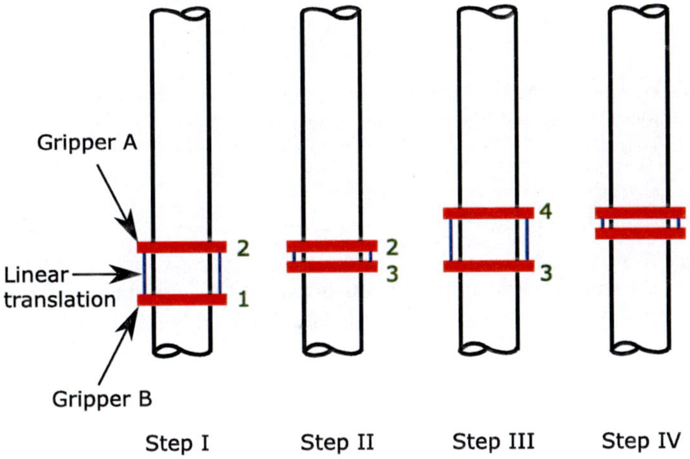

Fig. 9 Locomotion of an inchworm-type pipe climbing robot

tively. Gripper B is freed first, and then, a translational mechanism pulls it towards gripper A, up to point 3. Then, the gripper B is clamped at point 3 and gripper A is freed. The translational mechanism pushes gripper A away from gripper B up to point 4. Then, the gripper A is clamped at point 4 and the gripper B is again freed and pulled by the translational mechanism towards gripper A. This process repeats itself, thus resulting in climbing motion. The translational motion can be realized by an extensor contractor pneumatic artificial muscle, as shown in Fig. 10a or by using a stepper motor-driven lead screw, as shown in Fig. 10b. Information on pneumatic artificial muscle and extensor contractor pneumatic artificial muscle can be found in Refs. [37, 38], respectively.

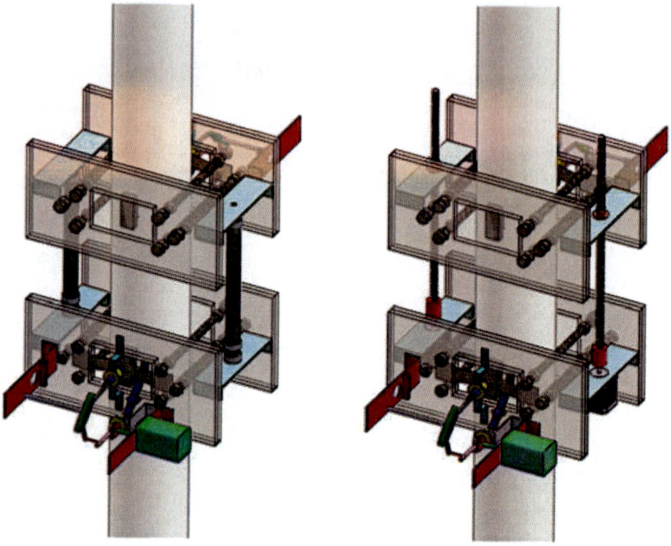

Fig. 10 Trimetric view of the 3D CAD model of the pipe climbing robot with **a** extensor contractor pneumatic artificial muscle and **b** stepper motor-driven lead screw

5 Conclusion

The paper proposes and designs a novel Scotch Yoke cum beam engine inspired mechanism for alternate actuation of a bistable buckled beam, used in a gripper for pipe climbing. The mechanism has been modelled and designed in Solidworks. After analysing the motion of the alternate actuation mechanism, the gripper is shown to be used in a pipe climbing robot by using either extensor contractor pneumatic artificial muscle or stepper motor-driven lead screw. Future work would be to carry out the kinetic analysis of the mechanism and to develop the pipe climbing robot.

References

1. Chen G, Gou Y, Zhang A (2011) Synthesis of compliant multistable mechanisms through use of a single bistable mechanism. J Mech Des 133(8)
2. Liu C-H, Huang G-F, Chen T-L (2017) An evolutionary soft-add topology optimization method for synthesis of compliant mechanisms with maximum output displacement. J Mech Robot 9(5)
3. Jensen B, Howell L, Salmon L (1999) Design of two-link, in-plane, bistable compliant micro-mechanisms. J Mech Des 121(3):416–423
4. Masters ND, Howell LL (2003) A self-retracting fully compliant bistable micromechanism. J Microelectromech Syst 12(3):273–280

5. Vangbo M (1998) An analytical analysis of a compressed bistable buckled beam. Sens Actuators A Phys 69(3):212–216
6. Cazottes P, Fernandes A, Pouget J, Hafez M (2009) Bistable buckled beam: modeling of actuating force and experimental validations. J Mech Des 131(10)
7. Qiu J, Lang JH, Slocum AH (2004) A curved-beam bistable mechanism. J Microelectromech Syst 13(2):137–146
8. Palathingal S, Ananthasuresh G (2017) Design of bistable arches by determining critical points in the force-displacement characteristic. Mech Mach Theory 117:175–188
9. Palathingal S, Ananthasuresh G (2016) Design of bistable pinned-pinned arches with torsion springs by determining critical points. In: Mechanism and Machine Science. Springer, pp 677–688
10. Charlot B, Sun W, Yamashita K, Fujita H, Toshiyoshi H (2008) Bistable nanowire for micromechanical memory. J Micromech Microeng 18(4):045005
11. Roodenburg D, Spronck J, Van der Zant H, Venstra W (2009) Buckling beam micromechanical memory with on-chip readout. Appl Phys Lett 94(18):183501
12. Diaconu C, Weaver P, Mattioni F (2007) Solutions for morphing airfoil sections using bi-stable laminated composite structures. In: 48th AIAA/ASME/ASCE/AHS/ASC Structures, Structural Dynamics, and Materials Conference, p 1719
13. Daynes S, Potter K, Weaver P (2008) Bistable prestressed buckled laminates. Compos Sci Technol 68(15–16):3431–3437
14. Pontecorvo ME, Barbarino S, Murray GJ, Gandhi FS (2013) Bistable arches for morphing applications. J Intell Mater Syst Struct 24(3):274–286
15. Cottone F, Gammaitoni L, Vocca H, Ferrari M, Ferrari V (2012) Piezoelectric buckled beams for random vibration energy harvesting. Smart Mater Struct 21(3):035021
16. Liu W, Badel A, Formosa F, Wu Y, Agbossou A (2013) Novel piezoelectric bistable oscillator architecture for wideband vibration energy harvesting. Smart Mater Struct 22(3):035013
17. Andò B, Baglio S, Bulsara A, Marletta V (2014) A bistable buckled beam based approach for vibrational energy harvesting. Sens Actuators A Phys 211:153–161
18. Palathingal S, Ananthasuresh G (2018) A bilateral relationship between stable profiles of pinned-pinned bistable shallow arches. Int J Solids Struct 143:183–193
19. Kim S-W, Koh J-S, Cho M, Cho K-J (2010) Towards a bio-mimetic flytrap robot based on a snap-through mechanism. In: 2010 3rd IEEE RAS & EMBS International Conference on Biomedical Robotics and Biomechatronics. IEEE, pp 534–539
20. Thomas S, Almanza M, Civet Y, Perriard Y (2018) Actuation displacement analysis of a self-switching shape memory alloy buckled beam. In: 2018 21st International Conference on Electrical Machines and Systems (ICEMS). IEEE, pp 1771–1776
21. Welsch F, Kirsch S-M, Motzki P, Schmidt M, Seelecke S (2018) Vacuum gripper system based on bistable SMA actuation. In: Smart Materials, Adaptive Structures and Intelligent Systems, vol 51944. American Society of Mechanical Engineers, p V001T04A014
22. Follador M, Conn A, Rossiter J (2015) Bistable minimum energy structures (BiMES) for binary robotics. Smart Mater Struct 24(6):065037
23. Jitosho R, Choi K, Foris A, Mazumdar A (2019) Exploiting bistability for high force density reflexive gripping. In: 2019 International Conference on Robotics and Automation (ICRA). IEEE, pp 1241–1247
24. Wang X, Khara A, Chen C (2020) A soft pneumatic bistable reinforced actuator bioinspired by venus flytrap with enhanced grasping capability. Bioinspir Biomim 15(5):056017
25. Schultz M, Hyer M (2004) A morphing concept based on unsymmetric composite laminates and piezoceramic MFC actuators. In: 45th AIAA/ASME/ASCE/AHS/ASC Structures, Structural Dynamics and Materials Conference, p 1806
26. Liu Y, Xu Q (2016) Design and analysis of a micro-gripper with constant force mechanism. In: 2016 12th World Congress on Intelligent Control and Automation (WCICA). IEEE, pp 2142–2147
27. Liu Y, Zhang Y, Xu Q (2016) Design and control of a novel compliant constant-force gripper based on buckled fixed-guided beams. IEEE/ASME Trans Mechatron 22(1):476–486

28. Huang H-W, Yang Y-J (2012) A MEMS bistable device with push-on-push-off capability. J Microelectromech Syst 22(1):7–9
29. Yang S, Xu Q (2017) Design and simulation of a passive-type constant-force MEMS microgripper. In: 2017 IEEE International Conference on Robotics and Biomimetics (ROBIO). IEEE, pp 1100–1105
30. Balakuntala MV, Palathingal S, Ananthasuresh G (2021) A passive universal grasping mechanism based on an everting shell. In: Mechanism and Machine Science. Springer, pp 601–608
31. Sarojini D, Lassche T, Herder J, Ananthasuresh G (2016) Statically balanced compliant two-port bistable mechanism. Mech Mach Theory 102:1–13
32. Uicker JJ, Pennock GR, Shigley JE, Mccarthy JM (2003) Theory of Machines and Mechanisms, vol 3. Oxford University Press, New York
33. Singh VP (1999) Theory of Machines. Dhanpat Rai and Co. (Pvt.) Ltd., Delhi
34. Ripin ZM, Soon TB, Abdullah A, Samad Z (2000) Development of a low-cost modular pole climbing robot. In: 2000 TENCON Proceedings. Intelligent Systems and Technologies for the New Millennium (Cat. No. 00CH37119), vol 1. IEEE, pp 196–200
35. Singh P, Ananthasuresh G (2012) A compact and compliant external pipe-crawling robot. IEEE Trans Robot 29(1):251–260
36. Tavakoli M, Marques L, de Almeida AT (2010) Development of an industrial pipeline inspection robot. Ind Robot Int J
37. Dutta SK, Sandeep Reddy B, Dwivedy SK (2022) Design of a two degrees of freedom actuator for rehabilitation robotic applications. In: Machines, Mechanism and Robotics. Springer, pp 1189–1196
38. Al-Fahaam H, Nefti-Meziani S, Theodoridis T, Davis S (2018) The design and mathematical model of a novel variable stiffness extensor-contractor pneumatic artificial muscle. Soft Robot 5(5):576–591

Modelling and Simulation of Car Suspension with Linear and Nonlinear Spring

Deepak Nigwal, Dinesh Kumar Pasi, and Manoj Chouksey

Abstract In this paper, a new approach is defined to change the linear stiffness of the spring element in the passive suspension system, instead of redefining the entire arrangement altogether. The linear cylindrical spring is replaced by a Nonlinear Conical spring, whose stiffness varies depending on the load-induced; and thus conical spring designs are studied and developed to fit in the application. The properties of this nonlinear spring are then imported into the car model using Combin39 Nonlinear Element in ANSYS, to check its response on vehicle dynamics such as bounce, roll, and pitch, under the road excitation of the speed bump, as compared to the linear rate spring working on the same range of vehicle loads. The car models are developed in ANSYS Workbench and MATLAB Simulink to compare with the analytical solution using the State-Space Model, for further validation. The results of Modal and Transient analysis showed some significant reduction in natural frequencies and amplitudes of sprung mass motions; thus concluding that conical springs do work and perform the best of both worlds, and fill the gap between passenger comfort and vehicle handling characteristics; without modifying the entire suspension arrangement.

Keywords Combin39 element · Modal analysis · Nonlinear conical spring · Transient analysis

1 Introduction

A good suspension system's job is to support the weight of the vehicle, provide a smooth and high level of ride quality to the vehicle, protect the structure of the vehicle from stresses by isolating from road surface irregularities, allow for rapid cornering and reduce body roll, keep the tires firmly in contact with the ground, reduce body squat and prevent body dive, allow the front wheels of the vehicle to steer and always maintain the correct alignment of the wheels. In recent years

D. Nigwal (✉) · D. K. Pasi · M. Chouksey
Shri G. S. Institute of Technology and Science, Indore 452003, India
e-mail: nigwal.deepak@yahoo.in

V. K. Gupta et al. (eds.), *Recent Advances in Machines and Mechanisms*, Lecture Notes in Mechanical Engineering, https://doi.org/10.1007/978-981-19-3716-3_28

authors have investigated more upon the human response subjected to the vibration excitation. Griffin, Michael et al. [1] investigated and concluded the range from 3 to 7 Hz to be the human body's fundamental natural frequency. Gillespie [2] described the stiffness ratio for the car suspension and suggested the ranges for stiffest and softest suspensions. The author said that a soft suspension isolates the ride, as for lower natural frequencies, the sprung mass acceleration is minimum; at 1 Hz natural frequency. Mitra et al., also studied the dynamic behavior but under the standard highway and city road profiles (top half sin wave) [3], and found the speed range of 5–10 kmph to be optimum to move over the bump without influencing the human tolerance of 0.315 m/s^2 to 0.625 m/s^2 as per ISO standard [4].

Modern studies have shifted toward full car models, as they allow the researchers to explore further into the vehicle dynamics. Jain [5] introduced a cabin to improve the ride comfort of the commercial vehicle using the ADAMS and reduced the natural frequencies to 1.8 to 2.2 Hz. Following this, Sharma et al., integrated the primary and cabin suspension, using a 7 and 11DOF car model with a 3DOF cabin on top [6]; suggesting improved vehicle performance and reduced cabin accelerations, thus improving ride comfort. In a similar study, Sharma et al. [7] gave the novel idea and determined the natural frequencies, mode shapes, and bode plots to better understand the effects of vibrations on the human body and comfort. The author thus suggested the use of a cabin for driver comfort.

Since passive suspensions were the most basic and least customizable, they are rather a compromise between passenger comfort and vehicle handling; giving rise to the controlled suspensions. This did solve major problems regarding acceleration and displacements of the body, but they were expensive for a daily driving passenger car. Some authors tried to find a different approach to make the ride dynamics more effective, leading to nontraditional methods; Mohite et al., developed a model having nonlinear spring and damper configuration for a quarter car [8]; and showed a significant decrease in the sprung mass motions. Shelke et al., defined a nonlinear variable stiffness spring and tire, by dividing its stiffness into two and three parts, respectively, i.e., linear, and nonlinear square and cubic stiffness [9]. Spring properties of linear and nonlinear spring properties were extracted experimentally using UTM and then fed into the models developed using MATLAB/Simulink and State-Space, and suggested the use of nonlinear spring to reduce the RMS acceleration for improved ride comfort.

Anubi et al. [10], also proposed a new design prototype of a passive suspension system incorporating the variable stiffness using the mechanical arrangement itself, consisting of simple horizontal and vertical struts mounted on a double-wishbone setup; where the vertical strut act as the primary suspension member and attached to a horizontal sliding mass connected to the horizontal strut (acting as a preloaded passive member). The design changed the stiffness of the system naturally with the loads and thus eliminating the need for any external control mechanism.

In the current study, a similar approach is investigated for varying the stiffness of the compression spring by considering a conical spring while replacing the linear rate spring, to achieve human comfort and ride quality. For this, the linear springs in the ANSYS model are converted using the Combin39 spring element. The models are simulated under the excitation of a speed bump and sprung mass motions are evaluated.

2 Mathematical, State-Space and ANSYS Models

Figure 1 shows the free body diagram of the sprung and unsprung mass; Fig. 2, shows the linear mathematical dynamics simulation model of the full car being studied and developed in this work. M_{usFL}, M_{usFR}, M_{usRL}, and M_{usRR} denotes the cubical

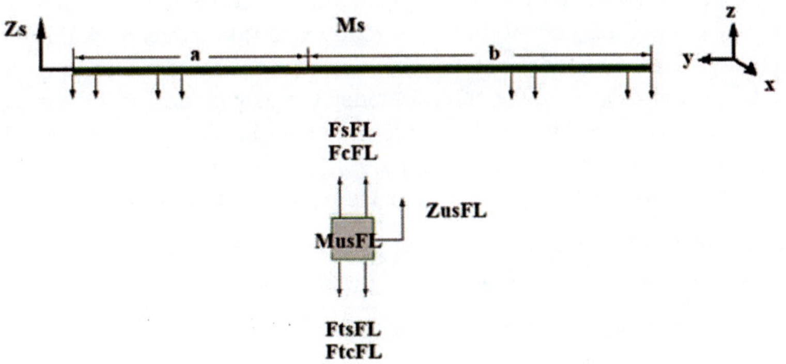

Fig. 1 Free body diagram of sprung and unsprung mass

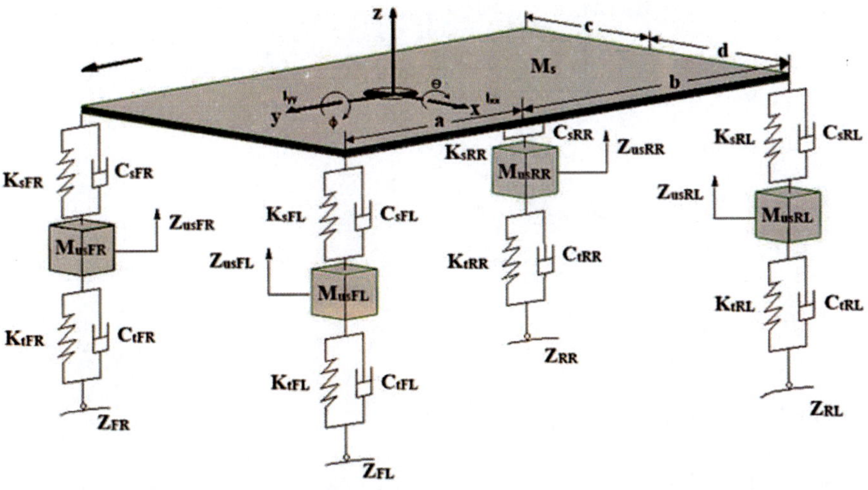

Fig. 2 7DOF Full Car Mathematical Model

shape unsprung masses of the car (M_{usFL} is the mass associated with the front left-hand side of the car), while the sprung mass having a large thin cuboidal shape is defined by M_s. Z_{usFL}, Z_{usFR}, Z_{usRL}, and Z_{usRR} denote the unsprung masses' vertical displacement, while Z_s denotes the vertical displacement of the sprung mass. θ and ϕ denote the degree of freedom corresponding to pitching and rolling, respectively. K_{sFL}, K_{sFR}, K_{sRL}, and K_{sRR}, and C_{sFL}, C_{sFR}, C_{sRL}, and C_{sRR} define the Suspension System stiffness and damping variables on all wheels individually. $K_{s\text{-}nFL}$, $K_{s\text{-}nFR}$, $K_{s\text{-}nRL}$, and $K_{s\text{-}nRR}$ defines the stiffness of all suspension springs having nonlinear characteristics. Similarly, K_{tFL}, K_{tFL}, K_{tRL}, and K_{tRR} (or K_t), and C_{tFL}, C_{tFR}, C_{tRL}, and C_{tRR} (or C_t), are the rigidity and damping variables of all four tires of the car. a and b indicates distance of the front and back axle from the sprung mass's center of gravity (CG); whereas c and d show distance from wheels to the center of gravity (CG). Z_{FL}, Z_{FR}, Z_{RL}, and Z_{RR} define the input road excitation induced upon each wheel respectively. Thus, the coordinates which describe the system are Zs, θ, ϕ, Z_{usFL}, Z_{usFR}, Z_{usRL}, and Z_{usRR}. For the current model, the yaw motion, roll bar effects, tire models, and steering angle of wheels are not taken into account. $F_{(s\text{-}c)FL}$, $F_{(s\text{-}c)FR}$, $F_{(s\text{-}c)RL}$, and $F_{(s\text{-}c)RR}$; and $F_{t(s\text{-}c)FL}$, $F_{t(s\text{-}c)FR}$, $F_{t(s\text{-}c)RL}$, and $F_{t(s\text{-}c)RR}$, are the spring and damper forces corresponding to the suspension and tyres, respectively.

Equations of motion: Newton's Law of motion is used to define the equations of motion for the suspension system of a full car, which is as follows:

Vertical Vibration of Front Left Unsprung Mass:

$$\begin{aligned} & M_{usFL}\ddot{Z}_{usFL} - C_{sFL}\left(\dot{Z}_S - a\dot{\theta} + c\dot{\varphi} - \dot{Z}_{usFL}\right) \\ & - K_{sFL}(Z_S - a\theta + c\varphi - Z_{usFL}) + C_t\left(\dot{Z}_{usFL} - \dot{Z}FL\right) \\ & + K_t(Z_{usFL} - ZFL) = 0 \end{aligned} \tag{1}$$

Similarly, equations of the rest of the unsprung masses can be defined to determine their motion, respectively.

Vertical Vibration Motion of Sprung Mass:

$$\begin{aligned} & M_S\ddot{Z}_S + C_{sFL}\left(\dot{Z}_S - a\dot{\theta} + c\dot{\varphi} - \dot{Z}_{usFL}\right) + K_{sFL}(Z_S - a\theta + c\varphi - Z_{usFL}) \\ & + C_{sFR}\left(\dot{Z}_S - a\dot{\theta} - d\dot{\varphi} - \dot{Z}_{usFR}\right) + K_{sFR}(Z_S - a\theta - d\varphi - Z_{usFR}) \\ & + C_{sRL}\left(\dot{Z}_S + b\dot{\theta} + c\dot{\varphi} - \dot{Z}_{usRL}\right) + K_{sRL}(Z_S + b\theta + c\varphi - Z_{usRL}) \\ & + C_{sRR}\left(\dot{Z}_S + b\dot{\theta} - d\dot{\varphi} - \dot{Z}_{usRR}\right) + K_{sRR}(Z_S + b\theta - d\varphi - Z_{usRR}) = 0 \end{aligned} \tag{2}$$

Rolling Motion of Sprung Mass:

$$\begin{aligned} & I_{YY}\ddot{\varphi} + c\left[C_{sFL}\left(\dot{Z}_s - a\dot{\theta} + c\dot{\varphi} - \dot{Z}_{usFL}\right) + K_{sFL}(Z_S - a\theta + c\varphi - Z_{usFL})\right] \\ & - d\left[C_{sFR}\left(\dot{Z}_S - a\dot{\theta} - d\dot{\varphi} - \dot{Z}_{usFR}\right) - K_{sFR}(Z_S - a\theta + d\varphi - Z_{usFR})\right] \\ & + c\left[C_{sRL}\left(\dot{Z}_S + b\dot{\theta} + c\dot{\varphi} - \dot{Z}_{usRL}\right) + K_{sRL}(Z_S + b\theta + c\varphi - Z_{usRL})\right] \end{aligned}$$

$$-d\left[C_{\text{sRR}}\left(\dot{Z}_S + b\dot{\theta} - d\dot{\varphi} - \dot{Z}_{\text{usRR}}\right) - K_{\text{sRR}}(Z_S + b\theta - d\varphi - Z_{\text{usRR}})\right] = 0 \quad (3)$$

Pitching Motion of Sprung Mass:

$$\begin{aligned} & I_{XX}\ddot{\theta} - a\left[C_{\text{sFL}}\left(\dot{Z}_S - a\dot{\theta} + c\dot{\varphi} - \dot{Z}_{\text{usFL}}\right) - K_{\text{sFL}}(Z_S - a\theta + c\varphi - Z_{\text{usFL}})\right] \\ & \quad - a\left[C_{\text{sFR}}\left(\dot{Z}_S - a\dot{\theta} - d\dot{\varphi} - \dot{Z}_{\text{usFR}}\right) - K_{\text{sFR}}(Z_S - a\theta + d\varphi - Z_{\text{usFR}})\right] \\ & \quad + b\left[C_{\text{sRL}}\left(\dot{Z}_S + b\dot{\theta} + c\dot{\varphi} - \dot{Z}_{\text{usRL}}\right) + K_{\text{sRL}}(Z_S + b\theta + c\varphi - Z_{\text{usRL}})\right] \\ & \quad + b\left[C_{\text{sRR}}\left(\dot{Z}_S + b\dot{\theta} - d\dot{\varphi} - \dot{Z}_{\text{usRR}}\right) + K_{\text{sRR}}(Z_S + b\theta - d\varphi - Z_{\text{usRR}})\right] = 0 \quad (4) \end{aligned}$$

2.1 State-Space Model

The linear state-space model and its analytical solution are used to validate the models developed in other tools. The general form of continuous state-space Eq. can be written as:

$$[\dot{x}] = [A][X] + [B][U] \quad (5)$$

$$[Y] = [C][X] + [D][U] \quad (6)$$

where $[\dot{x}]$ is the state of the system, $x = \frac{dx}{dt}$, $[A]$ is the system matrix, $[B]$ is the input matrix, $[C]$ is the output matrix, and $[D]$ is the direct transmission matrix, U is the input vector, and Y is the output vector. By following Eq. (1–7), the state, input, and output matrices are obtained and inserted in the general form (Eq. 8 and 9) to develop the state-space model. This model is then run in MATLAB program using parameters in Table 1, and further imported into Simulink environment to perform transient analysis.

2.2 ANSYS Model

ANSYS Workbench version 19.3 is used to create and analyze the 7 degrees of freedom full car model (developed by Sharma et al. [6]); consisting of 5 bodies, namely, a Sprung Mass and four Unsprung Masses, representing the body and wheel assembly masses of a car, respectively. The Sprung Mass geometry is slightly modified to account for the center of gravity, thus making it a front-biased body.

Speed bump: Hasbullah et al. [11] used the cosine function to define a single circular speed, Thus, the following input disturbances for road excitation on both wheels are

Table 1 Model parameters

Sr. No	Model parameter	Value	Unit
1	M_s	1243.4	kg
2	$M_{\mathrm{usFL}}, M_{\mathrm{usFR}}, M_{\mathrm{usRL}}, M_{\mathrm{usRR}}$	62.8	kg
3	$K_{\mathrm{sFL}}, K_{\mathrm{sFR}}$	36.26	N/mm
4	$C_{\mathrm{sFL}}, C_{\mathrm{sFR}}$	2.5	N-s/mm
5	$K_{\mathrm{sRL}}, K_{\mathrm{sRR}}$	28.49	N/mm
6	$C_{\mathrm{sRL}}, C_{\mathrm{sRR}}$	2.0	N-s/mm
7	$K_{\mathrm{s\text{-}nFL}}, K_{\mathrm{s\text{-}nFR}}$	30.12 +	N/mm
8	$K_{\mathrm{s\text{-}nRL}}, K_{\mathrm{s\text{-}nRR}}$	25.02 +	N/mm
9	$K_{\mathrm{tFL}}, K_{\mathrm{tFL}}, K_{\mathrm{tRL}}, K_{\mathrm{tRR}}$ (or K_{t})	200	N/mm
10	$C_{\mathrm{tFL}}, C_{\mathrm{tFR}}, C_{\mathrm{tRL}}, C_{\mathrm{tRR}}$ (or C_t)	0.5	N-s/mm
11	a	1.21	m
12	b	1.59	m
13	c, d	0.8	m
14	I_{XX}	940.85	Kg-m^2
15	I_{YY}	265.57	Kg-m^2

taken from the same, i.e.,

$$Z_{\mathrm{FR,FL}}(t) = \begin{cases} 0.5H_{\mathrm{bump}}(1-\cos(8\pi t)) & t_1 < t < t_2 \\ 0 & \text{otherwise} \end{cases} \tag{7}$$

where H_{bump} is the height of speed bump, 100 mm and Length, L 1000 mm; t_1 is 1 s.

By comparing the natural frequencies of the nonlinear model with that of the linear, a significant difference is observed; the first three frequencies are up to 73.11% less and the rest four frequencies are up to 6.45% less (Table 2).

Table 2 Natural frequencies of 7DOF full car linear and nonlinear model

Mode	Linear model (Hz)	Nonlinear model (Hz)	Difference, %
1	1.5007	0.86689	73.11
2	2.3736	1.4234	66.75
3	2.5879	1.5228	69.94
4	9.6193	9.2034	4.51
5	9.6328	9.2055	4.64
6	9.7981	9.2091	6.39
7	9.8042	9.2099	6.45

2.3 *Comparison of Linear State-Space and ANSYS Models*

Figure 3 shows the comparison of the linear model of both ANSYS and State-Space, but at 5 kmph, the ANSYS model show close agreement of the sprung mass bounce displacement with the state-space. However, the sprung mass bounce displacement differs significantly in the two models, as the 1st peak is at 33 mm, while the 2nd peak is at 24.7 mm, 1st peak and downward shoot of models, show 0.06% error.

3 Nonlinear Conical Spring Design

Wolansky, E. B. first derived the equations for buckling of the conical springs [12]. Wu et al. [13] gave the dynamic equations in partial differential forms and thus proposed an algorithm to find load–deflection relation. But the load–deflection curve for this work is defined using the MATLAB program from the formulations given by Rodriguez and Paredes [14]. The design parameters for the conical spring, to replace the linear one in the suspension application, are the pre-load and the suspension travel; thus the springs for both the axles are designed (Table 3). The properties of these developed conical springs are imported into the ANSYS 7DOF model using a nonlinear element.

COMBIN39 Nonlinear Spring Element: The default Combin14 Linear element in ANSYS is converted into the Combin39 element, referred from the APDL Element Reference [15]. This element supports nonlinear force–displacement behavior (inserted in the form of a user-defined curve) and is the reason for using it. Up to 20 coordinate points can be inserted in the form of real constants to define the curve in ascending order in 1st (compression) and moving towards the 3rd (tension)

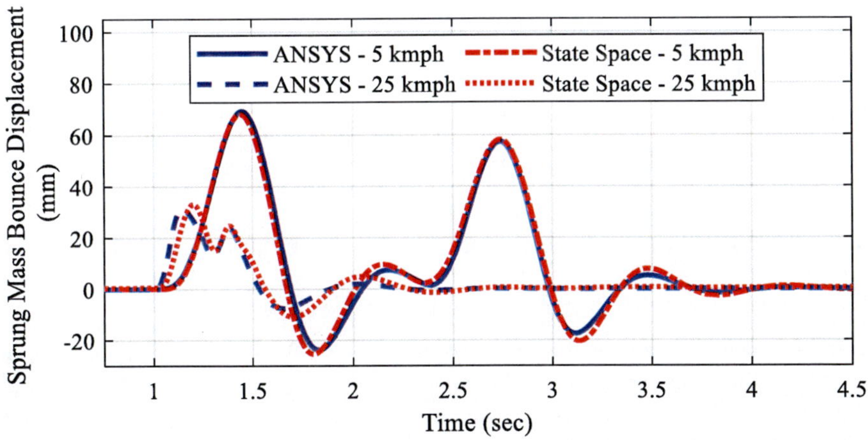

Fig. 3 Comparison of sprung mass bounce displacement

Table 3 Conical spring parameters

Symbol	Parameters	Front conical	Rear conical	Units
d	Wire diameter	10.5	9.5	mm
D_1	Mean coil diameter (Smallest)	60	49	mm
D_2	Mean coil diameter (Largest)	110	120.5	mm
N	Total number of Coils	8	6.5	–
n_a	Active number of coils	6	4.5	–
L_f	Free length	300	250	mm

quadrant. The values of the deflection are in ascending order and the minimum change in deflection is:

$$U_{i+1} - U_i > \Delta U_{\min} \tag{8}$$

where, $i \in [1; 19]$, U_i is the user input deflection, $\Delta U_{min} = (U_{max} - U_{min})/10^7$; The stiffness matrix, K_e and the load vector, F_e^{Sr} of this element is:

$$[K_e] = K^S \begin{bmatrix} 1 & -1 \\ -1 & 1 \end{bmatrix} \tag{9}$$

$$\{F_e^{Sr}\} = F_1 \begin{Bmatrix} 1 \\ -1 \end{Bmatrix} \tag{10}$$

where K^S is the slope of the active segment, and F_1 is the force in the element calculated from prior solved iteration. The commands used to develop a syntax are ET, KEYOPT, _sid, NLGEOM, ON, *R*, and RMORE; the syntax is pasted into the Commands (APDL) in the Mechanical module before simulation, to differentiate it into a nonlinear model.

4 Results

4.1 Comparison of Linear and Nonlinear Models in ANSYS

Figure 4 shows that the displacement values from the nonlinear model are significantly lower than that of the linear model as the vehicle passes over the speed bump. At the same time, the oscillations are more in the linear model and less in the nonlinear. The settling time on the other hand is more for the nonlinear model. The linear model showed 1st peak amplitude of 69.175 mm at 5 kmph, whereas the nonlinear model showed 59.385 mm. For the vehicle speed of 25 kmph, the nonlinear

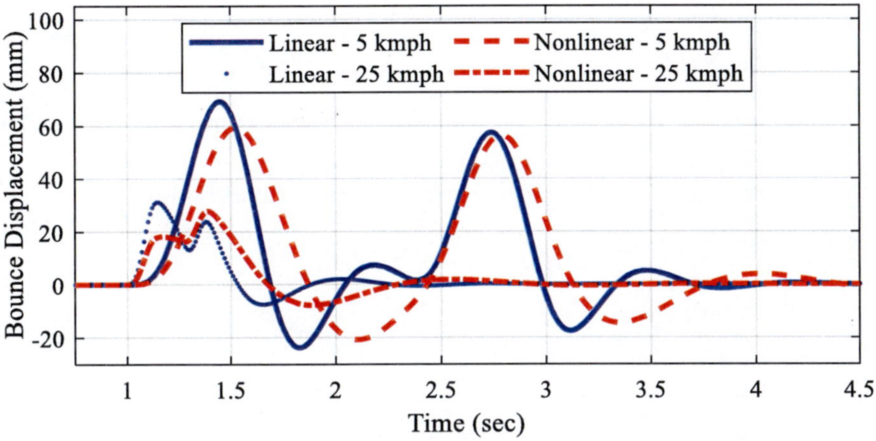

Fig. 4 Comparison of sprung mass bounce displacement (average)

model showed the 2nd peak to be higher than the first, at 18.188 mm, whereas the linear one peaked at 31.117 mm.

Figure 5 shows that the acceleration peaks tend to rise with the increase in vehicle speed as the vehicle passes over the speed bump. At 5 kmph, acceleration peaked at 2686.5 mm/s^2 and 2234.6 mm/s^2 (for 1st and 2nd peaks, respectively), and at 25 kmph, 7311.1 mm/s^2 and 7148.5 mm/^{2}, for 1st and 2nd peaks, respectively.

As evident from Fig. 6, unlike the linear model, the peak values are found to be lower, for the nonlinear model, at low as well as high speed, i.e., at 5 kmph, 1086 mm/s^2 and 829.05 mm/s^2, and 25 kmph, 4286.1 mm/s^2, and 4070.1 mm/s^2 (for 1st and 2nd peaks respectively). Figure 7 shows collectively percentage reduction in

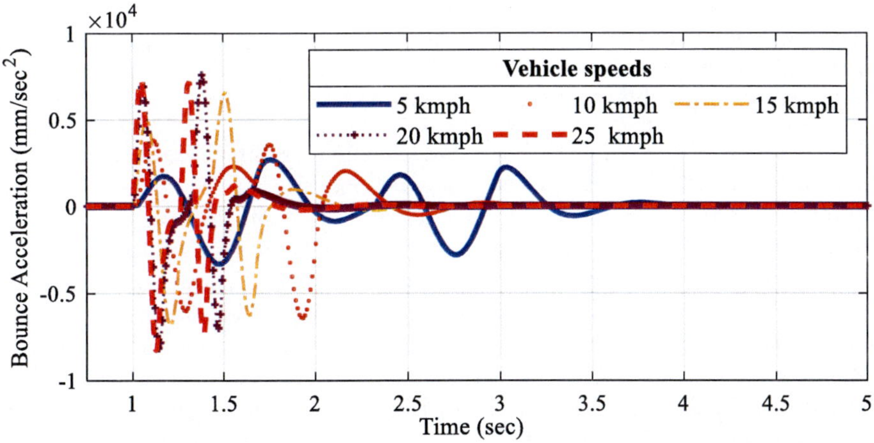

Fig. 5 Combined sprung mass bounce acceleration (average) for linear model

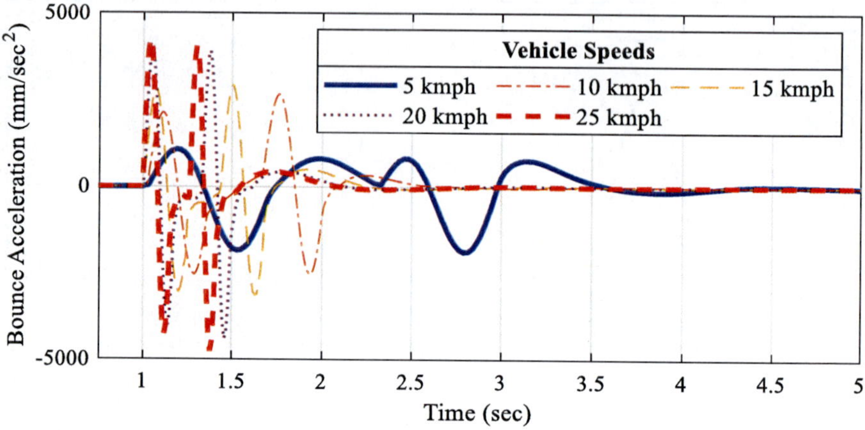

Fig. 6 Combined sprung mass bounce acceleration (average) for nonlinear model

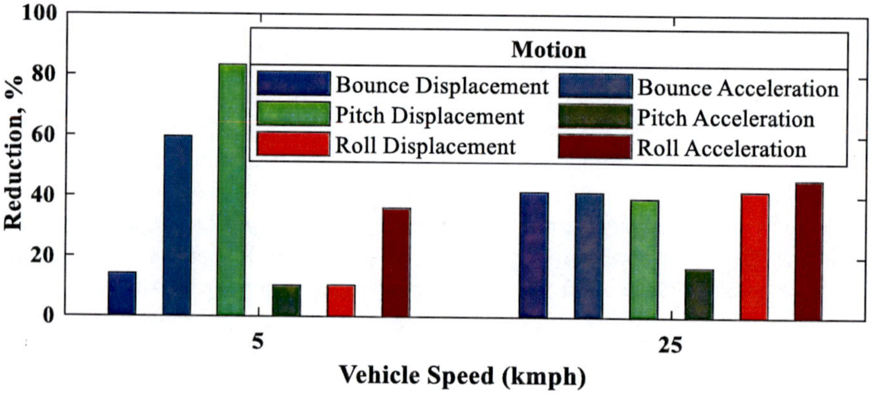

Fig. 7 Percentage reduction in nonlinear model's response as compared with the linear model's response, for sprung mass motions

vehicle's bounce/pitch displacement and accelerations at 5 kmph as well as 25 kmph. At a vehicle speed of 5 kmph, bounce displacement varied 14.15% (due to their soft stiffness nature) whereas acceleration varied up to 59.57%. Pitch values showed a maximum difference of 83.18% in displacement, but 10.32% in acceleration.

For roll motion, the percentage difference for both displacement and acceleration is 10.49% and 36.06%, respectively. Similarly, at 25 kmph, the difference in bounce displacement is 41.55% (due to the hardening effect caused by the increase in stiffness) and acceleration is 41.37%. The pitch motion's displacement ad acceleration difference is 39.37% and 16.65%, respectively; and for roll motion, the difference is 41.81% and 45.61%, respectively for displacement and acceleration. For human comfort, the vertical acceleration acting on the car body is compared with the literature, as Haniszewski et al. [16] suggested using ISO 2631-1, the values of sprung mass accelerations are under limits. Also, the nonlinear model's bounce and roll accelerations are 41.37% and 45.61% less as compared to the linear model, and thus suggesting even less discomfort.

5 Conclusion

1. The natural frequencies of the vehicle are found to be reduced with the use of conical i.e. a nonlinear spring.
2. There is a reduction in the overall bounce displacement response of the sprung mass for both linear and nonlinear stiffness springs with the increase in vehicle speed over the bump; vice versa for bounce acceleration.
3. Displacement and acceleration of sprung mass due to bounce, pitch, and roll are decreased for nonlinear stiffness spring as compared to linear stiffness spring.
4. At low vehicle speed, the reduction in nonlinear model's displacements is less when compared at the high vehicle speed, where this reduction is more.
5. The maximum sprung mass acceleration (average at CG) of the linear model is observed to be under the limits of human tolerances of comfort, which are defined in ISO 2631-1.
6. Passing over the speed bump at low vehicle speed is more desirable for both the linear and nonlinear models.

References

1. Griffin MJ, Erdreich J (1991) Handbook of human vibration. Acoustical Society of America
2. Gillespie TD (1992) Fundamental of vehicle dynamics, SAE International
3. Mitra A, Benerjee N, Khalane HA, Sonawane MA, Joshi DR, Bagul GR (2013) Simulation and analysis of full car model for various road profile on a analytically validated MATLAB/SIMULINK model. IOSR J Mech Civ Eng 22–33
4. ISO, I.S.O.J.I.S. (1997) Mechanical vibration and shock—Evaluation of human exposure to whole body vibration. Part 1: General requirements
5. Jain P (2007) Design and analysis of a tractor-trailer cabin suspension. SAE Technical Paper, pp 549–557
6. Sharma SK, Pare V, Chouksey M, Rawal BR (2016) Numerical studies using full car model for combined primary and cabin suspension. In: 3rd international conference on innovations in automation and mechatronics engineering: procedia technology, vol 23. Elsevier Ltd. pp 171–178
7. Sharma S, Chouksey M, Pare V, Jain P (2020) Modal and frequency response characteristics of vehicle suspension system using full car model. In: IOP conference series: materials science and engineering, p 810
8. Mohite AG, Mitra AC (2018) Development of linear and non-linear vehicle suspension model. Mater Today Proc 5(2):4317–4326
9. Shelke GD, Mitra AC, Varude VR (2018) Validation of simulation and analytical model of nonlinear passive vehicle suspension system for quarter car. Mater Today Proc 5(9):19294–19302

10. Anubi OM, Patel DR, Crane Iii CD (2013) A new variable stiffness suspension system: passive case. Mech Sci 4(1):139–151
11. Hasbullah F, Faris WF, Darsivan FJ, Abdelrahman M (2019) Ride comfort performance of a non-linear full-car using active suspension system with active disturbance rejection control and input decoupling transformation. Int J Heavy Veh Syst 26(2)
12. Wolansky EB (1996) Conical spring buckling deflection: how to derive the buckling deflection of conical springs with simply-supported or fixed ends. Springs 35:62–80
13. Wu MH, Hsu WY (1998) Modelling the static and dynamic behavior of a conical spring by considering the coil close and damping effects. J Sound Vib 214(1):17–28
14. Rodriguez E, Paredes M, Sartor M (2005) Analytical behavior law for a constant pitch conical compression spring. J Mech Des 128(6):1352–1356
15. Ansys (2017) ANSYS mechanical APDL element reference. Release 18.1 ed. 2017: Ansys Inc. 1480.
16. Haniszewski T, Michta A (2019) Preliminary studies of vertical acceleration of a passenger car passing through the speed bump for various driving speeds. J Transp Probl 14(1):12

Mechatronics and Algorithms to Analyse and Control the Laser Beam Spot Size for 3D Micro-printing Machine

Rajendra Kumar Arya, Ratnesha Bafna, Ujwal Pawar, and Prasanna Gandhi

Abstract Recent developments in digital manufacturing see widespread use of lasers of various kinds be it additive or subtractive methods. All laser beam-based technologies require control over the laser beam spot size to get the desired outcome. For systems requiring a frequent change of focusing optics or in-situ change of spot size while processing, a system having automatic control over laser spot size would be required. This paper proposes a mechatronic solution for the same (named as laser beam focusing system (LBFS)) specifically for an in-house micro-stereolithography (MSL) setup and proposes algorithms for detection of focus, estimation of spot size, and depth of focus (DOF) of laser. In this setup, a laser beam steered by mirrors is scanned over a photodiode positioned at a target plane or printing plane of MSL. This photodiode is masked partially with a sharp-edged metal sheet. Algorithms for scanning and processing the data obtained from photodiode are developed for automatic detection of the focal distance (i.e., focusing lens position) at the target position. Finally, the effectiveness of detection of focus spot is demonstrated fabricating high-resolution micro-features in MSL system.

Keywords Laser focusing · Spot size · Micro-stereolithography · Micro-feature · Mechatronics system

In today's technological world, lasers find their widespread use in several domains, may it be manufacturing, sensing, communication, and so on. Accurate control over the laser beam's spot size at the target plane is one of the basic requirements, especially in manufacturing domain. One of the applications considered here is use of laser technology to fabricate micro- and nano-scale components. Micro-stereolithography (MSL) is an example of advanced 3D micro-printing process to fabricate micro-products by laser curing of the photopolymerizable resin [1]. However, beam focusing and its alignment for getting minimum laser beam spot size at printing plane are one of the important pre-requirements of the process. The laser spot size control is

R. K. Arya (✉) · R. Bafna · U. Pawar · P. Gandhi
Mechanical Engineering Department, Indian Institute of Technology Bombay, Mumbai 400076, India
e-mail: aryarajendra10iitr@gmail.com

V. K. Gupta et al. (eds.), *Recent Advances in Machines and Mechanisms*, Lecture Notes in Mechanical Engineering, https://doi.org/10.1007/978-981-19-3716-3_29

essential for the MSL 3D printer to achieve high accuracy, better resolution, and repeatability on the printed micro-product [2]. This requires an appropriate method for estimating the laser spot size with tightly focused laser beams. Characterization of a printed object at different lens positions is a general method for getting the focused beam spot. This method gives sufficient information required for printing on MSL. However, it takes a lot of time and frequent change due to multiple characterizations of the printed object. An advanced method like laser beam profiling using a CCD camera is costly and requires sophisticated instruments.

Literature in optics domain presents various methods for estimating the laser spot size in the direction of propagation of the laser. However, each one has its own limitations. The knife-edge method has been frequently utilized to accomplish the purpose described above in various laser technology applications [3]. A laser beam focusing system with similar components (as in the knife-edge method) was used to determine focal distance and estimate the laser beam spot size in the present work. The system included a sharp-edged metal mask puts up on a photodiode (PD) while scanning the laser beam on PD, which produced photovoltaic signals that were further analyzed to get focal distance and beam spot size on printing plane at different lens positions. Eventually, the proposed method provided a rapid solution to autodetect the focal distance followed by controlling the laser spot size and estimated the depth of focus (DOF) in the MSL process. Finally, to ascertain the feasibility of the LBFS, some high-resolution 3D prints have been realized.

1 Material and Methods

1.1 Theoretical Considerations of the Laser Beam

A laser beam, in the direction of propagation, as it passes through a lens, converges (narrowing in focus) up to the focal position (within the focal distance) and then diverges (spreading out). Thus, this arrangement provides a tighter focus beam or minimum beam spot (also called beam wrist) at a focal distance on the target plane, shown in Fig. 1.

In literature, various types of laser beam profiles have been considered for their modeling, including Gaussian, multi-mode, tilted, at top and irregular beams [4]. Out of these, the Gaussian irradiance of laser been with TEM_{00} mode has been found most suitable for describing the laser beam's nature. Therefore, in the present work, the laser beam profile was considered as Gaussian. It is well known that the ideal Gaussian profile of a laser beam provides symmetrical irradiation (i.e., decreases as the distance from the axis increases) around its central axis. Similarly, an intensity profile of the Gaussian distributed laser beam can be described.

Various attenuation methods have been used to define the beam wrist, and all are shown in Fig. 2. Usually, the beam wrist (or beam spot radius) W is considered as a distance, where the irradiance is 1/e2 ($\approx$ 13.5%) of the beam's full intensity. Besides,

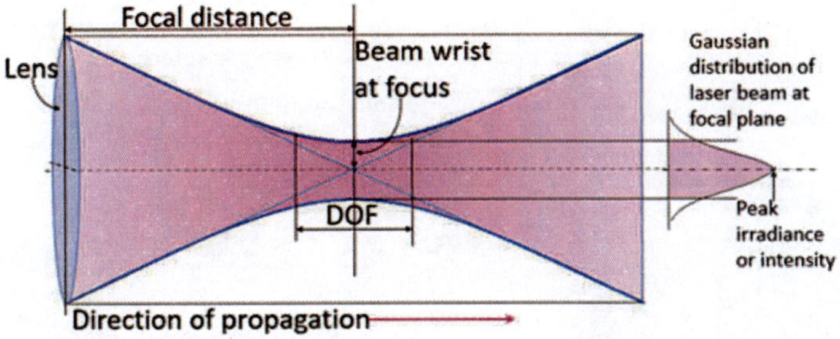

Fig. 1 Principle of laser beam

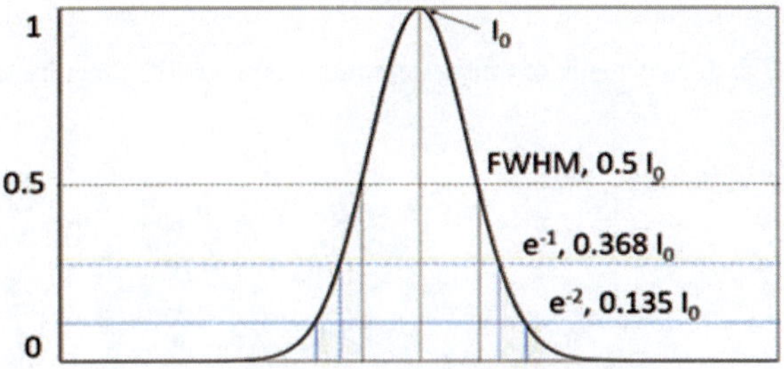

Fig. 2 Gaussian beam attenuation methods

sometimes, the laser spot size (beam wrist) has been estimated at the FWHM ($\approx 50\%$) considered to eliminate detraction and other environmental disturbances [5].

1.2 Experimental Setup

The micro-stereolithography (MSL) facility, which has been used in the present work, was developed in-house to fabricate micro-products [6]. The facility is capable of developing complex 3D products, though focusing the laser beam spot at the printing plane (VAT resin tank's surface) is a complicated and time-consuming task. Therefore, a mechatronics system facilitated to control over the laser spot size using photovoltaic signals was developed and utilized in the present work.

Figure 3 depicts an arrangement of the developed LBFS incorporated with the MSL facility and photographic view of micro-stereolithography 3D printer, and magnified picture of printing assembly followed by laser beam focusing system (LBFS) is shown in Fig. 4. In the current MSL facility, a laser source emits a laser beam with a 375 nm wavelength, which was directed to the printing plane by some mirrors and focused by a focusing lens. This provides polymerization of the photo resin, and eventually, 3D printing with the help of XYZ stages. Additionally, a LBFS with appropriate mechatronics selection was incorporated on MSL, actual picture LBFS can be seen in Fig. 4. In this system, a photodiode (PD) was employed on

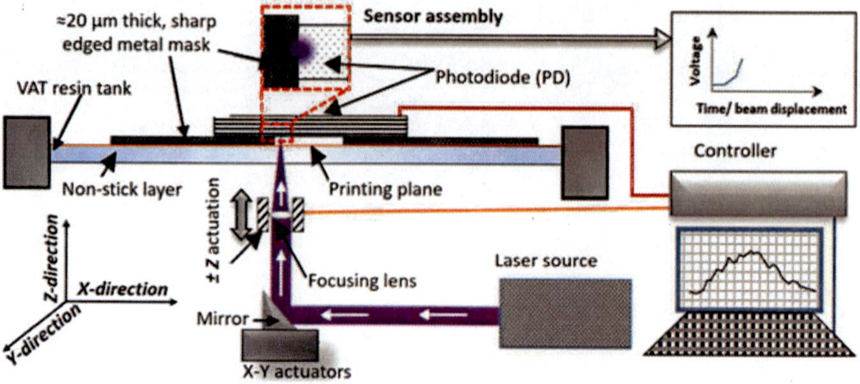

Fig. 3 Schematic of micro-stereolithography (MSL) facility with LBFS

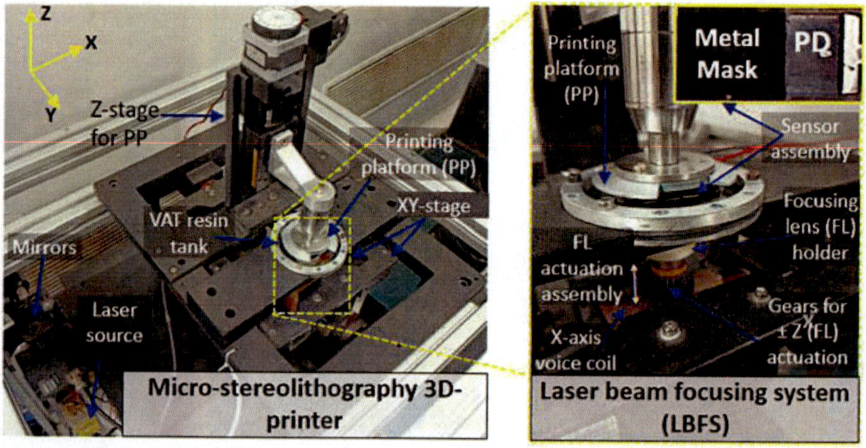

Fig. 4 Photographic view of micro-stereolithography 3D printer and magnified picture of printing assembly followed by laser beam focusing system (LBFS)

the printing plane to convert the laser light to the electric signal (i.e., photovoltaic signal). Also, the effective area of PD was slightly covered by a sharp-edged metal mask for getting different intensity signals during beam scanning on PD. Although, the aim was to estimate the spot size at the printing plane, which can be affected by placing the metal mask in between the PD and printing plane. Considering it and the depth of focus of the beam (i.e., ≈100 µm) a very thin metal mask of ≈20 µm was used. The mechatronics involved in controlling the lens's positions (in the laser's propagation axis, *Z*-axis) was provided by an actuator (i.e., connected to the lens), which was controlled by a controller. Eventually, the control over the laser spot size on the printing plane was accomplished by analyzing the obtained photodiode (PD) signals against the lens's positions. Besides, the controlling the lens's positions (in

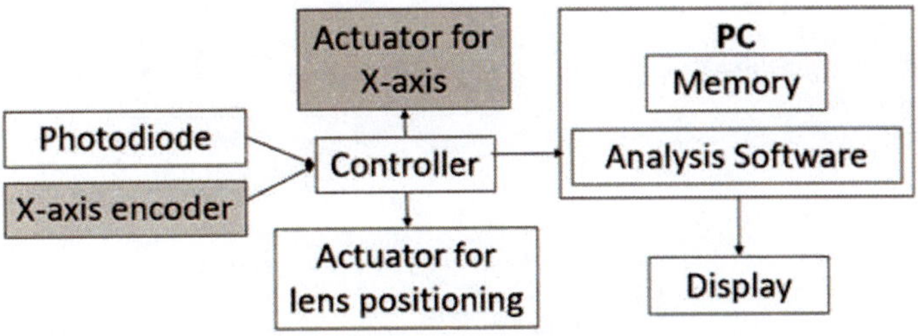

Fig. 5 Block diagram of the mechatronic involved in the LBFS

the laser's propagation axis, Z-axis) was provided by an actuator (i.e., connected to the lens through lens holder), which was controlled by a controller (dSPACE 1104 controller of MSL facility). The block diagram of the LBFS along with X-axis actuator and encoder (parts of MSL facility) is shown in Fig. 5. Here, it can be seen that the PD and encoder signal data were received by controller through data acquisition system (i.e., dSPACE 1104), which further stored and analyzed in the personal computer (PC). Subsequently, the actuator was instructed by using suitable algorithm for position the lens at focal distance to obtain minimum spot size.

Algorithm for data storage (i.e., V_n and X_n) in each scan for total m number of scans and analysis of stored data to determine the lens position at focal distance Z_0 and for autofocusing of lens at target plane is given in Fig. 6.

1.3 Methodology to Determine the Lens Potions at Focal Distance Followed by Estimation at the Laser Beam Spot Size

The purpose of the method used here is to determine lens positions in beam propagation (or Z coordinate) to get the smallest possible laser spot and depth of focus. The method used in the present work is similar to the most utilized method (i.e., knife-edge method) for laser beam analysis. As mentioned above, a sharp-edged metal mask was used on PD to filter out the laser light (100% filtration). Thus, as the laser beam passed through the mask, different intensity photovoltaic signals were generated at PD owing to the optical field distribution of beam spot.

The schematic of the scanning process and meantime captured signal through PD is shown in Fig. 7. It can also be seen from the figure that during the laser scanning on PD, the PD signals were minimum or nil when the beam's spot was entirely projected on the mask, and it raised when the beam's spot started projecting on the PD. Finally, as the beam's spot is entirely projected on PD, it provided maximum intensity signals (at particular laser power). It can easily understand that the obtained plot is a numerical integration of PD signal with respect to the beam displacement/position in scanning direction. Thus, it can be considered that the obtained value of PD

Fig. 6 Algorithm to determine the lens position at focal distance and autofocusing

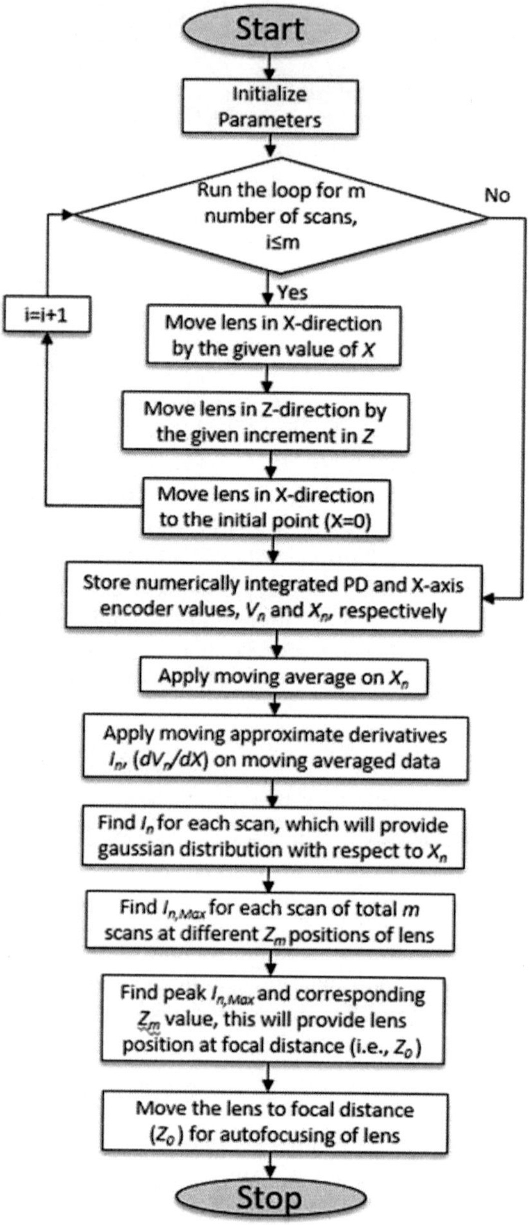

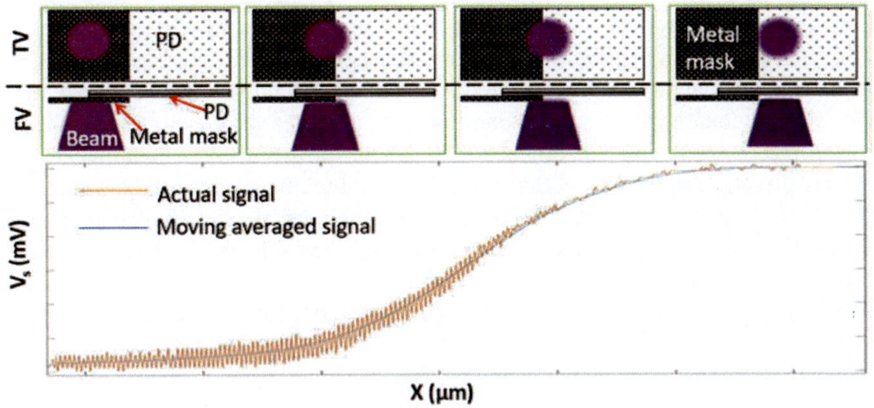

Fig. 7 Beam scanning on PD and obtained PD signals with beam displacement in scanning direction

signals V is the function of the beam displacement X in scanning direction. Also, the mathematical expression for the same can be written as in Eq. 1, similar to the knife-edge method;

$$V = k\int_{0}^{\infty} I\,\mathrm{dx} \tag{1}$$

where k is the PD sensitivity constant and I is the optical field distribution.

As mentioned above the laser beam spot follows Gaussian irradiance, thus the optical field distribution I can be found by derivation the V with respect to X,

$$I = -\frac{1}{k}\frac{\mathrm{dV}}{\mathrm{dx}} \tag{2}$$

Meanwhile, the steeper curve of numerically integrated PD signals provides higher value of derivative I, which can also be inferred as a focused position of lens (i.e., focal distance) with minimum beam spot size [7]. Figure 8 shows the numerical

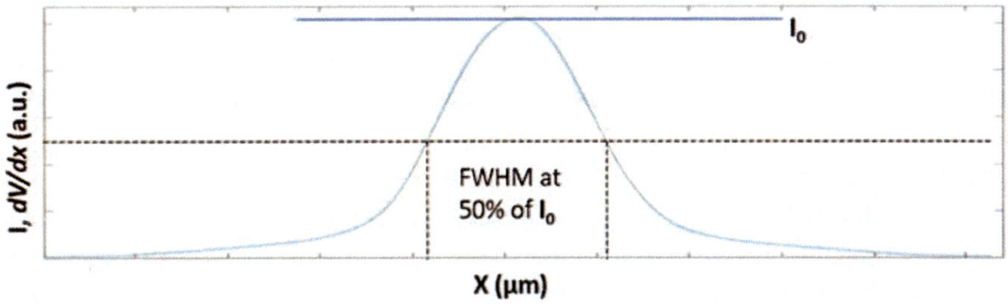

Fig. 8 Numerical derivative of PD signals versus with beam displacement in scanning direction

derivative of PD signals I versus with beam displacement in scanning direction. Also, the graph of numerical distribution follows the Gaussian fitting, hence it can be used to estimate the spot size of the laser beam. The full width at half maximum (FWHM) attenuation of the laser beam spot was considered for estimating spot size, owing to the elimination of diffraction and other environmental disturbances [5]. Eventually, the obtained focal distance with minimum bean spot can be used to 3D printing of high-resolution micro-parts. This method can also be adapted to estimate the depth of focus (DOF), which is a tolerance range established by the requirements for a particular application. Thus, for the micro-3D printing work, the allowable working range can be considered at the spot size under the minimum obtained spot size +1 μm.

2 Results and Discussion

Experiments were performed in two steps, initially to set the lens position for getting minimum laser beam's spot using the LBFS and estimating the DOF, this followed by 3D printing of high-resolution micro-feature on MSL.

2.1 *Estimation of Laser beam's Spot Size and Depth of Focus*

The experiments for scanning the laser beam on PD were performed at different positions of the focusing lens (in Z-direction), besides all the other parameters were fixed at a constant level. Subsequently, the obtained PD signals from scan were analyzed to estimate the laser beam spot size and DOF. Signal analysis procedure included (given in algorithm in Fig. 6), applying moving average filter on the raw PD signals followed by numerical differentiation. In this experiment, the beam was scanned at fixed lens position, after that the lens were moved in ±Z for a predefined value (i.e., 100 μm). Meanwhile, the PD signals and X-axis encoder data were recorded for further analysis. Thus, after multiple scans, the recorded data were analyzed. Subsequently, the lens position at focal distance was determined. The analyzed signal is also plotted for ±2600 μm lens displacement in Z-direction, as shown in Fig. 9. Here, figure shows the peak values of numerically differentiated signals (i.e., $I_{n,\,\mathrm{Max}}$) at different positions (i.e., ±Z) of lens in Z-direction. Also, Z_0 represents the lens position at focal distance; hence in this position, the beam spot will be minimum as compare to the other ±Z values. As can be seen in Fig. 9, maximum value of numerical differential was obtained at Z_0 owing to steep curve formation by numerical integration of PD signal.

Further, the spot size and DOC of the beam were estimated from the normalized derivative (i.e., I_n) values of individual scans at different lens positions (i.e., ±Z) using the method explained earlier (i.e., Sect. 2.3). Figure 10 shows the estimated

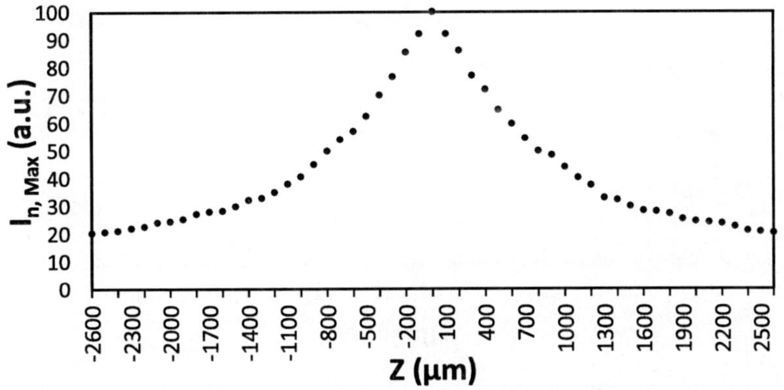

Fig. 9 Peak values of normalized derivatives obtained at different lens positions for course scan

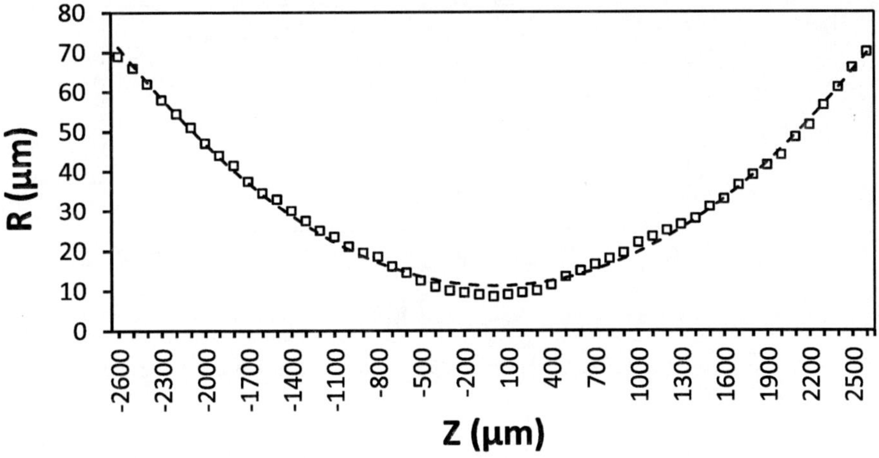

Fig. 10 Estimated spot size at different lens positions for course scan

spot size R (i.e., Spot diameter) for different lens positions $\pm Z$. Here, the convergent and divergent nature of the laser beam can be easily observed.

The spot size was estimated by shifting the lens position to 100 μm in each scan, and a range of lens positions was identified where the minimum spot size and DOP could be achieved. However, further refinement of lens position can provide tighten beam with smallest spot size. Therefore, in the next experiment, fine positioning (i.e., 10 μm sift) of the lens was performed to get minimum spot size and DOP. Here, the lens was moved within 200 μm (i.e., ±100 μm from previously obtained Z_0) with 10 μm shift in lens after each scan. Figure 11 shows $I_{n, \text{Max}}$ values obtained at various scans by moving lens in Z-direction. Here, the Z_0 was shifted to the new peak value of $I_{n, \text{Max}}$, thus, only ±100 μm Z positions are shown. This was because, in the previous experiment, the shift distance between two consecutive lens positions was

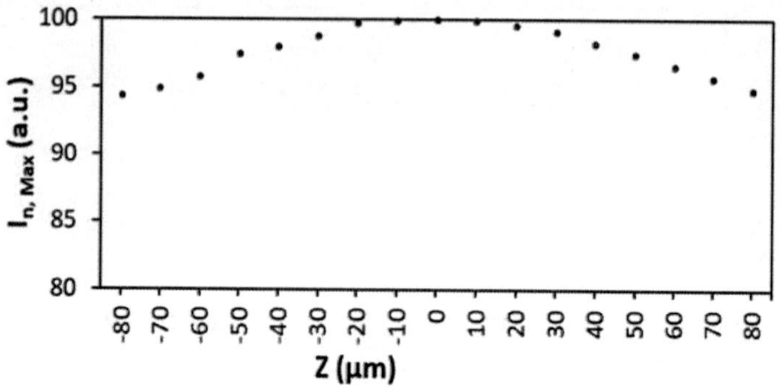

Fig. 11 Peak values of normalized derivatives obtained at different lens positions for fine scan

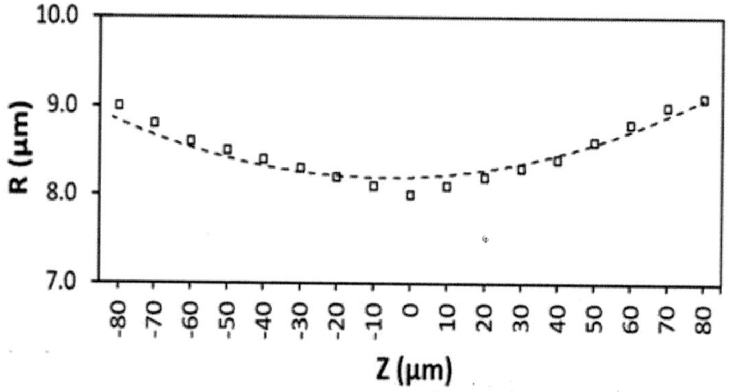

Fig. 12 Estimated spot size R (i.e., Spot diameter) at different lens positions Z for fine scan

high (i.e., 100 μm), which was 10 μm in this experiment. It can also be seen from the estimated spot size, shown in Fig. 12. Here, the minimum spot size was 8 μm. Therefore, the DOP can be estimated for the range of lens position, where the spot sizes are within 9 μm (i.e., minimum spot size + 1 μm).

2.2 *Validation of the Estimated Laser Spot Through 3D Printing Using Micro-stereolithography Process*

The experiments were performed for validation of previously estimated spot size at the printing plane. An in-house designed and developed facility of micro-stereolithography was used for printing micro-features by photopolymerization. The liquid photopolymer resin (i.e., acrylic-based) was photopolymerized by imposing the laser light at the printing plane through the focusing lens. It has been previously discussed that the spot size was controlled by adjusting the lens's position. Thus,

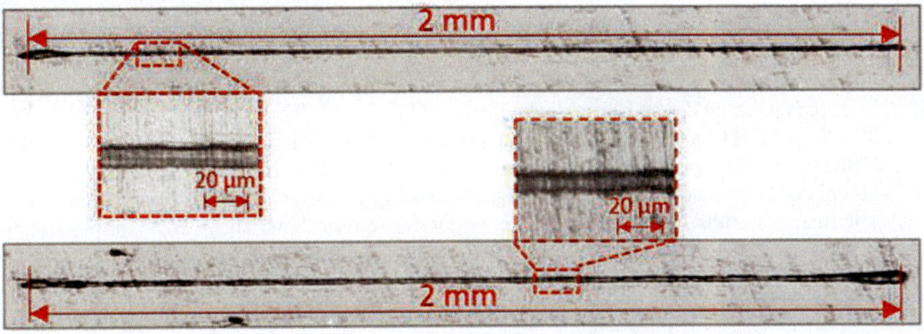

Fig. 13 Microscopic image of printed micro-feature (i.e., a straight line)

the printing was performed at lens position, which was identified for minimum spot size using above mentioned method. 2 mm line was printed by filling the liquid resin between the printing plan and a temporary plan with a gap of a few microns. The width of the line was measured at 10 points along the printed line, and the average of all the measured widths was considered as the final width. A microscopic image of printed micro-features can be seen from Fig. 13. Here, the repeatability in the process is shown. The average width of the micro-feature (i.e., a straight line) was 10.8 µm, which is close to the estimated spot size using the present method.

3 Conclusion

The present work demonstrated a method for estimating the laser spot size. A mechatronics-based laser beam focusing system (LBFS) was developed. The system has utilized a photodiode (PD) to sense the intensity of laser light at the printing plane. Further, the generated PD signal was used to determine the lens position at focal distance followed by estimation of laser spot size at different lens positions using an appropriate algorithm. The minimum spot size and desired depth of focus were obtained. Subsequently, the lens position for minimum spot size was identified and utilized for printing the micro-feature using the micro-stereolithography process. Eventually, a micro-feature (i.e., a straight line with 10.8 µm average width) was printed using an acrylic-based photopolymer.

References

1. Gandhi P, Deshmukh S, Ramtekkar R, Bhole K, Baraki A (2013) 'oN-AXIS' Linear focused spot scanning microstereolithography system: Optomechatronic design, analysis and development. J Adv Manuf Syst 12(1):43–68. https://doi.org/10.1142/S0219686713500030
2. Partanen JP (1996) Fabrication of parts containing small features using stereolithography. Annu Int Solid Free Fabr Symp 63–70, [Online]. Available: https://sffsymposium.engr.utexas.edu/Manuscripts/1996/1996-09-Partanen.pdf

3. Bilger HR, Habib T (1985) Knife-edge scanning of an astigmatic Gaussian beam. Appl Opt 24(5):686. https://doi.org/10.1364/ao.24.000686
4. Bonnett Del Alamo M, Soncco C, Helaconde R, Bazo J, Gago AM (2020) Laser spot measurement using cost-Affordable devices," *arXiv*
5. Chiu Y, Pan J-H (2007) Micro knife-edge optical measurement device in a silicon-on-insulator substrate. Opt Express 15(10):6367. https://doi.org/10.1364/oe.15.006367
6. Gandhi PS, Deshmukh S (2010) A 2D optomechanical focused laser spot scanner: Analysis and experimental results for microstereolithography. J Micromech Microeng 20(1). https://doi.org/10.1088/0960-1317/20/1/015035
7. Poce-Fatou JA, Martín J, Alcántara R, Fernández-Lorenzo C (2002) A precision method for laser focusing on laser beam induced current experiments. Rev Sci Instrum 73(11):3895. https://doi.org/10.1063/1.1511794

A Diamagnetically Levitated Actuator Capable of Independent In-Plane and Out-of-Plane Positioning

K. S. Vikrant and G. R. Jayanth

Abstract Precision positioning stages based on the principle of magnetic levitation find widespread use because of their large motion range and high positioning resolution. However, most of these stages are levitated by employing feedback control. Here, we propose a diamagnetically levitated actuator wherein the actuated magnetic platform is sandwiched between two identical printed circuit boards (PCBs). This configuration enables independently controlling the 3D electromagnetic forces acting on the platform and also increases the payload-carrying capacity of the platform. A mathematical model of the system is developed and is used to analyze the dependence of forces on the actuation currents flowing through the PCBs. Finally, the development of the actuator is discussed along with experiments demonstrating independent control of motion along *X*- and *Z*-axes.

Keywords Diamagnetic levitation · Precision positioning · Open-loop operation · Independent 3D motion

1 Introduction

With the advent of nanotechnology, demand for precision positioning stages has increased since they are employed for fabrication, characterization and manipulation of nanometer-scale samples. Magnetic levitation-based positioning stages employing electromagnetic actuation are suitable for developing positioners with high positioning accuracy and precision due to the absence of stiction and coulomb friction [1–4]. Active magnetically levitated stages employ electromagnets and feedback control to achieve precise positioning [5]. However, it is also possible to passively levitate a permanent magnet above a diamagnetic material and move it on the surface by means of current-carrying conductors. Since passive levitation is intrinsically

K. S. Vikrant · G. R. Jayanth (✉)
Indian Institute of Science, Bangalore 560012, India
e-mail: jayanth@iisc.ac.in

K. S. Vikrant
e-mail: vikrantsingh@iisc.ac.in

V. K. Gupta et al. (eds.), *Recent Advances in Machines and Mechanisms*, Lecture Notes in Mechanical Engineering, https://doi.org/10.1007/978-981-19-3716-3_30

stable, the control system is significantly simpler [6–8]. However, since the diamagnetic force is smaller in comparison with electromagnetic forces, passive stages have smaller payload capacity and are primarily suitable for samples whose sizes are about a millimeter or less.

Previously, permanent magnet arrays levitated on pyrolytic graphite, and actuated by current-carrying wires fabricated on a printed circuit board (PCB), have been employed as micro-robots to undertake a variety of manipulation tasks [9]. In particular, they have been shown to move over large distances on both flat and tilted surfaces [10–12]. By dividing a PCB into zones, a single array has been rotated in-plane, while multiple arrays have been moved independently [13]. Such arrays have been demonstrated to perform micromanipulation and assembly. The reported open-loop positioning precision is on the order of a few tens of nanometers and is limited only by thermal effects [12]. Further, to increase the payload capability, A. Hsu et. Al. have introduced a layer of ferrofluid between the magnetic platform and the graphite plate [14]. Although the reported systems are excellent as robotic platforms, the actuation strategy introduces limitations from the viewpoint of development of a positioning stage. In particular, the magnet experiences coupled in-plane and Z-forces which result in undesired Z-cross-axis motion during in-plane positioning [12]. This also reduces the payload-carrying capacity since the downward Z-force is often much larger than the weight of the magnet itself.

Here, we propose a new configuration where the magnet is sandwiched between two PCBs. Such an arrangement is shown to cancel the cross-axis Z-forces and to double the in-plane actuation forces when the currents in the two PCBs are identical. Conversely, they cancel the in-plane forces and double the Z-force when the currents are equal but opposite in sign. Consequently, independent actuation along the X- and Z-axes becomes possible. The payload capacity of the stage is also shown to increase significantly. The rest of the paper is divided as follows. Section 2 describes the principles of the actuator. The development and evaluation of the stage are presented in Sect. 3. Finally, conclusions are presented in Sect. 4.

2 Principles of the Actuator

2.1 *Overview*

The schematic representation of the proposed actuator is shown in Fig. 1a. The system comprises of two identical PCBs. The bottom PCB carries a pyrolytic graphite plate for diamagnetic levitation. The gap between the PCBs is so chosen that when no current is applied, the center of the levitating magnetic platform is positioned exactly between the two PCBs. Each PCB includes four independent current-carrying tracks $(I_{y1}, I_{y2}, I_{x1}, I_{x2})$ for actuating the levitated platform. The tracks I_{y1} and I_{y2} which are parallel to the Y-axis enable X-actuation, while the tracks I_{x1} and I_{x2} which are parallel to the X-axis enable Y-actuation. Each track comprises N straight conductors

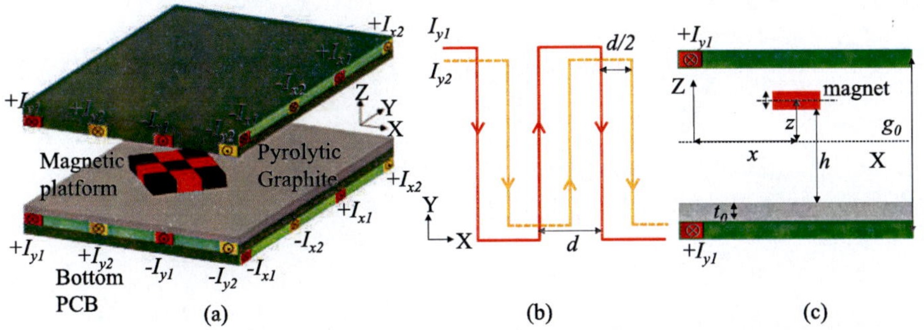

Fig. 1 Schematics showing **a** the complete actuator comprising top and bottom PCB, the pyrolytic graphite plate and the magnetic platform, **b** the tracks I_{y1} and I_{y2} of the PCB and (c) the cross-sectional view of the system with a single conductor for purposes of analysis

arranged in a meandering fashion and separated by a distance d (Fig. 1b). The two PCBs are aligned so that each track on one PCB is aligned with an identical track on the other. The magnetic platform is a chequerboard of identical smaller magnets, each of moment m, thickness t_m and mass m_m that are arranged such that the magnetic moments alternate between being upward and downward. During actuation, the in-plane force is proportional to the sum of the currents in the corresponding tracks of the top and bottom PCB, while the out-of-plane force is proportional to their difference.

Thus, in order to actuate the levitated platform in the XY-plane, identical currents are applied to the corresponding tracks on the two PCBs. This ensures that the in-plane forces applied by one PCB reinforce that applied by the other, while the out-of-plane forces cancel each other. In order to move the platform only along the Z-axis, the magnet is first trapped in-plane and additional equal but opposite currents are superposed on the currents that are already flowing through the corresponding tracks on the top and bottom PCBs. This causes a resultant force acting on the levitated platform along the Z-axis without destabilizing it along the in-plane axes and thus displaces the platform to a new Z-position.

Modeling and Analysis of the Actuator

Determination of the Levitation Height. By modeling a permanent magnet as a point dipole of moment m, the diamagnetic force experienced by the magnet whose base is at a distance h (Fig. 1c) from the plane of the pyrolytic graphite plate is given by

$$F_d(h) = \frac{3m^2|\chi|\mu_0}{64\pi}\frac{1}{(h + t_m/2)^4}, \tag{1}$$

where χ represents the magnetic susceptibility of the pyrolytic graphite plate. Thus, the levitation height h_0 obtained by equating the weight of the magnet $m_m g$ to $F_d(h)$ is found to be

$$h_0 = \frac{1}{2}\left(\sqrt[4]{\frac{3m^2|\chi|\mu_0}{4\pi m_m g}} - t_m\right). \tag{2}$$

If the thickness of the graphite plate is t_0, then the overall gap g_0 between the two PCBs is given by (Fig. 1c)

$$g_0 = 2(h_0 + t_0) + t_m. \tag{3}$$

Analysis of In-plane and Out-of-Plane Actuation Forces. To analyze the actuation forces on the magnetic platform, it is adequate to consider the forces applied on a single magnet in the platform by one pair of conductors of the first Y-track in both the PCBs, with one conductor of the pair being on the top PCB and the other being on the bottom PCB (Fig. 1c). It is also assumed that the two conductors are at the centers of their respective tracks. This analysis can subsequently be extended for the second Y-track and the two X-tracks.

Assuming that the conductors are much longer than the size of the platform, the force $\mathbf{F}(=\nabla(\mathbf{m}\cdot\mathbf{B}))$experienced by a single magnet in the platform at any point (x, z) is given by

$$\mathbf{F} = m\begin{bmatrix} \frac{\partial B_z}{\partial x} & 0 & \frac{\partial B_z}{\partial z} \end{bmatrix}^T. \tag{4}$$

The magnetic field B_z is the sum of the fields produced by the two conductors, each of which is proportional to the current through them. Thus, if I_{y1}^u, I_{y1}^ℓ represent the currents in the upper and lower PCBs, respectively, then B_z can be written as

$$B_z = b_z^u(x,z)I_{y1}^u + b_z^\ell(x,z)I_{y1}^\ell, \tag{5}$$

where $b_z^u(x,z)$, $b_z^\ell(x,z)$ are magnetic field per unit current due to the upper and the lower conductors, respectively, and are given by

$$\begin{aligned} b_z^u(x,z) &= -\frac{\mu_0 x}{2\pi(x^2 + (z - g_0/2)^2)} \\ b_z^l(x,z) &= -\frac{\mu_0 x}{2\pi[x^2 + (z + g_0/2)^2]}. \end{aligned} \tag{6}$$

Thus, the force can be written as

$$\mathbf{F} = \begin{bmatrix} f_x^u I_{y1}^u + f_x^\ell I_{y1}^\ell & 0 & f_z^u I_{y1}^u + f_z^\ell I_{y1}^\ell \end{bmatrix}^T, \tag{7}$$

where $f_x^i(x,z) = m\partial b_z^i / \partial x$, $f_z^i(x,z) = m\partial b_z^i / \partial z$ and $i = u, \ell$.

It is noted from Eqs. (6) and (7) that $f_x^u(x,0) = f_x^\ell(x,0)$ and $f_z^u(x,0) = -f_z^\ell(x,0)$. Thus, it is seen that if $I_{y1}^u = I_{y1}^\ell$, the magnetic moment experiences only in-plane (X) force from this pair, while if $I_{y1}^u = -I_{y1}^\ell$, the magnetic moment experiences only Z-force from this pair. Thus, while the former choice enables actuating along the X-axis with no cross-axis Z-motion, the latter choice enables actuating purely along the Z-axis with no cross-axis X-motion. However, to ensure that the platform is stable along the in-plane axes when it is actuated along the Z-axis, it is first trapped at a position x where it is in stable equilibrium and additional equal but opposite currents are superposed in the two PCBs on the currents that are employed to maintain stable equilibrium.

Since the first track comprises a meandered arrangement of conductors separated by distances d, the total X-force per unit current on a single magnet in the array is given by

$$F_x(x,0) = \sum_{k=-M}^{k=M} f_x(x-2kd,0) - \sum_{k=-M}^{k=M} f_x(x-(2k+1)d,0), \tag{8}$$

while the Z-force per unit current is given by

$$F_z(x,0) = \sum_{k=-M}^{k=M} f_z(x-2kd,0) - \sum_{k=-M}^{k=M} f_z(x-(2k+1)d,0). \tag{9}$$

where $M = (N-1)/4$.

It is worth noting that if the gradient in $F_z(x,0)$ along the Z-axis exceeds the gradient in diamagnetic force in the mid-plane between the two PCBs, viz., $\partial F_z(x,0)/\partial z\big|_{z=0} > \partial F_d/\partial z\big|_{h=h_0}$, then the magnet will become unstable along the Z-axis.

Analysis of the Payload-Carrying Capacity and Variation in the Levitation Height. The payload-carrying capacity of a single magnet is determined to be the extra load that gets the magnet's bottom surface to make contact with the top surface of pyrolytic graphite. In this scenario, $h = 0$ and the displacement of the center of the magnet from the plane of symmetry would be $-h_0$. Thus, assuming identical current I_{yl} is passed through the top and the bottom PCB traces, the payload-carrying capacity L per magnet is given by

$$L = F_d(0) + F_z(x,-h_0)I_{y1} - m_m g. \tag{10}$$

It is seen from Eq. (10) that the relationship between L and I_{y1} is linear at any given horizontal position x.

Figure 2a shows the dependence of normalized payload-carrying capacity $L/m_m g$ as a function of track current I_{y1} for the case of a single PCB and for two

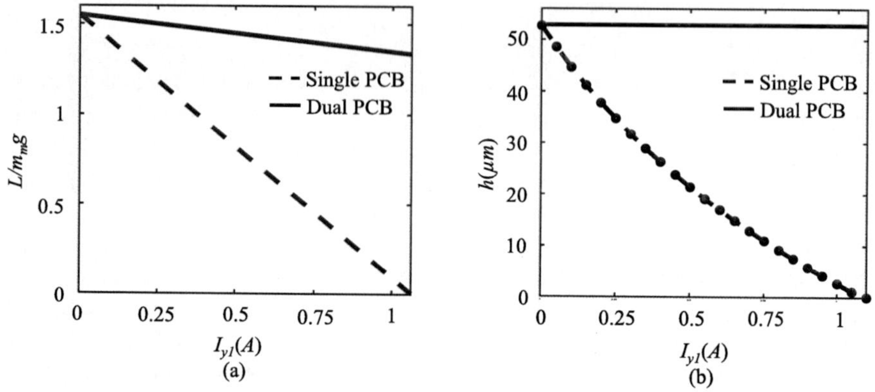

Fig. 2 Plots showing **a** the payload-carrying capacity of the magnet as a function of track current I_{y1} when x was chosen to be $-d/2$ so that the platform was in stable equilibrium, **b** the Z-position of the magnet as a function of track current I_{y1} for single and dual PCB configuration when x was chosen to be $-d/2$ so that the platform was in stable equilibrium. The values of I_{x1}, I_{x2} and I_{y2} used during the computations were 0. The values of m, m_m, χ, d, g_0 and N are 0.17×10^{-3} A/m^2, 5.78 mg, 4.1×10^{-4}, 1.27 mm, 1.28 mm and 24, respectively

PCBs and shows significant improvement for the latter case compared to the former. Figure 2b compares the height of levitation when the magnetic platform is situated symmetrically between the meanders for the cases when a single PCB is employed and when two PCBs are employed. It shows a significant reduction in height with the increased current for the former case due to cross-axis Z-force while for the latter, the height remains unaffected by the current I_{y1}. However, in the latter case, the current I_{y1} is limited to 1.34 A owing to the instability of the magnet platform beyond this value.

3 Development and Evaluation

To demonstrate the principles of the actuator, a magnetic platform, a pyrolytic graphite plate and two identical PCBs were arranged together in the proposed configuration. The magnetic platform is made of four NdFeB magnets of grade N40 arranged in a 2×2 chequerboard pattern. The dimensions of each magnet were 1.7 mm $\times$ 1.7 mm $\times$ 1.7 mm. The thickness of the employed pyrolytic graphite plate attached to the bottom PCB was 400 μm. The current-carrying tracks I_{y1}, I_{y2}, I_{x1}, I_{x2} were fabricated on a four-layer PCB. Each track consisted of multiple straight-line segments which were electrically connected in series. The straight segments of track I_{y1} were connected to each other by means of conducting traces present in layer 1 itself. The straight segments of track I_{y2} were connected to each other by means of vias that pass through layer 3. This ensured that the straight segments constituting tracks I_{y1} and I_{y2} were present in the same layer (layer 1) and thus possessed identical actuation gains. A similar strategy was employed to ensure that the straight segments of tracks I_{x1} and I_{x2} were both present in the same layer, viz., layer 2. The separation between layers 1 and 2 was 100 μm. The thickness and width of each track were 70 and 254 μm, respectively. The separation d between two consecutive straight segments of a given track was 1.27 mm. All these dimensions are specified in the datasheet of the PCBs with a maximum error less than ±10 μm.

The developed system was first employed to demonstrate that the levitation height does not change when the actuation current changes. To perform the experiment, the current through the single trace I_{y1} was varied in both the PCBs. For comparison, the same current I_{y1} was passed through just the bottom PCB alone, in order to mimic the situation with just a single PCB. The resulting motion of the magnet array is characterized by means of a modular side microscope (Cerna, Thorlabs) attached with a CMOS camera (MC050MG, Ximea). The side microscope employed a long working distance objective lens of magnification 5 and a tube lens of magnification 0.75 for characterizing the motion range and resolution of the positioner along the Z-axis. The position of the magnet array was measured using the digital image cross-correlation (DIC) technique [15] (Fig. 3).

Figure 3 plots the dependence of the levitation height on the current I_{y1} for the two cases and shows that the Z-position of the levitated platform varies by 32.2 μm for single PCB configuration. However, the Z-position of the levitated platform changes by only 2.76 μm for the dual PCB configuration as the actuation current I_{y1} is increased from 10 to 270 mA. Further, it was observed that the platform becomes unstable along the Z-axis when the actuation current was increased beyond 410 mA for the dual PCB configuration.

In the second experiment, the levitated magnetic platform was actuated along the X-axis first by varying track currents only in the bottom PCB. The images in

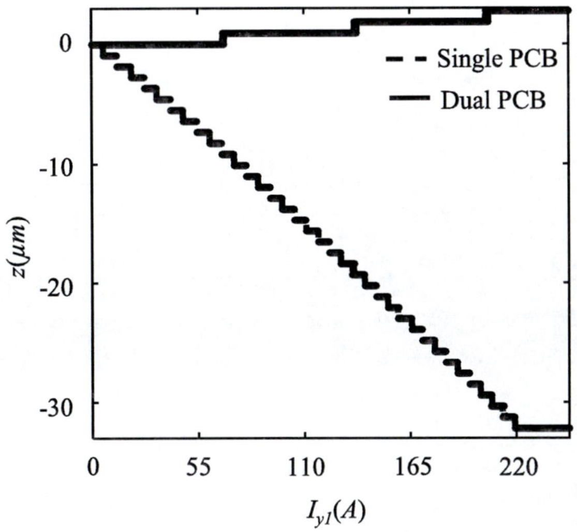

Fig. 3 Plot showing the Z-position of the magnetic platform as a function of track current I_{y1} for the cases of a single PCB and the dual PCB. The values of I_{x1}, I_{x2} and I_{y2} used during the experiment were 5 mA, 0 mA and 0 mA, respectively. The resolution of the employed position measurement system is 0.92 μm

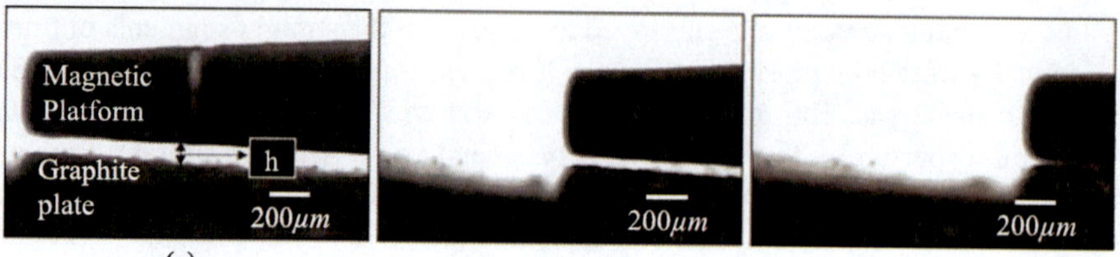

(a) (b) (c)

Fig. 4 Images showing the variation in the Z-position of the levitated magnetic platform as it moves along the X-axis when only bottom PCB is energized. The actuation current I_{y1} was fixed at 100 mA, while I_{y2} was varied between –100 and 100 mA

Fig. 4 show that the Z-position of the levitated magnetic platform varies as it moves along the X-axis when only the bottom PCB is used. Next, when both the PCBs are employed, the Z-position of the levitated magnetic platform is seen to remain at a constant height above the graphite plate as it moves along the X-axis (Fig. 5). On account of symmetry a similar result can also be obtained while actuating the platform to move along the Y-axis.

The final experiment was devoted to validating the ability to change the Z-height alone of the array. To perform this experiment, the current through the track I_{y1} of

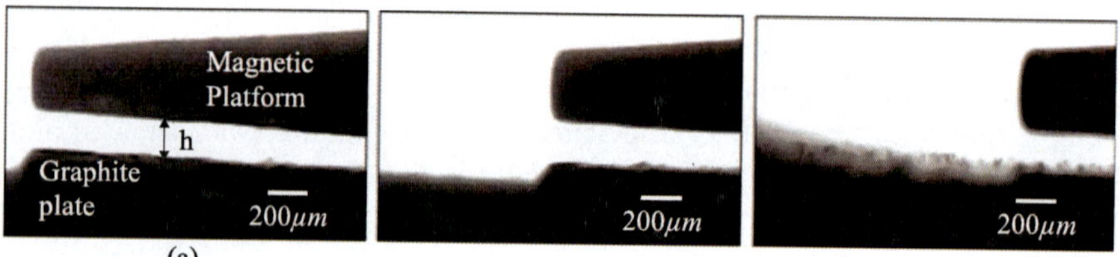

(a) (b) (c)

Fig. 5 Images showing the almost constant Z-position of the levitated magnetic platform as it moves along the X-axis when both PCBs are energized. The actuation current I_{y1} was fixed at 100 mA, while I_{y2} was varied between −100 and 100 mA

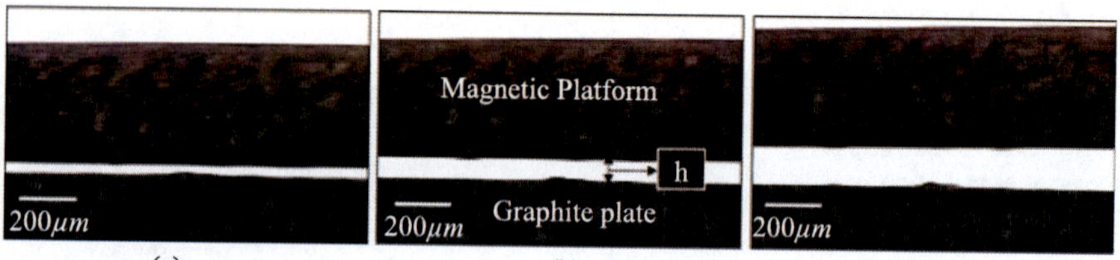

(a) (b) (c)

Fig. 6 Images showing the variation in only the Z-position of the levitated magnetic platform as both the PCBs are energized. While the current I_{y1} in top PCB is increased from 100 to 300 mA, the current in bottom PCB is kept constant at 300 mA, resulting in a travel range of approximately 80 μm the bottom PCB was kept constant and the current through the corresponding track on the top PCB was increased. This causes the levitated magnetic platform to move only along the *Z*-axis by about 80 μm as shown in Fig. 6.

4 Conclusion

This paper presented the design, development and evaluation of a diamagnetic levitation-based actuator. The goals of the developed system were to demonstrate independent motion along *X*, *Y* and *Z*-axis with an increased payload capacity of the levitated platform. The system comprised a moving platform made from permanent magnets, pyrolytic graphite plate for levitating the moving platform and two PCBs for actuating the platform. An analysis was performed to show that the motion along the *Z*-axis can be decoupled from the motion along the *X*- and the *Y*-axes by employing the symmetry present in the design. Further, analyses were performed to show that the payload-carrying capacity for the actuator is higher than that for the case where only a single PCB is used for actuation. Finally, the system was developed and experimentally demonstrated to enable independent motion along the *X*- and the *Z*-axes. Owing to symmetry the same independence can also be achieved between *Y*- and *Z*-axes. The developed actuator represents a building block which can be subsequently used for developing a multi-axis precision positioning stage with large range and high resolution.

Acknowledgements The authors wish to thank Mr. Rakesh Kumar (IAP, IISc Bangalore) and Mr. Kedar Badnikar (DESE, IISc Bangalore) for their assistance in the design of the PCB and milling of the pyrolytic graphite plate, respectively.

References

1. Verma S, Kim WJ, Gu J (2004) Six-axis nanopositioning device with precision magnetic levitation technology. IEEE/ASME Trans Mechatron 9(2):384–391
2. Kim WJ, Trumper DL (1998) High-precision magnetic levitation stage for photolithography. Precis Eng 22(2):66–77
3. Holmes M, Hocken R, Trumper DL (2000) The long-range scanning stage: a novel platform for scanned-probe microscopy. Precis Eng 24(3):191–209
4. Shan X, Kuo SK, Zhang J, Menq CH (2002) Ultra precision motion control of a multiple degrees of freedom magnetic suspension stage. IEEE/ASME Trans Mechatron 7(1):67–78
5. Earnshaw S (1839) On the nature of the molecular forces which regulate the constitution of the luminiferous ether. Trans Camb Phil Soc 7:97–112
6. Braunbek W (1939) Freischwebende Körper im elektrischen und magnetischen Feld. Z Phys 112:753–763
7. Waldron RD (1966) Diamagnetic levitation using pyrolytic graphite. Rev Sci Instrum 37:29–34
8. Simon MD, Heflinger LO, Geim AK (2001) Diamagnetically stabilized magnet levitation. Amer J Phys 69(6):702–713

9. Pelrine R et al (2012) Diamagnetically levitated robots: An approach to massively parallel robotic systems with unusual motion properties. In: 2012 IEEE international conference on robotics and automation. Saint Paul, MN, pp 739–744
10. Pelrine R, Wong-Foy A, Hsu A, McCoy B (2016) Self-assembly of milli-scale robotic manipulators: A path to highly adaptive, robust automation systems. In: 2016 international conference on manipulation, automation and robotics at small scales (MARSS). Paris, pp 1–6
11. Hsu A et al (2017) Automated 2D micro-assembly using diamagnetically levitated milli-robots. In: 2017 international conference on manipulation, automation and robotics at small scales (MARSS). Montreal, QC, pp 1–6
12. Hsu A, Chu W, Cowan C et al (2018) Diamagnetically levitated Milli-robots for heterogeneous 3D assembly. J Micro-Bio Robot 14:1–16
13. Pelrine R, Hsu A, Wong-Foy A (2019) Methods and results for rotation of diamagnetic robots using translational designs. In: 2019 international conference on manipulation, automation and robotics at small scales (MARSS). Helsinki, Finland, pp 1–6
14. Hsu A, Wong-Foy A, Pelrine R (2018) Ferrofluid levitated micro/milli-robots. In: 2018 international conference on manipulation, automation and robotics at small scales (MARSS). Nagoya, Japan, pp 1–7
15. Corporation I (2015) Itseez, Opencv 3.0 online documentation, 2015, [online] Available: http://docs.opencv.org

Control of Multiple Ferro-Bots for Steady Motion Using an Array of Electromagnets

Sudhanva Bhat and G. K. Ananthasuresh

Abstract Magnetic robots have been gaining attention due to their application in biomedical and other fields. Traditional magnetic robots are controlled using various methods, including powerful orthogonal fields with Helmholtz or Maxwell coils, moving permanent magnets, etc. Our work focuses on the control of multiple ferromagnetic objects, henceforth referred to as *ferro-bots,* using a planar magnetic field that changes spatially and temporally. The motions of the ferro-bots are thus constrained to a 2D plane. The system consists of a grid of solenoids, each standing vertically. The ferro-bots move on an acrylic sheet placed just above the solenoid grid. The ferro-bots are free bodies moving on a plane. When they are close to the solenoid, the magnetic force changes rapidly with distance from the centre of the solenoid. Controlling the ferro-bots in a region with rapidly varying forces, resisted only by nonlinear Coulomb friction, is the focus of this work. Computation of the magnetic force is complicated because of the induced magnetic field in the iron core of the solenoid and the ferro-bot: it leads to a transcendental equation in the magnetic field vector. Consequently, computing the magnetic force is also involved, as opposed to earlier work that used permanent magnets and an array of core-less coils. Therefore, a simplified numerical scheme is developed for computing the magnetic force on the ferro-bot based on coil currents and vice versa. There is another complication as there is a camera-based vision sensor to locate the positions of the ferro-bots. The frame rate of the camera is not sufficient for continuous control techniques. To overcome this, a strategy of pulsed motion, wherein current pulses are applied to get incremental stepped motion, is used. This control scheme is demonstrated by *Simscape* simulations in which the ferro-bot is moved along the prescribed trajectory. Simultaneous control of multiple ferro-bots is also demonstrated in the simulations. Preliminary experiments are performed to demonstrate the efficacy of the setup and the control algorithm.

S. Bhat (✉) · G. K. Ananthasuresh
Indian Institute of Science, Bengaluru, India
e-mail: sudhanvabhat@iisc.ac.in

G. K. Ananthasuresh
e-mail: suresh@iisc.ac.in

V. K. Gupta et al. (eds.), *Recent Advances in Machines and Mechanisms*, Lecture Notes in Mechanical Engineering, https://doi.org/10.1007/978-981-19-3716-3_31

Keywords Magnetic robots · Induced magnetic fields · Ferro-bots

1 Introduction

Robots enhance technological capacity in multiple ways, such as automating repetitive work, increasing precision and repeatability, working in hostile environments, and performing complex tasks. Traditional robotics involving actuated rigid-body mechanisms with multiple degrees of freedom (DoFs) and connected to a reference frame are most widely researched and used. Recently, untethered magnetic robots are finding applications in medical fields and microsystems. There is no physical contact between the source of the magnetic field and magnetic robots allowing for navigation of these robots inside covered environments, like the human body. Traditional techniques for forward and inverse dynamics and control do not work for magnetic actuation because of the inherent nonlinearity. Hence, developing suitable techniques has now become an important research topic.

The magnetic robots are controlled by current-carrying coils that produce the magnetic field. The magnetic force on the magnetic robot is proportional to the magnetic field gradient and magnetization. The magnetic torque is proportional to the magnetic field intensity. The work by Jeon et al. [1] showed the use of large Helmholtz coils, Maxwell coils, and gradient saddle coils. Helmholtz coils and Maxwell coils generate a uniform magnetic field. A combination of such coils arranged in multiple directions can generate the magnetic field of the required magnitude and direction. Multiple gradient saddle coils can be used to control the gradient of magnetic field. Kummer et al. [2] showed the use of a system of eight coils called the *Octomag* configured in an optimized arrangement for the control of a magnetic robot with five DoF. This arrangement of coils requires a relatively sophisticated mathematical modelling, but it is able to offer improved control. Diller et al. [3] demonstrated the use of a global magnetic field to control multiple heterogeneous objects having different magnetic properties. Chowdhury et al. [4] demonstrated the control of multiple permanent-magnet objects using a grid of printed circuit coils where the local magnetic field can be controlled. All these cases make use of permanent magnet robots that have fixed magnetization. Thus, there is no previous work on controlling ferromagnetic objects in the presence of locally varying magnetic fields. This paper focuses on the smooth control of multiple ferromagnetic robots (called *ferro-bots*) on a plane, using a grid of solenoids with ferromagnetic cores. Multiple permanent magnets stick to each other when they are brought together in close proximity. Separating them can be a challenging task. Ferromagnetic robots, on the other hand, can be separated from each other using electromagnets. Hence, having ferromagnet robots allows for relatively complex manoeuvres. The setup is designed to have the capacity to control meso-scale ferro-bots, whilst micro-scale ferro-bots are equally possible.

There are a few unique unsolved challenges addressed in this work. The ferromagnetism in the solenoid core and the ferro-bot poses an interesting modelling problem for which existing modelling in the literature cannot be used. The sizes of the bots are in the meso-scale, and hence, the field due to secondary induced magnetism, which is generally ignored, becomes significant. The Coulomb friction due to sliding of the ferro-bots on a surface and close proximity of the ferro-bots to the solenoids makes the ferro-bot behaviour complex. This complexity, along with the limitations of the vision-based hardware used for feedback, makes the control problem challenging.

We now place our work in the context of ongoing work in this field. Figure 1a shows the control of a magnetic robot using Octomag, and Fig. 1b shows the control of micro-robots made of permanent magnets using printed circuit coils. Former setup can control the field globally. The later setup has local control of the field. Figure 1c shows the solenoid grid setup along with meso-scale ferro-bots. Modelling of the magnetic field is straightforward with planar coils and permanent magnets as the objects, whereas it is not as easy when solenoids with ferromagnetic cores are used to move ferromagnetic objects. Ferro-bots are beneficial as they do not attract one another and hence are more easily navigable individually or in a swarm as compared to the permanent magnets. This benefit comes at the cost of complexity in modelling the magnetic field. This challenge is addressed in this paper.

The focus of this paper is on the design of a solenoid grid-based setup and development, implementation of the control strategy for independent, and steady control of multiple ferromagnetic robots on a 2D plane. The control strategy involves working out the forward and inverse problems for the designed magneto-mechanical system to compute the required current in each solenoid at any point in time to get prescribed ferro-bot trajectories.

The scope of this work includes modelling and design of the solenoid. A scheme for solving the forward problem of computing the force due to the coil currents in real-time is developed. The inverse problem of finding out the current required to get a particular force is also solved. The control strategy uses pulsed motion for ensuring steady motion along the required path. It is demonstrated with simulations. Basic actuation and independent control of multiple robots are demonstrated. Preliminary experiments are also described.

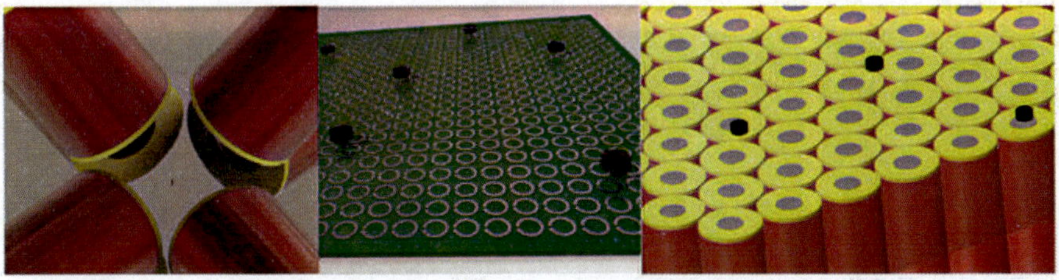

(a) (b) (c)

Fig. 1 Magnetic robots controlled using global and local magnetic fields **a** Octomag system controlling magnetic robots with global field [2], **b** an array of planar coils for moving permanent magnets [4], and **c** an array of cylindrical electromagnets to move ferromagnetic objects [this work]

2 Design of the Solenoid

Figure 2 shows a ferro-bot on a plane over a vertical solenoid. When current is applied to the solenoid, it becomes an electromagnet and attracts the ferro-bot towards it. As there is an acrylic sheet separating the electromagnet and the ferro-bot, the solenoid is a noncontact actuator. The nature of the solenoid is to be understood for developing a design suitable for the purpose of controlling multiple ferro-bots along prescribed trajectories. Beyzavi et al. [5] presented the modelling and optimization of planar micro-coils. The modelling of the solenoid with the iron core cannot be done analytically. We refer to the book by Inan and Inan [6], wherein the effect of induced magnetism is explained. The magnetic field in the presence of ferromagnetic objects is a transcendental vector equation. Thus, the magnetic force on a ferromagnetic object is complicated. We used a simplification as explained next.

The design parameters of the solenoid considered are length of the solenoid L, inner radius r_1, outer radius r_2, current in the solenoid I, number of turns in the solenoid n, distance of the iron object from the centre of the solenoid r, volume of the iron object v. The current density $\mathbf{J}$ is proportional to the current I and number of turns per unit length n/L. Based on the finite element analysis data, the force versus design parameters relation is assumed to be of the form (details are not included in this short paper):

$$F = (0.86 \times 10^{-7}) \frac{v^1 I^2 n^2 r_2 r_1^{1.7}}{r^{5.9}} \quad (1)$$

Note that Eq. (1) is a model developed based on approximate trends based on the observation of FEA data and not rigorous data fitting. For a given n/L ratio, the force is found to be proportional to L^2. Hence, L cancels off in the model.

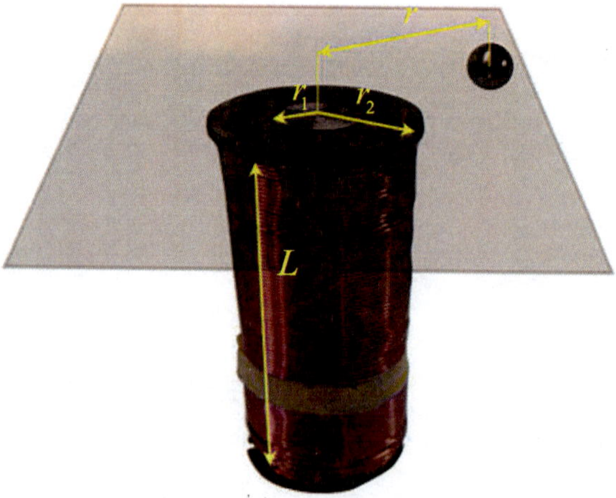

Fig. 2 Dimensions of importance in the solenoid and the ferro-bot interaction

The empirical model from Eq. (1) is used to design the solenoid. The idea here is to attract a ferro-bot in the presence of Coulomb friction from a distance little more than twice the solenoid radius. Consider a cylindrical ferro-bot of 8 mm diameter and 4 mm height made of mild-steel and having density 7850 kg/m^3. The mass of the ferro-bot will then be equal to 1.5 g. For a friction coefficient of $\mu = 0.25$ between ferro-bot and the sliding plane, the friction force will be $\mu mg = 0.0046$ N. The magnetic force has to be more than this. It is suitable to have thick copper wire for minimizing resistance and heating. A copper wire of 0.7 mm diameter was chosen as was available in the winding facility. It was found that for $r_2 = 0.015$ m, $r_1 = 0.008$ m, $L = 0.08$ m, a total of $n = 800$ can be accommodated. This wire can take up to 3.5 A current for short durations. The volume of the ferro-bot is $v = 2.01 \times 10^{-7}$ m^3. Though the ferro-bot we use is spherical, we design the solenoid to be able to attract a cylindrical button-shaped ferro-bot having friction coefficient of 0.3, so that the setup can be extended for the use of cylindrical ferro-bots later on. For $r = 0.03$ m and a current of $I = 3.3$ A, the force on the ferro-bot is found from Eq. (1) to be equal to $F = 0.0047$ N. This coil is able to attract the ferro-bot against the friction force from a distance equal to twice the outer radius of the coil. Because of the $1/r^{5.9}$ nature of the force, huge increments in the coil parameters like current and length only lead to minor improvement in the attraction range.

3 Experimental Setup

The experimental setup consists of a grid of solenoids arranged, as shown in Fig. 3. The hexagonal arrangement gives rise to dense packing. There are four main components of the setup: the solenoid actuator, the ferro-bot, the feedback camera sensor, and the control system. The solenoid grid is mounted on a table with holes for wire outlet at the bottom. An acrylic sheet is mounted on the solenoid grid to provide a surface for the ferro-bot to move on. The solenoids are sufficiently constrained so that they do not move when any of them are activated.

The ferro-bot here is a ferromagnetic ball of 6 mm dia. A plastic cap which has 7 mm ID and 8 mm OD is placed on the ball, as shown in Fig. 3. This plastic cap makes it easy to identify the position using a camera. The rolling effect of the ball, along with the sliding friction of the cap, brings the friction to a small value so that it is easier to control and yet stable from rolling due to slight inclines on the moving plane.

A Raspberry Pi camera is mounted on the top to measure the ferro-bot position. The camera has 640 $\times$ 480 resolution at 90 FPS. The camera is connected to a Raspberry Pi controller, which sends the 3 V control signals. The control signals then go to the pulse-width-modulation (PWM) generators that generate PWMs of 5 V. These PWM signals control switches to each of the coils, which are connected to a 12 V power source. The maximum current in the coil at this voltage is 3.3 A.

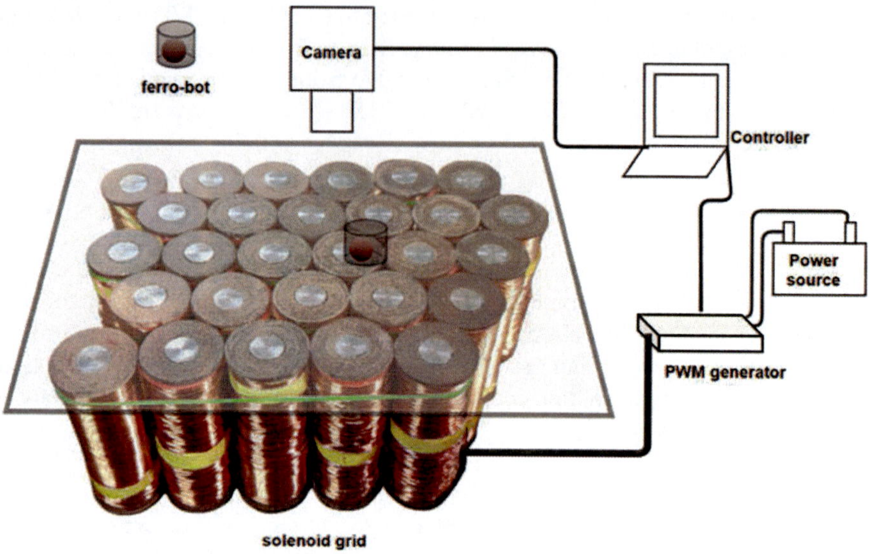

Fig. 3 Experimental setup

4 Relation Between Forces and Solenoid Currents

For the purpose of control, the currents in the solenoid that can give the required force is to be estimated. The force on an iron object is given by

$$\mathbf{F} = \int_V (\mathbf{M}.\nabla)\mathbf{B}dv = \frac{\chi_m}{\mu_0(1+\chi_m)} \int_V (\mathbf{B}.\nabla)\mathbf{B}dv \tag{2}$$

Here, μ_0 is the permeability of free space, $\mathbf{M} = (\chi_m \mathbf{B})/(\mu_0(1+\chi_m))$ is the magnetization or magnetic dipole moment per unit volume of the object of interest, χ_m is magnetic susceptibility of the ferromagnetic material, **B** is the magnetic field. **B** and gradient of **B** should be known for computing the force. COMSOL finite element analysis software (www.comsol.com) is used to generate the field data of a single solenoid with core of given design specifications carrying unit current. At any time, multiple coils are active. The field due to each solenoid at the point of interest can be summed up. There will be an additional field due to the magnetization of the ferro-bot. The additional field is proportional to magnetization and if found by FEA to be

$$\mathbf{B_M} = (8.36 \times 10^{-7})\mathbf{M} \tag{3}$$

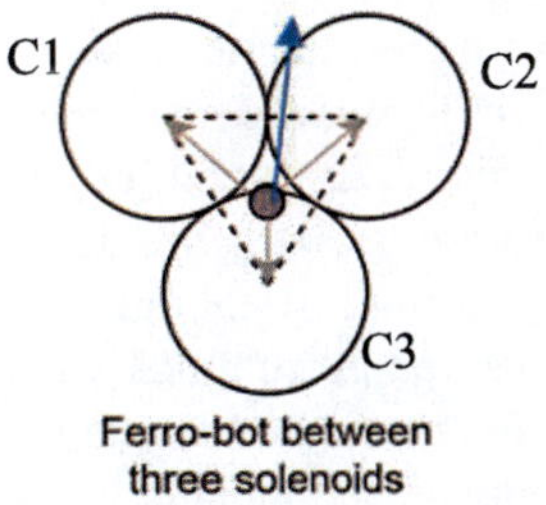

Fig. 4 Ferro-bot between three solenoids and the direction of applied force due to each solenoid

This additional field is added to the field from the coils. The gradient of field can be computed by applying finite difference to the FEA data. Hence, the force can be computed from Eq. (2). The magnetic field **B** is proportional to the coil currents I_i. So, the force equation in Eq. (2) can be written as

$$\mathbf{F} = \mathbf{C}_{ij} I_i I_j \tag{4}$$

$\mathbf{C}_{ij}$ are constants for a given position of the ferro-bot. By varying I_i and computing the resultant force, $\mathbf{C}_{ij}$ can be estimated. At any point, a ferro-bot is surrounded by three coils as shown in Fig. 4.

The coils along the either side of the direction of require force indicated by the blue arrow are shown in Fig. 4. Since $\mathbf{C}_{ij}$ are known, if a particular force has to be applied the currents can be estimated from Eq. (4). It is a quadratic equation in currents and can be solved symbolically. Hence, for a given force requirement, the solenoid currents can be estimated.

5 Control of Ferro-Bots to Follow Prescribed Trajectories

As described in the preceding sections, computing the magnetic force on the ferro-bots is an involved task. This complexity leads to challenges when multiple ferro-bots are to be moved along prescribed trajectories. [7–10] describe various control strategies for different kinds of micro-robot systems. The setup used in this work cannot directly use any of these strategies. The nonlinear dependence of the magnetic force on the distance from the centres of solenoids is one challenge. The limited reach of a solenoid to affect the motion of a ferro-bot to overcome Coulomb friction is another. Activating multiple solenoids at a time to have enough force in the right direction is necessary to move a ferro-bot along the required path. Sensing the position of the ferro-bot for feedback control is another challenge because we cannot mount any sensor on the ferro-bot. Hence, we use visual sensing with a camera mounted at the top. The frame rate of the camera limits the sensing rate. So, the velocity of the ferro-bot should also be controlled so that it does not go too far in a direction because the magnetic force rapidly rises in the inward radial direction of a solenoid.

All these challenges indicate that the standard linear control strategy might not work (in fact, it did not work in our trials). Hence, we use pulsed-signal strategy. That is, we activate a coil for a very short duration to give a small jerk to the ferro-bot so that it moves and stops due to Coulomb friction. We can control the small movement by passing just enough current in a solenoid or a set of solenoids. If the current pulse can be given quickly enough, it leads to a trajectory that stays with a band around the prescribed trajectory. This control strategy is described next, and its efficacy is demonstrated later in this section.

Figure 5 shows the schematic of the control system implemented. x_r, y_r are the reference position components at any given time. x_a, y_a are the actual position components. er_x, er_y are the position error components. F_x, F_y are the forces to be applied on the ferro-bot. v_x, v_y are the velocity components of the ferro-bot. I_i indicates the solenoid currents.

Even if there is a high-percentage error in computing the desired currents, the visual feedback can correct it in the next pulsed step. The position is measured and used to compute the position error as per the required trajectory. The forces are to be applied to the ferro-bot to correct the position error. The magnitude of the force to be applied and the pulse timing are decided based on friction and required velocity. The path travelled is more important in this experiment than the time taken.

The system of interest that includes multiple solenoids and ferromagnetic objects is simulated using Simscape in MATLAB. The scheme developed for computing the magnetic force is written as a function that takes in position, coil currents, and outputs

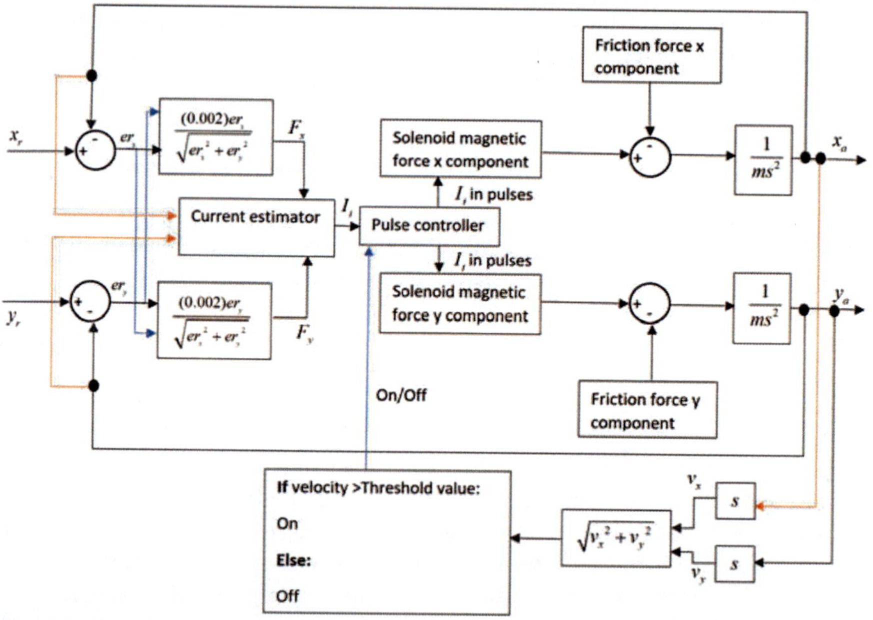

Fig. 5 Schematic of the control strategy

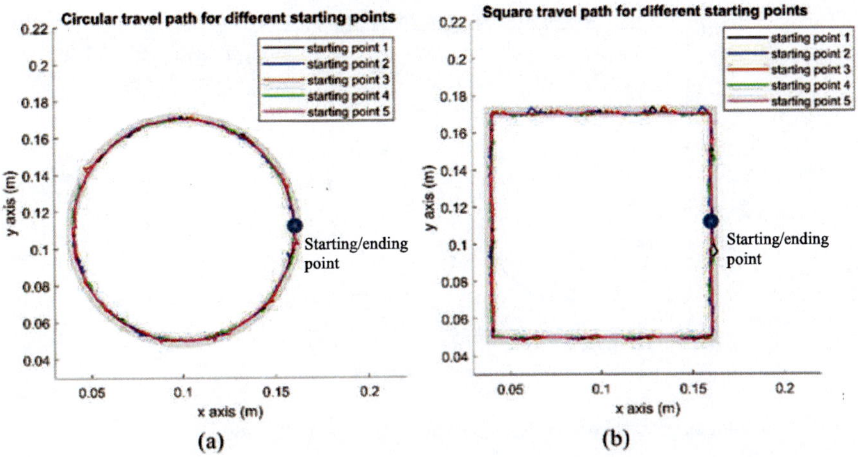

Fig. 6 Paths followed by the ferro-bots for various trajectories and starting points. **a** shows different circular trajectories overlapped and **b** shows square trajectories overlapped for comparison

force. This force is fed to Simscape. The Simscape selects the required ODE solver to solve and simulate the motion of the ferro-bot. There is a function for trajectory generation. Feedback control is used to control the ferro-bot, as illustrated in Fig. 6. The simulations are performed for moving a single object in various trajectories. The results are presented in Fig. 6. Initially, the ferro-bot is made to move along the anticlockwise circular trajectory starting at zero degrees. The same trajectory is repeated whilst offsetting the centre of the circle so as to change the starting point. Figure 6a shows multiple circular trajectories overlapped for comparison. Similarly, Fig. 6b shows multiple square trajectories with different centres overlapped for comparison.

The limitation in the inverse dynamics strategy can lead to an error of up to 3 mm from the intended path. Hence, our aim is to constrain the travel path within a band around the trajectory. The simulation also demonstrates the control of multiple objects, as can be seen in Fig. 7. Here, two ferro-bots are simultaneously moved along different paths.

6 Experimental Demonstration

The basic experiments are conducted to demonstrate the ability to control a disc ferro-bot. A small-scale setup with a grid of 19 solenoids was used. When a coil is switched on, any ferro-bot in the vicinity will move towards and come to rest over top of the solenoid. The smooth control scheme developed in this paper is yet to be experimented. The current experiment demonstrates the capacity of the solenoid to attract the ferro-bot. Hence, disc ferro-bots having higher friction are used for

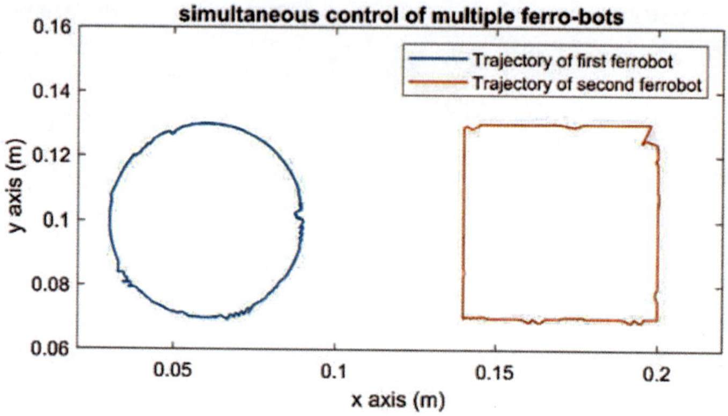

Fig. 7 Simultaneous control of two ferro-bots along different trajectories

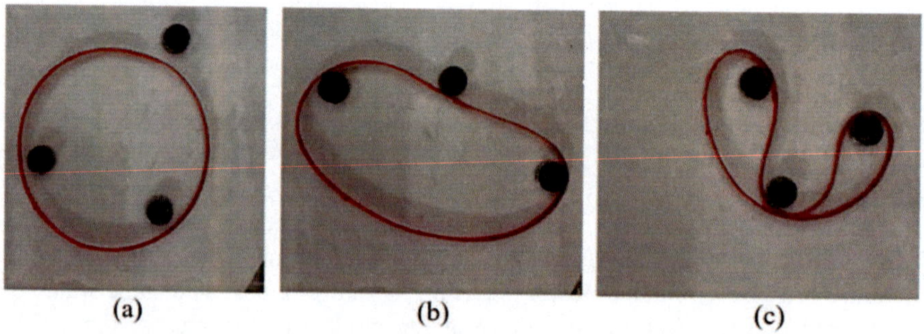

Fig. 8 Control of three ferro-bots for changing the shape of a compliant enclosure **a** initial position of the ferro-bots, **b** ferro-bots deforming a compliant enclosure, and **c** changing the shape of the compliant enclosure

demonstration. All of them are independently controlled. Figure 8 shows three disc ferro-bots. Figure 8a, b, and c show the ferro-bots at different configuration. Figure 8 shows the ferro-bots changing the shape of a compliant enclosure. The compliant enclosure is included to demonstrate the capacity of the ferro-bots to do work. The motion is not smooth as feedback control is not applied. This is a dynamic observation and is not visible in Fig. 8. The experiment demonstrates that the solenoid design is adequate. The smooth control is yet to be demonstrated.

7 Closure

This work's focus is the development of a solenoid grid-based setup for simultaneous individual control of multiple ferro-bots and the development of the control strategy. The solenoids were designed using FEA data. A setup is proposed based on the designed solenoid. A scheme is developed for estimating the currents in the coils so as to get the required force. The proposed scheme, along with feedback, is used to control the ferro-bot along a prescribed trajectory. A strategy of controlling the ferro-bot with pulsed motion is proposed. This control is demonstrated with a Simscape simulation. Preliminary experiments are performed to demonstrate crude control of ferro-bots so as to verify the designed solenoid capacity.

Acknowledgements This work was initiated with financial support from the Robert Bosch Centre for Cyber Physical Systems at the Indian Institute of Science towards the Cyber Gut project. Sudhanva Bhat is grateful for the Prime Minister's Research Fellowship (PMRF) provided by the Government of India.

References

1. Jeon S, Jang G, Choi H, Park S (2010) Magnetic navigation system with gradient and uniform saddle coils for the wireless manipulation of micro-robots in human blood vessels. IEEE Trans Magn 46(6), June 2010
2. Kummer MP, Abbott JJ, Kratochvil BE, Borer R, Sengul A, Nelson BJ (2010) OctoMag: an electromagnetic system for 5-DOF wireless micromanipulation. IEEE Trans Rob 26(6), Dec 2010
3. Diller E, Floyd S, Pawashe C, Sitti M (2012) Control of multiple heterogeneous magnetic micro-robots in two dimensions on non-specialized surfaces. IEEE Trans Rob 28(1), Feb 2012
4. Chowdhury S, Jing W, Cappelleri DJ (2015) Towards independent control of multiple magnetic mobile microrobots. Micromachines 7(1):3, 29 Dec 2015, https://doi.org/10.3390/mi7010003
5. Beyzavi A, Nguyen NT (2008) Modeling and optimization of planar microcoils. J Micromech Microeng 5 Aug 2008. https://doi.org/10.1088/0960-1317/18/9/095018
6. Inan US, Inan AS (1998) Engineering electromagnetics, 2nd edn. Addison-Wesley, USA
7. Ryan P, Diller E (2016) Five-degree-of-freedom magnetic control of micro-robots using rotating permanent magnets. In: IEEE international conference on robotics and automation (ICRA). https://doi.org/10.1109/ICRA.2016.7487317
8. Chowdhury S, Jing W, Cappelleri DJ (2015) An overview of multiple DoF magnetic actuated micro-robots. J Micro-Bio Rob 10:1–11. https://doi.org/10.1007/s12213-015-0083-6
9. Bouchebout S, Bolopion A, Abrahamians JO, Régnier S (2015) Controlling multiple micro-robots: recent progress and future challenges. J Micro-Bio Robot 10:1–11. https://doi.org/10.1007/s12213-015-0083-6
10. Wong D, Steager EB, Kumar V (2016) Independent control of identical magnetic robots in a plane. IEEE Rob Autom let 1(1), Jan 2016

Mechanism for Space Applications

Fold-line Mechanics for Ultra-Thin Membrane in Gossamer Space Structure

Parth K. Kamaliya and S. H. Upadhyay

Abstract Lighter mass and compact packaging of deployable space structures led to the advancement in folding patterns. Though crease(s) development is inevitable, they are preferred as it keeps thr fold in same location. However, it alters the compacted membrane's mechanical response leading to thru and post-deployment behavioral change in outer space. Thus, in this paper, the authors dealt with fold-line mechanics and its influence on stretched ultra-thin Kapton film. A connector-based model with integrated fold-line stiffness is suggested for membrane adopting Z-folding. Its selection depends on its appropriateness over pinned or fixed hinge connection techniques. At first, an experiment was carried out to measure the neutral angle. Later, a self-opened component profile is modeled in commercially available finite element package. Fold-lines are introduced using coupling constraint with assumed constant rotational stiffness. Authors have used an explicit time integration method for analyzing the deformed/fully deployed planar configuration. Uniform tensioning from opposite ends induce stresses in the connector and membrane region. The fold-line response over connection properties, axial displacement, and wrinkle formation in the structure is critical parameters for analysis. It was observed that mechanical response might vary depending on the form of the fold. Results also convey that the fold-line affects the stress transfer path, may cause stress concentration, and changes the wrinkling direction. A suggested technique is compared against a test case analyzed using a high number of smaller mesh elements with rotational stiffness for better estimation. Comparison conveys that the connector-based model requires less computational time at the cost of modeling complexity.

Keywords Deployable space structures · Creased membrane · Connecting element · Fold-line stiffness · Stress concentration

---

P. K. Kamaliya (✉) · S. H. Upadhyay
Smart Material and Structures Lab, Department of Mechanical and Industrial Engineering, Indian Institute of Technology Roorkee, Roorkee, Uttarakhand 247667, India
e-mail: pkamaliya@me.iitr.ac.in

S. H. Upadhyay
e-mail: sanjay.upadhyay@me.iitr.ac.in

V. K. Gupta et al. (eds.), *Recent Advances in Machines and Mechanisms*, Lecture Notes in Mechanical Engineering, https://doi.org/10.1007/978-981-19-3716-3_32

1 Introduction

Ultra-thin membrane becoming popular for solar sails with larger surface area, reflector portion in deployable space structure, and inflatable sunshield [1]. For their desire functioning and to carry an expected payload, membrane surface should be of size of a cricket field [2]. Thin in size and light-weight make it more suitable for compact packaging. This demand also arises from limited stowage volume of launch vehicle. Many folding patterns have been demonstrated for origami designs, but their implementation for thin-folded membrane is not as simple. Membrane thickness and selection of suitable folding pattern are two major concerns for in-space deployment [3]. This compaction creates a permanent crease, a plastic deformation to the membrane, which changes the material behavior as well as shape and size of membrane. Research investigated that the crease properties and fold-line stiffness has greater impact on final deployed shape.

The previous effort by scientists focused on crease characteristics, fold-line plasticity, and viscoelastic and elasto-plastic behavior of fold deployment. However, their emphasis limit to the single crease and fold deployment only. In designing the precise deployment, trial and error-based experimental testing is simply not preferred because of time, cost, and non-vacuum space environment. Hence, virtual simulations are more liable option in idealizing the conditions with accuracy and time. In formulating this, fold-mechanics play a key role. Papa and Pellegrino [4] introduced creases as a sharp edge in origami based folding method. Lee and Pellegrino [5] selected leaf folding pattern for CubeSats and examined the deployment force (with no additional external effort) for wrapped membrane. In both analyses, they assumed same material property for crease region and base membrane. Woo and Jenkins [6] incorporated crease as a wide strip and as a seam to analyze effect on wrinkling behavior of membrane. In recent years, researchers have considered either elastic, elasto-plastic, or plastic behavior of fold-line to investigate required folding force and to check deployment behavior [7–10]. However, these analyses are limited to thin-strip only, and have not been implemented in large size membrane.

Post-deployment effects such as wrinkling, multiple fold-lines, and intersected creases are untouched portions. Among the different materials, polyimide Kapton membrane is selected because of its property sustainability over extreme space environmental and thermal conditions. This paper focuses on understanding the effect of crease connection properties on deployed Kapton. A double-creased membrane which implies Z-folding approach is finalized as a part of larger sized reflector surface. Section 2 describes the selection of folding pattern and physical measurement of crease characteristics. Section 3 analyzes the different possible fold configurations and suggest a connector-based technique for finite element simulation. It is followed by implementation of technique in square membrane and discussion on the results obtained, in Sect. 4. Section 5 concludes the chapter.

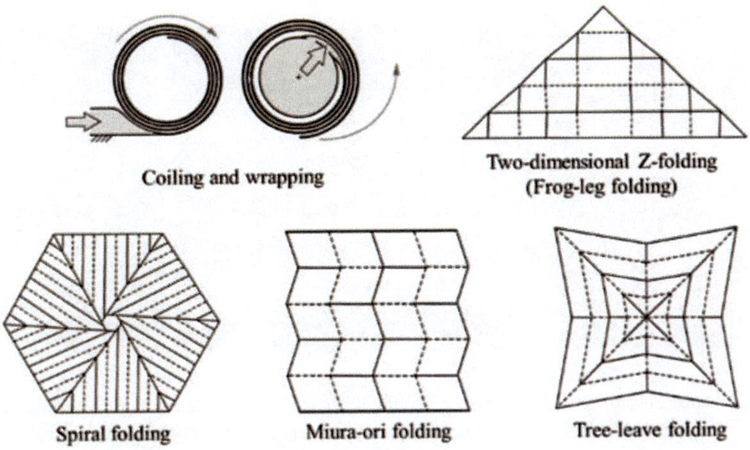

Fig. 1 Packaging approach for different shaped membrane with multiple fold-lines [11–13]

2 Characterizing the Crease in Membrane

2.1 *Fold-line Arrangement*

Biaxial compaction of membrane shapes, ranging from triangular to polygon and circular profile, can have different fold arrangements as shown in Fig. 1. Depending on that, fold-lines could follow Z-folding, zig-zag, spiral, coiling and wrapping or origami packaging [11–13]. Each pattern consists of many pick and valley profiles. For our reasearch, one small section is taken out from two-directional Z-folded membrane. This representative unit component consists of total four fold-lines (one peak and three valley) at 90-degree arrangement, shown in later part of the chapter (Fig. 7). Selection of this angle deepens on different folding pattern, usually 60-degrees in circumferential shape and 80–90 degrees in square/rectangular membrane.

2.2 *Experimental Measurement of Neutral Angle and Material Properties*

Experimental work focuses on the measurement of neutral angle and mechanical properties of creased membrane. Kapton strips of 125 mm × 25 mm with no visible defects are cut from large sheet. These specimens are lightly folded and pressed with roller. One-directional roller movement is such that a uniform compressive force is applied. Strips are kept in folded position for 24 hrs. to achieve a plastic deformation in the creased region. Viscoelastic nature of material causes it to unfold. Images were captured using a high-resolution camera and postprocessed to measure the

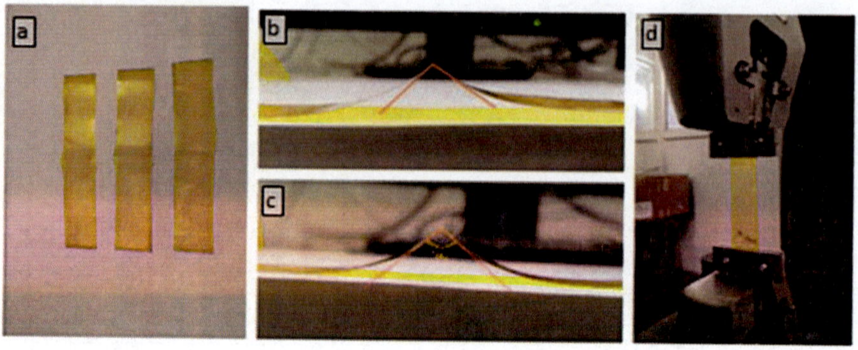

Fig. 2 **a** Kapton membrane strips with permanent crease, fold angle relaxation: **b** just after load removal and **c** after a day relaxation, **d** tensile test to obtain effective modulus of creased membrane

Table 1 Kapton membrane properties

Thickness (μm)	50
Density (kg/m^3)	1420
Creased membrane—Effective modulus (GPa)	2.104
Non-creased membrane—Young's modulus (GPa)	3.139
Yield stress (MPa)	
—Analytical	69.00
—Experimental	64.83
Creased region—Residual stress (MPa)	0.169
Poisson's ratio	0.34

equilibrium fold angle. Couple of samples were tested and observed the relaxation mechanism. Average value of neutral angle, just after load removal, found to be 87° ± 5° (Fig. 2b).

Non-linear behavior of crease substitutes the Young's modulus with effective crease modulus. The same creased strips, tested on INSTRON tensile testing machine with pulling load of 2 mm/min. Special jaw was made to grip the thin micron-size Kapton strips. Samples failed from middle were only in consideration. After thorough testing, obtained mechanical properties are given in Table 1.

3 Fold-line Simulation

Mechanical response of the fold provides neutral angle and fold-line stiffness properties. This section incorporates the same in finite element analysis. Figure 3 shows the two halves of rectangular Kapton strip, 62.5 mm × 25 mm each, modeled with 4-node quadrilateral shell element (S4R) with reduced integration.

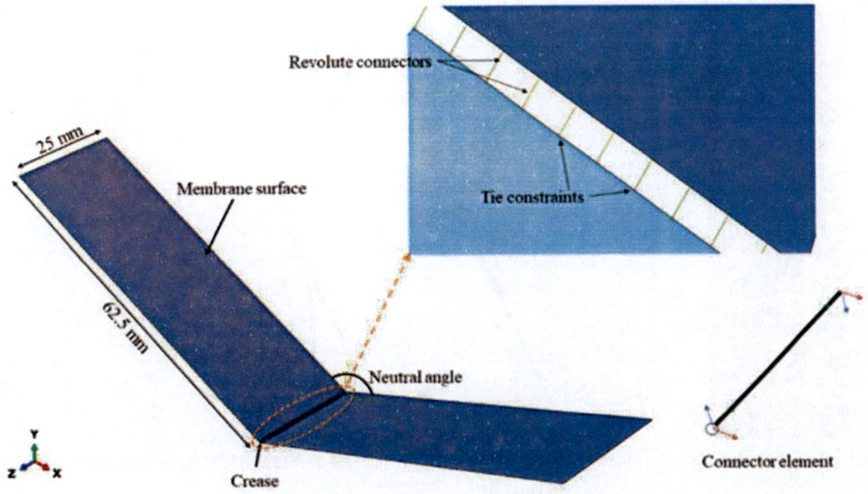

Fig. 3 Representation of shell membrane profile at neutral angle with creased region and zoomed-in connector elements

A Newton–Raphson time integration solver was used in typical finite element package. Computations were stabilized using *STATIC, STABILIZE option. Large deformations of shell elements are captured using *NL GEOM command. Two strips are apart at neutral angle, reason being avoiding the difficulty in idealizing from fully folded configuration. Central creased portion is modeled with different connection profiles, while the two extreme ends are subjected to shell edge loading to simulate the deployment process.

Three different techniques are used to idealize a crease in the membrane. The very first approach considered the mesh sweeping. Defining the finer mesh in the vicinity of crease and mesh density reduces as moving toward the end. This approach gives better accuracy in deployment results as the fold-line expect higher bending. A typical model consists of total 12,524 nodes with minimum length of 0.2 toward crease. Two halves are combined as a single entity, and typical boundary conditions were imposed. Center portion is kept fixed in translation and allowed for rotational movement in the direction normal to loading plane. These conditions create a fixed pin joint connection. Figure 4 shows the results of stress distribution and deployment from evaluation process. Second approach modifies the creased region properties. Experiments conducted in Sect. 2 measure the effective elastic modulus along with other mechanical properties. Membrane surface except fold-line has standard Kapton properties. In this case, external force for deployment is over and above the residual stress because of crease. Hence, requiring the higher than usual force for deployment. The last one contains the connector element with rotational stiffness. Here, two shell portions are connected with *TIE constraint and connector element. Rotational degree-of-freedom for connector element gives us the facility to define an individual DOF. In addition, rotational stiffness accounts for actual deployment sequence for

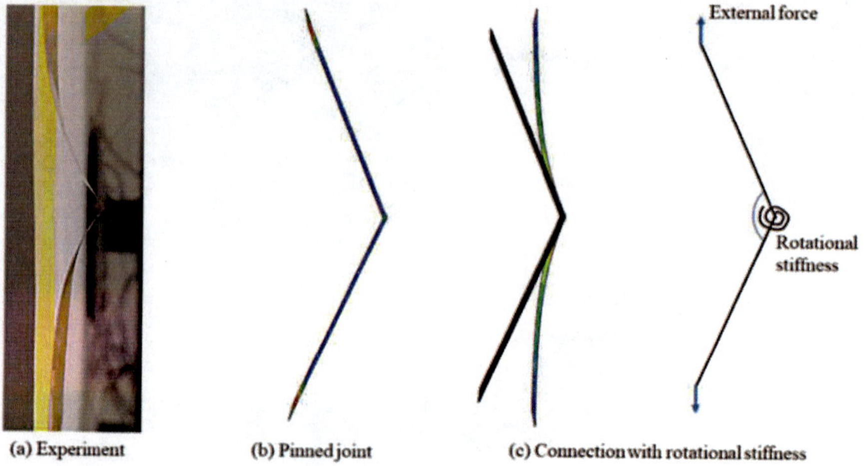

Fig. 4 Deformed profile comparison after applying external load

crease. This approach helps in idealizing the crease curvature, which should be implemented directly for membrane with multiple folds. Direct implementation of crease cusp radius makes the deployed membrane modeling and simulation easier.

4 Results and Discussion

Section 3 gives insight for crease profile accuracy. In that, connection with pinned joint has significant deviation from experimental outcome especially for crease curvature. The proposed model having rotational element generated the most realistic deployment of the crease profile and it stands out from the Fig. 4a and c. Talking about computational time, suggested approach of rotational stiffness outperforms hinge connection with requirement of over 900%. Deformed shapes of both methods were further analyzed for stress distribution, reaction forces-moments (Fig. 5). In pinned connection, membrane stresses are maximum at extreme ends and no/minimal stress profile observed for crease, see Fig. 5a. Contrarily, parallel alignment of connector elements transfers the stresses to its both end points. Hence, close gap is required between two connectors for uniform stress distribution. This seamless stress distribution is clearly captured in Fig. 5b. Reaction forces and moments (Fig. 5c and e) of magnitudes 3.501×10^{-3} N and 2.550×10^{-2} Nmm have significant impact on the rotational elasticity. Proposed connector element with rotational stiffness overcomes the residual stresses induced due to formation of the crease. Hence, reacting forces and moments are negligibly small (Fig. 5d and f).

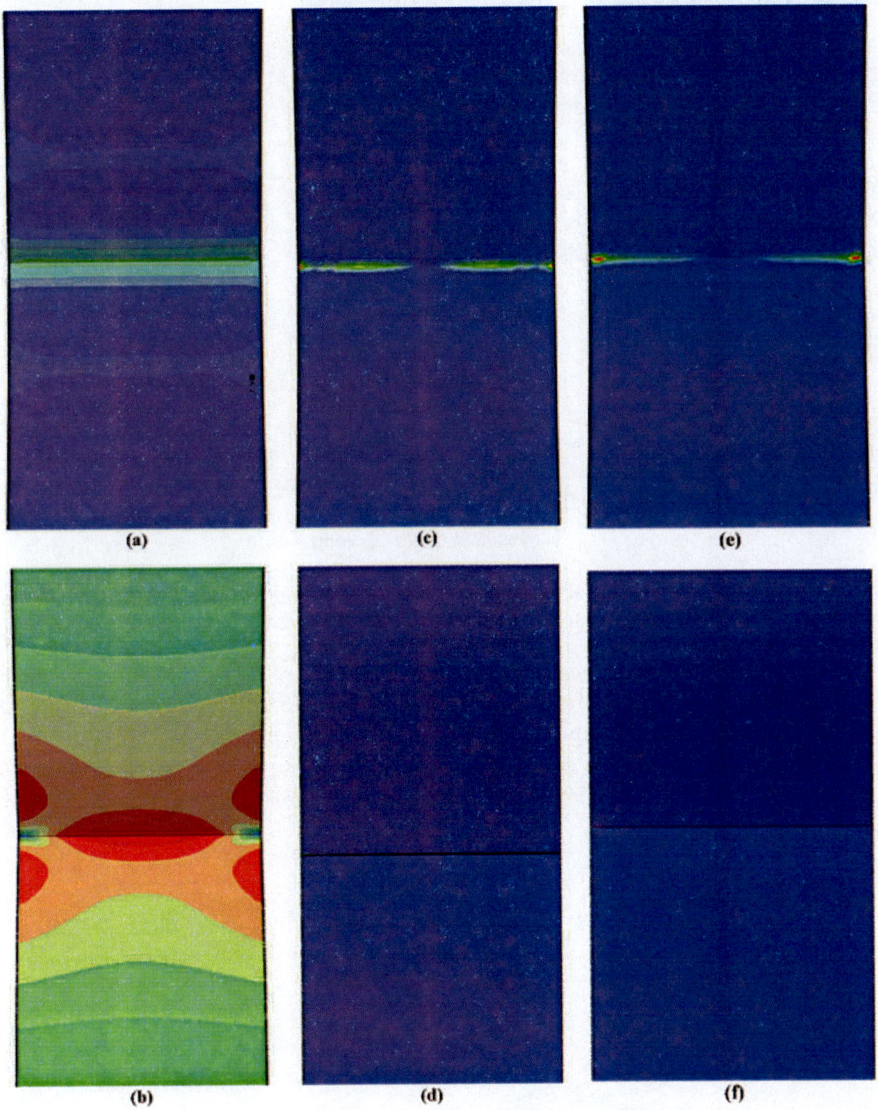

Fig. 5 Stress concentration at creased region for **a** pinned connection and **b** connector with rotational stiffness, followed by reaction forces **c-d** and moments **e-f** for both cases

In general, these forces and moments are related to the sequential membrane deployment. Figure 6 compares this membrane deployment calculation for numerical solutions against the experimental observations. Slight deviations are observed for proposed model, whereas the pinned connection is not even closure to the experimental data. Variations in the results are due to self-weight of the membrane and

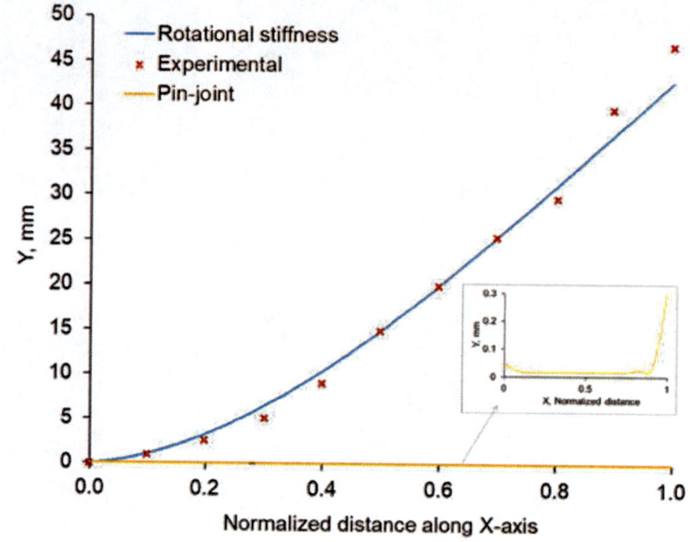

Fig. 6 Comparison of two different connector elements, replicating the crease, with experimental observations for 50 μm membrane

human error while extracting the data. This graph clearly conveys capability of proposed model for predicting the hinge stiffness and bending.

Fold-line properties and connection type finalized in Sect. 3 are implemented in two-directional Z-fold membrane reflector. A square membrane of size 0.5 m

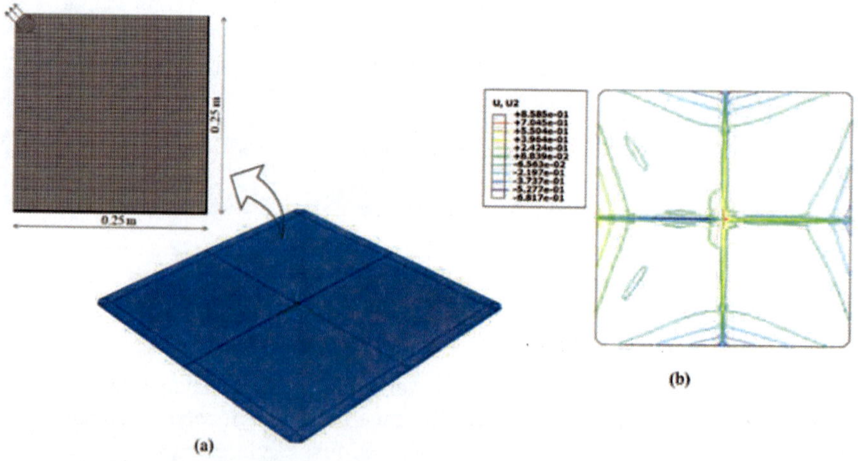

Fig. 7 **a** Double-creased corner loaded membrane and a quarter meshed region with relevant loading conditions, **b** stress distribution and wrinkle contours in the membrane loaded with equal force of 5 N at all four corners

× 0.5 m is shown in Fig. 7a with intersected crease line following the Z-fold approach. Each quarter portion of shell membrane geometry has 3057 elements totaling 12,254 for whole, out of which creased region is more refined with dense meshing (2144 elements). Except corners, where the triangular elements (S3) are introduced, remaining reflector surface has S4R quad elements. This mesh selection was finalized after mesh-sensitivity analysis. Sharp corners are avoided to observe the stress distribution and wrinkle formation in the membrane. Corners were truncated at an angle of 45 degrees and 15 mm wide. All four parts were assembled to form a deployed flat state. 2.5 mm of constant curvature of the crease was introduced from experimental observation of neutral angle and moment–angle response of crease. Then, connection is made using tie constraint with revolute connectors. These connectors provide rotational stiffness to the adjoining shell portion, especially crease. Extreme edges were assigned fixed boundary conditions allowing only in-plane movement. Specimen was pulled equally from all four corners. Magnitude of this external force is 5 N. Implemented non-linear post-buckling technique helps in getting the wrinkle behavior accurately.

Figure 7b presents the stress distribution and wrinkle contours of creased membrane. As seen, stresses are more likely to get scattered in the surface and only couple of stress lines remains. These stress lines are near boundaries and that is due the imposed fixed boundary conditions. Talking about the stress concentration, it is clearly visible near fold-lines. It is to note that, this also includes an effect of rotational stiffness of the crease. Comparing it with our non-creased membrane model, which has stress distribution over 50% of area. The wrinkles are also wide-spread over the membrane and covers up to 60% of surface compare to 35% in the creased profile. Thought the case for stress distribution seems promising, wrinkle amplitude is much higher in the crease vicinity (max.: 0.0874 mm and min.: −0.8206 mm) compare to maximum of 0.1546 mm and minimum of −0.0841 mm in non-creased profile. However, it should be compensated by achieved higher wrinkle free area, as creases are unavoidable for large size space antenna structures.

5 Conclusion

This chapter deals with fold-line mechanics and its influence on gossamer antenna reflector surface. The primary objective is to idealize the crease, an unavoidable entity for large membranes used for space applications. The selected approach is toward idealization of the fold-line with rotational stiffness and practical mechanical properties. This approach gives most realistic output compare to perfect hinge connection which is currently in practice. Offered possibility of this technique is implemented in the large size membrane reflector. Corner loaded two-dimensional Z-fold reflector surface shows stress concentration in the vicinity of crease. To eliminate irregular stress concentration, it is recommended to have more number of connectors. However, computational efficiency should be compromised.

Acknowledgements This work was supported by the Department of Science and Technology, Govt. of India under Indo-Sri Lanka joint research grant [No. DST/INT/SL/P-27/2016].

References

1. Liu ZQ, Qiu H, Li X, Yang SL (2017) Review of large spacecraft deployable membrane antenna structures. Chin J Mech Eng (English Ed. 30, 1447–1459 (2017). https://doi.org/10.1007/s10033-017-0198-x
2. Jenkins CHM (2006) Recent advances in gossamer spacecraft. American Institute of Aeronautics and Astronautics
3. Kamaliya PK, Upadhyay SH, Mallikarachchi HMYC (2021) Investigation of wrinkling behaviour in the creased thin-film laminates. Int J Mech Mater Des, In Press. https://doi.org/10.1007/s10999-021-09559-5
4. Papa A, Pellegrino S (2008) Systematically creased thin-film membrane structures. J Spacecr Rockets 45:10–18. https://doi.org/10.2514/1.18285
5. Lee N, Pellegrino S (2014) Multi-layered membrane structures with curved creases for smooth packaging and deployment. In: Spacecraft structures conference. AIAA SciTech, National Harbor, Maryland, pp 1–20
6. Woo K, Jenkins CH (2013) Effect of crease orientation on wrinkle-crease interaction for thin membranes. J Spacecr Rockets 50:1024–1034. https://doi.org/10.2514/1.A32183
7. Xia ZM, Wang CG, Tan HF (2018) Elastoplastic folding behavior of membrane ribbon based on plane strain beam theory. Int J Solids Struct 143:167–174. https://doi.org/10.1016/j.ijsolstr.2018.03.004
8. Torisaka A, Ogawa K, Miura S, Parque V, Miyashita T, Yamakawa H (2019) Study on in-plane and out-of-plane deformation considering elastic plasticity of membrane. In: AIAA SciTech forum. American Institute of Aeronautics and Astronautics, Inc., San Diego, pp 1–23
9. Jules T, Lechenault F, Adda-Bedia M (2019) Local mechanical description of an elastic fold. Soft Matter 15:1619–1626. https://doi.org/10.1039/c8sm01791c
10. Dharmadasa BY, McCallum MW, Mierunalan S, Dassanayake SP, Mallikarachchi CHMY, Jiménez FL (2020) Formation of plastic creases in thin polyimide film. J Appl Mech Trans ASME. https://doi.org/10.1115/1.4026559
11. Fu B, Sperber E, Eke F (2016) Solar sail technology—a state of the art review. Prog Aerosp Sci 86:1–19
12. Arya M, Lee N, Pellegrino S (2017) Crease-free biaxial packaging of thick membranes with slipping folds. Int J Solids Struct 108:24–39. https://doi.org/10.1016/j.ijsolstr.2016.08.013
13. Urbinati L (2020) Inflatable structures for space applications

Geometrical Analyses and Packaging Behavior of Foldable Cylinders with Bellow Pattern

Hemant Sharma and S. H. Upadhyay

Abstract Several folding methodologies have been presented for space-based inflatable booms based on different origami patterns. One dimensional expandable structures are essential components of any deployable space structure; some folding methodologies have been proposed for space-based cylindrical booms/masts based on several origami concepts. Noticeable work has been done in several technical areas related to deployable cylinders. However, the effect of pattern design parameters on packaging behavior is not studied. This paper is concerned with the geometry design and packaging behavior of cylindrical booms folded with bellow folding pattern. At first, the detailed geometry of the bellow pattern is discussed. Then, various geometry-based packaging parameters like the packaging efficiency, circumscribed radius of folded cylinder, and residual space after folding are calculated to investigate the packaging behavior. The influence of number of origami units and length ratio on the packaging behavior of cylindrical boom is discussed. The present research adds to the selection process of the folding patterns for inflatable cylindrical booms.

Keywords Bellow pattern · Geometric analyses · Packaging behavior

1 Introduction

Space-borne structures assembled with large sized membranes and deployable booms, often known as gossamer structure, have made revolutionary space missions possible by their exceptional capabilities in packaging and strength [1]. The demand for large space structures has attracted the interest of researchers in the use of membrane-based ultra-light assemblies.

It is found that under the action tension field, thin film membrane acts as a structural component, and this property can meet the requirements of a gossamer space

H. Sharma (✉) · S. H. Upadhyay
SMSL, MIED, IIT, Roorkee, Uttarakhand, India
e-mail: hsharma@me.iitr.ac.in

S. H. Upadhyay
e-mail: sanjay.upadhyay@me.iitr.ac.in

V. K. Gupta et al. (eds.), *Recent Advances in Machines and Mechanisms*, Lecture Notes in Mechanical Engineering, https://doi.org/10.1007/978-981-19-3716-3_33

structure. Inflatable booms are an important part of any inflatable space structure. The booms are compactly stowed before launching and quantified by the packaging efficiency, deformation caused during folding process, and the deployment ratio. The choice of folding pattern is an essential consideration in the design of the inflatable booms for space structures. Different type of folding patterns has been proposed to fold a cylindrical structure, such as spiral folding, Z-folding, telescopic folding, and origami folding methods [2]. The main origami folding methods include Yoshimura pattern, bellow pattern, Miura-ori folding pattern, Kresling pattern, and helically triangulated folding concept. Yoshimura [3] has analyzed the buckling behavior of a circular cylindrical shell under compression; the buckled surface settles into a localized form of a diamond-like pattern, named as Yoshimura pattern. Another folding concept similar to Yoshimura fold pattern was obtained from the torsional buckling of a thin cylindrical shell [4]. While solving the problem of biaxial folding of an infinite plane, Miura [5] discovered a concave polyhedral surface, which was composed of a repetition of four congruent parallelograms; this pattern was named Miura-ori folding pattern. Sogame and Furuya [6] presented a concept for cylindrical deployable space structures and examined their geometrical characteristics. Nojima [7] studied the folding methods of thin flat sheets and cylindrical shells by modifying the traditional Miura-ori pattern. Senda et al. [8] investigated several prototypes of inflatable cylindrical tubes packed with Miura-ori and Yoshimura pattern, and they found that in terms of local deformations and deployment straightness, Miura-ori folding with appropriate parameters is better than Yoshimura folding technique.

Guest and Pellegrio [9] studied the geometric formulation and folding properties of helically triangulated pattern for cylinders. Later, they have investigated the folding progression of triangulated cylinders with a well-known pin-jointed truss modeling concept [10] and examined the folding behavior experimentally [11]. Barker and Guest [12] analyzed the inflation and rigidization of thin shell triangulated cylinders. A unique type of helically triangulated pattern is called Kresling pattern [13], which is the buckling effect of a thin cylindrical shell under torsion. Cai et al. [14] investigated the geometric design and mechanical behavior and also addressed the bistable phenomenon cylinder with Kresling pattern. By splitting the degree-6 vertices on Yoshimura pattern, a classical bellow fold pattern can be obtained. Kane [15] presented mathematical optimization of family bellow fold patterns.

Among all the boom folding patterns presented in the literature, only a few can be folded and deployed without significant material strains. The design parameters of any folding pattern mainly affect the stowing and the deployment behavior of the boom. The boom can inflate unpredictably due to the air entrapped in the small empty spaces between the layers of the folded boom structure.

This unpredictable expansion of the residual gas led to several failures. The amount of residual gases can be reduced by analyzing the folded geometry of the boom. In the literature, no study presents the geometric analyses of the bellow folding pattern in order to obtain the empty space between the layers of the folded boom. The effect of different design parameters on the stowing behavior of the bellow folded boom has not been inspected in the literature.

In this study, the geometric design of the bellow folding pattern is discussed and various folding characteristics like packaging efficiency, the radius of packaged cylinder, and residual space at fully folded configuration are calculated. The influence of number of origami units and length ratio on the stowing behavior of the bellow folded boom is studied.

This paper is structured as follows. In Sect. 2, detailed geometry of the bellow folding pattern is presented, and mathematical expressions for various packaging parameters are obtained. In Sect. 3, the influence of design parameters on the packaging behavior of the bellow pattern is presented. Finally, Sect. 4 concludes the paper.

2 Geometry of Bellow Folding Pattern

A classic bellow pattern is shown in Fig. 1; the valley and mountain folds are denoted by dashed and solid lines, respectively. In a classical bellow pattern, a number of single reversal folds are organized along the periphery of the cylindrical boom, with the subsequent ring recurring along the length of the boom. In a strain-free folded configuration of the cylindrical boom, the consecutive reversal folds must develop a closed cross-section.

Figure 2 depicts a generalized form of a half-story segment unit cell of the bellow pattern that can be tessellated into a larger folding pattern (Fig. 3). The cylindrical closure condition for the resulting folded boom is given by

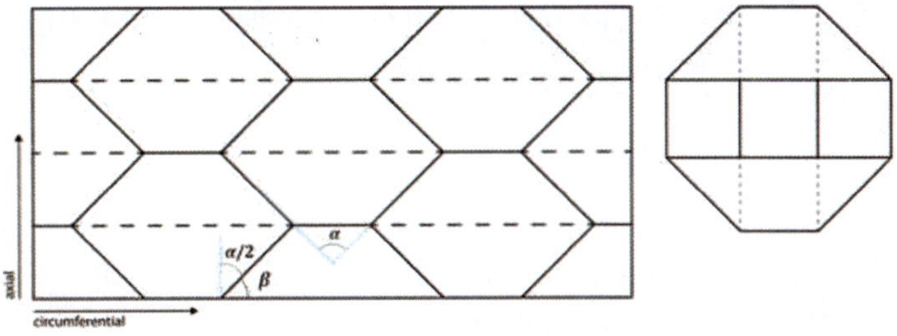

Fig. 1 Fold pattern and cross-section of a tetragonal bellow pattern

Fig. 2 Dimensions of a bellow unit

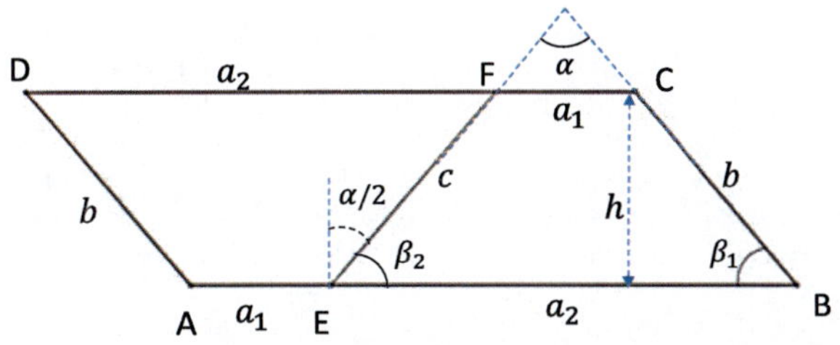

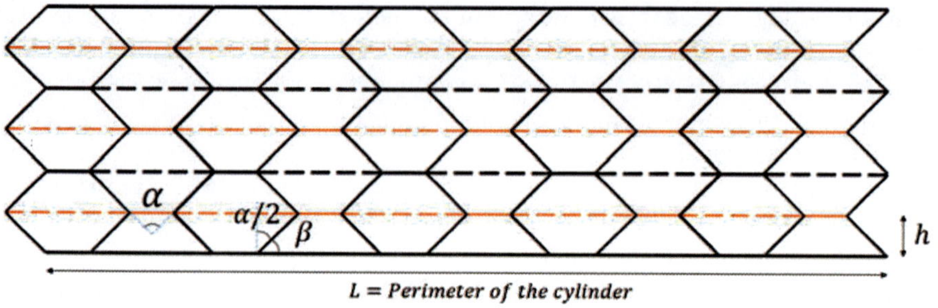

Fig. 3 Bellow folding pattern ($m = 4$)

$$m\alpha = (m-1)\pi \text{ or } m(\beta_1 + \beta_2) = \pi \tag{1}$$

where, $m =$ the number of bellow unit cells along the circumference.

α and $\beta =$ Angles between horizontal and diagonal fold lines.

In this paper, to achieve rotational symmetricity of folding polygons, the length of diagonal folds AD and EF is taken equal (i.e., b = c and $\beta_1 = \beta_2$.). The length of the total fold pattern L is equivalent to the circumference of the boom section as shown in Fig. 3, and each story of bellow pattern has m basic elements; then, the length of the single bellow unit can be written as

$$a = a_1 + a_2 = \frac{L}{m} \tag{2}$$

The ratio of the lengths of horizontal fold lines can be defined as length ratio (R) as follows

$$R = \frac{a_2}{a_1} \tag{3}$$

The value of length ratio will be greater than one for bellow pattern. The length of single unit (a), length ratio (R), and the angles α/β are taken as design parameters while designing a bellow folding pattern.

Using geometrical constraints of the folding pattern, all other dimensions can be defined in terms of design parameters as follows

$$a_1 = \frac{a}{(1+R)} \tag{4}$$

$$b = \frac{a}{2}\frac{(R-1)}{(R+1)}\sec\beta \tag{5}$$

$$h = \frac{a}{2}\frac{(R-1)}{(R+1)}\tan\beta \tag{6}$$

To investigate the stowing behavior of the bellow folding pattern having different configurations, several characteristics like the radius of packaged cylinder, stowing efficiency, and the residual volume inside the cylinder are defined. The expressions for these packaging parameters are established for the completely folded configuration of the cylindrical boom. Ideal folding condition (effect of thickness at the folds is neglected) is considered while obtaining the packaged height of the boom.

2.1 Radius of Packaged Cylinder

Figure 4 depicts the folded configuration of a cylinder based on bellow pattern ($m = 4$, $R = 3$); the radius of the circumscribed circle of the polygon, formed by the fold lines in fully stowed configuration, is called the radius of packaged cylinder (R_c). It is an essential parameter for evaluating the stowing behavior of the boom, and the expression for radius of bellow pattern folded cylinder can be obtained as follows

$$R_c = \frac{a_1}{2}\text{cosec}\frac{\theta_2}{2} = \frac{b}{2}\text{cosec}\frac{\theta_1}{2} \tag{7}$$

From the geometry

$$\frac{\sin\left(\frac{\theta_2}{2}\right)}{\sin\left(\frac{\theta_1}{2}\right)} = \frac{a_1}{b} \tag{8}$$

Fig. 4 Bellow polygon

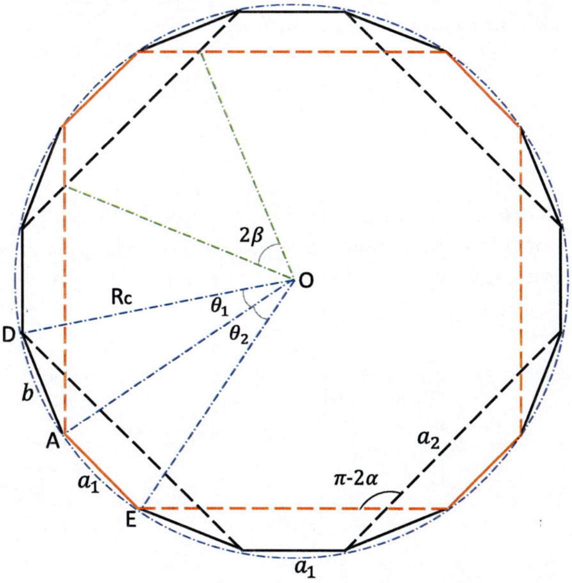

where

$$\theta_1 = \frac{\pi}{2m} + \theta; \theta_2 = \frac{\pi}{2m} - \theta \quad (9)$$

From Eqs. (9) and (5), (8) can be rewritten as

$$\frac{\sin\left(\frac{\pi}{4m} - \frac{\theta}{2}\right)}{\sin\left(\frac{\pi}{4m} + \frac{\theta}{2}\right)} = \frac{2\cos\beta}{R - 1} \quad (10)$$

Equation (10) is solved for θ using MATLAB for each combination of design parameters. Thus, θ_1 and θ_2 are obtained to calculate the radius R_c.

2.2 *Packaging Efficiency*

The stowage volume of the deployable structure is the main criterion while selecting the folding technique; it can be calculated from the geometry of the completely folded condition. The ability of packaging for any folding pattern can be described in terms of packaging/stowing efficiency. It can be written as follows

$$\text{PE} = \frac{\text{Vol}_{\text{Unfolded}} - \text{Vol}_{\text{folded}}}{\text{Vol}_{\text{Unfolded}}} \times 100 \quad (11)$$

Volume of folded configuration:

$$\text{Vol}_{\text{folded}} = \text{Area of outer polygon(A)} \times \text{Folded height}(\text{H}_\text{f}) \quad (12)$$

Area of Polygon (A)

In the bellow pattern, the fold polygon formed by outermost vertexes is an irregular polygon, which consists of two type of triangles as shown in Fig. 5. The complete fold polygon has n_1 number of such triangles ($n_1/2$ each) where n_1: number of sides of outer polygon = 4 m.

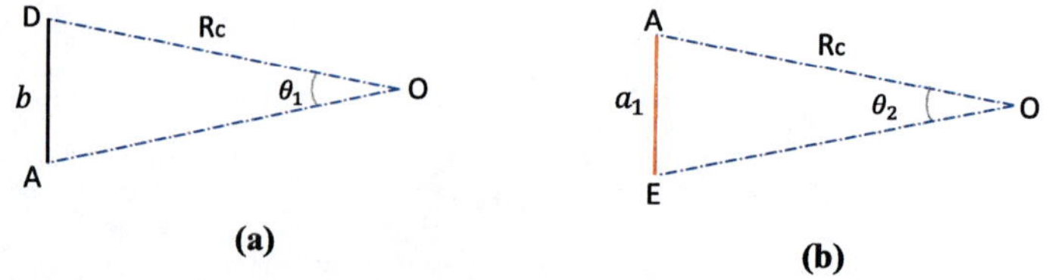

Fig. 5 Outer polygon triangles

The area of the irregular fold polygon can be expressed as

$$A = \frac{m}{2} b^2 \cot \frac{\theta_1}{2} + \frac{m}{2} a_1^2 \cot \frac{\theta_2}{2} \quad (13)$$

The value of θ_1 and θ_2 can be obtained from Eqs. (8)–(10).

Folded Height (H_f): The folded height of the cylinder cab be calculated by considering the total number of overlapping layers in stowed condition. It can be written as follows

$$H_f = 2 \times t \times S \quad (14)$$

where S = height of the deployed boom/single story height (2 h).

t = thickness of the sheet.

Volume of folded configuration:

$$\text{Vol}_{\text{folded}} = \pi \times \left(\frac{L}{2\pi}\right)^2 \quad (15)$$

Now, by using Eqs. (11)–(15), the packaging efficiency can be calculated for a particular folding pattern.

2.3 Residual Volume Calculations

The whole residual volume of completely folded boom includes inside residual volume $R_{\text{vol(inside)}}$ and residual volume between the layers $R_{\text{vol(b/w the layers)}}$.

Inside residual volume $R_{\text{vol(inside)}}$: The residual volume inside the folded boom can be obtained by multiplying the area of the polygon formed by innermost vertices and the folded height of the cylinder.

Inside residual area (A'): This inner polygon is a regular polygon, made-up of $n2$ same triangles (Δ POQ) as shown in Fig. 6, where $n2$: number of sides of inner polygon = 2 m. The inner polygon area can be expressed as follows

$$A' = \frac{m}{2} \text{PQ}^2 \cot \beta \quad (16)$$

where,

$$\text{PQ} = \left(a_1 + b \frac{\cos 2\beta}{\cos \beta}\right) \quad (17)$$

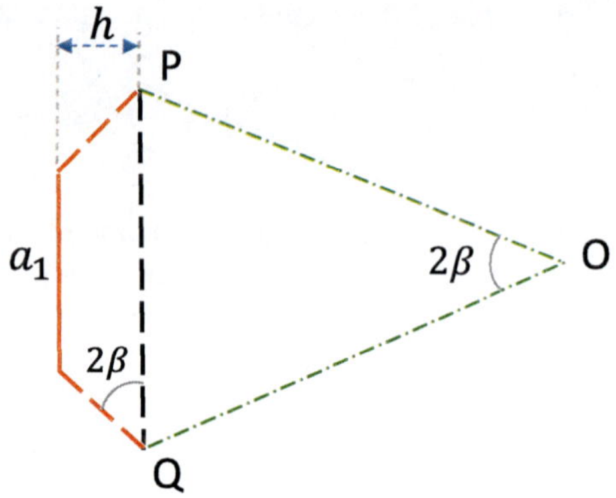

Fig. 6 Inside polygon triangle

$$R_{\text{vol(inside)}} = \text{Inside Residual area}(\text{A}') \times \text{Folded height}(\text{H}_\text{f}) \quad (18)$$

Residual volume in between layers

Residual area between layers: The folded configuration has trapezoidal (PQRS)-shaped free spaces (see Fig. 6) between the layers with height of two times of material thickness.

$$A'' = \text{PQ} \times b \sin \beta \quad (19)$$

$$R_{\text{vol(b/w the layers)}} = 2mtA''(2S - 1) \quad (20)$$

Total Residual Volume:

$$R_{\text{vol}} = R_{\text{vol(inside)}} + R_{\text{vol(b/w the layers)}} \quad (21)$$

3 Packaging Behavior

The knowledge of packaging behavior of folding patterns is essential while developing a packaging methodology. It includes the investigation of influence of design variables like number of origami units (m) and length ratio (R) on the packaging parameters like the radius of packaged cylinder (R_c), packaging efficiency (PE), and residual space inside the folded boom. The length of the boom and circumference are taken as 800 mm and 600 mm, respectively, throughout this paper.

3.1 Effect of the Number of Bellow Units

To examine the effect of the number of origami units on the stowing behavior of the boom, several bellow fold patterns were considered with different values of *m* while the length ratio (*R*) is kept constant, and all other dimensions were calculated using Eqs. (1)–(6). The variation of radius of stowed boom $R_{c,}$ and packaging efficiency with respect to *m* is shown in Figs. 7 and 8, respectively. Figure 9 displays the variational trends of residual volume curves with respect to *m*; it is clear that total and inside residual volume curves are almost equivalent; i.e., residual volume between the layers is significantly less compared to the inside residual volume. The change in the residual volume between folded layers with m is displayed in Fig. 10.

It is seen that as the number of origami units increases, initially R_c decreases rapidly, and at higher values of m, this decrement becomes minor. The packaging efficiency of the cylinder decreases significantly with the increment in m. Initially, the value of $R_{\text{vol(b/w the layers)}}$ increases rapidly, but at higher values of m, this increment become very less, whereas $R_{\text{vol(inside)}}$ increases continuously with increment in m.

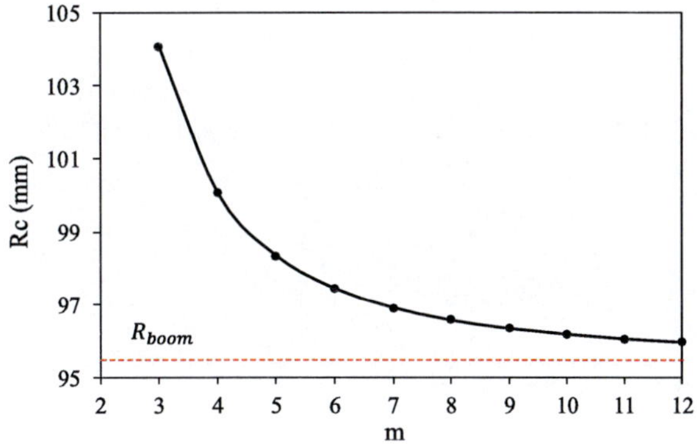

Fig. 7 Influence of *m* on R_c

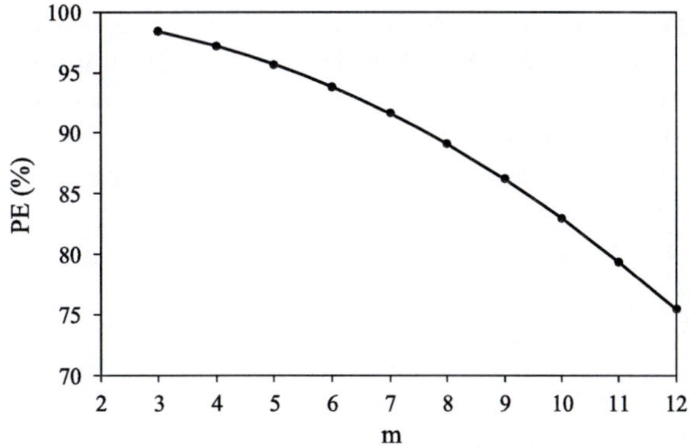

Fig. 8 Influence of *m* on PE

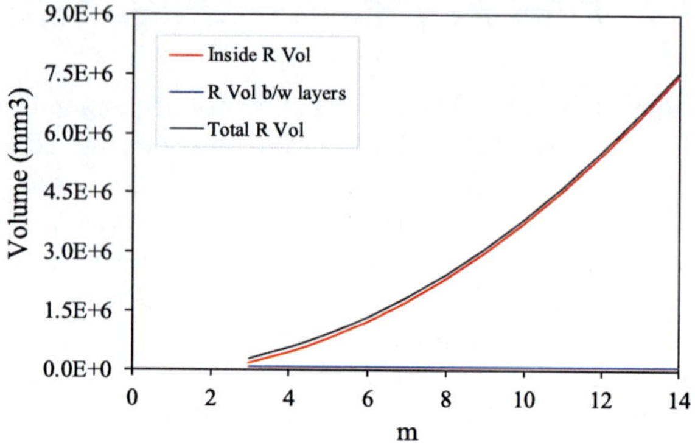

Fig. 9 Influence of m on Residual Vol

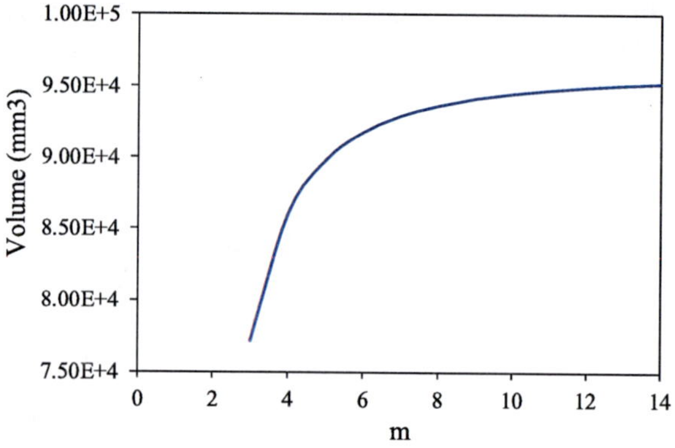

Fig. 10 Influence of m on R Vol b/w layers

3.2 *Effect of Length Ratio* (R)

A number of bellow fold patterns were designed having different length ratio (R) while keeping the number of origami units (m) constant, and all other dimensions change according to Eqs. (1)–(6). The change in the radius of stowed boom R_c and packaging efficiency with length ratio (R) for different values of m is presented in Figs. 11 and 12, respectively. Figure 13 shows the variation of the total residual volume with length ratio (R). It is seen that as the length ratio increases, R_c and PE increase but at higher values of R, there is not much increment. The residual volume decreases as the length ratio increases.

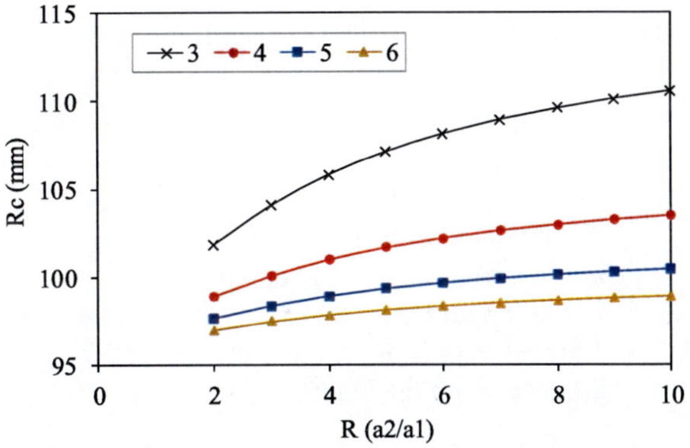

Fig. 11 Influence of length ratio (R) on R_c

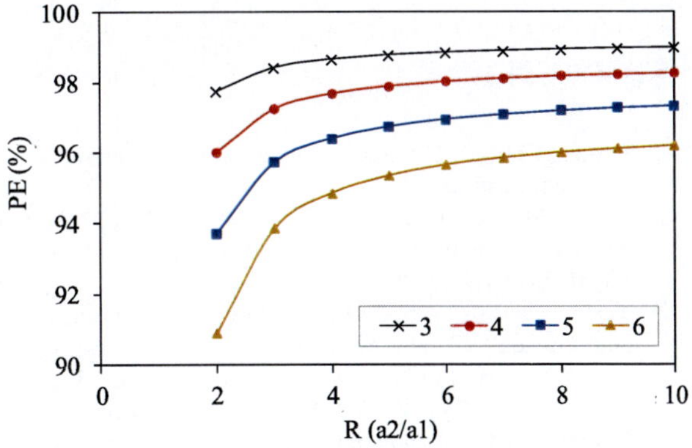

Fig. 12 Influence of length ratio (R) on PE

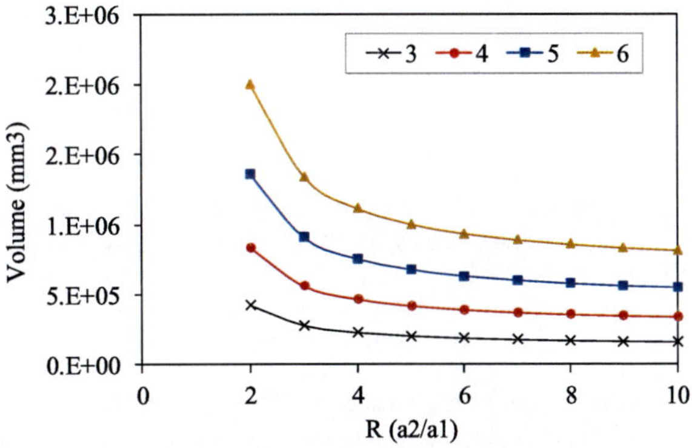

Fig. 13 Influence of length ratio (R) on total R Vol

4 Conclusion

The geometry of the packaged boom is analyzed to investigate the packaging behavior of the cylindrical boom. Furthermore, the influence of umber of origami units (m)

and length ratio (R) on the packaging behavior is investigated. The results indicate that:

1. As the m value increases, the radius of packaged cylinder reduces quickly at the starting; later on, the reduction becomes less significant. The packaging efficiency decreases with the increase in m value. Inside residual volume increases continuously with m, whereas increment of residual volume between layers becomes very less at higher values of m.
2. As the length ratio increases, R_c increases rapidly for lesser values of m, and at higher values of m, the increment of R_c is very less. Total residual volume decreases with the length ratio.

In the future, the deformation behavior of the bellow pattern will be studied.

References

1. Jenkins CH (2001) Gossamer spacecraft: membrane and inflatable structures technology for space applications
2. Schenk M, Viquerat AD, Seffen KA, Guest SD (2014) Review of inflatable booms for deployable space structures: packing and rigidization. J Spacecr Rocket 51(3):762–778. https://doi.org/10.2514/1.A32598
3. Yoshimura BY (1995) On the mechanism of buckling of a circular cylindrical shell under axial compression. National Advisory Committee for Aeronautics
4. Hunt GW, Ario I (2005) Twist buckling and the foldable cylinder: an exercise in origami. Int J Non-Linear Mech 40(6):833–843. https://doi.org/10.1016/j.ijnonlinmec.2004.08.011
5. Miura K (1985) Method of packaging and deployment of large membranes in space. Inst Space Astronaut Sci R 618:1–9. http://ci.nii.ac.jp/naid/110000029131/
6. Sogame A, Furuya H (2000) Conceptual study on cylindrical deployable space structures
7. Nojima T (1999) Modelling of folding patterns in flat membranes and cylinders by using origami. JSME Int J Ser C Mech Syst Mach Elem Manuf 1, 45:364–370
8. Senda K, Oda T, Ohta S, Igarashi Y, Watanabe A, Hori T, Ito H, Tsunoda H, Watanabe K (2006) Deploy experiment of inflatable tube using work hardening. No(May):3–8
9. Guest SD, Pellegrino S (1994) The folding of triangulated cylinders, part i: geometric considerations. J Appl Mech 61(4):773. https://doi.org/10.1115/1.2901553
10. Guest SD, Pellegrino S (1996) The folding of triangulated cylinders, part III: experiments. J Appl Mech 66(Dec 1994):778–783. https://doi.org/10.1115/1.2901554
11. Guest SD, Pellegrino S (1996) The folding of triangulated cylinders, part III: experiments. Trans-Am Soc Mech Eng J Appl Mech 63:77–83
12. Barker R, Guest S (2000) Inflatable triangulated cylinders. … Symposium on deployable structures: theory and …, 2000, 17–26
13. Abel JF, Cooke JR, Kresling B (2008) Natural twist buckling in shells : from the Hawkmoth ' s bellows to the deployable kresling-pattern and cylindrical miura-ori (May, 2008):1–4
14. Jianguo C, Xiaowei D, Ya Z, Jian F, Yongming T (2015) Bistable behavior of the cylindrical origami structure with kresling pattern. J Mech Des 137(6):061406. https://doi.org/10.1115/1.4030158
15. Kane NR (1995) Mathematically optimized family of ultra-low distortion bellow fold patterns

Evaluation of Dynamic System Characteristics of Payload Cover Mechanism for Spacecraft Applications

V. Sri Pavan Ravichand, Divesh Soni, Sanjay Gorur, Ghulam Sarwar, Shamrao, B. Lakshmi Narayana, S. Narendra, B. P. Nagaraj, and H. N. Suresha Kumar

Abstract Mechanisms for space applications are unique from those of ground applications from the perspective of their robust dynamic characteristics that are necessary to sustain launch vibration loads along with their kinematic characteristics. One such space-based mechanism is payload cover mechanism for spacecraft applications. In these missions, there are various scientific payloads that need to be protected during the ground, launch and initial on-orbit phase operations and positioned to functional configuration in orbit for scientific observations. These requirements are met by way of the payload cover mechanism with requisite dynamic and kinematic characteristics. Given the fact that each payload demands a unique payload cover mechanism, which has intricate geometry owing to the necessity of high stiffness and low mass constraints, it is very essential to not only estimate the frequency characteristics of mechanism using finite element analysis but also correlate the mathematical model of mechanism based on vibration test results. This helps in accurate prediction of system behavior at integrated spacecraft level as well as launch vehicle level. This paper presents the analysis results along with the correlation methodology adopted in case of a typical payload cover mechanism. The insights gained in the process of correlation have been highlighted. This includes the influence of contact area in determining the frequency characteristics as well as the load distribution in the system.

Keywords Deployment mechanism · Finite element analysis · Correlation · Payload cover mechanism

---

V. Sri Pavan Ravichand (✉) · D. Soni · S. Gorur · G. Sarwar · Shamrao · B. Lakshmi Narayana · S. Narendra · B. P. Nagaraj · H. N. Suresha Kumar
Spacecraft Mechanisms Group, U. R. Rao Satellite Centre, ISRO, Bengaluru 560017, India
e-mail: pavan.vsprc@gmail.com

V. K. Gupta et al. (eds.), *Recent Advances in Machines and Mechanisms*, Lecture Notes in Mechanical Engineering, https://doi.org/10.1007/978-981-19-3716-3_34

1 Introduction

Several deployable mechanisms are used to protect optical payloads from disturbances during launch and also during on-orbit maneuvers. These are deployed at appropriate time in-orbit for functional use. These mechanisms use revolute joints for deployment. Upon release, these mechanisms latch and change to structure. In addition, the mechanism needs to act as structure with requisite stiffness during launch phase so as to withstand dynamic loads. This is enabled using a hold-down unit that keeps both parts of the payload cover intact. Mechanisms used for space applications are designed with regard to stiffness owing to the necessity of catering to launch vibration loads. The configuration of a space-based mechanism is driven by analytical approach involving various iterative steps [1] that simulate launch to on-orbit phases. These steps including stiffness estimation based on normal mode analysis, need for test correlated mathematical model for spacecraft-level analysis and on-orbit body rate comparison with analytical predictions have been presented in reference paper [1]. An important step in this approach is correlation of mathematical model with experimental vibration test results. This paper presents the details on this step in case of payload cover mechanism. The hold-down state induces the elastic interaction effect [2] owing to the contact between the surfaces of two parts of the payload cover. This contact stiffness cannot be easily calculated from the geometry and material properties of the system. There have been many studies on contact stiffness. Some theoretical models in this direction have been proposed like the Hertz theory [3], the influence of surface topography on contact stiffness [4] and surface normal contact stiffness model [5]. In the present paper, contact stiffness is simulated using Rigid Body Element (RBE2) and analysis has been carried out using finite element analysis (FEA) by multiplying the unit contact stiffness with variable contact area. This contact stiffness has been correlated using experimental results from base excited vibration test of the mechanism hardware. This study provided insights on correlation methodology as well as on the influence of contact stiffness in the payload cover mechanisms used aboard satellites.

2 Analysis of Mechanism Using FEA

Finite element (FE) model of a typical payload cover mechanism used in space applications is shown in Fig. 1a.

FEA is carried out using commercial software packages MSC PATRAN and MSC NASTRAN. The FE model of the mechanism has been generated using TET10 elements so as to capture the stress distribution accurately of the intricate geometry. The four main constituents of the mechanism from functional perspective are hinge, hold down, top cover and bottom cover with baffle. The baffle hosting the payload is fixed on to the satellite structure.

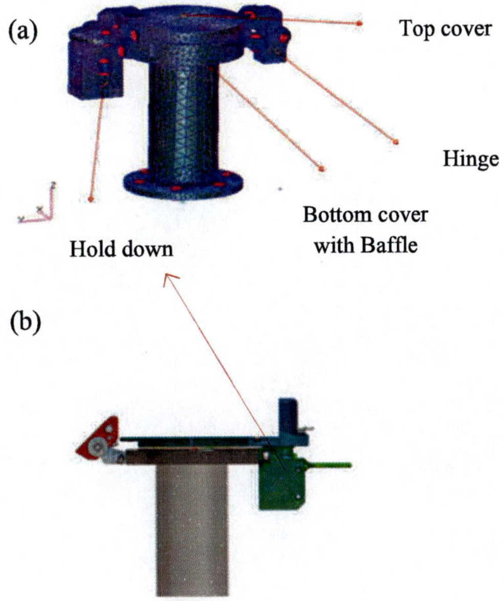

Fig. 1 **a** Finite element model of the payload cover mechanism. **b** Geometry model of the payload cover mechanism

The boundary constraint simulated in FEA is fixed boundary condition, wherein the interface of the baffle with satellite is constrained in all six directions. The hinge together with the hold down enables the system to act as structure during launch phase. The pre-compression due to the hold-down force has been modeled using RBE2 element. The pre-compression of hold down influences the contact area between the two covers in dynamic condition and thus influences the frequency characteristics. Because of this connectivity, the surfaces of the top and bottom covers assume a certain amount of contact. This contact adds stiffness to the system. In the initial phase of analysis, the contact stiffness is not considered as it helps in estimating the frequency results in a conservative manner. This approach enables to plan and conduct the vibration test in a cautious manner with regard to the sweep rate and notch sequencing.

Accordingly, following analyses, with fixed boundary condition, are carried out before the vibration test of hardware:

a. Normal mode analysis
The 1st mode is found to be 272 Hz with the mode shape as shown in Fig. 2. In the mode shape, it can be observed from the color contours (wherein red signifies maximum deformation, followed by yellow, green and blue) that the area near hinge is less stiffer compared to that near hold down, thereby pointing toward sensitivity analysis of hinge. This mode shape (Fig. 2) has been observed from the experimental data as well in the form of higher amplification at hinge location mounted accelerometer in comparison with that of hold-down location mounted accelerometer.

Fig. 2 First mode shape of payload cover mechanism

b. Frequency response for 1 m/s^2 of translational acceleration in all directions
The first resonance frequency is found to be around 270.5 Hz, with a structural damping coefficient of 4%.

Sensitivity analysis of hinge has been carried out using following two cases.

a. By disabling the hinge connectivity completely, wherein the system is held only by hold down. This gave a frequency of 103 Hz.
b. By making the hinge infinitely rigid so as to understand the upper limit of hinge contribution to stiffness. This gave a frequency of 440 Hz.

With this understanding of the system characteristics, the payload cover mechanism has been tested using base excited vibration. The test setup has been enabled with four accelerometers. The first mode in vibration test has been observed to be 542 Hz at the four locations, indicating a global mode. Subsequently, studies have carried out in order to correlate the FE model based on vibration hardware test data.

3 Correlation of FE Model Using Test Data

The mechanism can be considered into three sub-systems from the point of view of frequency characteristics. This categorization has been made specifically to understand the influence of various constituents of mechanism in understanding the stiffness characteristics of the system.

The three sub-systems are as follows:

a. Driving unit consisting of hinge, baffle, top cover and bottom cover
b. Hold-down unit
c. Contact area between the two covers after assembly of mechanism.

It can be observed from the configuration of the mechanism that the three sub-systems mentioned above are in parallel and add to the system-level stiffness. As mentioned earlier, the initial analytical estimations are carried out using only the first and second sub-systems without including the third sub-system, i.e., contact area between the two covers. Studies have been carried out using normal mode analysis to correlate the mathematical model of the payload cover mechanism with the vibration test data by varying the contact stiffness. This is done using unit contact stiffness [2], wherein the per unit area contact stiffness is assumed constant and the contact area is varied using RBE2 connectivity between interface surfaces of the two covers.

With regard to the contact stiffness, the mechanism primarily consists of connectivity at three locations, as shown in Fig. 3.

a. Location I—Between top and bottom covers
b. Location II—Between bottom cover and baffle
c. Location III—Between baffle and vibration table.

The studies have been carried out to understand the stiffness characteristics at these locations in a sequential manner so as to identify the influential location for correlation. Subsequently, the contact area at this connectivity is varied to correlate the frequency characteristics of FEA with that of vibration test data. These studies along with the first natural frequency results have been shown in Table 1.

It has been observed that the contact stiffness at interface of the two covers is influential in determining the 1st natural frequency of system.

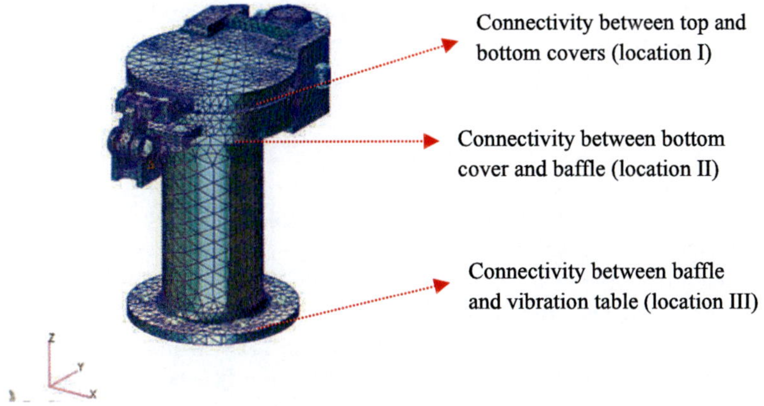

Fig. 3 Different connectivity locations for studies on contact stiffness

Table 1 Case studies

Case	Analysis studies carried out to understand influence of connectivity (location I/II/III of Fig. 3) w. r. t Case 1	1st Natural frequency (Hz)
1	Initial analysis with hinge stiffness and hold-down stiffness, without connectivity at any of the contacts in location I/II/III	272
2	Surface connectivity between cover and baffle, of full contact area (location II)	291
3	Surface connectivity between cover and baffle (location II) as well as baffle and vibration table (location III), of full contact areas	296
4	Surface connectivity between top and bottom covers (location I), of full contact area	733
5	Surface to surface connectivity between top and bottom covers (location I), of about 65% contact area	684
6	Surface to surface connectivity between top and bottom covers (location I), of about 46% contact area	536

a. From cases 2 and 3, it can be said that the influence of connectivity at cover-baffle interface and at baffle-vibration fixture is negligible.
b. From case 4, the above point can be further confirmed as the condition of full surface connectivity between top and bottom covers is resulting in a frequency of 733 Hz.
c. From sensitivity analysis of hinge presented in Sect. 2, it can be understood that there is significant influence of connectivity between the top and bottom covers that is causing the difference between the analysis results and the vibration data (542 Hz). This can be interpreted from the variation between 103 and 440 Hz, which point to the additional stiffness contribution apart from hold down.
d. In case of 5 and 6, the connectivity at the interface between the two covers has been studied by varying the contact area iteratively to understand the stiffness characteristics of the contact between the covers. It is found that for a contact area of about 46%, the FEA results are closer to the observed frequency data in vibration test.

Based on above cases, correlated FE model has been generated, and subsequent analysis [1] has been carried out on this model to estimate stiffness and strength parameters. These studies highlight the inter-related role of design, analysis and testing in mechanism development cycle.

It is found from the analysis of the correlated FE model that the frequency in X (536 Hz) and Y (510 Hz) directions is nearly close to test of around $542 \pm 10\%$ Hz, while in Z direction, the corresponding numbers are 1123 Hz in analysis and 906 Hz in test. It has been found that the contact stiffness at interface between the two covers is the influencing parameter. Thus, the methodology of correlation based on contact stiffness is found to yield good results in case of payload cover mechanism for space applications. This test correlated model, once integrated with mathematical model of full payload, helps in estimating accurately the dynamic characteristics of payload at sub-system level. Subsequently, the integrated model shall be used at spacecraft-level analysis activities.

4 Conclusion

This paper brings out a methodology for correlation of FE model with vibration test data based on influence of contact stiffness. The approach has been adopted in a logical manner by way of elimination of options as can be seen from the studies catering to the three connectivity locations in the payload cover mechanism. These studies not only catered to the contact stiffness perspective but also strengthened the approach by way of understanding the influence of two other major sub-units (hinge and hold down) on the stiffness characteristics.

Though the mechanism considered in the present paper is a typical payload cover mechanism used in satellites, the methodology and insights gained are equally applicable to any general preloaded mechanism, thereby paving way for further study in this direction. The paper models a fairly complex mechanism in its rigidized form, correlates the experimentally obtained first natural frequency with the frequency obtained from FE analysis and concludes that the extent of contact between the two covers is a major factor in the dynamic response. Different analysis estimations like normal mode frequency, stress and harmonic response have been presented for both the initial FE model and the correlated model. Finally, the paper highlights the inter-related role played by design, analysis and testing in successful development and realization of mechanisms. This is not only beneficial but also inevitable in case of space craft mechanisms, which need to cater to diverse requirements in terms of functionality as well as environment.

Acknowledgements The authors would like to thank Shri Alok Kumar Shrivastava, Deputy Director, MSA and Shri. M. Sankaran, Director, URSC for their constant support and encouragement.

References

1. Sri Pavan Ravi Chand V, Srivastava A, Kumar A, Kumar HNS, Keshavamurthy KA (2021) Analytical approach to develop a robust mechanism for on-orbit gimballing of satellite antenna. Spr J Adv Mech Eng June 2021
2. Kono D, Inagaki T, Matsubara A, Yamaji I (2013) Stiffness model of machine tool supports using contact stiffness. Elsevier J Precis Eng Jul 2013
3. Mindlin RD (1949) Compliances of elastic bodies in contact. J Appl Mech
4. Thornley RH, Connolly R, Barash MM, Koeningsberger F (1965) The effect of surface topography upon the static stiffness of machine tool joints. Int J Mach Tool Des Res
5. Yang H (2014) Relationship between normal and tangential contact stiffness of nominally flat surfaces. Int J Control Autom 7(6)

Rigidization Mechanism for Double-Layered Inflatable Circular Torus Structure

Vikas Rastogi, Sanjay H. Upadhyay, Sammir Sakhare, and Kripa S. Singh

Abstract Inflatable structures are widely using in space applications because of their unique properties such as high packaging efficiency, light in weight, highly flexible to fold in any configuration, and many more. Inflatable antennas require structural rigidity after deployment. This structural rigidity is provided by the support elements torus and booms. The support elements must rigidize after complete deployment of the inflatable antenna. This article is investigating the rigidity criterion for the support element torus. Here, the strain hardening method is used to rigidize the support element, i.e., torus. The analysis is made for the hollow, double-layered laminated circular torus. The layer-wise (LW) theory is implemented to calculate the stresses in the torus material. The rigidization pressure is found analytically and compared with the numerical analysis. Further, a parametric study is performed with different combinations of the laminate membrane thicknesses to find the optimum wall thickness of the support structure.

Keywords Gossamer structures · Strain hardening · Torus · Laminated membrane

1 Introduction

An extremely large size space antenna is desirable to increase the efficiency and working range. Many researchers have tried to explore the different methods to increase the size of the space antenna. Inflatable structures are the most attractive method to increase the size of space antennas. Inflatable structures are light in weight, required minimal energy to deploy, have high packaging efficiency, and have low cost

V. Rastogi (✉) · S. H. Upadhyay
Smart Materials and Structures Laboratory, MIED, IIT Roorkee, Roorkee, Uttarakhand 247667, India
e-mail: rvikas@me.iitr.ac.in

S. H. Upadhyay
e-mail: shumefme@iitr.ac.in

S. Sakhare · K. S. Singh
Space Application Centre-Indian Space Research Organization (SAC-ISRO), Ahmedabad, India

V. K. Gupta et al. (eds.), *Recent Advances in Machines and Mechanisms*, Lecture Notes in Mechanical Engineering, https://doi.org/10.1007/978-981-19-3716-3_35

in comparison with the conventional antennas [1]. Inflatable structures have many advantages but also come with some critical challenges. After complete deployment with the help of inflation gas, micrometeoroids may impact the surface of the antenna which leads to create micro holes. These micro holes will cause leakage of the inflation gas from the antenna, and the antenna will lose its erection strength that leads to deformation of the antenna. Rigidization of the antenna can allow the leakage of inflation gas without affecting its gained shape [2]. The researchers have studied many rigidization techniques for rigidize inflatable structures such as UV rigidization, foam rigidization, solvent boil-off rigidization, thermal rigidization, strain hardening rigidization, and many more [3]. Here, the strain hardening rigidization method is applied to rigidize the support structure torus. Researchers have studied the strain hardening method for boom and successfully rigidized the inflatable support element, boom [4]. Creases formed during folding are removed from the surface by applying the over-inflation pressure. This over-inflation pressure is slightly beyond the yield limit of the ductile material. Removal of the wrinkles from the surface of the boom enhances the structure health, i.e., structure's working life is extended.

The aim to present this work is to rigidize the torus structure by applying the strain hardening method. Here, optimum rigidization pressure is calculated analytically and compared with numerical analysis. Double layers of the torus structures are made of thin foil of aluminum and Kapton polyimide film. The thicknesses of the aluminum and Kapton membrane are ranging from 25 to 50 μm. The cross-sectional diameter of the torus is 100 mm, and the major diameter is 500 mm. The mechanical properties of the Al-Kapton laminated membrane are found through the uniaxial tensile test.

This article is arranged into five sections. The second section discusses the strain hardening method for torus structure. The third section explains the mathematical expression which is used to find the rigidization criterion of the structure. The fourth section is about the modeling and FEM analysis of the torus structure. In the last section, conclusion about the findings has been made.

2 Strain Hardening Method

A ductile thin film and a polyimide thin film are laminated to form the wall of the inflatable structure. The ductile material membrane acts as a strain storage source, while the polyimide film acts as a bladder to prevent the leakage of inflation gas. The inbuilt unique properties of both films are bound together to make them laminated membranes. The torus is inflated, which is made of ductile and polyimide films laminated membranes, slightly above the yield limit of the ductile membrane, and holds for a small time duration at the constant inflation pressure. Holding time serves as a transformation time for the ductile membrane to transform from elastic region to plastic region. After complete transformation, inflation gas supply is cut off. Elastic strain recovers while the plastic strain stores in the ductile membrane after cut off the inflation gas supply. The ductile membrane tries to expand the polyimide membrane, while the polyimide membrane resists this expansion and tries to recover its initial shape. Because of the different nature of both membranes, the ductile membrane experiences compressive stresses, while the polyimide membrane experiences tensile stresses, i.e., laminated membrane remains in the equilibrium state.

3 Rigidization Criterion

The membrane theory is also applicable for thin-walled toroidal structures when the shell experiences zero or negligible bending deformation [5]. Rigidization of the torus is based on von Mises yielding criterion. The rigidization pressure R_p has been found through

$$R_p = \sqrt{\frac{4}{3}} \cdot \frac{\sigma_y t}{r} \tag{1}$$

where σ_y is yield stress, t is the wall thickness, and r is the torus radius.

4 Rigidization of Torus

Computational analysis has been carried out for the rigidization of the torus structure to find the rigidization pressure with various wall thicknesses. Later, a comparison table has been created which shows the variation in results found through FEM analysis and mathematical analysis. A CAD model has been generated of the torus and analysis is carried out through a FEM tool. The torus wall is made of a laminated aluminum-Kapton polyamide film. In this analysis, total of six combinations have been analyzed of the wall thickness, i.e., Al-Kapton are 10.5–25, 10.5–50, 18–25, 18–50, 25–25, and 25–50. All thicknesses are in microns. A four-node quadrilateral mesh element has been generated for the surface distribution of the torus. The outer layer of torus is aluminum and inner layer is Kapton. The stress generation is obtained by using the layer-wise theory. Figure 1 is showing the Mises stress distribution on both layers. Both layers are in equilibrium and experiencing approximately same amount of stress. Figure 2 is showing cross-sectional view of laminated membrane.

5 Results and Discussion

Different graphs for different combinations are listed below. Each combination contains two plots. These plots are indicating rigidization criterion and rigidization pressure (Fig. 3).

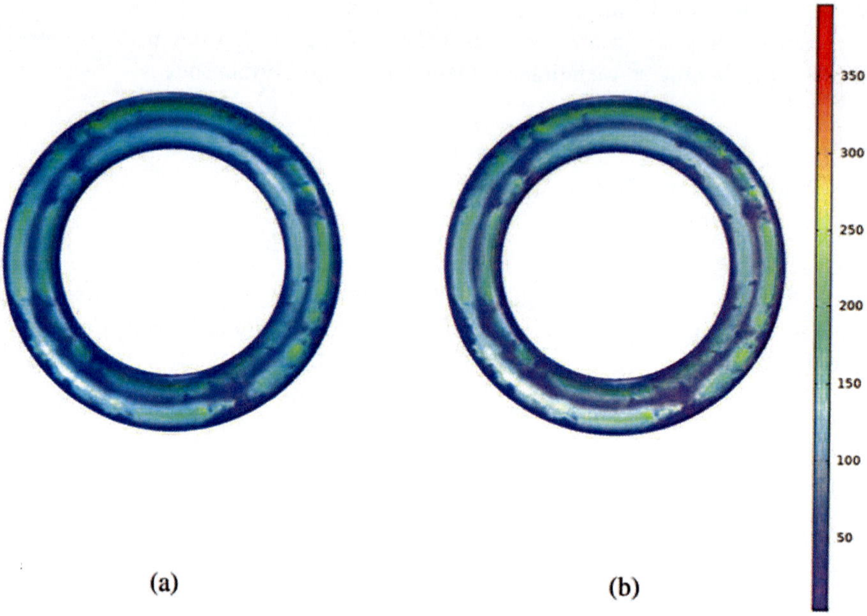

Fig. 1 Mises stress distribution on **a** outer layer, aluminum, **b** inner layer, Kapton

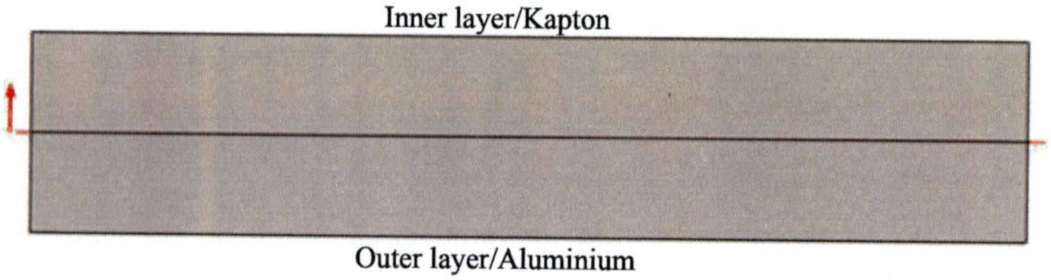

Fig. 2 Layers cross-sectional view

The hollow toroidal cavity has been filled through the inflation gas. As the volume of the cavity increases because of the filling of the inflation gas, elastic strain is stored in the structure. Storage of elastic strain in the structure continues until the torus material attains the yield limit. Inflation beyond the yield limit leads to plastic deformation of torus material. Holding inflation gas pressure, which created yield in the torus material, for some time so that material can take own time to flow in the yield direction. Elastic strain recovered as the inflation gas pressure is released but some amount of the plastic strain is remained stored in the torus wall. This permanently stored plastic strain is responsible for strain hardening of the torus wall material, i.e., rigidization of the torus is taking place. As shown in the aforementioned graphs, each combination of laminate is showing two graphs. The first graph is strain versus time and second is strain versus pressure. Strain vs time graph is showing the variation of

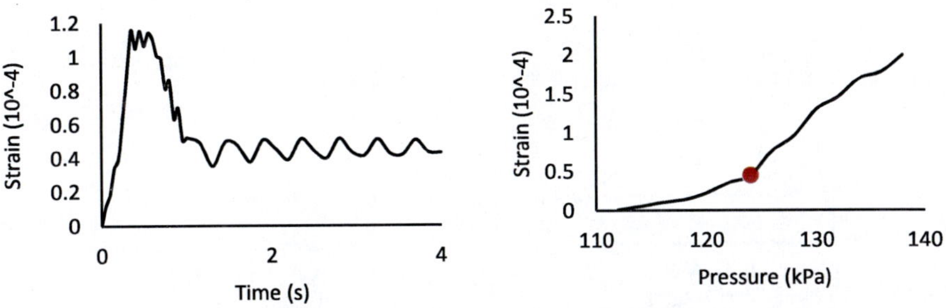

(a). T1. Thickness: - Aluminum – 10.5μm; Kapton – 25 μm

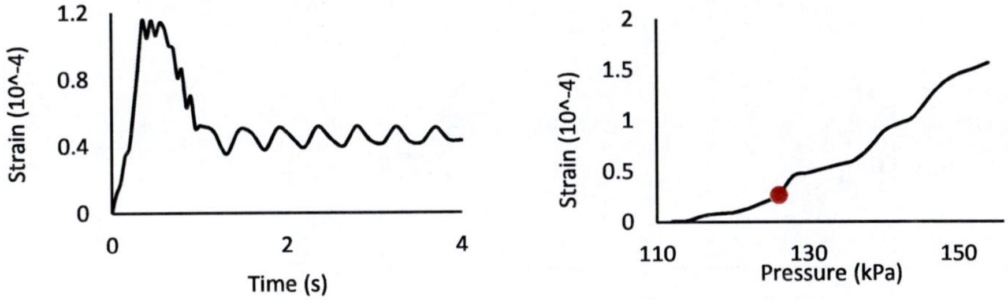

(b). T2. Thickness: - Aluminum – 10.5 μm; Kapton – 50 μm

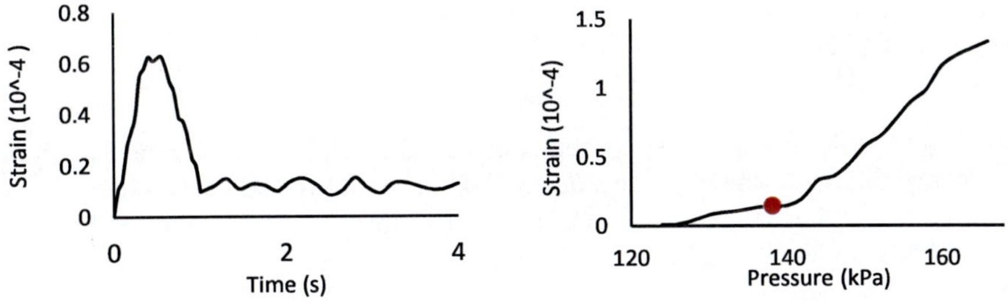

(c). T3. Thickness: - Aluminum – 18 μm; Kapton – 25 μm

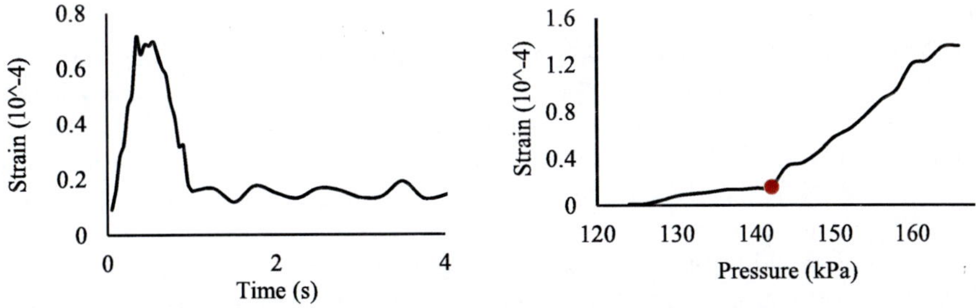

(d). T4. Thickness: - Aluminum – 18 μm; Kapton – 50 μm

Fig. 3 Plots for various combinations

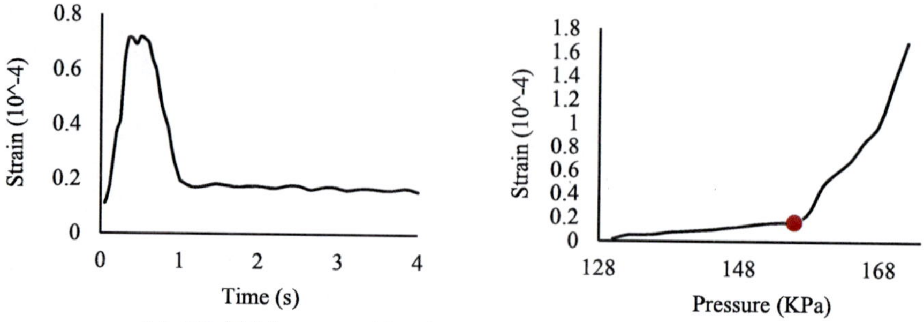

(e). T5. Thickness: - Aluminum – 25 μm; Kapton – 25 μm

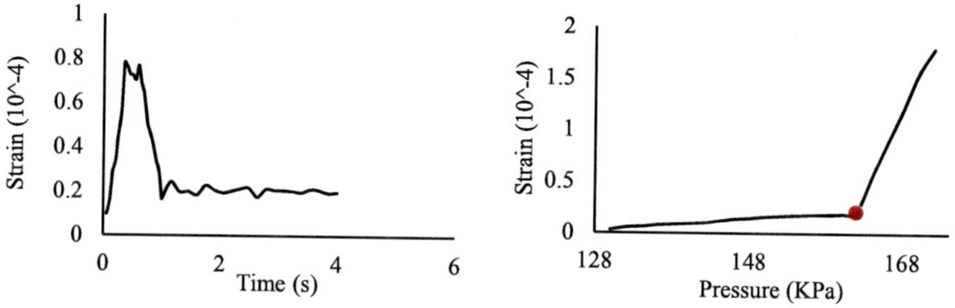

(f). T6. Thickness: - Aluminum – 25 μm; Kapton – 50 μm

Fig. 3 (continued)

strain w.r.t. time. All thickness combinations are T1, T2, T3, T4, T5, and T6. Strain vs time graph of T1 showing the storage and releases of the strain according to the inflation gas filling and releasing. The torus structure cavity is filled by the inflation gas slightly above the yield point of the torus wall material and hold for a small time period for elastic recovery. So, one can conclude that rigidization of the torus is taking place. Second plot is showing the pressure required for rigidization. After a certain pressure, strain increases drastically which showing a very high yielding of the torus wall material.

The results of computational analysis have been compared with the analytical approach. Equation (1) has been used to find the rigidization pressure analytically. Yield stress (σ_y) for each combination has been found through uniaxial tensile experiment. INSTRON UTS 5982 gives yield point of each membrane. Figure showing the experimental setup, shape and dimension of the specimen, and graph on stress–strain curve (Figs. 4 and 5).

After getting yield stress from the experiment for each combination, rigidization pressure is found through Eq. (1) and compared with computational results. Various data have been tabulated in the following Table 1:

The percentage error between computational results and analytical results is under 3% error. Hence, one can conclude that the computational and analytical results are in good agreement.

Fig. 4 Experimental setup (INSTRON UTS 5982)

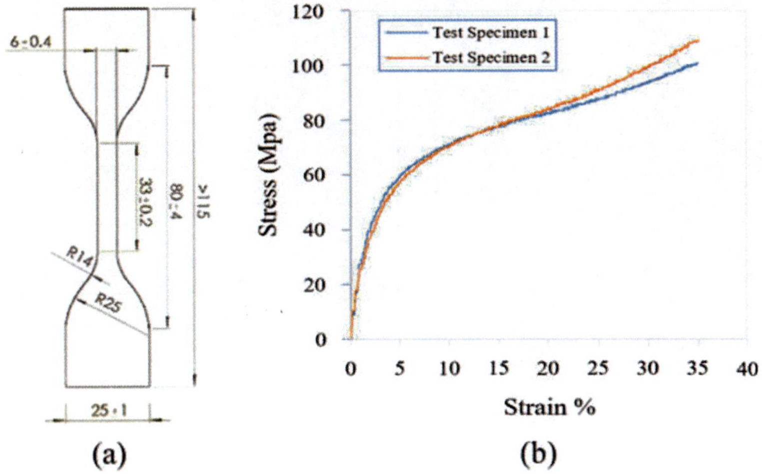

Fig. 5 **a** uniaxial tensile test specimen, **b** stress versus strain plot for Al-Kapton laminate

Table 1 Comparison between computational and analytical results

S. No	Combination name	Thickness		Yield stress (σ_y) (MPa)	Rigidization pressure (R_P)		% Error
		Al (μ)	Kapton (μ)		Computational (kPa)	Analytical (kPa)	
1	T1	10.5	25	153	124	125.5	1.19
2	T2	10.5	50	82.5	126	124.8	0.9
3	T3	18	25	130	138	135	2.2
4	T4	18	50	92.1	142	144.7	1.9
5	T5	25	25	134.8	156	155.7	0.19
6	T6	25	50	91.7	161	158.9	1.3

6 Conclusion

Inflatable structures are stowed in a canister before deployment so, wall thickness should be optimum. The thick wall will create a hurdle in folding and stowing, and high spring back force. A very thin wall will crack through fold lines and structure health will be compromised. T1 and T2 combinations have 10.5μ thin Al film which can lead to generate cracks in membrane during folding and unfolding. So, T1 and T2 combinations can be ruled out from structural health point of view. Out of T3, T4, T5, and T6 minimum rigidization pressure is required by T3 that is 138 MPa and its wall thickness is moderate. Crack will not propagate in T3 combination during folding and unfolding. Combinations T4, T5, and T6 can also be ruled out from wall thickness point of view. These combinations have thick wall thickness in comparison with T3 combination. Hence, combination T3 is providing optimum wall thickness with minimum rigidization pressure, i.e., T3 combination is the best available option out of six combinations.

Acknowledgements This research work carried out under the project grant no. ISR-1248-MID which is funded by the Indian Space Research Organization (ISRO). Smart Material and Structures Laboratory (SMSL) at Indian Institute of Technology, Roorkee (IITR), has been developed under this grant. We would like to acknowledge the ISRO scientist who gave their valuable feedback and resources to achieve the results.

References

1. Liu ZQ, Qiu H, Li X, Yang SL (2017) Review of large spacecraft deployable membrane antenna structures. Chin J Mech Eng 30(6):1447–1459
2. Babuscia A, Sauder J, Chandra A, Thangavelautham J, Feruglio L, Bienert N (2017) Inflatable antenna for CubeSat: a new spherical design for increased X-band gain. In: 2017 IEEE aerospace conference. IEEE, pp 1–10
3. Schenk M, Viquerat AD, Seffen KA, Guest SD (2014) Review of inflatable booms for deployable space structures: packing and rigidization. J Spacecr Rocket 51(3):762–778
4. Secheli G, Viquerat A, Lappas V (2015) An examination of crease removal in rigidizable inflatable metal-polymer laminate cylinders. In: 2nd AIAA spacecraft structures conference, p 0681
5. Fowler CP, Orifici AC, Wang CH (2016) A review of toroidal composite pressure vessel optimisation and damage tolerant design for high pressure gaseous fuel storage. Int J Hydrogen Energy 41(47):22067–22089

Mechanism for Biomedical Applications

Design of Jaw Rehabilitation Device for Patients with TMJ Disorder

Udit S. Parihar, Shreyas M. Patel, Suril V. Shah, Kaushal A. Desai, and Ankita Chugh

Abstract This paper presents a novel design of a rehabilitation device for patients suffering from jaw opening impairment due to temporomandibular joint (TMJ) disorder and oral fibrosis-inducing conditions. The design of the device is posed as a mechanism synthesis problem through path generation. The path generation is formulated as a multi-objective optimisation to minimise the error between the desired and actual jaw motion profiles and maximise the mechanical advantage. An iterative solution is proposed that follows the ideal jaw profile. A systematic selection of design variables is made, followed by formulating the objective function and identifying appropriate constraints. The optimisation routine uses the steepest descent gradient to find the feasible direction for the objective function and Lagrange multipliers for imposing constraint conditions. Limits on link lengths and coordinates of optimised linkages are imposed based on ergonomic considerations. The paper presents two solutions to achieve the desired objectives: a dual fourbar mechanism covering fourteen precision points and a single fourbar mechanism covering eight precision points. The designed mechanisms are compared for accuracy, and structural analysis is carried out to assess the strength. The solution consisting of a single 4-bar mechanism is found better from the strength and manufacturability point of view. A set of human trials would be required to examine the efficacy of the proposed design.

Keywords Synthesis of mechanism · Path generation · Optimisation · Jaw rehabilitation · TMJ disorder

U. S. Parihar · S. M. Patel · S. V. Shah (✉) · K. A. Desai
Indian Institute of Technology (IIT) Jodhpur, Jodhpur, India
e-mail: surilshah@iitj.ac.in

A. Chugh
All India Institute of Medical Sciences (AIIMS), Jodhpur, Jodhpur, India

V. K. Gupta et al. (eds.), *Recent Advances in Machines and Mechanisms*, Lecture Notes in Mechanical Engineering, https://doi.org/10.1007/978-981-19-3716-3_36

1 Introduction

The temporomandibular joint (TMJ) acts like a sliding hinge, connecting the jawbone to the skull, and controls the jaw movement. Injury to teeth or jaw, teeth grinding, arthritis, gum chewing, etc., are the leading cause of TMJ disorder [1–3]. There are some devices available for rehabilitation, as shown in Figs. 1, 2 and 3. However, these devices have a few drawbacks making them uncomfortable for the patients to use for a long duration. The major flaw with the current jaw opening devices is their inability to follow the natural path of the human jaw, as shown in Figs. 4 and 5. Studies show that the jaw path profile is concave inwards towards the mouth with a sharp bending point. However, the displacement analysis of existing devices shows that their path profiles are concave outwards in contrast. Other drawbacks of existing devices are: (i) Loading point is too narrow for teeth placement causing concentrated stresses at the point of contact, (ii) lack of proper scale to measure jaw opening for tracking improvements over time, (iii) lack of locking mechanism forces patient to hold the device at the uncomfortable position for a long time, and (iv) the devices are not designed for gaining maximum mechanical advantage without comprising the path precision and are also costly. So in this paper, a jaw opener as a fourbar mechanism is proposed with low cost and the ability to follow a profile similar to human jaw motion.

Fig. 1 Heister metal jaw opener [4]

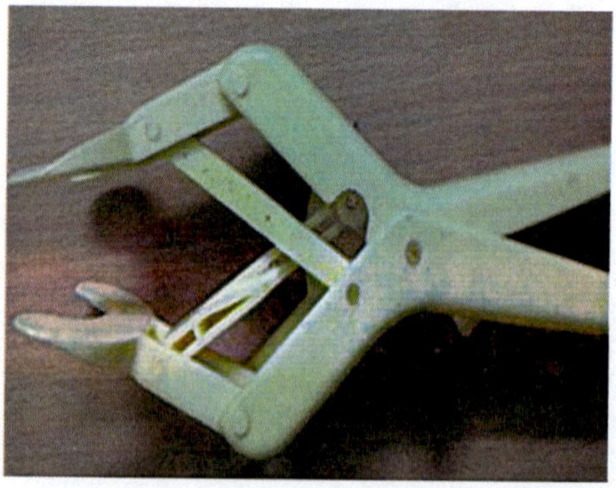

Fig. 2 OraStretch press rehab system [5]

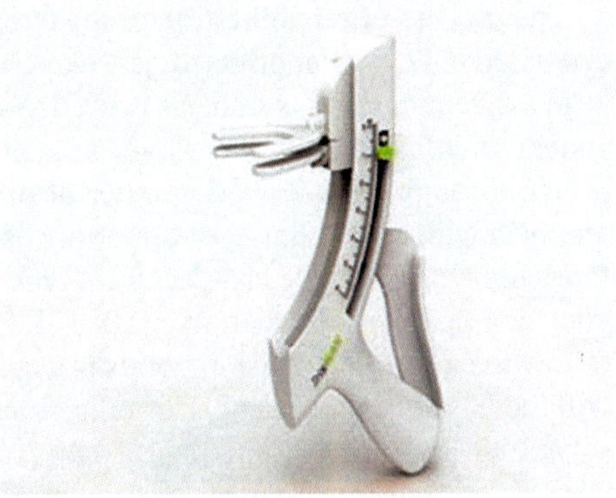

Fig. 3 TrisMax jaw rehabilitation tool [6]

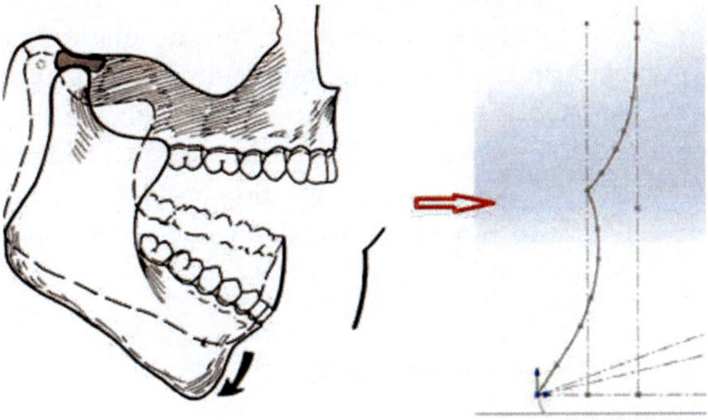

Fig. 4 Natural path of human jaw opening

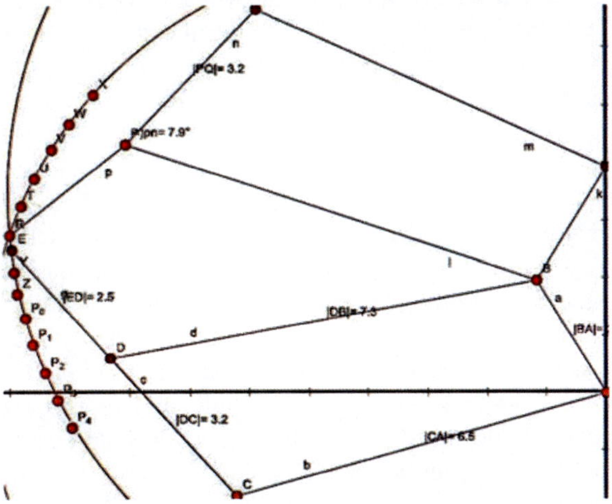

Fig. 5 Path of existing devices

The process of rehabilitation of the temporomandibular joint (TMJ) disorder was discussed in [7, 8]. For effective jaw exercises, its movement is needed to be traced by a device. Dimensional synthesis is used to solve this problem. More specifically, path generation of a linkage is employed to synthesis mechanisms to trace the desired jaw profile through a prescribed number of precision points [9]. Fourbar mechanism is one such widespread linkage commonly used in many applications. In the case of the fourbar mechanism, we can get an analytical solution to the problem if the number of precision points is less than five [10, 11]. Path generation of a fourbar mechanism to follow five or more precision points is a computationally heavy task, and complexity will increase with the number of precision points. Numerical methods were used to solve such a complex problem [12, 13]. A common approach is to solve the problem through optimisation subject to constraints [14] to find an optimal solution. There are many methods to solve the optimisation problem, as shown in [15].

In this paper, two different mechanism solutions are proposed to trace the motion of the human jaw profile. Design optimisation is used to obtain the solution for the proposed mechanisms' link lengths and relative angles. This paper is divided into five sections. Section 2 contains the design consideration for the synthesis problem. Section 3 formulates an optimisation problem to find a design solution. Results are shown in Sect. 4, whereas the conclusion of the paper is drawn in Sect. 5.

2 Design Considerations

The main goal of the proposed work is to design a device whose opening profile follows as closely as possible the human jaw motion profile with the minimum effort possible. This would ensure avoiding any unnecessary constraint forces at an angle to the jaw. We also target to develop a device that offers proper comfort to the patient by providing damping at the contact point between teeth and the device. To avoid continuous application of the force by the patient to hold the device, we have provided a locking mechanism to monitor the extent of jaw opening. A proper scale is also required in the device that would be helpful to track the patient's health.

We propose two solutions: (i) a dual fourbar mechanism comprised of two asymmetric fourbar mechanisms, each following different paths. Its advantage is that the resulting path is the sum of the individual paths, hence allowing us to track more precision points and, therefore, more accurate path. The upper fourbar follows six points, while the lower fourbar follows eight points. One connecting link between the two fourbars transfers the motion from the output link of one fourbar to the input link of the other fourbar and (ii) a single fourbar mechanism following eight precision points.

Our work is different from contemporary work by emphasising that the device's design follows the precise jaw path profile resulting in comfort during its usage. Experiments have been performed to calculate the mean path profile of a healthy human jaw, as shown in Fig. 4. Current medical devices, which do not follow precise path profiles, are ineffective in treating TMJ disorder. Our problem statement of following the jaw path is formulated as a path generation. We focus primarily on the fourbar mechanism and its novel variant dual fourbar mechanism for path generation of desirable profiles as proposed in the paper. A point on the coupler is chosen to follow the desired path. The link lengths and relative angles are obtained from the solution of the optimisation problem, as mentioned in the next section.

Another feature of the proposed design is the ability to lock down the mechanism between the minimum opening and its maximum opening. This is achieved by placing a nut and bolt between the ground and the crank link. This is also used to limit the maximum opening of the device. The design of the prototype also aims to provide an ergonomically better holding position, as shown in the final prototype in Fig. 12.

3 Formulation of Design Solution as an Optimisation Problem

The design of the jaw opener is posed as a design and synthesis of mechanism. The kinematic synthesis problem of determining link lengths and starting angles of the fourbar mechanism is converted into design optimisation. The objective function we want to minimise is the difference between precision points on a desired and calculated motion profiles. The constraints which should be maintained while achieving the required objective are adhering to the Grashoff condition, loop closure equation and limiting link lengths to their maximum and minimum values.

3.1 *Objective Function*

The input for the problem would be the coordinates of precision points on the required path. Minimising the distance between precision points is the same as minimising the square of the difference of x and y coordinates. To reduce the complexity of the objective function, we will minimise only the y coordinates of the points while keeping x coordinates equal for the given points and the optimised points. The objective function is defined as:

$$F = \sum_{i=1}^{n} (y_i - Y_i)^2 \tag{1}$$

where y_i is the discreet points on calculated curve and Y_i is the precision points on the input curve. n corresponds to the number of points optimised, e.g. for dual fourbar $n = 8$ for lower fourbar and $n = 6$ for upper fourbar, and F is value of objective function.

3.2 Loop Closure Equation

Various parameters used for the fourbar mechanism are shown in Fig. 6. The first equality constrained equation would be writing coupler endpoint vector **r** in terms of link vectors $\mathbf{Z}_6$, $\mathbf{Z}_2$ and $\mathbf{Z}_5$ and the second equation would be to force $\mathbf{Z}_1$, $\mathbf{Z}_2$, $\mathbf{Z}_3$ and $\mathbf{Z}_4$ to form a closed fourbar.

$$\mathbf{r} = \mathbf{Z}_6 + \mathbf{Z}_2 + \mathbf{Z}_5 \tag{2}$$

$$\mathbf{Z}_1 + \mathbf{Z}_4 = \mathbf{Z}_2 + \mathbf{Z}_3 \tag{3}$$

Rewriting vector equations into scalar form, one obtains:

$$X_i - Z_6 \cos\alpha - Z_2 \cos\gamma_i - Z_5 \cos(\angle Z_{3i} + \lambda) = 0 \tag{4}$$

$$y_i - Z_6 \sin\alpha - Z_2 \sin\gamma_i - Z_5 \sin(\angle Z_{3i} + \lambda) = 0 \tag{5}$$

$$Z_1 \cos\beta + Z_4 \cos(\angle Z_{4i}) - Z_3 \cos(\angle Z_{3i}) - Z_2 \cos\angle\gamma_i = 0 \tag{6}$$

$$Z_1 \sin\beta + Z_4 \sin(\angle Z_{4i}) - Z_3 \sin(\angle Z_{3i}) - Z_2 \sin\angle\gamma_i = 0 \tag{7}$$

Each of the above four equations corresponds to the single point on the curve for $i = 1, 2, \ldots, n$; hence for 14 points, we would have $4 \times 14 = 56$ equality constraint equations.

3.3 Limiting Values of Design Variables

The device's manufacturing requirements and ergonomics aspect would require the link lengths to be greater than a certain value and smaller than the certain maximum value. Hence, the inequality constraint equations are introduced as:

$$K_1 \geq Z_1 \geq \min;\ K_2 \geq Z_2 \geq \min;\ K_3 \geq Z_3 \geq \min \tag{8}$$

$$K_4 \geq Z_4 \geq \min;\ K_5 \geq Z_5 \geq \min;\ K_6 \geq Z_6 \geq \min \tag{9}$$

Here, K_i, where $i = 1, 2, \ldots, 6$, are the upper values of link lengths and min is the lower value of link lengths, which is different for different links. In order to avoid

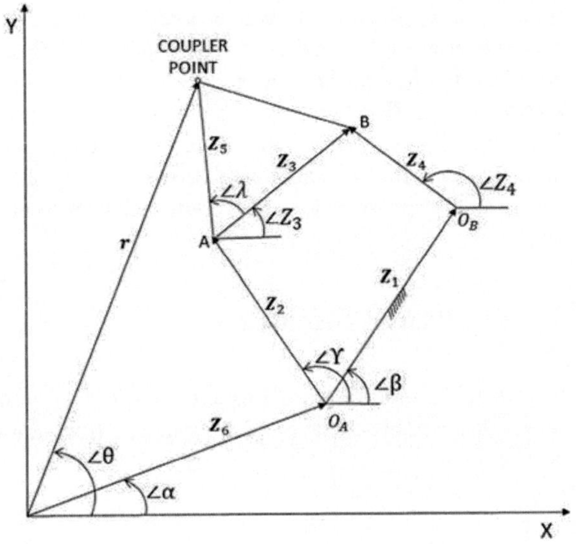

Fig. 6 Definition of various parameters of a fourbar mechanism

overlapping of successive crank angle obtained from optimisation, there should be a minimum amount of angle difference, K_7, between them, as shown below:

$$\gamma_{i+1} - \gamma_i \geq K_7 \tag{10}$$

where $i = 1, 2, \ldots, n - 1$ for discreet points. The other inequality constraint is required to ensure minimum starting angular difference, K_8, between ground (β) and crank (γ). The maximum size of the device would give us another inequality between angles of vectors Z_6 and r as follows:

$$\gamma_i - \beta \geq K_8 \tag{11}$$

$$\theta_i - \alpha \geq K_9 \tag{12}$$

The design variables $Z_1, Z_2, Z_3, Z_4, Z_5, Z_6, \alpha, \beta, \lambda, y_i, \gamma_i, \angle Z_{3i}, \angle Z_{4i}$ can be varied to satisfy the above-mentioned equalities and inequalities while minimising the objective function. So, we have $9 \times 1 + 4 \times 14 = 65$ design variables which need to be optimised. The input variables are the constants K_i, where $i = 1, 2, \ldots, 9$, and the coordinates of the required curve X_i, Y_i where $i = 1, 2, \ldots, n$ and n is the number of precision points on the required curve.

The mathematical solver known as *fmincon* is used to solve the nonlinear objective function with equality and inequality constraints. The solver works by calculating the hessian and gradient of the objective function at numerous points and directing the solution to the minimum value of the objective function at each step. The iterative process stops when the objective function reaches a specific minimum value given by the user at the beginning of the programme. *fmincon* starts the calculation from an initial point, and the convergence time and value are heavily dependent on the starting point. So to provide a better starting point, we used another linear solver *linprog* with the value of the objective function as zero while forcing the design variables as mentioned above equality and inequality constraints. Due to the linear nature of the solver, it does not require the user to give an initial point.

4 Results and Discussion

Results of optimisation are presented in this section in the form of two design solutions. Both solutions have been compared from the point of view of strength and accuracy.

4.1 Design 1: Dual Fourbar Mechanism

The dual fourbar mechanism is comprised of two asymmetric fourbar mechanisms, each following two different paths. The upper fourbar follows six precision points, while the lower fourbar follows eight precision points. One connecting link between the two fourbars transfers the motion from the output link of one fourbar to the input link of the other fourbar. The solution of optimisation gives us values of the design variables. The link lengths and initial angles are extracted from the solution to form a dual fourbar linkage.

Figure 7 depicts the length of links obtained through optimisation. The values of the objective functions for the upper and lower fourbar are given by $F_{\text{upper}} = 0.0121$ and $F_{\text{lower}} = 0.00027$ at the end of optimisation. Figure 8 shows the conceptualised prototype, the dual fourbar, with a connecting link. Despite providing a higher path accuracy, dual fourbar fails to withstand jaw forces for a long duration of time, as seen in Fig. 9. This is mainly due to the mechanism's complexity resulting in smaller and thinner links and joints. It was also observed that the resulting design is also challenging from the point of view of manufacturability. Figure 10 show the path traced by the fourbar, which is the same as the required path shown in Fig. 7 in red dots.

4.2 Design 2: Single Fourbar Mechanism

The single fourbar mechanism-based device traces eight points. Figure 11 shows link lengths of optimised fourbar. The value of objective function is $F = 0.2946$ at the end. Figure 12 shows the conceptualised prototype of a single fourbar mechanism.

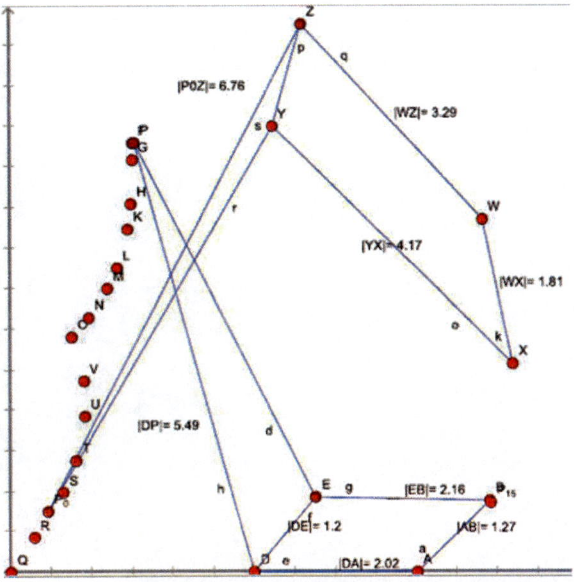

Fig. 7 Optimised link lengths of dual fourbar mechanism

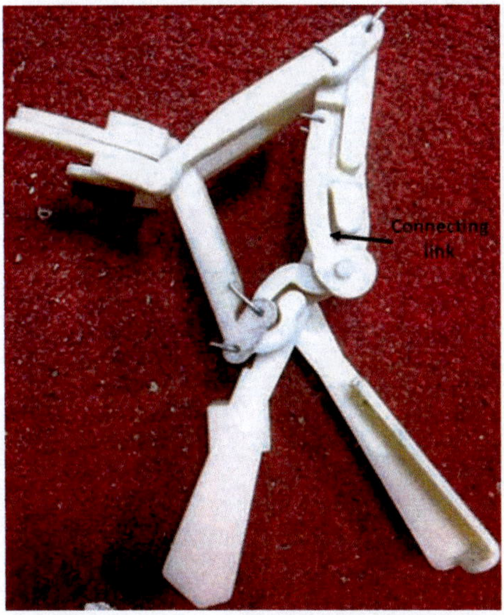

Fig. 8 Prototype of dual fourbar mechanism

The lower mouthpiece is connected to the coupler, which traces the jaw path, while the upper mouthpiece is connected to the ground. The larger link lengths also help in achieving a bigger mechanical advantage. The mechanical advantage increases with increasing crank angle, making it easier to hold the device during the maximum opening condition. The locking and scaling system can be easily integrated into the proposed design. The single fourbar allows us to have a larger link length with a fewer

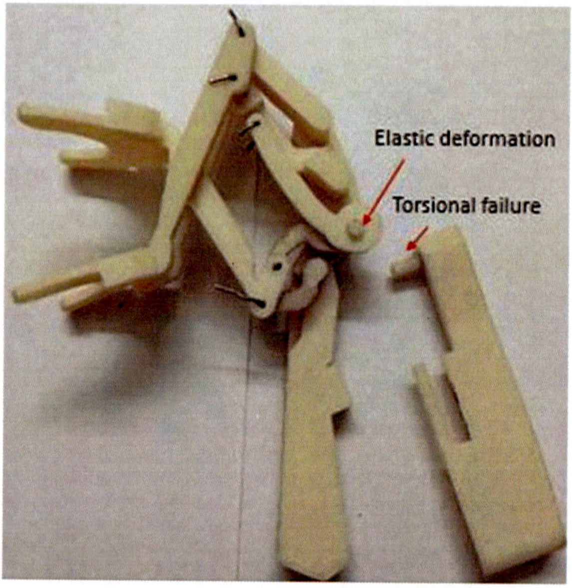

Fig. 9 Torsion failure of 3D printed prototype of dual fourbar

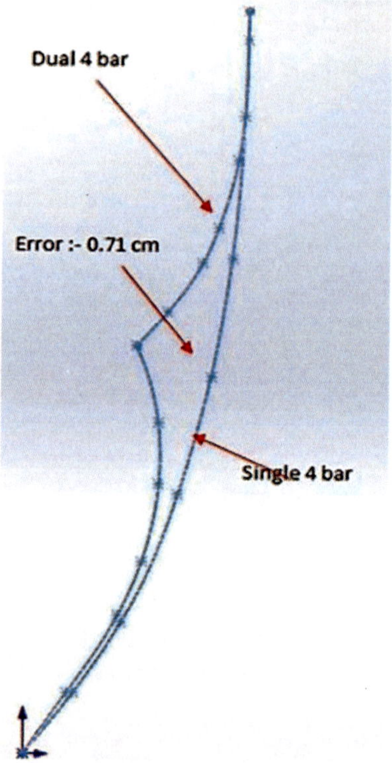

Fig. 10 Path traced by single and dual fourbar

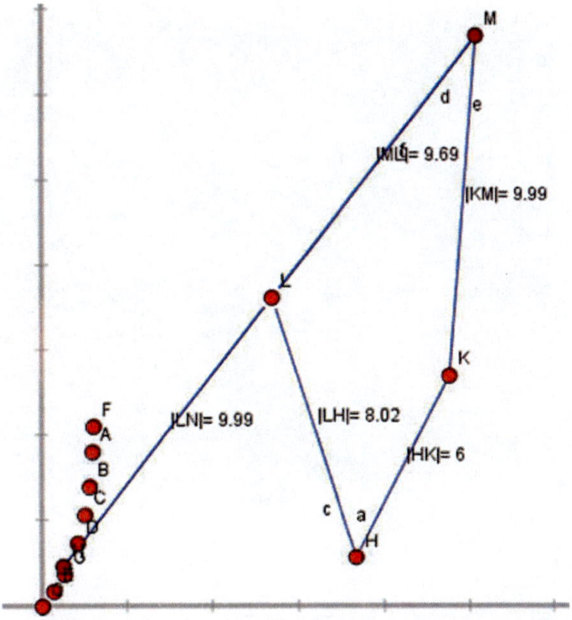

Fig. 11 Optimised link lengths of single fourbar mechanism

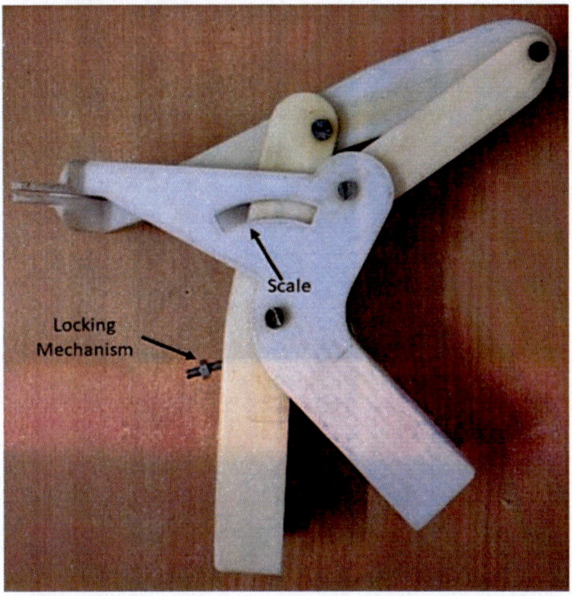

Fig. 12 Prototype of single fourbar mechanism

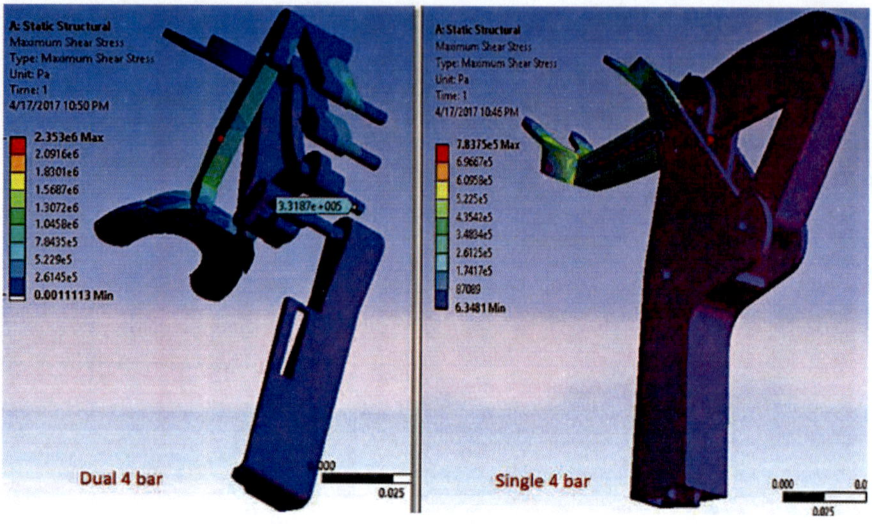

Fig. 13 Stress analysis of dual and single fourbar mechanisms

number of links. However, it will induce an error of 0.71 cm^2 from its natural path (Fig. 10). This error can be compensated by the fact that each person has different jaw paths; hence, a small error can be average out over large people. The single fourbar mechanism-based design was also found to be easier and cheaper to manufacture.

Structural analysis of both designs in Fig. 13 shows that stresses induced in a single fourbar are much less than dual fourbar mechanism. This indicates that the single fourbar outperforms from the strength point of view of strength. The larger link lengths in the single fourbar also enable us to use metallic pins instead of ABS pins, unlike the dual fourbar.

5 Conclusion

Two novel designs of jaw openers are proposed in this work for a patient suffering from TMJ disorder. Both designs can trace human jaw motion profiles. We also investigated the accuracy and strength of both designs as a trade-off. With the help of the dual fourbar mechanism, it was possible to trace higher precision points resulting in better accuracy. However, it required smaller and thinner links that could not withstand the jaw forces, resulting in lower strength. The single fourbar-based mechanism helped increase the strength of the device considerably while slightly compromising the path's accuracy. This is not of significant concern because it is averaged out by variations in the jaw motion profile from one person to another. In future, systematic human trials will be performed to make more improvements based on the feedback received from patients. Further, five or more links mechanisms will be explored for the design of the Jaw opener.

References

1. Buescher JJ (2007) Temporomandibular joint disorders. Am Fam Phys 76(10):1477–1482
2. Yildirgan K, Zahir E, Sharafi S, Ahmad S, Schaller B, Ricklin ME, Exadaktylos AK (2017) Mandibular fractures admitted to the emergency department: data analysis from a Swiss level one trauma centre. Emerg Med Int 2016
3. Haggerty CJ, Laughlin RM (2015) Atlas of operative oral and maxillofacial surgery. Wiley
4. Heister jaw opener. http://www.sksurgicals.net/heister-jaw-opener-mouth-gag.htm
5. TrisMax. http://www.ncbi.nlm.nih.gov
6. OraStretch press rehab system. https://www.craniorehab.com/orastretch-information
7. Murphy I (2019) Temporomandibular disorders: manual therapy, exercise and needling. Br Dent J 226(1):6
8. Andrews JR, Harrelson GL, Wilk KE et al (2012) Physical rehabilitation of the injured athlete: expert consult—online and print. Elsevier Health Sciences
9. Norton RL, Han J (1999) Design of machinery, vol 2. McGraw-Hill, New York
10. Khorshidi M, Soheilypour M, Peyro M, Atai A, Panahi MS (2011) Optimal design of four-bar mechanisms using a hybrid multi-objective GA with adaptive local search. Mech Mach Theory 46(10):1453–1465
11. Freudenstein F (2010) Approximate synthesis of four-bar linkages. Resonance 15(8):740–767
12. McLarnan C (1963) Synthesis of six-link plane mechanisms by numerical analysis
13. Roth B, Freudenstein F (1963) Synthesis of path-generating mechanisms by numerical methods
14. Fox R, Willmert K (1966) Optimum design of curve-generating linkages with inequality constraints. In: Mechanical engineering, vol 88. ASME—American Society of Mechanical Engineers, p 63
15. Singiresu S et al (1996) Engineering optimization: theory and practice. Wiley

Design and Development of Double Air Suction Resuscitation Device Using Scotch Yoke Mechanism

Shivdayal Patel and Tanuja Sheorey

Abstract Aim was to make a low cost, lightweight, portable, automated cardiopulmonary resuscitation (CPR) device, which can be used while shifting patients, or emergency situation and cater to patient's breathing requirement. Looking at the situations likes Covid-19 or natural disaster, two point air supplies to cater to two patients simultaneously is planned. Proposed design will provide customized air supply with variation in pressure as well as volume of air for maintaining breathing comfort of different age group of patients. The device will be fabricated mainly in three parts, namely actuating mechanism, arrangement for variable air supply pressure, and flow volume and feedback control loop for automated customized operation. Scotch Yoke mechanism has been used to convert rotary motion of DC motor to linear motion. One-point rotary motion is converted to two-point linear motion. Hence, the device may be used to provide CPR to two patient's simultaneously. Variable speed is attained through PWM that may change current supply which in turn gets stabilized via inbuilt switching IC. The hardware has been incorporated. Similarly, to control flow volume, position of crank pin will be altered by using guide ways via feedback control.

Keywords Resuscitation device · Scotch Yoke mechanism · Ambu-bag · Sensor

1 Introduction

Aim was to make a low cost, lightweight, portable, automated CPR device, which can be used while shifting patients, or emergency situation and cater to patient's breathing requirement. Looking at the situations likes Covid-19 or natural disaster, two point air supplies to cater to two patients simultaneously is planned. Proposed design will provide customized air supply with variation in pressure as well as volume of air for maintaining breathing comfort of different age group of patients. Double air suction resuscitation device can be used in all Govt. PHC's (especially villages)

S. Patel (✉) · T. Sheorey
PDPM Indian Institute of Information Technology, Design and Manufacturing, Jabalpur, Madhya Pradesh 482005, India
e-mail: shivdayal@iiitdmj.ac.in

V. K. Gupta et al. (eds.), *Recent Advances in Machines and Mechanisms*, Lecture Notes in Mechanical Engineering, https://doi.org/10.1007/978-981-19-3716-3_37

to cater at the time of need patient such as snake bite, shifting patient from PHC to referral hospital, at the time of natural disaster, Covid-19 patients, operation theaters of hospitals: During the operation, oxygen supply is required. The device can be used in ambulances and all the residential educational institutions, etc. In 1940–1950, positive pressure device only controlled the volume. Soar et al. [1] developed the modified CPR device which had reduced complications that can lead to further negative outcomes after ROSC. The resuscitation-related rib and sternum injure are the most common complications; almost, 60% of patients sustain rib fractures during M-CPR. Lee et al. [2] developed the CPR device for clinically vulnerable people such as pregnant women and infants. Maternal cardiac arrest remains a concern as very little has been done toward this problem and can be devastating, often resulting in the death of the mother. Tan et al. [3] investigated the modified CPR system to increase the efficiency of CPR system. A chest compression test is used to minimize the risk of the patient during an oxygen supply. This system is involving a boar model, an end-tidal CO2-guided computerized robot CPR system. Maertens et al. [4] developed the mechanical chest compressor cardiopulmonary resuscitation system to recover the cardiac arrest patients. This CPR device is basically used for the survival and neurological recovery, but this device was not used for the pregnant women or infants [3–6].

The basic aim was to develop a low cost, lightweight, portable, automated CPR device, which can be used while shifting patients, or emergency situation and cater to patient's breathing requirement. Scotch Yoke mechanism has been used to convert rotary motion to linear motion. One-point rotary motion is converted to two-point linear motion. Hence, the device may be used to provide CPR to two patient's simultaneously.

2 Methodology of Double Air Suction Resuscitation Device

Devices available in this category internationally have also used Ambu-bag for air supply. However, different mechanisms are used for conversion of rotary motion of a motor to linear or oscillatory motion required for compression. Some devices have incorporated built-in air flow meter for air supply pressure and volume measurement with analog indicators for display. For changing supply parameters, microcontroller, variable resistor, etc., have been used.

2.1 *Actuating Mechanism*

Four bar mechanism, i.e., rotary to oscillatory motion, slider crank, rack and pinion, and cam follower mechanism have been used to get desirable motion to operate Ambu-bag with respect to pressure and suction creation. In our proposed device, Scotch Yoke mechanism has been used to convert rotary motion of DC motor to linear motion. One-point rotary motion is converted to two-point linear motion. Hence, the device may be used to provide CPR to two patient's simultaneously.

2.2 *Variable Air Supply Pressure and Flow Volume*

In order to have customized air supply as per the patient's need, variable motor speed and voltage is required. Techniques used in other such devices are variable resistor to sense position of pressing arm, variable voltage supply, etc. Similarly, to get variable flow volume, different controls have been used to vary the angle by which oscillatory motion is performed to press Ambu-bag. In our proposed device, variable speed is attained through PWM that may change current supply which in turn gets stabilized via inbuilt switching IC. The hardware has been incorporated. Similarly, to control flow volume, position of crank pin will be altered by using guide ways via feedback control.

2.3 *Feedback Control Loop*

In the absence of expert technician, it is difficult to know how much and at what rate the air supply need of the patient is. So far, we didn't get this feature in the available devices under our category. They are being operated manually to bring any change. In our proposed design, automation of air supply pressure and volume would be achieved based on feedback control. Motor speed and oxymeter reading is fed to the control loop, providing correction in either of the air supply pressure or volume or both. The same will be displayed on screen attached to the device.

2.4 *Principle or Operation*

Scotch Yoke mechanism is a very basic mechanism used to convert rotary motion to linear motion as shown in Fig. 1. A simple DC motor may be used for the purpose. With the additional link, rotary motion is converted to two-point linear motion, 180° apart, i.e., opposite to each other.

3 Working of Double Air Suction Resuscitation Device

For varying volume and supply pressure of air through the device, as per the need to the patient, simple two-knob arrangements are provided that can be handled by any person. The device can provide air–oxygen mixture at high, intermediate, and low

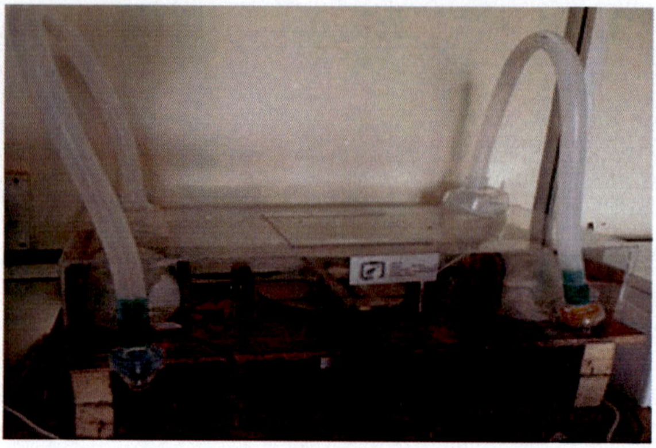

Fig. 1 Prototype model of double air suction resuscitation device

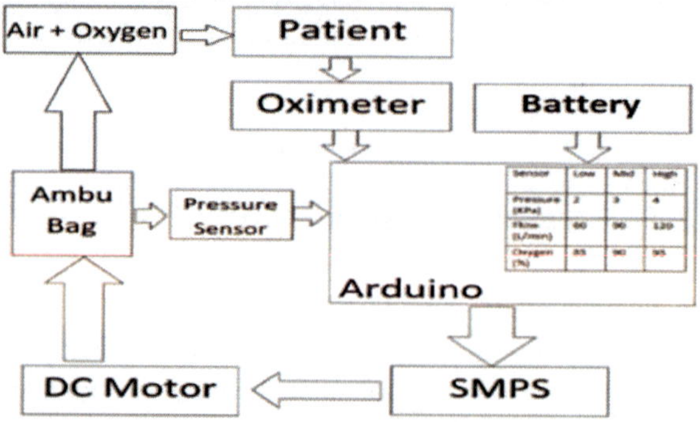

Fig. 2 Flowchart of the double air suction resuscitation device

volume. Similarly, it can provide the mixture at low, intermediate, and high pressure. At the time of emergency situation like natural disaster and gas leak, the device caters to two patients simultaneously. Pressure sensor and display LED are used to measure the air pressure inside resuscitator bag and display volume values, respectively, as shown in Fig. 2. Developed prototype provides customized air supply with variation in pressure as well as volume of air for maintaining breathing comfort of different age group of patients. The variation in these parameters is brought by simple knob arrangement. Attached LCD screen provides the supply parameter values for monitoring purpose.

4 Conclusions

Scotch Yoke mechanism has been used to convert rotary motion to linear motion. One-point rotary motion is converted to two-point linear motion. Hence, the device may be used to provide CPR to two patient's simultaneously. For varying volume and supply pressure of air through the device, as per the need to the patient, simple two-knob arrangements are provided that can be handled by any person. The device can provide air–oxygen mixture at high, intermediate, and low volume. Similarly, it can provide the mixture at low, intermediate, and high pressure. At the time of emergency situation like natural disaster gas leak, the device caters to two patients simultaneously. Pressure sensor and display LED are used to measure the air pressure inside resuscitator bag and display volume values, respectively.

References

1. Soar J, Nolan JP, Böttiger BW et al (2015) European resuscitation council guidelines for resuscitation 2015: section 3. Adult advanced life support. Resuscitation 95:100–147
2. Lee HM, Cho KH, Choi YH, Yoon SY, Choi YH (2008) Can you deliver accurate tidal volume by manual resuscitator? EMJ 25(10): 632–634
3. Tan D, Xu J, Shao S, Fu Y, Sun F, Zhang Y et al (2017) Comparison of different inspiratory triggering settings in automated ventilators during cardiopulmonary resuscitation in a porcine model. PLoS ONE 12:e0171869
4. Maertens VL, De Smedt LE, Lemoyne S et al (2013) Patients with cardiac arrest are ventilated two times faster than guidelines recommend: an observational prehospital study using tracheal pressure measurement. Resuscitation 84(7):921–926
5. Aufderheide TP, Sigurdsson G, Pirrallo RG et al. (2004) Hyperventilation-induced hypotension during cardiopulmonary resuscitation. Circulation 109(13):1960–1965
6. Aufderheide TP, Lurie KG (2004) Death by hyperventilation: a common and life-threatening problem during cardiopulmonary resuscitation. Crit Care Med 32(Suppl):S345–S351

Design of Mechanism for Actuating Piezoelectric-Based Sector Micropump

Bittu Kumar Singh, Tanuja Sheorey, and Vijay Kumar Gupta

Abstract In recent years, the micropump has been investigated by many researchers as drug delivery and disease diagnostic device. These micropumps are designed to control a small, specified amount of fluid. Piezoelectric-based micropump is one of these devices which is used to collect the required amount of blood from the human body for diagnostic. As the size of these devices is very small, therefore, handling and controlling these devices are a challenging task.

One such device has been designed by Haldkar et al. (J Mech Sci Technol 31(6):1–9 (2017)). The device consists of seven micropumps, each of 45° circular sectors, arranged circularly with separate microneedles attached to them. For collecting the blood sample, the microneedle needs to be pressed down by 1 mm so that it can penetrate human skin and extract blood for testing. This work aims to design a mechanism as an integral part of the wearable device, to operate the piezoelectric-based micropump. The cam-follower mechanism has been used for pressing the microneedle and pulling back after blood extraction. This paper discusses the design and analysis of the mechanism.

Keywords Cam · Micropump · Microneedle · Gear

1 Introduction

A number of microdevices have been developed for various applications. Some of the applications include medical devices, biochemical analysis, and microelectronics cooling systems [1–3]. In the biomedical field, these micropumps have been used for

B. K. Singh · T. Sheorey (✉) · V. K. Gupta
Discipline of Mechanical Engineering, Indian Institute of Information Technology, Design and Manufacturing, Jabalpur, Madhya Pradesh, India
e-mail: tanush@iiitdmj.ac.in

B. K. Singh
e-mail: 1813205@iiitdmj.ac.in

V. K. Gupta
e-mail: vkgupta@iiitdmj.ac.in

V. K. Gupta et al. (eds.), *Recent Advances in Machines and Mechanisms*, Lecture Notes in Mechanical Engineering, https://doi.org/10.1007/978-981-19-3716-3_38

blood testing to detect various diseases or as drug delivery systems [4–6]. According to the available literature, modeling and flow analysis of a piezoelectric-based sector micropump has been done by Haldkar et al. [7]. The micropump geometry is a 45° sector with a 6 mm radius. The purpose of the development of this micropump is to use it as a glucometer for the measurement of glucose level present in the blood of a diabetic patient. A microneedle extracts blood from the body of a patient and collects it in a pump chamber. It eventually comes in contact with the biosensor placed inside the pump chamber. After diagnostic, the biosensor shows the reading of glucose level. For proper suction, the microneedle needs to be inserted in the human skin and to be held for some time and pull back. In this paper, a mechanism for pushing and timing of the microneedle inside the skin is proposed.

1.1 Micropump

The micropump developed by [7, 8] is shown in Fig. 1. It consists of a pump chamber, piezoelectric bimorph, and a microneedle. A piezoelectric bimorph is attached at the top of the micropump with a diaphragm. A microneedle is attached at the base of the micropump for the suction of blood from the human body. When a sinusoidal voltage is applied across the piezoelectric, bimorph oscillates creating suction inside the pump chamber, and blood is sucked through the microneedle. A biosensor is placed at a specified location from the microneedle as shown. The blood sucked through the microneedle is collected at the biosensor for diagnosis. A circular section of device consists of seven such sectors as shown in Fig. 2. An angular space of 6.42° is provided between two micropumps to avoid frictional resistance at the time of operation. At the time of operation, only, one micropump gets operated. Figure 3 shows the single pump (sector) with microneedle and follower mechanism used for timing.

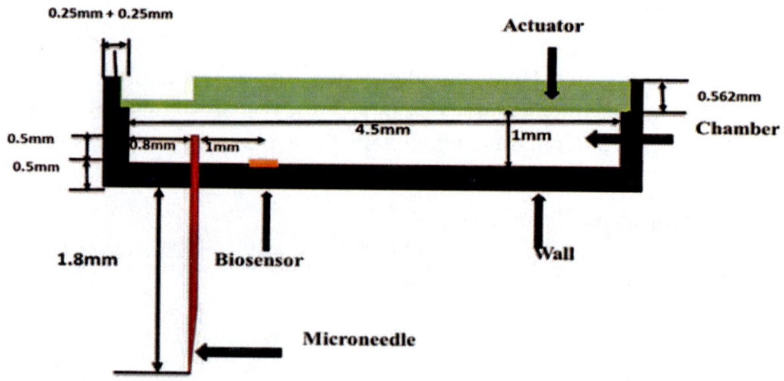

Fig. 1 Micropump model by Haldkar et al. [7]

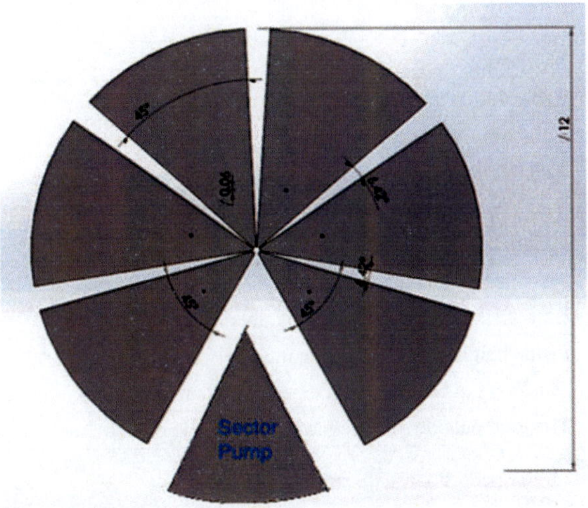

Fig. 2 Top view of 7-sector micropump [9]

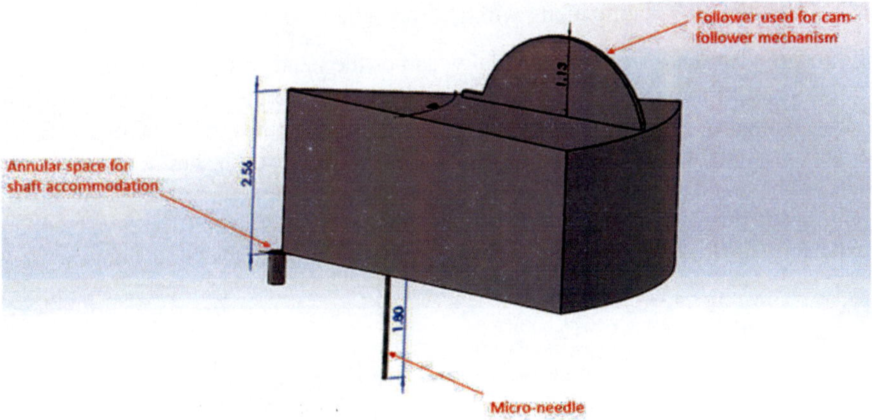

Fig. 3 Single-sector micropump with follower at the top

The dimensions of the micropump are considered the same as taken by Haldkar et al. [7] and are summarized in Table 1.

2 Micropump Timer Mechanism

For the blood extraction, following sequence of operations are performed:

a. First, the pump along with microneedle is moved downward by 1 mm so that needle is inserted inside the skin to extract blood.

Table 1 Dimensions of micropump [7]

Parameters	Dimension
Radius of the micropump, R	6 mm
Height of the micropump, h	2 mm
Angular space between two micropumps, θ_2	6.42°
Diameter of the microneedle, d	60 μm
Length of the microneedle, l	1.8 mm
Thickness of the piezo-bimorph, t_1	0.562 mm
Base thickness of the pump chamber, t_2	0.50 mm
Inside height of the pump chamber, t_3	1 mm
Thickness of the cover plate on the top surface of the pump chamber, t_4	0.5 mm
The total outside thickness of the pump chamber	2.56 mm

b. After pressing to required depth, the needle is held inside the skin for 3 s while pump is extracting the blood [9].
c. Next, the micropump along with microneedle is to be pulled back.

To perform the operations manually, as per the need of the timing of the blood sampling, cam and follower mechanism has been considered as one of the alternative. The cam rotation is controlled by a knob mounted outside the circle as shown in Fig. 4. A gear arrangement is used to transfer the rotation of the knob shaft to the camshaft. The knob is used to control two motions, (i) rotation of the sector micropump and (ii) pulling and pushing of micropump with the microneedle. The knob has two positions to perform the tasks which can be set by pushing and pulling the knob. For the current

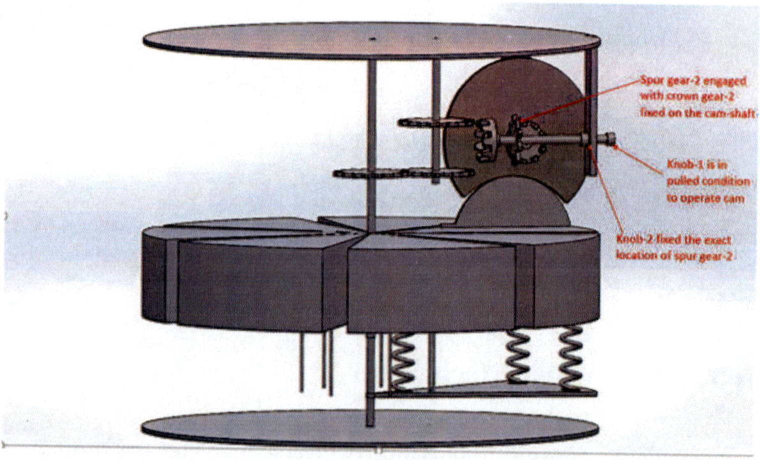

Fig. 4 Knob arrangement for the microneedle operation

timer motion, the knob is pulled outside as shown in Fig. 4. In this position of the knob, the spur gear-2 engages with crown gear-2; therefore, the rotational motion of the knob gets transmitted to the camshaft.

When the crown gear mounted on the knob is in mesh with the spur gear mounted on the compound gear shaft, the rotation of the sectors is achieved by the rotation of the knob. The gear ratio between the crown gear and the compound spur gear is two; therefore, when the micropump is required to rotate by a certain angle of rotation, the knob is required to rotate by twice the angle of rotation of the micro-pump. When the rotation of the micropump is done, the next sector micropump is placed at the location to conduct the next testing.

In this work, it has been assumed that a human can rotate the knob at a speed of 15 rpm approximately.

2.1 *Cam Design*

For conducting test, cam needs to be operated through the knob, so the knob shaft is pulled back to engage the worm wheel with a worm gear to transfer rotational motion between the knob shaft and camshaft. The total operation needs to be completed in 4 s with one revolution.

Based on the requirement of the micropump, the cam can be divided in the following intervals:

(a) Rise interval: During this interval, the microneedle is pressed down and penetrates the human skin. The rise angle is very low; therefore, the time taken by the needle to penetrate the skin is very less. The angle is selected as 45° so that penetration completed in 0.5 s (1/8th of total cycle).
(b) Dwell interval: During this interval, the microneedle does not move and stays stationary in the penetrated condition, and suction of blood takes place through the microneedle. The interval is chosen as 270° (3/4th of the total duration). The microneedle will remain penetrated for 3 s.
(c) Return interval: During this interval, the microneedle moves back to the original position. The return motion is completed in 0.5 s (1/8th of total cycle).

The cam design parameters are summarized in Table 2, and designed cam is shown in Fig. 5.

2.2 *Theoretical Calculations*

Input rotational speed applied to the knob, N_1=15 rpm.

The rotational speed of cam, $N_2 = 15\text{ rpm} = \frac{15}{60} = 0.25$ rps (revolution per seconds),

Table 2 Cam design parameters

Parameters	Values
Cam maximum radius, r_1	2 mm
Cam minimum radius, r_2	1 mm
Cam thickness, t	0.2 mm
Follower stroke-length, x	1 mm
Outward angle, θ_{rise}	45°
Return angle, θ_{return}	45°
Dwell angle, θ_{dwell}	270°

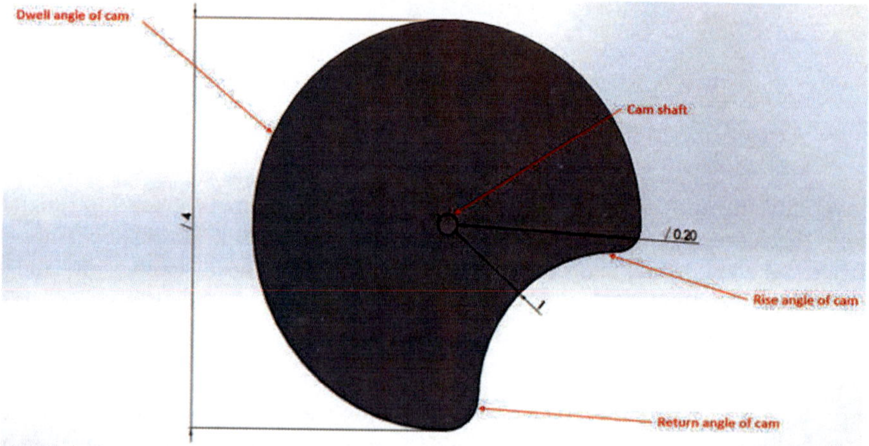

Fig. 5 Cam model its not looking circular

For a blood sample test, the cam needs to rotate about its axis by 360° or one revolution.

So, the time required for cam to complete one revolution,

$$t = \frac{\text{Number of revolution required}}{\text{Rotational speed of cam } (N_2)}$$

$$t = \frac{1}{0.25} = 4 \text{ s}$$

2.3 *Simulation Results from Solidworks*

The whole cam along with the gear mechanism is modeled in Solidworks for further analysis and simulated. Figure 6 shows the initial condition of the micropump; the

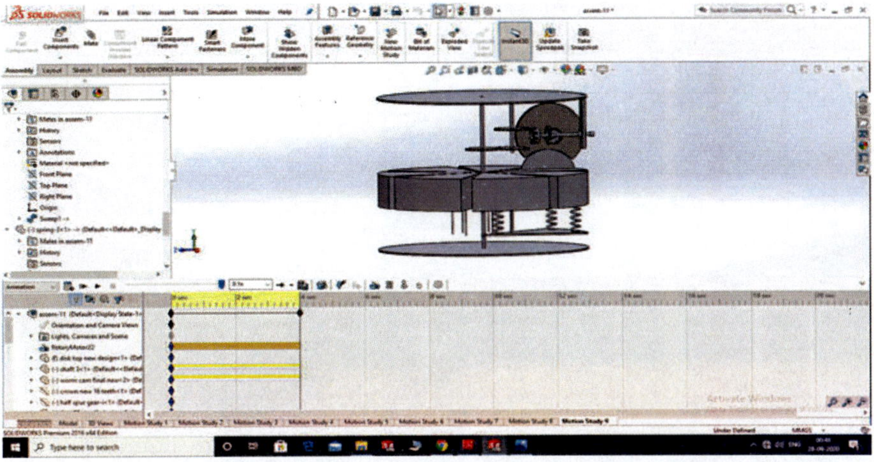

Fig. 6 Initial condition of micropump (at 0 s)

knob is engaged to the cam with the help of crown gear arrangement. Figure 7 shows the pressed condition of the micropump, and the microneedle can be seen to be pressed down. During the rise angle, micropump is pressed down, and the needle penetrates the skin. During the dwell time, the pump is operated, and the suction of blood takes place. During the fall angle, microneedle comes out to its original position. The total cycle time of 4 s can be seen clearly in Fig. 8.

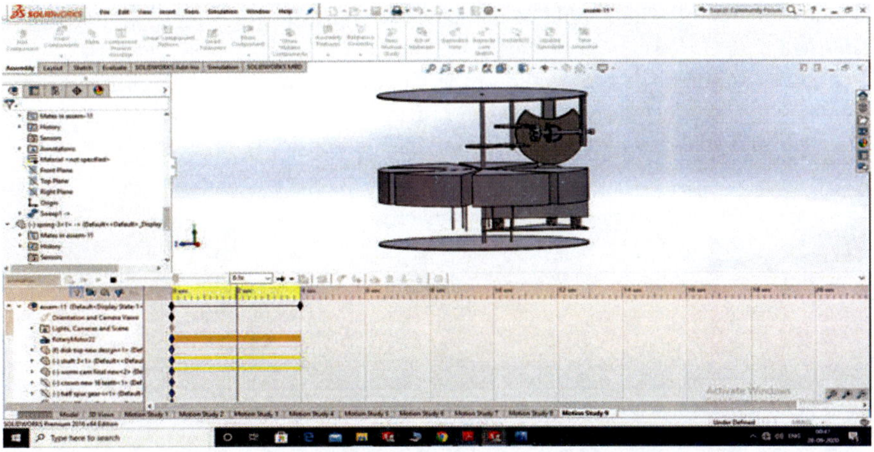

Fig. 7 Pressed condition of micropump (at 2 s)

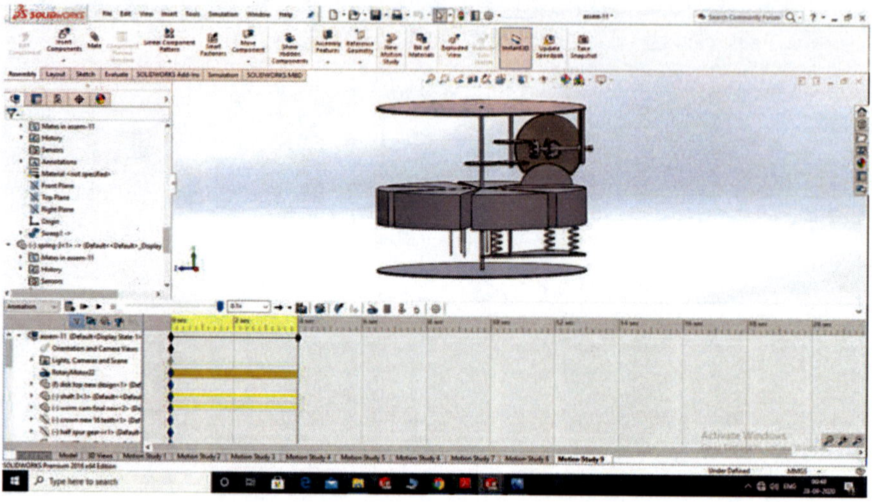

Fig. 8 Final condition of micropump (at 4 s)

3 Cam and Follower Analysis Using ADAMS Analysis Software

To find the displacement, velocity, and acceleration variation of microneedle during cam rotation in timer mechanism, cam and follower simulation has been carried out using ADAMS. 2D model of cam and follower mechanism as designed in ADAMS is shown in Fig. 9. The profile of the cam is developed using multiple B-spline in ADAMS^{C}.

Figure 10a, b shows the profile of the cam as generated in ADAMS and Solidworks. The cam designed in ADAMS is irregular in shape compared to the cam designed in Solidworks because B-spline has been used to draw the curve of cam

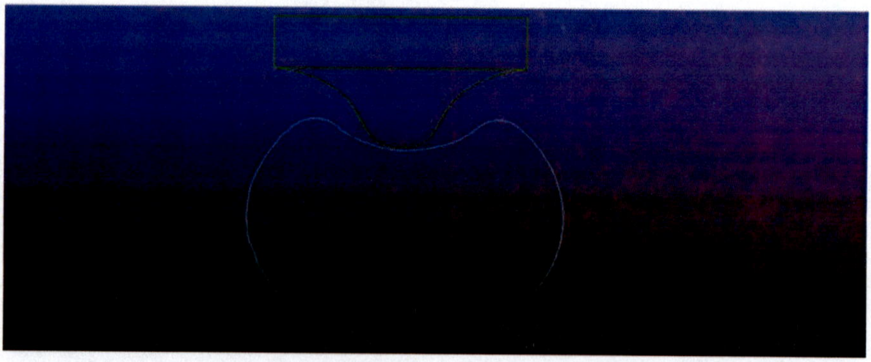

Fig. 9 2D-model of cam and follower drawn in ADAMS

(a) Irregular cam surface

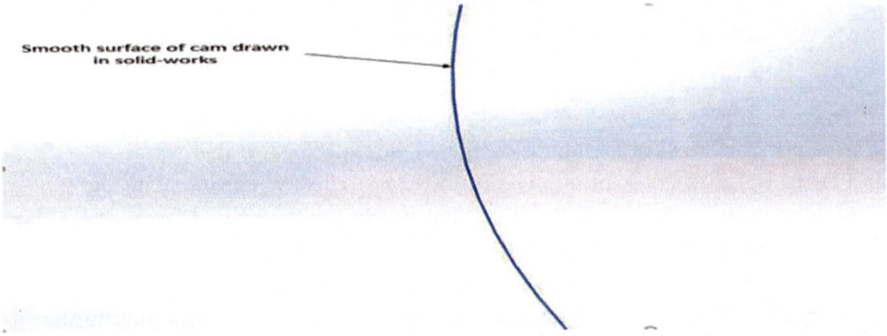

(b) Smooth cam surface

Fig. 10 Comparison of surface of cam model drawn in ADAMS and Solidworks. **a** Irregular cam surface, **b** smooth cam surface

profile in ADAMS. From the figures, it is clear that the cam model designed in Solidworks has smooth surfaces without any sharp corners or irregularities; therefore, it can be assumed that the motion of the microneedle is smooth and constant without any acceleration or jerk.

3.1 Results and Discussion

Figure 11 shows the displacement, velocity, and acceleration profile for the follower motion as simulated in ADAMS. It can be seen that the follower takes 0.5 s to rise linearly during cam rotation, and for the next 3 s follower, displacement is approximately constant, while cam rotates, then for the last 0.5 s, the follower returns linearly to its initial position.

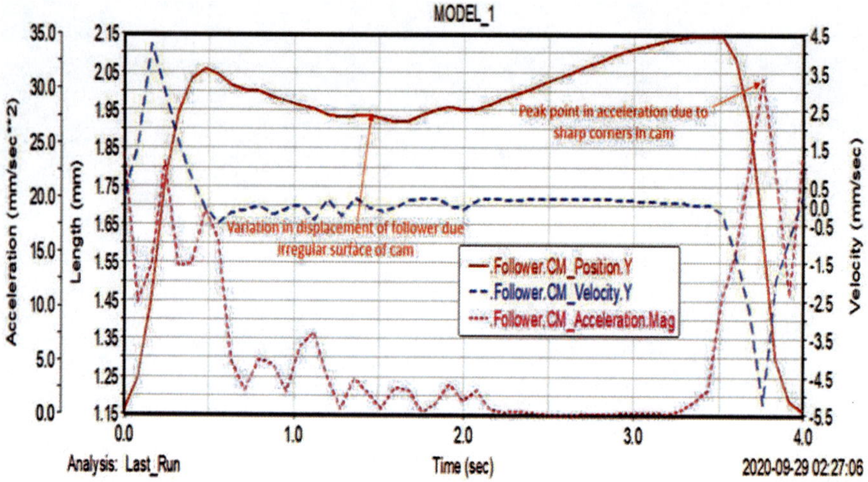

Fig. 11 Variation of displacement, velocity, and acceleration of follower

The displacement pattern of the follower in the graph is observed to be irregular. This irregularity can be attributed to the irregular surface of the cam profile drawn in ADAMS. The acceleration of the follower has some peak points during the rise and return states of the follower. These peak points have occurred due to some sharp corners in the cam profile. From the graph, it has been observed that the acceleration is minimum during constant displacement time. When the cam starts rotating, the velocity of the follower linearly increases and decreases with time during the rise and return stroke of the follower. The velocity of the follower is zero during the rise condition of the follower as shown in the graph.

4 Stress Analysis of Gear and Shaft

This device is designed to operate a micropump to extract blood from human skin. When the cam rotates with the help of a knob, it presses the needle down and forces the microneedle to penetrate the human skin. Ramasubramanian et al. [10] have investigated that when a microneedle of diameter equals to labium of a mosquito is penetrated inside human skin then 14.7 mN (milli-newton) force is required. Due to this force, gears and shafts of the actuating mechanism are subjected to stresses. The stresses are calculated in this section to ensure that the device can operate without failure.

5 Torque and moment calculations

Force applied on the circumference of the cam during penetration of microneedle inside human skin,

$$F = 14.7\ \text{mN} = (14.7 \times 10^{-3})\ \text{N},$$

The radius of cam, $R = 2$ mm,

From the definition of torque, torque applied on camshaft is given by,

$T = (F * R)$ Nm.

Therefore, $T = (14.7 \times 10^{-3}) \times (2 \times 10^{-3}) = (29.4 \times 10^{-6})$ Nm.

So, the torque (T) is required to apply at the knob shaft to penetrate the microneedle inside human skin, and the same is used to calculate the stress condition of components.

(a) Stress analysis of knob shaft

The stress analysis of the knob shaft is performed on SOLIDWORKS by applying the input torque value obtained from theoretical calculations.

Torque applied on the shaft, $T = (29.4 \times 10^{-6})$ Nm,

Diameter of knob shaft, $d = 0.2$ mm,

Length of knob shaft, $l = 3.5$ mm,

Here, polyethylene terephthalate (PET) material has been considered to design shaft and gear components; properties of the material have given below:

Tensile strength, $\sigma_t = 61.67$ MPa,

Compressive strength, $\sigma_c = 92.9$ MPa,

The stress and strain condition of the knob shaft is shown in Figs. 12 and 13 and summarized in Table 3.

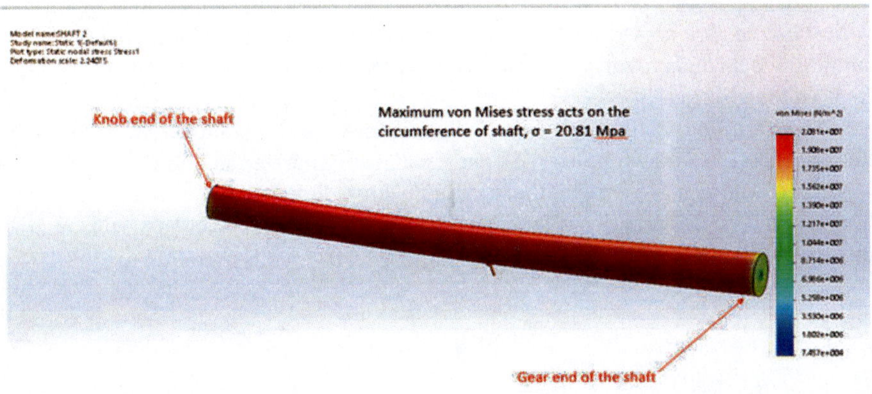

Fig. 12 Stress analysis of shaft

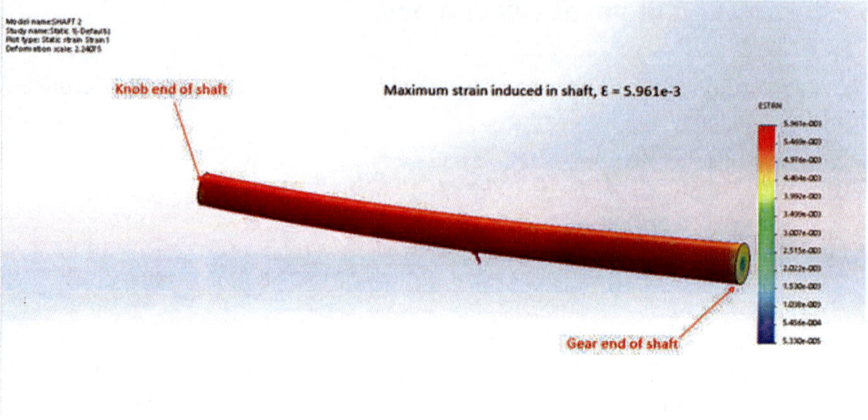

Fig. 13 Strain analysis of shaft

Table 3 Results of stress analysis of shaft

Parameters	Values
Maximum stress induced in the shaft, σ	20.81 MPa
Yield stress of shaft material, σ_y	61.67 MPa
Maximum strain induced in the shaft, ε	5.961e–03

From Table 3, it is clear that the stress induced in the shaft is less than the yield strength of the material; therefore, shaft design is safe and can be operated for infinite times.

(b) **Stress analysis of spur gear-2**

A small spur gear is used to transmit motion from the input shaft to the camshaft with the help of a crown gear mounted on the camshaft. As the dimensions of the small spur gear and crown gear are the same; therefore, torque and motion transmitted by both the gears are the same.

Torque applied on the gear, $T = 29.4\text{e–}06$ Nm. The stress and strain condition of the spur gear-2 is shown in Figs. 14 and 15 and summarized in Table 4.

From Table 4, it is clear that the maximum stress induced in the small spur gear is less than the yield strength of the material; therefore, this gear can be used to operate this mechanism for infinite cycles.

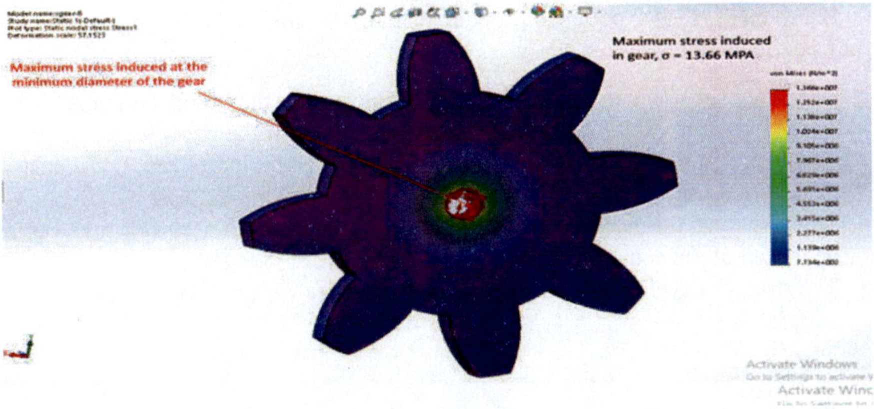

Fig. 14 Result of stress analysis of spur gear-2

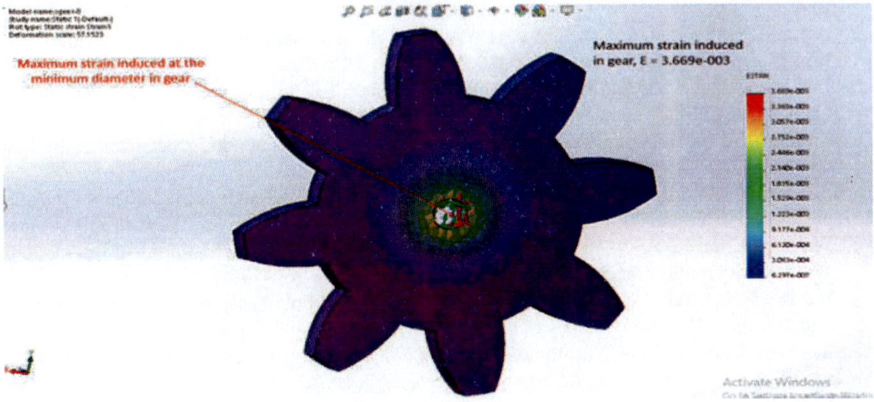

Fig. 15 Result of strain analysis of spur gear-2

Table 4 Results of stress analysis of spur gear-2

Parameters	Values
Maximum stress induced in gear, σ	13.66 MPa
Yield stress of gear material, σ_y	61.67 MPa
Maximum strain induced in gear, ε	3.699e–03

6 Conclusions

In this work, actuating mechanisms for a piezoelectric-based micropump have been developed and analyzed. Spur gear and crown gear have been used to transmit motion and torque between two perpendicular shafts. To operate a micropump, a cam has been designed.

Based on the results obtained for both the mechanisms, the following conclusions may be drawn.

- The designed timer mechanism can be used to time the blood sampling in 4 s. The simulation result matches with the theoretical calculations without any error.
- The designed cam-follower mechanism is capable of providing required displacement of 1 mm for 3 s of the follower.
- The actuating mechanism is not subjected to jerk at any instant of time during operation of the cam-follower.
- The stresses developed in knob gear and shafts are within the allowable value.

References

1. Laser DJ, Santiago JG (2004) A review of micro-pumps. J Micromech Microeng 14(6):R35–R64
2. Abhari F, Jaafar H, Md Yunus NA (2012) A comprehensive study of micropumps technologies. Int J Electrochem Sci 7(10):9765–9780
3. Das PK, Hasan ABMT (2017) Mechanical micropumps and their applications: a review. AIP Conf Proc 1851
4. Smits JG (1990) Piezoelectric micropump with three valves working persistaltically. Sens Actuators A 21(1–3):203–206
5. Spencer WJ, Corbett WT, Dominguez LR, Shafer BD (1978) An electronically controlled piezoelectric insulin pump and valves. IEEE Trans Sonics Ultrason 25(3):153–156
6. Wang X, Zhou Z, Ye X, Li Y, Zhang W (1998) A PZT-driven micropump. In: International symposium on micromechatronics and human science, pp 269–272
7. Haldkar RK, Gupta Vk, Sheorey T (2017) Modeling and flow analysis of piezoelectric based micropump with various shapes of microneedle. J Mech Sci Technol 31(6):1–9
8. Tsuchiya K, Naoyuki N, Yasutomo U, Eiji N (2005) Development of blood extraction system for health monitoring system. Biomed Microdev 7(4):347–353. https://doi.org/10.1007/s10544-005-6077-8
9. Haldkar RK (2019) Design, simulation and analysis of piezo actuated micropump for blood sampling. PhD Thesis, PDPM Indian Institute of Information Technology, design and manufacturing, Jabalpur, India
10. Ramasubramanian MK, Barham OM, Swamynathan V (2008) Mechanics of mosquito bite with applications to microneedle design. Bioinspirat Biomimet. https://doi.org/10.1088/1748-3182/3/4/046001

Robotics

An Optimization-Based Design of Open-Chain Manipulator Arm: Incorporating Dimensional Uncertainty

Saurabh Chaudhary, Virendra Kumar, and Soumen Sen

Abstract We present an approach to design an open-chain manipulator arm (OCMA) addressing dimensional uncertainty. A constrained design optimization problem (DOP) is framed, in that design parameters (DP) and constraints cast in the interval domain. The design of the OCMA addresses the criteria that the arm achieves high dynamic manipulability with maximized reachable space. Hence, a multi-objective (MO) DOP is formulated considering two objective functions, namely *Global Dynamic Manipulability* and the *Kinematic Workspace*. It considers the inertial and geometric properties of the links. Optimization of the interval domain provides a solution in terms of a nominal value of the DP with a so-called tolerance. The design constraints are imposed on the maximum allowable tool-point deflection and maximum developed stress. A 2R OCMA moving under gravity and with a payload is considered to show the method's effectiveness, giving rise to a tolerance design of the manipulator.

Keywords Robot design · Interval method · Dynamic manipulability

1 Introduction

A manipulator is designed to an assured performance, quantified through some performance metrics. Generally, a design optimization problem is formulated using the performance criteria. This is associated with several design constraints, including strength and allowable deflection, joint range, and some feasible limits for the dimensions of the link geometries. A conventional design can only result in nominal values of the design parameters. However, unmodeled uncertainties in designing and manufacturing the components and their assemblies (including off-the-shelf components)

S. Chaudhary (✉)
Indian Institute of Engineering Science and Technology, Howrah, India
e-mail: saurabhc783@gmail.com

V. Kumar · S. Sen
CSIR-Central Mechanical Engineering Research Institute, Durgapur, India

V. K. Gupta et al. (eds.), *Recent Advances in Machines and Mechanisms*, Lecture Notes in Mechanical Engineering, https://doi.org/10.1007/978-981-19-3716-3_39

are always associated with variations or tolerances. These variations result in deviations of performance from the desired one. A preferred approach to deal with the uncertainties is to bring in the tolerable perturbations in performance as the system's specification at the design level itself and obtain the variations in design parameters (such as dimensions) as design tolerances. The interval method offers a powerful tool to deal with these parameter and performance uncertainties and is efficiently used in the design optimization of different machines and their parts, considering the uncertainties in the system. A design parameter has a nominal center value and a radius (tolerance) in the interval method [1].

In the robotics literature, there are instances of interval methods used to analyze robot workspace, robot design, and inverse kinematics. The interval Newton method [2, 3, 5, 6] and Krawczyk method [4] perform the inverse kinetics redundant manipulator. The optimum dimensions of the manipulator have been found in [7], where the robot has to reach some predefined configurations. A parallel manipulator robot's design is optimized in [8], and tolerances in the design parameters of Stanford's arm are determined in [9]. On another side, various works in literature are present in which performance measures are used to design manipulators. Several performance metrics were presented in the literature and showed their effectiveness in the designing of manipulators. A few performance measures such as manipulability, dynamic manipulability, isotropy, conditioning index, and generalized inertia ellipsoids keep the manipulator away from singular configurations. The manipulability measure proposed in [10] presents an ability to maneuver the manipulator uniformly within its workspace. Dynamic manipulability [11] is an addition to manipulability, where the inertia of the manipulator is considered in this performance measure which gives the ability to accelerate uniformly. Normalized Jacobian's condition number, i.e., isotropy, is used to examine the conditioning (ill-conditioned or not) of the Jacobian matrix, which also inspects how equally the joint rates contribute to the tool-point velocity. Generalized inertia ellipsoid is discussed in [12], which measures the task space inertia of the manipulator. Global conditioning index (GCI) [13] is defined as the weighted average of the inverse of the condition number of the Jacobian matrix computed over the entire robot. Using GCI, Kumar. V et al. [14] found the optimum link lengths of a serial link redundant manipulator. Three performance measures, i.e., structural length index, global conditioning index, and modified dynamic conditioning index [15], are used to find the optimal length of a 7-DOF serial link manipulator. A multi-objective optimization approach [16] is used to find the optimal link lengths.

The studies cited above use performance measures in manipulator design, but the consequence of uncertainties was not effectively attended. In reality, the existence of integral uncertainties causes deviation of the performance of the robot manipulator from the designed desired performance. In this article, an approach is presented, which considers the uncertainties present in the geometry of the links. The proposed approach uses the manipulator's *dynamic manipulability* and *workspace area* as the objective function in global and some primary constraints for the multi-objective optimization problem. One of the constraints enforces a condition on developed stress so that it is always kept below the allowable strength of the chosen material.

The second constraint considers the deflection at the tip of the end effector (*in worst-case configuration*), which should be kept within a specified tolerable value. The optimization is formulated by taking design parameters and constraints in the interval (mid-value and radius). The interval form results gives the output nominal values and the design parameters' tolerances as radii of the intervals. A 2R OCMA moving under gravity and with a payload is considered to show the method's effectiveness, giving rise to a tolerance design of the manipulator.

2 Interval Method and Manipulator Performance Metric

The fundamental principles and the laws of interval methods used in this work are presented [1]. An interval A is written as (i) using lower and upper bound $A = [\underline{A},\ \overline{A}]$ and (ii) by specifying the center and its radius $A = \{\langle x,\ y\rangle\}$ where $\underline{A}$ is lower and $\overline{A}$ is upper bound, and x = center and y = radius. Two objective functions are chosen to formulate the design optimization problem, namely the dynamic manipulability and the workspace area are discussed below.

2.1 *Dynamic Manipulability*

Dynamic manipulability (DM) characterizes the maneuverability of the robot manipulator taking into account the arm kinematics and dynamics. It deals with the degree to which a manipulator can modify the end-effector acceleration in all directions for a given joint torque constraint. The dynamic manipulability [11] is described

$$\eta_{\mathrm{d}} = \sqrt{\det\left(J\left(M^{\mathrm{T}}M\right)^{-1}J^{\mathrm{T}}\right)} \tag{1}$$

where $M = M(\theta)$ is mass inertia matrix (configuration dependent), $J(\theta)$ is the Jacobian matrix. If the manipulator is non-redundant, then the DM shrinks to

$$\eta_{\mathrm{d}} = \frac{\det(J)}{\det(M)} \tag{2}$$

DM of a 2R OCMA shown in Fig. 1 is defined by Eq. (2), and the elements of the mass inertia and Jacobian matrix are.

$$J = \begin{bmatrix} -L_1S_1 - L_2S_{12} & -L_2S_{12} \\ L_1C_1 + L_2S_{12} & L_2C_{12} \end{bmatrix}$$

M

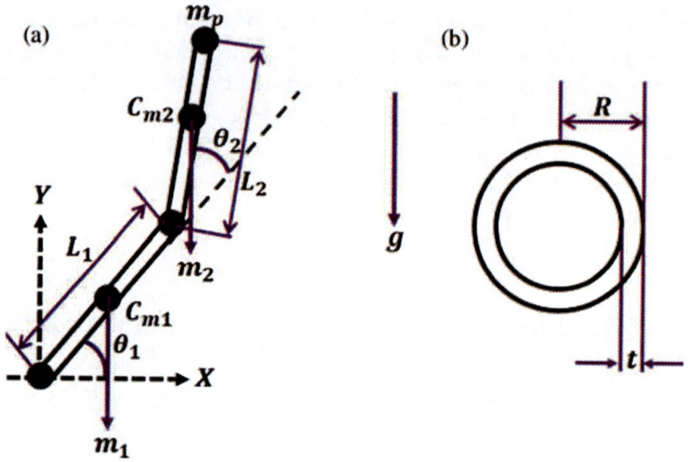

Fig. 1 **a** Schematic diagram, **b** cross-sectional view of 2D of manipulator

$$= \begin{bmatrix} I_{ZZ1} + I'_{ZZ2} + m_1 C_{m1}^2 + m'_2\left(L_1^2 + C'^2_{m2} + L_1 Cm'_2 C_2\right) & I'_{ZZ2} + m'_2\left(C'^2_{m2} + L_1 C'_{m2} C_2\right) \\ I'_{ZZ2} + m'_2\left(C'^2_{m2} + L_1 C'_{m2} C_2\right) & I'_{ZZ2} + m'_2 C'^2_{m2} \end{bmatrix}$$

where C_1, C_{12}, S_1, S_{12} are $\cos(\theta_1)$, $\cos(\theta_1 + \theta_2)$, $\sin(\theta_1)$ and $\sin(\theta_1 + \theta_2)$, L_1, L_2, m_1, m'_2, θ_1 and θ_2 are the lengths, mass, and joint angles of the first link and second link (mass with payload), Cm_1 and Cm'_2 are center of mass first and the second link (with payload), I_{zz1} and I'_{zz2} are the mass moment of inertia of the first link and the second link (with payload). The above-defined DM is configuration dependent and known as a local performance measure. The global variant of the DM is obtained by taking the average over the entire workspace and given as.

$$\eta_{d_g} = \frac{\int_w \eta_d \mathrm{d}w}{\int_w \mathrm{d}w} \quad (3)$$

where $\int_w \mathrm{d}w$ is the volume of the entire workspace. Numerically, for the global variant of DM computation, the whole workspace is distributed into several points (scattered over the grid). If N is the total points on the grid, η_{d_i} is the DM (configuration dependent); at any point, DM is defined as

$$\eta_{d_g} = \frac{1}{N}\sum_{i=1}^{N} \eta_{d_i} \quad (4)$$

2.2 Workspace Area

The second objective function is chosen as the robot manipulator's workspace area. The shape of the workspace of the 2R OCMA is not entirely circular, but it is an annular shape. The outer radius $A_{r_1}^2$ of the annular workspace is equal to the sum of the link lengths, and the inner radius $A_{r_2}^2$ is equal to the difference in the link lengths. The area is given as

$$W_A = \pi A_{r_1}^2 - \pi A_{r_2}^2 \tag{5}$$

3 Problem Formulation

A multi-objective maximization optimization problem has been framed for 2R OCMA, moving vertically against gravity and with a payload at the end effector. The problem is formulated using two objective functions, namely global dynamic manipulability and the workspace area, as defined in Sect. 2. Objective functions determine link geometries (length and radius) under certain structural strength constraints. A hollow right circular cylinder has been considered the links, parametrized by outer radius and wall thickness. The structural strength constraints are imposed by keeping the maximum bending stress developed below the allowable material strength (with a factor of safety) and maintain the maximum deviation at the end effector within a specified limit in the worst-case configuration possible. Moreover, the end effector is required to reach all points distributed across a predefined grid in the workspace into this optimization problem. With a specified tolerance, the end effector should reach inside the predefined grid area. The inverse kinematics of the manipulator examines the possibility of reaching a point taking into account the median point of the link lengths and grid points. The forward kinematics of the manipulator checks the tolerance at grid points by taking the link lengths (in interval form). As a design parameter, the link length ratio is considered to reduce the degree of complexity of the problem. The optimization problem is cast in the interval domain by considering the design variables and other constant material factors as intervals. The objective function and the constraint relations are also expressed and are computed in the intervals. The multi-objective optimization problem 2R OCMA is framed by merging both the objective functions by taking a weighted sum. To cope with the unit mismatch issue, the objective metrics are normalized in formulating the multi-objective function. The optimization problem is thus framed as.

Cost Function:

$$\underset{\eta_{d_g}, W_A}{\text{maximize}}\ f = \omega_1 \times \eta_{d_g} + \omega_2 \times W_A$$

Subjected to

$$\rho_{\min} \leq \rho \leq \rho_{\max}$$

$$\delta_{\text{end - effector}} < \delta_{\text{allow}}$$

$$\sigma_i \leq \sigma_{\text{allow}}$$

$$\theta_i \leq \theta_i \leq \theta_{i_{\max}}$$

$$R_{i_{\min}} \leq R_i \leq R_{i_{\max}}$$

$$\in_{i_{\min}} \leq \in_i \leq \in_{i_{\max}}$$

$$r_{i_{\min}} \leq r_i \leq r_{i_{\max}}$$

where $i = 1, 2$ ω_1, ω_2 are the weighting factor, ρ is the link length ratio $\left(\rho = \frac{L_1}{L_2}\right)$, σ_i, and σ_{allow} are the maximum bending stress at each joint and allowable bending stress of the material (yield stress with FOS), $\delta_{\text{end - effector}}$ and $\delta_{\text{allowable}}$ are end-effector's deflection and maximum allowable end-effector's deflection, R_i is the link radius of each link, $\in_i$ and r_i are the tolerance in each link length and radii, respectively. In this optimization, problem aluminum is taken as the material of the links. The properties of the aluminum are taken in the interval domain, i.e., density is [2640, 2810] kg/m^3, Young's modulus of elasticity is [68.7, 70] N/m^2, yield strength of the material is 240×10^6 N/m^2 with FOS as 2. The manipulator's parameters are also taken in the interval domain, i.e., g is [9.8, 9.81] m/s^2, link thickness t is [1.75, 2.25] mm, the payload is [4.9, 5.1] kg, δ_{allow} is 2 mm, end-effector tolerance is ±5 mm, and total length is 1 m. The ρ is varied from 0.2 to 2; tolerances in length are varied from 50 µm to 6 mm; link radii and tolerances are varied from 10 mm to 50 mm and 10 µm to 50 µm, respectively; the joint angles vary from 0 to π rad; the manipulator is operating only in the first two quadrants. The optimization is solved using the grid search method, and the steps are given below. In which, the objective functions and the constraints are expressed in intervals.

(i) Initialize with material properties, and constraints, geometric parameters.
(ii) Set the bounds (lower and upper) of design parameters.
(iii) Start the iteration from the lower to the upper bound of the chosen DP and check the constraints.
(iv) If all the constraints are satisfied, calculate the objective function.
(v) Change the design value and repeat the step (iii) and (iv) until the upper bound of the design variables is reached.

Table 1 Optimized design parameters with varying weights

Weights		Optimized design variables				
ω_1	ω_2	ρ	$\in_1$ (mm)	R_1 (mm)	R_2 (mm)	$\in_2$ (μm)
0.1	0.9	1	2.7	36.7	18.9	50
0.2	0.8	1.2	2.7	36.7	18.9	50
0.3	0.7	1.2	2.7	36.7	18.9	50
0.4	0.6	1.4	3.4	36.7	18.9	50
0.5	0.5	1.4	3.4	36.7	18.9	50
0.6	0.4	1.4	3.4	36.7	18.9	50
0.7	0.3	0.2	3.4	36.7	27.8	50
0.8	0.2	0.2	3.4	36.7	27.8	50
0.9	0.1	0.2	3.4	36.7	27.8	50

(vi) Find the maximum objective function value.
(vii) Select the optimum design parameters corresponding to the maximum objective function.

4 Results and Discussion

The formulated problem is numerically solved through the algorithm in Sect. 3. The solution provides the optimum link length ratio and link radii of the manipulator with tolerances. The multi-objective optimization problem is formulated by combining the objective functions in a single objective function using a weighted sum of the normalized individual functions. The objective function is normalized as $f_{\text{normal}} = \frac{f}{\text{mid}(f_{\max}) - \text{mid}(f_{\min})}$ and calculated in the interval domain, and $\text{mid}(f_{\max})$ and $\text{mid}(f_{\min})$ are the midpoint of the maximum and minimum interval values of the objective function, respectively. The optimization problem is solved in two parts. In the first part, equal weights are used, and in the second part, different weights are used such that their sum is always equal to 1. When the weights are equal, the optimal link length ratio is 1.4, the tolerance in link length is ±3.4 mm; the link radii are 36.7 and 18.9 mm, respectively, with tolerance as ±50 μm. The optimized design variables using different weights are arranged in Table 1, and values are not rounded off. All the calculations are performed on MATLAB© using INTLAB© [17].

5 Conclusion

The article presents an interval mathematics-based method for designing OCMA, providing results in interval valued design parameters (nominal values and tolerances). The proposed methodology addresses the inherent uncertainties (errors due to manufacturing and other errors) in designing an OCMA. The problem is formulated in a multi-objective optimization framework, and the solution is obtained by the nterval method. The proposed optimization method in the interval domain is applied to design a 2R OCMA, moving vertically under gravity. Global dynamic manipulability and workspace area are chosen as the objective function and solved in the interval domain. This leads to a tolerance design of the links for given variations in performance. Future work will focus on the algorithm part of the proposed methodology and explore an improved algorithm than a basic grid search method.

References

1. Moore RE, Kearfott RB, Cloud MJ (2009) Introduction to interval analysis. Soc Indus Appl Mathe
2. Castellet A, Thomas F (1997) Towards an efficient interval method for solving inverse kinematic problems. In: Proceedings of international conference on robotics and automation. vol 4(4), pp 3615–3620
3. Kumar V, Sen S, Roy SS, Das S, Shome SN (2015) Inverse kinematics of redundant manipulator using interval newton method. Int J Eng Manuf 5(2):19–29
4. Rao RS, Asaithambi A, Agrawal SK (1998) Inverse kinematic solution of robot manipulators using interval analysis. J Mech Des 120(1):147
5. Pac MR, Popa DO (2012) Interval analysis for robot precision evaluation. In: Proceedings-IEEE internatonal conference robotics automation, pp 1087–1092
6. Pac MR, Rakotondrabe M, Khadraoui S, Popa DO, Lutz P (2013) Guaranteed manipulator precision via interval analysis of inverse kinematics. In: 15th international conference adv veh. technology 10th international conference des. education 7th international conference micro-nanosystem, vol 1 (9). p V001T09A017
7. Eric Lee JPM (2003) Constantinos mavroidis, geometric design of spatial RRR manipulators using interval analysis. NSF Des Manuf Grantees Conf 6–9
8. Hao F, Merlet JP (2005) Multi-criteria optimal design of parallel manipulators based on interval analysis. Mech Mach Theory 40(2):157–171
9. Wu W, Rao SS (2007) Uncertainty analysis and allocation of joint tolerances in robot manipulators based on interval analysis. Reliab Eng Syst Saf 92(1):54–64
10. Yoshikawa T (1984) Analysis and control of robot manipulators with redundancy. In: Robotics research the first international symposium, pp 735–747
11. Yoshikawa T (1985) Dynamic manipulability of robot manipulators. In: Proceedings 1985 IEEE international conference on robotics and automation, vol 2. pp 1033–1038
12. Asada H (1983) A geometrical representation of manipulator dynamics and its application to arm design. J Dyn Syst Meas Control 105(3):131
13. Gosselin C, Angeles J (1991) A global performance index for the kinematic optimization of robotic manipulators. J Mech Des 113(3):220
14. Kumar V, Sen S, Roy SS, Har C, Shome SN (2014) Design optimization of serial link redundant manipulator: an approach using global performance metric. Proc Technol 14:43–50
15. Hwang S, Kim H, Choi Y, Shin K, Han C (2017) Design optimization method for 7 DOF robot manipulator using performance indices. Int J Precis Eng Manuf 18(3):293–299
16. Lim H, Hwang S, Shin K, Han C (2010) Design optimization of the robot manipulator based on global performance indices using the grey-based taguchi method. IFAC Proc 43(18):285–292
17. Rump SM (1999) INTLAB—interval laboratory. In: Developments in reliable computing. Springer, pp 77–104

Optimum-Blended Trajectory Generation of ABB SCARA Robot to Minimize Travel Time and Jerk with Dynamic Motion Analysis

Kaustav Ghar, Bhaskar Guin, Nipu Modak, and Tarun Kanti Naskar

Abstract The ABB SCARA robot is widely used in industries as an assembly-line robot. This paper aims at presenting an optimized trajectory so that robotic manipulation can be achieved in minimum time while ensuring adequate motion smoothness, i.e., minimizing jerk. Joint trajectories can be interpolated using polynomial and trigonometric blending functions. This paper uses two types of blending trajectories—a linear segment with parabolic blends (LSPBs) and a linear segment with sinusoidal blends (LSSBs). An optimization problem is formulated to minimize travel time and jerk using G.A. The maximum tool center point (TCP) velocity and acceleration are considered optimization constraints. Graded weight functions are introduced that facilitate converting a multi-objective optimization problem into a single-objective optimization one. Lagrangian mechanics is used for dynamic analysis. The validation of the work is done using SolidWorks®.

Keywords Optimum joint trajectory · LSPB · LSSB · GA · TCP · Minimum travel time · Minimum jerk · Lagrangian dynamics

1 Introduction

Since the development of the first industrial robot in 1959, there has been continuous development in the field of robotics [1]. Robots are now equipped with advanced microcontrollers that enable humans to control a robot's motion using software [2]. SCARA robot is one of the most commonly used assembly-line robots [3]. Recently, a modular design of a traditional SCARA robot, namely M-SCARA, has been presented in [4]. Reducing the joint jerk is desired for smooth performance. The lower is the jerk greater is the degree of smoothness of its motion. The jerk of a joint depends on the type of polynomial used as trajectory. Joint jerk minimization using G.A. has been presented in [5]. It uses joint angle, velocity, angular acceleration, and torque as optimization constraints. A smooth trajectory has been generated using

K. Ghar · B. Guin · N. Modak (✉) · T. K. Naskar
Department of Mechanical Engineering, Jadavpur University, Kolkata, India
e-mail: nmechju@gmail.com

V. K. Gupta et al. (eds.), *Recent Advances in Machines and Mechanisms*, Lecture Notes in Mechanical Engineering, https://doi.org/10.1007/978-981-19-3716-3_40

the parametric NURBS curve [6]. The approach uses gradient-based optimization to minimize the jerk. A time-optimal smooth joint path planning is presented in [7], where piecewise cubic and B-spline polynomial curves are used for trajectory. A smooth trajectory planning is presented in [8] using a fifth-order B-spline.

The paper aims at presenting a smooth trajectory of the ABB SCARA robot for minimum possible travel time, which would be beneficial for industrial application. Different polynomial and trigonometric joint interpolation functions are constructed keeping in view the workspace. The effect of minimizing the jerk and travel time has been compared using MATLAB®. The kinematic and dynamic models of the robot are validated for those trajectories using Solidworks® motion analysis.

2 Mathematical Modeling

The ABB SCARA robot is a 4-axis R-R-P-R robot. The moment of inertia of a body about the longitudinal axis is less than any other axis. Hence, the torque required for roll motion of the end effector is negligible compared to other joints. That is why it is not considered in the analysis.

The forward kinematic analysis obtains the workspace. The trajectory in the Cartesian space is chosen such that it lies within the reachable workspace. Three precision points define the trajectory—start (A), stop (C), and via point (B), as shown in Fig. 1a. The joint actuation required to allow the end effector to trace the trajectory is obtained through the inverse kinematic model.

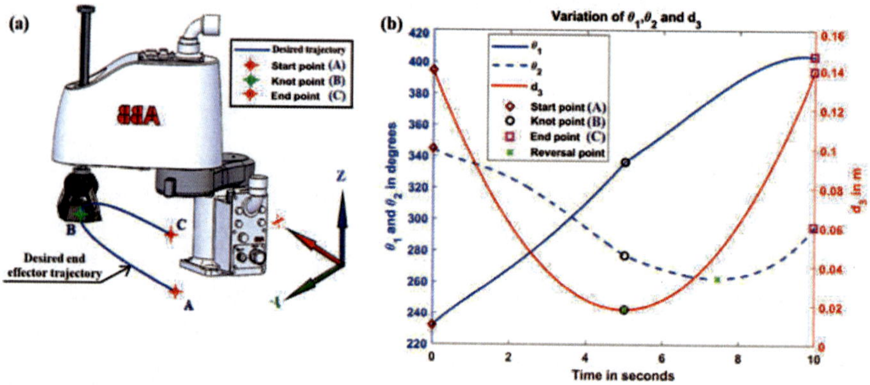

Fig. 1 **a** End-effector trajectory in Cartesian space, **b** variation of θ_1, θ_2, and d_3

2.1 Inverse Kinematics Model

The variation of the joint actuation has been computed using inverse kinematics [9]. Coordinates at any point on the trajectory are defined by $[x_e, y_e, z_e]^T$ in Cartesian space. The actuation parametersθ_1, θ_2, and d_3 can be obtained as shown in Eq. 1,

$$\begin{cases} \theta_2 = \pm\cos^{-1}\left(\frac{x_e^2+y_e^2-L_1^2-L_2^2}{2L_1L_2}\right) \\ \theta_1 = \tan^{-1}\left(\frac{-(L_2\sin\theta_2)x_e+(L_1+L_2\cos\theta_2)y_e}{(L_1+L_2\cos\theta_2)x_e+(L_2\sin\theta_2)y_e}\right) \\ d_3 = L_{01} - Z_e \end{cases} \quad (1)$$

where L_1, L_2 be the lengths of link 1 and link 2, respectively. L_{01} is the offset distance between link 1 and link 2. The quadrant correction has to be implemented to select the desired value of the joint actuation after inversion; for this purpose, the multiple solutions obtained in the current step can be compared with the value of the previous step to obtain the required solution.

2.2 Joint Trajectory Interpolation

From the variations of the joint actuation, it has been observed that there may be a point in the joint trajectory at which the actuator reverses the direction of its motion. This point is termed the reversal point. The reversal point divides the joint trajectory into two segments. Depending on its position, the knot point may lie in the 1st or 2nd segment, or it may coincide with the reversal point. If the knot point coincides with the reversal point, then it is treated as a reversal point. So, apart from the start and endpoint, the trajectory has a knot point and sometimes a reversal point.

The joint trajectory can be interpolated using a linear segment with polynomial [10] or trigonometric blends. The general equation of a linear segment with a polynomial blend is represented in Eq. 2.

$$\theta(t) = \begin{cases} \sum_{i=0}^{n_1} A_i t^i; & 0 \le t \le t_a \\ \sum_{i=0}^{n_2} B_i t^i; & t_a \le t \le t_d \\ \sum_{i=0}^{n_3} C_i t^i; & t_d \le t \le t_f \end{cases} \quad (2)$$

where t_a and t_d are the blend times for acceleration and deceleration, respectively. If the knot point lies in the 1st blend segment, then $n_1 = n + 1, n_2 = 1, n_3 = n$. If the knot point lies in the 2nd blend segment, then $n_1 = n, n_2 = 1, n_3 = n + 1$. If the knot point lies in the linear segment, then $n_1 = n, n_2 = 2, n_3 = n$. Here, n = 2 for a linear segment with parabolic blends (LSPB). A linear joint trajectory can also be blended using sinusoidal segments [11]. The normalized equation for a linear segment with sinusoidal blends is shown in Eq. 3.

$$\overline{\theta}(t) = \begin{cases} \dfrac{\frac{2t_a}{\pi}\left(\sin\left(\frac{\pi t}{2t_a} - \frac{\pi}{2}\right) + 1\right)}{f} & 0 \le t \le t_a \\ \dfrac{\frac{2t_a}{\pi} + t - t_a}{f} & t_a \le t \le t_d \\ \dfrac{\frac{2t_a}{\pi} + t_d - t_a + \frac{2(1-t_d)}{\pi}\sin\left(\frac{\pi(t-t_d)}{2(1-t_d)}\right)}{f} & t_d \le t \le 1 \end{cases} \quad (3)$$

where $f = \frac{2t_a}{\pi} + t_d - t_a + \frac{2(1-t_d)}{\pi}$ and t_a, t_d are normalized blend times for acceleration and deceleration, respectively. $\overline{\theta}(t)$ is the normalized joint displacement.

LSPB and LSSB trajectories with or without reversal points can be constructed using Eqs. 2 and 3, respectively. The LSSB trajectory is interpolated such that the knot point always lies in the linear segment since constructing a cycloid profile passing through three assumed points is a complex mathematical procedure. However, for the LSPB trajectory, if the knot point lies in any segment, the corresponding segment is interpolated using a polynomial of the next higher order. The joint temporal behavior includes joint velocity $\dot{\theta}$, acceleration $\ddot{\theta}$, and jerk $\dddot{\theta}$ computed using the central difference method [12].

2.3 *Joint Dynamic Behavior*

The dynamic equation of a robotic manipulator consists of inertia, centrifugal, Coriolis, friction, and gravity terms. Lagrangian dynamic formulation [13] has been used to formulate the dynamic equation of the robotic manipulator. Equation 4 shows the generalized equation of motion for a robotic manipulator:

$$M(q)\{\ddot{q}\} + H(q, \dot{q}) + F(\dot{q}) + G(q) = \tau \quad (4)$$

where q is the joint variable, $\dot{q}$ is the generalized velocity, $\ddot{q}$ is the generalized acceleration, and τ is the generalized torque vector. M is the inertia terms; H is the Coriolis and centripetal force terms; F is the viscous and Coulomb friction force term, and G is the gravity term. The inverse dynamics [14] approach has been used to compute the joint torques from their respective temporal behavior.

2.4 *Optimization*

Minimum travel time

The objective of this problem is to minimize the travel time. The constraints taken into consideration are maximum allowable velocity and acceleration of the tool center point (TCP). A constant value of blending time is considered for constructing the LSPB and LSSB joint trajectories. The total travel time is reduced at each iteration until constraints are violated. Hence, minimum travel time for a particular blend time can be obtained.

Jerk minimization for constant travel time

In this problem, the travel time is assumed to be constant, while the jerk for each joint trajectory segment is minimized. The joint jerk is a function of the blend times; hence, the blending time for each joint trajectory is considered as the optimization variable. The constraints are similar to those used previously for minimizing travel time. As the joints are actuated simultaneously, the overlapping joint segments must also be optimized simultaneously, considering it a problem of multi-objective optimization. Such problems are faster and more efficient to be solved by converting them to a single-objective problem by adding the objective functions pre-multiplied by weight functions. In the case of robotic manipulators, joints are coupled, and hence, a concept of *graded weight function* is introduced. *Graded weight functions* mimic the joint coupling and temporal behavior of respective joints. The nature and process of selection of the graded weight functions are discussed below:

Selection of weights based on the degree of joint coupling (k_i^d)

More weight is given to that joint of the robot whose degree of coupling is higher, i.e., whose actuation has a significant effect on the overall kinematic and dynamics of the robot. The weights of this category are designated as k_i^d, where i is the joint number of the robot. In this problem, joint 3 can be considered to be independent of joint 1 and joint 2, and so, it carries the least weightage, whereas joint 1 is highly coupled and therefore allotted the highest weightage. So, the weight functions are graded based on the degree of joint coupling.

Selection of weights based on the time duration of each segment (k_j^t)

The smaller the travel time, the greater is the expected joint jerk; hence, more weightage is given to that segment of the joint trajectory which has a smaller duration of travel time. The weights of this category are designated as k_j^t, where j is the segment number of the robot joint trajectory.

Overall graded weight factor (k_j^o)

The overall weights for each joint trajectory segment are the products of k_i^d and k_j^t. The overall graded weight factor k_j^o is shown in Eq. 5. The overall joint jerk is the sum of the maximum jerk for each joint trajectory segment multiplied by its corresponding graded weight factor as shown in Eq. 6.

$$k_j^o = k_i^d * k_j^t \tag{5}$$

$$J = \sum_{j=0}^{n} k_j^o * \max\left|\text{jerk}_j\right| \tag{6}$$

where n is the no. of segments in the joint trajectory, j is the segment of the joint trajectory, and J is the peak joint jerk. For such an optimization problem, the following can be defined:

- *Objective function:* minimize *peak joint jerk* (J).
- *Optimization variable:* blending time t_a and t_d. For simplification, $t_d = (1 - t_a)$, i.e., equal blend times for acceleration and deceleration.
- *Constraints:* Limiting values of tool center point (TCP) velocity and acceleration denoted by v_{limit}, a_{limit}, respectively.
- *Constants:* t_{traj} (total time taken to traverse the trajectory)

Thus, the optimization problem can be stated as:

$$\text{Minimize } J(t)$$

$$\text{Subject to} : v \leq v_{limit} \text{ and } a \leq a_{limit}$$

The optimization has been carried out using MATLAB® optimization toolbox. The solver selected for this purpose is the genetic algorithm (GA) which can minimize constrained, nonlinear, single-objective multivariable problems. Hence, the optimum blend times for each segment can be obtained.

Combined travel time and jerk minimization

In this problem, the travel time is minimized along with joint jerk. The optimization problem remains the same as in the case of jerk minimization for constant travel time, discussed previously. However, the travel time t_{traj}, is not constant but reduced at each iteration as long as the constraints are satisfied. Hence, it is possible to minimize the combined travel time and joint jerk.

3 Results and Discussion

Based on the problem description, the mathematical model was developed and simulated by programming in MATLAB® software. The kinematic and dynamic equations are solved, and validation of joint torques has been done using SolidWorks® motion simulation. The details are enumerated below. The desired trajectory of the end effector is constructed within the workspace region of the robot through three precision points A (−0.3248, −0.3059, −0.18) m, B (0.1131, −0.3195, −0.058) m, and C (0.3794, 0.05048, −0.18) m as shown in Fig. 1a. These points have been interpolated in the x–y and x–z planes using parabolic curves as shown in Fig. 2.

Since joint trajectories are interpolated using different polynomials, a deviation inevitably occurs at points other than A, B, and C. The combined plot of the desired trajectory and interpolated end-effector trajectory for LSPB and LSSB joint trajectory having minimized travel time and jerk along with its validation with SolidWorks®

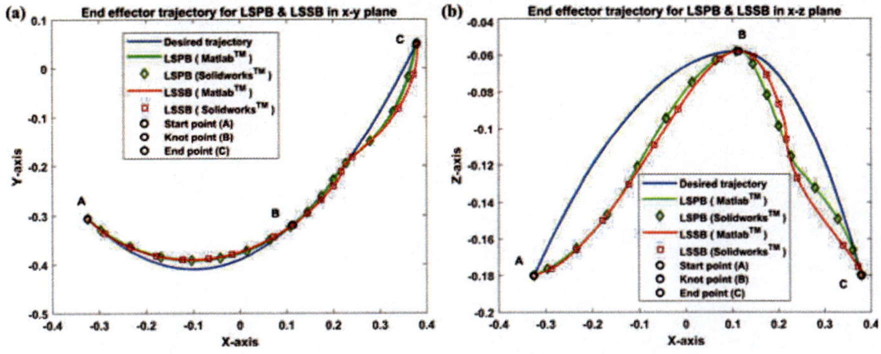

Fig. 2 Projection of end-effector trajectory for LSPB and LSSB in **a** x–y plane, **b** x–z plane

motion analysis has been shown in Fig. 2. The robot end effector gripped with a 3 kg payload is moved along the desired trajectory from start to end through the point (knot point) as shown in Fig. 1a.

Using the inverse kinematics model, the variation of θ_1, θ_2, and d_3 has been obtained as shown in Fig. 1b. It has been found that the joint variable θ_1 did not have a reversal point, while θ_2 and d_3 have a reversal point. The reversal point in d_3 coincides with the knot point; hence, the knot point is treated as the reversal point in this case. Initially, a total travel time (t_{traj}) is assumed to be 10 s, and normalized blend time $t_a = 0.2$ has been considered to interpolate the LSPB and LSSB joint trajectories for each joint; thereafter, optimization is performed to minimize the total travel time and jerk.

3.1 Effect of Blend Times on Actuator Jerk and Torque/Force

Based on the simulation results, a comparison is made for LSPB and LSSB joint trajectories for different cases.

- Case A—constant travel time and assumed blend time (variable jerk).
- Case B—constant travel time and minimized jerk.
- Case C—minimized travel time and assumed blend time (variable jerk).
- Case D—minimized travel time and minimized jerk.

Table 1 shows the variation of jerk and torques for LSPB and LSSB joint trajectories for case A and case B as obtained from simulation. The results of jerk minimization show that there has been a decrease in the R.M.S. value of jerk and also a reduction in the R.M.S. value of joint torques for each joint trajectory segment as compared to the non-optimized case (Case A). This is because minimizing the jerk smoothens the joint motion for a particular joint trajectory; hence, load on the actuator also decreases.

Table 1 Variation of the joint jerk and joint torque for case A and case B

Sl. No.	Type of Polynomial	t_{traj}	Case	Parameters	Joint 1	Joint 2		Joint 3		CM_ TCP Vel	CM_ TCP Acc
						Segment 1	Segment 2	Segment 1	Segment 2		
1	LSPB	10	A	BT	0.2	0.2	0.2	0.2	0.2	0.17	0.25
				RJ	33.08	37.67	48.97	0.03	0.03		
				T/F	0.40	0.26		30.51			
		10	B	BT	0.11	0.45	0.44	0.46	0.45	0.18	0.2
				RJ	20.85	15.11	32.63	0.02	0.02		
				T/F	0.48	0.30		30.51			
2	LSSB	10	A	BT	0.2	0.2	0.2	0.2	0.2	0.17	0.46
				RJ	70.62	82.56	52.78	0.029	0.029		
				T/F	0.60	0.39		30.51			
		10	B	BT	0.09	0.5	0.5	0.49	0.49	0.18	0.30
				RJ	46.65	15.42	22.09	0.01	0.01		
				T/F	0.57	0.35		30.51			

Here, B.T., R.J., T/F represent normalized blend time, R.M.S. value of jerk, and R.M.S. values of joint torques or forces, respectively. The unit of R.J. for joints 1 and 2 is degree/s^3, while for joint 3, its unit is m/s^3. The unit of T/F for joints 1 and 2 is Nm, while for joint 3, its unit is N CM_TCP Vel is the calculated maximum TCP velocity in m/s CM_TCP Acc is the calculated maximum TCP acceleration in m/s^2.

Table 2 shows the variation of jerk and torques for LSPB and LSSB joint trajectories for case C and case D. It has been observed that minimizing the total travel time increases the jerk. For combined travel time and jerk minimization (case D), lowering the jerk lowers the peak acceleration and velocity for a given travel time, so it is possible to reduce the travel time further. Hence, the optimum travel time obtained in this case is even lower than that in case C. The R.M.S. value of jerk is higher in this case as the travel time and jerk are two mutually conflicting optimization parameters. This high value of jerk would ideally not affect the system as the velocity and acceleration constraints are within limits. However, there is a deviation observed in torque of joint 2 for case D where the LSPB joint trajectory has a higher R.M.S. value of joint torque than the LSSB joint trajectory. Since joint 3 is independent, the force required for its actuation remains constant for all cases.

3.2 *Effect of Joint Jerk Minimization*

Figure 3 shows the variation of jerk of joint 1 for LSPB and jerk of joint 3 for LSSB joint trajectories. It has been observed that LSSB trajectories have a smoother jerk profile as it has continuous higher-order derivatives; however, the torque required is also higher than LSPB for most cases. Since sinusoidal curves are infinitely derivable, the jerk at start and end is always zero in the case of the LSSB trajectory. This is the advantage of using a sinusoidal blend as compared to polynomial blends, making it the ideal choice for use in high-speed applications.

3.3 *Variation of Joint Torques Due to Minimization of the Joint Jerk*

Figure 4 shows the plot of torque of joint 1 for LSPB and force of joint 3 for LSSB joint trajectories. It has been observed that the torque of joint 1 has more fluctuations, whereas the force of joint 3 is more uniform. This is because there exists a coupling between joint 1 and joint 2. Also, the torque/force required to actuate a joint interpolated with an LSSB is higher than that of an LSSB. For a constant travel time, minimizing the jerk reduces the joint torques/forces for each type of joint trajectories. The torque of joint 1 is higher for the combined minimum jerk and travel time except for joint 3 in the case of LSSB, where there is no significant increase in actuator force.

Table 2 Variation of the joint jerk and joint torque for case C and case D

Sl. No.	Type of Polynomial	t_{traj}	Case	Parameters	Joint 1	Joint 2		Joint 3		CM_ TCP Vel	CM_ TCP Acc
						Segment 1	Segment 2	Segment 1	Segment 2		
1	LSPB	8.73	C	BT	0.2	0.2	0.2	0.2	0.2	0.19	0.32
				RJ	49.65	56.54	73.49	0.4	0.0.4		
				T/F	0.52	0.35		30.51			
		6.49	D	BT	0.07	0.11	0.07	0.46	0.11	0.2	1.47
				RJ	90.57	205.23	536.78	0.10	0.19		
				T/F	1.27	0.89		30.51			
2	LSSB	8.60	C	BT	0.2	0.2	0.2	0.2	0.2	0.2	0.62
				RJ	110.67	129.37	82.71	0.045	0.045		
				T/F	0.81	0.54		30.51			
		7.24	D	BT	0.03	0.06	0.43	0.49	0.49	0.2	1.16
				RJ	192.41	167.27	90.28	0.02	0.02		
				T/F	1.29	0.75		30.51			

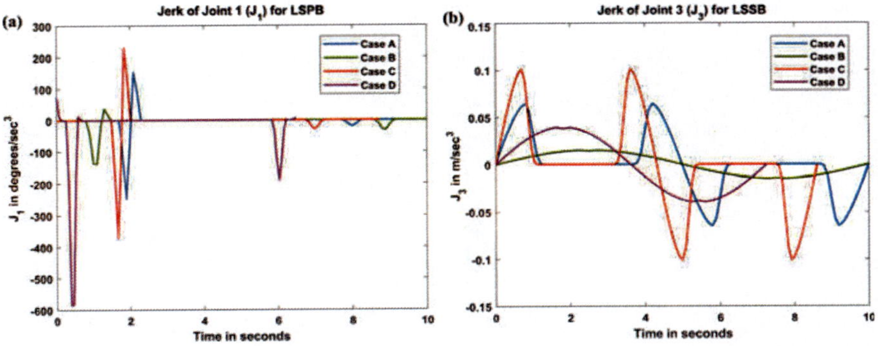

Fig. 3 Joint jerk for blended joint trajectory polynomial, **a** jerk of joint 1 for LSPB, **b** jerk of joint 3 for LSSB

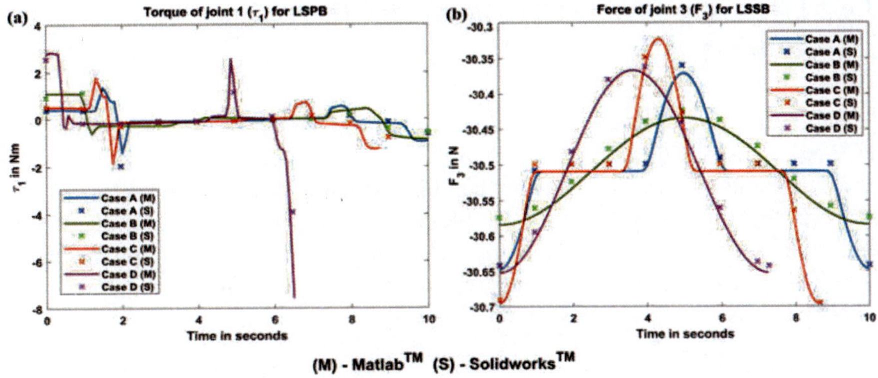

Fig. 4 Joint torques/forces for blended joint trajectory polynomial, **a** troque of joint 1 for LSPB, **b** force of joint 3 for LSSB

The validation of the joint torque and forces is done using SolidWorks® motion analysis. The initial position was set such that the end effector is located at position A; then, the joint trajectories are fed to the joints such that it may trace the path from A to C via B. Various kinematics and dynamic parameters such as velocity, acceleration, joint torque/force, and the trace of end-effector trajectory were obtained as a result of the motion simulation. It has been observed that the MATLAB® results are compliant with the results obtained from SolidWorks®, the former being conservative compared to the latter.

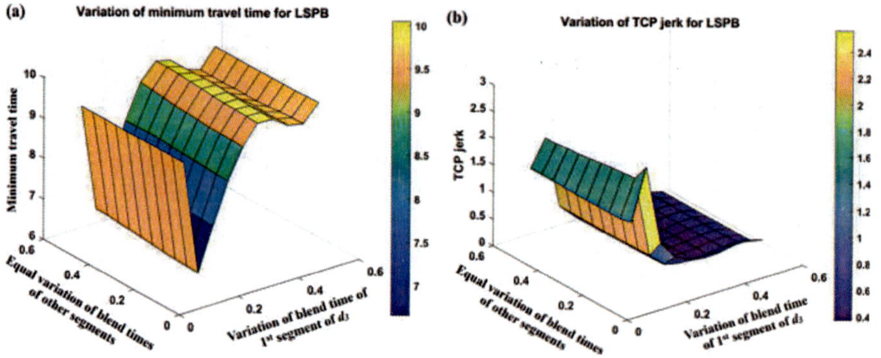

Fig. 5 Variation of, **a** minimum travel time, **b** TCP jerk, with blending time for LSPB joint trajectory

3.4 Effect of Combined TCP Jerk and Travel Time Minimization

An algorithm similar to joint jerk minimization has been followed to minimize the TCP jerk for an LSPB joint trajectory. Figure 5a, b shows the variation of minimum travel time and TCP jerk for different combinations of blend times, respectively. It is observed that the two surfaces are complementary, i.e., with the gradual decrease in minimum travel time; the TCP jerk increases, and there occurs an overall maximum peak jerk corresponding to the overall minimized travel time. Similar results can also be obtained in the case of LSSB joint trajectory.

4 Conclusion

In this paper, the objective of obtaining the optimal trajectory has been successfully obtained by minimizing jerk and travel time both independently and simultaneously using G.A. A realistic approach has been used to develop the mathematical model for the ABB SCARA robot. The paper's novelty is the use of graded weight functions, which ensure fast convergence of multi-objective optimization problems. It can be concluded that it is possible to further lower the minimum travel time by simultaneous optimization of jerk and travel time. For a particular joint trajectory, minimizing the jerk also reduces the torque/force required for its actuation, while independent minimization of travel time increases the jerk and also torque/force. The effect of using LSPB and LSSB joint trajectories has been explored. The variation of joint jerk and force is smoother for LSSB joint trajectory, and hence, it is recommended for all practical applications. The success of the proposed model has been ensured by comparing the calculated results with the dynamic response obtained from SolidWorks® motion simulation.

References

1. Singh B, Sellappan N, Kumaradhas P (2013) Evolution of industrial robots and their applications. Int J Emerg Technol Adv Eng 3(5):763–768
2. Pan Z, Polden J, Larkin N, Van Duin S, Norrish J (2012) Recent progress on programming methods for industrial robots. Robot Comput Integr Manuf 28(2):87–94
3. Suri S, Jain A, Verma N, Prasertpoj N (2018) SCARA industrial automation robot. In: 2018 international conference on power energy, environment and intelligent control (PEEIC), pp 173–177
4. Chen D, Zeng Z, Guan Y, Zhu H, Zhang T (2020, October.) SCARA robots developed with modular method. In: 2020 IEEE international conference on mechatronics and automation (ICMA), pp 454–459, IEEE
5. Huang P, Chen K, Yuan J, Xu Y (2007). Motion trajectory planning of space manipulator for joint jerk minimization. In: 2007 international conference on mechatronics and automation pp 3543–3548, IEEE
6. Hashemian A, Hosseini SF, Nabavi SN (2017) Kinematically smoothing trajectories by NURBS reparameterization–an innovative approach. Adv Robot 31(23–24):1296–1312
7. Cao B, Dodds GI (1994) Time-optimal and smooth joint path generation for robot manipulators. In: 1994 international conference on control-control'94, vol 2, pp 1122–1127, IET
8. Gasparetto A, Zanotto V (2007) A new method for smooth trajectory planning of robot manipulators. Mech Mach Theory 42(4):455–471
9. KuCuk S, Bingul Z (2004) The inverse kinematics solutions of industrial robot manipulators. In: Proceedings of the IEEE international conference on mechatronics, 2004. ICM'04, pp 274–279
10. Kunz T, Stilman M (2011) Turning paths into trajectories using parabolic blends. Georgia Institute of Technol
11. Parent R (2012) Interpolating values. Computer animation algorithms and techniques, 3rd (ed), Morgan Kaufmann, pp 81–83
12. Balaguruswamy E (1999) Numerical differentiation. In: Numerical-methods, 1st (edn.), TataMcGraw-Hill Education, pp 347–351
13. Roulstone I (2003) Lagrangian dynamics. Encycl Atmos Sci 1098–1104. https://doi.org/10.1016/b0-12-227090-8/00194-9
14. Lauß T, Oberpeilsteiner S, Sherif K, Steiner W (2019) Inverse dynamics of an industrial robot using motion constraints. In: 2019 20th international conference on research and education in mechatronics (REM), pp 1–7, IEEE

Tension Adjustment in Cable-Driven Robots Used for MIS

Amanpreet Singh and Jitendra P. Khatait

Abstract Nowadays, robotic manipulators act as the essential members in various domains to perform a wide range of tasks. In medical/surgical domain, integration of minimally invasive surgery (MIS) and robotic manipulators has given significant benefits to both, surgeons as well as to patients. In literature, notable contributions have been reported by various researchers for the design and development of MIS robots. For the transmission design of these manipulators, cables are the key elements. Being unilateral elements, they work in tension only. Therefore, for their efficient operation, pre-tension is the prerequisite. To achieve the same, different ways have been introduced in past. However, in the current work, it is noticed that design of such manipulators needs to be refined further. In this direction, current work introduces a novel tension adjustment mechanism for cable-driven robotic manipulator and presents its kinematic and force analyses. Further, various kinematic parameters and forces characterizing the proposed concept are plotted in the MATLAB environment. From these plots, it is learnt that derived formulation is in good agreement with physical behavior of the devised mechanism. To demonstrate the introduced concept, it is proposed to implement the same to a double-parallelogram-based RCM mechanism. Typically, this mechanism is used for MIS applications. For the realization of the embodiment, development of a prototype is in progress. Moreover, the proposed mechanism has potential to provide precise and accurate cable tension with mere standard manufacturing tolerances and assembly methods.

Keywords Tension adjustment · Cable-driven robot · Surgical robot · RCM mechanism · MIS · Pre-tension

1 Introduction

Mechanisms play a vital role in the wide range of applications. Broadly speaking, these include defense, space, underwater exploration and medical/surgical-related

A. Singh (✉) · J. P. Khatait
Department of Mechanical Engineering, Indian Institute of Technology Delhi, Hauz Khas, New Delhi 110016, India
e-mail: amanpreet.singh@mech.iitd.ac.in

V. K. Gupta et al. (eds.), *Recent Advances in Machines and Mechanisms*, Lecture Notes in Mechanical Engineering, https://doi.org/10.1007/978-981-19-3716-3_41

applications. In literature, a variety of motion/power transmission methods exist for the actuation of various DOFs of a given mechanism [1]. Mainly, they are of two types. First method includes diverse forms of gear drives and trains, bar linkages, etc. Second method presents the variety of tendon-drives. The implementation of the former method may degrade the system dynamics. Due to this, base-mounted actuation is preferred over joint mounted actuation [1–3]. Such actuation employs tendons and pulleys as the main transmission members. Lee [4] and Camarillo [5] have studied the mechanics and control of tendon-drives. The key demand of a tendon-drive is to have adequate initial tension as a tendon element works in tension only. Madhani et al [6] have developed an 8-DOFs cable-driven RCM manipulator for MIS applications. The required initial tension was adjusted with provision of pulley movement. On the basis of this system, Intuitive Surgical Inc. [7] has introduced state-of-the-art da Vinci Surgical System. Another 8-DOFs robotic manipulator has been reported by Chen et al [8]. The developed system was equipped with force-feedback and employs cables to drive its arm mechanism. Further, to address the limitation (large foot-print) of mechanical RCM mechanisms for MIS, Liu et al [9] have introduced a cable-based RCM linkage. The cable tension was determined using a novel methodology which was based on constraints analysis. Cavusoglu et al [10] have proposed a telerobotic surgical workstation for suturing and knot tying. The patient-side manipulator was actuated using DC servo motors via cable transmission. The RAVEN a 7-DOFs cable-driven robot was presented by King et al [11, 12]. Later, Hannaford et al. [13] presented the second version of this system, denoted as RAVEN-II, which has improved cable-routing, geometry, and tension adjustment. Another MIS robot has been reported by Li et al [14]. In the proposed system, the RCM feature based on double-parallelogram mechanism was realized using cable-pulley arrangement. A tendon-driven wearable exoskeleton for elderly and patients has been introduced by Kong and Jeon [15]. To achieve required tendon tension, springs were employed in the tendon loops. Basile and Haase [16] have presented a system for tension compensation in mechanical control cables. The proposed concept used springs for intended purpose. Segreti et al [17] have described a technique to approximate tendon length in an underactuated snake-like manipulator. Li et al [2] have proposed a novel RCM mechanism. The other tendon-driven systems have been reported in [18–20].

Several techniques of tension adjustment have been presented in past. This paper introduces a novel tension adjustment mechanism for cable-driven robots used for MIS. The proposed concept offers the better features as compared to the existing tension adjustment methods. The organization of this manuscript is as follows. Sect. 2 describes the proposed mechanism. Sects. 3 and 4 present its kinematic and force analyses, respectively. Sect. 5 develops the relation between input and output variables, and Sect. 6 discusses the variation of various mechanism parameters against the applied input. Thereafter, based upon the current findings, conclusion is drawn.

2 Mechanism Description

This section describes the devised tension adjustment mechanism (TAM) (see Fig. 1a) for cable-driven robotic manipulators. It comprised of a base link, a lever, cables and pulleys, etc. An extension spring is used to generate the pre-tension force. In addition, it compensates for the variation in tension either due to the manufacturing or assembly errors, or both. Depending upon the spring characteristics, a dash-pot mechanism may be employed. In this paper, the embodiment of the TAM is coined as primary mechanism while, TAM itself is called as secondary mechanism. On the similar basis, each of pulleys and cables utilized in these mechanisms is prefixed by words "primary" and "secondary", respectively. Figure 1a describes the different members of the TAM, whereas Fig. 1b represents the preferred embodiment of the same. It is a double-parallelogram RCM mechanism which is commonly used for MIS applications.

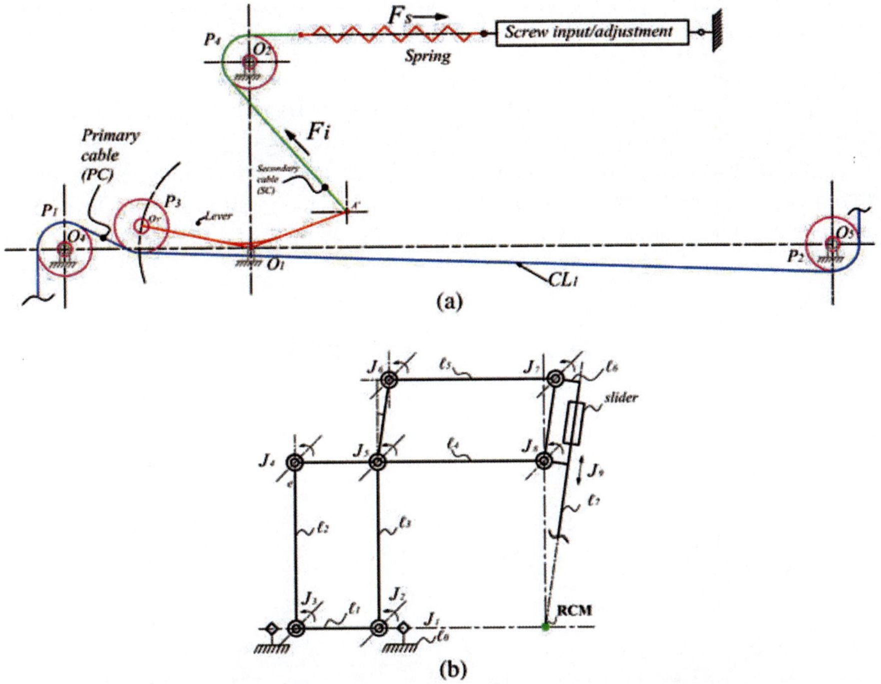

Fig. 1 Schematics of **a** concept of tension adjustment mechanism (TAM) and **b** preferred embodiment of TAM

3 Kinematic Analysis

This section derives the kinematic formulation of the proposed TAM. It is characterized by the primary cable angles (β and γ), secondary cable angle (α), lever angle (ϕ), and mechanism geometric attributes. The secondary cable angle denotes the assumed revolute joint between secondary cable and lever pin. It is formulated as a function of lever angle (ϕ), pulley radius (r_p), radius of rotation of lever (r_i), and d_{12} which is the distance between points O_1 and O_2. Figure 2 represents the two poses of TAM, first, when there is no input and second (in dashed line), after some input displacement (d) is given. In this paper, planer case is considered. All the parameters of interest are referenced with respect to frame (x, y). Further, the expression for angle α is given by Eq. (1a). To obtain the expression for initial value of this angle, i.e. α_i, put $\phi = \phi_i$.

$$\alpha = 2\tan^{-1}\left\{\frac{\left(\sqrt[2]{\left(d_{12}^2 - 2r_i d_{12}\sin(\phi) + r_i^2 - r_p^2\right)} + r_i\cos(\phi)\right)}{\left(d_{12} - r_p - r_i\sin(\phi)\right)}\right\} \tag{1a}$$

Next task is to derive the formulation for the primary cable angles β and γ. As stated earlier for the initial value of secondary cable angle (α), β_i, and γ_i corresponds to the initial values of primary cable angles. Figures 3a through c depict the initial and changed geometry of the primary cable, when some input is given. The current analysis assumed that primary cable, at point T_P, is tangent to the pulley P_3. The

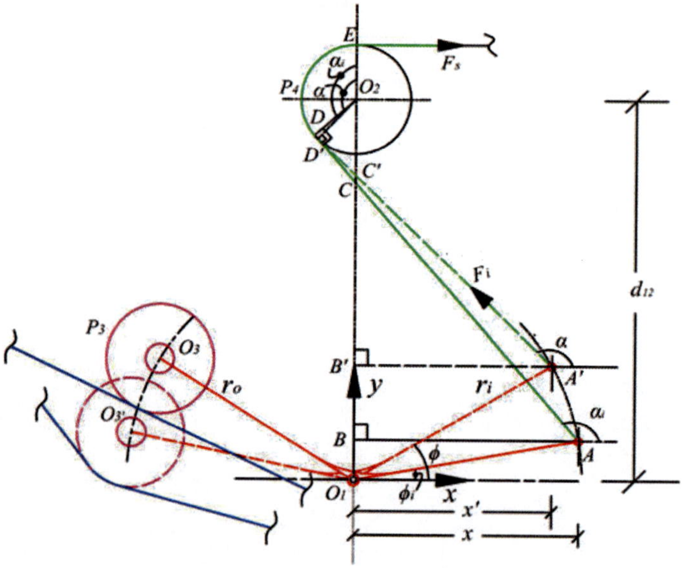

Fig. 2 Schematic representation of angles α_i, α and ϕ

distance d_{14} is expressed in terms of geometric properties to construct Eq. (2a). Put $\phi = \phi_i$ to obtain β_i.

$$2\tan^{-1}\left\{\frac{\left(d_{14} + r_o\cos(\psi) + \sqrt{\left(d_{14}^2 + r_o^2 - 4r_p^2 + 2d_{14}r_o\cos(\psi)\right)}\right)}{\left(2r_p - r_o\sin(\psi)\right)}\right\} \tag{2a}$$

where $\psi = (\phi + \theta_l)$.

Now, the formula of second primary cable angle γ is to be obtained (see Figs. 3a through d). In Fig. 3a, it is noticed that angles $(\pi - \beta_i)$ and $(2\pi - \gamma_i)$ are alternate

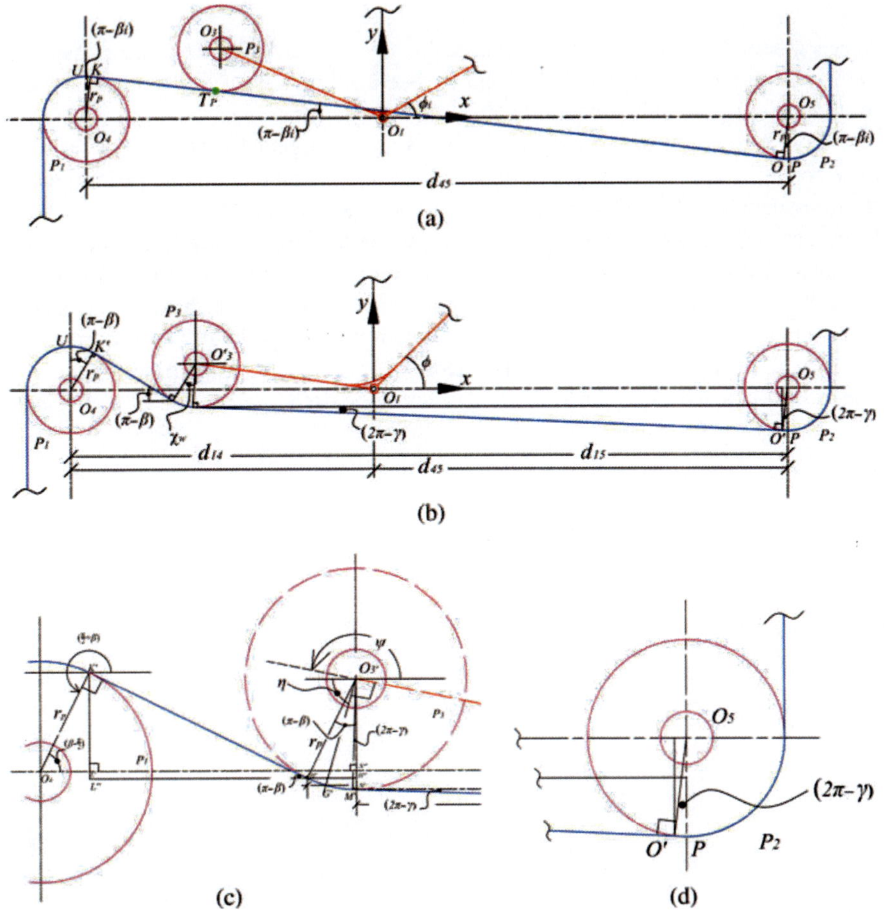

Fig. 3 Schematic representation of **a** lever, primary cable, pulley P_3, and associated geometry when $d = 0$, **b** lever, primary cable, pulley P_3, and associated geometry when $d > 0$, **c** magnified view of pulleys P_1 and P_3, **d** magnified view of pulley P_2

when there is no input. Therefore, angle γ_i is obtained following Eq. (2a) as well. The pulley P_2 is magnified (see Fig. 3d). After analyzing these figures, implementing mathematical identities and necessary rearrangements, Eq. (3a) is developed which provides the expression of angle γ. In addition to Eq. (2a), the Eq. (3a) is utilized, by substituting $\phi = \phi_i$, to get the expression of γ_i.

$$\gamma = 2\pi - \left\{ \tan^{-1}\left(\frac{(r_0 \sin(\psi))}{(d_{15} - r_0 \cos(\psi))} \right) \right\} \tag{3a}$$

4 Force Analysis

This segment deals with force analysis of the presented TAM. As the input displacement d is applied, the force F_i at point A' becomes nonzero and the same is transmitted to primary cable through lever-pulley (P_3) arrangement (see Fig. 4a). After this force transmission, primary cable is no more tangent to the pulley P_3, rather, have some nonzero wrap of primary cable. Further, T_1 and T_2 are the primary cable tensions developed by the force F_0. Figure 4b depicts the force triangle of these forces, at point G'. In this analysis, it is assumed that pulleys are friction-free, and power remains same at input and output. The following Eqs. (4a) through (4d) are formulated. It is clear from Eq. (4d) that both tensions are equal (say T). The tension (T) is obtained from tension–deformation relation of the primary cable, as given by Eq. (5a) To account for both, initial tension (T_0), if any, and tension (T_s) generated due to implementation of TAM, Eq. (5b) should be used.

$$T_1 \cos(\beta) + T_2 \cos(\gamma) = F_0 \cos(\eta) \tag{4a}$$

$$T_1 sin(\beta) + T_2 sin(\gamma) = F_0 \sin(\eta) \tag{4b}$$

$$T_2 r_p - T_1 r_p = 0 \tag{4c}$$

$$\boxed{T_1 = T_2 = T} \tag{4d}$$

$$T_s = \left\{ \frac{E_{pc} A_{pc}\left(\delta_{pc}(\phi)\right)}{\mathcal{L}^i_{pc}} \right\} \tag{5a}$$

$$T = T_0 + T_s \tag{5b}$$

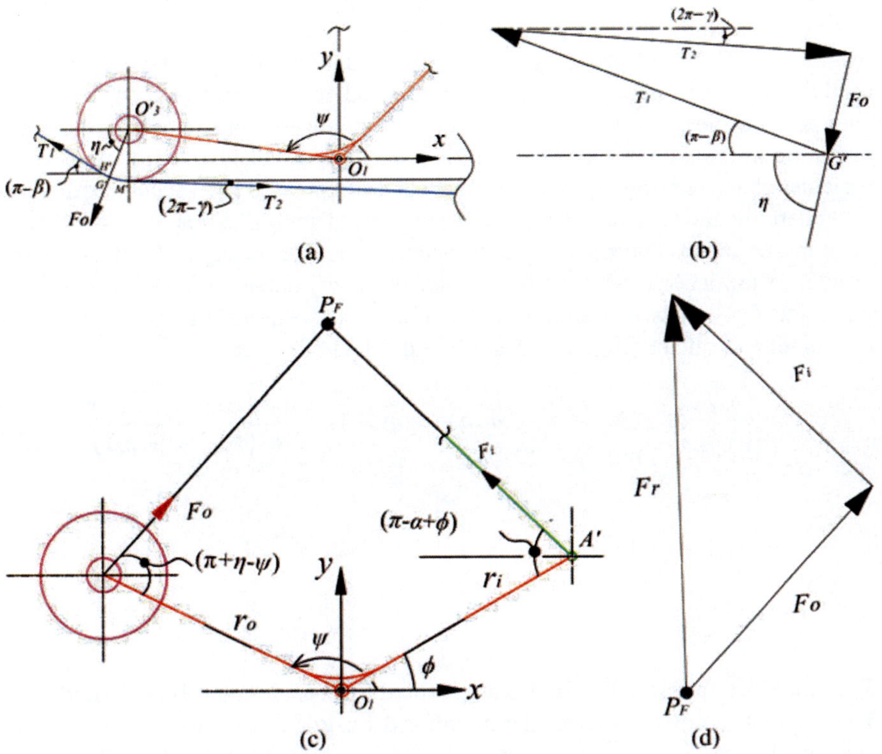

Fig. 4 **a** Schematic of principal forces on primary cable at point G' in TAM, **b** Force triangle at point G', **c** Forces, F_i and F_o, at points A' and O_3' of lever $O_3' O_1 A'$, **d** force triangle at point P_F

Next objective is to derive the formulation of input and spring forces which are assumed to be equal in magnitude. These are derived in terms of primary and secondary cables angles, lever angle, output tension, and other geometric properties. Further, force F_i is resolved at point of its action which is A (see Fig. 4c). The expression of force $F_i(\phi)$ is obtained by taking moment balance about z-axis at O_1 and after suitable rearrangements, it is presented in Eq. (10).

$$F_i = \left\{ \left(T_0 + \frac{E_{\mathrm{pc}} A_{\mathrm{pc}} \left(\delta_{\mathrm{pc}}(\phi) \right)}{\mathcal{L}^i_{\mathrm{pc}}} \right) \frac{(\cos(\gamma) + \cos(\beta))(\sin(\psi - \eta))}{\cos(\eta)\sin(\alpha - \phi)} \left(\frac{r_{\mathrm{o}}}{r_i} \right) \right\} \quad (6)$$

5 Input Displacement and Lever Angle

This section relates the input displacement (d) and lever angle (ϕ) of the proposed mechanism. Here, the objective is to determine angle ϕ which denotes the rotation of lever when some input is given by means of a screw arrangement, however, not restricted to. The prescribed tangent constraint and included angle of the lever (θ_l)are used to calculate its initial value. On the other hand, it may be set as a constant design parameter of the lever while retaining the stated tangent constraint and, therefore, obtaining included angle of lever. In this work, ϕ_i is taken as zero. As of now, variables characterizing the proposed concept are expressed as a function of angle ϕ. After deriving the required relations and performing simplification, the Eq. (7) gives the required relation between input displacement and lever angle. To obtain ϕ, this problem is produced in MATLAB environment. The f solve subroutine is utilized to minimize $f(\phi)$. Once ϕ is obtained, thereafter, it is substituted back into the derived formulation to obtain different variables of the TAM.

$$f(\phi) = \left\{\left(\frac{F_s}{k_{\text{sp}}} + r_p(\alpha_i - \alpha + \tan(\alpha) - \tan(\alpha_i)) + \frac{x'}{\cos(\alpha)} - \frac{x}{\cos(\alpha_i)}\right) - d\right\} \tag{7}$$

6 Discussion

The kinematics proposed in Sect. 3 are verified by considering cables as rigid links. The function formulated in Sect. 5 is minimized, by dropping the force term, to obtain lever angle (ϕ). To achieve it, required data is presented in Table 1. Thereafter, substituting for ϕ in the presented kinematics. The geometrical configuration of proposed mechanism, corresponding to each value of lever angle, is produced in solid works environment from where primary and secondary cables angles are measured. It has been noticed that both results are in good concurrence with each other. Figures 5a through Fig. 5d present the plots of the angles ϕ, α, β, and γ, respectively. It is noted that lever angle is surging right from the starting while the secondary cable angle initially rises, however, falls later on. Both are in good agreement with motion of the proposed mechanism. Further, angle β declines first, however, rises afterward while angle γ surges right from the beginning. The variation of these angles is in accordance with the motion of TAM. Furthermore, Fig. 6 shows the limiting cases of TAM, first with no input and second at limiting values of previously mentioned four angles.

Now, to investigate the dependence of primary cable tension on various angles and forces of the devised TAM, spring deflection is accounted for as given in Eq. (7) and it is minimized again to find out ϕ. The variation of force (F_i) at point A and primary cable tension (T) is plotted (see Figs. 7a and b) as the input displacement is increased from 0 to 15 mm at step-size of 0.1 mm. It is noticed that F_i is zero when $d = 0$, whereas tension T is not zero. It is the initial tension (T_0) prior to the application of TAM to primary mechanism. As d turns out to be nonzero, F_i becomes nonzero which in turn increases the tension. Furthermore, the surge in this force and generated tension are primarily restricted by TAM parameters which include the load

Table 1 Details of the parameters for the implementation of TAM

S. No.	Parameter		Value	S. No.	Parameter	Value
1	d_{12}	(mm)	75	10	$E_{pc}\left(\mathrm{N}/\mathrm{mm}^2\right)$	187.5×10^3
2	d_{14}		75	11	$k_{sp}\left(\mathrm{N}/\mathrm{mm}\right)$	20.1
3	d_{15}		150	12	ϕ_i (rad)	0
4	d_{45}		225	13	θ_l (rad)	2.58
5	d_{26}		150	14	T_0(N)	25
6	r_o		40	15	F_{br}^{pc} (N)	544
7	r_i		40	16	δ_{max}^{sp} (mm)	8.3
8	r_p		12.885	17	d_{max}(mm)	60
9	d_{pc}		0.813			

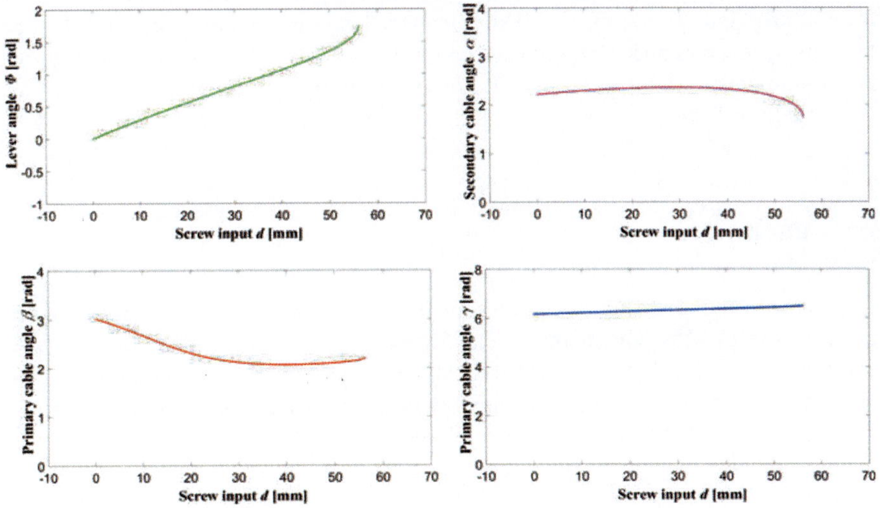

Fig. 5 (left to right) plots of angles ϕ, α, β, and γ, respectively against input d

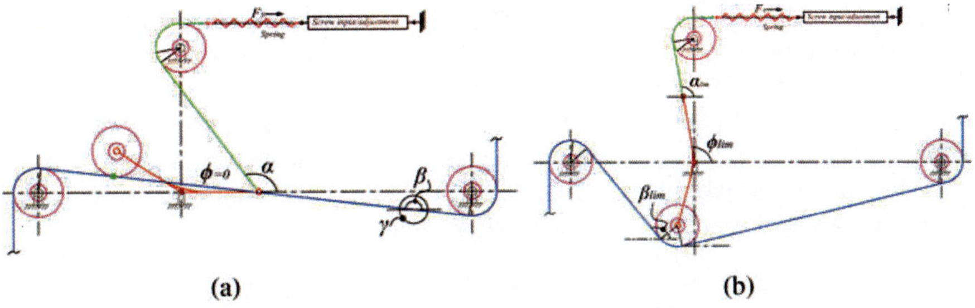

Fig. 6 Schematics of proposed TAM **a** initial pose and **b** limiting case

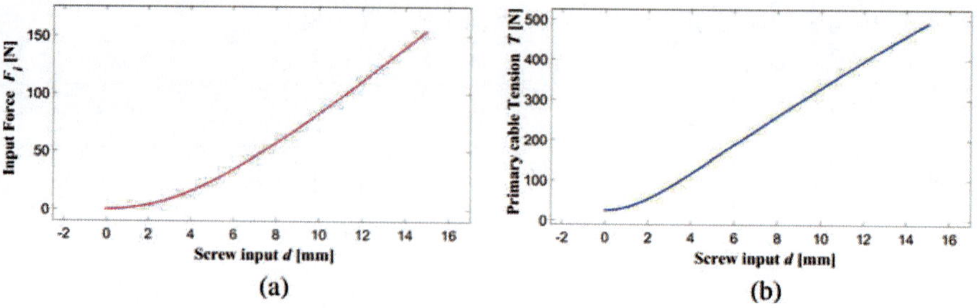

Fig. 7 Plots of **a** force F_i and **b** tension (T) generated in primary cable against input d

carrying capacities of primary and secondary cables, accessible input d, peak spring deformation, and lever angle (ϕ_{lim}). It is worth to add, the energy required to pre-tension primary cable is reduced notably by choosing the ratio of radii of rotation of lever appropriately. In turn, it will significantly lower the related cost. In this work, this ratio is taken as one. The proposed concept is applicable in several configurations to provide tension adjustment which are under study, and a provisional patent has been filed for TAM.

7 Conclusion

This paper introduced a novel tension adjustment mechanism for cable-driven robotic manipulator for MIS. The mathematical formulation has been derived for kinematic and force analyses. The proposed mechanism takes care for the variation in cable tension due to manufacturing and assembly inaccuracies. It does not demand precise and accurate installation of associated elements such as spring and cables. In addition, it offers precise tension adjustment according to the need of a given drive. The current version of the devised mechanism operates as per screw motion of input device which gives quite precise and accurate force regulation in the absence of any electronic force sensor for surgical applications. Further, the rotating lever delivers mechanical advantage which considerably decreases the specifications of springs and other required components. Also, instead of pre-stressed cables, standard cables are used which are relatively cheaper. It is due to the fact that effect of constructional stretch, presents in the standard cables, is eliminated through the adjustment offered by the proposed concept. Moreover, during assembly of mechanism it does not need any actuator, to stretch/release spring, and provides better safety to surrounding occupants. Thus, presented mechanism offers an economic and safe design.

References

1. Tsai LW (1995) Design of tendon-driven manipulators. J Mech Des 117(B):80–6
2. Li J, Wang S, Wang X, He C (2010) Optimization of a novel mechanism for a minimally invasive surgery robot. Int J Med Robot Comput Assist Surg 6(1):83–90
3. Baser O, Konukseven EI (2010) Theoretical and experimental determination of capstan drive slip error. Mech Mach Theor 45(6):815–827
4. Lee JJ (1991) Tendon-driven manipulators: analysis, synthesis, and control. Doctoral dissertation, University of Maryland
5. Camarillo DB (2008) Mechanics and control of tendon driven continuum manipulators. Doctoral dissertation, Stanford University
6. Madhani AJ (1998) Design of tele-operated surgical instruments for minimally invasive surgery. Doctoral dissertation, Massachusetts Institute of Technology
7. Lee WJ, Robot S, Center MI (2016) Ten-year experience of the da Vinci robotic surgery at Severance Yonsei University Hospital in Korea Hanyang. Med Rev 36(4):215–224
8. Chen X, Xin X, Zhao B, He Y, Hu Y, Liu S (2017) Design and analysis of a haptic master manipulator for minimally invasive surgery. In: Proceedings of IEEE, international conference on information and automation (ICIA), pp 260–265
9. Liu S, Chen B, Caro S, Briot S, Harewood L, Chen C (2016) A cable linkage with remote centre of motion. Mech Mach Theor 105:583–605
10. Çavuşoğlu MC, Williams W, Tendick F, Sastry SS (2003) Robotics for telesurgery: second generation Berkeley/UCSF laparoscopic telesurgical workstation and looking towards the future applications. Ind Robot Int J
11. King HH, Tadano K, Donlin R, Friedman D, Lum MJ, Asch V, Wang C, Kawashima K, Hannaford B (2009) Preliminary protocol for interoperable telesurgery. In: International conference on advanced robotics, pp 1–6
12. Lum MJ, Friedman DC, Sankaranarayanan G, King H, Fodero K, Leuschke R, Hannaford B, Rosen J, Sinanan MN (2009) The Raven: design and validation of a telesurgery system. Int J Robot Res 28(9):1183–1197
13. Hannaford B, Rosen J, Friedman DW, King H, Roan P, Cheng L, Glozman D, Ma J, Kosari SN, White L (2012) Raven-II: an open platform for surgical robotics research. IEEE Trans Biomed Eng 60(4):954–959
14. Li J, Zhou N, Wang S, Gao Y, Liu D (2012) Design of an integrated master–slave robotic system for minimally invasive surgery. Int J Med Robot Comput Assist Surg 8:77–84
15. Kong K, Jeon D (2006) Design and control of an exoskeleton for the elderly and patients. IEEE/ASME Trans Mechatron 11(4):428–432
16. Basile P, Haase C (2008) Tension compensating assembly for mechanical control cables, US Patent No: 7,469,617 B2
17. Segreti SM, Kutzer MD, Murphy RJ, Armand M (2012) Cable length estimation for a compliant surgical manipulator. In: Proceedings of IEEE international conference of on robotics and automation (ICRA), pp 701–708
18. Tzemanaki A, Fracczak L, Gillatt D, Koupparis A, Melhuish C, Persad R, Rowe E, Pipe AG, Dogramadzi S (2016) Design of a multi-DOF cable-driven mechanism of a miniature serial manipulator for robot-assisted minimally invasive surgery. In: 6th international proceedings on biomedical robotics and biomechatronics (BioRob), pp 55–60
19. Yousef BF, Aiash FM (2013) A mechanism for surgical tool manipulation. In: 9th Asian control conference (ASCC), pp 1–5
20. Kim SK, Shin WH, Ko SY, Kim J, Kwon DS (2008) Design of a compact 5-DOF surgical robot of a spherical mechanism: CURES. In: IEEE/ASME international conference on advanced intelligent mechatronics, pp 990–995

Kinematic and Dynamic Analysis of a Six-Bar Aerial Gripper Mechanism

V. S. Rajashekhar and Debasish Ghose

Abstract Unmanned aerial vehicles (UAV) are being used for commercial and rescue operations. A gripper attached to the UAV would help in grasping, holding and placing the object of interest during these operations. In this work, we propose and analyze one degree of freedom gripper mechanism that has six links and seven revolute joints. The kinematic analysis is done by treating the gripper mechanism as two four-bar mechanisms that form two loops. A two-stage vector loop method is used to determine the joint angles. In order to statically balance the gripper mechanism, the coordinates of all the joints in the mechanism at various positions of the links are found out. The method of principal vectors is used for the complete balancing of the mechanism. In this method, each vector is directed along with one of the links, and the position of the center of mass of the entire mechanism is given by the addition of these vectors in series. The coefficients of these principal vectors are determined by summing the mass-weighted position vector of every center of mass of the link. Counterweights are added to the links for statico-dynamic balancing of the gripper mechanism. The dynamic analysis is done using the Newtonian solution method. There are 15 equations with 15 unknown contact forces and torque acting in the gripper mechanism. Solving these equations, we obtain the forces acting on the links, and torque required for driving the gripper mechanism. The gripper mechanism was then modeled using a CAD software. It was fabricated by 3D printing the parts and assembling them together. The gripper was finally mounted below a hexacopter and tested in the outdoor environment by carrying and dropping a spherical object.

Keywords Aerial gripper · Gripper mechanism · Six-bar mechanism · Unmanned aerial vehicles · Balancing · Gripper kinematics · Gripper dynamics

---

V. S. Rajashekhar (✉) · D. Ghose
GCDSL, Department of Aerospace Engineering, Indian Institute of Science, Bangalore, India
e-mail: vsrajashekhar@gmail.com

D. Ghose
e-mail: dghose@iisc.ac.in

V. K. Gupta et al. (eds.), *Recent Advances in Machines and Mechanisms*, Lecture Notes in Mechanical Engineering, https://doi.org/10.1007/978-981-19-3716-3_42

1 Introduction

Unmanned aerial vehicles (UAV) are being used for grasping, transporting and placing payloads for commercial purposes [13]. To conduct this operation, the UAV platforms are installed with gripper and manipulator mechanisms [16]. A gripper is usually light in weight and is installed near the center of mass of the overall system in order to easily grasp objects. The task of grasping, transporting and placing can be effectively achieved using the two main states in a gripper mechanism: grasp and release. The UAV dynamics have limited impacts when grippers are used (in contrast to the manipulators) since there is no relative motion of the payload after being grasped. However, the grippers have to be designed for a specific purpose based on the properties of the payload. A survey on aerial manipulation [17] shows the following types of aerial grippers that exist in the literature: impactive gripper, ingressive gripper, mechanical compliant gripper, magnetic gripper and vacuum gripper. Among them, mechanical grippers offer a high shape compatibility without any material restriction. It grasps a target object by holding it without causing any destruction to it. A survey on payload transportation using UAVs [24] shows that grasping is recommended for individual object transportation in any environment (cluttered or uncluttered) where the maneuverability is less and payload is low. The survey also recommends grasping for cooperative transportation in uncluttered environments where the maneuverability is medium and the payload is heavy. This was the motivation for proposing a mechanical gripper to grasp an object moving in space.

The need for multiple kinematic loop mechanisms arises when the operating workspace is small and the actuator needs to be positioned far away from the operating workspace. They are also used in applications where there is a need for high mobility and speed, high stiffness, high load carrying capacity and precise positioning [5]. This scenario is common for grippers and manipulators used in unmanned aerial vehicles (UAVs). The more number of links in a mechanism leads to more possible trajectories, but it would cause linkage interference [28]. When compared to the four-bar mechanism which is often used in gripping applications [18], the chosen six-bar mechanism also has one degree of freedom but offers better grasping using two jaws [12].

Grippers that use six-bar mechanisms offer the advantage of aligning and re-orienting objects [4]. Six-bar mechanisms are commonly used in various applications such as positioning and orientating bottles from one conveyor to another [4], leg mechanisms in bio-inspired robots like biped walking [6], biped robot running on water surface [27], locust inspired jumping robot [29], bionic leg mechanism [26], hand rehabilitation [7], walking machine [23], sewing machine [20], computing apparatus [1], flapping wing mechanism [2, 9, 25], knee-ankle-foot orthosis [8] and mobility assistance device [21].

In this work, we present the kinematic and dynamic analysis of a six-bar aerial gripper mechanism that was synthesized earlier [12]. Section 2 explains the gripper mechanism design and working principle along with the link dimensions. In Sect. 3, kinematic analysis along with a numerical example is presented. Section 4 presents the balancing of the mechanism using the method of principal vectors. Section 5 presents the dynamic analysis of the mechanism using the Newtonian solution method. The fabrication and testing of the mechanism is presented in Sect. 6 before the concluding remarks given in Sect. 7.

2 The Gripper Mechanism

2.1 Mechanism Design

A 1-DoF aerial gripper mechanism was dimensionally synthesized [12]. Figure 1 shows the development of the gripper. In Fig. 1a, a four-bar mechanism is formed. In Fig. 1b, a six-bar gripper mechanism is formed by adding two binary links to the four-bar mechanism. In this mechanism, there are six links (four are binary and two are ternary) and seven revolute joints (joints 1, 3 and 5 are binary; joints 2 and 4 are ternary). The fixed link is between the joints 1 and 4.

2.2 Working Principle

The working of the gripper mechanism is shown in Fig. 2. In Fig. 2a, the gripper is in open position. In Fig. 2b, the gripper is in close position. When the UAV is searching for the target object, the gripper would be in open position. Once the target object is reached and comes with in the workspace of the gripper, it would grasp and be in the close position.

2.3 Dimensions of the Links

The dimensional synthesis of the gripper mechanism was done using geometric programming [22] and is presented in Laishram et al. [12]. The dimensions of the

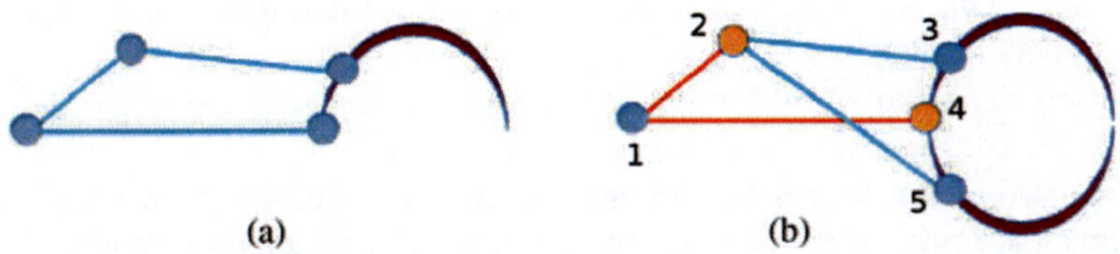

Fig. 1 Development of the mechanism. **a** A four-bar mechanism is formed. **b** A six-bar gripper mechanism is formed by adding two binary links

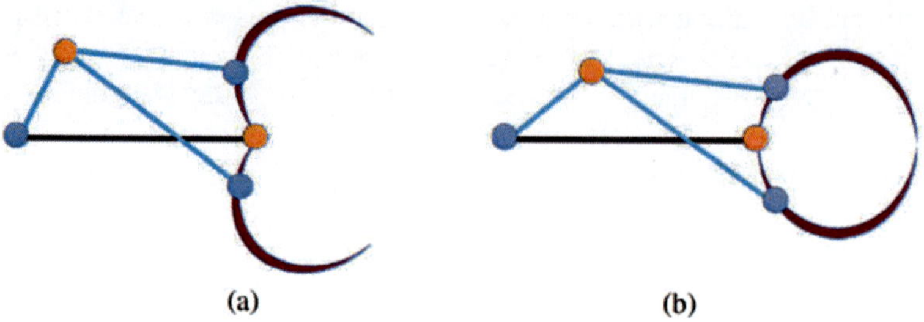

Fig. 2 The working of gripper mechanism **a** open state, **b** close state

Fig. 3 The notations used to represent the dimensions of the gripper mechanism

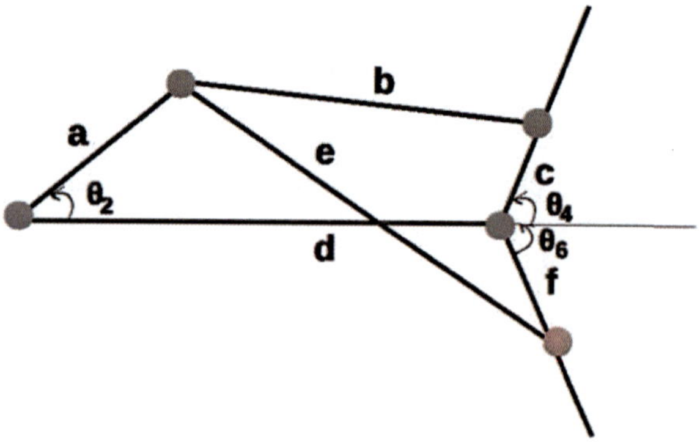

Table 1 The dimensions of links in the gripper mechanism

Link	Length (in mm)	Width (in mm)	Thickness (in mm)
a	100	26.66	27.47
b	396	26.66	15.84
c	20	26.66	12.58
d	300	26.66	17.70
e	396	26.66	15.84
f	20	26.66	12.58

links are found out by predetermining the ratio of the actuated angle θ_2 to the half grasping angle θ_4. The results are briefly presented here for the sake of ease of understanding the proposed mechanism. Figure 3 shows the simplified representation of the gripper mechanism.

The length, width and thickness of the links in the gripper mechanism, found using geometric programming, are presented in Table 1.

After deriving the kinematic and dynamic equations for the gripper mechanism, the numerical solutions are obtained using the values presented in Table 1.

3 Kinematic Analysis of the Mechanism

The kinematic analysis is done by treating the mechanism as two four-bar mechanism forming two loops. The vector loop equations are formed and solved for various angles as mentioned in Norton [19].

The two loops in the gripper mechanism is shown in Fig. 4.

The vector loop diagram of the gripper used for aerial grasping is shown in Fig. 5.

The two vector loop equations are as follows:

$$R_2 + R_3 - R_4 - R_1 = 0 \tag{1}$$

$$R_2 + R_5 - R_6 - R_1 = 0 \tag{2}$$

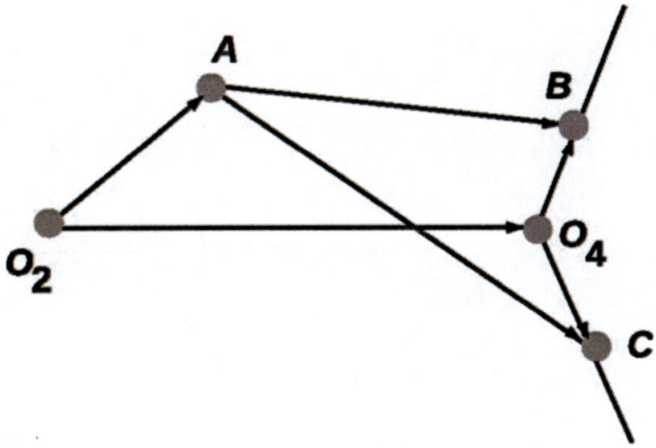

Fig. 4 The vector loop diagram of the gripper mechanism. Loop 1: O_2–A–B–O_4–O_2; loop 2: O_2–A–C–O_4–O_2

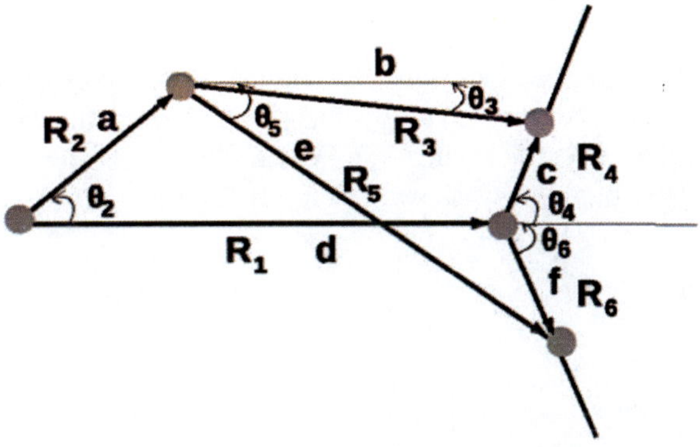

Fig. 5 The vector loop diagram of the gripper mechanism with angles and scalar notations labeled

Here, θ_2 is the input angle and θ_4 and θ_6 are the output angles. The other unknown angles are θ_3 and θ_5. In loop 1,

$$\theta_4 = 2\arctan\left(\frac{-B + \sqrt{B^2 - 4AC}}{2A}\right) \tag{3}$$

$$\theta_3 = 2\arctan\left(\frac{-E - \sqrt{E^2 - 4DF}}{2D}\right) \tag{4}$$

where,
$A = \cos\theta_2 - \frac{d}{a} - \frac{d\cos\theta_2}{c} + \frac{a^2+c^2+d^2-b^2}{2ac}$
$B = -2\sin\theta_2$
$C = \frac{d}{a} - (\frac{d}{c} + 1)\cos\theta_2 + \frac{a^2+c^2+d^2-b^2}{2ac}$
$D = \cos\theta_2 - \frac{d}{a} - \frac{d\cos\theta_2}{b} + \frac{c^2-d^2-a^2-b^2}{2ab}$
$E = -2\sin\theta_2$
$F = \frac{d}{a} - (\frac{d}{b} + 1)\cos\theta_2 + \frac{c^2-d^2-a^2-b^2}{2ab}$.

In the loop 2,

$$\theta_6 = 2\arctan\left(\frac{-H - \sqrt{H^2 - 4GI}}{2G}\right) \tag{5}$$

$$\theta_5 = 2\arctan\left(\frac{-K - \sqrt{K^2 - 4JL}}{2J}\right) \tag{6}$$

where
$G = \cos\theta_2 - \frac{d}{a} - \frac{d\cos\theta_2}{f} + \frac{a^2+f^2+d^2-e^2}{2ac}$
$H = -2\sin\theta_2$
$I = \frac{d}{a} - (\frac{d}{f} + 1)\cos\theta_2 + \frac{a^2+f^2+d^2-e^2}{2af}$
$J = \cos\theta_2 - \frac{d}{a} - \frac{d\cos\theta_2}{e} + \frac{f^2-d^2-a^2-e^2}{2ae}$
$K = -2\sin\theta_2$
$L = \frac{d}{a} - (\frac{d}{e} + 1)\cos\theta_2 + \frac{f^2-d^2-a^2-e^2}{2ae}$.

A Numerical Example The lengths of the links that were presented in Table 1 were substituted from (3) to (6). The graph showing the values of θ_3 and θ_5 with reference to the driver θ_2 is shown in Fig. 6a. The graph showing the values of θ_4 and θ_6 for a particular θ_2 is as shown in Fig. 6b.

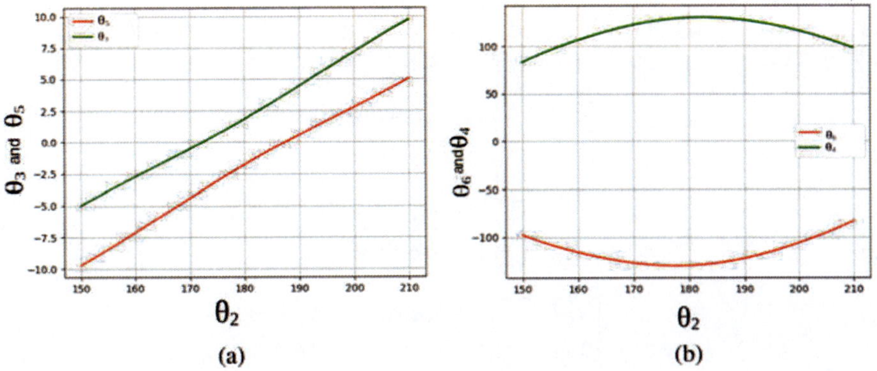

Fig. 6 **a** θ_3 and θ_5 with reference to the driver θ_2, **b** θ_4 and θ_6 with reference to the driver θ_2

Fig. 7 The joint and coupler positions of the gripper represented as points

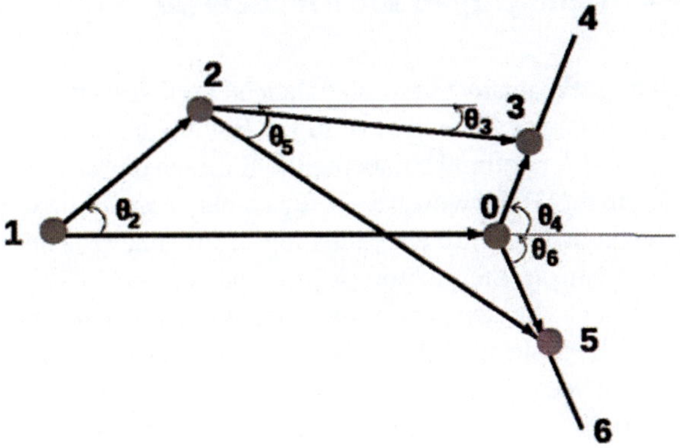

3.1 *Position of Joints and Couplers*

The various joint and coupler positions of the gripper mechanism are represented as points in Fig. 7.

The coordinates of all the joints and the coupler points present in the gripper mechanism is as mentioned in the equations below.

$$\begin{bmatrix} x_0 \\ y_0 \end{bmatrix} = \begin{bmatrix} 0 \\ 0 \end{bmatrix} \tag{7}$$

$$\begin{bmatrix} x_1 \\ y_1 \end{bmatrix} = \begin{bmatrix} l_1 \cos \pi \\ l_1 \sin \pi \end{bmatrix} \tag{8}$$

$$\begin{bmatrix} x_2 \\ y_2 \end{bmatrix} = \begin{bmatrix} x_1 + l_2 \cos \theta_2 \\ y_1 + l_2 \sin \theta_2 \end{bmatrix} \tag{9}$$

$$\begin{bmatrix} x_3 \\ y_3 \end{bmatrix} = \begin{bmatrix} l_3 \cos\theta_4 \\ l_3 \sin\theta_4 \end{bmatrix} \tag{10}$$

$$\begin{bmatrix} x_4 \\ y_4 \end{bmatrix} = \begin{bmatrix} x_3 + l_4 \cos(\theta_4 - \pi/4) \\ y_3 + l_4 \sin(\theta_4 - \pi/4) \end{bmatrix} \tag{11}$$

$$\begin{bmatrix} x_5 \\ y_5 \end{bmatrix} = \begin{bmatrix} l_5 \cos\theta_6 \\ l_5 \sin\theta_6 \end{bmatrix} \tag{12}$$

$$\begin{bmatrix} x_6 \\ y_6 \end{bmatrix} = \begin{bmatrix} x_5 + l_6 \cos(\theta_6 + \pi/4) \\ y_5 + l_6 \sin(\theta_6 + \pi/4) \end{bmatrix} \tag{13}$$

4 Balancing of the Mechanism

The mechanism has to be balanced so that it does not cause any disturbance to the drone during the period of flight. For a planar six-bar mechanism, genetic algorithm is used for optimal balancing to minimize the variations in shaking force and shaking moment [3]. However, a survey on balancing of linkages [14] show that there are five methods for complete balancing of the mechanism. Among them, for our proposed mechanism, the method of principal vectors [15] is chosen since it is efficient and simple to implement. In this method, each vector is directed along one of the links and the position of the center of mass of the entire mechanism is given by the addition of the vectors in series. The coefficients of these principal vectors are determined by summing the mass-weighted position vector of every center of mass of the link. The parameters of the mechanism used for analysis is shown in Fig. 8.

The position of the center of mass of the mechanism is given by the following equation.

$$\vec{r_s} = \frac{1}{\mu}(m_1\vec{r_{s1}} + m_2\vec{r_{s2}} + m_3\vec{r_{s3}} + m_4\vec{r_{s4}} + m_5\vec{r_{s5}}) \tag{14}$$

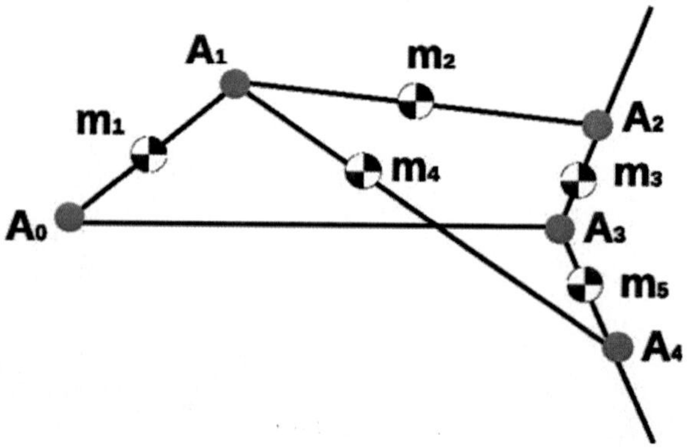

Fig. 8 The balancing of the mechanism with clearances considered at joints

where,
μ—the total mass of the moving links
m_i—the mass of the individual links
r_{si}—the vector which describes the position of center of mass of the link
n_i—the unit vector of the link.

The position vectors of the link centers of mass is given as follows,

$$
\begin{aligned}
\overrightarrow{r_{s1}} &= b_1\overrightarrow{n_1} \\
\overrightarrow{r_{s2}} &= a_1\overrightarrow{n_1} + b_2\overrightarrow{n_2} \\
\overrightarrow{r_{s3}} &= a_1\overrightarrow{n_1} + a_2\overrightarrow{n_2} + b_3\overrightarrow{n_3} \\
\overrightarrow{r_{s4}} &= a_1\overrightarrow{n_1} + b_4\overrightarrow{n_4} \\
\overrightarrow{r_{s5}} &= a_1\overrightarrow{n_1} + a_4\overrightarrow{n_4} + b_5\overrightarrow{n_5}
\end{aligned} \tag{15}
$$

Substituting (15) in (14), the following is obtained.

$$
\begin{aligned}
\overrightarrow{r_s} = \frac{1}{\mu}(&m_1b_1\overrightarrow{n_1} + m_2a_1\overrightarrow{n_1} + m_2b_2\overrightarrow{n_2} + m_3a_1\overrightarrow{n_1} + m_3a_2\overrightarrow{n_2} + m_3b_3\overrightarrow{n_3} \\
&+m_4a_1\overrightarrow{n_1} + m_4b_4\overrightarrow{n_4} + m_5a_1\overrightarrow{n_1} + m_5a_4\overrightarrow{n_4} + m_5b_5\overrightarrow{n_5})
\end{aligned} \tag{16}
$$

From the above equation, the grouping is done as follows.

$$
\begin{aligned}
h_1 &= \frac{1}{\mu}(m_1b_1 + m_2a_1 + m_3a_1 + m_4a_1 + m_5a_1) \\
h_2 &= \frac{1}{\mu}(m_2b_2 + m_3a_2) \\
h_3 &= \frac{1}{\mu}(m_3b_3) \\
h_4 &= \frac{1}{\mu}(m_4b_4 + m_5a_4) \\
h_5 &= \frac{1}{\mu}(m_5b_5)
\end{aligned} \tag{17}
$$

Equation (16) can be written as,

$$\overrightarrow{r_s} = h_1\overrightarrow{n_1} + h_2\overrightarrow{n_2} + h_3\overrightarrow{n_3} + h_4\overrightarrow{n_4} + h_5\overrightarrow{n_5} \tag{18}$$

or, alternatively, as

$$\overrightarrow{r_s} = \overrightarrow{h_1} + \overrightarrow{h_2} + \overrightarrow{h_3} + \overrightarrow{h_4} + \overrightarrow{h_5} \tag{19}$$

where, $\overrightarrow{h_i}$ is the principal vector.

Table 2 The numerical values considered for balancing of mechanism

Link	*h*	*m* (g)	*a* (cm)	*b* (cm)
1	8.1818	30	10	5
2	12.0871	80	39.6	19.8
3	0.1224	20	2	1
4	12.0871	80	39.6	19.8
5	0.1224	20	2	1

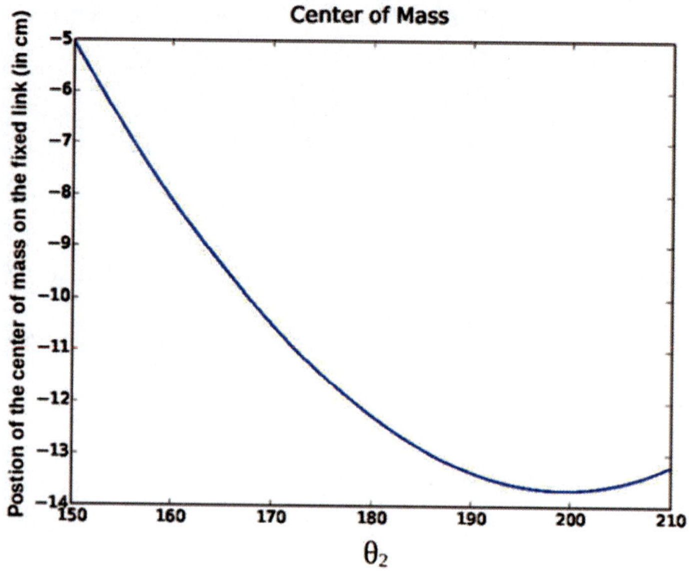

Fig. 9 The results of dynamic balancing where the θ_2 ranges between 150° and 180°

A Numerical Example A numerical example is shown for the manipulator that was manufactured and mounted on the drone. The values of *h*, *m*, *a* and *b* are given in Table 2. The values of θ_2 ranges from 150 to 180°. The values of θ_2, θ_3, θ_4 and θ_5 were taken from the kinematic analysis. The equations were framed taking the above conditions into account and graph as shown in Fig. 9 was obtained. The graph shows the variation of the center of mass with respect to the change in value of θ_2.

4.1 *Addition of Counter Weights for Statico-Dynamic Balancing*

The static, dynamic and statico-dynamic balancing of a four-bar mechanism are discussed in Kamenskii [11]. By observing the symmetry along the horizontal plane

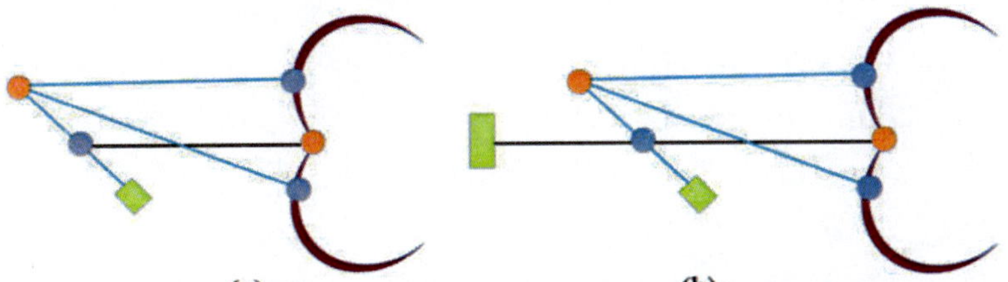

Fig. 10 **a** Stage 1 of dynamic balancing where the mechanism is balanced completely. **b** Stage 2 of dynamic balancing where the center of gravity is bought to the back

in the gripper during the initial position of θ_2 (180° of the crank), it can be concluded that the center of mass lies on the fixed link and is statically balanced. When θ_2 moves to 150°, the gripper is in closed position. There is an imbalance in the mechanism which makes the center of mass to go off the fixed link. This creates a dynamic balancing problem. This problem is solved in two stages.

In the first stage, as done in Kamenskii [10], a counter mass is fixed to the crank at the bottom as shown in Fig. 10a. This makes the center of mass to lie on the fixed link throughout the period of motion of the crank.

In the second stage, the center of mass is brought back near the crank so that the entire center of mass is always below the drone and not offset from it. It is as shown in Fig. 10b. Hence the gripper mechanism is statico-dynamically balanced.

5 Dynamic Analysis of the Mechanism

The Newtonian solution method is used to perform the dynamic analysis of the mechanism. The dynamic analysis is done using Newtonian solution method as mentioned in Norton [19]. The contact forces and torque (T_{12}) are represented in the free body diagram which is shown in Fig. 11.

Link 2:

$$\begin{aligned} &F_{42x} + F_{12x} + F_{32x} = m_2 a_{G2x} \\ &F_{42y} + F_{12y} + F_{32y} = m_2 a_{G2y} \\ &T_{12} + (R_{12x}F_{12y} - R_{12y}F_{12x}) + (R_{32x}F_{32y} - R_{32y}F_{32x}) \\ &\quad + (R_{42x}F_{42y} - R_{42y}F_{42x}) = I_{G2}\alpha_2 \end{aligned} \tag{20}$$

Link 3:

$$\begin{aligned} &F_{53x} - F_{32x} = m_3 a_{G3x} \\ &F_{53y} - F_{32y} = m_3 a_{G3y} \\ &(R_{53x}F_{53y} - R_{53y}F_{53x}) - (R_{23x}F_{32y} - R_{23y}F_{32x}) = I_{G3}\alpha_3 \end{aligned} \tag{21}$$

Link 4:

$$\begin{aligned} F_{64x} - F_{42x} &= m_4 a_{G4x} \\ F_{64y} - F_{42y} &= m_4 a_{G4y} \\ (R_{64x} F_{64y} - R_{64y} F_{64x}) - (R_{42x} F_{42y} - R_{42y} F_{42x}) &= I_{G4} \alpha_4 \end{aligned} \quad (22)$$

Link 5:

$$\begin{aligned} F_{15x} - F_{53x} &= m_5 a_{G5x} \\ F_{15y} - F_{53y} &= m_5 a_{G5y} \\ (R_{15x} F_{15y} - R_{15y} F_{15x}) - (R_{53x} F_{53y} - R_{53y} F_{53x}) &= I_{G5} \alpha_5 \end{aligned} \quad (23)$$

Link 6:

$$\begin{aligned} F_{16x} - F_{64x} &= m_6 a_{G6x} \\ F_{16y} - F_{64y} &= m_6 a_{G6y} \\ (R_{16x} F_{16y} - R_{16y} F_{16x}) - (R_{64x} F_{64y} - R_{64y} F_{64x}) &= I_{G6} \alpha_6 \end{aligned} \quad (24)$$

Equations (20) to (24) can be converted into a matrix form $Ab = c$, where A is a 15×15 matrix containing the known perpendicular distances and b is 15×1 vector containing the unknown forces and torque. The vector b is found out by $b = A^{-1}c$.

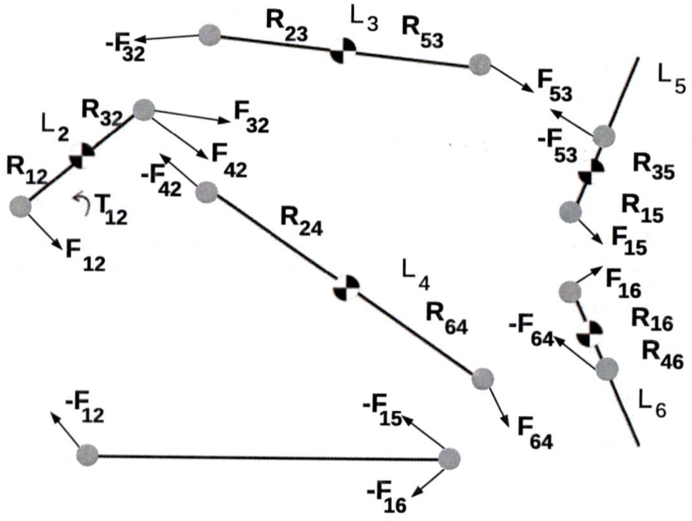

Fig. 11 The free body diagram used for the dynamic analysis of the gripper mechanism

5.1 A Numerical Example

The objective of the numerical analysis is to compute the forces at the joints and the torque required to drive the mechanism. The mechanism without the actuator is as shown in Fig. 12 and is considered for the numerical analysis.

The values of the lengths of the links are as follows: $L_1 = 300\,\text{mm}$, $L_2 = 100\,\text{mm}$, $L_3 = 396\,\text{mm}$, $L_4 = 396\,\text{mm}$, $L_5 = 20\,\text{mm}$ and $L_6 = 20\,\text{mm}$. The mass of the links are as follows: $m_2 = 0.011\,\text{kg}$, $m_3 = 0.064\,\text{kg}$, $m_4 = 0.064\,\text{kg}$, $m_5 = 0.045\,\text{kg}$, $m_6 = 0.045\,\text{kg}$.

Based on the kinematic analysis presented in Sect. 3, the vectors of the link are $R_{32} = R_{12} = 50\,\text{mm}$ @ $165°$, $R_{53} = R_{23} = 198\,\text{mm}$ @ $-1.25°$, $R_{64} = R_{24} = 198\,\text{mm}$ @ $-6°$, $R_{46} = R_{16} = 10\,\text{mm}$ @ $-125°$, $R_{15} = R_{35} = 10\,\text{mm}$ @ $125°$.

The mass moment of inertia values of the links is as follows: $I_{G2} = 1.343\,\text{kg mm}^2$, $I_{G3} = 4.635\,\text{kg mm}^2$, $I_{G4} = 4.635\,\text{kg mm}^2$, $I_{G5} = 3.606\,\text{kg mm}^2$ and $I_{G6} = 3.606\,\text{kg mm}^2$. The angular acceleration values of the links are as follows: $\alpha_2 = 2.0944\,\text{rad/s}^2$, $\alpha_3 = 1.0472\,\text{rad/s}^2$, $\alpha_4 = 1.0472\,\text{rad/s}^2$, $\alpha_5 = 3.49066\,\text{rad/s}^2$ and $\alpha_6 = 3.49066\,\text{rad/s}^2$. The acceleration of the center of mass of all the links is taken to be $1\,\text{mm/s}^2$. The above values are substituted in the matrix formed by Eqs. (20) to (24).

Solving $b = A^{-1}c$, we get the unknown forces (N) and torque (Nm) as follows: $F_{12x} = -1.8375\,\text{N}$, $F_{12y} = 0.1529\,\text{N}$, $F_{32x} = -0.6148\,\text{N}$, $F_{32y} = -0.0011\,\text{N}$, $F_{42x} = 2.4524\,\text{N}$, $F_{42y} = -0.1517\,\text{N}$, $F_{53x} = -0.6147\,\text{N}$, $F_{53y} = -0.0011\,\text{N}$, $F_{64x} = 2.4524\,\text{N}$, $F_{64y} = -0.1516\,\text{N}$, $F_{15x} = -0.6147\,\text{N}$, $F_{15y} = -0.0011\,\text{N}$, $F_{16x} = 2.4525\,\text{N}$, $F_{16y} = -0.1516\,\text{N}$, $T_{12} = 0.1125\,\text{Nm}$.

The magnitude of these forces is used to determine the size of the links and pivot pins, and selection of pivot bearings. The driving torque is used to select the motor needed to actuate the mechanism. These forces and torques are also used as a basis for calculating stress limits in order to prevent failure during the required operating life. Although the sizes of the links and pivot pins were determined previously using geometric programming [12] and presented in Table 1, it is the Newtonian solution method that provides a good estimate of the torque acting in the mechanism.

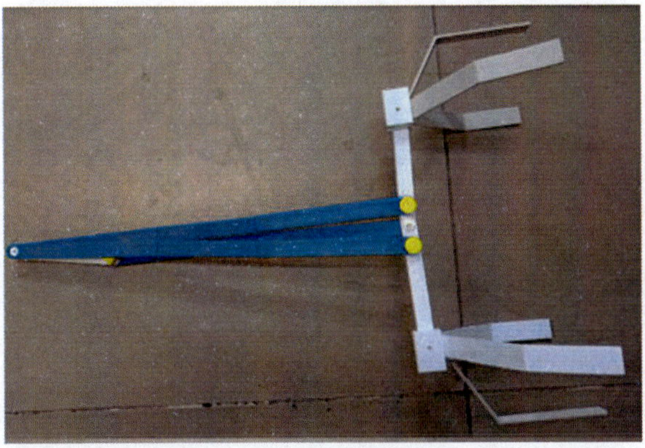

Fig. 12 The prototype of the gripper mechanism without the actuator

6 Fabrication and Testing the Mechanism

6.1 Modeling and Fabrication of the Gripper Mechanism

Based on the obtained driving torque, a DC motor with encoder was chosen. The parts of the gripper mechanism were modeled and assembled in a computer-aided design (CAD) software as shown in Fig. 13a. The parts were 3D printed and assembled with the DC motor as shown in Fig. 13b. The fixed link of the gripper mechanism was then mounted on a carbon fiber rod (about 1 m length). The carbon fiber rod had a rack and pinion mechanism that enabled the gripper to move front and back in case the target object is slightly out of reach. This setup was fixed below a hexacopter. It is shown in Fig. 14a.

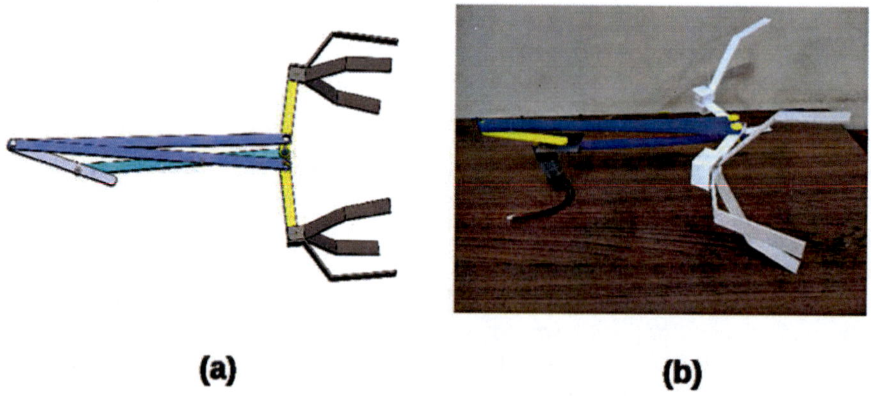

Fig. 13 The aerial gripper mechanism. **a** The computer-aided design of the gripper. **b** The prototype of the gripper with synthesized dimensions

Fig. 14 The gripper mechanism mounted on the UAV and tested without and with the payload. **a** The gripper mechanism mounted on the UAV. **b** The UAV in flight without payload. **c** The UAV in flight with a payload

6.2 Testing Conditions

The opening and closing of the gripper mechanism was done before and after mounting on the hexacopter. The experiments were carried out in the outdoor environment. The signal to the DC motor was given through the *Arduino Nano* Microcontroller which received commands from the *NVIDIA Jetson TX2 Development board* mounted on the drone. This board received open and close commands for the gripper from the *Robot Operating System (ROS)* that was installed in a off-board computer. The hexacopter was initially flown without payload to check the working of the gripper mechanism. It is shown in Fig. 14b. Then it was flown with a spherical object of mass 0.15 kg and 150 mm in diameter. It is shown in Fig. 14c. It was observed that system was stable in both the conditions.

7 Conclusions

In this work, a six-bar aerial gripper mechanism was proposed and analyzed. The kinematic analysis using a two-stage vector loop method was done for the gripper mechanism. The statico-dynamic balancing of the gripper mechanism was done to maintain stability during operation. The Newtonian solution method was used for dynamic analysis, which yielded the forces acting on the links and torque required for driving the mechanism. The claw of the gripper was designed in order to grasp a spherical object. The mechanism was fabricated and tested on a hexacopter in an outdoor environment. The analysis presented in this work will be used for controller design.

Acknowledgements The authors would like to acknowledge partial support from Robert Bosch Center for Cyber-Physical Systems (RBCCPS), Indian Institute of Science, Bengaluru, India, and Khalifa University, Abu Dhabi.

References

1. Antonin S (1944) Mechanism for use in computing apparatus. US patent 2,340,350
2. Bae SY, Koh JS, Jung GP (2021) A miniature flapping mechanism using an origami-based spherical six-bar pattern. Appl Sci 11(4):1515
3. Belleri BK, Kerur SB (2020) Balancing of planar six-bar mechanism with genetic algorithm. J Mech Energy Eng 4(4):303–308
4. Cepolina EE, Zaplana I, D'Imperio M, Gagliardi R, Baizid K, Scaccia M, Dai J, Muscolo G, Cannella F (2021) Scalable six bar linkage mechanism for re-orienting and aligning objects: design methodology. Procedia CIRP 97:66–70
5. Dasgupta B, Mruthyunjaya T (2000) The Stewart platform manipulator: a review. Mech Mach Theory 35(1):15–40
6. Funabashi H, Ogawa K, Honda I, Iwatsuki N (1985) Synthesis of leg-mechanisms of biped walking machines: part II, synthesis of foot-driving mechanism. Bull JSME 28(237):544–549

7. Gezgin E, Chang PH, Akhan AF (2016) Synthesis of a watt II six-bar linkage in the design of a hand rehabilitation robot. Mech Mach Theory 104:177–189
8. Ghosh S, Robson NP, McCarthy JM (2020) Kinematic design and evaluation of a six-bar knee-ankle-foot orthosis. J Eng Sci Med Diagn Ther 3(2)
9. Guo Z, Guo Z, Zhao L, Zhou S, Chen W (2018) A new type of flapping-wing robot based on the Stephenson II six-bar mechanism. In: 2018 IEEE international conference on mechatronics, robotics and automation (ICMRA). IEEE, pp 145–149
10. Kamenskii V (1968a) On the problem of the number of counterweights in the balancing of plane linkages. J Mech 3(4):323–333
11. Kamenskii V (1968b) On the question of the balancing of plane linkages. J Mech 3(4):303–322
12. Laishram R, Das K, Ghose D et al (2020) Synthesis of a six-bar gripper mechanism for aerial grasping. arXiv preprint http://arxiv.org/abs/2012.06724
13. Liew CF, DeLatte D, Takeishi N, Yairi T (2017) Recent developments in aerial robotics: a survey and prototypes overview. arXiv preprint http://arxiv.org/abs/1711.10085
14. Lowen G, Berkof R (1968) Survey of investigations into the balancing of linkages. J Mech 3(4):221–231
15. Lowen G, Tepper F, Berkof R (1983) Balancing of linkages—an update. Mech Mach Theory 18(3):213–220
16. Meng X, He Y, Han J (2020) Survey on aerial manipulator: system, modeling, and control—corrigendum. Robotica 38(7):1343
17. Mohiuddin A, Tarek T, Zweiri Y, Gan D (2020) A survey of single and multi-UAV aerial manipulation. Unmanned Syst 8(02):119–147
18. Mumcuoğlu ME, Yüksel I, Altuğ E (2020) Design of an automatic item pick-up system for unmanned aerial vehicles. Celal Bayar Univ J Sci 16(1)
19. Norton RL (2011) Kinematics and dynamics of machinery. McGraw-Hill Higher Education
20. Ogawa K (1962) An application of six-bar linkage to the thread take-up lever in a sewing machine. Bull JSME 5(19):554–560
21. Purwar A, Jadhav A (2017) A novel six-bar mechanism for executing motion trajectories of sit-to-stand transformation in a multifunctional mobility assist device. In: ASME 2017 international design engineering technical conferences and computers and information in engineering conference. American Society of Mechanical Engineers digital collection
22. Rao A (1979) Synthesis of 4-bar function-generators using geometric programming. Mech Mach Theory 14(2):141–149
23. Shieh WB, Tsai LW, Azarm S (1997) Design and optimization of a one-degree-of-freedom six-bar leg mechanism for a walking machine. J Robot Syst 14(12):871–880
24. Villa DK, Brandão AS, Sarcinelli-Filho M (2019) A survey on load transportation using multirotor UAVs. J Intell Robot Syst 1–30
25. Wang PL, McCarthy JM (2017) Design of a spatial six-bar flapping wing mechanism for combined control of swing and pitch. In: International design engineering technical conferences and computers and information in engineering conference, vol 58189. American Society of Mechanical Engineers, p V05BT08A027
26. Xu K, Liu H, Zhu X, Song Y (2019) Kinematic analysis of a novel planar six-bar bionic leg. In: IFToMM world congress on mechanism and machine science. Springer, pp 13–21
27. Xu L, Cao K, Wei X, Shi Y (2013) Dynamics analysis of fluid-structure interaction for a biologically-inspired biped robot running on water. Int J Adv Robot Syst 10(10):373
28. Zhang Y, Liu S, Mo X, Yang Y, Ge W (2019) Optimization and dynamics of six-bar mechanism bionic knee. In: 2019 WRC symposium on advanced robotics and automation (WRC SARA). IEEE, pp 91–96
29. Zhang Z, Chang B, Zhao J, Yang Q, Liu X (2020) Design, optimization, and experiment on a bioinspired jumping robot with a six-bar leg mechanism based on jumping stability. Math Probl Eng 2020

Bond Graph Modelling and Simulation of Pneumatic Soft Actuator

Garima Bhandari, Pushparaj Mani Pathak, and Jung-Min Yang

Abstract This paper presents the design and dynamic modelling of a soft pneumatic actuator that can be used to mimic snake or worm-like locomotion. The bond graph technique is used to derive the dynamics of the actuator. To validate the accuracy of the derived dynamic model, we conduct numerical simulations using 20-sim® software. Experimental results demonstrate that the soft actuator achieves bidirectional bending and linear displacement, which is essential for mimicking snake or worm-like locomotion.

Keywords Bond graph · Control · Dynamics · Snake robots · Soft robots

Nomenclature

μ	Transformer coefficient in bond graph model
ρ	Density of air supplied by air pump
ω	Angular frequency for desired extension/contraction
A	Area of elastic packet
C_d	Orifice coefficient of discharge
D	Diameter of orifice in elastic packet for air supply/removal
k	Stiffness of elastic packet
k_d	Derivative gain constant
k_p	Proportional gain constant
m	Mass of elastic packet
R	Gas constant

Supported by organization DST, India and NRF, South Korea.

G. Bhandari (✉) · P. M. Pathak
Indian Institute of Technology Roorkee, Haridwar, Uttarakhand 24766, India
e-mail: gsoharu@me.iitr.ac.in

P. M. Pathak
e-mail: pushparaj.pathak@me.iitr.ac.in

J.-M. Yang
Kyungpook National University, 80, Daehak-ro, Buk-gu, Daegu, South Korea
e-mail: jmyang@ee.knu.ac.kr

V. K. Gupta et al. (eds.), *Recent Advances in Machines and Mechanisms*, Lecture Notes in Mechanical Engineering, https://doi.org/10.1007/978-981-19-3716-3_43

R_b	Damping in elastic packet
T	Temperature of air supplied
x	Amplitude of desired reference trajectory

1 Introduction

In nature, many limbless reptiles with slender bodies use their body's flexibility and frictional properties to move around and overcome obstacles in environment. This has inspired new field of soft robots which mimic the motions of natural creatures like a snake, elephant trunk, etc. The rigid robots will require an infinite number of joints to realize the flexibility to mimic the behaviour of snake locomotion, which is not an energy-efficient solution. Whereas soft robots have an infinite passive degree of freedom (DoFs), enabling them to be passively deformed when interacting with environments under simple actuation. Another advantage of soft robots over conventional rigid robots is safe interaction with humans and adaptability to complex and uncertain environment at low cost [15].

Several soft robots perform various movements as bending, rotation, extending/contracting, and twisting using multiple techniques and materials, including shape memory alloys, shape morphing polymers, dielectric elastomers, tendon drive, piezoelectric actuation, and fluid power [1, 8, 9, 23]. Amongst these, pneumatically driven is found most advantageous because of their large deformation/force, good power-to-weight ratio, and low manufacturing cost [18]. They have gained popularity as a fibre-reinforced soft bending actuator (FRSBA) [6], but the use of viscoelastic material makes soft robots a highly nonlinear system. The nonlinearity of the dynamics of the soft robots increases the complexity of control [3], due to which most control strategies are open-loop in the case of soft robots [13].

Few studies have attempted to implement close loop control by dynamically controlling the fluid pressure or mass inside the robot body [7, 11]. Some have ignored dynamics completely and tried model-free control methods or training methods [19, 22]. But ignoring dynamics does not guarantee accurate control performance. Further, few studies implemented theoretical kinematics and dynamics models for controller design. These studies included various approaches, including finite element method (FEM), constant curvature method, concentrated mass method, and Euler–Lagrange method [10, 17, 21]. The dynamics provided by before mentioned methods are ambiguous, complex, and difficult to analyse, thus making controller design difficult. Further, few studies implemented empirical methods such as Jacobian method [16], visual servo control [12], neural network, sliding mode control, and adaptive control [14, 20]. These studies provide a simpler approach towards system modelling and controller design than the strategies involving more complex theoretical models. However, the nonlinear behaviours are not perfectly described, and the validities are limited to the specified experimental conditions [4].

The paper presents a simple soft actuator design based on bidirectional bending and stretching/contracting capabilities due to which it can have snake or worm-like

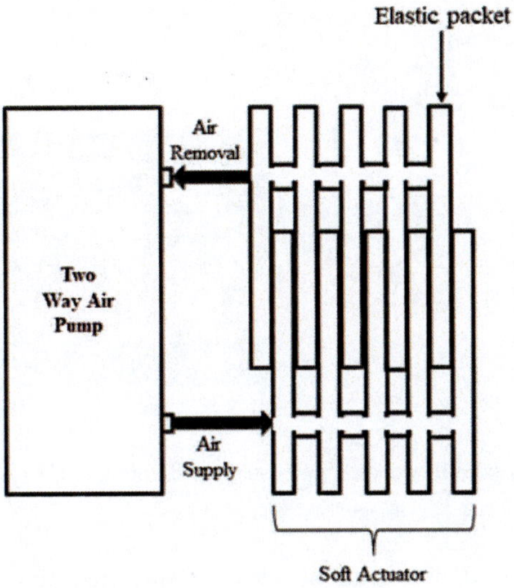

Fig. 1 Schematic representation of working concept

movements. The actuator can be used in various applications as a manipulator or mobile robot. Further, a dynamic model of the soft actuator has been derived from the bond graph modelling approach. The proposed modelling formalism can accommodate major dynamic parameters of the soft actuator and system nonlinearities. The robot has been controlled using PD controller. The performance of the soft actuator modelling is validated through numerical simulations, where the soft robot emulates bidirectional bending and extension/contraction as desired by the controller input.

The rest of this paper is structured as follows. Section 2 presents the design and modelling of a soft actuator in framework of bond graph technique. Section 3 presents detailed numerical simulation results to validate proposed dynamics with close loop PD control implemented. Finally, Sect. 4 concludes the paper.

2 Dynamic Modelling

We will discuss the dynamics of the soft actuator in this section. To understand the dynamics, it is important to understand the working and design of the soft actuator. The following sections explain in detail the design and bond graph modelling of the actuator.

2.1 Design of Soft Actuator

The design of the soft actuator is bellow-like. The soft actuator has packets made of elastic material joined together, as shown in Fig. 1.

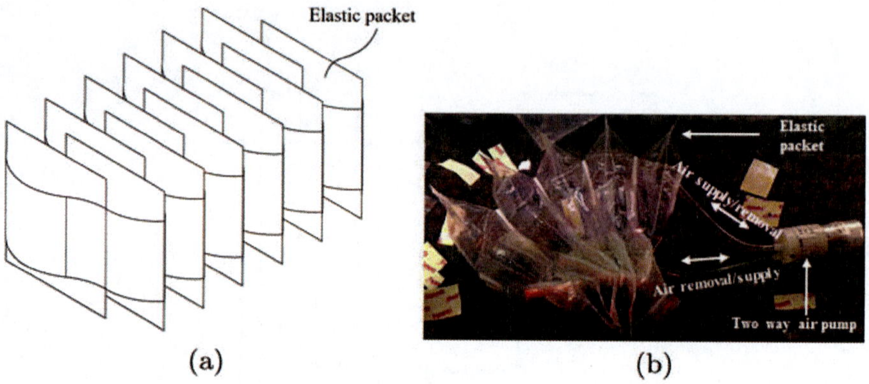

Fig. 2 Soft actuator's **a** solid model and **b** pictorial representation of its working concept

The elastic packets inflate and deflate like balloons when air is supplied or removed from them. The two-way air pump can supply and remove air; the connection of air pump to soft actuator is as shown in Fig. 1. The bidirectional bending is emulated by alternating supplying and removing air from each side of the soft actuator. The actuator bends in the direction of the side from where the pump removes air.

The microcontroller controls the alternating air supply and removal of air for bidirectional bending. We can attach various sensors to the soft actuator to measure the pressure, extension, and bending angle and employ them to control the soft actuator. Figure 2a, b shows solid model and the pictorial representation of the working concept of the soft actuator. Now, with a clear understanding of working and control of soft actuator, we will model the bond graph to derive the dynamics in the next section.

2.2 *Bond Graph Modelling of Soft Actuator*

The bond graph modelling technique reduces mathematical complexities related to highly nonlinear system, such as soft actuator presented in this paper. Since our actuator design is bellow-like, so we model each elastic packet as a bellow. For the basic understanding of bond graph model, we explain Fig. 3 in detail.

The source of effort element (**SE**) is the pressure (P_1) supplied to the elastic packet from the air pump. The resistance element (**R**) is the resistance the pneumatic flow faces when it enters through the orifice in the elastic packet. We calculate value of resistance as:

$$R = \frac{\sqrt{P_1 - P_2}}{C_d D} \tag{1}$$

where C_d is the orifice coefficient of discharge and D is diameter of orifice.

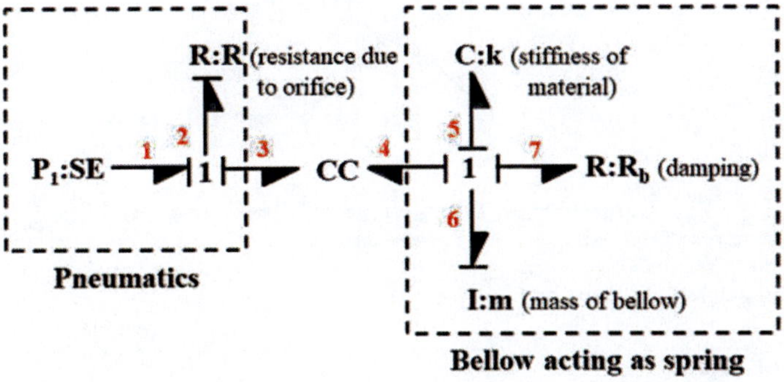

Fig. 3 Bond graph of the bellow

We have assumed extension and contraction of soft actuator synonymous with that of spring for the ease of modelling; with this assumption, we infer that the effort from the pneumatic domain is causing displacement in the mechanical part. We use C-field element (**CC**) in cases like these where the efforts and displacements are in a different domain but interrelated. Thus, we have two capacitance values in this system, C_1, due to change in volume of the elastic packet and C_2 due to compressibility of air [2]. We can calculate their value as follows:

$$C_1 = \frac{A^2}{k}$$
$$C_2 = \frac{Ax}{\rho RT} \tag{2}$$

where A and k are the area and stiffness of elastic packet respectively, x is the extension in the elastic packet after air supply, ρ is the density of air, R is gas constant, and T is the temperature of air supplied.

The resistance element (**R**) represented as R_b is damping in elastic packet and inductance element (**I**) is its mass (m). The dynamic equations we get after bond graph modelling are:

$$R(C_1 + C_2)\frac{dP_2}{dt} + P_2 = P_1 \tag{3}$$

$$m\ddot{x} + R_b\dot{x} + kx = P_2A \tag{4}$$

where P_2 is the pneumatic pressure inside the elastic packet, which is given by effort in bond number 3 in Fig. 3.

The bond graph model of elastic packet is then used to make the both sides of the actuator as shown in Fig. 4. We have implemented half car model for bidirectional bending of the actuator. Since we want to alternately supply and remove air from each side of actuator, the pressure P_1 is provides as:

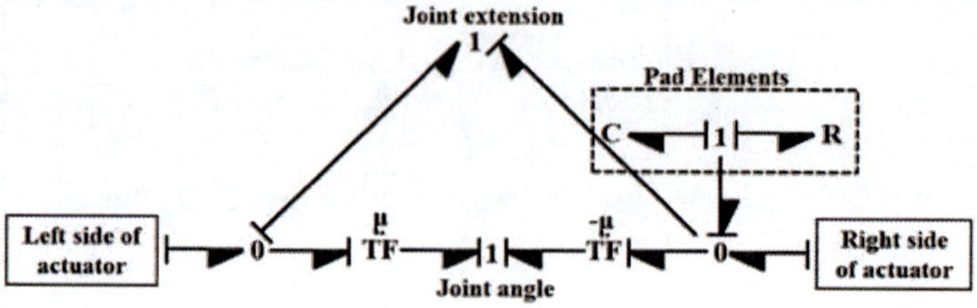

Fig. 4 Bond graph of the actuator

Left side of actuator

$$P_1 = k_p(x_{Lref} - x_L) + k_d(\dot{x}_{Lref} - \dot{x}_L)$$
$$\text{where, } x_{Lref} = x\sin(wt) \tag{5}$$

Right side of actuator

$$P_1 = k_p(x_{Rref} - x_R) + k_d(\dot{x}_{Rref} - \dot{x}_R)$$
$$\text{where, } x_{Rref} = x\sin(wt + \pi) \tag{6}$$

k_p and k_d are proportional and derivative control gains, whereas x_L and x_R are actual extension or contraction of left and right side of the actuator.

From Eqs. 5–6, we infer that we alternately supply air to the left side of the actuator and remove air from its right side, thus making it bend in both directions. We can also achieve simple extension and contraction by providing air supply as in Eq. 5 on both sides of the actuator.

3 Numerical Simulations

After creating bond graph model of actuator, we conduct numerical simulations to validate its accuracy in the framework of 20-sim® (ver. 4.7) [5]. Table 1 summarizes the system parameters of actuator used in the numerical experiment.

Table 1 System parameters used in numerical experiment

Parameter	Value	Parameter	Value
m	0.015 kg	x	0.3 m
R_b	0.4 kg/s	ρ	1.225 kg/m^3
k	350 N/m	R	8.31451 J/K mol
C_d	0.8	T	300 K
D	0.008 m	k_p	40
A	0.0096 m^2	k_d	10
μ	2.5	ω	1 rad/s

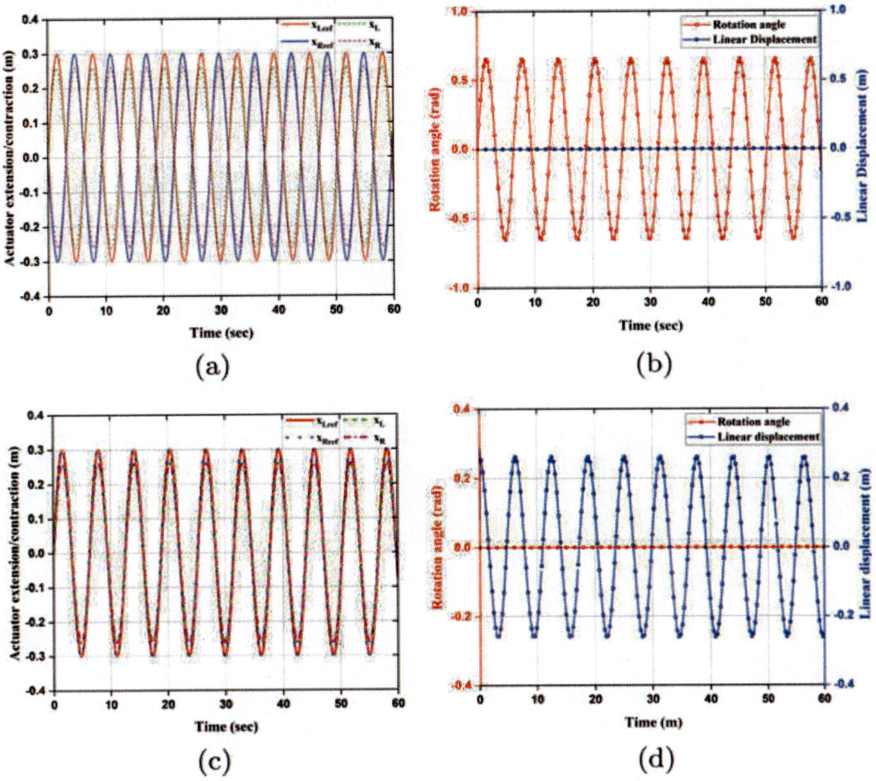

Fig. 5 Behaviour of actuator when **a** air is removed and supplied alternately, **b** rotation and extension due to alternate air supply, **c** air is removed and supplied simultaneously, and **d** rotation and extension due to simultaneous air supply

Figure 5a, b shows results for bidirectional bending of actuator. We can see in, Fig. 5a, each side of actuator extends and contracts alternately according to x_{ref} for each side. Figure 5b shows the result for bidirectional bending and rotation angle achieved. We can see there is no linear displacement in this case.

The extension and contraction of both sides of the actuator for linear displacement of the actuator are drawn in Fig. 5c. In this case, we can see both sides of the actuator are expanding and contracting simultaneously. Figure 5d demonstrates that the actuator has only linear displacement when each side expands and contracts simultaneously.

As explained in Sect. 2.2, controller output is pressure, P_1, supplied to actuator (see Eqs. 5–6). Figure 6 shows the results for rotation angle, linear displacement, and torque generated by actuator for the controlled pressure supply.

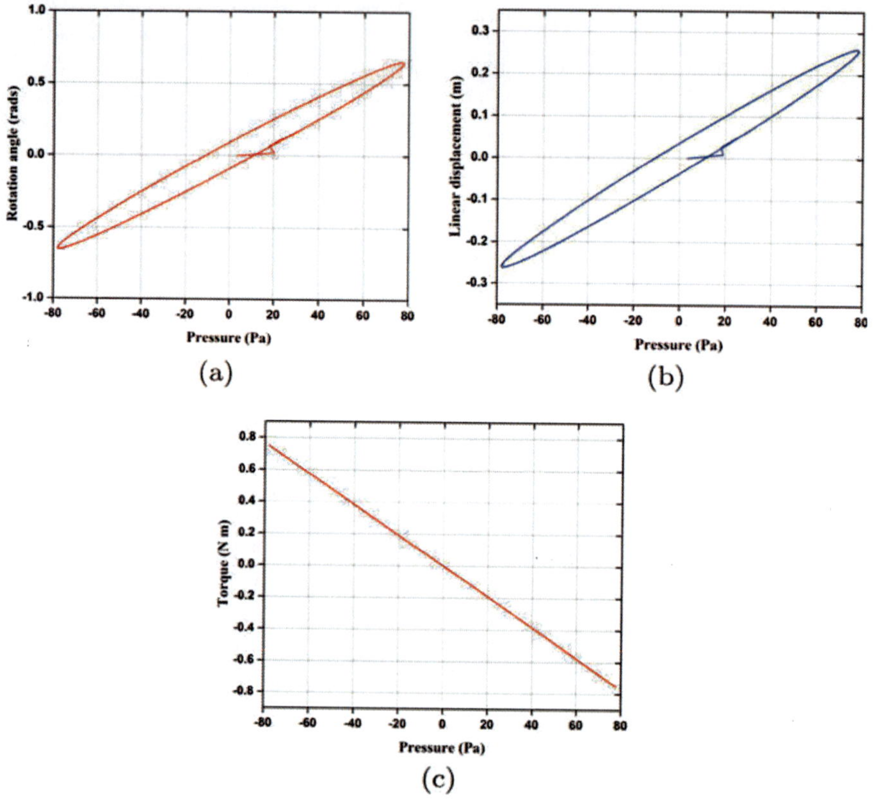

Fig. 6 Performance of actuator in terms of **a** rotation, **b** linear displacement, and **c** torque obtained for the controlled output

4 Conclusion

The paper presented a design and dynamic modelling of the soft actuator, which can be used in the varied application. The dynamics developed in this paper will be the basis for the complete dynamic modelling of the manipulator or robot created using our actuator. The conventional PD control is used in this paper, but we can use superior control methods since we have an accurate dynamic model. It would be interesting for future studies to achieve other important objectives such as path planning and obstacle avoidance for a soft robot made from the soft actuator.

Acknowledgements This work was supported in part by Indo-Korea JNC Program of Department of Science and Technology (DST), Government of India (INT/Korea/JNC/Robotics, dated 23-03-2018), in part by the National Research Foundation of Korea (NRF-2017K1A3A1A68072072).

References

1. Blumenschein LH, Gan LT, Fan JA, Okamura AM, Hawkes EW (2018) A tip-extending soft robot enables reconfigurable and deployable antennas. IEEE Robot Autom Lett 3(2):949–956. https://doi.org/10.1109/LRA.2018.2793303
2. Bolton W (1999) Mechatronics: electronic control systems in mechanical engineering. Addison Wesley Longman, Harlow, Essex, England
3. Boyraz P, Runge G, Raatz A (2018) An overview of novel actuators for soft robotics. Actuators 7(3). https://doi.org/10.3390/act7030048. https://www.mdpi.com/2076-0825/7/3/48
4. Chen C, Tang W, Hu Y, Lin Y, Zou J (2020) Fiber-reinforced soft bending actuator control utilizing on/off valves. IEEE Robot Autom Lett 5(4):6732–6739. https://doi.org/10.1109/LRA.2020.3015189
5. Controllab Products (2018) 20sim. https://www.20sim.com
6. Deimel R, Brock O (2013) A compliant hand based on a novel pneumatic actuator. In: 2013 IEEE international conference on robotics and automation, pp 2047–2053. https://doi.org/10.1109/ICRA.2013.6630851
7. Deimel R, Radke M, Brock O (2016) Mass control of pneumatic soft continuum actuators with commodity components. In: 2016 IEEE/RSJ international conference on intelligent robots and systems (IROS). IEEE Press, pp 774–779. https://doi.org/10.1109/IROS.2016.7759139
8. Elango N, Faudzi AAM (2015) A review article: investigations on soft materials for soft robot manipulations. Int J Adv Manuf Technol 80(5):1433–3015. https://doi.org/10.1007/s00170-015-7085-3
9. Elango N, Razif MR, Faudzi AAM, Palanikumar K (2020) Evaluation of a suitable material for soft actuator through experiments and FE simulations. Int J Manuf Mater Mech Eng 10(2):64–76. https://doi.org/10.4018/IJMMME.2020040104
10. Falkenhahn V, Hildebrandt A, Neumann R, Sawodny O (2017) Dynamic control of the bionic handling assistant. IEEE/ASME Trans Mechatron 22(1):6–17. https://doi.org/10.1109/TMECH.2016.2605820
11. Gerboni G, Diodato A, Ciuti G, Cianchetti M, Menciassi A (2017) Feedback control of soft robot actuators via commercial flex bend sensors. IEEE/ASME Trans Mechatron 22(4):1881–1888. https://doi.org/10.1109/TMECH.2017.2699677
12. Greer JD, Morimoto TK, Okamura AM, Hawkes EW (2017) Series pneumatic artificial muscles (spams) and application to a soft continuum robot. In: 2017 IEEE international conference on robotics and automation (ICRA), pp 5503–5510. https://doi.org/10.1109/ICRA.2017.7989648
13. Huang H, Lin J, Wu L, Fang B, Sun F (2019) Optimal control scheme for pneumatic soft actuator under comparison of proportional and PWM-solenoid valves. Photon Netw Commun 37(2):153–163. https://doi.org/10.1007/s11107-018-0815-3
14. Hyatt P, Wingate D, Killpack MD (2019) Model-based control of soft actuators using learned non-linear discrete-time models. Front Robot AI 6:22. https://doi.org/10.3389/frobt.2019.00022
15. Laschi C, Mazzolai B, Cianchetti M (2016) Soft robotics: technologies and systems pushing the boundaries of robot abilities. Sci Robot 1(1). https://doi.org/10.1126/scirobotics.aah3690
16. Li M, Kang R, Branson DT, Dai JS (2018) Model-free control for continuum robots based on an adaptive Kalman filter. IEEE/ASME Trans Mechatron 23(1):286–297. https://doi.org/10.1109/TMECH.2017.2775663
17. Marchese AD, Tedrake R, Rus D (2016) Dynamics and trajectory optimization for a soft spatial fluidic elastomer manipulator. Int J Robot Res 35(8):1000–1019. https://doi.org/10.1177/0278364915587926
18. Mohd Nordin INA, Muhammad Razif MR, Faudzi AAM, Natarajan E, Iwata K, Suzumori K (2013) 3-D finite-element analysis of fiber-reinforced soft bending actuator for finger flexion. In: 2013 IEEE/ASME international conference on advanced intelligent mechatronics, pp 128–133. https://doi.org/10.1109/AIM.2013.6584080

19. Polygerinos P, Wang Z, Overvelde JTB, Galloway KC, Wood RJ, Bertoldi K, Walsh CJ (2015) Modeling of soft fiber-reinforced bending actuators. IEEE Trans Robot 31(3):778–789. https://doi.org/10.1109/TRO.2015.2428504
20. Skorina EH, Luo M, Tao W, Chen F, Fu J, Onal CD (2017) Adapting to flexibility: model reference adaptive control of soft bending actuators. IEEE Robot Autom Lett 2(2):964–970. https://doi.org/10.1109/LRA.2017.2655572
21. Thor Morales B, Frederick L, Alexandre K, Zhongkai Z, Rochdi M, Christian D (2018) Finite element method-based kinematics and closed-loop control of soft, continuum manipulators. Soft Robot 5(3):348–364. https://doi.org/10.1089/soro.2017.0079
22. Wu P, Jiangbei W, Yanqiong F (2018) The structure, design, and closed-loop motion control of a differential drive soft robot. Soft Robot 5(1):71–80. https://doi.org/10.1089/soro.2017.0042
23. Zhang J, Sheng J, O'Neill CT, Walsh CJ, Wood RJ, Ryu JH, Desai JP, Yip MC (2019) Robotic artificial muscles: current progress and future perspectives. IEEE Trans Robot 35(3):761–781. https://doi.org/10.1109/TRO.2019.2894371

Design of Pneumatically Actuated Soft Robotic Gripper for Gripping Cylindrical Objects of Varying Diameters

Monalisa Sharma and Shubhashis Sanyal

Abstract Soft robotic grippers enable safe interactions with objects due to excellent compliance and ability to absorb energy arising in case of sudden impact, unlike their rigid counterparts. To effectuate an automated laboratory enabling remote experimentations without risking human operators, a soft robotic gripper is designed for handling fragile and delicate laboratory equipment. Constant curvature kinematics have been explored, proposing two types of gripping mechanisms for cylindrical objects-planar gripping mechanism and coiled gripping mechanism. The range of object diameters that a given gripper can handle has been obtained based on the choice of gripping mechanism.

Keywords Soft robotic gripper · Constant curvature kinematics · Gripping mechanism

1 Introduction

For several years, the field of robotics has been dominated by robots made of rigid links, joints and actuators, showing remarkable speed, accuracy and heavy load handling capacity. However, they become quite inadequate and unsafe in real-life scenarios like an unknown surrounding consisting of damageable goods as well as people in the vicinity.

Recently, scientists have begun exploring bio-inspired robotic architectures predominantly composed of soft materials, like elastomers, to impart compliance and adaptability to robots. These are called soft robots. From initial hyper redundant and hard continuum structures [1, 2], to soft universal gripper [3] or multifunctional tentacles [4, 5], soft robots have achieved hitherto unrealistic movements like stretching, twisting, multidirectional bending, squeezing, coiling, etc. They have diverse methods of actuation like use of pressurized air (or fluid) [5, 6], controlling cable lengths [1, 7], use of electroactive polymers which show deformation upon

M. Sharma (✉) · S. Sanyal
Department of Mechanical Engineering, National Institute of Technology Raipur, Raipur, Chhattisgarh, India
e-mail: monalisa.sharma204@gmail.com

V. K. Gupta et al. (eds.), *Recent Advances in Machines and Mechanisms*, Lecture Notes in Mechanical Engineering, https://doi.org/10.1007/978-981-19-3716-3_44

voltage supply, shape memory alloys which can deform and then revert back to its original shape upon heating, electrorheological or magnetorheological fluids which undergo stiffness variation when subjected to electric or magnetic fields, principle of particle jamming [3], etc.

Since soft materials can conform to the shape of an object and enable a smooth jerk-free contact, they are the most suitable candidates for grasping delicate objects or sensitive equipment. Particularly when handling apparatus containing infectious or hazardous samples, soft grippers are far more potent than traditional hard grippers. They can manipulate safely in an unspecified environment and pose no risk to humans. Automated laboratories are the need of the hour, which could enable remote experimentations involving infectious samples, without endangering human operators. Seeking for effective grippers to aid in this purpose, a soft gripper is designed with novel channel structure, and gripping mechanisms are proposed for cylindrical objects of varying diameters, since commonly used laboratory equipment like glass vials, beakers, flasks, test tubes, etc., are cylindrical in shape. This gripper is designed using a fluidic elastomer actuator [4, 8], actuated through pressurized air. The novelty of this work lies in the proposed soft gripper design and kinematic analysis of two different types of gripping mechanisms specific to cylindrical objects.

Gripper's design is described in Sect. 2. Section 3 introduces kinematics for a bending soft robot, while in Sect. 4, these kinematic calculations are implemented in the proposed gripper with two types of gripping methods for handling cylindrical objects of varying diameters. The range of object diameters that a specified gripper length can handle has been identified as per the gripping mechanism, which is a uniqueness of this work.

2 Gripper Design

Soft pneumatically actuated gripper designs span from simple two-fingered grippers [6] to six-membered starfish shaped gripper [9], from tentacle-inspired gripping structures [4, 5] to anthropomorphic grippers [10]. Most of these designs combine an expanding elastomer with an inextensible strain-limiting layer, resulting in a bending motion [5, 6, 8, 9, 11, 12]. The proposed gripper design is unique as it is a homogeneous elastomeric single-finger design whose bending relies on the channel's design. This design is mechanically simple and easy to manufacture, as shown in Fig. 1. The choice of dimensions was based on multiple trials and is given in Table 1.

The compliant elastomer body contains four hollow longitudinal channels shaped like a quadrant of a circle. When a channel is pressurized with compressed air, the thin outer walls being the least stiff region expand easily, while the thick inner walls acting as chamber separations are more resistant and remain comparatively unaltered. This variation in dimensions causes the structure to bend.

Some distinguishable features of this design are described here, which make the proposed gripper stand out amongst others. While some morphologies utilize thread windings to prevent excessive radial expansion of the channels, the thick

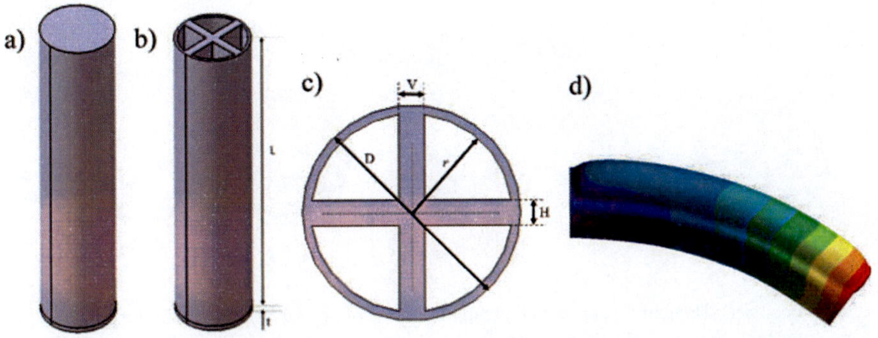

Fig. 1 Single-finger gripper model. **a** gripper model is closed at both ends with endcaps. **b** the top end is open. **c** gripper cross-section **d** actuated gripper's bent shape obtained from ANSYS simulation at 45 kPa pressure

Table 1 Gripper dimensions

Parameter	Dimensions (mm)
Outer Diameter, *D*	50
Chamber Radius, *r*	23
Thickness of vertical chamber separation, *V*	6
Thickness of horizontal chamber separation, *H*	6
Length of Gripper, *L*	200
Thickness of end-caps, *t*	3

compartmental separations in this design keep the transverse expansion in check. Being composed of a single elastomer material, this design can be moulded using 3D printed moulds, in a single step, minimizing its overall fabrication time. This also eliminates any risk of delamination of separate material layers at higher pressures. Provision of a rotating base plate will enable rotating the gripper and allow it to bend along any angular direction. The gripper surface is enhanced with gecko adhesive [6] to provide a firm grasp.

2.1 Bending Capabilities of Proposed Gripper

This gripper is capable of bending in four planes—two orthogonal planes and two diagonal planes, attaining both positive and negative curvatures in each plane. Single chamber actuation results in diagonal bending, such that the plane containing gripper's curved configuration lies at an angle of $\emptyset = \pm 45°$ with respect to the vertical

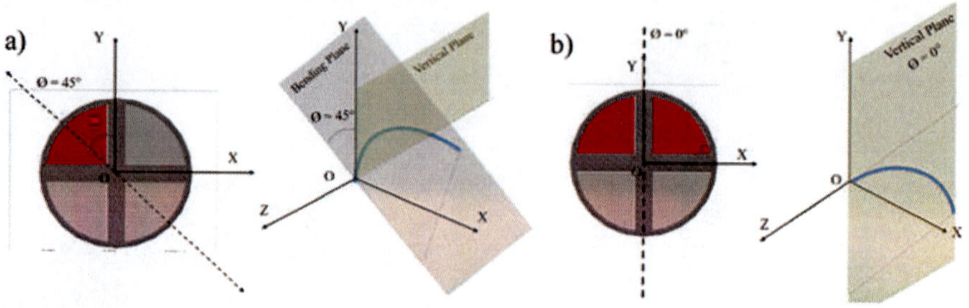

Fig. 2 Position of actuated chamber and resultant plane of arc for **a** single chamber actuation, and **b** two adjacent chamber actuation. Here, the blue curve represents the soft gripper and shows its bending direction

plane. Pressurizing two adjacent chambers causes bending in two orthogonal planes, i.e. vertical or horizontal planes.

Figure 2 depicts bending caused by single chamber and dual chamber actuation, respectively. Single actuated chamber results in bending of gripper along a plane making 45° with the vertical plane. Two adjacent actuated chambers cause bending of gripper in the vertical plane. Thus, the choice of actuated chamber decides the bending plane and curvature direction.

The nature of bending of the gripper has been verified through simulations in ANSYS software as shown in Fig. 1d. A hyperelastic incompressible Yeoh material model, with two parameters, was used to capture the nonlinear material behaviour of hyperelastic material of the gripper.

$$C_1 = 0.11MPa$$

$$C_2 = 0.02MPa$$

$$D_1 = 0$$

$$D_2 = 0$$

These coefficients fit the soft material Elastosil and are taken from [12]. The bent configuration obtained from simulations at different supply pressure helps to obtain various arc parameters required for kinematic analysis discussed in the following sections.

3 Constant Curvature Kinematics

Kinematics enables us to control and predict the robot's motion for practical applications. Due to the bending behaviour of our gripper, conventional DH parameters valid for rigid link robots become ineffectual. Instead, constant curvature kinematic formulations [1, 2, 7, 13] are used. These formulations treat a bending-type robot as a series of constant curvature arcs joined together such that they have a common tangent at their joints. The curves formed by bending soft robots are described by its curvature (κ), angle of plane containing the arc, or plane angle (Ø) and arc length l (or $s \in [0, l]$, where s is the arc length at any point lying along the curve), as shown in Fig. 3a. Bending angle θ is related to arc length and curvature as: $\theta = \kappa l$.

3.1 *Forward Kinematics*

Forward kinematics is the process to obtain the position (x, y and z) and orientation of the curve tip (or any other point located along the curve) in the base frame, when arc parameters, κ, Ø and l, are given. Several literature resources are available detailing the forward kinematic calculations. The resultant transformation matrix varies with the choice of coordinate frames, hence, it is important to follow the same rules of frame assignment at each step. It has been shown that multiple approaches essentially produce the same results [2].

Frame Assignment and Transformation Matrix: For the base frame, positive Z-axis is assigned to the tangent at the base of arc as shown in Fig. 3. Positive Y-axis is the direction about which the arc bends when $Ø = 0$. Positive X-axis is the final position taken by the arc if bending angle is assumed to be 180°, as shown in Fig. 3.

Fig. 3 Forward and inverse kinematic relations between **a** arc parameters (κ, Ø, l) and **b** position coordinates (x, y, z) in reference frame for a single segment constant curvature robot

$$T = \begin{bmatrix} \cos(\emptyset)\cos(\kappa l) & -\sin(\emptyset) & \cos(\emptyset)\sin(\kappa l) & \cos\emptyset\left(\frac{1-\cos(\kappa l}{\kappa}\right) \\ \sin(\emptyset)\sin(\kappa l) & \cos(\emptyset) & \sin(\emptyset)\sin(\kappa l) & \sin\emptyset\left(\frac{1-\cos(\kappa l)}{\kappa}\right) \\ -\sin(\kappa l) & 0 & \cos(\kappa l) & \frac{1}{\kappa}\sin(\kappa l) \\ 0 & 0 & 0 & 1 \end{bmatrix} \tag{1}$$

Equation (1) gives the transformation matrix. Here, the end frame is aligned such that its *Z*-axis is tangent to the curve and *X*-axis points towards the centre of the curve. The position coordinates of the curve tip (*l*) or any desired point on curve ($s \in [0, l]$), with respect to base frame, are given by first three elements of column 4. The first three elements of column 1 are unit vectors giving the principal direction of the X-axis of the tip frame described with respect to the base frame.

$$\vec{X}_{\text{tip}} = \{\cos(\varnothing)\cos(\kappa l)\}\hat{i} + \{\sin(\varnothing)\sin(\kappa l)\}\hat{j} + \{-\sin(\kappa l)\}\hat{k} \tag{2}$$

Similarly, the unit vectors giving principal directions of Y-axis and Z-axis of tip frame are given by first three elements of column 2 and 3, respectively. The above matrix can be used for a planar curve by putting plane angle $\emptyset = 0$.

3.2 Derivation of Inverse Kinematics from Forward Kinematics

Inverse kinematics involves calculating the arc parameters (κ, Ø and *l*) required to position the curve tip at a specified location (*x*, *y* and *z*) in the base frame. Although inverse kinematic formulae have been derived geometrically [14], in this work analytical calculations have been used, similar to conventional method of deriving inverse kinematics from forward kinematic transformation matrix obtained from DH tables. This derivation is presented below. Substituting $\kappa = 1/r$ and $\kappa l = \theta$ in the position coordinates of the curve tip obtained from Eq. (1),

$$x = r\cos\emptyset(1 - \cos\theta) \tag{3}$$

$$y = r\sin\emptyset(1 - \cos\theta) \tag{4}$$

$$z = r\sin\theta \tag{5}$$

Squaring, adding and rearranging these equations,

$$(1 - \cos\theta) = \frac{2(x^2 + y^2)}{x^2 + y^2 + z^2} = \frac{x^2 + y^2 + z^2}{2r^2} \tag{6}$$

By rearranging Eq. (6) to get an expression for r, curvature is obtained as

$$\kappa = \pm \frac{2\sqrt{x^2 + y^2}}{x^2 + y^2 + z^2} \tag{7}$$

Substituting the value of r in Eq. (5) to get $\sin\theta$, and by taking the value of $\cos\theta$ from Eq. (6), bending angle is obtained as

$$\theta = \tan^{-1}\left(\frac{2z\sqrt{x^2 + y^2}}{z^2 - x^2 - y^2}\right) \tag{8}$$

From Eqs. (4) by (3), plane angle Ø is obtained as

$$\emptyset = \tan^{-1}\left(\frac{y}{x}\right) \tag{9}$$

And, arc length can be obtained using, $l = \theta/\kappa$.

4 Gripping Mechanism and Parameterization

Two types of gripping mechanisms have been proposed for cylindrical objects which are distinctive to this work-planar gripping and coiled gripping, with elaborated kinematics shown through a case study. For a chosen mechanism, the range of object diameters possible to handle can be identified which is another unique contribution of this work.

4.1 Planar Gripping Mechanism

In this method, the gripper surrounds a cylindrical object in a circular grip, contacting the object along its entire length. For this type of grip, gripper's curvature is equal to the object's curvature. To form a grip, it is assumed the gripper must encircle at least 2/3rd of the object circumference, i.e. the circular arc subtends a bending angle of 240°. Clearly, the maximum extent of this grip is when the gripper completely surrounds the circumference of the object.

An intuitive assumption has been made regarding the angle subtended by the gripper for gripping the object because the main aim here is to develop a gripping mechanism and incorporate continuum kinematics. Further study needs to be done in order to validate the gripping angle for a sturdy grip. The angle will depend on the gripper material as well as the material being gripped. Force analysis was beyond the

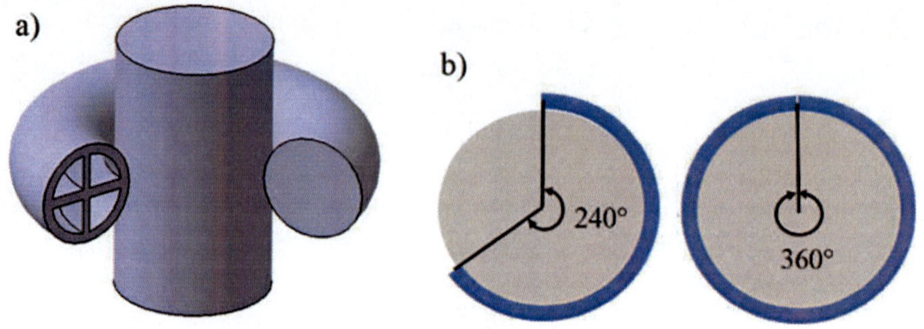

Fig. 4 **a** Gripper surrounding an object in planar grip, **b** minimum and maximum angles subtended by gripper (blue) surrounding an object (grey) in a planar grip

scope of this work which mainly intends to elaborate kinematics of gripping; hence, the angle values are suggested on the assumption that frictional forces will be able to hold the object. These values are neither fixed nor final (Fig. 4)

Fixed Gripper Length: This is a commonly encountered case where gripper has a fixed length l, while the cylindrical objects vary in size, like small test tubes, large beakers, glass vials, etc. Using $\kappa l = \theta$, the range of possible gripper curvatures is

$$\kappa_{\min} = \theta_{\min}/l = 4\pi/3l \text{ and } r_{\max} = 1/\kappa_{\min} \tag{10}$$

$$\kappa_{\max} = \theta_{\max}/l = 2\pi/l \text{ and } r_{\min} = 1/\kappa_{\max} \tag{11}$$

Here, $r_{\min}$ and $r_{\max}$ are the minimum and maximum object radius, respectively, that a gripper can handle using planar gripping mechanism.

Fixed Object Radius: This is a special case where a gripper is designed in accordance with the object's size. Say, the object radius is r. Then gripper should have a curvature of $\kappa = 1/r$. The range of gripper length required to encompass at least 2/3rd of object circumference and at most the full circumference is estimated using $l = \theta/\kappa$,

$$l_{\min} = (4\pi/3)/\kappa \tag{12}$$

$$l_{\max} = 2\pi/\kappa \tag{13}$$

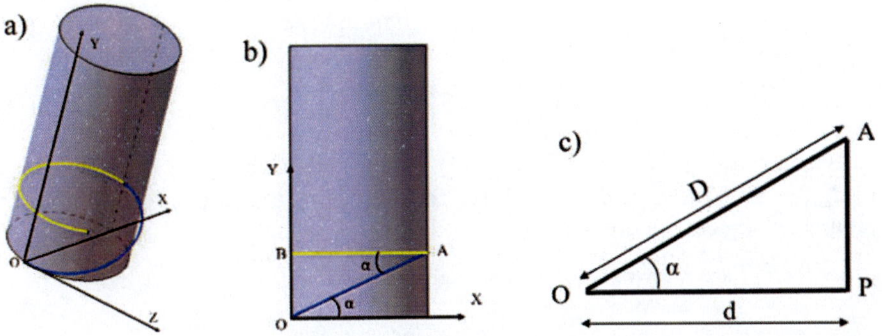

Fig. 5 **a** Coiled gripping mechanism by a two-segment gripper, **b** Segment-1 (blue) forms semi-circle at angle α along plane OA, and segment-2 (yellow) forms planar grip along plane AB and **c** geometric parameters for coiled grip are shown in a line diagram

4.2 *Coiled Gripping Mechanism*

Coiling is natural phenomenon found in plants and even used by some animals in nature to form a grip. This mechanism tries to mimic the natural coiling in order to get more contact area and thus improve the load bearing capacity of the gripper. A rotating base plate attached at the base of each gripper segment enables it to bend along a plane lying at any angle Ø. It is proposed to form a coil using two gripper segments per coil winding, as shown in Fig. 5a. Multiple gripper segments can be joined to obtain desired number of coil windings.

Segment 1 (blue) forms a semicircular arc along plane OA, but the cylinder's surface profile along plane OA is elliptical. Hence, it contacts the object only at points O and A. Segment 2 (yellow) bends in plane AB parallel to the cylinder's base surface, forming a planar grip with full contact along entire gripper length. This gripping mechanism is explained using a case study.

Case Study: Each gripper segment is of length 20 cm. A single-winding coiled grip is desired around an object of diameter 10 cm.

Target object diameter = $d = 10$ cm, gripper segment length = $l = 20$ cm, segment 1 forms a semicircle with diameter D, therefore bending angle $\theta_1 = 180°$. This arc's plane is rotated at an angle α i.e. $Ø_1 = \alpha$.

From geometry in Fig. 5c, $D = OA = d/\cos\alpha$. Therefore, curvature of arc-1 is

$$\kappa_1 = \frac{2}{D} = \frac{2\cos\alpha}{d} = \frac{\theta_1}{l} \tag{14}$$

The value of α can vary from 0° to 45°. The maximum limit is 45°, beyond this value the gripper won't reach around the cylindrical object, but intersect the object at some point on the front face itself. So, this becomes the deciding factor in determining range of objects. From Eq. (14), $d = (2\cos\alpha)l/\theta$. Substituting the limits of α, range of object diameters is obtained as

$$d_{\max} = (2\cos 0°)l/\theta = 2l/\pi \tag{15}$$

$$d_{\min} = (2\cos 45°)l/\theta = (\sqrt{2}l)/\pi \tag{16}$$

By substituting $l = 20$ cm in Eqs. (15) and (16), object diameter lies in the range of $d_{\min} = 9.003$ cm and $d_{\max} = 12.732$ cm. Therefore, this gripper can form a coil around an object with d = 10 cm.

Since θ and l are known, curvature of segment-1, is

$$\kappa_1 = \theta/l = \pi/20 = 0.157$$

Value of α is obtained from Eq. (14) as, $\alpha = 38.24°$. Substituting these curve parameters in Eq. (1), transformation matrix for segment 1 is obtained as

$$^{0}T_1 = \begin{bmatrix} -0.7854 & -0.6189 & 0 & 10.0054 \\ -0.6189 & 0.7854 & 0 & 7.8848 \\ 0 & 0 & -1 & 0 \\ 0 & 0 & 0 & 1 \end{bmatrix}$$

Column 3 gives position coordinates of tip of segment-1. This tip frame {1} has its Z-axis tangent to the curve, X-axis pointing towards the centre of the curve. This tip frame is the base frame for segment 2.

Arc parameters for Segment 2: For planar grip, gripper's curvature = curvature of object, i.e.

$$\kappa_2 = 2/d = 0.2$$

Plane angle with respect to segment 1 is, $\emptyset_2 = \alpha = 38.24°$, arc length l is same as before. Bending angle is $\theta_2 = \kappa_2 l = 0.2 \times 20 = 4\,\text{rad} = 229.183°$. Substituting these in Eq. (1), transformation matrix for segment 2 is obtained as

$$^{1}T_2 = \begin{bmatrix} -0.5133 & -0.6189 & -0.5944 & 6.4940 \\ -0.4684 & 0.7854 & -0.4684 & 5.1176 \\ 0.7568 & 0 & -0.6536 & -3.7840 \\ 0 & 0 & 0 & 1 \end{bmatrix}$$

Total transformation for coil's tip with respect to the absolute reference frame {0} located at base of gripper segment 1, is obtained as

$$^{0}T_2 = {}^{0}T_1\,{}^{1}T_2 \tag{17}$$

$$
{}^{0}T_{2} = \begin{bmatrix} 0.6930 & 0 & 0.7567 & 1.7377 \\ -0.0501 & 0.9998 & 0.0000 & 7.8850 \\ -0.7568 & 0 & 0.6536 & 3.784 \\ 0 & 0 & 0 & 1 \end{bmatrix}
$$

This case study gives a method for precise positioning and manipulating of a soft gripper. Position coordinates obtained above agree with the geometrically obtained position. Thus, two gripping mechanisms with predictable positioning for soft grippers have been presented.

5 Conclusion

The unparalleled compliance and shape adaptability enable soft grippers to be used for grasping objects based on operator's intuition instead of depending on complex positioning algorithms. However, in certain applications, prediction of exact position is important, as described in the present work. The proposed gripping mechanisms mainly elaborate the kinematics involved in predicting the gripper's position. Their implementation and application require further experimentation and consideration of force analysis which gives scope for more research.

Due to complete contact between the gripper and object, planar grip is very firm and secure, which makes handling delicate test tubes containing contagious samples extremely safe. To grip heavier loads, coiled grip can be used with multiple windings.

Though the range of object size that can be handled by a gripper of fixed length is very small, the benefits of soft robotics outweigh these limitations and encourage seeking for methods of improvement. There is much scope for exploring new gripper capabilities. Instead of searching for abilities to suit an application, applications that can benefit from a certain ability should be sought to make soft robots a part and parcel of our daily life.

Acknowledgements The valuable assistance and support from the Department of Mechanical Engineering, National Institute of Technology, Raipur, has made this work possible and is gratefully acknowledged.

References

1. HannanMW, Walker ID (2003) Kinematics and the Implementation of an Elephant's Trunk Manipulator and Other Continuum Style Robots," 2003. [Online]. Available: www.interscience.wiley.com
2. Webster RJ, Jones BA (2010) Design and kinematic modeling of constant curvature continuum robots: a review. Int J Robot Res 29(13):1661–1683. https://doi.org/10.1177/0278364910368147
3. Brown E et al (2010) Universal robotic gripper based on the jamming of granular material. Proc Natl Acad Sci U S A 107(44):18809–18814. https://doi.org/10.1073/pnas.1003250107

4. Marchese AD, Rus D (2016) Design, kinematics, and control of a soft spatial fluidic elastomer manipulator. Int J Robot Res 35(7):840–869. https://doi.org/10.1177/0278364915587925
5. Martinez RV et al (2013) Robotic tentacles with three-dimensional mobility based on flexible elastomers. Adv Mater 25(2):205–212. https://doi.org/10.1002/adma.201203002
6. Glick P, Suresh SA, Ruffatto D, Cutkosky M, Tolley MT, Parness A (2018) A Soft Robotic Gripper with Gecko-Inspired Adhesive. IEEE Robot Autom Lett 3(2):903–910. https://doi.org/10.1109/LRA.2018.2792688
7. Jones BA, Walker ID (2006) Kinematics for multisection continuum robots. IEEE Trans Robot 22(1):43–55. https://doi.org/10.1109/TRO.2005.861458
8. Marchese AD, Katzschmann RK, Rus D (2015) A recipe for soft fluidic elastomer robots. Soft Robot 2(1):7–25. https://doi.org/10.1089/soro.2014.0022
9. Ilievski F, Mazzeo AD, Shepherd RF, Chen X, Whitesides GM (2011) Soft robotics for chemists. Angew Chemie Int Ed 50(8):1890–1895. https://doi.org/10.1002/anie.201006464
10. Deimel R, Brock O (2013) A compliant hand based on a novel pneumatic actuator. In: Proceedings of IEEE international conference on robotics and automation, pp 2047–2053. https://doi.org/10.1109/ICRA.2013.6630851
11. Deimel R, Brock O A compliant hand based on a novel pneumatic actuator
12. Polygerinos P et al (2015) Modeling of soft fiber-reinforced bending actuators. IEEE Trans Robot 31(3):778–789. https://doi.org/10.1109/TRO.2015.2428504
13. Hannan MW, Walker ID (2000) Novel kinematics for continuum robots
14. Neppalli S, Csencsits MA, Jones BA, Walker I (2008) A geometrical approach to inverse kinematics for continuum manipulators. In: 2008 IEEE/RSJ internatinal conference on intelligent robots system IROS, pp 3565–3570. https://doi.org/10.1109/IROS.2008.4651125

Estimation of Internal Joint Forces and Resisting Torques for Impact of Walking Robot Model

K. Ramachandra and Sourav Rakshit

Abstract Quadruped motion in legged locomotion provides an outlook to explore both static and dynamic gaits. The leg contact of the quadruped with the environment is an impulsive contact with a non-smooth interaction. This impulsive force is resisted by the torques applied at the leg joints of the model, and hence, the realistic estimate of these resisting torques is vital for the stable operation of the robot. In the current study, the dynamic model of a planar two-legged mobile robot is formulated for an impact problem by utilizing the external impact model as proposed by Lee et al. (Modeling and analysis of internal impact for general classes of robotic mechanisms. In: Proceedings. IEEE/RSJ international conference on intelligent robots and systems, vol. 3, pp. 1955–1962, 2000 [1]). The impulse encountered during contact of the leg foot with the ground is utilized as the basis to compute the resisting forces at the joints in the model. These resisting joint forces due to impact can be used to decide the configuration of mobile robot leg during landing on the ground. From the simulation, it is established that the impact force on the joints is greatest when the orientation of the lower link of the leg is perpendicular to the contact environment.

Keywords Impact model · Legged robot · Unilateral contact

1 Introduction

Quadruped motion which is a major area in legged locomotion provides an outlook to explore both static and dynamic locomotion. The leg contact of the quadruped with the ground/external obstacle constrains the allowable motion of the body. Further, this contact is an impulsive contact with a non-smooth interaction and hence is one of the sources of energy loss during locomotion. During contact, the impulsive force acting on the body would be resisted by the torques applied at the leg joints of the

K. Ramachandra (✉) · S. Rakshit
Department of Mechanical Engineering, IIT Madras, Chennai, India
e-mail: ramachandra.kodi@gmail.com

S. Rakshit
e-mail: srakshit@iitm.ac.in

V. K. Gupta et al. (eds.), *Recent Advances in Machines and Mechanisms*, Lecture Notes in Mechanical Engineering, https://doi.org/10.1007/978-981-19-3716-3_45

model, and hence, a realistic estimate of these resisting torques is vital for the stable dynamic operation of the robot. The impact effect encountered during an unilateral contact of a floating body with the ground includes several factors such as the elastic properties of both bodies, velocities during contact, and direction of impact.

The robot performing a dynamic gait, when comes in contact with the ground experiences external impact force at the leg foot and in turn, propagates as internal impulsive forces on the joints of the robot. The external impulse model for estimation of impulse for a serial type of robot was introduced by Walker [2], where he also proposed a method to reduce it by self-motion in case of a redundant manipulator. Zheng and Hemami [3] derived the internal impulse model at the joints, but their model was confined to the serial-type manipulators. Lee et al. [1] extended the study of the internal impulse model to a general class of mechanism by using Newton–Euler method to obtain the internal impulsive forces. Lee et al. [4] suggested an external impulse model and an internal impulse measure for sawing task to show its efficiency by using a dual-arm. Bowling [5] utilized the impact model to determine the effect of impact forces on legged robot agility, wherein he computes the performance curves describing how well the legged system can use the impact forces to accelerate itself. Karssen et al. [6] suggested the swing-leg retraction model to improve the disturbance rejection and reduce impact energy loss during contact. He further describes the effect of horizontal velocity of leg on the optimal swing-leg retraction.

In the current study, a planar two-legged mobile robot was formulated and subjected to external impulsive contact with the ground. The impulsive force obtained during contact of the leg foot with the ground was utilized as the basis to compute the resisting forces on the leg joints of the model. This computation of contact forces requires only the state of the robot before impact. The resisting torques on the leg joints obtained due to external impact together with the internal resisting joint forces can be utilized as a factor to decide the configuration of the robot for minimal impact during contact on the ground.

2 Dynamic Model of Planar Robot

In this section, the dynamic model of a planar two-legged mobile robot is formulated using augmented Lagrangian formulation by considering the joint constraints in the model. The planar robot configuration as shown in Fig. 1 consists of two legs, each with two degrees of freedom (DoF) links and a floating base to which these legs are connected. The system is represented by seven independent system coordinates given as $\mathbf{q_i} = [x_1, y_1, \theta_1, \theta_2, \theta_3, \theta_4, \theta_5]^{\mathrm{T}}$. Of these seven system coordinates, three system coordinates (x_1, y_1, θ_1) are of the under-actuated base and the four coordinates $(\theta_2, \theta_3, \theta_4, \theta_5)$ are of the actuated revolute joints. In a planar formulation, each revolute joint imposes two translation constraints, one in horizontal and the other in the vertical direction of the links connected to the joint. Thus, the four joints of the system impose eight joint constraints, which show up as internal resisting forces acting on the joints in the inertial coordinate frame. The coordinates of these eight constraints form the dependent coordinates given as: $\mathbf{q_d} = [x_2, y_2, x_3, y_3, x_4, y_4, x_5, y_5]^{\mathrm{T}}$.

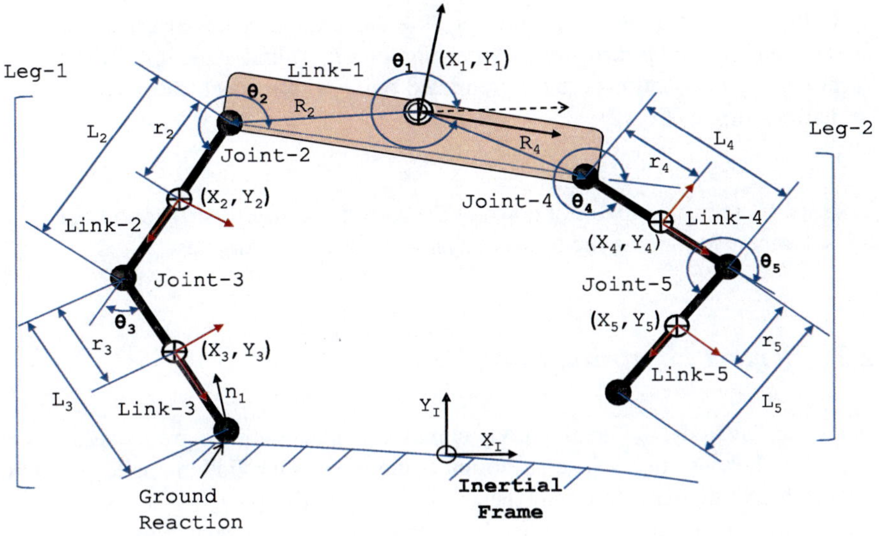

Fig. 1 Schematic of planar robot

In Fig. 1, x_i and y_i (i varies from 1 to 5) are x- and y-components of position vectors locating the center of mass (CoM) of link-i with respect to inertial frame. The angles, θ_2 to θ_5 are orientation of link-2 to link-5, respectively, with respect to the previous link. Further, $\mathbf{n_1} = (nx_1, ny_1)$ indicates the unit normal vectors at the contact point of leg-1 with the ground.

2.1 *Joint Constraint Equations*

The system consists of four actuated revolute joints. Each revolute joint of system imposes two constraints in translation between the two connected links and is given by Eq. (1) as:

$$\begin{aligned}
Jx_{12} &= x_1 + R_2\cos(\delta_2 + \theta_1) - r_2\cos(\theta_1 + \theta_2 + \pi) - x_2 = 0\\
Jy_{12} &= y_1 + R_2\sin(\delta_2 + \theta_1) - r_2\sin(\theta_1 + \theta_2 + \pi) - y_2 = 0\\
Jx_{23} &= x_2 + (L_2 - r_2)\cos(\theta_1 + \theta_2) - r_3\cos(\theta_1 + \theta_2 + \theta_3 + \pi) - x_3 = 0\\
Jy_{23} &= y_2 + (L_2 - r_2)\sin(\theta_1 + \theta_2) - r_3\sin(\theta_1 + \theta_2 + \theta_3 + \pi) - y_3 = 0\\
Jx_{14} &= x_1 + R_4\cos(\delta_4 + \theta_1) - r_4\cos(\theta_1 + \theta_4 + \pi) - x_4 = 0\\
Jy_{14} &= y_1 + R_4\sin(\delta_4 + \theta_1) - r_4\sin(\theta_1 + \theta_4 + \pi) - y_4 = 0\\
Jx_{45} &= x_4 + (L_4 - r_4)\cos(\theta_1 + \theta_4) - r_5\cos(\theta_1 + \theta_4 + \theta_5 + \pi) - x_5 = 0\\
Jy_{45} &= y_4 + (L_4 - r_4)\sin(\theta_1 + \theta_4) - r_5\sin(\theta_1 + \theta_4 + \theta_5 + \pi) - y_5 = 0
\end{aligned} \quad (1)$$

In Eq. (1), considering Jx_{12} and Jy_{12}, the subscript indicates that the constraints are applied in x- and y-directions on the joint formed by link-1 and link-2. The above eight constraint equations are differentiated twice with respect to time and written in matrix[1] form as Eq. (2):

$$\mathbf{C_q}\ddot{\mathbf{q}} + (\mathbf{C_q}\dot{\mathbf{q}})_\mathbf{q}\dot{\mathbf{q}} = 0 \tag{2}$$

Here, $\mathbf{C_q}$ is 8×15 matrix, known as the *Joint Constraint Matrix* and $(\mathbf{C_q}\dot{\mathbf{q}})_\mathbf{q}\dot{\mathbf{q}}$ is 15×1 vector containing the terms of system velocity coordinates.

2.2 Contact Constraint Equations

In the current study, the floating body comes in contact with the ground at only one leg tip as shown in Fig. 1. The constraint is modeled as a rigid kinematic constraint which ensures non-penetration in the ground. Considering a friction-less surface, the constraint is imposed in the normal direction of contact. The constraint is computed as a dot product of the position vector from inertial frame to the leg tip contact point with the unit contact normal ($\mathbf{n_1}$) and given as:

$$\begin{aligned} Lp_{n1} &= nx_1(x_1 + R_2\cos(\delta_2 + \theta_1) + L_2\cos(\theta_1 + \theta_2) + L_3\cos(\theta_1 + \theta_2 + \theta_3)) \\ &\quad + ny_1(y_1 + R_2\sin(\delta_2 + \theta_1) + L_2\sin(\theta_1 + \theta_2) + L_3\sin(\theta_1 + \theta_2 + \theta_3)) \end{aligned} \tag{3}$$

The ground is considered as a flat surface with the inertial frame on top of the surface. In Eq. (3), the loop value of Lp_{n1} would be zero whenever the kinematic constraint is imposed, which thereby indicates the contact of leg tip with the ground. Writing the statement as Eq. (4):

$$Lp_{n1} \leq 0 \longrightarrow Gl_{n1} \text{ is True} \tag{4}$$

Here Gl_{n1} is the variable indicating contact of leg-1 with ground if condition is true. To obtain the contact constraints consistent with the acceleration of system coordinates, the above constraints are differentiated twice with respect to time to obtain the equation[2] as Eq. (5):

$$\mathbf{J_a}\ddot{\mathbf{q}} + (\mathbf{J_a}\dot{\mathbf{q}})_\mathbf{q}\dot{\mathbf{q}} = 0 \tag{5}$$

Here, $\mathbf{J_a}$ is 1×15 matrix, known as the the *Contact Jacobian matrix* and $(\mathbf{J_a}\dot{\mathbf{q}})_\mathbf{q}\dot{\mathbf{q}}$ is 1×1 matrix containing the terms of system velocity coordinates.

---

[1] The details of matrices for $\mathbf{C_q}$ and $(\mathbf{C_q}\dot{\mathbf{q}})_\mathbf{q}\dot{\mathbf{q}}$ are given in Appendix.

[2] The matrix obtained is given in Appendix.

2.3 System Lagrangian Formulation

The system Lagrangian formed is summation of kinetic energy, potential energy, joint constraints, and contact constraints in the system and is given as Eq. (6).

$$\begin{aligned}\text{Lag} = \text{KE} - \text{PE} + (Jx_{12}\lambda x_{12} + Jy_{12}\lambda y_{12} + Jx_{23}\lambda x_{23} + Jy_{23}\lambda y_{23}\\ + Jx_{14}\lambda x_{14} + Jy_{14}\lambda y_{14} + Jx_{45}\lambda x_{45} + Jy_{45}\lambda y_{45}) + (Lp_{n1}\lambda_{n1})\end{aligned} \tag{6}$$

Here, KE and PE are the summation of kinetic and potential energy of each link in the model, λ_{n1} is Lagrange multiplier representing the contact force generated due to the kinematic constraint of the leg-1 tip with the ground during contact in the normal direction, λx_{ij} and λy_{ij} are the Lagrange multipliers indicating the internal joint forces in x- and y-directions, respectively, at the joint formed by ith and jth link.

The equation of motion for the system[3] is:

$$\mathbf{M\ddot{q}} + \mathbf{G_q} - \mathbf{C_q^T}\boldsymbol{\lambda}_{\mathbf{ij}} - \mathbf{J_a^T}\boldsymbol{\lambda}_{\mathbf{n1}} = \tau \tag{7}$$

Here, $\mathbf{M}$ is the mass matrix of the system, $\mathbf{G_q}$ is vector of gravity terms, $\mathbf{C_q}$ is the *Joint Constraint Matrix*, as obtained in Eq. (2), τ is vector of actuator joint torques and λ_{ij} is vector of Lagrange multipliers i.e., $\boldsymbol{\lambda}_{\mathbf{ij}} = [\lambda x_{12}; \lambda y_{12}; \lambda x_{23}; \lambda y_{23}; \lambda x_{14}; \lambda y_{14}; \lambda x_{45}; \lambda y_{45}]$. Augmenting Eq. (7) with joint and contact constraint matrices (Eqs. 2 and 5) and writing the equations in matrix form as:

$$\begin{bmatrix} \mathbf{M} & -\mathbf{C_q^T} & -\mathbf{J_a^T} \\ \mathbf{C_q} & 0 & 0 \\ \mathbf{J_a} & 0 & 0 \end{bmatrix} \begin{bmatrix} \mathbf{\ddot{q}} \\ \boldsymbol{\lambda}_{\mathbf{ij}} \\ \boldsymbol{\lambda}_{\mathbf{n1}} \end{bmatrix} = \begin{bmatrix} \tau - \mathbf{G_q} \\ -(\mathbf{C_q\dot{q}})_\mathbf{q}\mathbf{\dot{q}} \\ -(\mathbf{J_a\dot{q}})_\mathbf{q}\mathbf{\dot{q}} \end{bmatrix} \tag{8}$$

Here the terms $\mathbf{C_q}$, $\mathbf{J_a}$ and $\mathbf{G_q}$ are function of time.

3 External Impact Force on System

Considering the collision of the rigid body against a fixed surface, wherein the collision is partially elastic, the change in velocity of the rigid body normal to the surface of contact is given by:

$$(\mathbf{v_1} + \Delta\mathbf{v_1})\mathbf{n} = -e(\mathbf{v_1})\mathbf{n} \tag{9}$$

Here $\mathbf{v_1}$ is the velocity vector of the body before the collision, $\Delta\mathbf{v_1}$ is the vector of velocity increment of the rigid body due to collision, $\mathbf{n}$ is the unit vector normal to the surface of contact, and e is coefficient of restitution indicating a partially elastic collision ranging from $0 < e < 1$. Rearranging Eq. (9) in terms of velocity increment of the rigid body:

[3] The details of the matrices, $\mathbf{M}$ and $\mathbf{G_q}$ obtained are given in Appendix.

$$(\Delta \mathbf{v_1})\mathbf{n} = -(1-e)(\mathbf{v_1})\mathbf{n} \tag{10}$$

Writing Eq. (10) in terms of velocity increment normal to the surface of collision:

$$\Delta v_{n1} = -(1-e)v_{n1} \tag{11}$$

where $v_{n1} = (\mathbf{v_1})\mathbf{n}$ and $\Delta v_{n1} = (\Delta \mathbf{v_1})\mathbf{n}$.

To evaluate the magnitude of external impulse on the system, consider leg-1 of the planar robot coming in contact with the ground and is subjected to impact due to the kinematic constraints with the surface. Leg-2 is in the air during this impact. The equation of motion of the system Eq. (7) during this instant is modified as proposed by Lee et al. [1] as:

$$\tau = \mathbf{M}\ddot{\mathbf{q}} + \mathbf{G_q} - \mathbf{C_q^T}\lambda_{\mathbf{ij}} - \mathbf{J_a^T} F_{n1} \tag{12}$$

Here F_{n1} is the external impulsive force experienced at contact point in normal direction of contact, $\mathbf{G_q}$ is vector of gravity terms, $\mathbf{J_a}$ is the contact Jacobian matrix detailed in Eq. (5).

In order to find a relation between the external impulse acting on the system to the velocity increment during impact, Eq. (12) is integrated over the time interval (t_0 and $t_0 + \Delta t$) of contact. Since position, velocities and actuation torque are finite all the time during impact, the non-impulsive terms such as $\mathbf{C_q}$, $\mathbf{G_q}$ and τ drop out from Eq. (12). The resulting equation is:

$$\mathbf{M}(\dot{\mathbf{q}}(t_0 + \Delta t) - \dot{\mathbf{q}}(t_0)) = \mathbf{J_a^T} P_{cn1} \tag{13}$$

Here, $\dot{\mathbf{q}}(t_0 + \Delta t) - \dot{\mathbf{q}}(t_0)$ is the velocity increment of joint variables due to impact, P_{cn1} is the external impulse at contact point in the normal direction of contact, given as $P_{cn1} = \int_{t_0}^{t_0+\Delta t} F_{n1}\, dt$. Rearranging Eq. (13) in terms of velocity increments:

$$\Delta\dot{\mathbf{q}} = \mathbf{M}^{-1}\mathbf{J_a^T} P_{cn1} \tag{14}$$

where, $\Delta\dot{\mathbf{q}} = \dot{\mathbf{q}}(t_0 + \Delta t) - \dot{\mathbf{q}}(t_0)$ is the velocity increment due to impact. The incremental velocity of leg tip before contact, normal to the contact surface is given as:

$$\Delta v_{n1} = \mathbf{J_a}\Delta\dot{\mathbf{q}} \tag{15}$$

Substituting for Δv_{n1} and $\Delta\dot{\mathbf{q}}$ from Eqs. (11) and (14), respectively, in Eq. (15) as:

$$-(1-e)v_{n1} = \mathbf{J_a}\mathbf{M}^{-1}\mathbf{J_a^T} P_{cn1} \tag{16}$$

Rearranging Eq. (16) in terms of P_{cn1}, the magnitude of external impulse on the foot due to the impact is given as:

$$P_{cn1} = \frac{-(1+e)v_{n1}}{(\mathbf{J_a M^{-1} J_a^T})} \tag{17}$$

With the external impulse of contact (P_{cn1}) value known for leg-1, the ground contact force normal to the ground is given as:

$$\lambda_{n1} = P_{cn1}/\Delta t \tag{18}$$

where Δt is the time step of numerical iteration. In the current model, a time step of 0.0001 s is used for numerical iteration. The actuator wrench acting on the joints of leg due to ground contact force is given as:

$$\tau_{\text{react}} = (\mathbf{J_a}^{\mathbf{T}} \lambda_{n1}) \tag{19}$$

To evaluate the generated impulsive force due to contact, the planar model is subjected to gravity free-fall along various touchdown[4] angles of link-3 (lower link of the leg-1). A typical configuration of the planar robot with indication of the touchdown angle is shown in Fig. 2. The magnitude of the external impulsive force for various touchdown angles is plotted in Fig. 3. It is seen from the plot that link-3 is subjected to maximum impact force when the orientation of link-3 is perpendicular to the ground. The link parameters considered for the evaluation of the system from link-1 to link-5 as shown in Fig. 1 are with mass of [8.5, 1.12, 1.17, 1.12, 1.17] kg and with link lengths of [0.8, 0.4, 0.4, 0.4, 0.4] m, respectively.

4 Internal Impact Forces on Joints

Due to impact of the leg foot with the ground, internal impulsive forces act on each joint of the leg. This impulsive forces acting on each joint provide scope of orientation of leg links during impact to minimize the effect of impact forces. To obtain an equation to evaluate the internal impulsive forces, from Eq. (2),

$$\mathbf{C_q \ddot{q}} = \mathbf{Q_c} \quad \text{where} \quad \mathbf{Q_c} = -(\mathbf{C_q \dot{q}})_{\mathbf{q}} \mathbf{\dot{q}} \tag{20}$$

and from Eq. (7),

$$\mathbf{M\ddot{q}} + \mathbf{G_q} - \mathbf{C_q^T} \lambda_{\mathbf{ij}} - \mathbf{J_a^T} \lambda_{\mathbf{n1}} = \tau$$

Rearranging Eq. (7), in terms of $\mathbf{\ddot{q}}$ as:

$$\mathbf{\ddot{q}} = \mathbf{M}^{-1}(\tau + \mathbf{C_q^T} \lambda_{\mathbf{ij}} + \mathbf{J_a^T} \lambda_{\mathbf{n1}} - \mathbf{G_q}) \tag{21}$$

[4] Touchdown angle is the angle formed between the lower link of the leg (link-3) with reference to inertial frame during contact.

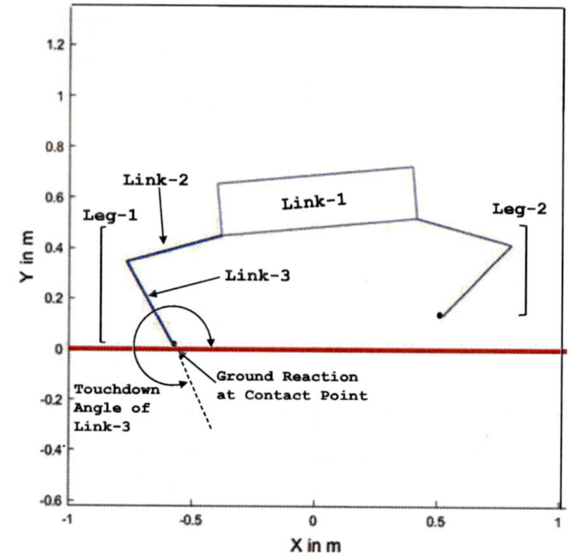

Fig. 2 Typical configuration of planar robot indicating the touchdown angle of link-3

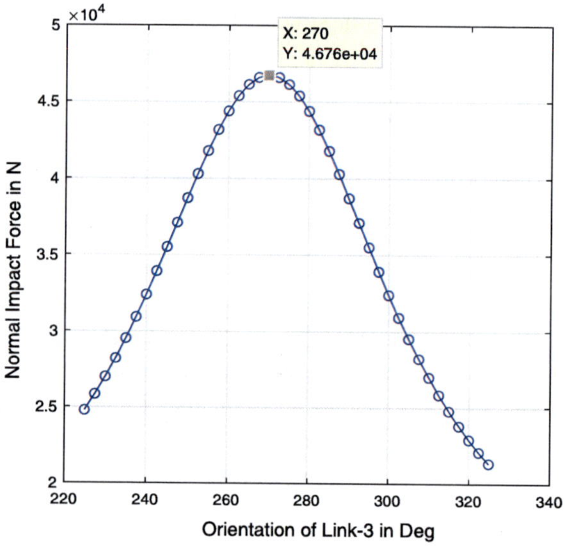

Fig. 3 External impulsive force at various touchdown angles of link-3

Substituting Eq. (21) in terms of $\ddot{\mathbf{q}}$ in Eq. (20),

$$\mathbf{C_qM^{-1}}(\tau + \mathbf{C_q^T}\lambda_{\mathbf{ij}} + \mathbf{J_a^T}\lambda_{\mathbf{n1}} - \mathbf{G_q}) = \mathbf{Q_c} \tag{22}$$

Rearranging the terms:

$$\mathbf{C_qM^{-1}}(\tau + \mathbf{J_a^T}\lambda_{\mathbf{n1}} - \mathbf{G_q}) + \mathbf{C_qM^{-1}C_q^T}\lambda_{\mathbf{ij}} = \mathbf{Q_c}$$

Shifting the term $\mathbf{C_qM^{-1}}(\tau + \mathbf{J_a^T}\lambda_{\mathbf{n1}} - \mathbf{G_q})$ to RHS of the equation:

$$\mathbf{C_qM^{-1}C_q^T}\lambda_{\mathbf{ij}} = \mathbf{Q_c} - \mathbf{C_qM^{-1}}(\tau + \mathbf{J_a^T}\lambda_{\mathbf{n1}} - \mathbf{G_q}) \quad (23)$$

Rearranging the terms in Eq. (23) to obtain the vector of joint resisting forces:

$$\lambda_{\mathbf{ij}} = (\mathbf{C_qM^{-1}C_q^T})^{-1}(\mathbf{Q_c} - \mathbf{C_qM^{-1}}(\tau + \mathbf{J_a}^{\mathbf{T}}\lambda_{\mathbf{n1}} - \mathbf{G_q})) \quad (24)$$

To evaluate the model, the planar model with leg links having constant joint angles was subjected to free fall under gravity for varying orientation of link-3 (lower link of leg-1). Figure 4 shows the configuration of the system at varying orientation of link-3 impacting the ground. The velocity of the model in the vertical direction was held constant (at 3 m/s in the simulation) during the impact of model with ground for all orientation of link-3. The magnitude of impulsive force at various touchdown angles of link-3 and its resisting forces at Joint-2 and Joint-3 were computed using Eq. (24) and were plotted as in Fig. 5. The orientation of the body was inclined by 5° and the upper link of leg-1 was at 220° to the body. During the simulation, the contact happened only at leg-1. It can be observed from the plot that the internal impact forces were high when the orientation of link-3 is perpendicular to the surface of contact. Further, Fig. 6 shows the reaction joint torques at the instant of impact for various touchdown angle of link-3. It is seen from the figure that the reaction torque at Joint-3 of leg-1 is zero when the link-3 is perpendicular to the ground and increases when moved away from this orientation, whereas the reaction joint torque in upper joint (Joint-2) reduces as the angle orienting the link-3 increases.

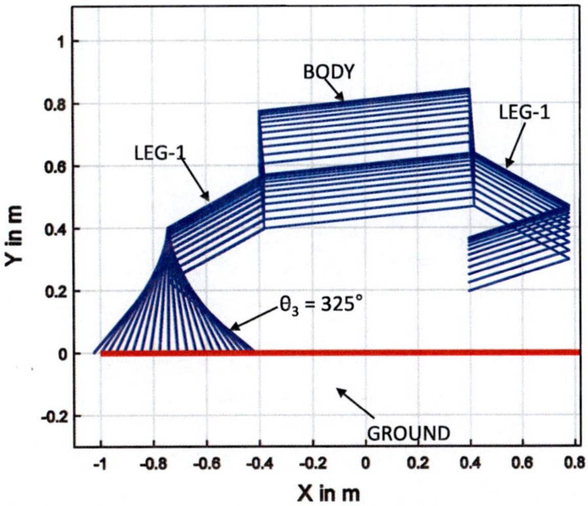

Fig. 4 Configuration of planar robot at various touchdown angle of link-3

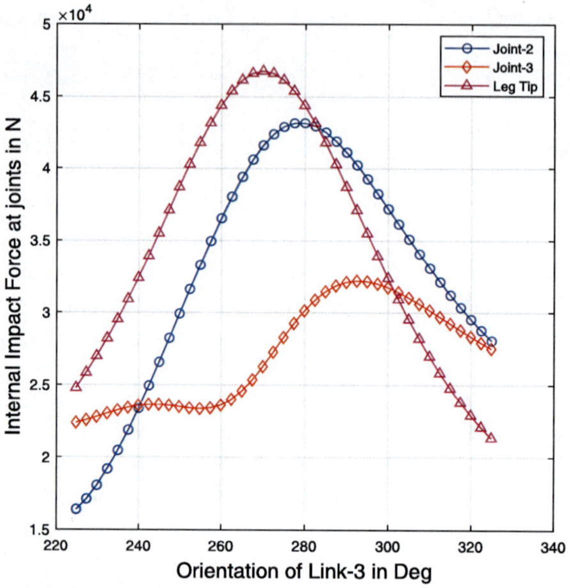

Fig. 5 Internal impulsive forces at joints at various touchdown angle of link-3

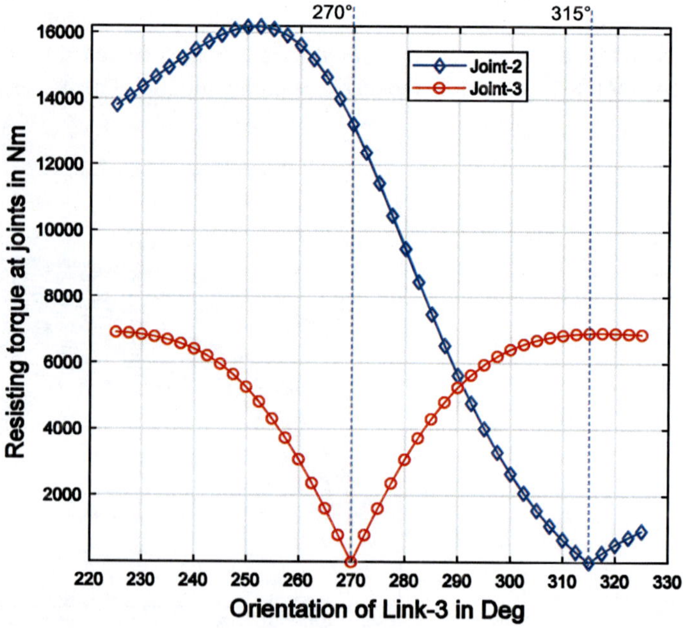

Fig. 6 Magnitude of reaction torque at joints against various touchdown angle of link-3

5 Comparison with Simulation Results Obtained from Multi-body Dynamics Software

To compare the results obtained from the equations of resisting joint forces, Fig. 5 and joint reaction torques, Fig. 6, the robot model parameters were utilized to construct a simulation model in ADAMS software with identical masses and dimensions. Using the restitution model with a penalty factor of 1E4, the coefficient of restitution as 0.8 and with no friction acting at the contact point with ground was used as a contact model in the simulation. The simulation was performed using a WSTIFF integrator, with a time step of 0.001 s. The model was simulated for a gravity fall, with a vertical velocity of 3 m/s at the instance of impact with the ground. The resisting joint forces and reactive joint torques obtained were plotted in Figs. 7 and 8, respectively, for various touchdown angles of link-3.

Comparing the plots of resisting joint forces from Figs. 5 and 7, the plot obtained from Adams software simulation shows a gradual decrease in impulsive force from touchdown angle of 230 to 320°, whereas the simulation plot from the current algorithm shows that the impulsive force at leg tip is highest when the orientation of link-3 is normal to the ground. Further from Figs. 6 and 8, when comparing the reactive joint torques due to impulse, the plot for reaction torque of joint-3 from the current algorithm has zero magnitude when the touchdown angle is 270°, whereas it has finite value in the plot obtained from Adams simulation.

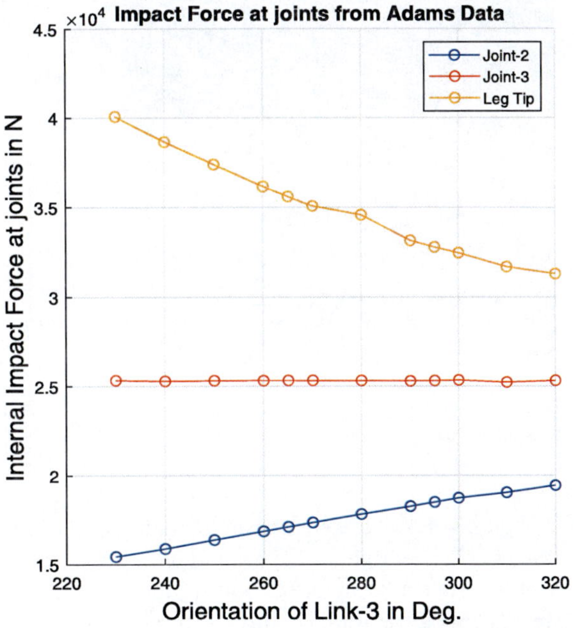

Fig. 7 Internal impulsive forces at joints at various touchdown angles of link-3 obtained from simulation in Adams software

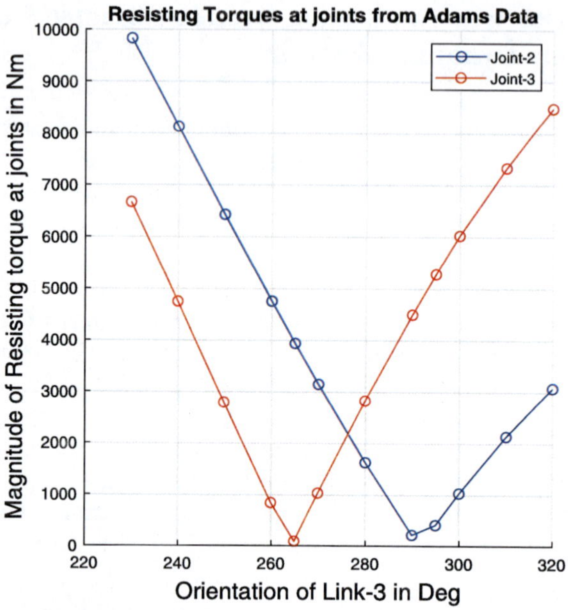

Fig. 8 Magnitude of reaction torque at joints against various touchdown angle of link-3 obtained from simulation in Adams software

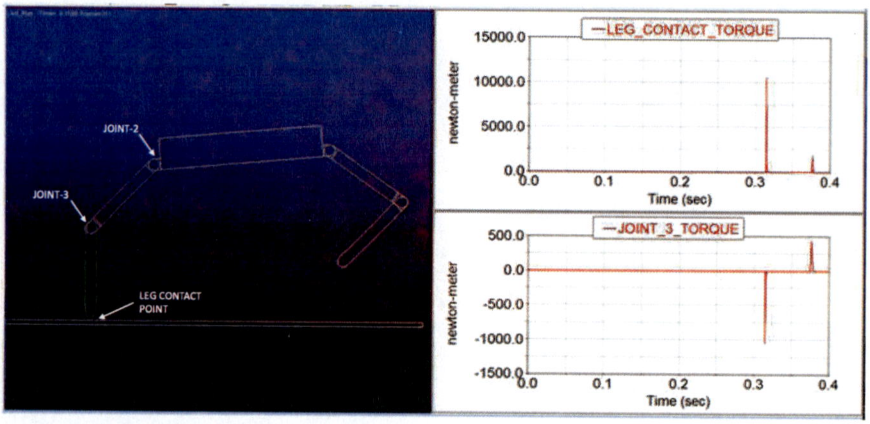

Fig. 9 Torque acting at joint-3 and at the leg contact point in Adams simulation when the touchdown angle is normal to ground

Figure 9 shows the simulation model at the instance of impact where the touchdown angle is normal to the ground with the condition of no friction at the contact. At this instance, the plot of joint torque at joint-3 indicates a torque of 1043 Nm. Since, being a friction-less surface and link-3 being normal to the ground, the horizontal component of joint impulsive force at joint-3 is zero, the torque at this joint

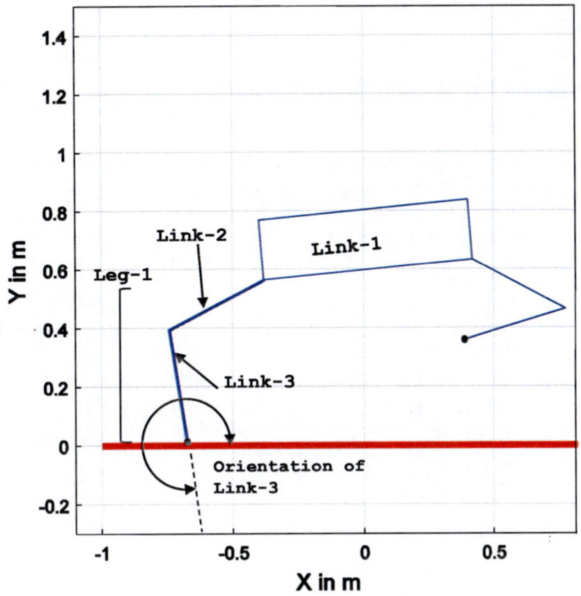

Fig. 10 Touchdown angle during impact to minimize impulse force on joints

should have been zero. Further, when measured, there also exists torque at the point of contact of the leg tip (Fig. 9) with the ground even though being a friction-less surface.

6 Conclusion and Future Work

In this work, first the floating body dynamic model of planar robot was formulated including the kinematic constraint of the leg with the ground, then the external impulse model was applied to obtain the impact force on one leg when in contact with the ground. Further, the internal joint impulsive model was formulated using the joint constraint equations of the system. It is observed from Fig. 3 that the external impulsive force plot and shows high value of impact force when the orientation of link-3 is normal to the ground. From Fig. 5, it is seen that the internal joint resisting forces plot, joint-2 and joint-3 show high value of force when the orientation of link-3 is beyond normal to the ground. Also from Fig. 6 of reaction torques, it is observed that the effect of impulse forces on joints can be reduced and also the torque required be minimized by orienting link-3 beyond 280° with respect to the inertial frame as shown in Fig. 10 at the instance of impact.

This formulation currently worked on a planar case is proposed to be extended on a full 3D quadruped robot model wherein multiple legs come in contact with the environment.

Appendix

Joint Constraint Equations

The joint constraint equations for the system (described in Fig. 1) as given in Eq. (1) and are written in matrix form as:

$$\mathbf{C_q}\ddot{\mathbf{q}} + (\mathbf{C_q}\dot{\mathbf{q}})_\mathbf{q}\dot{\mathbf{q}} = 0$$

where

$$\mathbf{C_q} = \begin{bmatrix} 1 & 0 & Cq_{13} & Cq_{14} & 0 & 0 & 0 & -1 & 0 & 0 & 0 & 0 & 0 & 0 & 0 \\ 0 & 1 & Cq_{23} & Cq_{24} & 0 & 0 & 0 & 0 & -1 & 0 & 0 & 0 & 0 & 0 & 0 \\ 0 & 0 & Cq_{33} & Cq_{34} & Cq_{35} & 0 & 0 & 1 & 0 & -1 & 0 & 0 & 0 & 0 & 0 \\ 0 & 0 & Cq_{43} & Cq_{44} & Cq_{45} & 0 & 0 & 0 & 1 & 0 & -1 & 0 & 0 & 0 & 0 \\ 1 & 0 & Cq_{53} & 0 & 0 & Cq_{56} & 0 & 0 & 0 & 0 & 0 & -1 & 0 & 0 & 0 \\ 0 & 1 & Cq_{63} & 0 & 0 & Cq_{66} & 0 & 0 & 0 & 0 & 0 & 0 & -1 & 0 & 0 \\ 0 & 0 & Cq_{73} & 0 & 0 & Cq_{76} & Cq_{77} & 0 & 0 & 0 & 0 & 1 & 0 & -1 & 0 \\ 0 & 0 & Cq_{83} & 0 & 0 & Cq_{86} & Cq_{87} & 0 & 0 & 0 & 0 & 0 & 1 & 0 & -1 \end{bmatrix}$$

here,

$Cq_{13}: -\sin(\theta_1+\theta_2)r_2 - \sin(\delta_2+\theta_1)R_2$

$Cq_{14}: -\sin(\theta_1+\theta_2)r_2$

$Cq_{23}: \cos(\theta_1+\theta_2)r_2 + \cos(\delta_2+\theta_1)R_2$

$Cq_{24}: \cos(\theta_1+\theta_2)r_2$

$Cq_{33}: -\sin(\theta_1+\theta_2)(L_2-r_2) - \sin(\theta_1+\theta_2+\theta_3)r_3$

$Cq_{34}: -\sin(\theta_1+\theta_2)(L_2-r_2) - \sin(\theta_1+\theta_2+\theta_3)r_3$

$Cq_{35}: -\sin(\theta_1+\theta_2+\theta_3)r_3$

$Cq_{43}: \cos(\theta_1+\theta_2)(L_2-r_2) + \cos(\theta_1+\theta_2+\theta_3)r_3$

$Cq_{44}: \cos(\theta_1+\theta_2)(L_2-r_2) + \cos(\theta_1+\theta_2+\theta_3)r_3$

$Cq_{45}: \cos(\theta_1+\theta_2+\theta_3)r_3$

$Cq_{53}: -\sin(\theta_1+\theta_4)r_4 - \sin(\delta_4+\theta_1)R_4$

$Cq_{56}: -\sin(\theta_1+\theta_4)r_4$

$Cq_{63}: \cos(\theta_1+\theta_4)r_4 + \cos(\delta_4+\theta_1)R_4$

$Cq_{66}: \cos(\theta_1+\theta_4)r_4$

$Cq_{73}: -\sin(\theta_1+\theta_4)(L_4-r_4) - \sin(\theta_1+\theta_4+\theta_5)r_5$

$Cq_{76}: -\sin(\theta_1+\theta_4)(L_4-r_4) - \sin(\theta_1+\theta_4+\theta_5)r_5$

$Cq_{77}: -\sin(\theta_1+\theta_4+\theta_5)r_5$

$Cq_{83}: \cos(\theta_1+\theta_4)(L_4-r_4) + \cos(\theta_1+\theta_4+\theta_5)r_5$

$Cq_{86}: \cos(\theta_1+\theta_4)(L_4-r_4) + \cos(\theta_1+\theta_4+\theta_5)r_5$

$Cq_{87}: \cos(\theta_1+\theta_4+\theta_5)r_5$

and

$$(\mathbf{C_q\dot{q}})_\mathbf{q}\mathbf{\dot{q}} = \begin{bmatrix} r_2(\omega_1+\omega_2)^2\cos(\theta_1+\theta_2) + R_2\omega_1^2\cos(\delta_2+\theta_1) \\ r_2(\omega_1+\omega_2)^2\sin(\theta_1+\theta_2) + R_2\omega_1^2\sin(\delta_2+\theta_1) \\ (\omega_1+\omega_2)^2\cos(\theta_1+\theta_2)(L_2-r_2) + r_3(\omega_1+\omega_2+\omega_3)^2\cos(\theta_1+\theta_2+\theta_3) \\ (\omega_1+\omega_2)^2\sin(\theta_1+\theta_2)(L_2-r_2) + r_3(\omega_1+\omega_2+\omega_3)^2\sin(\theta_1+\theta_2+\theta_3) \\ r_4(\omega_1+\omega_4)^2\cos(\theta_1+\theta_4) + R_4\omega_1^2\cos(\delta_4+\theta_1) \\ r_4(\omega_1+\omega_4)^2\sin(\theta_1+\theta_4) + R_4\omega_1^2\sin(\delta_4+\theta_1) \\ (\omega_1+\omega_4)^2\cos(\theta_1+\theta_4)(L_4-r_4) + r_5(\omega_1+\omega_4+\omega_5)^2\cos(\theta_1+\theta_4+\theta_5) \\ (\omega_1+\omega_4)^2\sin(\theta_1+\theta_4)(L_4-r_4) + r_5(\omega_1+\omega_4+\omega_5)^2\sin(\theta_1+\theta_4+\theta_5) \end{bmatrix}$$

Contact Constraint Equation

The loop equation, Eq. (3) is differentiated twice with respect to time to obtain the equation in matrix form as:

$$\mathbf{J_a\ddot{q}} + (\mathbf{J_a\dot{q}})_\mathbf{q}\mathbf{\dot{q}} = 0$$

where $\mathbf{J_a}$ is given as:

$$\mathbf{J_a} = \left[nx_1, ny_1, J_{a13}, J_{a14}, J_{a15}, 0, 0, 0, 0, 0, 0, 0, 0, 0, 0\right]$$

here,

$$\begin{aligned} J_{a13} &= (-\sin(\theta_1+\theta_2)L_2 - \sin(\theta_1+\theta_2+\theta_3)L_3 - \sin(\delta_2+\theta_1)R_2)nx_1 \\ &\quad + (\cos(\theta_1+\theta_2)L_2 + \cos(\theta_1+\theta_2+\theta_3)L_3 + \cos(\delta_2+\theta_1)R_2)ny_1 \\ J_{a14} &= (\cos(\theta_1+\theta_2)L_2 + \cos(\theta_1+\theta_2+\theta_3)L_3)ny_1 - (\sin(\theta_1+\theta_2)L_2 \\ &\quad + \sin(\theta_1+\theta_2+\theta_3)L_3)nx_1 \\ J_{a15} &= \cos(\theta_1+\theta_2+\theta_3)L_3ny_1 - \sin(\theta_1+\theta_2+\theta_3)L_3nx_1 \end{aligned}$$

and $((\mathbf{J_a\dot{q}})_\mathbf{q}\mathbf{\dot{q}})$ is given by:

$$\begin{aligned} (\mathbf{J_a\dot{q}})_\mathbf{q}\mathbf{\dot{q}} = &-L_2nx_1(\omega_1+\omega_2)^2\cos(\theta_1+\theta_2) - L_3nx_1(\omega_1+\omega_2+\omega_3)^2 \\ &\cos(\theta_1+\theta_2+\theta_3) - L_2ny_1(\omega_1+\omega_2)^2\sin(\theta_1+\theta_2) \\ &- L_3ny_1(\omega_1+\omega_2+\omega_3)^2\sin(\theta_1+\theta_2+\theta_3) \\ &+ nx_1R_2\omega_1^2(-\cos(\delta_2+\theta_1)) - ny_1R_2\omega_1^2\sin(\delta_2+\theta_1) \end{aligned}$$

System Lagrangian Equations

The system Lagrangian formed is summation of kinetic energy, potential energy, joint constraints, and contact constraints in the system. The kinetic energy (KE) of the system is determined as a summation of KE of each body in the model, given by:

$$\mathrm{KE} = K_1 + K_2 + K_3 + K_4 + K_5$$

where:

K_1 to K_5: Kinetic energy of Link-1 to Link-5 respectively given as

$K_1 = 0.5m_1(\dot{x}_1^2 + \dot{y}_1^2) + 0.5\mathrm{MI}_1\omega_1^2$

$K_2 = 0.5m_2(\dot{x}_2^2 + \dot{y}_2^2) + 0.5\mathrm{MI}_2(\omega_1 + \omega_2)^2$

$K_3 = 0.5m_3(\dot{x}_3^2 + \dot{y}_3^2) + 0.5\mathrm{MI}_3(\omega_1 + \omega_2 + \omega_3)^2$

$K_4 = 0.5m_4(\dot{x}_4^2 + \dot{y}_4^2) + 0.5\mathrm{MI}_4(\omega_1 + \omega_4)^2$

$K_5 = 0.5m_5(\dot{x}_5^2 + \dot{y}_5^2) + 0.5\mathrm{MI}_5(\omega_1 + \omega_4 + \omega_5)^2$

Here m_1–m_5 indicates the masses of link-1 to link-5, respectively, as shown in Fig. 1, MI_1 to MI_5 are the moment of inertia about the CoM of link-1 to link-5, respectively, ω_1 to ω_5 are the angular velocities of link-1 to link-5, respectively. The potential energy (PE) of the system is given by:

$$\mathrm{PE} = gm_1y_1 + gm_2y_2 + gm_3y_3 + gm_4y_4 + gm_5y_5$$

The Lagrangian obtained for the system is given as Eq. (6). The Lagrangian thus obtained is evaluated for each system coordinates. The equation of motion for the system is obtained as:

$$\mathbf{M}\ddot{\mathbf{q}} + \mathbf{G}_{\mathbf{q}} - \mathbf{C}_{\mathbf{q}}^{\mathbf{T}}\lambda_{\mathbf{ij}} - \mathbf{J}_{\mathbf{a}}^{\mathbf{T}}\lambda_{\mathbf{i}} = \tau$$

Here, mass matrix (**M**) is a square positive definite matrix obtained as:

$$\mathbf{M} = \begin{bmatrix}
m_1 & 0 & 0 & 0 & 0 & 0 & 0 & 0 & 0 & 0 & 0 & 0 & 0 & 0 & 0 \\
0 & m_1 & 0 & 0 & 0 & 0 & 0 & 0 & 0 & 0 & 0 & 0 & 0 & 0 & 0 \\
0 & 0 & M_{33} & M_{34} & M_{35} & M_{36} & M_{37} & 0 & 0 & 0 & 0 & 0 & 0 & 0 & 0 \\
0 & 0 & M_{34} & M_{34} & M_{35} & 0 & 0 & 0 & 0 & 0 & 0 & 0 & 0 & 0 & 0 \\
0 & 0 & M_{35} & M_{35} & M_{35} & 0 & 0 & 0 & 0 & 0 & 0 & 0 & 0 & 0 & 0 \\
0 & 0 & M_{36} & 0 & 0 & M_{36} & M_{37} & 0 & 0 & 0 & 0 & 0 & 0 & 0 & 0 \\
0 & 0 & M_{37} & 0 & 0 & M_{37} & M_{37} & 0 & 0 & 0 & 0 & 0 & 0 & 0 & 0 \\
0 & 0 & 0 & 0 & 0 & 0 & 0 & m_2 & 0 & 0 & 0 & 0 & 0 & 0 & 0 \\
0 & 0 & 0 & 0 & 0 & 0 & 0 & 0 & m_2 & 0 & 0 & 0 & 0 & 0 & 0 \\
0 & 0 & 0 & 0 & 0 & 0 & 0 & 0 & 0 & m_3 & 0 & 0 & 0 & 0 & 0 \\
0 & 0 & 0 & 0 & 0 & 0 & 0 & 0 & 0 & 0 & m_3 & 0 & 0 & 0 & 0 \\
0 & 0 & 0 & 0 & 0 & 0 & 0 & 0 & 0 & 0 & 0 & m_4 & 0 & 0 & 0 \\
0 & 0 & 0 & 0 & 0 & 0 & 0 & 0 & 0 & 0 & 0 & 0 & m_4 & 0 & 0 \\
0 & 0 & 0 & 0 & 0 & 0 & 0 & 0 & 0 & 0 & 0 & 0 & 0 & m_5 & 0 \\
0 & 0 & 0 & 0 & 0 & 0 & 0 & 0 & 0 & 0 & 0 & 0 & 0 & 0 & m_5
\end{bmatrix}$$

where

$M_{33} = \mathrm{MI}_1 + \mathrm{MI}_2 + \mathrm{MI}_3 + \mathrm{MI}_4 + \mathrm{MI}_5$

$M_{34} = \mathrm{MI}_2 + \mathrm{MI}_3$

$M_{35} = \mathrm{MI}_3$

$M_{36} = \mathrm{MI}_4 + \mathrm{MI}_5$
$M_{37} = \mathrm{MI}_5$
and $\mathbf{G_q}$ is a vector containing the gravity terms with respect to each system coordinates and is given as:

$$\mathbf{G_q} = \left[0,\ gm_1,\ 0,\ 0,\ 0,\ 0,\ 0,\ 0,\ gm_2,\ 0,\ gm_3,\ 0,\ gm_4,\ 0,\ gm_5\right]^{\mathrm{T}}$$

References

1. Lee SH, Kim SH, Kwak YK (2000) Modeling and analysis of internal impact for general classes of robotic mechanisms. In: Proceedings. IEEE/RSJ international conference on intelligent robots and systems, vol 3, pp 1955–1962. https://doi.org/10.1109/IROS.2000.895257
2. Walker ID (1990) The use of kinematic redundancy in reducing impact and contact effects in manipulation. In: Proceedings. IEEE international conference on robotics and automation, vol 1, pp 434–439. https://doi.org/10.1109/ROBOT.1990.126016
3. Zheng Y-F, Hemami H. Mathematical modeling of a robot collision with its environment. J Robot Syst 2(3):289–307. https://doi.org/10.1002/rob.4620020307
4. Byung SR, Lee JH, Yi BJ, Kim W (2005) Robot motion generation considering external and internal impulses. In: IEEE/RSJ international conference on intelligent robots and systems, pp 948–953. https://doi.org/10.1109/IROS.2005.1545208
5. Bowling A (2011) Impact forces and agility in legged robot locomotion. J Vib Control 17(3):335–346. https://doi.org/10.1177/1077546309357855
6. Karssen JGD, Haberland M, Wisse M, Kim S (2011) The optimal swing-leg retraction rate for running. In: IEEE international conference on robotics and automation, pp 4000–4006. https://doi.org/10.1109/ICRA.2011.5980168

Motion Planning and Control of Two Quadcopters with Cable-Suspended Point Mass Payload

Pratik Prajapati and Vineet Vashista

Abstract Considering the low payload carrying capacity for a single quadcopter, one option to increase the payload carrying capacity is the use of more than one quadcopter. This work focuses on trajectory planning and control of two quadcopters with a cable-suspended point mass payload system using a leader–follower scheme. For safe and stable transportation of suspended payload, the quadcopters should be controlled to not lead to cable slackness. Accordingly, wrench closure analysis is carried out for the follower quadcopter with respect to the leader quadcopter, which helps design desired trajectory generation for the quadcopters. The performance of the proposed motion planning strategy is verified by conducting simulation in SIMSCAPE multibody package in MATLAB.

Keywords Aerial transportation · Multiple quadcopters · Cable-suspended payload · Wrench closure workspace

1 Introduction

Quadcopters have a wide range of applications in the both military and civil sectors, because of which they have been a subject of research for many research groups. Some of the applications are aerial photography, agriculture use, aerial surveillance, humanitarian operations in disaster situations, etc. In many of these applications, the quadcopter is required to carry some specific type of payload.

Various methodologies are presented in the literature to transport payloads from one place to another, for example, using robotic arms/grippers [1–3] or by suspending the payload using cables [4–11] or direct attachment to the chassis [12, 13]. However, the agility of the quadcopter retains while transporting payload by suspending it through cables because it decouples the attitude dynamics of the quadcopters. To increase the payload carrying capacity, one has to increase the capacity

P. Prajapati · V. Vashista (✉)
Human-Centered Robotics Lab, Indian Institute of Technology Gandhinagar, Gandhinagar 382355, India
e-mail: vineet.vashista@iitgn.ac.in

V. K. Gupta et al. (eds.), *Recent Advances in Machines and Mechanisms*, Lecture Notes in Mechanical Engineering, https://doi.org/10.1007/978-981-19-3716-3_46

of motors and need to change the overall structure of the chassis. On the other hand, another alternative to increase the payload carrying capacity is by using multiple quadcopters.

A motion planning approach using a transition-based rapidly exploring random tree algorithm was implemented in [14] for the system consisting of three aerial robots transporting a rigid payload to ensure the wrench-feasibility constraint in the cables. A wrench set analysis technique from the cable-driven parallel robots (CDPR) was implemented in [15] for aerial cable towed system (ACTS) consisting of the quadcopters which can be used for the cable-tension distribution and motion planning. In [16], cooperative transportation of a cable-suspended payload using multiple quadcopters was demonstrated in which reconfigurable parallel robot techniques were used to avoid specifying the prior forces in the cables and get rid of the tension distribution algorithms. Geometric nonlinear control for the multiple quadcopters with cable-suspended payload system was implemented in the [17] by simulations and experiments. All of these methods are subjected to higher computational costs to run trajectory planning and control algorithms. Moreover, some methods cannot be directly implementable in outdoor environment settings.

The present work focuses on trajectory planning and control for cable-suspended point mass payload using two quadcopters. The governing dynamical equation of motion of the system is derived using the Newton–Euler equation, considering cable remains taut during transportation. A leader–follower scheme is considered in which the leader quadcopter is autonomously controlled to direct the motion of the entire system, and the follower quadcopter is commanded for safe and stable transportation of the payload while keeping positive tension in the cables. Accordingly, wrench closure workspace analysis is carried out for the follower quadcopter, ensuring positive tension in the cables. This workspace analysis is further extended for trajectory generation for the follower quadcopter, which results in tracking the prescribed trajectory of the payload. Finally, the SIMSCAPE multibody dynamic toolbox of SIMULINK (in MATLAB) is used to check the performance of the developed motion planning and control strategy.

2 Method

The line diagram of two quadcopters with a cable-suspended point mass payload system is shown in Fig. 1. The inertial frame of reference, $\{I\}$, is represented as three orthonormal vectors $[a_1, a_2, a_3]^{\mathrm{T}}$ where a_3 is taken in the vertically upward direction. Body frame of reference to each quadcopter $\{B_i\}$, where $i = \{1, 2\}$ is considered using three orthogonal unit vectors, $[b_{1_i}, b_{2_i}, b_{3_i}]^{\mathrm{T}}$ where $i = \{1, 2\}$, attached to its center of gravity (CG) where first body axis, i.e., b_{1_i}, is pointed toward the middle of first and second motors as shown in Fig. 1, and third axis, i.e., b_{3_i}, is pointed perpendicular to plane of the quadcopter. The mass and moment of inertia about CG of ith quadcopter are denoted as m_i, J_i respectively. It is assumed that point mass payload, m_0, is suspended using inextensible cables to the quadcopter's CG.

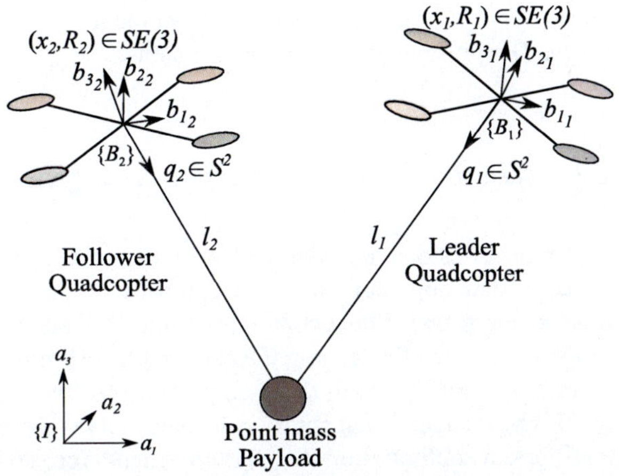

Fig. 1 Line diagram of the two quadcopters with a cable-suspended point mass payload system

The translational position of the payload is denoted as $X_0 \in \mathbb{R}^3$. The length of the cable from the quadcopter 1 and 2 to payload is denoted as l_1 and l_2, respectively. The translational position and attitude of the quadcopter in frame $\{I\}$ are denoted as $X_i = [x_i, y_i, z_i]^{\mathrm{T}} \in \mathbb{R}^3$ and $R_i \in \mathrm{SO}(3)$, respectively. Standard *ZXY* Euler angle parameterization is used to represent the attitude of the quadcopters given as in Eq. (1), where $(\phi_i, \theta_i, \psi_i)$ represents roll, pitch, and yaw angle of the quadcopter [18]. The angular position of each cable is measured from a_1 and a_2 axes denoted as ϕ_{p_i} and θ_{p_i}, respectively, which also can be represented in two sphere as $q_i = \{a \in \mathbb{R}^3 : ||a|| = 1\}$. Corresponding expression for q_i is given in Eq. (2). The body angular velocity of ith quadcopter is denoted as $\omega_i = [\omega_{1_i}, \omega_{2_i}, \omega_{3_i}]^{\mathrm{T}}$.

$$R_i = \begin{bmatrix} c\psi_i c\theta_i - s\phi_i s\psi_i s\theta_i & -c\phi_i s\psi_i & c\psi_i s\theta_i + c\theta_i s\phi_i s\psi_i \\ s\psi_i c\theta_i + c\psi_i s\phi_i s\theta_i & c\phi_i c\psi_i & s\psi_i s\theta_i - c\theta_i s\phi_i c\psi_i \\ -s\theta_i c\phi_i & s\phi_i & c\phi_i c\theta_i \end{bmatrix} \tag{1}$$

$$q_i = \begin{bmatrix} c\theta_{p_i} & 0 & s\theta_{p_i} \\ s\phi_{p_i} s\theta_{p_i} & c\phi_{p_i} & -c\theta_{p_i} s\phi_{p_i} \\ -c\phi_{p_i} s\theta_{p_i} & s\phi_{p_i} & c\phi_{p_i} c\theta_{p_i} \end{bmatrix} \tag{2}$$

where $c(\cdot) = \cos(\cdot)$, $s(\cdot) = \sin(\cdot)$. Using Newton–Euler equations, the equation of motion for two quadcopters with cable-suspended point mass payload system considering the cable remains taut is given in Eqs. (3)–(5) as derived in [19].

$$m_i \ddot{X}_i = f_i R_i a_3 - m_i g a_3 + T_i q_i \tag{3}$$

$$J_i \dot{\omega}_i + \omega_i \times J_i \omega_i = M_i \tag{4}$$

$$m_0 \ddot{X}_0 = -\sum T_i q_i - m_0 g a_3 \tag{5}$$

where, f_i, M_i is the thrust force and moment generated by ith quadcopter respectively. T_i represents tension in the ith cable.

2.1 Wrench Closure Workspace Analysis

This work considers a leader–follower scheme where one quadcopter is considered a leader quadcopter autonomously controlled to direct the system's motion. Another quadcopter, follower quadcopter, is commanded such that it tracks desired trajectory of the payload while keeping the cable taut. As preliminary work, a wrench closure workspace (WCW) analysis is carried out for the follower quadcopter (indicated as 2) when leader quadcopter (indicated as 1) is fixed at the origin. Further, static equilibrium configuration (SEC) is considered for WCW analysis. At SEC, follower quadcopter hovers at a particular point in space with zero linear and rotational velocities. From Eqs. (3)–(5), at SEC, the attitude of quadcopter, forces, and tension in each cable can be found using Eqs. (6)–(8).

$$f_1 R_1 a_3 = m_1 g a_3 - T_1 q_1 \tag{6}$$

$$f_2 R_2 a_3 = m_2 g a_3 - T_2 q_2 \tag{7}$$

$$T_1 q_1 + T_2 q_2 = -m_0 g a_3 \tag{8}$$

Using parameter values given in Table 1, wrench closure workspace for follower quadcopter is shown in Fig. 2a and corresponding cross-sectional view of WCW from YZ plane passing through origin O is shown in Fig. 2b.

The volume contained by the magenta sphere 1 with the volume subtracted from the other two magenta spheres (2 and 3) in Fig. 2a indicates wrench closure workspace for the follower quadcopter. WCW indicates the region in which the cable tension remains positive all the time. The surface of blue sphere 4 indicates a possible region for the payload position. As shown in Fig. 2b, an instance when cable 1 is at an angle of α from the Y-axis is shown, and the highlighted arc shows the corresponding allowable position for follower quadcopter. Further, at every angular position of cable 1, i.e., from 0 to 360° corresponding allowable position for the follower quadcopter is shown by colored arcs. Using wrench closure workspace analysis, the following two corollaries are extracted at SEC.

Table 1 Parameters used for simulation

Description	Notation	Value
Mass of follower quadcopter	m_2	1.2 kg
MOI of follower quadcopter	J_2	diag(0.016, 0.017, 0.032) kgm^2
Mass of payload	m_0	0.5 kg
Length of cables	l_1, l_2	1 m

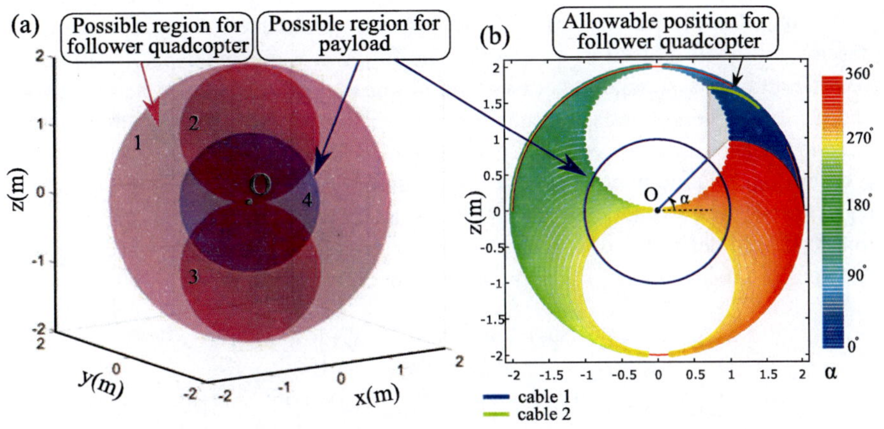

Fig. 2 **a** Static wrench closure workspace for follower quadcopter with respect to leader quadcopter. **b** Cross-sectional view of WCW from YZ plane passing through origin O

1. If the translational position of the follower quadcopter is known with respect to leader quadcopter, then there exists only one feasible solution for the position of payload.
2. For the given angular position of the leader quadcopter's cable, we require only one parameter to define the system configuration completely. (This one parameter could be the translational position of the follower quadcopter or angular position of the follower quadcopter's cable.)

The trajectory for the follower quadcopter can be planned using the above corollaries to track the desired payload's trajectory. As the trajectory is planned based on the WCW, the resulting motion of the quadcopter ensures the positive tension in the cables.

2.2 Controller Design for Follower Quadcopter

Consider a case when the leader quadcopter is fixed at origin, and we want to track the desired trajectory for payload denoted as X_{p_d}. Using WCW analysis, we can define the corresponding trajectory for the follower quadcopter, which lets the payload track desired trajectory, X_{p_d}, while making sure the cables not get slacked. From the desired trajectory of the payload, cable 1's angular position can be found out, i.e., $q_{1_d} = X_{p_d}/l_1$. Using, WCW and corollary 2 if we know the angular position of the cable 2, i.e., q_{2_d} corresponding to q_{1_d}, we can calculate the desired position for follower quadcopter as given in Eq. (9).

$$X_{2_d} = l_1 q_{1_d} - l_2 q_{2_d} \tag{9}$$

To track the desired position for the follower quadcopter, a linear controller as described in [20] is used in this work. The follower quadcopter has to orient at a specific attitude and generate corresponding thrust force to balance its own weight and payload's weight at desired position X_{2_d}. Using Eqs. (6)–(8), the attitude of the quadcopter and thrust force at SEC can be found which is denoted as ϕ_{2_e}, θ_{2_e}, and f_{2_e}, respectively. The moment M_2 is calculated using PID controller as given in Eq. (10) and desired roll and pitch angle, i.e., ϕ_{2_d}, θ_{2_d} are calculated using Eqs. (11)–(12). In this work, the desired yaw angle is taken as zero, i.e., $\psi_{2_d} = 0$. The thrust force f_2 is calculated using Eq. (13) which controls the desired altitude z_{2_d}.

$$M_2 = \begin{bmatrix} k_{P_\phi}(\phi_{2_d} - \phi_2) + k_{I_\phi}\int(\phi_{2_d} - \phi_2)dt + k_{D_\phi}(\dot{\phi}_{2_d} - \dot{\phi}_2) \\ k_{P_\theta}(\theta_{2_d} - \theta_2) + k_{I_\theta}\int(\theta_{2_d} - \theta_2)dt + k_{D_\theta}(\dot{\theta}_{d_i} - \dot{\theta}_2) \\ k_{P_\psi}(\psi_{2_d} - \psi_2) + k_{I_\psi}\int(\psi_{2_d} - \psi_2)dt + k_{D_\psi}(\dot{\psi}_{2_d} - \dot{\psi}_2) \end{bmatrix} \tag{10}$$

$$\phi_{2_d} = \phi_{2_e} + \frac{-1}{g}(\ddot{y}_{2_d} - K_{\dot{y}_2}(\dot{y}_2 - \dot{y}_{2_d}) - K_{y_2}(y_2 - y_{2_d})) \tag{11}$$

$$\theta_{2_d} = \theta_{2_e} + \frac{1}{g}(\ddot{x}_{2_d} - K_{\dot{x}_2}(\dot{x}_2 - \dot{x}_{2_d}) - K_{x_2}(x_2 - x_{2_d})) \tag{12}$$

$$f_2 = f_{2_e} + m_2(\ddot{z}_{2_d} - K_{\dot{z}_2}(\dot{z}_2 - \dot{z}_{2_d}) - K_{z_i}(z_2 - z_{2_d})) \tag{13}$$

3 Results and Discussion

The performance of the proposed motion planning and control strategy is simulated using SIMSCAPE multibody dynamics software in SIMULINK. The parameters used to conduct simulations are listed in Table 1. Initially, both the cables are kept inclined at 45° from vertical plane. Hence, the angular position of the cables' are $q_{1_0} = [-0.707, 1, -0.707]^T$, $q_{2_0} = [0.707, 1, -0.707]$. For the initial three seconds, the system reaches to static equilibrium configuration from the defined initial configuration. After that, two different cases are considered for desired payload trajectory as given below.

1. Case-1: Horizontal payload trajectory
   $x_{P_d}(t) = -0.5\cos(\frac{t-3}{2})$ m, $y_{P_d}(t) = 0.5\sin(\frac{t-3}{2})$ m, $z_{P_d}(t) = -0.866$ m
2. Case-2: Vertical payload trajectory
   $x_{P_d}(t) = -0.866$ m, $y_{P_d}(t) = 0.5\sin(\frac{t-3}{2})$ m, $z_{P_d}(t) = -0.5\cos(\frac{t-3}{2})$ m

Payload's desired and simulated translational position and follower quadcopter's desired and simulated translational position for tracking of 0.5 m horizontal circular payload trajectory is shown in Fig. 3 and for tracking of 0.5 m vertical circular trajectory is shown in Fig. 4. For the first three seconds, the system converges to SEC. After the third second, the follower quadcopter is commanded to track the desired

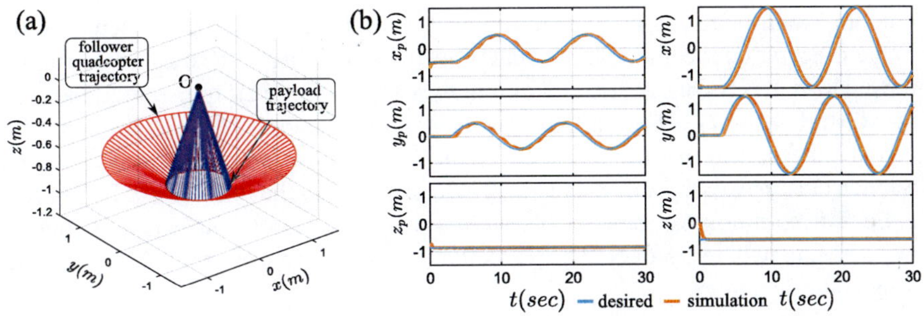

Fig. 3 Simulation results to track 0.5 m horizontal payload trajectory. **a** 3D plots for payload and follower quadcopter trajectory. **b** Translational position of payload and follower quadcopter. Sky blue color plot shows the desired trajectory, and orange color plots show simulated trajectory

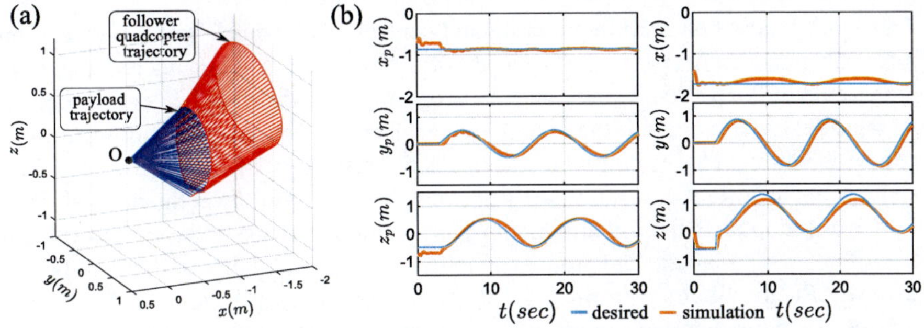

Fig. 4 Simulation results to track 0.5 m vertical payload trajectory. **a** 3D plots for payload and follower quadcopter trajectory. **b** Translational position of payload and follower quadcopter. Sky blue color plot shows the desired trajectory, and orange color plots show simulated trajectory

trajectory X_{2_d} according to track the payload desired trajectory up to 30 s. Sky blue color plots show the desired trajectory to be tracked, and orange color plots show simulated trajectories.

In case 1, root mean square (RMS) and standard deviation (STD) of error between desired and simulated trajectory for quadcopter's translational position are (0.1514 ± 0.1514 m, 0.1563 ± 0.1559 m, 0.0461 ± 0.0456 m) and for payload's translational position are (0.0557 ± 0.0556 m, 0.0565 ± 0.0563 m, 0.0101 ± 0.0101 m) along x, y, z-axes, respectively. In case 2, RMS and STD of error between desired and simulated trajectory for quadcopter's translational position are (0.0870 ± 0.0543 m, 0.0939 ± 0.0899 m, 0.1231 ± 0.1078 m) and for payload's translational position are (0.0551 ± 0.0543 m, 0.0649 ± 0.0632 m, 0.1097 ± 0.1032 m) along x, y, z-axes, respectively. The small values of the STD in payload and follower quadcopter indicate the presented controlled tracks the desired trajectory accurately.

The developed controller requires feedback of follower quadcopter's position with respect to leader quadcopter; by placing onboard cable attitude measurement devices as demonstrated in [21], the presented control strategy can be extended for the outdoor experiments easily. From the qualitative analysis, it is inferred that the motion of the follower quadcopter did not lead to accurate tracking of the desired position of the payload. It is because of oscillations of the payload while the transportation generates disturbances. However, as demonstrated in the literature [22, 23], the controller can be modified to incorporate minimization of the payload oscillations. Dynamic wrench closure workspace analysis will be carried out in the future work from which agile transportation can be possible without slacking cables. Furthermore, outdoor experiments will be conducted to check the feasibility of the presented motion planning control strategy.

4 Conclusion

This work focuses on motion planning and control of two quadcopters with a point mass cable-suspended payload. The leader–follower scheme is considered to control the motion of the payload where one quadcopter is considered as leader quadcopter, which directs the motion of the system, whereas another quadcopter, follower quadcopter, is commanded such that it tracks desired trajectory of the payload. Further, static wrench closure workspace analysis is carried out for follower quadcopter when leader quadcopter is fixed while helping to generate slackness-free trajectories for the follower quadcopter. The proposed motion planning modality is simulated using SIMSCAPE multibody dynamics software to check its performance. The simulation results demonstrated the feasibility of the proposed methodology in enabling slackness-free motion of the payload. The trajectory generation in this work is less computational costly.

Acknowledgements This work is supported by Core Research Grant (CRG/2020/004990) from SERB India.

References

1. Thomas J, Loianno G, Polin J, Sreenath K, Kumar V (2014) Toward autonomous avian-inspired grasping for micro aerial vehicles. Bioinspir Biomim 9(2):025010
2. Mellinger D, Lindsey Q, Shomin M, Kumar V (2011) Design, modeling, estimation and control for aerial grasping and manipulation. In: 2011 IEEE/RSJ international conference on intelligent robots and systems. IEEE, pp 2668–2673
3. Kim S, Choi S, Kim HJ (2013) Aerial manipulation using a quadrotor with a two DOF robotic arm. In: 2013 IEEE/RSJ international conference on intelligent robots and systems. IEEE, pp 4990–4995
4. Sreenath K, Michael N, Kumar V (2013) Trajectory generation and control of a quadrotor with a cable-suspended load—a differentially-flat hybrid system. In: 2013 IEEE international conference on robotics and automation. IEEE, pp 4888–4895

5. Foehn P, Falanga D, Kuppuswamy N, Tedrake R, Scaramuzza D (2017) Fast trajectory optimization for agile quadrotor maneuvers with a cable-suspended payload. In: Robotics: science and systems, pp 1–10
6. Tang S, Kumar V (2015) Mixed integer quadratic program trajectory generation for a quadrotor with a cable-suspended payload. In: 2015 IEEE international conference on robotics and automation (ICRA). IEEE, pp 2216–2222
7. Goodarzi FA, Lee T (2016) Stabilization of a rigid body payload with multiple cooperative quadrotors. J Dyn Syst Meas Control 138(12)
8. Sreenath K, Lee T, Kumar V (2013) Geometric control and differential flatness of a quadrotor UAV with a cable-suspended load. In: 52nd IEEE conference on decision and control, pp 2269–2274
9. Dai S, Lee T, Bernstein DS (2014) Adaptive control of a quadrotor UAV transporting a cable-suspended load with unknown mass. In: 53rd IEEE conference on decision and control, pp 6149–6154
10. Gassner M, Cieslewski T, Scaramuzza D (2017) Dynamic collaboration without communication: vision-based cable-suspended load transport with two quadrotors. In: 2017 IEEE international conference on robotics and automation (ICRA). IEEE, pp 5196–5202
11. Lee T, Sreenath K, Kumar V (2013) Geometric control of cooperating multiple quadrotor UAVs with a suspended payload. In: 52nd IEEE conference on decision and control. IEEE, pp 5510–5515
12. Vergouw B, Nagel H, Bondt G, Custers B (2016) Drone technology: types, payloads, applications, frequency spectrum issues and future developments. In: The future of drone use. Springer, pp 21–45
13. Augugliaro F, Lupashin S, Hamer M, Male C, Hehn M, Mueller MW, Willmann JS, Gramazio F, Kohler M, D'Andrea R (2014) The flight assembled architecture installation: cooperative construction with flying machines. IEEE Control Syst Mag 34(4):46–64
14. Manubens M, Devaurs D, Ros L, Cortés J (2013) Motion planning for 6-D manipulation with aerial towed-cable systems. In: Robotics: science and systems (RSS), 8 p
15. Erskine J, Chriette A, Caro S (2019) Wrench analysis of cable-suspended parallel robots actuated by quadrotor unmanned aerial vehicles. J Mech Robot 11(2)
16. Masone C, Bülthoff HH, Stegagno P (2016) Cooperative transportation of a payload using quadrotors: a reconfigurable cable-driven parallel robot. In: 2016 IEEE/RSJ international conference on intelligent robots and systems (IROS). IEEE, pp 1623–1630
17. Goodarzi FA, Lee D, Lee T (2014) Geometric stabilization of a quadrotor UAV with a payload connected by flexible cable. In: 2014 American control conference. IEEE, pp 4925–4930
18. Michael N, Mellinger D, Lindsey Q, Kumar V (2010) The grasp multiple micro-UAV testbed. IEEE Robot Autom Mag 17(3):56–65
19. Sreenath K, Kumar V (2013) Dynamics, control and planning for cooperative manipulation of payloads suspended by cables from multiple quadrotor robots. rn 1(r2):r3
20. Michael N, Mellinger D, Lindsey Q, Kumar V (2010) The grasp multiple micro-UAV testbed. IEEE Robot Autom Mag 17(3):56–65
21. Prajapati P, Parekh S, Vashista V (2021) On-board cable attitude measurement and controller for outdoor aerial transportation. Robotica 1–15
22. Prajapati P, Parekh S, Vashista V (2019) Collaborative transportation of cable-suspended payload using two quadcopters with human in the loop. In: 2019 28th IEEE international conference on robot and human interactive communication (RO-MAN). IEEE, pp 1–6
23. Prajapati P, Parekh S, Vashista V (2020) On the human control of a multiple quadcopters with a cable-suspended payload system. In: 2020 IEEE international conference on robotics and automation (ICRA), pp 2253–2258

Robust Path Following Control of Autonomous Underwater Vehicle Using Combined Time Delay Estimation and Backstepping Method

Diwakar Gurung, C. S. Kumar, and Vishwanath Nagarajan

Abstract The present work addresses the nonlinear control problem for the path following of an autonomous underwater vehicle (AUV) by using the combined backstepping method and time delay estimation (TDE). The proposed controller guides the AUV along a predefined geometric path whilst handling the modelling uncertainties and external disturbances. The control system is developed at two stages: kinematic and dynamic level. At a kinematic level, path following errors are formulated using the path frame, and the desired reference velocities are derived to reduce these errors. At the dynamic level, the actual control force and the moment are obtained to track the desired AUV velocities. The dynamic controller incorporates the time delay estimation method to estimate the unmodelled dynamics and external disturbances. Using Lyapunov stability analysis, the proposed control architecture is proven to be stable, and all the control loop signals are shown to be uniformly ultimately bounded. Finally, numerical simulations are presented to exhibit the controller performance in the face of disturbances and model inconsistencies. The proposed controller performance is compared with the traditional feedback linearisation (FL) control design which lacks robustness property.

Keywords Autonomous underwater vehicle · Time delay estimation · Path following

1 Introduction

In recent times, underwater vehicles have become an important subject of research because of their broad usage in the military, oil and gas industries and many more.

D. Gurung (✉) · C. S. Kumar
Department of Mechanical Engineering, Indian Institute of Technology Kharagpur, Kharagpur, West Bengal, India
e-mail: diwakargurung@iitkgp.ac.in

V. Nagarajan
Department of Ocean Engineering and Naval Architecture, Indian Institute of Technology Kharagpur, Kharagpur, West Bengal, India

V. K. Gupta et al. (eds.), *Recent Advances in Machines and Mechanisms*, Lecture Notes in Mechanical Engineering, https://doi.org/10.1007/978-981-19-3716-3_47

Many underwater tasks like surveying and seafloor mapping demand the robust path following capabilities of the AUV with effective control and guidance strategies. A typical path following problem deals with the development of a robust control law for an AUV to reach and follow the predefined path without any temporal constraints. The AUV dynamics is highly nonlinear, and the dynamic model parameters like added mass and hydrodynamic damping coefficients are not precisely known. Moreover, the AUV is also subjected to external disturbances like ocean waves and currents. To address the path following problem, several seminal control and guidance techniques have been proposed and implemented. These methods include model-based PID and backstepping control [1–4], adaptive/sliding mode control [5–10], intelligent-based control [11–15] and many more. The methods have their own set of advantages and drawbacks. In the recent past, several efforts have been made to design the partially model-free robust control scheme. Active disturbance rejection control (ADRC), being one of effective choices, has been successfully implemented in [16, 17]. In the ADRC method, an observer is designed to estimate the hydrodynamic uncertainties and disturbances online and is compensated in the existing control law. Another method for estimating unknown systems is the time delay estimation method and is effectively used for the robot manipulator in [18–20]. The TDE technique is simple and uses only the previous system input and output information to estimate the unknown dynamics. The concept has been used for the AUV trajectory tracking problem as mentioned in [21–23]. In this paper, a control law based on the backstepping and TDE method is designed for the path following of an underactuated AUV subjected to parameter perturbation and disturbances. The stability of the control law is proven using Lyapunov stability criteria, and the boundedness of TDE error is proven. In this work, only, the AUV motion in a horizontal plane is considered, and the AUV kinematics and dynamics are described in the next section.

2 Kinematics and Dynamics

Considering Fig. 2, the body frame $\{b\}$ with axes $(\hat{x}_b, \hat{y}_b, \hat{z}_b)$ is attached with the AUV at point O. The body frame axes coincide with the AUV principal axes with $\hat{z}_b$ pointing into the page. The kinematic model of an AUV motion in a horizontal plane is represented by the following equation:

$$[\dot{x}, \dot{y}, \dot{\psi}]^{\mathrm{T}} = R_i^b [u, v, r]^{\mathrm{T}} \tag{1}$$

$$R_i^b = \begin{bmatrix} \cos(\psi) & -\sin(\psi) & 0 \\ \sin(\psi) & \cos(\psi) & 0 \\ 0 & 0 & 1 \end{bmatrix} \tag{2}$$

where (x, y, ψ) designates the AUV positions and orientation (yaw angle) in the inertial frame $\{i\}$, (u, v) are the AUV surge and sway velocities, and r is the vehicle yaw rate, and R_i^b is the rotation matrix from frame $\{b\}$ to $\{i\}$. The dynamics of an

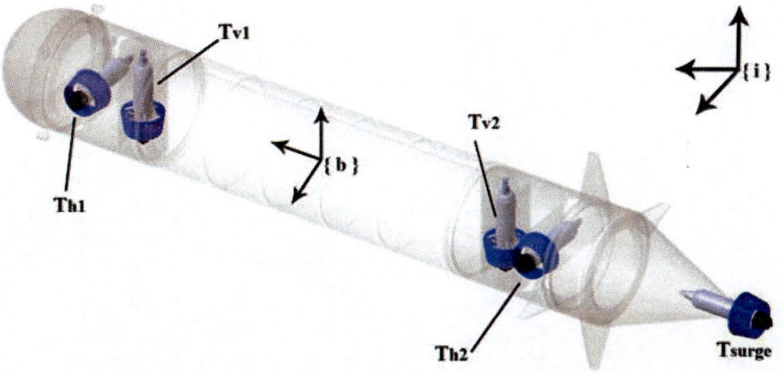

Fig. 1 AUV thruster distribution

underactuated AUV is represented as follows:

$$\dot{u} = \frac{1}{(m - X_{\dot{u}})}\left[-X_{|u|u}|u|u - (Y_{\dot{u}} - m)vr + \tau_{\mathrm{u}} + X_{\mathrm{ext}}\right] \tag{3a}$$

$$\dot{v} = \frac{1}{(m - Y_{\dot{v}})}\left[-Y_{|v|v}|v|v - Y_{|r|r}|r|r - (m - X_{\dot{u}})ur + Y_{\mathrm{ext}}\right] \tag{3b}$$

$$\dot{r} = \frac{1}{(I_z - N_{\dot{r}})}\left[-N_{|v|v}|v|v - N_{|r|r}|r|r - (X_{\dot{u}} - Y_{\dot{v}})uv + \tau_{\mathrm{r}} + N_{\mathrm{ext}}\right] \tag{3c}$$

Here, m is the AUV mass; I_z is the inertia of AUV about vertical axis; $X_{\dot{u}}$, $Y_{\dot{v}}$ and $N_{\dot{r}}$ are the added masses, and $X_{u|u|}$, $Y_{v|v|}$, $Y_{r|r|}$, $N_{v|v|}$, $N_{r|r|}$ are the hydrodynamic damping coefficient. τ_{u} and τ_{r} denote control inputs, i.e. surge force and yaw moment, respectively, and X_{ext}, Y_{ext} and N_{ext} represent external disturbances acting on the AUV. The AUV dynamics doesn't have control input in the sway direction which makes the AUV underactuated.

In the present study, the model of the AUV-Test bed developed by IIT-Kharagpur is used for the study [24]. The AUV is entirely thruster driven and is primarily used for slow-speed manoeuvring. The underwater vehicle has five thrusters (developed by Tecnadyne) that are arranged to generate the necessary forces/moment. The distribution of these thrusters is shown in Fig. 1. Two thrusters T_{H1} and T_{H2} generate the moment τ_{r} responsible for heading or yaw motion. Besides these thrusters, the AUV also has another two thrusters T_{V1} and T_{V2} for heave motion. T_{surge} is the main thruster generating surge force τ_{u} for the forward motion.

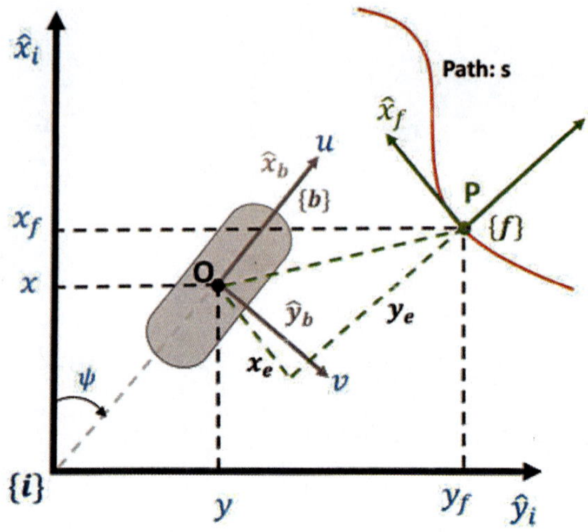

Fig. 2 Path following of AUV

3 Problem Formulation

To address the path following control problem, the kinematics of AUV is derived in the Serret–Frenet (S-F) frame assigned to a virtual moving point in the desired spatial path [1]. As depicted in Fig. 2, the AUV must follows the virtual target point P that moves on a geometric path parameterized by a path variable s. Here, s is the curvilinear abscissa of the moving point on the path. To represent the virtual point movement, a Serret–Frenet frame $\{f\}$ is attached with the target point P with its x-axis ($\hat{x}_f$) pointed towards the path tangent. The inertial position of P is given by $[x_f(s), y_f(s)]^T$ and the S-F frame orientation is given by ψ_f with respect to the inertial frame $\{i\}$. Let $[x_e, y_e]^T$ be the distance between the AUV point O and the path point P defined in the frame $\{f\}$ and is expressed by the following relation:

$$[x_e, y_e]^T = R_f^i[(x - x_f(s)), (y - y_f(s))]^T \tag{4}$$

$$R_f^i = \begin{bmatrix} \cos(\psi_f) & \sin(\psi_f) \\ -\sin(\psi_f) & \cos(\psi_f) \end{bmatrix} \tag{5}$$

x_e and y_e are known as along-track and cross-track error. Here, R_f^i is the rotation matrix, mapping the frame $\{i\}$ to $\{f\}$ parameterised by the S-F frame orientation ψ_f. Where,

$$\psi_f = \arctan\left(\frac{y_f'(s)}{x_f'(s)}\right) \tag{6}$$

Let $c(s)$ be the path curvature, $v_{\mathrm{t}} = \sqrt{u^2 + v^2}$ be the vehicle total velocity, and $\beta = \arctan(v/u)$ be the side slip angle. Recalling $\dot{\psi}_{\mathrm{f}} = c(s)\dot{s}$, the error dynamics is given by:

$$\dot{x}_{\mathrm{e}} = -\dot{s}(1 - c(s)y_{\mathrm{e}}) + v_{\mathrm{t}} \cos \psi_{\mathrm{e}} \tag{7a}$$

$$\dot{y}_{\mathrm{e}} = -\dot{s}c(s)x_{\mathrm{e}} + v_{\mathrm{t}} \sin \psi_{\mathrm{e}} \tag{7b}$$

where $\psi_{\mathrm{e}} = \psi + \beta - \psi_{\mathrm{f}}$. The rate of evolution of the variable s will be derived in the controller design section. The path following problem for an underactuated AUV in a horizontal plane can be described as follows:

Given a geometric path parameterised by the virtual target point P and the Serret–Frenet frame $\{f\}$ attached with the point, develop robust control laws (surge force τ_{u} and yaw moment τ_{u}) for an underactuated AUV to reduce the along-track (x_{e}) and cross-track (y_{e}) errors to a small neighbourhood of zero, whilst its surge velocity is regulated to the desired speed and the total velocity vector aligns with the path tangent.

4 Controller Design

The overall architecture of the proposed controller is shown in Fig. 3. Since the underactuated AUVs are highly nonlinear and are also subjected to disturbances and modelling uncertainties, the combined backstepping and TDE-based nonlinear control method is adopted to address the path following control problem. The control design aims to reduce the path following error, i.e. the cross-track and along-track error whose dynamics is given by Eq. (7).

4.1 *Kinematic Controller*

To start off with the control formulation, consider the following Lyapunov candidate function:

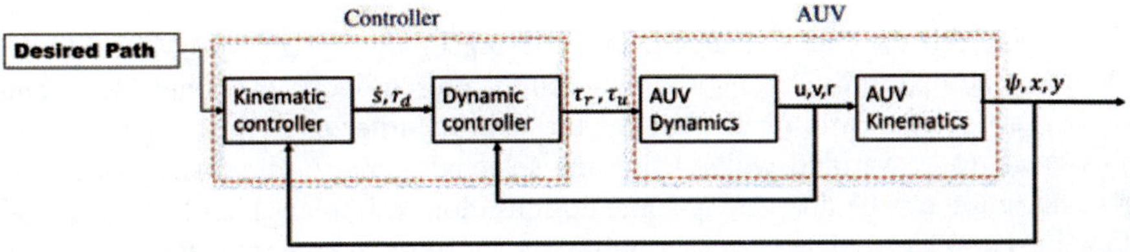

Fig. 3 AUV controller

$$V_1 = \frac{1}{2}x_e^2 + \frac{1}{2}y_e^2 \tag{8}$$

Differentiating (8) along the equation of $\dot{x}_e$ and $\dot{y}_e$ in (7):

$$\dot{V}_1 = x_e\dot{x}_e + y_e\dot{y}_e = x_e(-\dot{s}(1 - c(s)y_e) + v_t \cos\psi_e) + y_e(-\dot{s}c(s)x_e + v_t \sin\psi_e) \tag{9}$$

To get the above Lyapunov function derivative to be negative definite, the variable ψ_e and $\dot{s}$ is selected as:

$$\begin{cases} \psi_e = \psi_{LOS} = -\sin^{-1}\left(\frac{y_e}{\sqrt{y_e^2+\Delta_{ye}^2}}\right) \\ \dot{s} = v_t \cos(\psi_e) + k_1 x_e \end{cases} \tag{10}$$

The Lyapunov function derivative in (9) becomes:

$$\dot{V}_1 = -k_1 x_e^2 - \frac{y_e v_t}{\sqrt{\Delta_{ye}^2 + y_e^2}} \tag{11}$$

where $\Delta_{ye} > 0$ and $k_1 > 0$ are the kinematic control parameter. Assuming the perfect heading tracking, i.e. $\psi_e = \psi_{LOS}$ and surge speed $u \geq u_{min} > 0$ making $v_t > 0$, Lyapunov function derivative $\dot{V}_1$ will be negative definite, and the position error (x_e, y_e) will converge to zero.

To regulate the heading error $\tilde{\psi}_e = \psi_e - \psi_{LOS}$ to near zero, another Lyapunov function is considered as follows:

$$V_2 = \frac{1}{2}\tilde{\psi}_e^2 \tag{12}$$

The time derivative of V_2 becomes:

$$\dot{V}_2 = \dot{\tilde{\psi}}_e(r + \dot{\beta} - c(s)\dot{s} - \dot{\psi}_{LOS}) \tag{13}$$

Thus, the desired yaw velocity to regulated the heading error is given as:

$$r_d = -k_2(\tilde{\psi}_e) - \dot{\beta} + c(s)\dot{s} + \dot{\psi}_{LOS} \tag{14}$$

where $k_2 > 0$ is the control gain and thus making $\dot{V}_2 = -k_2(\tilde{\psi}_e)^2 \leq 0$. This implies that the heading error is Lyapunov stable. In the present work, the control law gains (k_1, k_2) are selected arbitrarily although the stability criterion of the Lyapunov analysis needs to be verified whilst using the selected gains. Further, analysis on the gain selection can be done using some optimisation techniques based on some cost functions and the vehicle state constraints. It is also to be noted that the gain Δ_{ye} is the line of sight distance and is taken to be twice the length of the AUV [4]. Small Δ_{ye} value makes the kinematic control law aggressive generating larger yaw angle command. Thus, the effect of the selected control parameter on the vehicle state has to be evaluated and analysed during the overall manoeuvring process in simulation study.

4.2 Dynamic Controller

As shown in Fig. 3, the kinematic controller serves as a reference subsystem, providing the command velocity signal to the dynamic controller. Using the backstepping method, the command velocity signals can be extended to the AUV dynamics and derive the actual control surge force and yaw moment.

Consider the desired constant surge velocity u_d and the reference yaw rate r_d, the dynamic control is designed to regulate the AUV surge and yaw rate (u, r) to the desired values by following steps:

The dynamic equation of surge and yaw rate (3a) and (3c) can be simplified as:

$$\begin{cases} m_u\dot{u} = N_u + \tau_u + D_u \\ m_r\dot{r} = N_r + \tau_r + D_r \end{cases} \tag{15}$$

where, $m_u = m - X_{\dot{u}}, m_r = m - X_{\dot{r}}$, $N_u = -X_{|u|u}|u|u - (Y_{\dot{u}} - m)vr$ and $N_r = -N_{|v|v}|v|v - N_{|r|r}|r|r - (X_{\dot{u}} - Y_{\dot{v}})uv$ denote the known dynamic quantities. D_u and D_r define the unknown dynamics and external disturbances. Again, rearranging the simplified dynamic equation (15) as:

$$\begin{cases} \bar{m}_u\dot{u} = N_u + h_u + \tau_u \\ \bar{m}_r\dot{r} = N_r + h_r + \tau_r \end{cases} \tag{16}$$

where $\bar{m}_u$ and $\bar{m}_r$ are the inertial gains to be used in the dynamic control law. $h_u = -m_u\dot{u} - \bar{m}_u\dot{u} + D_u$ and $h_r = -m_r\dot{r} - \bar{m}_r\dot{r} + D_r$ are the nonlinear terms which incorporate the lumped dynamic disturbances. Consider the velocity tracking error as:

$$\tilde{u} = u_d - u, \tilde{r} = r_d - r \tag{17}$$

The time derivative of velocity errors $\tilde{u}$ and $\tilde{r}$ is given by:

$$\begin{cases} \dot{\tilde{u}} = \dot{u}_d - \frac{h_u}{\bar{m}_u} - \frac{N_u}{\bar{m}_u} - \frac{\tau_u}{\bar{m}_u} \\ \dot{\tilde{r}} = \dot{r}_d - \frac{h_r}{\bar{m}_r} - \frac{N_r}{\bar{m}_r} - \frac{\tau_r}{\bar{m}_r} \end{cases} \tag{18}$$

To regulate the velocity errors, the actual dynamic control input is selected as:

$$\begin{cases} \tau_u = N_u + \hat{h}_u + \bar{m}_u\xi_u \\ \tau_r = N_r + \hat{h}_r + \bar{m}_r\xi_r \end{cases} \tag{19}$$

where $\xi_u = k_3\tilde{u}$ and $\xi_r = k_4\tilde{r} + \dot{r}_d$ and $k_3 > 0$ and $k_4 > 0$ are the control gain. $\hat{h}_u$ and $\hat{h}_r$ are the estimated value of h_u and h_r obtained by using the TDE method. TDE concept is based on the assumption that a continuous signal or variation of dynamics is negligible during a small time period, and a control action can estimate the unknown dynamics based on past observations of system dynamics [19, 20]. Using Eq. (16), the estimated values are obtained as follows:

$$\hat{h}_u = h_{u(t-\Delta t)} = -\tau_{u(t-\Delta t)} + \bar{m}_u \dot{u}_{(t-\Delta t)} - N_{u(t-\Delta t)} \tag{20}$$

$$\hat{h}_r = h_{r(t-\Delta t)} = -\tau_{r(t-\Delta t)} + \bar{m}_r \dot{r}_{(t-\Delta t)} - N_{r(t-\Delta t)} \tag{21}$$

where Δt denotes delay and is set as the sampling time and the quantity with subscript $(.)_{(t-\Delta t)}$ is the previous time value. Using the control efforts (19) complimented by the estimated value (20) and (21), the velocity error dynamics reduces to the following:

$$\dot{\tilde{u}} = -k_3\tilde{u} + \epsilon_{h_u} \tag{22}$$

$$\dot{\tilde{r}} = -k_4\tilde{r} + \epsilon_{h_r} \tag{23}$$

where $\epsilon_{h_u} = \frac{\hat{h}_u}{\bar{m}_u} - \frac{h_u}{\bar{m}_u}$ and $\epsilon_{h_r} = \frac{\hat{h}_r}{\bar{m}_r} - \frac{h_r}{\bar{m}_r}$ represents the estimation error in the estimates $\hat{h}_u$ and $\hat{h}_r$. The estimation errors ϵ_{h_u} and ϵ_{h_r} at time t are given by

$$\epsilon_{h_u(t)} = (1 - \bar{m}_u m_{u(t)}^{-1})\epsilon_{h_u(t-\Delta t)} + \Gamma_{h_u 1(t)} + \Gamma_{h_u 2(t)} \tag{24}$$

where, $\Gamma_{h_u 1(t)} = (1 - \bar{m}_u m_{u(t)}^{-1})(\xi_{u(t)} - \xi_{u(t-\Delta t)})$ and $\Gamma_{h_u 2(t)} = m_{u(t)}^{-1}[N_{u(t)} - N_{u(t-\Delta t)} + (m_{u(t)} - m_{u(t-\Delta t)})\dot{u}_{(t-\Delta t)}]$
and,

$$\epsilon_{h_r(t)} = (1 - \bar{m}_r m_{r(t)}^{-1})\epsilon_{h_r(t-\Delta t)} + \Gamma_{h_r 1(t)} + \Gamma_{h_r 2(t)} \tag{25}$$

where $\Gamma_{h_r 1(t)} = (1 - \bar{m}_r m_{r(t)}^{-1})(\xi_{r(t)} - \xi_{r(t-\Delta t)})$ and $\Gamma_{h_r 2(t)} = m_{r(t)}^{-1}[N_{r(t)} - N_{r(t-\Delta t)} + (m_{r(t)} - m_{r(t-\Delta t)})\dot{r}_{(t-\Delta t)}]$.

Since the velocity error dynamics is stimulated by the estimation errors, the dynamic control stability is proven through boundedness of the estimation error signals. Assuming $\Gamma_{h_u 1(t)}, \Gamma_{h_u 2(t)}, \Gamma_{h_r 1(t)}$ and $\Gamma_{h_r 2(t)}$ to be the bounded forcing functions, the error signals are proven to be bounded by satisfying the following conditions:

$$|1 - \bar{m}_u m_{u(t)}^{-1}| < 1, |1 - \bar{m}_r m_{r(t)}^{-1}| < 1 \tag{26}$$

To satisfy the above stability criterion, the inertial gains $\bar{m}_u$ and $\bar{m}_r$ are selected as $0 < \bar{m}_u < 2m_{u(t)}$ and $0 < \bar{m}_r < 2m_{r(t)}$. Selecting the inertial gain value as the vehicle rigid body inertia term will satisfy the stability criterion of the time delay control [22].

5 Simulation Study and Comparative Assessment

A numerical simulation is performed to verify the proposed controller's performance in the face of external disturbances and model uncertainties. The TDE-based controller is evaluated with traditional feedback linearisation control at the dynamic level as follows:

$$\begin{cases} \tau_{\mathrm{u}} = N_{\mathrm{u}} + \bar{m}_{\mathrm{u}}\xi_{\mathrm{u}} \\ \tau_{\mathrm{r}} = N_{\mathrm{r}} + \bar{m}_{\mathrm{r}}\xi_{\mathrm{r}} \end{cases} \tag{27}$$

The circular path, parameterised by $\omega(s)$, is considered for the simulation test as follows:

$$x_{\mathrm{f}}(\omega) = 70 + 50\sin\omega; \qquad y_{\mathrm{f}}(\omega) = 70 - 50\cos\omega \tag{28}$$

where ω is estimated by $\dot{\omega} = \frac{\dot{s}}{\sqrt{[x_{\mathrm{f}}'(\omega)]^2+[y_{\mathrm{f}}'(\omega)]^2}}$. The starting point of the target point on the circular is $(x_{\mathrm{f}}(0), y_{\mathrm{f}}(0))$ at $\omega = 0$.

The AUV model parameters for the simulation are provided in Table 1. The AUV initial position and orientation and the control gains are provided in Table 2. The desired surge speed is $u_{\mathrm{d}} = 1.5$ m/s, and the disturbances of $d_{\mathrm{u}}(t) = -5(1 + \sin(0.1t))$ and $d_{\mathrm{r}}(t) = -5(1 + \sin(0.1t))$ are applied in surge and yaw directions, respectively. For the model inconsistency, the dynamic model parameter in Table 1 is reduced by 30%. As shown in Fig. 4, the AUV with TDE control law follows the circular path precisely, whilst the feedback linearisation controller produces the undulating path trajectory. The cross-track and along-track error reduces to zero in around 25 s for TDE-based controller. However, for the FL controller, the path error doesn't reduce to zero and takes a long time to follow the path orientation as shown in Fig. 5. For speed regulation, the TDE-based control design achieves the desired surge speed of 1.5 m/s whilst the other controller produces some steady error. The irregularities found in the FL controller are due to its inability to compensate for the

Table 1 AUV parameters

$m = 66$ kg	Length = 1.77 m
$X_{\dot{u}} = 2.43$ kg	$Y_{\dot{u}} = 61.25$ kg
$N_{\dot{r}} = 8.60$ kg m^2/rad	$X_{\lvert u\rvert u} = 2.43$ kg/m
$Y_{\lvert v\rvert v} = 245.59$ kg/m	$Y_{\lvert r\rvert r} = -7.25$ kg m/rad^2
$N_{\lvert v\rvert v} = -14.50$ kg	$N_{\lvert r\rvert r} = 33$ kg m^2/rad^2

Table 2 Control parameters

$k_1 = 1, k_2 = 0.5, k_3 = 4, k_4 = 3$
$\Delta_{\mathrm{ye}} = 2 \times$ length of AUV, $\bar{m}_{\mathrm{u}} = 80, \bar{m}_{\mathrm{r}} = 50$
$x(0) = 70, y(0) = 0, \psi(0) = 0$

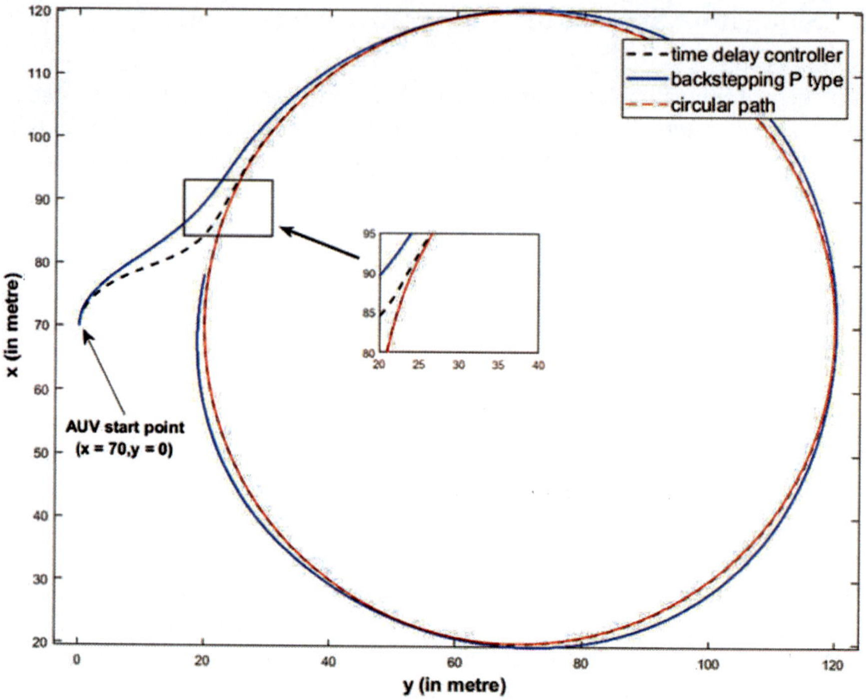

Fig. 4 AUV trajectory

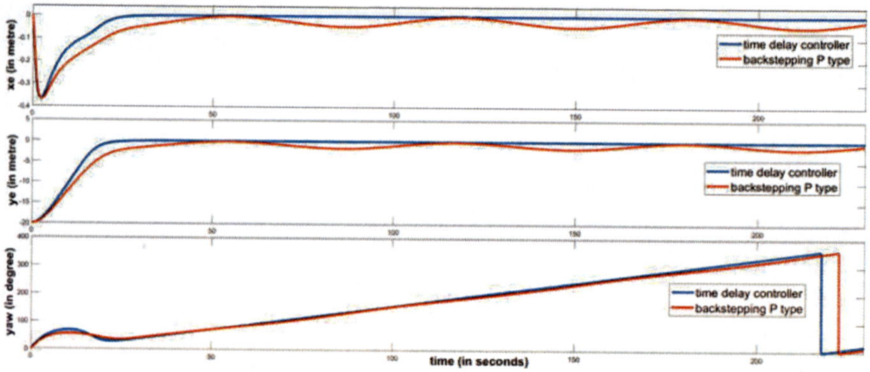

Fig. 5 Path error and yaw response

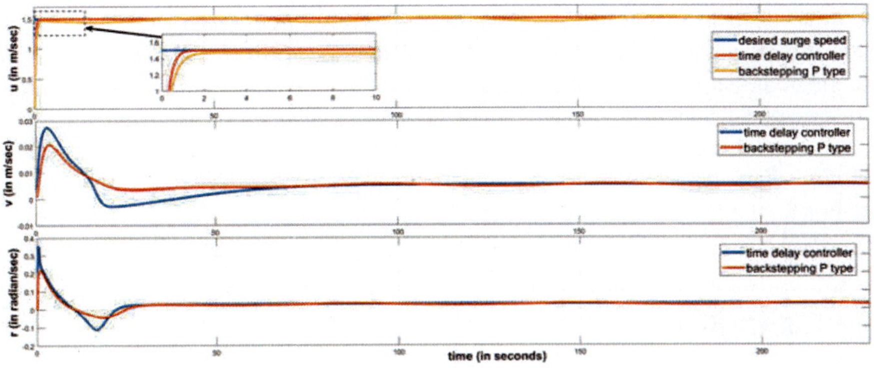

Fig. 6 AUV velocities

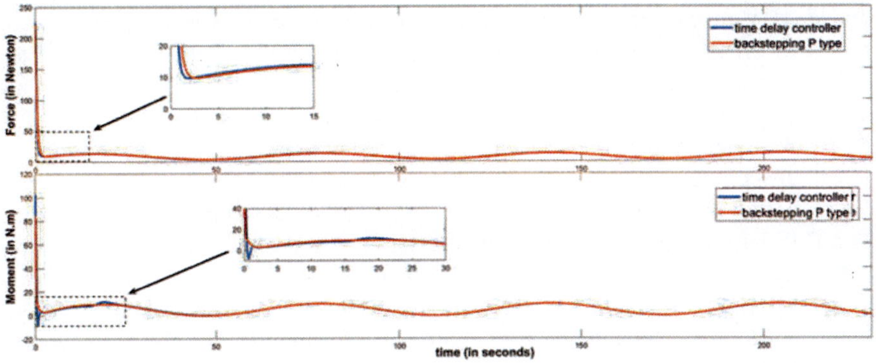

Fig. 7 Surge force and yaw moment

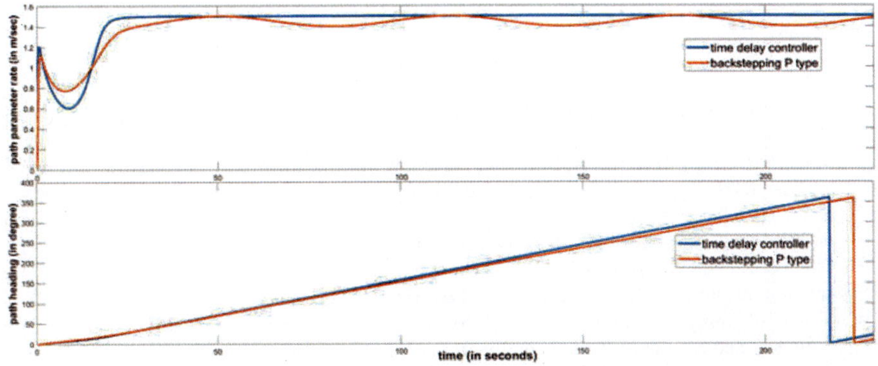

Fig. 8 Serret Frenet frame trajectory

external disturbances and model uncertainties. Sway velocity and yaw rate response are also provided in Fig. 6. The control efforts, i.e. the surge force and the yaw moment are shown in Fig. 7 respectively. Initially, the control efforts are high due to large path errors and reduce to small value as the vehicle steers towards the path. Both controllers exert almost equivalent control efforts. The Serret–Frenet frame motion characterised by the path orientation and the rate of path length $\dot{s}$ is provided in Fig. 8.

6 Conclusion

The paper describes the implementation of a combined backstepping and time delay estimation-based control strategy for the path following operation of an underactuated AUV in a horizontal plane. The control strategy aims at reducing the path following errors; generating the desired vehicle velocities and the dynamic control law is obtained to regulate the vehicle velocities to the desired ones. The convergence of the AUV to the desired path has been illustrated in detail, and all the closed-loop feedback signals in the control structure are proven to be bounded. The simulation test results are obtained using the parameters of the AUV testbed to follow the geometric path with the proposed control strategy and show satisfactory performance. The TDE-based controller performs well, attaining the desired velocities values even in the face of external disturbance and AUV uncertainties. The performance of the proposed control strategy compared with the feedback linearization control design. Future work will extend the derived control law for the 3D path following control with experimental validations.

References

1. Lapierre L, Soetanto D (2007) Nonlinear path-following control of an AUV. Ocean Eng 34(11–12):1734–1744
2. Maurya P, Desa E, Pascoal A, Barros E, Navelkar G, Madhan R, Mascarenhas AAMQ, Prabhudesai S, Afzulpurkar S, Gouveia A et al (2006) Control of the Maya AUV in the vertical and horizontal planes: theory and practical results. In: Proceedings of the 7th IFAC conference on manoeuvring and control of marine craft, pp 20–22
3. Duc Do K, Pan J, Jiang Z-P (2004) Robust and adaptive path following for underactuated autonomous underwater vehicles. Ocean Eng 31(16):1967–1997
4. Xiang X, Lapierre L, Jouvencel B (2015) Smooth transition of AUV motion control: from fully-actuated to under-actuated configuration. Robot Auton Syst 67:14–22
5. Lapierre L, Jouvencel B (2008) Robust nonlinear path-following control of an AUV. IEEE J Ocean Eng 33(2):89–102
6. Zeng J, Wan L, Li Y, Dong Z, Zhang Y (2017) Adaptive line-of-sight path following control for underactuated autonomous underwater vehicles in the presence of ocean currents. Int J Adv Robot Syst 14(6):1729881417748127

7. Borhaug E, Pettersen KY (2005) Adaptive way-point tracking control for underactuated autonomous vehicles. In: Proceedings of the 44th IEEE conference on decision and control. IEEE, pp 4028–4034
8. Yoerger D, Slotine J (1985) Robust trajectory control of underwater vehicles. IEEE J Ocean Eng 10(4):462–470
9. Ruiz-Duarte JE, Loukianov AG (2015) Higher order sliding mode control for autonomous underwater vehicles in the diving plane. IFAC-PapersOnLine 48(16):49–54
10. Elmokadem T, Zribi M, Youcef-Toumi K (2017) Terminal sliding mode control for the trajectory tracking of underactuated autonomous underwater vehicles. Ocean Eng 129:613–625
11. Xiang X, Caoyang Y, Zhang Q (2017) Robust fuzzy 3D path following for autonomous underwater vehicle subject to uncertainties. Comput Oper Res 84:165–177
12. Loebis D, Naeem W, Sutton R, Chudley J, Tetlow S (2007) Soft computing techniques in the design of a navigation, guidance and control system for an autonomous underwater vehicle. Int J Adapt Control Signal Process 21(2–3):205–236
13. Bian X, Zhou J, Yan Z, Jia H (2012) Adaptive neural network control system of path following for AUVs. In: 2012 proceedings of IEEE SoutheastCon. IEEE, pp 1–5
14. Guo J, Chiu F-C, Huang C-C (2003) Design of a sliding mode fuzzy controller for the guidance and control of an autonomous underwater vehicle. Ocean Eng 30(16):2137–2155
15. Zhang J, Xiang X, Zhang Q, Li W (2020) Neural network-based adaptive trajectory tracking control of underactuated AUVs with unknown asymmetrical actuator saturation and unknown dynamics. Ocean Eng 218:108193
16. Miao J, Wang S, Zhao Z, Li Y, Tomovic MM (2017) Spatial curvilinear path following control of underactuated AUV with multiple uncertainties. ISA Trans 67:107–130
17. Yan Z, Liu Y, Zhou J, Wu D (2014) Path following control of an AUV under the current using the SVR-ADRC. J Appl Math 2014
18. Youcef-Toumi K, Ito O (1990) A time delay controller for systems with unknown dynamics
19. Cho GR, Chang PH, Park SH, Jin M (2009) Robust tracking under nonlinear friction using time-delay control with internal model. IEEE Trans Control Syst Technol 17(6):1406–1414
20. Hsia TC, Gao LS (1990) Robot manipulator control using decentralized linear time-invariant time-delayed joint controllers. In: Proceedings. IEEE international conference on robotics and automation. IEEE, pp 2070–2075
21. Cho GR, Li J-H, Park D, Jung JH (2020) Robust trajectory tracking of autonomous underwater vehicles using back-stepping control and time delay estimation. Ocean Eng 201:107131
22. Prasanth Kumar R, Dasgupta A, Kumar CS (2007) Robust trajectory control of underwater vehicles using time delay control law. Ocean Eng 34(5–6):842–849
23. Kumar RP, Kumar CS, Sen D, Dasgupta A (2009) Discrete time-delay control of an autonomous underwater vehicle: theory and experimental results. Ocean Eng 36(1):74–81
24. Prasanth Kumar R, Sarath Babu S, Srilekha Y, Kumar CS, Sen D, Dasgupta A (2006) Test-bed for navigation and control of a thruster based AUV. In: OCEANS 2006-Asia Pacific. IEEE, pp 1–4

Effect of Passive Springs on Taskspace Stiffness of a Cable-Driven Serial Chain Manipulator

N. S. S. Sanjeevi and Vineet Vashista

Abstract Understanding the end-effector stiffness of a manipulator is an effective tool to interpret manipulators' interaction capabilities in the taskspace environment. A manipulator having the ability to alter its stiffness characteristics provides an advantage to make it suitable for versatile applications. Recently, cable-driven serial chain manipulators, CDSMs, have emerged as an important robotic architecture for various applications due to their promising features, such as low moving inertia, large payload handling capacity, and flexibility in altering its system architecture. Owing to the unidirectional force application property of cables, these systems are redundantly actuated. The flexibility in architecture modulation within a CDSM implies the possibility of varying system performance. Accordingly, system parameters of CDSM can be altered to modulate the stiffness behavior of the manipulator. In this work, we attempt to modify the stiffness characteristics of CDSMs through the addition of passive springs. Two different modalities of passive spring attachments were considered, and preliminary results on their effect on taskspace stiffness are presented.

Keywords Cable-driven systems · Taskspace stiffness · Passive stiffness

1 Introduction

Cable-driven manipulators (CDMs) are robots where motor-actuated cables apply external forces on links to control the end-effector motion. They inherit the advantage of a high load capacity, small inertia and ability to alter the workspace through varying system parameters [1, 2]. Accordingly, they show a great attraction for various applications, such as camera systems [3], cargo handling [4, 5], rehabilitation [6, 7], and radio telescopes [8]. Since actuators are fixed on the base, and the weight of cables is almost negligible, the motion inertia is reduced significantly. However, for CDMs, a critical issue is the unidirectional nature of forces exerted by cables and

N. S. S. Sanjeevi · V. Vashista (✉)
Indian Institute of Technology Gandhinagar, Gandhinagar, Gujarat 382355, India
e-mail: vineet.vashista@iitgn.ac.in

V. K. Gupta et al. (eds.), *Recent Advances in Machines and Mechanisms*, Lecture Notes in Mechanical Engineering, https://doi.org/10.1007/978-981-19-3716-3_48

should remain in tension while performing tasks that demand actuation redundancy. Thus, in order to fully control a CDM with n degrees-of-freedom (DOF), a minimum of $n + 1$ cables are needed.

Manipulators with serial chain architecture have been widely used in the robotics community due to their unique properties such as high repeatability, large workspace, and ease of control. Historically, their usage has been mainly in industrial operations, such as pick-and-place, painting, and welding [9, 10]. Recently, cable-driven serial chain manipulators, CDSMs, have emerged as an important robotic platform due to their unique properties of low moving inertia and flexibility in altering system architecture. Among the various performance measures of a robotic manipulator, stiffness plays a vital role in robustly and safely interacting with unknown environments, including industrial and human-in-the-loop applications. A manipulator's stiffness regulates its rigidity, positioning, and trajectory tracking accuracy during any physical interaction task. Notably, CDSMs, providing flexibility in altering the system architecture, present an advantage in stiffness modulation. Accordingly, the literature on cable-driven manipulators has reported works on understanding stiffness and modulating it by changing various CDM system parameters like cable stiffness [11, 12], actuator locations [13].

The literature shows that cable tensions are used to regulate the stiffness of CDMs when cables with linear stiffness springs are used. Yu et al. [14] performed stiffness control on CDMs by optimizing the tension distribution of the cables. The stiffness of the end-effector is controlled to enhance the accuracy of trajectory tracking and disturbance rejection performance. Changes in the stiffness of the CDMs through varying stiffness across cables and cable tensions are reported in [15]. Cables with nonlinear stiffness characteristics are used to achieve changes in the stiffness of the CDM when manipulating the cable tensions in [16]. Osada et al. [17] employed a series of fixed and movable pulleys to produce devices with nonlinear stiffness characteristics. Another way to alter the stiffness is by adding passive stiffness elements to the manipulator, which is explored in this work.

Notably, usage of passive springs to alter the manipulator performance of CDMs is reported in the literature. Usage of the passive springs to alter the performance of the system is reported in the literature. Intuitively, springs may keep cables in tension with properly chosen placement parameters. A CDM to perform sub-millimeter operations where a compression spring was used to keep cables in tension is presented in [18]. A planar translational cable direct driven robot (CDDR) with a linkage mechanism was tensioned with two torque springs to prevent slack in the cables [19]. The introduction of springs to alter the workspace of single or multi-body cable-driven robots is presented in [20–22].

Accordingly, this work presents some preliminary analysis on the effect of passive springs on the taskspace stiffness characteristics of CDSMs. Two different modalities of passive spring attachments were considered. Firstly, passive springs are attached to the links from different ground points. Secondly, passive springs are routed across the pulleys positioned over the joints of the manipulator. Thus, it varies stiffness across that joint. Finally, stiffness indices were used to analyze the resulting stiffness changes in the CDSM.

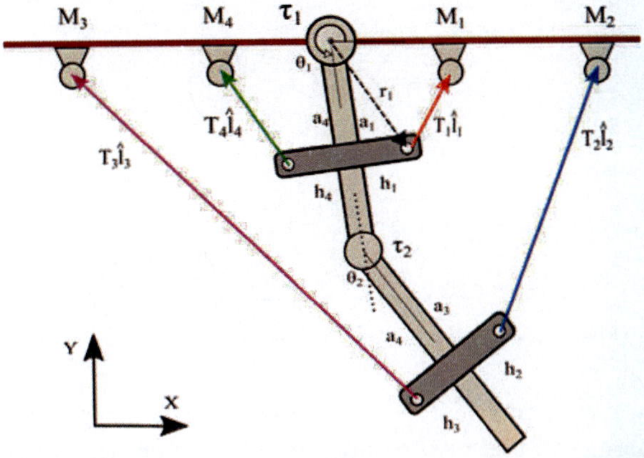

Fig. 1 Illustration of the considered 2 DOF CDSM model and respective cable configuration. O represents origin, and M_i represents motor positions w.r.t O. T_i and l_i represents cable tension and cable length. D_i represents link lengths, r_i and h_i represent cable attachment points and offset position on links, and their respective values are given in Table. 1. τ_i denotes joint torques

2 Architecture

In this work, a planar two-link cable-driven serial chain manipulator (CDSM) is considered for analysis. Figure 1 shows the schematic of the CDSM, where the rigid links are connected with revolute joints and motor actuated cables are connected to the links to apply external forces on the links to generate the required motion. However, as a cable can only apply a pulling force, actuator redundancy is required to control a CDSM [20]. Specifically, a minimum of three cables is required to control the considered two degrees-of-freedom (DOF) manipulator. In this work, four actuated cables, i.e., two actuators redundancy, are considered as it facilitates larger workspace [2, 23]. The link lengths are denoted as D_i, and the generalized position and force coordinates of the single DOF revolute joint as θ_i and τ_i, respectively. The cable attachment point on each link is at a distance a_i from the joint center and at an offset h_i from the link axis, and M_i denotes the corresponding motor position from O.

The dynamic model of the considered two DOFs serial chain manipulator can be formulated using Lagrange's method. With L, the Lagrangian and θ_i and τ_i being the generalized joint variable and joint torque values, respectively, the equations of motion can be written as presented in Eq. (1).

$$\frac{d}{dt}\left(\frac{\partial L}{\partial \dot{\theta}_i}\right) - \frac{\partial L}{\partial \theta_i} = \tau_i, i = 1, 2 \tag{1}$$

In the current work, the actuated cables is modeled as a source of pure force at the attachment point on the links to generate the joint torques. In general for a n DOFs system with m actuated cables, the joint torques can be expressed as $\tau_i = \tau_i^c$.

$$\tau_i^c = \sum_{j=1}^{m} \left(T_j \hat{l}_j \cdot \frac{\partial \mathbf{r}_j}{\partial \theta_i} \right) \tag{2}$$

$$\underbrace{\begin{bmatrix} \tau_1 \\ \tau_2 \end{bmatrix}}_{\tau_{(2\times1)}} = \underbrace{\begin{bmatrix} \hat{l}_1 \frac{\partial \mathbf{r}_1}{\partial \theta_1} & \hat{l}_2 \frac{\partial \mathbf{r}_2}{\partial \theta_1} & \hat{l}_3 \frac{\partial \mathbf{r}_3}{\partial \theta_1} & \hat{l}_4 \frac{\partial \mathbf{r}_4}{\partial \theta_1} \\ \hat{l}_1 \frac{\partial \mathbf{r}_1}{\partial \theta_2} & \hat{l}_2 \frac{\partial \mathbf{r}_2}{\partial \theta_2} & \hat{l}_3 \frac{\partial \mathbf{r}_3}{\partial \theta_2} & \hat{l}_4 \frac{\partial \mathbf{r}_4}{\partial \theta_2} \end{bmatrix}}_{\mathbf{A}_{(2\times4)}} \underbrace{\begin{bmatrix} T_1 \\ T_2 \\ T_3 \\ T_4 \end{bmatrix}}_{T_{(4\times1)}} \tag{3}$$

Accordingly, torque contribution at both the joints, due to all cables, is modeled. Further, rearranging all these expressions in matrix from results in Eq. (3). Here, $\mathbf{A} \in \mathbb{R}^{2\times4}$ denotes the structure matrix which represents the linear mapping. The parameters r_i and l_i in A are vectors defining cable attachment points on the links and actuated cable directions, respectively, as shown in Fig. 1. Essentially, the mapping A is a function of system geometry and captures the effect of cable routing architecture of a CDSM.

3 Stiffness Formulation

Stiffness is defined as the amount of resistance to an unit deformation. In particular, taskspace stiffness implies the manipulator's resistance to unit deformations in the end-effector's positions for unit force. For a multi-DOFs system, it is formulated mathematically using a matrix. It can be interpreted using stiffness ellipsoids where ellipsoids shape, axes direction, and radii magnitude is used to perform the stiffness analysis.

3.1 *Taskspace Stiffness*

Taskspace stiffness for a planar manipulator can be interpreted as the resistance offered by a manipulator for a small change in position for a unit force applied to end-effector at a quasi-static condition. For small forces, $\delta F \in \mathbb{R}^{2\times1}$, changes in position, $\delta X \in \mathbb{R}^{2\times1}$, are evaluated using taskspace stiffness matrix,$\mathbf{K}_X \in \mathbb{R}^{2\times2}$.

$$\delta F = \mathbf{K}_X \delta X \tag{4}$$

Using the principle of virtual work and the definition of the Jacobian, the relationship between the joint torques and the force applied on the end-effector can be written as in Eq. (5).

$$\tau = J^T F \tag{5}$$

$$\delta\tau = (\delta J^T)F + J^T(\delta F) \tag{6}$$

Similar to unit forces in taskspace, at a quasi-static condition, for small joint torques, $\delta\tau \in \mathbb{R}^{2\times1}$, variations in the joint variables, $\delta\theta \in \mathbb{R}^{2\times1}$, are computed using the joint space stiffness matrix, $\mathbf{K}_\theta \in \mathbb{R}^{2\times2}$, as presented in Eq. (7).

$$\delta\tau = \mathbf{K}_\theta \delta\theta \tag{7}$$

$$\mathbf{K}_\theta = \frac{\delta J^T}{\delta\theta} F + J^T \mathbf{K}_X \frac{\delta X}{\delta\theta} \tag{8}$$

Replace the torque and force expression in terms of joint space and taskspace stiffness matrices. Also, the manipulator Jacobian matrix, *J*, relates the differential joint displacements to infinitesimal movement of the end-effector, i.e., $\delta X = J\delta\theta$. Notably, Jacobian, *J*, is a function of manipulators link length parameters and joint angles. Noting the joint space stiffness relation, Eq. (7) and the kinematic relationship between taskspace and joint space, the expression for $\mathbf{K}_X$ is written in Eq. (9).

$$\mathbf{K}_X = (J^T)^{-1} \left[\mathbf{K}_\theta - \underbrace{\frac{\delta J^T}{\delta\theta} F}_{\mathbf{K}_g} \right] (J^{-1}) \tag{9}$$

$$\mathbf{K}_X = (J^T)^{-1} \mathbf{K}_\theta (J^{-1}) \tag{10}$$

$$\mathbf{K}_\theta = \frac{\delta A}{\delta\theta} T - A k_c A^T \tag{11}$$

The term $\mathbf{K}_g$ represents the effect of differential Jacobian under an applied external load *F*. Accordingly, under no load condition, Eq. (9) reduces to Eq. (10). For CDSM, joint space stiffness, $\mathbf{K}_\theta$ is dependent on cable tension distribution, cable routing, i.e., system architecture and cable stiffness as presented in Eq. (11) [24].

4 Stiffness Alteration

Equation (9) establishes the dependency of manipulators taskspace stiffness, $\mathbf{K}_X$, on joint space stiffness, $\mathbf{K}_\theta$ which is dependent on system architecture, *A*, and cable stiffness, k_c. Thus, any structural alterations of CDSM by adding passive stiffness elements alter the joint space stiffness, thus overall taskspace stiffness behavior. In particular, two different modalities of passive spring attachments were considered in this work.

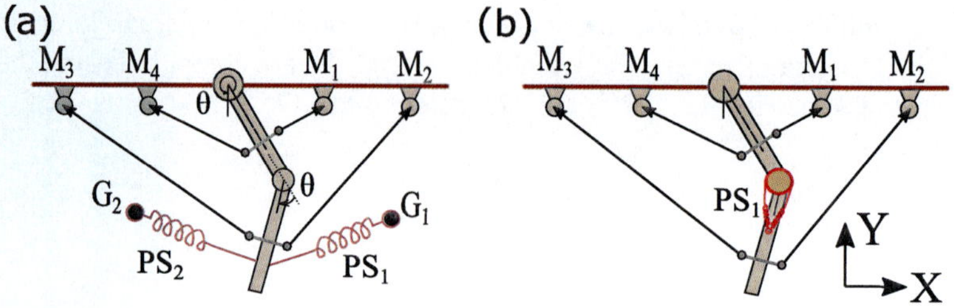

Fig. 2 Passive stiffness configurations **a** Springs attached to distal link **b** springs routed along joint 2 through pulley to provide external joint stiffness only at joint 2

4.1 Passive Springs Attached to Links

In this case, one end of the passive spring is connected to the links, while the other end is connected to a fixed ground point, shown in Fig. 2a. Essentially, these springs generate resisting or assisting forces on the links when CDSM moves along the trajectory in accordance with their ground and cable attachment positions. Notably, the addition of springs alters the torque-tension mapping, Eq. (12). Here, τ_{PS} is the torque generated due to the external forces on the links generated through the passive spring attachments. It can lead to assisting or resisting torque. Thus, to generate the desired torque, τ_{des}, the torque τ_{CDSM}, that should be generated by CDSM, will vary accordingly. Further, suppose the linear springs are elongated and are in tension. In that case, they will act as extra cables in the cable-driven system, which are to be modeled through structure matrix, A_{PS}, using the formulation presented in Sect. 2. The contribution of the additional springs to the overall joint stiffness is presented in Eq. (13). Now, using this updated $\mathbf{K}_{\theta_{New}}$, one can calculate the taskspace stiffness matrix, $\mathbf{K}_X$, using Eq. (10). Notably, any changes in the passive spring stiffness and their attachment positions will alter the taskspace stiffness characteristics.

$$\tau_{des} = \tau_{CDSM} + \tau_{PS} \tag{12}$$

$$\tau_{des} = \underbrace{AT}_{\text{CDSM}} + \underbrace{A_{PS}T_S}_{\text{PS}}$$

$$\mathbf{K}_{\theta_{New}} = \mathbf{K}_{\theta_{CDSM}} + \mathbf{K}_{\theta_{PS}} \tag{13}$$

4.2 Passive Spring Routed Across Joint

In this case, passive springs were positioned on the links and routed along the joint through a pulley mechanism as shown in Fig. 2b. Essentially, these springs create additional joint stiffness on the joint, which is joint 2 in Fig. 2b. Accordingly, this alters the joint stiffness characteristics as presented in Eq. (14). Here, the additional

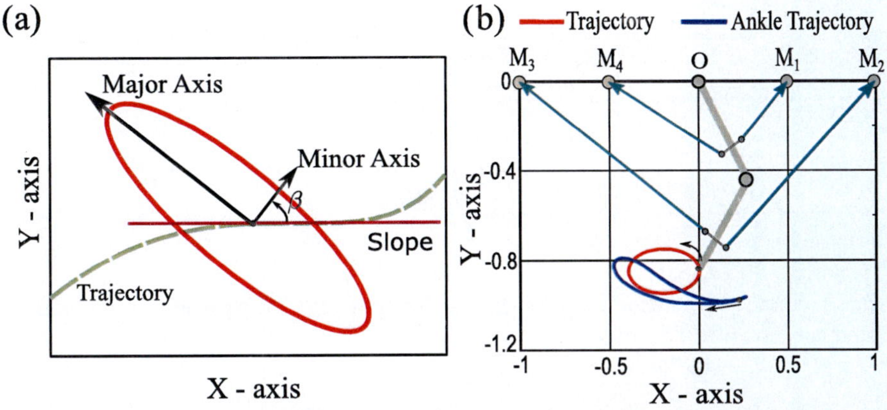

Fig. 3 **a** Stiffness ellipse **b** chosen taskspace trajectories to alter stiffness, circular and human ankle trajectory

passive joint stiffness, $k_{\theta_{\mathrm{PS}}} = 2r_{\mathrm{p}}^2 k_{\mathrm{ps}}$ [25], is a function of passive spring stiffness, k_{ps} and the radius of the pulley, r_{p}; it is routed across on the joint. Further, the taskspace stiffness matrix, $\mathbf{K}_X$, using Eq. (10).

$$\mathbf{K}_{\theta_{\mathrm{New}}} = \underbrace{\begin{bmatrix} k_{\theta_{11}} & k_{\theta_{12}} \\ k_{\theta_{21}} & k_{\theta_{22}} \end{bmatrix}}_{\mathrm{CDSM}} + \underbrace{\begin{bmatrix} 0 & 0 \\ 0 & k_{\theta_{\mathrm{PS}}} \end{bmatrix}}_{\mathrm{PS}} \tag{14}$$

Performance Parameters: To analyze the stiffness behavior of a cable-driven system, typically indices based on eigen values of $\mathbf{K}_X$ are used [24]. Condition number (CN) is one such index which is the ratio of on the maximum and minimum eigenvalues. CN value is always equal or greater than 1, where $CN = 1$ denotes an isotropic stiffness behavior.

$$CN = \frac{\lambda_{\max}}{\lambda_{\min}}$$

Further, a stiffness ellipse can be defined for the $\mathbf{K}_X$. Here, the eigenvectors of the stiffness matrix, $\mathbf{K}_X$, are directed along the axes of the ellipse such that the minor axis of the stiffness ellipse is directed along the eigen vector with larger eigenvalue. Similarly, the major axis of the ellipse is directed along the eigen vector with smaller eigenvalue. Essentially major axis of ellipse implies direction of maximum resistance to disturbance and minor axis implies least resistance. Figure 3a shows a stiffness ellipse at a point on the trajectory. The knowledge of minor axis orientation can be used to examine the robot trajectory tracking performance. In particular, angle β is defined to record the orientation of minor axis with the slope of taskspace trajectory such that when a CDSM is used a zero value of angle β represents a case where the end-effector experience least stiffness along the taskspace trajectory. In this work, all

Table 1 System parameters of CDSM. Both the links are assumed to be of equal lengths of 50 cm

	Motor Positions $M_i = [X_i, Y_i]$ (in M)	Attachment Positions (in *cm*)
CDSM	$M_1 = [0.175, 0.02]$ $M_2 = [0.25, -0.02]$ $M_3 = [-0.25, -0.02]$ $M_4 = [-0.175, 0.02]$	$a_1 = a_4 = 34$ $a_2 = a_3 = 34$ $h_1 = h_4 = 6.25$ $h_2 = h_3 = 6.25$

these indices, *CN*, stiffness ellipse and orientation angle, were used for the stiffness analysis.

5 Results and Discussion

This section presents the simulation results for taskspace stiffness analysis of two DOFs, two redundancy CDSMs. Link lengths (D_i) are assumed to be of 50 *cms*. Further, the CDSM system parameters, namely motor positions (M_i) and cable attachment positions (a_i) and (h_i) used for this analysis, are listed in Table 1. Stiffness across all the cables, k_{c_i}, is assumed to be of 1500 N/m. For the stiffness analysis, two different trajectories are chosen in the taskspace as shown in Fig. 3b. A circular trajectory considers CDSM as an industrial manipulator, and a human ankle trajectory considers CDSM as lower-limb exoskeleton assisting the human lower-limb motion. Further, stiffness indices were evaluated and compared for CDSM with both the considered passive spring attachment modalities.

For a CDSM, end-effector stiffness characteristics that resist deflections from the trajectory path due to undesired external forces are desired. The prospect of stiffness index β can be interpreted as its value along the trajectory should be minimum. Essentially, the minimum value of β means the manipulator will not resist deflections along the trajectory. From a stiffness ellipse perspective, the ellipse's minor axis, which represents the least resistance to external forces, should be oriented along the trajectory. Under such conditions, manipulator resits deflections in all other directions, with maximum resistance being offered along the direction perpendicular to the trajectory. Further, *CN* should be small, as higher *CN* represents skewed ellipse which means resistance is maximum only in one direction. Thus, optimal index values would be smaller β and *CN* and the minor axis of the ellipse oriented along the trajectory.

Figure 4a–c presents the orientation angle, β, *CN*, and stiffness ellipse variations across the circular trajectory for the three cases, without passive spring, with passive springs, and torsional spring at joint 2. These indices which were evaluated using the taskspace stiffness matrix are computed from Eq. (10). All three cases reported significant variations of stiffness indices along the trajectory. From the stiffness ellipse orientation, it is observed that the major axis is oriented closely toward the y-axis. It means the manipulator resists deflection along y-direction compared to

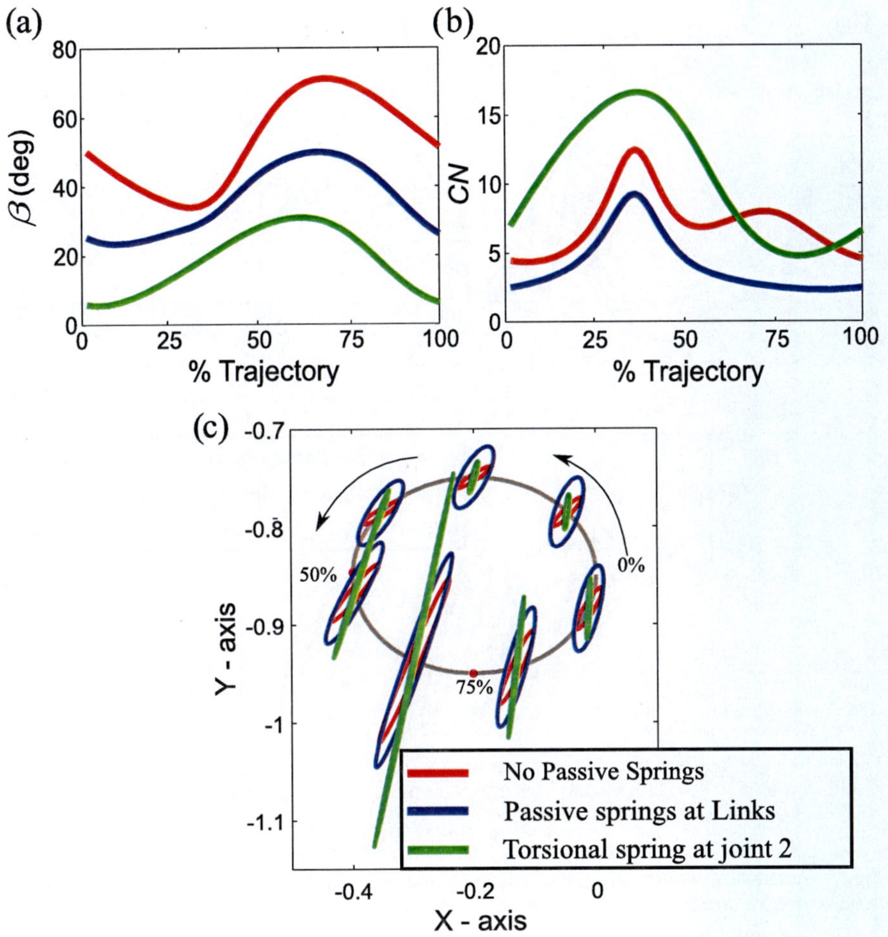

Fig. 4 Variation of **a** orientation angle (β) **b** *CN* and **c** stiffness ellipses along the circular trajectory in taskspace for different passive spring attachment conditions

x-direction for most of the trajectory. Notably, adding a torsional spring at joint two resulted in a reduction in the β value when compared with the other two cases, Fig. 4a. But, it made stiffness ellipse skewed as reflected in Fig. 4c and also can be inferred from an increase in the *CN* values, Fig. 4b. Adding external passive springs to the links did not have much effect on β and orientation of ellipses but reduced the *CN* values. Further, stiffness ellipse for external passive springs case encompassed the ellipse of no springs cases. It essentially means the stiffness of the system is larger. It is intuitive, as the addition of springs increases stiffness. Further, the addition of torsional spring resulted in an increase in the stiffness along the *y*-direction in taskspace as reflected in ellipses from Fig. 4c.

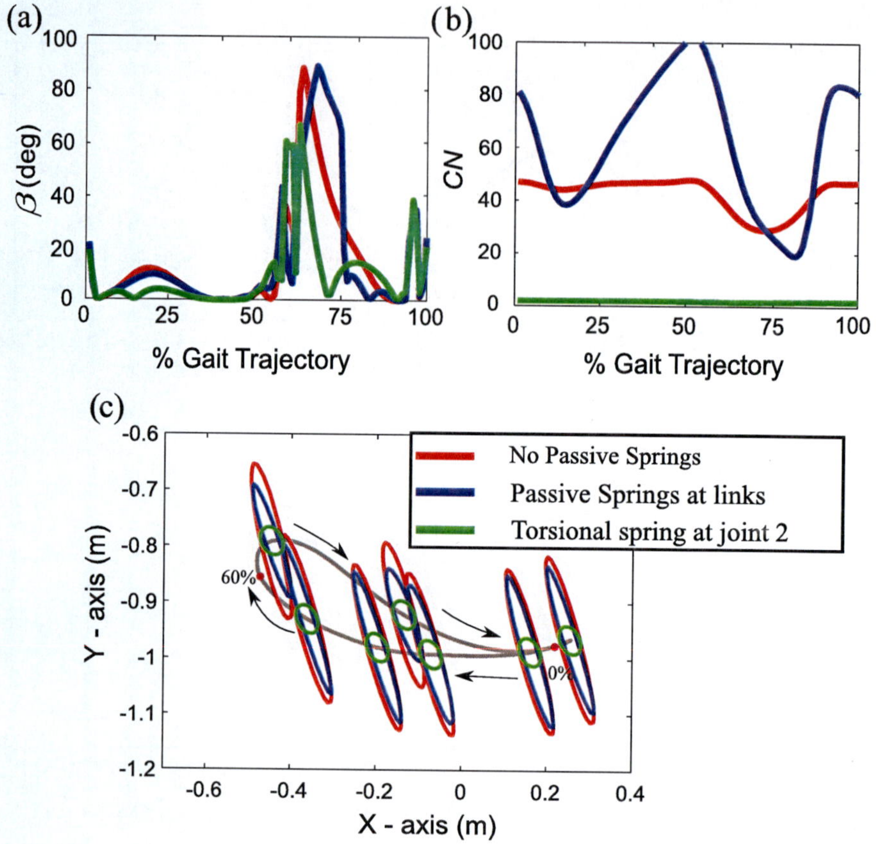

Fig. 5 Variation of **a** orientation angle (β) **b** *CN* and **c** stiffness ellipses along the ankle trajectory in taskspace for different passive spring attachment conditions

Figure 5a–c presents the orientation angle, β, *CN*, and stiffness ellipse variations across the human ankle trajectory. Here, CDSM is considered a lower-limb exoskeleton that assists the lower limb in executing the human ankle motion. Exoskeletons are devices that assist the motion and provide rehabilitation training. For an exoskeleton to provide effective rehabilitation, it should facilitate the wearer to move along the desired trajectory and resist deviations from the desired ankle trajectory. Thus, the desired indices would be minimal $beta$, the minor axis of stiffness ellipse along the ankle trajectory, and smaller *CN*. A minimum value of β represents a case where the exoskeleton imposes minimum stiffness along the desired ankle trajectory. Having minimum stiffness along the trajectory means the ankle gets a resistance to deviate from the desired trajectory.

Stiffness indices were computed for all three modalities. Orientation angle, β, is minimum along major portions of the ankle trajectory, except for 50–75%, Fig. 5a. *CN* value is larger for the without and with passive spring cases, Fig. 5b. Notably, *CN* values reduced significantly for the torsional spring case. It is also reflected in the stiffness ellipses, as they are less skewed compared to other cases Fig. 5c. Notably, the addition of torsional spring improved the β value too to some extent during 50–75% of ankle trajectory. However, it resulted in the reduction of the overall stiffness magnitude. It can be inferred from stiffness ellipses where ellipses for torsional spring case become smaller. However, this behavior can be changed by changing the spring stiffness values. Notably, for the presented analysis, the spring positions and lengths were chosen so that they were elongated throughout the trajectory. However, one can optimize their ground positions so that springs will activate only during some portions of the trajectory.

Overall, these preliminary results highlight that the changes result from the addition of external passive springs on the stiffness characteristics of the manipulator. For the presented results, passive spring stiffness values and their initial positions were chosen randomly. However, one can formulate the optimization schemes based on these indices that solve for optimal stiffness and mounting positions of passive springs to achieve the desired stiffness characteristics along the desired trajectory.

6 Conclusion

This work presents some preliminary analysis of passive springs' effect on the taskspace stiffness characteristics of cable-driven serial chain manipulators, CDSMs. Two different modalities of passive spring attachments, where passive springs are attached to the links from different ground points and passive springs are routed through the pulleys positioned over the joints of the manipulator, were considered for the analysis. The resulting changes in the stiffness characteristics were analyzed through stiffness indices, namely stiffness ellipse and condition number. Simulation results emphasize that the addition of passive springs results in significant alterations in the stiffness characteristics of CDSMs. Future work would be to develop optimization schemes based on stiffness indices to identify optimal passive spring stiffness and mounting positions that provide desirable stiffness characteristics along the planned trajectory.

Acknowledgements This work was supported in part by the Inspire funding from the Department of Science and Technology and in part by the Scheme for Promotion of Academic and Research Collaboration (SPARC) program, Ministry of Human Resource Development, Government of India (Grant SPARC/2018-2019/P481/SL).

References

1. Merlet JP (2008) Kinematics of the wire-driven parallel robot MARIONET using linear actuators. In: 2008 IEEE international conference on robotics and automation. IEEE, 2008
2. Gouttefarde M, Gosselin CM (2006) Analysis of the wrench-closure workspace of planar parallel cable-driven mechanisms. IEEE Trans Robot 22(3):434–445
3. Cone LL (1985) Skycam-an aerial robotic camera system. Byte 10(10):122
4. Dagalakis NG et al (1989) Stiffness study of a parallel link robot crane for shipbuilding applications, 183–193
5. Schindele D, Aschemann H (2011) Fast nonlinear MPC for an overhead travelling crane. IFAC Proc 44(1):7963–7968
6. Mao Y, Agrawal SK (2012) Design of a cable-driven arm exoskeleton (CAREX) for neural rehabilitation. IEEE Trans Robot 28(4):922–931
7. Sanjeevi NSS, Vashista V (2021) Stiffness modulation of a cable-driven leg exoskeleton for effective human-robot interaction. Robotica: 1–21
8. Duan X et al (2011) Motion prediction and supervisory control of the macro-micro parallel manipulator system. Robotica 29(7):1005
9. Pandilov Z, Dukovski V (2014) Comparison of the characteristics between serial and parallel robots. Acta Tech Corvininesis-Bull Eng 7(1)
10. Summers M (2005) Robot capability test and development of industrial robot positioning system for the aerospace industry. SAE Trans: 1108–1118
11. Zhou X, Jun SK, Krovi V (2015) A cable based active variable stiffness module with decoupled tension. J Mech Robot 7(1)
12. Yeo SH, Yang G, Lim WB (2013) Design and analysis of cable-driven manipulators with variable stiffness. Mech Mach Theory 69:230–244
13. Abdolshah S et al (2017) Optimizing stiffness and dexterity of planar adaptive cable-driven parallel robots. J Mech Robot 9(3)
14. Yu K et al (2010) Enhanced trajectory tracking control with active lower bounded stiffness control for cable robot. In: 2010 IEEE international conference on robotics and automation. IEEE, 2010
15. Kawamura S (1995) Development of an ultrahigh speed robot FALCON using wire drive system°. Robot Autom: 215–220
16. Gialias N, Yoky M (2004) Muscle actuator design for the ACT Hand. In: Proceedings on IEEE international conference on robotics and automation, vol 4. ICRA'04. 2004. IEEE, 2004
17. Osada M et al (2010) Realization of flexible motion by musculoskeletal humanoid "Kojiro" with add-on nonlinear spring units. In: 2010 10th IEEE-RAS international conference on humanoid robots. IEEE, 2010
18. Russell RA (1994) A robotic system for performing sub-millemetre grasping and manipulation tasks. Robot Auton Syst 13(3):209–218
19. Trevisani A, Gallina P, Williams RL (2006) Cable-direct-driven robot (CDDR) with passive SCARA support: theory and simulation. J Intell Robot Syst 46(1):73–94
20. Mustafa SK, Agrawal SK (2011) On the force-closure analysis of n-DOF cable-driven open chains based on reciprocal screw theory. IEEE Trans Robot 28(1):22–31
21. Taghavi A et al (2013) Workspace improvement of two-link cable-driven mechanisms with spring cable. Cable-driven parallel robots. Springer, Berlin, pp 201–213
22. Duan Q, Vashista V, Agrawal SK (2015) Effect on wrench-feasible workspace of cable-driven parallel robots by adding springs. Mech Mach Theory 86:201–210
23. Sanjeevi NSS, Vashista V (2019) Effect of cable co-sharing on the workspace of a cable-driven serial chain manipulator. In: Proceedings of the advances in robotics 2019. 2019, pp. 1–6
24. Sanjeevi NSS, Vashista V (2017) On the stiffness analysis of a cable driven leg exoskeleton. In: 2017 International conference on rehabilitation robotics (ICORR). IEEE, 2017
25. Rao P, Deshpande AD (2018) Analyzing and improving Cartesian stiffness control stability of series elastic tendon-driven robotic hands. In: 2018 IEEE international conference on robotics and automation (ICRA). IEEE, 2018

Mechanism for Agriculture Applications

Design and Development of Multi-crop Fibre Extracting Machine

Deepak Mahapatra and Jagpal Singh Bal

Abstract Design and development of machinery for handicraft products is generally done by the artisan himself based on his/her experience, intuition and experimentation mostly by trial and error process. Growing demand for products like handloom-based textiles has created a requirement of production in large quantities which has generated demand for raw materials such as yarn of different types. Application of natural fibre derived from biomass is increasing in handloom-based textile processing. This has created a need for fibre extraction from biomass and several machines are developed in different parts of the world. A review of existing machinery and associated mechanisms is presented in this paper to address the need for research in development of sustainable support systems to handloom-based artisans. A human powered mechanism is suggested for extraction of fibre from biomass.

Keywords Fibre crops · Extraction machines · Sisal fibre · Flax fibre · Banana fibre

1 Introduction

In the recent past the demand for natural fibres has grown steadily for the textile industry as well as for their use in the form of additives to make natural composites. The reason for this rise in demand owe to the fact that natural polymers are low cost, biodegradable, ease of availability and also are human friendly thus leading to sustainable development. The rise in demand accounts for steady increase in production of natural fibres and hence search for newer and efficient methods of extraction of natural fibres has been attracting the researchers in a global platform. Extraction and processing of traditional natural fibres like cotton, jute, etc. has been

D. Mahapatra (✉)
College of Food Technology, IGKV, Raipur, CG, India
e-mail: medeepakmahapatra@gmail.com

J. S. Bal
Directorate of Technical Education, Raipur, CG, India

V. K. Gupta et al. (eds.), *Recent Advances in Machines and Mechanisms*, Lecture Notes in Mechanical Engineering, https://doi.org/10.1007/978-981-19-3716-3_49

sufficiently researched and the government is promoting new diversified innovations in obtaining fibres from other sources as well [1].

Handicraft industry has been primarily based on traditional methods of extraction of fibres as most of the artisans belong to the marginal to cottage level. They are financially weak as well as ignorant to technological developments. Hence they are not able to sustain competition to the textile requirements of the country. However, to view the other way round, it is also observed that the machinery that they use have evolved out of their years of experience, and experimentation and intuition that they have developed in course of time. Most of their machinery is low cost manually operated devices that come up with gradual trial and error approaches.

It can also be noticed that the handloom industry has survived and continues to show its existence despite the tough environment framed by the textile giants. It is largely felt that there is a need to bring forward the traditional experiences of the handloom artisans and assimilate them with new innovative ideas so as to develop simple, cost effective natural fibre extraction and processing machines.

Growing demand for products like handloom-based textiles has created a requirement of production in large quantities which has generated demand for raw materials such as yarn of different types. Application of natural fibre derived from biomass is increasing in handloom-based textile processing. This has created a need for fibre extraction from biomass and several machines are developed in different parts of the world. The present report is an attempt to review the existing machinery and associated mechanisms so as to address the need for research in development of sustainable support systems to handloom-based artisans. The conclusion of the review must pave the way for developing a low cost manually operated multi-crop fibre extraction unit.

2 Review of Machinery for Plant Fibre Extraction

Natural fibres are seen as alternatives for catering the needs of the industry in future as they can be used in textile industry, composites as well as medical and biotech industries as well. In the present era, natural fibres have definitely been looked forward to for their degradable nature and more importantly they are also renewable.

A number of varieties of biomass exist in nature which contains fibres that can be used for various applications. There are thousands of species of plants that can yield natural fibres and in India itself we have around 800 fibre yielding plants [2]. However the potential for commercial utilization in terms of their availability, fibre properties, etc. of natural fibres exist in only a few particular varieties. The most important types of fibre yielding varieties are sisal, banana, flax, kenaf, hemp, pineapple leaf, ramie, etc. Fibre is extracted from a particular part of the plant for example the leaves of sisal, pine apple, palm, etc. are used whilst skin of banana, hemp, jute are found suitable for extraction; similarly stem of banana, coconut, bamboo, etc. are utilized for fibre extraction, whereas fruits of cotton plant yield fibre.

Apart from the industrial use, the above fibre plants are also looked forward to by the handloom sector so that obtained fibres may be utilized in single form or in union with other established fibres like jute and cotton [3]. For obtaining suitable quality of the yarn (single or in union) it is necessary to know the properties of the fibres and fibre yielding plants, however this is a herculean task as the properties of plants vary according to geographical and climatic conditions. Still any relevant information will help in improvising the state of art techniques of fibre extraction.

Hence a compilation of the works related to the extraction techniques of natural fibres is necessary. In the present paper review has been made to compare the existing methods of fibre extraction with an insight of the traditional knowledge in the handloom sector. In the successive paragraphs a brief discussion on the various machines and related mechanisms of fibre extraction from biomass is discussed. The intention is to develop manually operated low cost machinery for fibre extraction suitable for handloom artisans and also to create awareness for exploring the possible commercial applications of other available bio-vegetation.

2.1 *Brief History of Machinery Related to Fibre Extraction*

Yusof and Adam [4] have reported that inventions of machines for natural fibre extraction date back to the 1900s with names like decorticator, abstractor, defibrator, etc. And according to published review, the inventions and patents corresponding to the fibre extraction were in large numbers in the first half of the twentieth century. However after 1950, the corresponding number of publications has been decreasing till 2000, probably due to the growth in the utilization of synthetic fibres. With the emerging area of natural composites the fibre extraction and processing techniques have rejuvenated in the research.

An exhaustive review on the variety of fibre extracting machines developed in the time span of 1920–2012 has been reported in [4]. It is remarked that in most of the machines the extraction is accompanied by scratching the plant material with sharp tools like knife, blade, shredder after which a few include crushing and bending. According to Gardener, [5] that crushing and crumping leads to clogging of the decorticated material leading to separating problems which was the reason for not including crushing in his design. Whilst not many agree on the verdict of Gardener and have the opinion that crushing breaks the rubbery waxy layer thus making it easier to extract the fibre [6]. The idea behind is that by providing a pre-crushing action on the leaves of the plant using a series of rollers accompanied by washing leads to less stressing of fibres, eventually leads to lesser damage. In 1998, Leduc et al. [7] came up with a design that contained only rollers and did not contain any type of blade or knives as an extracting tool (Fig. 1). In a slightly different design Cook [8] has used the fluid power to rotate the crushing and grinding wheels and in addition fluid washes the fibres and tool also. The various machinery and mechanisms relevant to the natural fibre extraction in global platforms may be referred from Table 1.

An excellent overview of the available machines and currently used by small scale artisans in India has been reported in [3]. The extraction of different fibres like Jute,

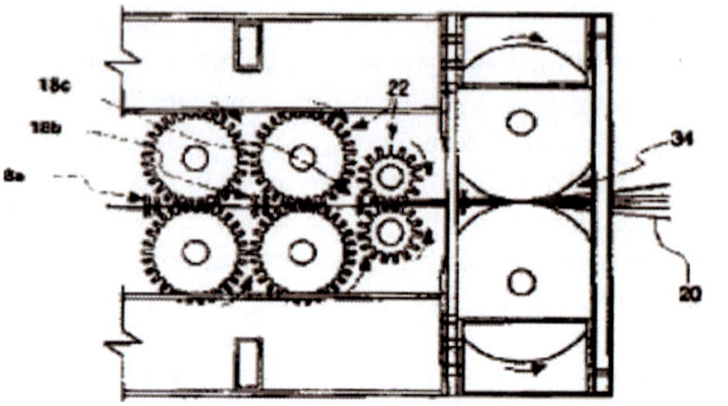

Fig. 1 Fibre extracting machine patented by Ludec et al. [7]

cotton, sisal, flax, pineapple leaf and coir has been discussed in terms of capacity and output. However the mechanism of nearly all these machines is based on scratching through knives, with only a few machines having arrangements for pre-crushing of plants.

2.2 *Traditional Methods of Fibre Extraction*

Before proceeding this topic one must know the steps to be followed in extraction of the fibre. However each variety of fibre requires a specific process to be followed, for brevity let us classify the fibres broadly into two categories-leaf fibres (sisal) and stem or bast (flax) fibres. A flow chart of each category is shown in Fig. 2.

Retting is a traditional method of extracting the fibres by soaking it in water or other chemicals so as to remove lignin, pectin and other substances. Retting may also be done by keeping the plant products in atmospheric air; however this method takes more time and is also liable to be inefficient in case of sudden changes in environment.

Manual scratching or stripping is done with the help of a broken plate or broken coconut shell to extract the fibre. After this the fibres are washed and dried in air. The fibres are then coated with wax to remove any entanglements and then they are knotted to form yarns. However this entire process takes too much time and is labour intensive. But the quality of fibres obtained is better as compared to mechanical methods [10].

The small scale artisans follow mostly the processes by manual techniques. Most of the operations like harvesting of crops, retting, scotching, spinning, decortication, etc. are carried out manually. Even if a few tools or machinery are used, then they have been developed from intuition after years of experience gained through working under challenging environments. In Figure 3 traditional methods of extraction of fibres and crafting it into useful products has been shown.

In Chhattisgarh, the district of Janjgir-Champa is known for its skilled handicraft art based on Kosa silk. However due to lack of availability of raw materials the local

Table 1 Technological innovations in fibre extraction machinery in last century [4, 8]

Patent/Journal	Blade/edged blade/knives/shredder	Crusher/bender
Fibre removing machine for flax, hemp, ramie and other textile plants (1925)	√	√
Fibre removing machine for flax, hemp, ramie and other textile plants (1925)	√	√
Fibre-decorticating machine (1926)	√	√
Decorticating apparatus for treatment of sisal and other fibrous leaves and stalks (1927)	√	√
Decorticating and deliberating machine (1934)	√	√
Fibre separating machine (1937)	√	–
Method of and machine for abstracting and preparing fibres (1939)	√	–
Method and machine for the abstraction and preparation of fibres (1939)	√	√
Decorticating machine (1940)	√	–
Apparatus for abstracting and preparing fibres from fibre-bearing plants (1942)	–	√
Machine for cleaning and separating flax or other fibrous plants (1948)	√	√
Apparatus for obtaining fibres from plant leaves (1955)	√	√
Self-propelled fibre harvesting and decorticating machine (1960)	–	√
Method for decorticating plant material (1998)	√	√
Plant material processing system (2000)	√	√
Plant material processing system (2002)	√	√
Utilization of pineapple leaf agro waste for extraction of fibre and the residual biomass for Vermi-composting (2011)	√	√
Extraction device for fibre in plant stems and leaves (2012)	√	√

artisans are looking for alternatives to get fusion with silk. It has been reported that sisal fibre can act as a suitable match [1], hence fibre extraction techniques may be quite relevant in this area. The local artisans in this area use customized tools for yarn making and loom; the local crafting techniques show remarkable experience and aptitude to carve very fine quality handloom products from natural fibres as well. Hence it is observed that there is a big scope of intervention of mechanization to support the local artisans to increase their productivity and in turn their living standards.

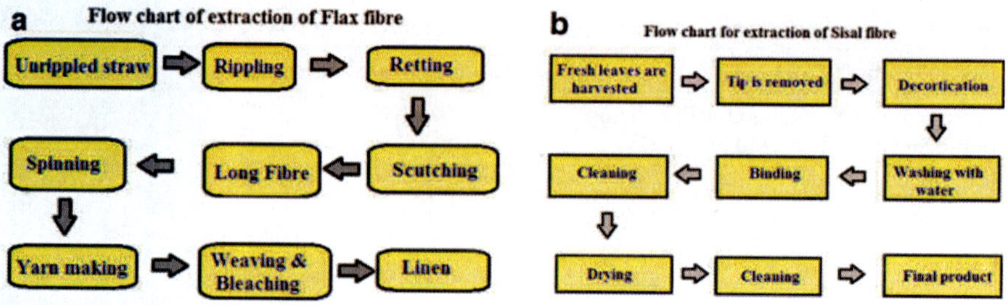

Fig. 2 Flow chart of fibre extraction **a** flax, **b** sisal

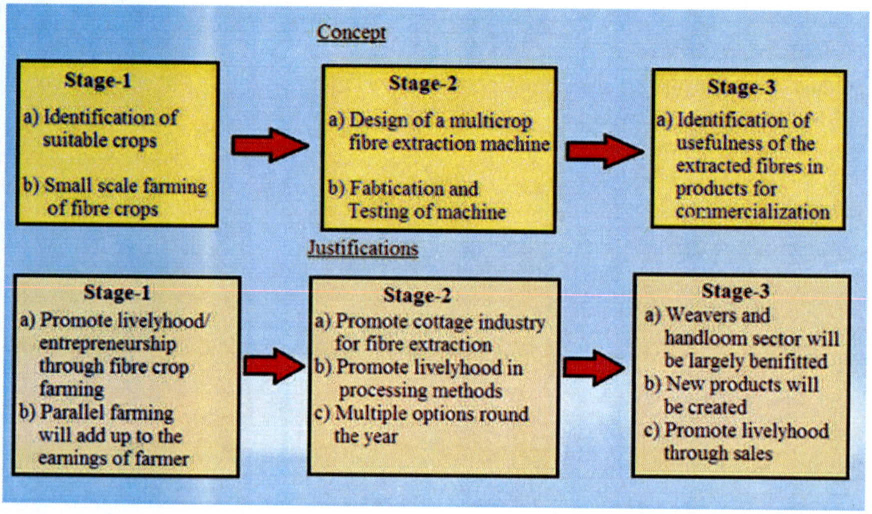

Fig. 3 Conceptual idea of the research work

3 Background and Motivation

It is an established fact that Agave (Sisal) fibre can be used as a replacement for the silk fibre [1]. It is also presumed that the properties of many other natural fibres can serve similar purposes with comparable effectiveness. In addition to this at present most of the fibre extraction methods are manually operated leading to low production rates. Hence it is foreseen that this area needs to be further explored to assess the commercial viability at regional level.

There are many factors which led us to the development of this innovative idea. Some of these factors are listed below.

- Available methods of fibre extraction are traditional (manual).
- Manual methods of extraction lead to non-uniformity and reduced outputs in production.

- Higher costs of machinery (for small scale processing).
- Limited technical intervention to promote sustainable production.
- High demand of Kosa silk and other fine yarns (sisal fibre, linen).
- Presence of a large number of weavers and artisans in the region with strong traditional weaving/handloom experience.
- There is a scope for sustainable natural fibre made finished products (handlooms for example) to be made comparable to synthetic contemporaries.
- Studies pertinent to developing multiple fibre crop options for the local weavers are limited in the state.

The conceptual idea has been presented in three stages as shown in Fig. 3.

It is observed that a separate study on the agricultural aspects is relevant to the accomplishment of the above concept. The current authors are also working on the agro-climatic study of fibre crop cultivation and its promotion amongst the rural community. This being a slightly contrast area is not included here and a detailed report on those observations is to be published separately.

4 Design Features of a Manually Operated Fibre Extractor

Before proceeding to the design parameters some end conditions must be fixed that must contain the broad objectives of the idea. Some of them are listed below.

- Low cost, portable, efficient, pedal operated fibre decorticator
- Production capacity of quality fibre must be 1 kg per hour (approx.)
- Machine must run with manual/animal power only. The input power must be minimal through suitable speed reduction.
- Weight must be low so that even women may operate it easily.
- Maintenance must be easy, which means machinery should be simple, with ease of assembly and dismantling.
- It must have provision for incorporation of new attachments, so that possibility of fibre extraction of a new crop may also be handled.

In this section design of a portable fibre extraction unit for (leaf fibre extraction) is discussed briefly. As discussed in Sect. 2.1, the machinery required must contain a blade or knife operating in cyclic operation that must scratch the waxy layer from the top and bottom surface of the leaf. A pre-crushing arrangement must be fitted prior to scratching to decrease the effort on the roller. The actuation of both the rotating elements must be provided with the same input. The number of blades may be between 12 and 16, depending on the size of the drum. A schematic of the mechanism is shown in Fig. 4.

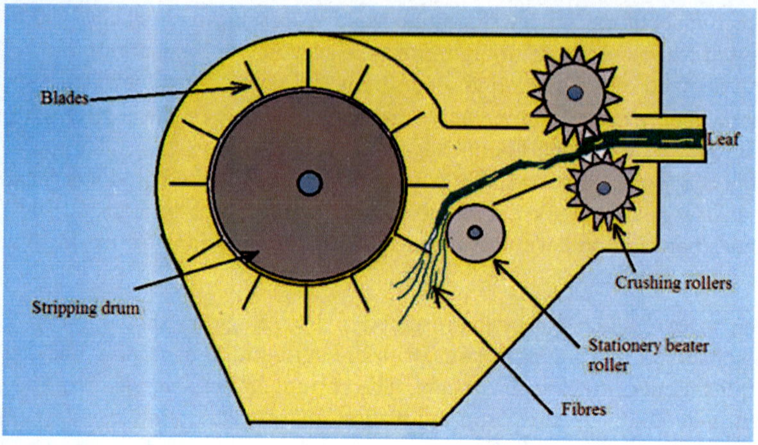

Fig. 4 Schematic diagram of mechanism of fibre extracting machine

4.1 *Design of a Prototype of Fibre Extractor*

The primary components of the extracting unit are: a pair of crushing roller to crush the leaf, grip it and feed it towards the rotating drum fitted with blades. The blades scrap the crushed leaf against the static beater roller to separate the fibres from the remaining scalp, which is then washed and dried to obtain plant fibre.

Figure 5 shows the power transmission unit for a manually operated extracting unit. It is similar to that provided in a sewing machine in which the operator is moving the pedal with his foot whilst in sitting position, the pedal in turn operates a belt-pulley through a crank fitted near the pedal that ultimately drives the drum and the crushing rollers. Hence the manual effort applied at the pedal must be sufficient to drive the two rollers and drum, simultaneously to scrap the leaf. Let us try to assess the power calculations for sisal fibre decortication numerically.

The Young's modulus, E of sisal leaf is assumed to be 10 GPa. An average length of leaf can be taken as 500 mm, whilst its cross section may be assumed as 150 mm and a thickness of around 10 mm. (The maximum leaf thickness of sisal plant can be even 3.5 cm [11]). Let the leaf be crushed and bend by 5 mm, then the force required can be obtained from the following equation:

$$y = \frac{Fl^3}{48\text{EI}} \quad (1)$$

which gives $F = 240$ N. Hence the torque required at crushing rollers will be

$$T_c = F \times r_c$$

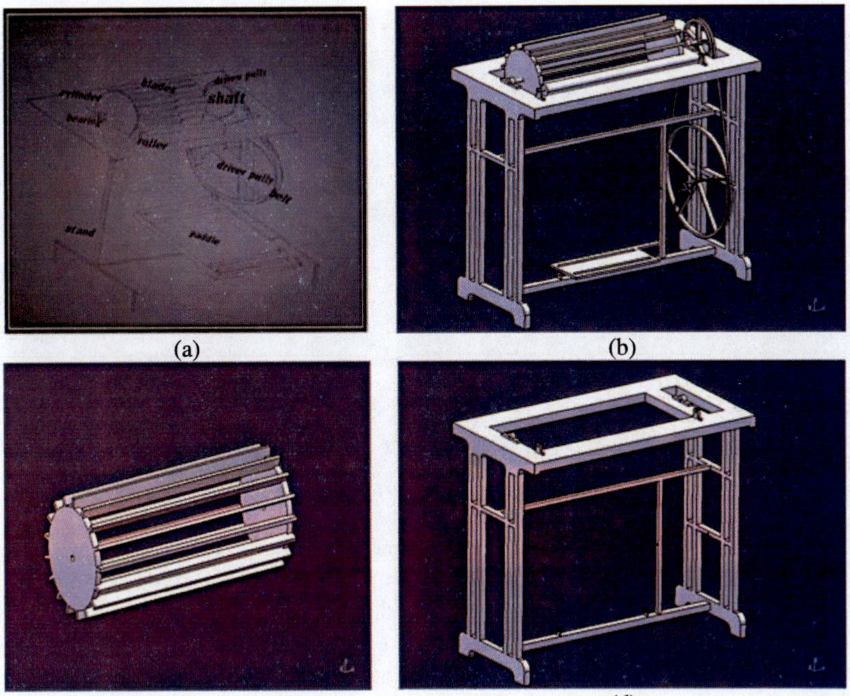

(a) (b) (c) (d)

Fig. 5 **a** Drawing of the manual decorticator, **b** isometric view of assembly drawn in Creo 2.0, **c** isometric view of the drum, **d** isometric view of the mounting table

If the diameter of the rollers is 100 mm then torque required will be 12 N m.

Let F_1 be the resistance offered by the leaves for decortication at each blade in drum then the torque required to rotate the drum will amount to $T_d = F_1 * r_d$. Hence if the tearing resistance of the leaf is assumed to be 400 N and the diameter of drum is 250 mm, then the torque required will be 50 N m. Hence total torque is 62 N m.

Now let us assume that the all the rollers are rotating at 100 rpm, then the power required can be calculated from the equation:

$$P = \frac{2\pi \mathrm{NT}_{\mathrm{net}}}{60} \quad (2)$$

This gives $P = 648$ W.

As we have neglected friction and other effects hence a slightly higher value of power input may be required. It can be noted that the power required is quite appreciable if a manually operated machine is to be developed (however it is also felt that the value of tearing resistance that is assumed above is on a higher side). Hence it is suggested to hybrid the system with a small motor (around 1 HP) that is attached

(a) (b)

Fig. 6 **a** Prototype of sisal fibre extracting machine, **b** sisal fibre obtained from the developed machine

to the pedal which will help to overcome the inertia effects and support during excess power requirements. In [12] authors have developed portable sisal fibre extractor for small scale farmers, however the motor used in the study is of 5HP considerably higher than the present proposal. The other components like shafts, bearings, belt drive, etc. can easily be designed by conventional methods in standard texts and hence it is not discussed in this study. The design may be further optimized to reduce the weight and cost.

Further modifications and innovative designs may be successively thought off to promote the above concept. Apart from extraction of fibre, portable units to make yarn from fibre, mechanization of the loom, etc. may be further explored. New research in collaboration between the agriculture sector and engineering sector must come up to provide sustainable solutions to the rural community in developing wider livelihood opportunities.

In Fig. 6a, a prototype of the manual sisal fibre extracting machine is presented. It can be observed that the frame of a sewing machine is utilized for providing the thrust. The operator will use his feet to provide the required torque. In Fig. 6b corresponding fibre obtained from the machine is shown. The design is further subjected to improvisations for better output.

5 Conclusions

Review of research in the field of natural fibre extraction and associated machinery is presented. A flow chart of the fibre extraction process is prepared and technical intervention needed to support artisans is discussed. Idea of a human powered mechanism suitable for natural fibre extraction from different biomass is coined. Preliminary design of mechanism is presented. Development of a prototype is carried out and tested with different types of fibres. Results are encouraging.

The research and development in this direction may also lead to sustainable handicraft product innovation.

References

1. Anonymous (2009) Report by Ministry of Textiles, Govt. of India
2. Sahu SC, Pattnaik SK, Dash SS, Dhal NK (2013) Fibre-yielding plant resources of Odisha and traditional fibre preparation knowledge—an overview. Indian J Nat Prod Resour 4(4):339–347
3. Das PK, Nag D, Debnath S, Nayak LK (2010) Machinery for extraction and traditional spinning of plant fibres. Indian J Trad Knowledge 9(2):386–393
4. Yusof Y, Adam A (2013) Review on PALF extraction machines for natural fibers. Adv Mater Res 781–784:2699–2703. https://doi.org/10.4028/www.scientific.net/AMR.781-784.2699
5. Gardner FP (2293056) (1942) Apparatus for abstracting and preparing fibers from fiber bearing plants
6. James M (US2722039) (1955) Apparatus for obtaining fibers from plant leaves
7. Leduc PJ (Humboldt, CA), Hill LG (Humboldt, CA), Kelly DH (Humboldt, CA), Stratton Mark A (Saskatoon, CA) (1998) Method for decorticating plant material
8. Adam A (2016) Design and development of pineapple leaf fiber machine 1(PALF M1) for small scale industry, Master's thesis, UTHM University, Malasiya
9. Saravanabavana D, Mohan Kumar GC (2011) Conceptual design features and eco methods for the extraction of natural fibers in the materialistic earth. Research into design—supporting sustainable product development, Research publishing, pp 461–471 ISBN: 978-981-08-7721-7
10. Jose S, Salim R, Ammayappan L (2016) An overview on production, properties, and value addition of pineapple leaf fibers (PALF). J Nat Fiber 13(3):362–373. https://doi.org/10.1080/15440478.2015.1029194
11. Naik RK, Dash RC, Goel AK (2013) Mechanical properties of sisal (A. *sisalana*) relevant to harvesting and fibre extraction. Int J Agricult Eng 6(2):423–426
12. Naik RK, Kar G (2020) Development and testing of a portable sisal fibre extractor for small scale growers Int J Curr Microbiol Appl Sci 9(10):2467–2472

Design Modification in Transplanting Mechanism of a Manual Rice Transplanter for Improved Performance

Deepak Mahapatra, Vikky Kumhar, and Manoj Verma

Abstract An improvement in the transplanting mechanism is suggested to reduce losses of the seedlings during the operation. A four bar mechanism incorporated with slight modification so as to control the speed in an appropriate way leading to better control and stability.

Keywords Rice transplanter · Four bar mechanism · Manually operated

Abbreviations

CFB Mechanism	Conventional four bar mechanism
QR Mechanism	Quick return mechanism
L_1–L_4	Link lengths of CFB mechanism
L_P, ψ	Coupler length parameters of CFB
L_0, δ	Parameters for QR
θ_0, θ_2	Input angles for CFB and QR respectively
ω_0, ω_2	Angular velocities of CFB and QR

1 Introduction

Agriculture is the backbone of Indian economy. In terms of timely farm operations, reduced losses and better management of inputs, agricultural mechanization holds

D. Mahapatra (✉)
College of Food Technology, IGKV, Raipur, CG, India
e-mail: medeepakmahapatra@gmail.com

V. Kumhar
SSTC, Bhilai, CG, India

M. Verma
UTD, CSVTU, Bhilai, CG, India

V. K. Gupta et al. (eds.), *Recent Advances in Machines and Mechanisms*, Lecture Notes in Mechanical Engineering, https://doi.org/10.1007/978-981-19-3716-3_50

the key to sustainable growth. However, there is a large gap in actual utilization of the available mechanization techniques only to a small fraction of farming community. One of the reasons is the poor living standards of farmers who cannot afford costlier equipment's and other, lack of awareness in using available techniques. A large part of the farming community still uses manual or animal power in farm mechanization and hence improved techniques of manual or animal power techniques are required to be developed.

Rice cultivation is a major portion of agricultural production and most of the associated processes are labor oriented. Depending on the availability of water, sowing of paddy/rice can be done directly or by using transplanting. Seeds are spread directly to a wet or dry field in direct seeding technique; however, seedlings are first cultivated in a seeded nursery and uprooted for transplantation either manually or automatically in transplanting method. In direct sowing method, the weeds are controlled by chemical application however mechanical techniques can be used in transplanting approach. The author deliberates a number of advantages of the transplanting approach [1].

2 Review of Literature

Mechanical rice transplantation goes back to 1955, when Taiwan created and tested a gravity type-hand type operational transplanter. In 1956, China tested the first manually controlled six row transplanter. Despite the fact that these robots produced roughly twice as much as manual transplanting, they failed owing to uneven planting and a high labor need [2]. Miura [3] has reported the development of rice transplanters in Japan. In the early seventies, a manual rice transplanter with trade name, 'Annapoorna' was developed. It had 10 rows and root washed seedlings were used. This machine can cover about 0.16 h/day [4].

Kohli and Agrawal [4] offered an article for the optimization and dynamic analysis of transplanting unit of a paddy transplanter. Two types of mechanisms are worth noting in order to acquire a cycle of planting operation while keeping in mind the needed route and simplification of the mechanism: a four bar mechanism and a six-bar mechanism, which is a modified version of the four bar itself. The six-bar mechanism has one link that acts as a pusher and is best suited for mat seedlings. However, no grasping was supplied for traditional type seedlings; therefore this mechanism is frequently dismissed.

According to Anonymous [2] and Thomas [5], the majority of power controlled transplanter planting devices is four bar linkage crank and rocker mechanisms. The seedlings are picked from tray by a planting finger and sowed in the soil in an upright position. To place the seedlings efficiently in position it is necessary to guide the finger through a desired path (optimized path of finger) in a cyclic manner so that it must not disturb the seedlings while return. Hence a detailed kinematic analysis of the planting mechanisms will be worth in improving the new designs.

The design synthesis of mechanism of rice transplanter must include the determination of link lengths, coupler extension and orientation of finger and fixed link.

In [5] Thomas has synthesized a mechanism for transplanter by assessing numerous trial values for the various parameters to arrive at optimized path required for transplanting operation. However, it is assessed that the velocity and acceleration components (point no. iii, v and ix in [4]) are not addressed satisfactorily in the reports [4, 5]. The present work is an effort to bridge this gap by introducing a quick return mechanism for the input crank so as to regulate the speed in the forward and backward motions of the finger in the mechanism.

3 Materials and Methods

A cycle of planting operation includes picking of a desired number of seedlings from a tray by mechanical fingers then sowing them into a puddled field planting them hill to hill and row to row in precision spacing at a predetermined depth and nearly upright position, subsequently the finger must return to its original path without effecting the tray of other seedlings, thus completing the cycle. A machine fitted with such mechanism may be move continuously or intermittently, depending on its design and requirement [6].

In conventional transplanter design a planar four bar linkage owing to its simplicity, is used for obtaining the desired path of the finger as discussed above. The input is given to the crank in a continuous and rotational manner through a wheel. As a result, the mechanism has one degree of freedom, and a coupler point capable of constructing a loop can be included. The coupler point will be where the planting finger will be connected. A detailed study of working procedure may be found in [4, 5].

The present work study assumes summation of two linkages, the conventional four bar (CFB) mechanism [5] and a quick return (QR) mechanism (which is also an inversion of the four bar linkage) that are connected in such a way that the QR mechanism's output becomes input to the CFB mechanism by making the CFB crank a slotted link (Fig. 1). Thus the angular velocity of the crank of CFB is controlled using a constant angular velocity input to the QR mechanism. The parameters that define the CFB mechanism are selected from [5] while the additional parameters for QR mechanism are:

Parameters for QR mechanism (Fig. 1).

- Length of crank, L_0
- Position of the axis of rotation of crank, δ
- Angular velocity of crank, ω_0

Initially the speed of travel is neglected as it is assumed to be very small as compared to planting velocity. The angle of inclination of the fixed link is also assumed to be zero initially. For brevity, the synthesis of CFB mechanism has been adopted and the values of parameters L_1–L_4 and L_F and the extension angle (ψ) has been considered from the literature [5]. The angular velocity of the crank of CFB

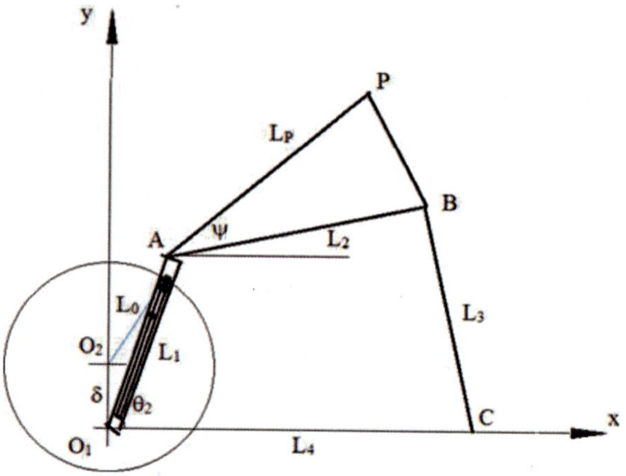

Fig. 1 A quick return mechanism provides input to four bar mechanism in rice transplanter mechanism

mechanism (ω_1) is a variable that needs to be calculated with respect to the input angular velocity (ω_0) of the QR mechanism.

A Matlab code has been developed to simulate and plot the path of the coupler point *P* and subsequently its velocity in case of CFB Mechanism. Figure 2a represents the simulation of the CFB mechanism while Fig. 2b shows the coupler curve of the finger with the velocity of finger at given points. It can be observed that the velocity of the finger in the forward motion, when it picks the seedling, is very high resulting in severe losses.

Here it is proposed to introduce a quick return mechanism as shown in Fig. 3 to regulate the speed of the finger during planting operation and also increase the speed

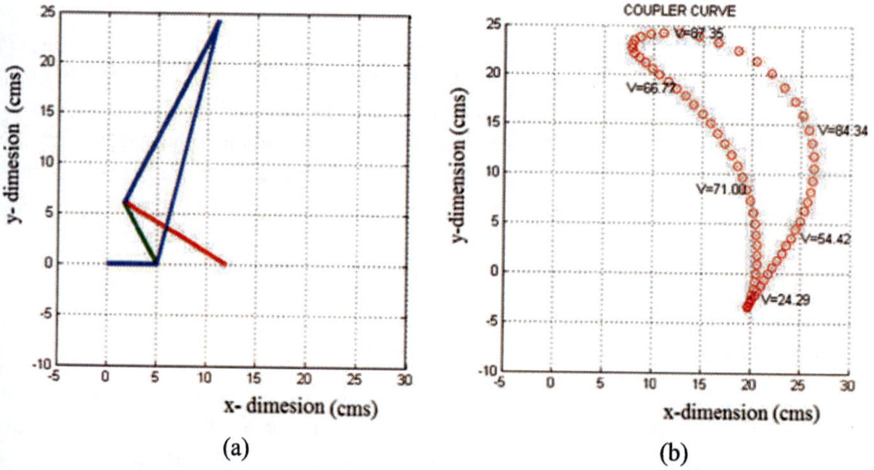

Fig. 2 **a** Simulation of CFB mechanism in Matlab, **b** coupler curve of the path of finger with instantaneous velocit

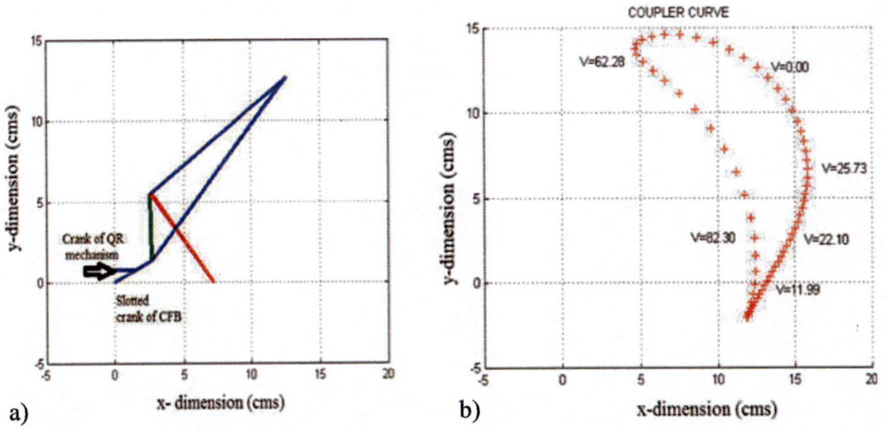

Fig. 3 **a** Simulation of modified mechanism in Matlab, **b** Coupler curve of the path of finger with instantaneous velocity

when it return back to complete the cycle. The crank of CFB mechanism is replaced with a slotted link of same dimension.

The length of the slot is kept constant as 80% of the length of the link. A slider is made to slide on this slotted link whose movement is thus governed by the crank of QR mechanism. Mathematically the angular velocity of crank of CFB can be related easily with the angular velocity of the QR crank. After knowing variation of the angular velocity of CFB crank, the velocity of the finger attached to it can be easily calculated. The composite mechanism and resultant velocity of finger are shown in Fig. 3a and b.

4 Results and Discussion

It can be observed by comparing Fig. 3b with Fig. 2b that the speed of the finger before catching the seedling has been considerably reduced in the modified transplanting mechanism.

Figure 4 shows the tip velocity with crank angle as input for CFB mechanism and modified mechanism with QR input. It can be observed from the figure that using the QR input, the transplanting speed during the downward movement is reduced appreciably and it is increased during the return motion. This improvement will certainly aid in reducing the seedling loss and increase the efficiency of the transplanter mechanism. In this context the appropriate size of the link lengths of QR mechanism (crank and its axis with respect to the axis of the crank of CFB mechanism) must be calculated in an optimum way. The optimum dimensions of the QR mechanism can be synthesized using a trial and error method however that discussion is not described here for brevity.

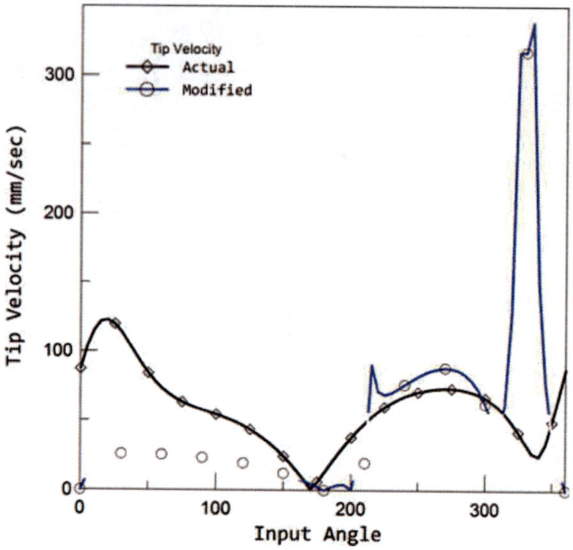

Fig. 4 Tip velocity (finger velocity) with respect to crank rotation

A prototype of the above modified form of transplanting mechanism to obtain smooth transmission of seedlings was fabricated to test the effectiveness of the theoretical study. It was then tested in laboratory and then in farm of BRSM CAET and RS, IGKV, Mungeli CG (Fig. 5).

5 Conclusions

In this work it is proposed to include a quick return input link that may control the speed of the crank of the transplanter thereby regulating the speed of the finger. The crank of transplanter machine is slotted so as to contain the slider of quick return input.

It is observed that using an additional link to regulate the speed has shown remarkable control in speed of the finger. The speed of finger is considerably reduced during the downward motion of transplanting operation while at the same time it return with a high speed to the initial position. In this process a synthesis of the new links is also approached by trial and error method. A relationship between the input crack speed and finger speed has also been developed.

As a future scope the following points can be suggested. Optimum synthesis using precision techniques may be undertaken to calculate the link parameters of quick return input mechanism. An elaborate evaluation of the performance in real field conditions with a sturdy/ rigid transplanter is also needed in future to validate the results obtain in this work with respect to efficiency in handling the seedlings.

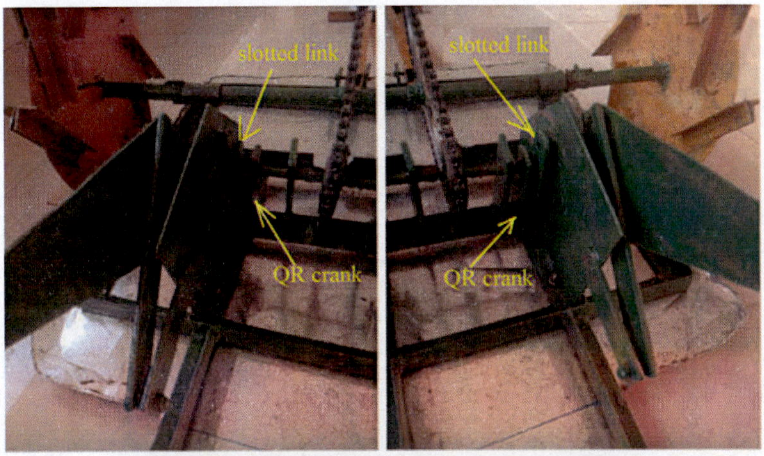

(a)

(b)

Fig. 5 **a** Prototype of modified transplanter, **b** testing in real field conditions

Acknowledgements The authors are enthusiastically thankful to Chhattisgarh Swami Vivekanand Technical University, Bhilai (TEQIP III, Project No. CSVTU/CRP/TEQIP-III/32) for providing financial support and of BRSM CAET and RS, IGKV, Mungeli CG for providing infrastructural support for carrying out the present research work.

References

1. Dixit A, Khurana R, Singh J, Singh G (2007) Comparative performance of different paddy transplanter. Agric Rev 28(4):262–269
2. Anonymous (1979) Latest Japanese Paddy Transplanter. Yanmar Argriculture Equipment Company, Ltd., Osaka, Japan.
3. Miura (1966) Rice transplanting machine. Japan Agricult Res Quarterly 1(3):19–22
4. Mahapatra AP (1973) Transplanter, Indian Farm 23(8):11–13

5. Kohli SS, Agrawal VP (1993) Analysis of a four-bar linkage for a proposed transplanter. Trans Am Soc Agric Eng 36(2):317–320
6. Thomas (2002). Development of a mechanism for transplanting rice seedlings. Mech Mach Theory 37:395–410

Miscellaneous

Gear Meshing Visualization for Effective Education Using MechAnalyzer Software

Nilabro Saha, Rajeevlochana G. Chittawadigi, and Subir Kumar Saha

Abstract Teaching and learning of Theory of Machines can be made more effective with the usage of physical or virtual models of mechanical components. Physical models require material, manufacturing, and assembly, etc., whereas virtual models only require a software to visualize the model. Animation of the same will improve the understanding further. Out of the many such visualization software, the second and third authors have been involved with the development of MechAnalyzer software. Many of the existing modules in the software have already been reported in the literature elsewhere. In this paper, the new developments related to the drawing of gear tooth profiles, the animation of meshing and an interface to solve typical textbook problems have been reported. In addition, implementation of gears with complicated working principles, such as bevel gears, planetary gears, gear trains, etc. is also reported. All these have easy to use interface which allows teachers and students to vary their input parameters and observe the model and the animation of its working.

Keywords Gears · Animation · Education · Mechanisms · Bevel gears · Gear tooth profile

1 Introduction

A comprehensive understanding of different mechanisms and linkages in a theory of machines course is the foundation for advanced subjects. However, in a classroom setting, it is difficult for teachers to convey a visual understanding of the motions

N. Saha
Department of Mechanical Engineering, National Institute of Technology Durgapur, Durgapur, India

R. G. Chittawadigi (✉)
Department of Mechanical Engineering, Amrita School of Engineering, Bengaluru, Amrita Vishwa Vidyapeetham, Bengaluru, India
e-mail: rg_chittawadigi@blr.amrita.edu

S. K. Saha
Department of Mechanical Engineering, Indian Institute of Technology Delhi, New Delhi, India

V. K. Gupta et al. (eds.), *Recent Advances in Machines and Mechanisms*, Lecture Notes in Mechanical Engineering, https://doi.org/10.1007/978-981-19-3716-3_51

of different mechanisms to the students. Such classroom teaching can be enhanced by the use of computer simulations. Several such software have been reviewed and compared with MechAnalyzer, a software developed by the second and third authors, earlier versions of which have been reported in [1].

MechAnalyzer (hereon referred to as MA) is a 3D model-based learning software, developed as a Windows desktop application, with 5 modules: the *Core Module* contains 16 pre-loaded mechanisms with variable input parameters, animation, and kinematic analysis plots; the velocity and acceleration diagram (VAD), instantaneous center (IC), and static force analysis (SFA) modules cater to kinematic analysis using graphical methods; the *Cam Module* focuses on the cam-follower mechanism; the *Gear Profile Module* is to learn about gear profile curves; and the *Gear Meshing Module* allows the user to animate mating involute gears and solve basic textbook problems in the software.

There exist commercial programs for simulation and analysis of gear trains, e.g., *KISSsoft* [6], *GearTeq* and *GearTrax* [5], and *Gears App* [4]. Other programs with noteworthy teaching value include *Gear Template Generator Program* [3], *Gear Generator* [2], and *Planet Gear Simulator* [8]. In MA, the newly added simple and compound gear trains, the planetary gearset, the standard open differential gearset, and the *Gear Meshing Module* show the novelty of MA more on the educational value than in the research content.

2 The Simple and Compound Gear Trains

A parallel shaft gear train comprises meshing gears mounted on multiple parallel shafts. Here in MA, helical gears have been implemented. A 0° helix angle corresponds to a spur gear. The *GearMechanism* interface is shown in Fig. 1i–iv. In the simple gear train (loaded by default), the intermediate gears (*idler gears*) only determine the distance through which motion is transmitted (observe by tweaking the number of teeth), or change the direction of output motion. In MA, the two idler gears ensure that the sense of output motion is always opposite to the input. The compound train in MA has the two intermediate gears which are mounted on a single shaft. Their effect is easily seen by varying their teeth count.

3 The Planetary Gearset

The planetary gearset in MA aims for an understanding of the planetary gear unit in an automatic transmission system. It has four components: a sun gear, a ring gear, a planet gear, and a planet carrier. The user can vary the number of teeth in the planets and the sun. A particular velocity ratio may be obtained by designating one of the components as input and one as output, while fixing a third. Four modes of operation

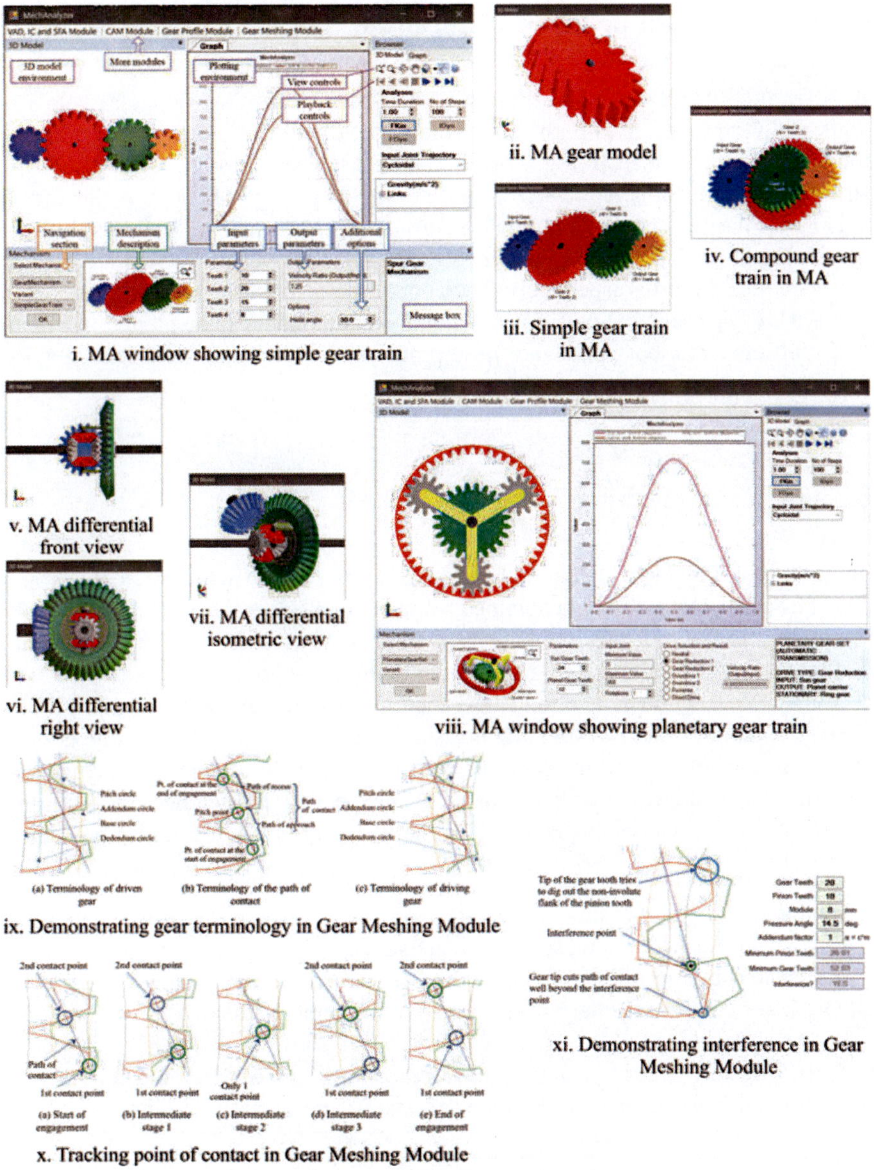

Fig. 1 New gear trains and the new gear meshing module in MA

are possible: neutral, reduction drive, overdrive, and reverse drive. Figure 1viii shows the MA window with the planetary gearset loaded. The angular velocities of the components about their own axes is governed by the Willis equation [7].

4 The Standard or Open Differential

A differential gearbox lets the driving wheels rotate at different angular velocities while delivering power to both. The ordinary automotive differential with straight tooth bevel is presented in Fig. 1v–vii. The angular velocities of the left and right side gears are given by $\omega_l = -2\lambda\omega_c$ and $\omega_r = 2(1-\lambda)\omega_c$, where ω_c is the angular velocity of the crown gear, and λ is the *turn direction parameter* which is available as user input. $\lambda = 0, 0.5,$ and 1 correspond to a sharp left turn, forward motion, and a sharp right turn, respectively. The angular velocity of the spider gears is $\omega_s = (2\lambda - 1)\omega_c$. The kinematic parameters of each component can be plotted in MA for different input motion types, and these plots can also be exported in the required format.

5 The Gear Meshing Module

The gear meshing module (reported earlier in [1]) is a tool to demonstrate the properties of involute gears. The data obtained from its three side-pane tabs and the plotting section can be put to numerous uses. With the new feature additions in MA Version 6, a few ways to use this module have been highlighted here.

Being able to animate and to change the input parameters quickly makes it convenient to demonstrate how the features shown in Fig. 1i behave when one or more inputs change. In Fig. 1ii, the point of contact is tracked through one engagement period. It moves along the path of contact, as is expected. Interference can be demonstrated in MA, as shown in Fig. 1iii. The methods to remedy it can also be demonstrated thus:

- By *increasing the pressure angle* and noting the values in the *Results* tab, it can be shown how this method relieves interference.
- Given the maximum allowable addendum in the *Results* tab, the suitable *addendum coefficient* to prevent interference can be calculated. Reducing the addendum is known as *stubbing*.
- *Increasing the number of teeth* can decrease interference. That the contact ratio also increases is an added advantage and can be shown in this module.

6 Conclusion

Earlier papers in the literature have elucidated how MA has been designed with the goal of aiding in a holistic study of theory of machines. In this paper, MA's application in learning about gears and gear trains—machine components that are ubiquitous in this field—has been highlighted. The MA gear meshing module and the gear trains in MA core module have been designed to aid in an in-depth study of the subject. It is hoped that a skilled teacher and an inquisitive student can draw far more value from MechAnalyzer than has been described in this paper.

References

1. Dikshithaa R, Jain S, Swaminathan J, Chittawadigi RG, Saha SK (2018) MechAnalyzer: software to teach kinematics concepts related to cams, gears, and instantaneous center. In: Mechanism and machine science. Springer, pp 135–149
2. Gear generator. Available: https://geargenerator.com/
3. Gear template generator program. Available: https://woodgears.ca/gear/
4. Gears app by drivetrain hub. Available: https://drivetrainhub.com/gears/
5. Geartrax. Available: https://camnetics.com/geartrax/index.htm
6. Kisssoft. Available: https://www.kisssoft.com/en
7. Muller HW, Mannhardt WG, Glover JH (1982) Epicyclic drive trains: analysis, synthesis, and applications. Wayne State University Press
8. Planet gear simulator. Available: http://www.thecatalystis.com/gears/

Role of Profile Parameters on the Sensitivity of Cantilever Sensor: A Numerical Analysis

Shivanku Chauhan and Mohd. Zahid Ansari

Abstract Presented work is an effort to find the effective way to improve the sensitivity of the cantilever sensing devices utilizing different profile modification technique. The effect of size scaling, mass reduction holes (MRH), and stepping the profile is observed on the sensitivity of the cantilever sensor. A rectangular profile of the cantilever is used to analyze the three effects on the resonant frequency of the cantilever. The numerical simulation is performed to ascertain the optimum number of MRH and optimized position of them with respect to the maximum resonant frequency of the cantilever. The analysis of size scaling, MRH size effect, and stepped profile effect shows that these techniques proved beneficial to enhance the sensitivity of the cantilever sensor. The size scaling technique is found more effective than the two other.

Keywords Cantilever · Sensitivity · Size scaling · Mass reduction holes

1 Introduction

Cantilever sensing devices has the plenty scope in the various field such as mass sensing, detection of explosive, chemicals, myopathy, and other online monitoring application [1–5]. The cantilever sensor can be operated in the static and dynamic mode basically. For sensing, the static mode of operation employs its deflection because of the differential surface stresses while the dynamic mode of operation employs the change in its resonant frequency because of some external means. Dynamic mode of operation of the cantilever sensor is extensively used in the online monitoring [6–9]. The mass sensing can be done by a cantilever sensor by detecting its resonant frequency shift with and without the mass applied on it. Many efforts have been made to enhance the sensitivity and limit of detection of these cantilever

S. Chauhan · Mohd. Zahid Ansari (✉)
MEMS and Microfluidics Lab, Discipline of Mechanical Engineering, PDPM Indian Institute of Information Technology, Design and Manufacturing, Airport Road, Jabalpur, MP 482005, India
e-mail: 1823602@iiitdmj.ac.in

V. K. Gupta et al. (eds.), *Recent Advances in Machines and Mechanisms*, Lecture Notes in Mechanical Engineering, https://doi.org/10.1007/978-981-19-3716-3_52

sensors. Aboelkassem et al. [10] did the effort for sensitivity enhancement of a biomass cantilever sensor by utilizing the higher mode of its vibration. An impedance-based model of piezoelectric cantilever sensor connected with an impedance circuit was proposed by Zhao et al. [11]. It is an effective way for the sensitivity enhancement. Weigert et al. [12] showed the effect of dimensions of cantilever on its resonant frequency which directly affects its sensing abilities. Canavese et al. [13] developed a more sensitive design of a cantilever sensor by making a hole at its free end. This model significantly improved the resonant frequency of the cantilever compared to the base model and hence provided the higher sensitivity. Zhao et al. [14] found the effect of the shape of the cantilever and the mode of vibration on the accuracy of the fluid property measurement. For the sensitivity improvement Gao et al. [15] suggested an optimized profile cantilever sensor in which the profile of the cantilever was made stepped. This is an effective method for the sensitivity enhancement without using the cantilever sensor at higher mode of vibration.

Thus the cantilever design should have good deflection abilities (decreased bending stiffness) with sufficient frequency characteristics and the geometric dimension reduction and configuration modification is a promising sensitivity improving method. This presented work provides an effective mean to improve the sensitivity of the cantilever sensor by different profile modification. Size scaling, mass reduction holes and stepping method are analyzed to maximize the resonant frequency of the cantilever which directly affects its sensitivity.

2 Numerical Modeling and Analysis

The numerical modeling and analysis in this work is done in two sections basically. The modeling is inspired from the fact that the hole at the free end of a cantilever reduced the vibrating mass of the cantilever and in turns enhanced its resonant frequency. The model analysis is performed in Ansys Workbench 18.1 and all the cantilevers are discretized using solid 45 type of elements. As the cantilevers profile is symmetric and straight, so the simple line meshing is done for all the models. Also, the mesh convergence test is performed by successively refining the mesh. This convergence test confirms the accuracy of the used numerical model. In the first analysis a simple steel cantilever of dimensions 25 mm $\times$ 5 mm $\times$ 1 mm is modeled and analyzed while in the second analysis a steel cantilever with the piezoelectric patch is modeled with different dimension as given in the Fig. 2. Both the steel and PZT-5*H* are considered linear isotropic in the modeling and the properties of these material utilize in the modeling and analyses are given in the Table 1. The material damping is ignored in all the cantilever models for the simplicity of the analysis and it does not affect the intended purpose of the work presented.

The numerical modal analysis can provide the Eigen frequencies of the particular structure corresponding to the different mode of vibration. In the presented work, the resonant frequencies corresponding to the first six mode of the vibration are obtained

Table 1 Material Properties

Materials	Density (kg/m^3)	Young's modulus (MPa)	Poisson's ratio
Steel	7800	2 × 105	0.31
PZT-5H	7500	6.4 × 104	0.30

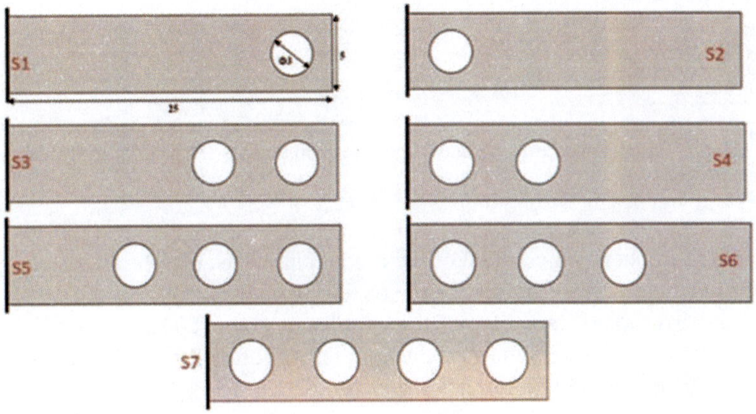

Fig. 1 Rectangular cantilever profiles with different number of MRH at different position

through the model analysis and the first mode of the vibration is used here for the study.

2.1 *Analysis-I*

In the first analyses, 1–4 mass reduction hole (MRH) are provided over the rectangular profile of the cantilever during the modeling of the sensor and the numerical simulation is performed to discover the optimum number of holes and optimized position of the MRH with respect to the maximum resonant frequency of the cantilever. The diameter of each MRH is kept 3 mm. Figure 1 shows the models of the cantilever with different number of MRH over their profile.

2.2 *Analysis-II*

In the second phase of the numerical analyses, the steel-piezo cantilever is modeled and simulated involving the piezo patch stacked over it. For the interfacing between the PZT-5H patch and steel plate, perfectly bonded connection is provided at the contact region of both the material. This study analyzes three different methods of

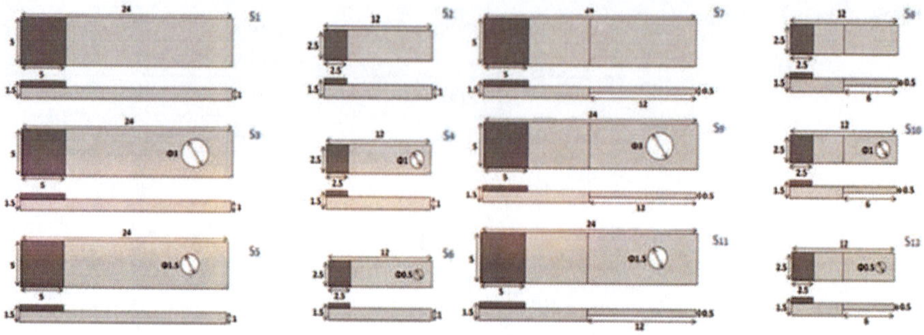

Fig. 2 Steel-piezo rectangular (1–6) and stepped-rectangular (7–12) cantilever profiles with size scaling and MRH

the resonant frequency increment of the model named as; (1) The size scaling effect on the fundamental resonant frequency of the cantilever, (2) Steeping effect on the fundamental resonant frequency of the cantilever, and (3) The MRH effect on the fundamental resonant frequency of the cantilever. Total 12 profile of the cantilever are modeled shown in Fig. 2.

3 Results and Discussion

The measurement sensitivity of a cantilever sensor generally improves on its fundamental resonant frequency increment. The higher resonant frequency of the cantilever improves its signal to noise ratio which directly improves the sensitivity. So, it is important to design a cantilever sensor with the higher resonant frequency and without affecting its deflection abilities so much. This study is an effort to improve the sensitivity of the cantilever sensor and the two finite element analyses are done to get the higher sensitivity model of the cantilever sensor.

3.1 Analysis-I

The optimum number of MRH holes and their optimized positions with respect to the maximum resonant frequency of the rectangular cantilever are found in this analysis. The result obtained from the model analysis for the different model of the cantilever is shown in the Table 2.

It is clear from the Fig. 3 that the MRH near to the fixed end of the cantilever results in the decrement in the resonant frequency because of the stiffness decrement. Although, the hole also reduces the mass of the cantilever but the overall ratio of the stiffness and mass decreases which in-turn decreased the resonant frequency of the cantilever. This analysis suggests to provide the MRH near to the free end of the

Table 2 Modal analysis results for rectangular profile cantilevers with different MRHs on profile

Model	Dimensions ($L \times B \times T$) mm	Number of MRH and position	f (Hz)
S1	25 × 5 × 1	1-Near far end	1405
S2	25 × 5 × 1	1-Near fixed end	1131
S3	25 × 5 × 1	2-Near far end	1416
S4	25 × 5 × 1	2-Near fixed end	1080
S5	25 × 5 × 1	3-Near far end	1330
S6	25 × 5 × 1	3-Near fixed end	1092
S7	25 × 5 × 1	4-Throughout span	1168

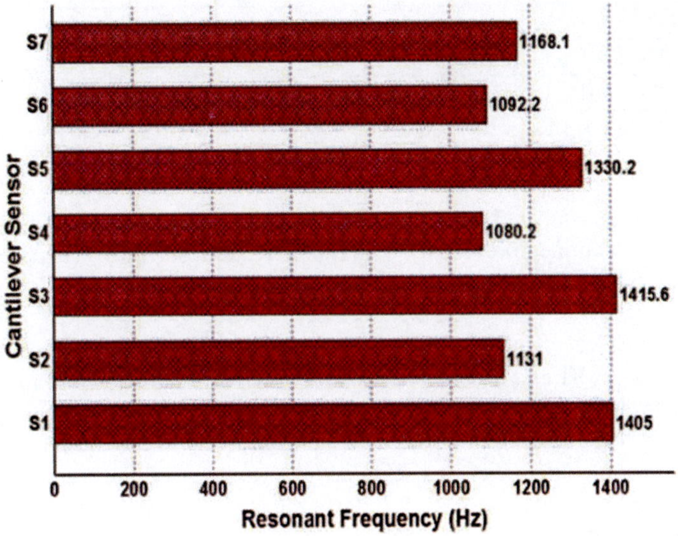

Fig. 3 Comparison of the resonant frequency of the different MRHs cantilever

cantilever because this design increase the overall ratio of stiffness to mass hence enhanced the resonant frequency. Further, 2 MRH at the free end side are optimum for the maximum resonant frequency in the present model and more increment in the number of MRH decreases the resonant frequency.

3.2 Analysis-II

The results of analysis-1 show that the optimum position of MRH is at the free of the cantilever. Considering this fact, further investigation is done about the effect of size scaling, steeping and the MRH size on the fundamental resonant frequency of the

Table 3 Modal analysis results for Steel-piezo cantilever sensor

Model	Dimensions ($L \times B \times T$) mm	MRH diameter (mm)	Step dimension ($L \times T$) mm	f (Hz)
S1	24 × 5 × 1	–	–	1691
S2	12 × 2.5 × 1	–	–	6628
S3	24 × 5 × 1	φ3	–	1824
S4	12 × 2.5 × 1	φ1	–	6845
S5	24 × 5 × 1	φ1.5	–	1721
S6	12 × 2.5 × 1	φ0.5	–	6679
S7	24 × 5 × 1	–	12 × 0.5	1886
S8	12 × 2.5 × 1	–	6 × 0.5	7330
S9	24 × 5 × 1	φ3	12 × 0.5	2032
S10	12 × 2.5 × 1	φ1	6 × 0.5	7578
S11	24 × 5 × 1	φ1.5	12 × 0.5	1925
S12	12 × 2.5 × 1	φ0.5	6 × 0.5	7405

cantilever. Model analysis is done for the total 12 profile of the cantilever involving the three effects mentioned previously. The fundamental resonant frequencies obtained for all the models of cantilever are shown in Table 3.

Size Scaling Effect

Cantilever model S1 and S7 are two models which are scale down to model S2 and S8, respectively, by 0.5%. Scale down the cantilever size by half (scaling factor = 0.5) shows a huge increment in the resonant frequency of the model. For a simple rectangular profile cantilever this scale down has increased the resonant frequency by 292% approximately (S1–S2). Similarly the resonant frequency has increase by 289% approximately in case of stepped profile cantilever (S7–S8). Figure 4 shows the size scaling effect on the resonant frequency of the cantilever.

Steeping Effect

The six different profiles of the cantilever S1, S2, S3, S4, S5, and S6 are made stepped to S6, S7, S8, S9, S10, S11, and S12, respectively. This Steeping of the cantilever profile also increased the resonant frequency of the cantilever models but the increment is lesser compared to the scaling effect. This increment in the resonant frequency of the models because of the steeping the profile is approximately 10–11% for each model. Figure 5 shows the stepping effect on the resonant frequency of the cantilever.

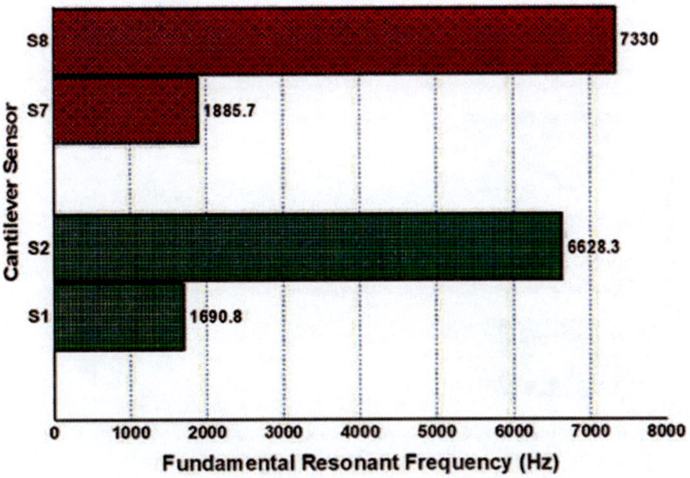

Fig. 4 Size scaling effect on the fundamental resonant frequency of the steel-piezo cantilever

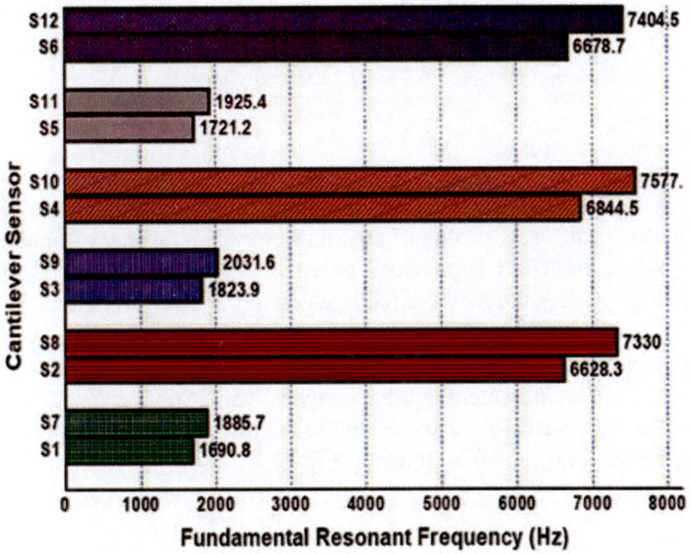

Fig. 5 Steeping effect on the fundamental resonant frequency of the steel-piezo cantilever

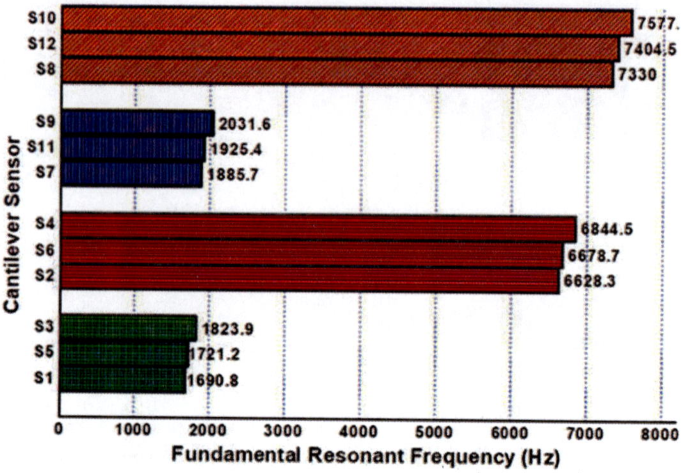

Fig. 6 MRH effect on the fundamental resonant frequency of the steel-piezo cantilever

The increment in the resonant frequency is observed because of the steeping effect is; S1 → S7 = 11.53%, S2 → S8 = 10.59%, S3 → S9 = 11.39%, S4 → S10 = 10.71%, S5 → S11 = 11.86%, S6 → S12 = 10.86%.

MRH Effect

The effect of MRH is already analyzed in the Analysis-I which suggested to provide the MRH at free end of the cantilever. Here basically different MRH are modeled at the free end of different profile of the cantilever. The resonant frequency is also increasing when the MRH is provided near to the free end of the cantilever. The increment on providing the MRH at the free end of the cantilevers is as; S1 → S5 = 1.8, S2 → S6 = 0.8, S7 → S11 = 2.1, S8 → S12 = 1.0%. Further, the increment in the diameter of the MRH by two times again increases the resonant frequency of the each model. This increment is as; S5 → S3 = 5.9, S6 → S4 = 2.4, S11 → S9 = 5.5, S12 → S10 = 2.3%. The increment in the resonant frequency because of the MRH effect is as follows and is shown in Fig. 6.

4 Conclusion

The role of the profile parameters to decide the resonant frequency of the cantilever sensor is studied in the work. The resonant frequency of the cantilever is directly related to its sensitivity. Providing MRH near to the free end of the cantilever is proved beneficial in respect to the resonant frequency increment and the 2-MRH at the free end are found optimum in the presented model. Scaling down the cantilever size by a factor of 0.5 increases the resonant frequency by 292% and 289% for the steel-piezo rectangular and stepped-rectangular cantilever, respectively. Stepping the cantilever profile has not much affect the resonant frequency compared to the scaling effect and the increment in the resonant frequency is only 10 to 11%. Increasing the MRH size improves the resonant frequency of each model a bit more.

References

1. Zhao J, Zhang Y, Gao R, Liu S (2015) A new sensitivity improving approach for mass sensors through integrated optimization of both cantilever surface profile and cross section. SensActuators B Chem 206:343–350
2. Korayem AK, Taghizade M, Korayem MH (2018) Sensitivity analysis of surface topography using the submerged non uniform piezoelectric micro cantilever in liquid by considering inter atomic force interaction. J Mech Sci Technol 32(5):2201–2207
3. Park Y, Kim S, Park J (2018) Mass measurement method based on resonance frequency of the capacitive thin membrane sensor. J Mech Sci Technol 32(7):3263–3271
4. Na Y, Lee H, Park J (2018) A study on piezoelectric energy harvester using kinetic energy of ocean. J Mech Sci Technol 32(10):4747–4755
5. Johnson BN, Mutharasan R (2012) Biosensing using dynamic-mode cantilever sensors: a review. Biosens Bioelectron 32(1):1–18
6. Chauhan S, Ansari MZ, Sahu S, Husain A (2021) sensitivity improvement of piezoelectric mass sensing cantilevers through profile optimization. Adv Manuf Ind Eng 1007–1014
7. Gruszkiewicz MS, Rother G, Wesolowski DJ, Cole DR, Wallacher D (2012) Direct measurements of pore fluid density by vibrating tube densimetry. Langmuir 28:5070–5078
8. Khan MF, Schmid S, Larsen PE, Davis ZJ, Yan W, Stenby EH, Boisen A (2013) Online measurement of mass density and viscosity of pL fluid samples with suspended microchannel resonator. Sens Actuat B Chem 185:456–461
9. Burg TP, Manalis SR (2003) Suspended microchannel resonators for biomolecular detection. Appl Phys Lett 83:2698–2700
10. Aboelkassem Y, Nayfeh AH, Ghommem M (2010) Bio-mass sensor using an electrostatically actuated microcantilever in a vacuum microchannel. Microsyst Technol 16(10):1749–1755
11. Zhao J, Liu S, Huang Y, Gao R (2017) A new impedance based sensitivity model of piezoelectric resonant cantilever sensor. World congress on advances in structural engineering and mechanics (ASME17). Korea
12. Weigert S, Dreier M, Hegner M (1996) Frequency shifts of cantilevers vibrating in various media. Appl Phys Lett 69(19):2834–2836
13. Canavese G, Ricci A, Gazzadi GC, Ferrante I, Mura A, Marasso SL, Ricciardi C (2016) Resonating behaviour of nanomachined holed microcantilevers. Sci Rep 5:17837
14. Zhao L, Hu Y, Wang T, Ding J, Liu X, Zhao Y, Jiang Z (2016) A MEMS resonant sensor to measure fluid density and viscosity under flexural and torsional vibrating modes. Sensors 16(830):1–15
15. Gao R, Huang Y, Wen X, Zhao J, Liu S (2017) Method to further improve sensitivity for high order vibration mode mass sensors with stepped cantilevers. Sensors 17(14):4405–4411

Position Control Using a Physics-Based Model for Biomimetic Underwater Propulsor Actuated by IPMC

Ankur Gupta, Satyendra K. Prajapati, and Sujoy Mukherjee

Abstract In this paper, an effort is made to provide a better position control model using the physics-based model for biomimetic underwater propulsor actuated by ionic polymer-metal composite (IPMC). The biomimetic underwater propulsor considering the body caudal fin-type locomotion is undertaken where the oscillating fin tail of the original fish is replaced by a fin tail comprising an active IPMC beam. The supported cantilevered IPMC beam model is analyzed and the relations for displacement and velocity are studied using physics-based modeling in a fluidic environment. The robotic swimming velocity response due to the active IPMC tail fin deflection under applied electric potential has been observed. Considering the Laplace transform of the mathematical model, the transfer function relating the tail tip deflection and input voltage has been generated and proportional, integral and derivative (PID) closed-loop controller has been developed to control the deflection from IPMC in an underwater environment. The controlled and efficient desired displacement response of 1 mm through simulations was achieved. The PID gains are tuned using the Ziegler–Nichols tuning method to make the output response quick and robust.

Keywords Position control · Physics-based model · PID control · IPMC · Underwater propulsor

1 Introduction

In the fast-growing field of biomimetic underwater swimming robots, the overview of fish swimming [1] with a variety of morphological and structural features for underwater propulsion with efficiency and maneuverability has been presented. After many years of evolution, the aquatic animals mainly the fish are considered to be

S. K. Prajapati · S. Mukherjee
Department of Mechanical Engineering, PDPM Indian Institute of Information Technology, Design and Manufacturing, Dumna Airport Road, Jabalpur, M.P 482005, India

A. Gupta (✉)
Department of Mechanical Engineering, Lingaya's Vidyapeeth, Nachauli, Jasana Road, Faridabad, Haryana 121002, India
e-mail: 1613601@iiitdmj.ac.in

V. K. Gupta et al. (eds.), *Recent Advances in Machines and Mechanisms*, Lecture Notes in Mechanical Engineering, https://doi.org/10.1007/978-981-19-3716-3_53

the best underwater swimmers [2] and based on biomimetics the aquatic robots are developed and their kinematics and hydrodynamics have been studied. Many underwater swimming robots have been developed using motors as actuators but are heavier in size, noisy and need large power so, in recent years researchers are working on small size underwater robots actuated by smart materials. Ionic polymer–metal composite (IPMC) is the smart material that comes under the category of flexible electroactive polymers (EAP) which shows actuation characteristics under a low applied voltage and can be used as artificial muscle for underwater robotics [3]. The working principle of IPMC is described as, when a voltage is applied to the material there is the movement of water molecules with hydrated cations and the associated electromechanical interaction causes the bending effect which is explored as actuator and reported as IPMC-based underwater robotic fish by various groups [4–6]. Most of the work [7–11] investigated the bending response of IPMC through Euler Bernoulli beam modeling, fabrication of IPMC-based underwater robot for generating complex flapping motion, molding IPMC actuator into boot material to get complex deformation and the ability of IPMC to generate various locomotion types for biomimetic robots but the fundamental physics of IPMC bending that covers the movement of ions and also the investigation on free-swimming robot actuated by IPMC based on physics model is missing.

The ionic polymer–metal composite as an actuator has been used by various researchers for various applications like microgrippers, manipulators, propulsors, etc. [12–15] and its control model also has been developed [16–18]. The control of the IPMC-based underwater swimmer has been developed [19]. The position control of IPMC-based underwater biomimetic propulsor has been developed using the Euler Bernoulli beam theory dynamic equation [20]. But, the control of IPMC actuated underwater biomimetic swimmer based on the physics-based model has very limited literature. A dedicated model is desirable for an ideal design and control of the biomimetic underwater robotic swimmer actuated by IPMC. A physics-based model for biomimetic underwater robotic fish actuated by IPMC [21] acting as caudal fin has been explored for actuation characteristics.

In this paper, we provide a better position control model using the physics-based model for biomimetic underwater propulsor actuated by ionic polymer–metal composite (IPMC). The biomimetic underwater propulsor considering the body caudal fin-type locomotion is undertaken where the oscillating fin tail of the original fish is replaced by a fin tail comprising of an active IPMC beam. The supported cantilevered IPMC beam model with actuation dynamics is analyzed and the relations for displacement and velocity are studied using a physics-based model in a fluidic environment. The robotic swimming velocity response due to the active IPMC tail fin deflection under applied electric potential has been observed. Considering the Laplace transform of the mathematical model, the transfer function relating the tail tip deflection and input voltage has been generated and proportional, integral and derivative (PID) closed-loop controller has been developed to control the deflection from IPMC in an underwater environment to get the controlled and efficient output response through MATLAB Simulink tool. The PID gains are tuned using the Ziegler–Nichols tuning method to make the output response quick and robust. A more detailed account of the approach follows.

2 Mathematical Model

The mathematical model for the biomimetic underwater swimmer actuated by IPMC consists of two parts firstly, the slender body propulsion theory in which the swimmer body is considered as a slender body means the cross-section area of the body changes with length. Secondly, the underwater robotic fish actuated by IPMC acting as the caudal fin is modeled considering the physics behind the material actuation.

2.1 *Slender Body Propulsion*

Lighthill's slender body propulsion theory [21] is considered for thrust as the robotic structure considered to have an slender body and the relation is shown below:

$$T = \left[\frac{m}{2}\left(\left(\frac{\delta w(x,t)}{\delta t}\right)^2 - V^2\left(\frac{\delta w(x,t)}{\delta x}\right)^2\right)\right]_{x=L} \tag{1}$$

where $x = L$ denotes the end of the tail. We are considering the mean value of thrust, m denotes the virtual mass density at the tail tip.

$$m = \frac{1}{4}\pi W^2 \rho \beta \tag{2}$$

Here, W is the width of the tail at the end, ρ is the density of the fluid and β is a non-dimensional parameter close to 1 [15].

$$F = \frac{C\rho V^2 A}{2} \tag{3}$$

where A is the wetted surface area, C is the drag coefficient. The mean thrust calculated is balanced by the drag force at the steady-state, from which the cruising speed V can be written as

$$V = \left[\sqrt{\frac{m\left(\frac{\delta w(x,t)}{\delta t}\right)^2}{C\rho A + m\left(\frac{\delta w(x,t)}{\delta x}\right)^2}}\right]_{x=L} \tag{4}$$

2.2 *Model of Robotic Fish Motion with IPMC Tail as a Caudal Fin*

Considering Fig. 1, which shows a bending deflection of an IPMC beam in a cantilevered configuration under the application of applied electric potential. The transfer function relating deflection $w(x, s)$ and applied electric potential $\varphi(s)$ obtained by Laplace transformation is shown as a product of the bending moment dynamics $M(x, s)$ and the viscoelastic dynamics of the IPMC beam $D(s)$ as presented in [22]:

$$TF(x, s) = \frac{w(x, s)}{\varphi(s)} = M(x, s).D(s) \tag{5}$$

The bending moment $M(x, s)$ is expressed as shown in [23]

$$M(x, s) = \frac{\alpha W K k(\gamma - 1)}{E_q I(\gamma s + K)}\left(\frac{X(x, s)}{1 + r_2\theta(s)}\right) \tag{6}$$

where α is the stress charge coupling constant, W is the width of IPMC, k is the dielectric constant of the base polymer, $I = \frac{2Wh^3}{3}$ is an inertial moment and r_2 is electrode resistance per unit length.

E_q is the equivalent Young's modulus of IPMC in the fluid [22] and expressed as

$$E_q = \frac{E}{\left(1 + \frac{\pi\rho W}{8\rho_{\text{IPMC}} h}\right)} \tag{7}$$

where ρ_{IPMC} is the density of IPMC, E is Young's modulus of IPMC in air. K and γ are constant terms expressed as

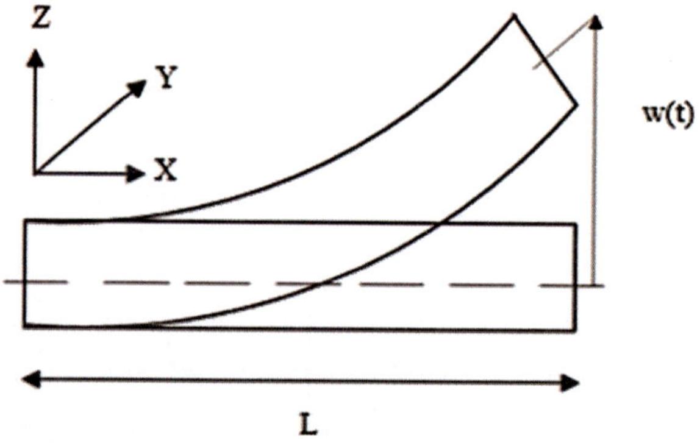

Fig. 1 IPMC beam as a caudal tail fin in a cantilevered configuration

$$K = \frac{F^2 \mathrm{d}C^-}{k\mathrm{RT}}\left(1 - C^- \Delta v\right) \text{ and } \gamma = \sqrt{\frac{K}{d}} h \tag{8}$$

where F is the Faraday's constant, d denotes ionic diffusivity, C^- denotes anion concentration, R is the gas constant, T denotes absolute temperature and Δv denotes volumetric change.

The functions defined in the above equation can be defined as:

$$\theta(s) = \frac{Wks\gamma(s + K)}{h(s\gamma + K)} \tag{9}$$

$$X(x, s) = \frac{1 + (\sinh(B(s)x) - B(s)x)\tanh(B(s)L) - \cosh(B(s)x)}{B(s)^2} \tag{10}$$

$$B(s) = \sqrt{r_1\left(\frac{\theta(s)}{(1 + r_2\theta(s))} + \frac{2}{R_p}\right)} \tag{11}$$

where r_1 denotes electrode resistance per unit length and R_p denotes polymer resistance.

The relative low actuation bandwidth of IPMC cause viscoelastic dynamics and can be expressed as:

$$D(s) = \frac{\omega_n^2}{s^2 + 2\xi\omega_n + \omega_n^2} \tag{12}$$

where ω_n denotes the natural frequency of IPMC in fluid and ξ denotes damping ratio.

The natural frequency ω_n in terms of beam dimensions and mechanical properties can be expressed as [22]:

$$\omega_n = \frac{C_1^2}{L^2}\sqrt{\frac{E_q I}{2\rho_{\mathrm{IPMC}} W h}} \tag{13}$$

where C_1 denotes the constant associated with beam oscillation.

Now to couple viscoelastic dynamics to hydrodynamics, a transfer function is generated relating the slope of the beam to the input electric potential.

$$\mathrm{TF}_c(x, s) = \frac{\frac{\delta w(x,s)}{\delta x}}{\varphi(s)} = M_c(x, s)D(s) \tag{14}$$

$$M_c(x, s) = \frac{\psi_s WKk(\gamma - 1)}{E_q I(\gamma s + K)} \frac{(\cosh(B(s)x) - 1)\tanh(B(s)L) - \sinh(B(s)x)}{B(s)(1 + r_2\theta(s))} \tag{15}$$

The input electric potential $\varphi(t) = A\sin(\omega t)$, the displacement and the slope of the IPMC beam at the end can be expressed as:

$$\begin{aligned} w(L, t) &= A|TF(L, j\omega)|\sin(\omega t + \text{phase angle}) \\ \frac{\delta w(L, t)}{\delta x} &= A|TF_c(L, j\omega)|\sin(\omega t + \text{phase angle}) \end{aligned} \tag{16}$$

The cruising speed can be rewritten with respect to deflection and slope under the applied electric potential as

$$V = \sqrt{\frac{\text{mA}^2\omega^2|\text{TF}(L, j\omega)|^2}{2C_\rho A + \text{mA}^2|\text{TF}_c(L, j\omega)|^2}} \tag{17}$$

The various parameter values used for evaluating velocity of robotic swimmer, tail tip deflection and solving transfer function is given in Table 1.

Table 1 Various parameter values for dynamic model [22]

Parameters	Definition	Value	Unit
F	Faraday's constant	96,487	C/mol
R	Gas constant	8.31431	J/mol·K^2
T	Absolute temperature	297	K
R_p	Through-polymer resistance per unit	34	Ω/m
E	Young's modulus of IPMC	571	MPa
r_1	Resistance per unit lenght in x direction	2219	Ω/m
r_2	Resistance per unit length in y direction	8.2^{-3}	Ω/m
d	Ionic diffusivity	3.08×10^{-7}	m^2/s
C^-	Anion concentration	1091	mol/m^3
κ	Effective dielectric constant	1.48×10^{-6}	F/m
α	Coupling constant	0.05	J/C
C_1	First mode oscillation constant	1.8751	
ρ	Water density	1000	kg/m^3
ξ	Damping ratio	0.225	
ρ_{IPMC}	IPMC density	1600	kg/m^3
L	IPMC tail length	33	mm
W	Width of IPMC	14.7	mm
h	height	175	µm

3 Results

The numerical actuation model is explored through simulations for the first mode of bending for the IPMC under the applied electric potential using parametric values given in Table 1. The transfer function considering the displacement and input electric potential has been used to plot the Bode plot as shown in Fig. 2, which suggests the system cut-off frequency at around 3.8 Hz.

Figure 3, shows the swimming velocity of the underwater biomimetic robot actuated by IPMC and suggests that the swimming speed is related to the actuation frequency. Also, there is one optimal frequency at which the swimming speed attains the highest speed which depends on the dimension of the IPMC strip.

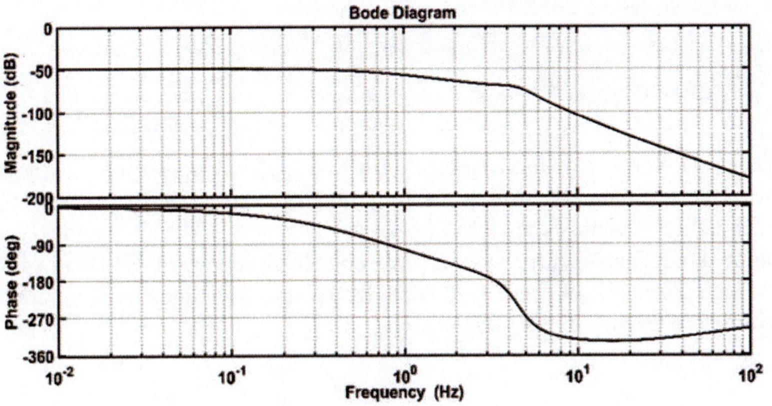

Fig. 2 Bode plot for IPMC tail fin showing cut-off frequency

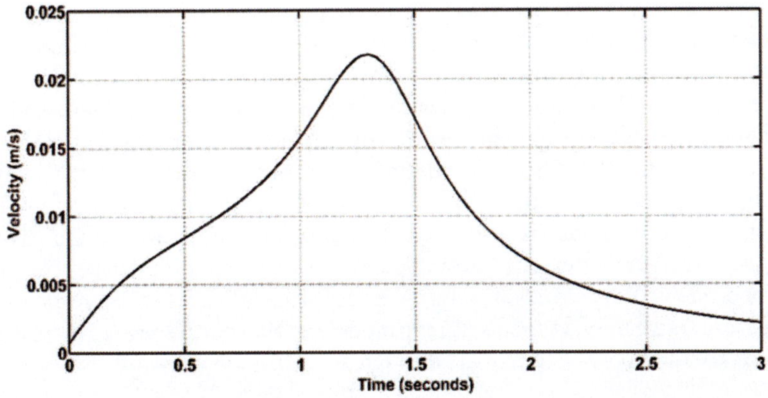

Fig. 3 Velocity versus time of the biomimetic robotic fish

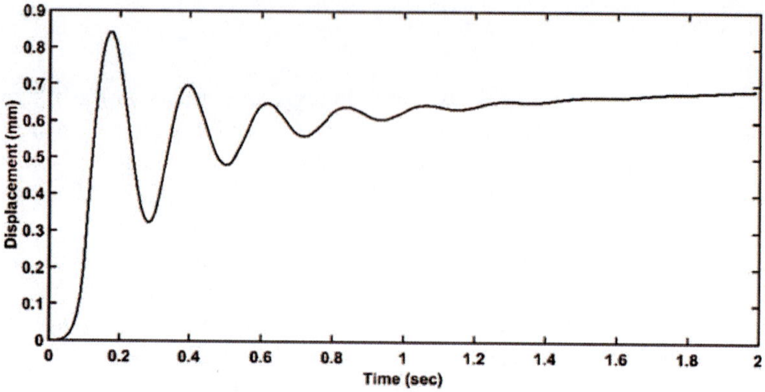

Fig. 4 Open-loop response of IPMC underwater actuation

3.1 *Open-Loop Response*

The transfer function relating IPMC bending and applied electric potential have been used and the simulated open-loop response under the step electric potential applied to the actuation model is shown in Fig. 3.

Figure 4, suggests that the system has a cut-off frequency of 3.8 Hz as suggested in the Bode plot as shown in Fig. 2. Also, the natural frequency in water conditions is identified to be $\omega_1 = 23.8$ rad/s.

3.2 *Closed-Loop PID Control Response*

The control of a system is necessary in order to get desired response from the system for a particular application. The closed-loop control is used to get the desired response with the error signal adjusting the input electric potential. The proportional, integral and derivative (PID) control has been applied to the transfer function relating the IPMC displacement and applied electric potential as shown in Fig. 5. As the name suggests each term plays its role in controlling the output response like Proportional term means the control response is proportional to the error signal so if an error is large the control response is also large. The integral term integrates the past error values to diminish the residual error and the derivative term anticipates the current trend of error so as to give a better control effect to achieve the desired response [24].

Figure 6, suggests that the closed-loop model can achieve the desired displacement of 1 mm after the observation made through an open-loop response.

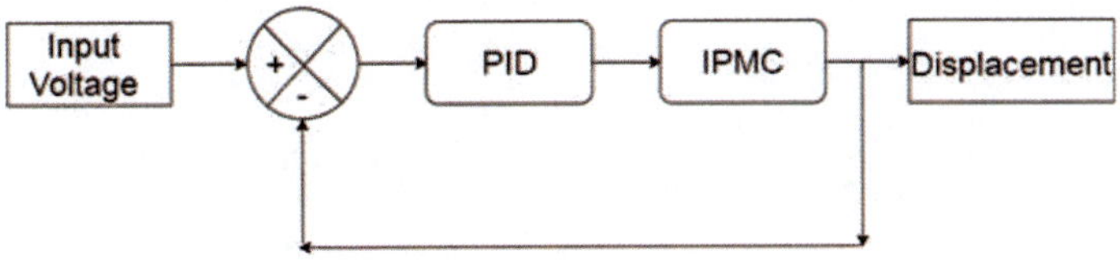

Fig. 5 Block diagram of closed-loop position control

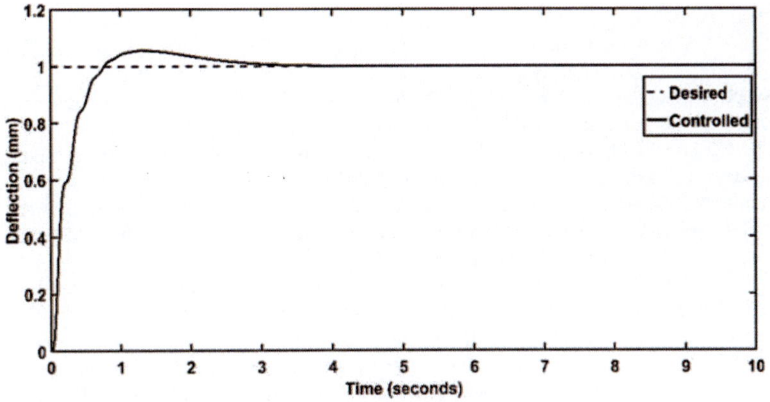

Fig. 6 Closed-loop controlled response for desired 1 mm displacement

4 Conclusion

In this paper, physics-based modeling of the steady-state forward locomotion was explored for an underwater robotic structure actuated by an ionic polymer-metal composite. The model considers the physics behind the IPMC actuation dynamics and also the hydrodynamics related to the active IPMC beam interacting with the fluidic environment. The biomimetic underwater propulsor considering the body caudal fin-type locomotion is undertaken where the oscillating fin tail of the original fish is replaced by a fin tail comprising an active IPMC beam. The cruising velocity of the robotic swimmer and the open-loop deflection of the active IPMC beam has been plotted which suggests the robotic forward locomotion and also the natural frequency of IPMC beam in the fluidic environment has been identified as 23.8 rad/s. Considering the Laplace transform of the mathematical model, the transfer function relating the deflection and input voltage has been generated and proportional, integral and derivative (PID) closed-loop controller has been developed to control the deflection from IPMC in an underwater environment. The controlled and efficient desired displacement response of 1 mm through simulations was achieved. The controlled model has been provided to make the output response quick and robust. This work will help researchers in modeling a biomimetic underwater robotic swimmer actuated by an IPMC beam as an active caudal fin tail.

References

1. Sfakiotakis M, Lane DM, Davies JBC (1999) Review of fish swimming modes for aquatic locomotion. IEEE J Ocean Eng 24:237–252
2. Lauder GV, Anderson EJ, Tangorra J, Madden PGA (2007) Fish biorobotics: kinematics and hydrodynamics of self-propulsion. J Exp Biol 210:2767–2780
3. Shahinpoor M, Kim KJ (2001) Ionic polymer metal composites: I. Fundamentals Smart Mater Struct 10:819–833
4. Shahinpoor M, Kim KJ (2005) Ionic polymer metal composites: IV. Ind Med Appl Smart Mater Struct 14:197–214
5. Kim KJ, Yim W, Paquette JW, Kim D (2007) Ionic polymer-metal composites for underwater operation. J Intell Mater Syst Struct 18:123–131
6. Aureli M, Kopman V, Porfiri M (2010) Free-locomotion of underwater vehicles actuated by ionic polymer metal composites. IEEE/ASME Trans Mechatron 15:603–614
7. Chen Z, Um TI, Bart-Smith H (2011) Ionic polymer-metal composite enabled robotic manta ray. Electroact Polym Actuators Dev 7976:797637
8. Palmre V, Hubbard JJ, Fleming M, Pugal D, Kim S, Kim KJ, Leang KK (2013) An IPMC-enabled bio-inspired bending/twisting fin for underwater applications. Smart Mater Struct 22. doi:https://doi.org/10.1088/0964-1726/22/1/014003
9. Yang T, Chen Z, Member I (2015) Development of 2D maneuverable robotic fish propelled by multiple ionic polymer—metal composite artificial fins. Proceeding of IEEE international Conference on Robots. Biomimetics, pp 255–260
10. Trabia S, Shen Q, Stalbaum T, Hunt R, Hwang T, Kim K () Numerical and experimental investigation of a biomimetic robotic jellyfish actuated by ionic polymer-metal composite. In: Proceeding of 13th international conference ubiquitous robot. ambient intelligent. URAI, pp 204–205
11. Salazar R, Campos A, Fuentes V, Abdelkefi A (2019) A review on the modeling, materials, and actuators of aquatic unmanned vehicles. Ocean Eng 172:257–285
12. Jain RK, Datta S, Majumder S, Dutta A (2011) Two IPMC fingers based micro gripper for handling. Int J Adv Robot Syst 8(1):1–9
13. Jain RK, Khan A, Inamuddin Abdullah MA (2019) Design and development of non-perfluorinated ionic polymer metal composite (IPMC) based flexible link manipulator for robotic assembly. Polymer Compos 40(7):2582–2593
14. Gupta A, Mukherjee S (2021) Dynamic modeling of biomimetic undulatory ribbon fin underwater propulsor actuated by IPMC. Mater Today Proc 44(1):1086–1089. https://doi.org/10.1016/j.matpr.2020.11.183
15. Shen Q, Wang T, Liang J, Wen L (2013) Hydrodynamic performance of a biomimetic robotic swimmer actuated by ionic polymer metal composite. Smart Mater Struct 22(7):075035
16. Richardson RC, Levesley MC, Brown MD, Hawkes JA, Watterson K, Walker PG (2003) Control of Ionic polymer metal composites. IEEE/ASME Trans Mechatron 8:245–253
17. Chen X (2018) Discrete-time adaptive control design for ionic polymer-metal composite actuators. IEEE Access. 6:28114–28121
18. Zakeri, E., Moeinkhah, H (2019) Digital control design for an IPMC actuator using adaptive optimal proportional integral plus method: simulation and experimental study. Sens Actuators A Phys 298:111577
19. Shen Q, Wang T, Wen L, Liang J (2013) Modelling and fuzzy control of an efficient swimming ionic polymer-metal composite actuated robot. Int J Adv Robot Syst 10:1–13
20. Gupta A, Mukherjee S (2021) Position control of a biomimetic IPMC underwater propulsor. J Inst Eng India Ser C 102(4):1031–1040
21. Lighthill MJ (1960) Note on the swimming of slender fish. J Fluid Mech 9:305–317
22. Mbemmo E, Chen Z, Shatara S, Tan X (2008) Modeling of biomimetic robotic fish propelled by an ionic polymer–metal composite actuator. In: Proceedings of IEEE international conference on robotics and automation, pp. 689–694. Pasadena, USA

23. Chen Z, Tan X (2007) A control-oriented, physics-based model for ionic polymer-metal composite actuators. In Proceedings of the 46th IEEE conference on decision and control, pp 590–595. New Orleans, LA
24. Ogata K (2010) Modern control engineering, 5th edn. Prentice-Hall, New Jersey, pp 568–590

Composites

Effect of Interface on Elastic Properties and Vibration Characteristics of CNT-Reinforced Composites

Surendra Kumar, Saurabh Mishra, and Amit Kumar

Abstract Interface between embedded carbon nanotube (CNT) and its surrounding matrix has an important role in predicting efficiency of CNT as reinforcement in CNT-reinforced polymer nanocomposites (CNTRCs). This paper presents effect of interface on the effective elastic properties and vibration characteristics of CNTRCs using an analytical model and finite element method. The effective elastic properties (elastic moduli and Poisson's ratios) of the CNTRCs in the presence of an interface are predicted using finite element modeling of a three-dimensional representative volume element. Parametric studies are carried out to investigate the effects of elastic properties and the thickness of the interface on the effective elastic properties of these composites. Interphase is also characterized to have a varied modulus through its thickness. Subsequently, the vibration frequencies of these composites having different interface characteristics are predicted using an analytical (closed form) model developed based on first order shear deformation theory of laminated plate and also using a finite element modal analysis.

Keywords CNT · Nanocomposites · Mechanical properties · Finite element method · Vibration

1 Introduction

Carbon nanotubes (CNTs), because of their unprecedented mechanical properties, have become favorable and extensively used nanofillers for polymer composites for structural applications. Despite ample revelations in form of improvement in the mechanical properties and many experimental and theoretical studies made to characterize and quantify the functionality of the nanocomposites, there are several limitations that are being investigated and need to be resolved.

The load transfer between matrix and CNTs is not only characterized by atomic structure at nano scale but also rely upon the interfacial region, thus impacting the

S. Kumar (✉) · S. Mishra · A. Kumar
CSIR-Central Mechanical Engineering Research Institute, Durgapur 721309, India
e-mail: surend_kr@yahoo.com

V. K. Gupta et al. (eds.), *Recent Advances in Machines and Mechanisms*, Lecture Notes in Mechanical Engineering, https://doi.org/10.1007/978-981-19-3716-3_54

effective mechanical properties of nanocomposites. Han and Elliot [1] have shown that interfacial effects cannot be ignored while predicting the effective properties of CNTRCs having strong interfacial interactions, and thus the general macroscopic rule of mixtures, which is mainly based on the CNT volume fraction, does not satisfactorily predict composite modulus. Wan et al. [2] studied the effect of length of SWCNTs with hard and soft interphase inside the matrix and reported that the critical fiber length was reduced up to 46% by the existence of hard interphase, however the effective moduli was insensitive to interphase. Hammerand et al. [3] performed analysis of hollow fiber nanocomposites by assuming an interphase layer between the fiber and the matrix. Simulation using finite element method (FEM) showed that in presence of a hard interphase (having Young's modulus ten times that of the polymer) with 10% volume fractions of fiber and interphase each, effective axial modulus had increased about 3.75%. In order to investigate the influence of interface at nanoscale in CNT-Polyethylene (PE) nanocomposites, Li and Seidel [4] carried out parametric studies based on molecular dynamics (MD) simulation and characterized the non-functionalized interface in terms of length and numbers of polymer chains, grip position, and temperature dependency. Rafiee et al. [5] studied influence of the interphase resembling normal CNTs and functionalized CNTs nanocomposites and reported that the functionalization had reduced the Young's modulus of RVE in nanoscale but also predicted that it could be improved in meso scale due to good quality of dispersion resulting from cross-linked covalent bonds between the CNTs and the polymer. Banerjee et al. [6] studied the role of interphase on mechanical properties of SWCNT composites using FEM. They reported that the matrix/CNT interphase can be characterized in terms of a weak boundary layer (WBL) and low value of fracture stress at WBL is the main reason for real-time low value of elastic moduli of CNT-reinforced composites.

Choi et al. [7] used MD-FEM linked multiscale model of SWNT-epoxy based nanocomposites to evaluate mechanical properties and their dependence on CNT size. During this study, a thin but highly concentrated deformation energy region was found around the CNT, identified as the ductile/soft interfacial region, responsible for inefficient load transfer and weakened interfacial properties. Results revealed rise in interface stiffness with decrease in the diameter of CNT. Malagù et al. [8] proposed a computational methodology on the basis of MD simulations and equivalent continuum mechanics model to estimate size effects in non-functionalized SWCNT polymer composites. MD simulations had revealed that solely interphase is responsible for the mechanical response of nanocomposites. Maghsoudlou et al. [9] conducted finite element analysis on the CNT-epoxy nanocomposites for different cylindrical RVEs by incorporating curvature, interfacial interaction, and agglomeration effects. By incorporating the interfacial effects, the discrepancy between numerically and experimentally obtained moduli was only 8.4%, and later, accounting the agglomeration with interphase, it was reduced to 6.8% at 0.1 wt% of SWCNT. Amraei et al. [10] proposed a closed-form mathematical model to predict the transverse isotropic elastic properties of interphase by assuming the interfacial elastic properties varying radially. The developed three-phase micromechanical model was found to be accurate in prediction of longitudinal elastic modulus for 4 vol% (8, 8)

SWCNT/polycarbonate nanocomposite with only 0.5% deviation as compared to the prediction of MD simulation.

Most studies in the literature on CNTRCs have concentrated on their material properties. So far as study of structural responses (vibration, etc.) of the actual structural bodies made of CNTRC materials is concerned, different computational mechanics approaches have been used to determine the vibration responses of CNTRC structures and functionally graded CNTRC structures in the form of beams [11], plates [12–16], and shells [17, 18]. However, to the best of authors' knowledge, the influence of interphase characteristics on the vibration responses of CNTRCs is not reported in literature.

The present study is thus intended to analyze the effect of interfacial region on the effective elastic properties as well as vibration characteristics of the CNTRCs using an analytical model and finite element method. A continuum mechanics based three-dimensional finite element analysis is performed on a three-phase representative volume element (RVE) comprising embedded CNT, interphase and the polymer matrix. Parametric analyzes are carried out by changing the thickness and elastic modulus of the interphase. Subsequently, vibration responses of the CNTRCs having different interfacial characteristics are predicted using an analytical (closed form) model developed based on first order shear deformation theory (FSDT) of laminated plate. A finite element modal analysis of the CNTRC plate is also carried out for the sake of comparison. The novelty of this paper is reflected in the study of the effect of interface having a varied modulus (ranging from that of CNT to that of polymer) through its thickness.

2 Elastic Properties of a Three-Phase CNTRC

The approach used to predict effective elastic properties of CNTRCs is based on a so-called representative volume element (RVE) of square cross section. The RVE based modeling reduces the complexity in nanoscale structures and computational efforts. In three-phase composite model, it consists of three components: (a) carbon nanotubes, (b) interfacial region, and (c) surrounding polymer matrix. In this work, the non-functionalized SWCNT is converted into an equivalent solid fiber having diameter same as the outer diameter of the SWCNT, and the interphase is converted into either a single-layer or a multilayer concentric cylindrical shell surrounding the SWCNT. The schematic of the three-phase composite model subjected to a longitudinal strain is shown in Fig. 1.

The material system taken is (10, 10) SWCNT having mean diameter as 1.356 nm, thickness (t_{CNT}) as 0.34 nm and length (L_{CNT}) as 10 nm embedded in polymethyl methacrylate (PMMA) polymer matrix. Interphase of varied thickness from 0.5 t_{CNT} to 2.0 t_{CNT}, i.e. from 0.17–0.68 nm is considered for parametric analyzes. Due to its ambiguous structure, dimensional properties of interphase are still unclear, but literature has suggested that the thickness of interphase particularly lies in this range. Dimensions of the RVE are calculated based on the volume fractions of different

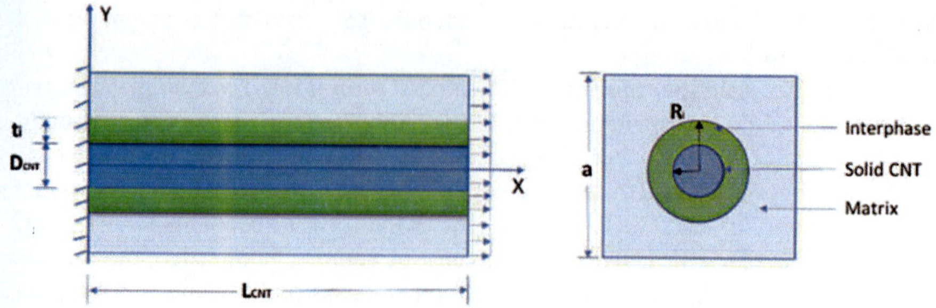

Fig. 1 Schematic of a three-phase composite model subjected to a longitudinal strain

phases. The corresponding volume fractions of the three phases are calculated by Eq. (1)

$$V^{\mathrm{nt}} = \frac{\pi R_{\mathrm{CNT}}^2}{a^2}, \quad V^{\mathrm{int}} = \frac{\pi \left(R_i^2 - R_{\mathrm{CNT}}^2\right)}{a^2} \quad \text{and} \quad V^m = 1 - V^{\mathrm{nt}} - V^{\mathrm{int}} \tag{1}$$

in which R_{CNT} is the radius of the equivalent solid CNT, R_i the outer radius of the interphase, and a the side length of the square RVE.

2.1 Finite Element Modeling

A three-dimensional finite element (FE) model of the three-phase RVE was generated in ANSYS software by discretizing the CAD model into high order structured hexahedral elements optimally. In order to obtain five independent elastic constants (elastic moduli and Poisson's ratios) of the CNTRC, the FE model of the RVE is subjected to certain loads and boundary conditions. These loading conditions are categorized as uniaxial tension, lateral uniform tension, and shear loading, and are depicted in Fig. 2.

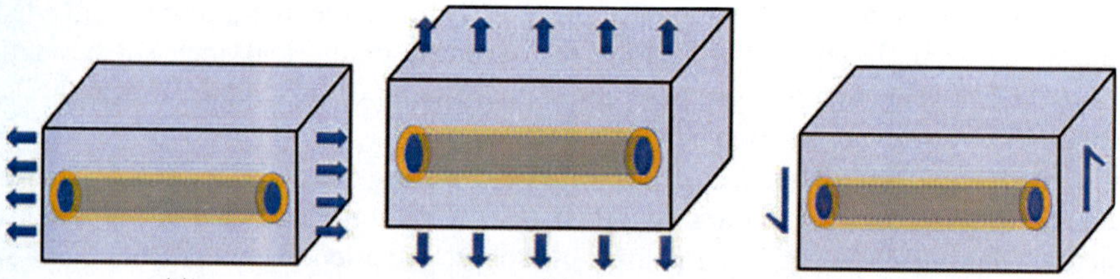

Fig. 2 RVE loading conditions: **a** uniaxial tension, **b** lateral uniform tension, and **c** shear loading

A symmetrical quarter FE model is considered for axial and lateral loadings by taking advantages of symmetry, whereas shear loading is performed on a full FE model. In case of uniaxial tension, the RVE is subjected to 1% longitudinal strain as displacement boundary condition on one end and is restrained on the other end. Uniform lateral loading and shear loading have similarly been implemented. Appropriate symmetrical boundary conditions were applied on symmetric surfaces.

The average stress and strains corresponding to different boundary conditions are computed at the middle section of the RVE model. Since the CNT is aligned and continuous along the x-direction inside the polymer matrix, the resulting nanocomposite will have transversely isotropic elastic properties having strain–stress relations given by Eq. (2) from which the five independent elastic constants can be evaluated.

$$\begin{Bmatrix} \epsilon_x \\ \epsilon_y \\ \epsilon_z \\ \gamma_{yz} \\ \gamma_{xz} \\ \gamma_{xy} \end{Bmatrix} = \begin{bmatrix} \frac{1}{E_x} & -\frac{\nu_{xy}}{E_x} & -\frac{\nu_{xy}}{E_x} & 0 & 0 & 0 \\ -\frac{\nu_{xy}}{E_x} & \frac{1}{E_y} & -\frac{\nu_{yz}}{E_y} & 0 & 0 & 0 \\ -\frac{\nu_{xy}}{E_x} & -\frac{\nu_{yz}}{E_y} & \frac{1}{E_y} & 0 & 0 & 0 \\ 0 & 0 & 0 & \frac{2(1+\nu_{yz})}{E_y} & 0 & 0 \\ 0 & 0 & 0 & 0 & \frac{1}{G_{xy}} & 0 \\ 0 & 0 & 0 & 0 & 0 & \frac{1}{G_{xy}} \end{bmatrix} \begin{Bmatrix} \sigma_x \\ \sigma_y \\ \sigma_z \\ \tau_{yz} \\ \tau_{xz} \\ \tau_{xy} \end{Bmatrix} \tag{2}$$

3 Vibration Analysis of the CNTRC

An analytical (closed-form) model based on FSDT of a laminated plate is developed to predict free vibrational characteristics of a CNTRC rectangular plate embedded with uniformly distributed, aligned, and straight CNTs (Fig. 3). The effect of interphase is included in the model by assigning material properties obtained from the three-phase RVE analysis to the CNTRC plate.

The equations of motion for the CNTRC plate can be found using Hamilton's principle as in Eq. (3).

Fig. 3 Configuration of a CNTRC plate with uniformly distributed CNTs

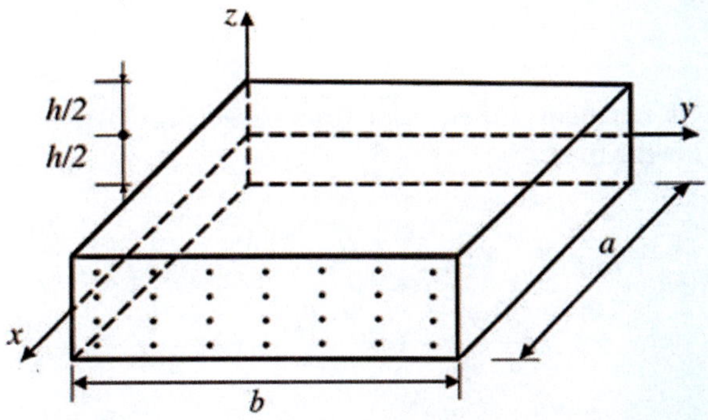

$$\frac{\partial N_x}{\partial x}+\frac{\partial N_{xy}}{\partial y}=I_1\ddot{u}_0+I_2\ddot{\psi}_x,\quad \frac{\partial N_{xy}}{\partial x}+\frac{\partial N_y}{\partial y}=I_1\ddot{v}_0+I_2\ddot{\psi}_y,$$
$$\frac{\partial Q_x}{\partial x}+\frac{\partial Q_y}{\partial y}+q=I_1\ddot{w}_0,\quad \frac{\partial M_x}{\partial x}+\frac{\partial M_{xy}}{\partial y}-Q_x=I_2\ddot{u}_0+I_3\ddot{\psi}_x\quad \text{and}$$
$$\frac{\partial M_{xy}}{\partial x}+\frac{\partial M_y}{\partial y}-Q_y=I_2\ddot{v}_0+I_3\ddot{\psi}_y \tag{3}$$

In Eq. (3), N_x, N_y and N_{xy} are in-plane (membrane) stress resultants (membrane forces per unit length); M_x, M_y and M_{xy} are moment resultants (moments per unit length); and Q_y and Q_x are transverse shear stress resultants (shear forces per unit length). These are defined by Eq. (4).

$$\begin{Bmatrix} N_x \\ N_y \\ N_{xy} \end{Bmatrix}=h\begin{bmatrix} C_{11} & C_{12} & 0 \\ C_{12} & C_{22} & 0 \\ 0 & 0 & C_{66} \end{bmatrix}\begin{Bmatrix} \frac{\partial u_0}{\partial x} \\ \frac{\partial v_0}{\partial y} \\ \frac{\partial u_0}{\partial y}+\frac{\partial v_0}{\partial x} \end{Bmatrix},$$
$$\begin{Bmatrix} M_x \\ M_y \\ M_{xy} \end{Bmatrix}=\frac{h^3}{12}\begin{bmatrix} C_{11} & C_{12} & 0 \\ C_{12} & C_{22} & 0 \\ 0 & 0 & C_{66} \end{bmatrix}\begin{Bmatrix} \frac{\partial \psi_x}{\partial x} \\ \frac{\partial \psi_y}{\partial y} \\ \frac{\partial \psi_x}{\partial y}+\frac{\partial \psi_y}{\partial x} \end{Bmatrix}\quad \text{and}$$
$$\begin{Bmatrix} Q_y \\ Q_x \end{Bmatrix}=\alpha h\begin{bmatrix} C_{44} & 0 \\ 0 & C_{55} \end{bmatrix}\begin{Bmatrix} \frac{\partial w_0}{\partial y}+\psi_y \\ \frac{\partial w_0}{\partial x}+\psi_x \end{Bmatrix} \tag{4}$$

In Eq. (3), $(I_1, I_2, I_3)=\int_{-h/2}^{h/2}\rho(1, z, z^2)dz$, ρ being the density of the CNTRC plate. In Eq. (4), $u_0(x, y)$ and $v_0(x, y)$ are the mid-surface in-plane displacements, and $w_0(x, y)$ is the mid-surface transverse displacement. ψ_x and ψ_y are the rotations of a line normal to the mid-surface. In Eq. (4), the reduced stiffnesses C_{ij} are given as

$$C_{11}=\frac{E_x}{1-\nu_{xy}\nu_{yx}},\quad C_{12}=\frac{\nu_{xy}E_y}{1-\nu_{xy}\nu_{yx}}=\frac{\nu_{yx}E_x}{1-\nu_{xy}\nu_{yx}},\quad C_{22}=\frac{E_y}{1-\nu_{xy}\nu_{yx}},$$
$$C_{44}=\frac{E_y}{2(1+\nu_{yz})}\quad \text{and}\quad C_{55}=C_{66}=G_{xy}$$

On expanding the equations of motion (Eq. (3)) using Eq. (4), we get Eq. (5) in which t is the time.

$$C_{11}\frac{\partial^2\psi_x}{\partial x^2}+C_{66}\frac{\partial^2\psi_x}{\partial y^2}+(C_{12}+C_{66})\frac{\partial^2\psi_y}{\partial x\partial y}-\frac{12\alpha C_{55}}{h^2}\left(\psi_x+\frac{\partial w_0}{\partial x}\right)=\rho\frac{\partial^2\psi_x}{\partial t^2},$$
$$(C_{12}+C_{66})\frac{\partial^2\psi_x}{\partial x\partial y}+C_{66}\frac{\partial^2\psi_y}{\partial x^2}+C_{22}\frac{\partial^2\psi_y}{\partial y^2}-\frac{12\alpha C_{44}}{h^2}\left(\psi_y+\frac{\partial w_0}{\partial y}\right)=\rho\frac{\partial^2\psi_y}{\partial t^2},$$
$$\alpha C_{55}\left(\frac{\partial \psi_x}{\partial x}+\frac{\partial^2 w_0}{\partial x^2}\right)+\alpha C_{44}\left(\frac{\partial \psi_y}{\partial y}+\frac{\partial^2 w_0}{\partial y^2}\right)+\frac{q}{h}=\rho\frac{\partial^2 w_0}{\partial t^2} \tag{5}$$

For the simply-supported CNTRC plate, the boundary conditions are given by

$$w_0 = \frac{\partial \psi_x}{\partial x} = 0 \quad \text{at } x = 0, a \quad \text{and} \quad w_0 = \frac{\partial \psi_y}{\partial y} = 0 \quad \text{at} \quad y = 0, b \tag{6}$$

The solutions to these governing equations are determined by using appropriate Fourier series expansions for the loads, displacements, and rotations. Fourier series expansions that satisfy the boundary conditions, are given by

$$\begin{aligned} \psi_x &= \sum_m \sum_n \Psi^x_{\text{mn}}(t) \cos \frac{m\pi x}{a} \sin \frac{n\pi y}{b}, \\ \psi_y &= \sum_m \sum_n \Psi^y_{\text{mn}}(t) \sin \frac{m\pi x}{a} \cos \frac{n\pi y}{b} \quad \text{and} \\ w_0 &= \sum_m \sum_n W_{\text{mn}}(t) \sin \frac{m\pi x}{a} \sin \frac{n\pi y}{b} \end{aligned} \tag{7}$$

and the load function is represented by $q(x, y, t) = \sum_m \sum_n Q_{mn}(t) \sin \frac{m\pi x}{a} \sin \frac{n\pi y}{b}$. Neglecting the rotary inertia, the governing equations (Eq. 5) can be simplified using the fourier series expansions for each set of integers m and n into

$$\begin{bmatrix} H_{11} & H_{12} & H_{13} \\ H_{12} & H_{22} & H_{23} \\ H_{13} & H_{23} & H_{33} \end{bmatrix} \begin{Bmatrix} \Psi^x_{\text{mn}}(t) \\ \Psi^y_{\text{mn}}(t) \\ W_{\text{mn}}(t) \end{Bmatrix} = \frac{12}{h^3} \begin{Bmatrix} 0 \\ 0 \\ Q_{\text{mn}}(t) - \rho h \ddot{W}_{\text{mn}}(t) \end{Bmatrix} \tag{8}$$

The elements of the symmetric matrix H_{ij} are

$$\begin{aligned} H_{11} &= C_{11}\left(\frac{m\pi}{a}\right)^2 + C_{66}\left(\frac{n\pi}{b}\right)^2 + \frac{12\alpha}{h^2} C_{55}, \quad H_{12} = (C_{12} + C_{66})\left(\frac{m\pi}{a}\right)\left(\frac{n\pi}{b}\right), \\ H_{13} &= \frac{12\alpha}{h^2} C_{55}\left(\frac{m\pi}{a}\right), \quad H_{22} = C_{66}\left(\frac{m\pi}{a}\right)^2 + C_{22}\left(\frac{n\pi}{b}\right)^2 + \frac{12\alpha}{h^2} C_{44}, \\ H_{23} &= \frac{12\alpha}{h^2} C_{44}\left(\frac{n\pi}{b}\right) \quad \text{and} \quad H_{33} = \frac{12\alpha}{h^2} C_{55}\left(\frac{m\pi}{a}\right)^2 + \frac{12\alpha}{h^2} C_{44}\left(\frac{n\pi}{b}\right)^2 \end{aligned} \tag{9}$$

Substitution of first and second equations of Eq. (8) into its last (third) equation results in Eq. (10)

$$(H_{13} K_A + H_{23} K_B + H_{33}) W_{\text{mn}}(t) = \frac{12}{h^3}\left[Q_{\text{mn}}(t) + \rho h \omega^2_{\text{mn}} W_{\text{mn}}(t)\right], \tag{10}$$

in which $K_A = \frac{H_{12}H_{23}-H_{13}H_{22}}{H_{11}H_{22}-H_{12}^2}$ and $K_B = \frac{H_{12}H_{13}-H_{11}H_{23}}{H_{11}H_{22}-H_{12}^2}$. For free vibrations, the loading is zero. Thus, we have

$$\omega_{mn}^2 = \frac{h^2(H_{13}K_A + H_{23}K_B + H_{33})}{12\rho}. \tag{11}$$

Square root of Eq. (11) gives the fundamental frequencies of the CNTRC plate.

4 Results and Discussion

To study the influence of the interphase on the elastic properties and the vibration response of the CNTRC, the analyzes are performed for different interface thicknesses as well as different elastic moduli of the interphase categorized into soft and stiff interphases. Soft interphase refers to the interphase having elastic modulus less than that of the polymer, while a stiff interphase has elastic modulus higher than that of the polymer. Interphase is also modeled to have a varied modulus through its thickness, whose value ranges from a high value equal to the CNT modulus at the CNT-interphase junction to a low value equal to the polymer modulus at the interphase-polymer junction. Although any kind of variation can be considered, a linear variation with respect to the interphase thickness is used here. The material properties of the equivalent SWCNT and PMMA matrix are taken as reported in [1], and the interphase is assigned varied elastic properties. These properties are listed in Table 1. In this work, the volume fraction of CNT (V^{nt}) is kept constant at 12%.

4.1 *Prediction of Elastic Properties*

Finite element analysis is performed for the CNT/PMMA nanocomposite having interfacial regions of different elastic and geometrical properties. Two types of interphases (a) soft interphase ($E^{int} = 0.1E^m$) and (b) stiff interphase ($E^{int} = 10E^m$), both having thicknesses as 0.17, 0.34, 0.51, and 0.68 nm, are taken.

Table 1 Material properties of different phases of the RVE

Phase	Young's modulus (GPa)	Poisson's ratio	Density (kg/m^3)
CNT	$E_L^{nt} = 600$ (longitudinal), $E_T^{nt} = 10$ (transverse)	$\nu^{nt} = 0.175$	1400
Matrix	$E^m = 2.5$	$\nu^m = 0.34$	1150
Interphase	$E^{int} =$ 0.1 E^m, 10 E^m	$\nu^{int} = 0.34$	1150

Table 2 Mesh sensitivity analysis for the prediction of elastic properties of the CNTRC having hard interphase ($t_i = 2\,t_{CNT}$)

Mesh element size (nm)	Number of elements	Longitudinal modulus (GPa)
0.25	3440	80.115
0.1	50,200	80.27
0.07	143,572	80.285
0.045	536,984	80.295
0.038	876,744	80.30
0.032	1,464,214	80.30

The analysis is conducted on a three-phase RVE discretized into high-order structured hexahedral elements optimally. As a first step, mesh sensitivity analysis is also performed. For a typical case of the three-phase FE model having hard interphase with thickness $t_i = 2t_{CNT}$, the convergence of results (in the form of effective longitudinal modulus) with mesh refinement is depicted in Table 2. The comparative results of different elastic moduli of CNT/PMMA with different moduli and thicknesses of interphase are depicted in Table 3, along with the results predicted using two-phase model without consideration of any interphase. It can be observed from the table that the presence of interphase has significantly influenced the elastic properties of the CNT/PMMA nanocomposite. Soft interphase region has weakening effect on the elastic modulus and increase in its thickness causes decrease in the elastic modulus of the composite; however, opposite trend is observed in case of the stiff interphase. Moreover, the soft interphase has very small influence on the longitudinal elastic modulus as compared to the stiff interphase. Effects of interphase thickness on the longitudinal and transverse moduli of the nanocomposite are plotted in Figs. 4a and b for the two types of interphase. Shown in this figure is also the longitudinal modulus of the nanocomposite predicted using a three-phase rule of mixtures (ROM) as given by Eq. (12). ROM prediction for longitudinal modulus is in consistent with the finite element result.

$$E_x = V^{\text{nt}} E^{\text{nt}} + V^{\text{int}} E^{\text{int}} + V^m E^m \tag{12}$$

Table 3 Effective material constants of CNTRC with and without interphase

Material constants	Two-phase model	Effect of interphase			
		Soft ($E^{\text{int}} = 0.25$ GPa)		Stiff ($E^{\text{int}} = 25$ GPa)	
		($t_i = 0.17$ nm)	($t_i = 0.68$ nm)	($t_i = 0.17$ nm)	($t_i = 0.68$ nm)
Ex (GPa)	74.2	74.1	73.6	75.4	80.3
Ey (GPa)	3.13	2.43	1.23	3.34	4.42
νxy	0.31	0.32	0.33	0.32	0.33
Gxy (GPa)	1.19	0.84	0.54	1.30	1.94

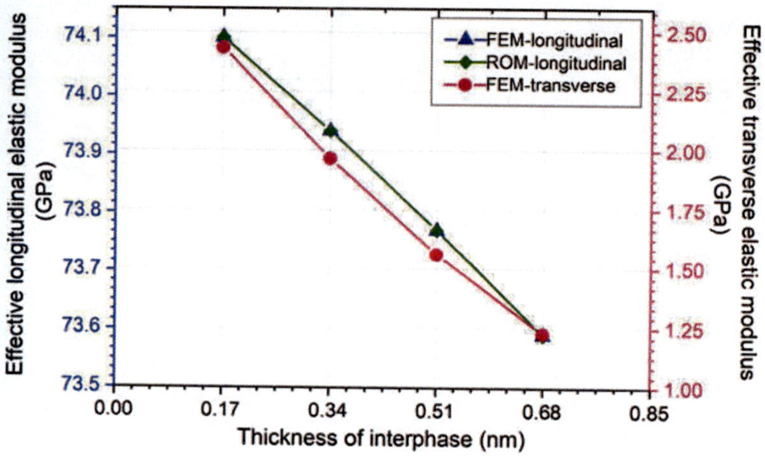

(a)

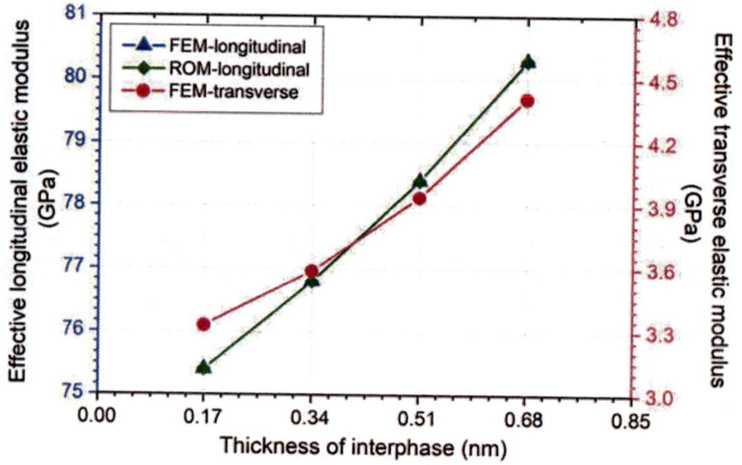

(b)

Fig. 4 Effect of interphase on elastic moduli (longitudinal and transverse) of the CNTRC: **a** soft interphase ($E^{\text{int}} = 0.25$ GPa) and **b** stiff interphase ($E^{\text{int}} = 25$GPa)

In case of the stiff interphase, the increase in effective longitudinal modulus has improved from 1.6 to 8.2% with increase in thickness from 0.17 to 0.68 nm. A significant rise is also observed for the effective transverse elastic modulus from 6.6 to 41% with increase in thickness. On the other hand, increasing the interphase thickness from 0.17 to 0.68 nm in case of the soft interphase, a drastic decrease in effective transverse elastic modulus takes place spanning from 22 to 61%.

The effective elastic properties of the CNTRC having interphase with varied modulus through its thickness for different thicknesses of the interphase are depicted

Table 4 Predicted elastic properties of CNT/PMMA composite having interphase with varied modulus and comparison with MD simulation results [1]

Interphase thickness (nm)	Material constants			
	E_x (GPa)	E_y (GPa)	νxy	Gxy (GPa)
0.17	89.5	3.41	0.34	1.27
0.25	97.3	3.52	0.34	1.35
0.34	106.5	3.66	0.34	1.42
MD simulation [1]	94.6	2.9	–	–

in Table 4. Table 4 also compares the predicted results with MD simulation results reported in [1]. A comparison of MD simulation result with the result listed in Table 3 depicts that the two-phase RVE model underestimates the longitudinal modulus of the nanocomposite by about 22%, while incorporation of the interphase (with varied modulus and having a thickness in the range of CNT thickness) has a stiffening effect on the nanocomposite and results in a much lesser deviation from the MD prediction.

4.2 Prediction of Fundamental Frequencies

Effect of interphase on the free vibration response of a CNT/PMMA nanocomposite plate is investigated using the developed analytical model. A square plate of side $a = 20$ mm and thickness $h = 2$ mm and subjected to a simply-supported boundary condition is considered, and fundamental frequencies are predicted. The analytical results for the two-phase RVE model (without consideration of the interphase) are validated using results available in literature. For each case, a finite element modal analysis is also done in ANSYS software, and the results are compared. The frequencies calculated are non-dimensionalized ($\overline{\omega}$) for the sake of comparison using the following formula: $\overline{\omega} = \omega\left(\frac{a^2}{h}\right)\sqrt{\frac{\rho^m}{E^m}}$. The dimensionless frequencies for the simply-supported nanocomposite plate with and without interphase are summarized in Table 5.

The analytical results match with those reported in [12, 17] as well as those calculated by finite element method, indicating that the proposed analytical model can quickly provide closed-form solutions for vibration frequencies of CNTRCs. Effects of interphases with two different moduli and various thicknesses on the first six dimensionless frequencies of the CNTRC plate subjected to simply-supported boundary condition are plotted in Fig. 5.

It can be deduced from the results that the interphase has also significant influence over the vibrational response of the CNTRC plate. Presence of stiff interphase considerably enhances the frequencies of vibration and the increase in frequencies is more prominent for higher modes of vibration. In contrast, soft interphase, particularly that with higher thickness of interfacial region, reduces the frequencies of vibration of the CNTRC.

Table 5 Effect of interphase on dimensionless frequencies of the simply-supported CNT/PMMA nanocomposite plate

Interphase properties	Mode	Dimensionless frequency			
		Analytical	FEM	Wang et. al. [12]	Aragh et. al. [17]
Two-phase model	(1, 1)	12.68	12.50	12.27	12.26
	(1, 2)	17.96	17.17	16.81	16.70
	(2, 1)	32.06	32.13	29.44	29.42
Soft interphase (t^i = 0.17 nm)	(1, 1)	11.95	11.89	–	–
	(1, 2)	16.45	16.18	–	–
	(2, 1)	29.35	29.69	–	–
Soft interphase (t^i = 0.68 nm)	(1, 1)	10.47	10.37	–	–
	(1, 2)	13.98	12.99	–	–
	(2, 1)	24.42	24.79	–	–
Stiff interphase (t^i = 0.17 nm)	(1, 1)	13.40	13.27	–	–
	(1, 2)	19.19	18.68	–	–
	(2, 1)	34.69	34.88	–	–
Stiff interphase (t^i = 0.68 nm)	(1, 1)	14.79	14.70	–	–
	(1, 2)	21.22	21.32	–	–
	(2, 1)	40.06	40.22	–	–

The first six dimensionless frequencies of the CNTRC having interphase with varied modulus through its thickness, and subjected to simply-supported boundary condition are plotted in Fig. 6 for different thicknesses of the interphase. It can be observed that the incorporation of the interphase (with varied modulus and having a thickness in the range of CNT thickness) results in a considerable enhancement in the frequencies of vibration of the CNTRC plate.

5 Concluding Remarks

This paper presents the influence of interphase (interfacial region between CNT and polymer) on the effective elastic properties and vibration characteristics of a CNT-reinforced PMMA polymer nanocomposite (CNTRC). Continuum mechanics based finite element analysis is performed on a three-phase RVE for evaluating effective elastic properties of the CNTRC having an interphase. Vibration frequencies of the three-phase CNTRC are calculated using an analytical (closed form) model developed based on first order shear deformation theory for a laminated plate, and the results obtained are also compared with those calculated using a finite element modal analysis of the CNTRC plate. Parametric analyzes are carried out by changing the thickness as well as elastic modulus of the interphase. The interphase is found to significantly influence the elastic properties of the CNTRC. Stiff interphase has

Fig. 5 Effect of interphase on the dimensionless frequencies of the CNTRC plate: **a** soft interphase ($E^{\text{int}} = 0.25$ GPa) and **b** stiff interphase ($E^{\text{int}} = 25$ GPa)

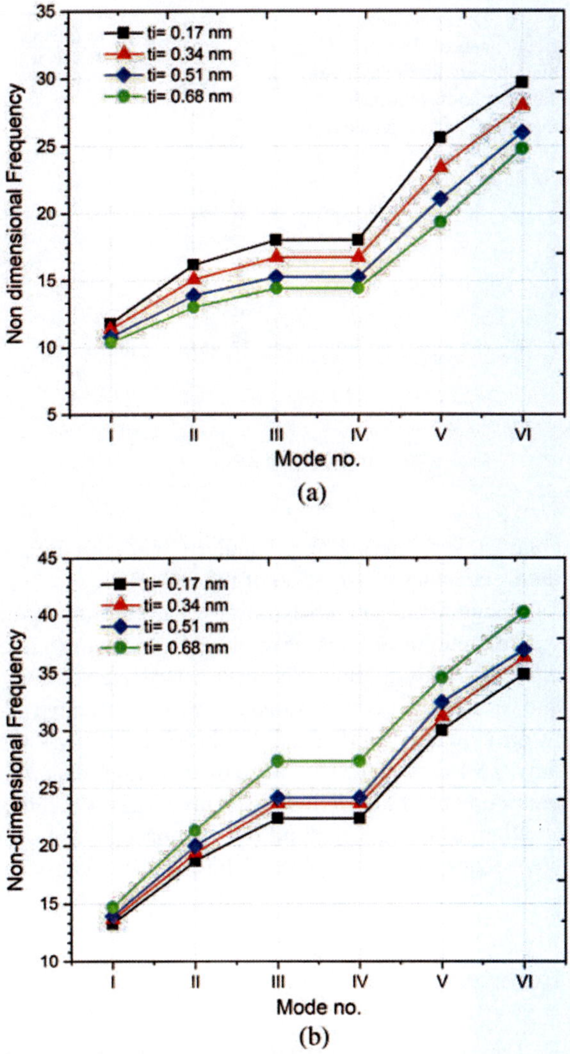

strengthening effect on the effective elastic modulus of the CNTRC and increase in its thickness causes increase in the elastic modulus; however, opposite trend is observed in case of the soft interphase. The soft interphase has very small influence on the longitudinal elastic modulus as compared to the stiff interphase. However, both have significant influences on the transverse elastic modulus of the CNTRC.

The developed analytical (closed form) model was found to quickly and accurately determine vibration frequencies of the CNTRC plates. It was found that the interphase has significant influence over the vibrational response of the CNTRC as well. Presence of stiff interphase considerably enhances the frequencies of vibration, and the increase in frequencies is more pronounced for higher modes of vibration. In

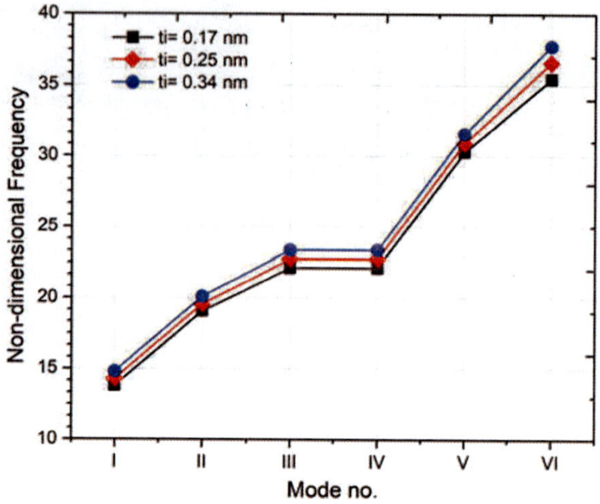

Fig. 6 Dimensionless frequencies of the CNTRC plate having interphase with varied modulus through its thickness, and subjected to a simply-supported boundary condition

contrast, soft interphase, particularly that with higher thickness, drastically reduces the frequencies of vibration of the CNTRC.

A significant finding is observed by the incorporation of an interphase having varied modulus through its thickness and having a thickness in the range of the CNT thickness. This interphase is observed to have a stiffening effect on the nanocomposite and resulted in a considerable increase in effective elastic properties of the CNTRC when compared with those predicted using two-phase RVE model consisting of only CNT and the matrix. Moreover, a significant decrease in deviation of the FEM prediction for the effective elastic properties was found when compared with the MD simulation result. This kind of interphase is also observed to appreciably enhance the frequencies of vibration of the CNTRC plate.

References

1. Han Y, Elliott J (2007) Molecular dynamics simulations of the elastic properties of polymer/carbon nanotube composites. Comput Mater Sci 39(2):315–323
2. Wan H, Delale F, Shen L (2005) Effect of CNT length and CNT-matrix interphase in carbon nanotube (CNT) reinforced composites. Mech Res Commun 32(5):481–489
3. Hammerand DC, Seidel GD, Lagoudas DC (2007) Computational micromechanics of clustering and interphase effects in carbon nanotube composites. Mech Adv Mater Struct 14(4):277–294
4. Li Y, Seidel GD (2014) Multiscale modeling of the effects of nanoscale load transfer on the effective elastic properties of unfunctionalized carbon nanotube-polyethylene nanocomposites. Model Simulation in Mater Sci Eng 22(2)
5. Rafiee R, Pourazizi R (2015) Influence of CNT functionalization on the interphase region between CNT and polymer. Comput Mater Sci 96:573–578
6. Banerjee D, Nguyen T, Chuang TJ (2016) Mechanical properties of single-walled carbon nanotube reinforced polymer composites with varied interphase's modulus and thickness: a finite element analysis study. Comput Mater Sci 114:209–218

7. Choi J, Shin H, Cho M (2016) A multiscale mechanical model for the effective interphase of SWNT/epoxy nanocomposite. Polymer 89:159–171
8. Malagù M, Goudarzi M, Lyulin A, Benvenuti E, Simone A (2017) Diameter-dependent elastic properties of carbon nanotube-polymer composites: emergence of size effects from atomistic-scale simulations. Compos B Eng 131:260–281
9. Maghsoudlou MA, Isfahani RB, Saber-Samandari S, Sadighi M (2019) Effect of interphase, curvature and agglomeration of SWCNTs on mechanical properties of polymer-based nanocomposites: experimental and numerical investigations. Compos Part B: Eng 175(12):107–119
10. Amraei J, Jam JE, Arab B, FirouzAbadi RD (2019) Modeling the interphase region in carbon nanotube-reinforced polymer nanocomposites. Polym Compos 40(S2):E1219–E1234
11. Ke LL, Yang J, Kitipornchai S (2010) Nonlinear free vibration of functionally graded carbon nanotube-reinforced composite beams. Compos Struct 92(3):676–683
12. Wang Z, Shen HS (2011) Nonlinear vibration of nanotube-reinforced composite plates in thermal environment. Comput Mater Sci 50:2319–2330
13. Zhu P, Lei Z, Liew K (2012) Static and free vibration analyses of carbon nanotube-reinforced composite plates using finite element method with first order shear deformation plate theory. Compos Struct 94(4):1450–1460
14. Huang B, Guo Y, Wang J, Du J, Qian Z, Ma T, Yi L (2017) Bending and free vibration analyses of antisymmetrically laminated carbon nanotube-reinforced functionally graded plates. J Compos Mater 51:3111–3125
15. Ahmadi M, Ansari R, Rouhi H (2018) Free vibration analysis of carbon fiber-carbon nanotube polymer matrix composite plates by a finite element-based multiscale modeling approach. J Multiscale Model 9(2):1850(002)
16. Safaei B, Ahmed N, Fattahi A (2019) Free vibration analysis of polyethylene/CNT plates. Eur Phys J Plus 134(271)
17. Aragh BS, Barati A, Hedayati H (2012) Eshelby-Mori-Tanaka approach for vibrational behavior of continuously graded carbon nanotube-reinforced cylindrical panels. Compos B Eng 43(4):1943–1954
18. Pouresmaeeli S, Fazelzadeh S (2016) Frequency analysis of doubly curved functionally graded carbon nanotube-reinforced composite panels. Acta Mech 227:2765–2794

A Review on Recent Techniques and Current Challenges in Identifying Defects in Additively Manufactured Metal Components

Vivek V. Bhandarkar and Puneet Tandon

Abstract Additive manufacturing (AM) or 3D printing is a process of successively joining layers of materials to fabricate a replica of a 3D CAD model. Capability to manufacture complex geometries and reduced material wastage are some of the key advantages of the 3D printing process over conventional subtractive manufacturing technologies. The technology is being evolved across most of the industrial sectors like biomedical, aerospace, and construction. However, the process is not as much extensively adopted as the traditional manufacturing techniques due to certain flaws and defects that are associated with the parts being fabricated. The defects like porosity, balling, cracking, and residual stresses in the part may hinder the application of a 3D printed part in a device or machine. The defective parts sometimes may lead to the failure of the machine. Hence, it becomes very crucial to identify such defects in real-time while the part is being fabricated and analyze the defect causing process parameters so that they can be eliminated in the final part. A comprehensive review of various types of defects in additively manufactured metal parts, possible defect causing process parameters, and the recent techniques to identify those defects is carried out in this study. In addition, the paper investigates several challenges being faced by researchers in implementing the defect identifying techniques for various 3D printing processes.

Keywords Additive manufacturing · 3D printing · Defects · Defects identification techniques

1 Introduction

Additive manufacturing (AM) creates physical objects from virtual 3D models by depositing material in a layer-by-layer manner with the help of digitally controlled tools. The layer-wise manufacturing approach can produce precise objects with complex geometries that are impossible to produce by conventional manufacturing

V. V. Bhandarkar · P. Tandon (✉)
deLOGIC Lab, Discipline of Mechanical Engineering, Pandit Dwarka Prasad Mishra Indian Institute of Information Technology, Design and Manufacturing, Jabalpur, MP 482005, India
e-mail: ptandon@iiitdmj.ac.in

V. K. Gupta et al. (eds.), *Recent Advances in Machines and Mechanisms*, Lecture Notes in Mechanical Engineering, https://doi.org/10.1007/978-981-19-3716-3_55

processes [1]. AM was first introduced in 1987 by the name Rapid Prototyping (RP) with Stereo lithography (SL) technology for prototyping non-metallic objects. SL involves the joining of thin layers of liquid polymers with the help of a laser source. A generalized flow process of AM involves creating a desired 3D CAD model, converting the CAD model into STL file format, G-code generation by slicing the 3D object and communicating the codes from the computer to the 3D printer [2]. American Society for Testing and Materials (ASTM) standard classifies AM processes into seven broad categories: Binder Jetting, Material Extrusion, VAT Photopolymerization, Material Jetting, sheet lamination, directed energy deposition (DED), and powder bed fusion (PBF). These broad categories are further classified into various sub-categories that can print a wide range of materials like polymers, metals, ceramics, and composites [3]. The innovative applications of AM are persistently being developed for the past couple of decades for the manufacturing of direct parts in the major industrial sectors like aerospace, automotive, military, medical, architectural, electronics, and various other areas. However, the sectors like automotive, aerospace, and industrial machines use these processes mainly for fabricating metallic products [4]. Binder Jetting, Sheet Lamination, PBF, and DED can produce metallic parts along with other innovative materials parts [5]. Nowadays, most industrial applications rely on the precision of mechanical properties and surface finish of the additively manufactured products, but AM alone is not capable of printing parts within the acceptable tolerances because of the occurrence of certain internal defects while printing [6, 7]. In the case of Selective Laser Melting (SLM), a lot of internal defects like balling, porosity, cracks, powder accumulation, powder vaporization, and thermal stresses occur while the part is being printed. These defects eventually influence the mechanical properties and quality of the final additively manufactured products [8]. Although AM technology is successfully adopted in almost every sector as compared to the traditional manufacturing methods, defects and flaws hinder its application in the functioning machinery. Therefore, certain techniques have been widely evolved in recent years to identify the defects and flaws in the additively manufactured components to improve their geometrical accuracy and mechanical properties. These techniques detect a particular defect and obtain the information related to the defect like its type, depth, contour, and size [9]. Low cost, real-time detection, capable of detecting different defects, detecting defects in complex geometries are some of the key prerequisites for advanced defects identification techniques in AM [10]. Identification of defects in real-time in AM facilitates modification in its parameters to eliminate the defects or stop the process to reduce material wastage and time. Defects identifying techniques in AM are classified into two groups, i.e., traditional techniques which are non-destructive, and techniques based on machine learning and deep learning [11].

The proposed study reviews the various categories of defects that occur in additively manufactured metallic components and encapsulates the techniques for identifying the associated defects. The study is divided into six modules: Section 1 provides brief introduction of the study. Section 2 summarizes the various kinds of defects in metal AM. Section 3 illustrates traditional methods of defects identification in AM. The defects identifying techniques in AM with the help of machine learning and deep learning technology are depicted in Sect. 4, while Sect. 5 discusses different challenges for implementing the advanced defects identifying techniques. At last, Sect. 6 discusses the conclusion.

2 Types of Defects in Metal Additive Manufacturing (MAM)

A lot of research is carried out in metal additive manufacturing (MAM) over recent years; however, it is still in the developing stage. Several process parameters can be optimized for minimizing discontinuity in printing materials. Defects in AM parts can take place due to non-optimized process parameters and these defects leads to the failure of parts that hinder its application in various industries. Despite many advantages of AM, the defects restrict the process in real-life usage due to unacceptable accuracy and mechanical properties. This study reviews the widely occurring defects in the case of MAM. The most common defects in the parts being printed in case of MAM are porosity, balling, cracking, and residual stresses [12].

2.1 Porosity

In MAM techniques like powder bed fusion (PBF) and directed energy deposition (DED), porosity occurs very frequently. Pores introduced due to a lack of powder particles fusion and confined gases that affect the density and mechanical properties of the fabricated parts [13]. Although there are some applications where controlled porosity is maintained to achieve desired properties of the parts like for scaffolds fabrication in biomedical applications [14], porosity is considered as a defect in most industrial applications. Hence it becomes essential to develop certain techniques to overcome porosity by reducing the pores. Günther et al. studied the defects on titanium alloy Ti-6Al-4 V in two different methods of MAM: Electron beam melting (EBM) and selective laser sintering (SLS) [15]. Figure 1 illustrates the scanning electron microscopic (SEM) images of porosity defects in these two MAM methods.

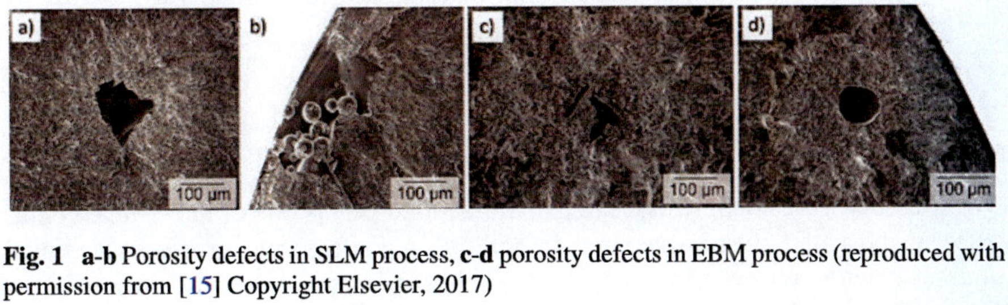

Fig. 1 **a-b** Porosity defects in SLM process, **c-d** porosity defects in EBM process (reproduced with permission from [15] Copyright Elsevier, 2017)

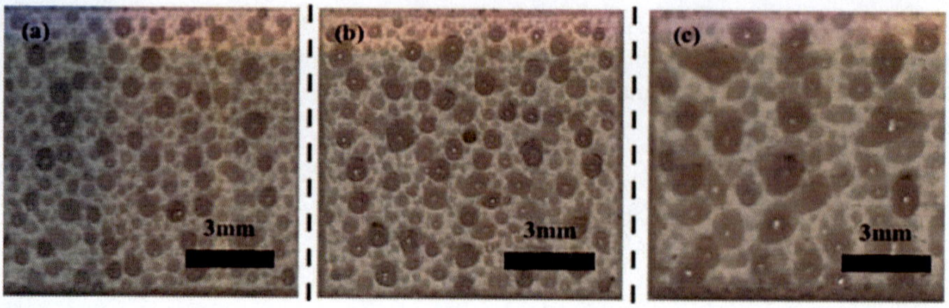

Fig. 2 Optical images of balling phenomenon on alumina material in SLM with different hatch spaces: **a** 0.15 mm; **b** 0.10 mm; **c** 0.05 mm (reproduced with permission from [17] Copyright Elsevier, 2020)

2.2 *Balling*

Balling defects are nothing but the formation of melt metal balls in the case of MAM which arises when the solidification of molten metal takes place in the form of spheres instead of the desired solidified layers. This defect is responsible for the improper interconnection of the subsequent layers that eventually leads to the geometrical inaccuracy, undesired surface roughness, and unacceptable mechanical properties of the final part [16]. Qiu et al. conducted SLM on alumina and probed the various defects including the balling defect that occurred during the process [17]. Figure 2 illustrates the optical images of balling occurrence in the AM parts.

2.3 *Cracking*

Cracks are commonly occurring defects in the parts manufactured by most of the metal AM processes. These cracks are originated primarily due to the thermal stresses and suppress the quality and mechanical properties of the parts [18]. Gao et al. investigated the crack growth rates, surface microstructures, and fracture mechanisms of TC4-DT alloy components produced by wire-arc additive manufacturing [19]. Figure 3 shows the propagation of cracks in wire-arc additively manufactured components on TC4-DT alloy once it originates on its grain boundaries.

2.4 *Residual Stresses*

The presence of residual stresses in a manufactured part is one of the critical conditions in most of industrial sectors. In additive manufacturing processes, residual stresses occur due to varying temperature fields and trapped thermal stresses in the fabricated parts. When the confined stresses within the part are rescued suddenly,

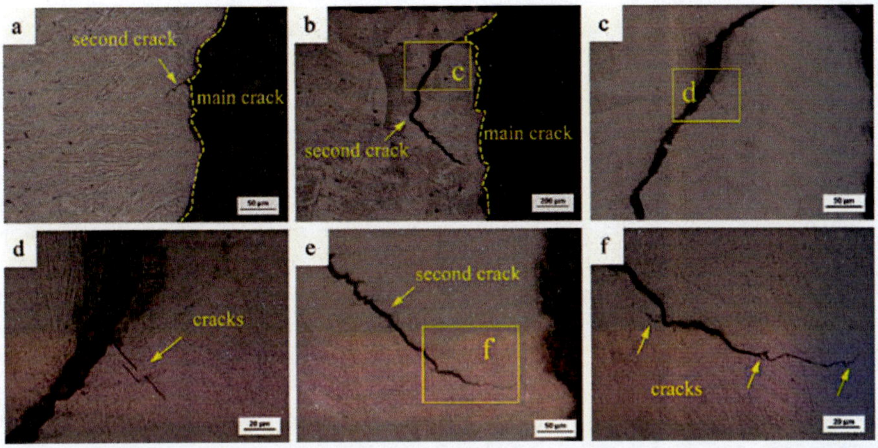

Fig. 3 SEM images of cracking phenomenon and its growth in TC4-DT alloy parts fabricated by wire-arc additive manufacturing. (reproduced with permission from [19] Copyright Elsevier, 2021)

warpage and cracking occur on the part surface or inside the surface that directly deforms the part geometry and affects the part quality [20].

3 Traditional Techniques of Defects Identification in AM

The traditional techniques of identifying or detecting defects in AM are non-destructive in nature and the most used such techniques are ultrasonic defects identification, penetration defects identification, defects identification using infrared imaging, and with the help of eddy current.

3.1 Ultrasonic Defects Identification Technique

Ultrasonic waves are used in this technique for identifying the internal defects in the metal parts with the help of a transmitting probe that releases the waves toward the surface of the metallic component and then inside the component. Based on these propagated waves, the probe gets back the signals reflected from different interfaces of the component and identifies the defect by comparing the time difference between these reflected signals. The defect characteristics like its size, position, and nature are evaluated by analyzing the height and position of the reflected signals on the display screen [21]. Millon et al. used a pulsed laser to emit ultrasonic waves on the additively manufactured part surface for detecting cracks in the part [22].

3.2 *Penetration Defects Identification Technique*

It is a non-destructive inspecting technique that uses a capillary concept for identifying the defects on the additively manufactured parts. Here, a penetrant is used for applying fluorescent or colored agents to the part surface. The penetrant then penetrates and settles into the defects on surfaces with the help of the capillary phenomenon. Finally, the effect of the light source is used to identify the associated surface defects. Performance of penetrant, fluorescent agents and the associated defects are the factors that decide the sensitivity of this technique. Glenn Research Center of NASA utilizes this technique to investigate the parts of rocket engine that are fabricated by SLM and EBM processes [23]. The limitation of this technique is that it is not suitable for the porous parts manufactured by AM.

3.3 *Defects Identification Using Infrared Imaging*

This technique uses the intensity of thermal radiations of the additively manufactured parts to identify the shape and contour of defects. The infrared images depict defects by analyzing the difference of thermal radiations between the defects and the neighboring materials. Thermal imaging cameras are generally used in this technique to inspect the temperature of the surfaces of additively manufactured parts for analyzing the thermal behavior and identification of defects location. Schwerdtfeger et al. identified the positions of defects based on the intensity of radiations by conducting layer-wise infrared imaging in the electron beam selective melting process and found that the defects possess higher radiation intensity [24].

3.4 *Defects Identification Using Eddy Current*

The principle of electromagnetic induction works in this non-destructive testing technique to identify defects in additively manufactured conductive material parts by analyzing the changes in induced eddy currents. The technique conveys acceptable results for the defects like cracks and non-fusion pores. Du et al. investigated the eddy current defects identifying technique on 3D printed parts of composites [25].

4 Machine Learning and Deep Learning-Based Defects Identification Techniques

Of late, machine learning (ML) technology is used by most of the scholars for identifying defects in 3D printed parts due to the numerous benefits it offers over traditional techniques. Some of such benefits include the technique being highly automated and better detection rate and accuracy. These technologies use a neural network consisting of numerous convolutional layers which are used to learn and provide precise output features. Then the data is classified based on the extracted output features through its powerful learning capability. ML and deep learning (DL)-based techniques are carried out in two steps. The first step involves capturing and collecting the images of AM parts and in the second step, defects are detected by analyzing various features. The most widely used such techniques are discussed in this section.

4.1 Defects Identification Using Convolutional Neural Network (CNN)

CNN-based defect identification technique involves creating a complicated CNN structure with different layers and then identifying image-based defects through end-to-end training of the obtained image data features from various layers [26]. The biggest benefit of CNN for defects identification is that it can provide high level abstract image data features by learning from large input data. However, the computational complexity of CNN increases as the depth of the network increases. Using a reasonable camera system, AM process images are captured as raw data in this technique. Then the image data is subsequently utilized to train several CNN architectures for defect classification, resulting in the development of an automated process assessment and documentation system. Westphal and Sietz offer complicated transfer learning (TL) techniques based on CNN for automatically classifying powder bed defects in the SLS process utilizing relatively little datasets [27].

4.2 Defects Identification Using Autoencoder Network Method

A feedforward network with one or more linked hidden layers is known as an Autoencoder neural network architecture. It employs a nonlinear mapping function to transform the original data into learnt characteristics by encoding and decoding the fully interconnected neurons. Encoding and decoding are the two steps involving in this technique. In the first step, features are extracted from the input data by converting them into encoded signals. Then in the second step (i.e., decoding step), reconstructed signals are obtained from the extracted features by adjusting weight and offset with a minimum or negligible error [28]. Defects in additively manufactured components are then identified based on these reconstructed signals. The Autoencoder network maps between the input space and feature space for reducing reconstruction error from the extracted features and identifies the associated defect [29].

4.3 *Defects Identification Using Deep Residual Neural Network*

Residual neural network (ResNet) uses skip connections (also known as shortcut connections) to link input from one layer to the next without modifying the input. We deliberately allow stacked layers to fit a residual mapping rather than connecting them to match a chosen underlying mapping. It is feasible to have a deeper network and gain greater performance by skipping the connection. In this technique of defects identification, the residual optimization phenomenon is used to increase the number of network layers in a continuous manner when network structure complexity increases giving the convolutional layer output in residual unit with the same dimensions as the input data [30]. This technique enhances the performance of classification by lowering the convergence and preventing over-fitting and hence it can be used for identifying defects in additively manufactured parts through part images classification [31].

4.4 *Defects Identification Using Recurrent Neural Network (RNN)*

A typical CNN framework uses convolution and pooling tasks to extract features from input layers of test data. Whereas RNN framework extract input data layer features by performing the recurrent operation and hence it eliminates the pooling task. Hence, this technique can be applied for extracting essential features where the input sample testing data is limited. This technique minimizes the loss of data in the pooling operation. The limitation of this technique is that the RNN model may exhibit over-fitted behavior when the number of iterations during network training increases [32]. In this technique, a dataset of additively manufactured components images is generated which is then divided into training set and testing set. Depending on the image features, the RNN model classifies the images with defects. For instance, void defect exists on an image as black areas and the existence of more than expected black color in the image shows a deviation and this deviation is marked by RNN model as a defect.

5 Challenges for Implementing Advanced Defects Identifying Techniques

Many research scholars have concluded that machine learning-based defects identification techniques are more suitable as compared to the traditional techniques in the case of additively manufactured components due to their ability to adapt complex data structures and conduct efficient feature extraction. Koeppe and Hernandez Padilla performed prediction of stress in lattice structures with the help of a trained ML framework and found that the framework took only 0.47 s whereas finite element method (FEM) simulation takes 5–10 h to complete the prediction [33]. However, processing and training a large amount of AM data can be time-consuming, costly, and challenging. The absence of accurate, accessible, and comprehensive datasets on AM processes, products, and materials is a key challenge to the development of implementing ML-based defects identification techniques in AM. While each build of a part in AM process might create terabytes of data, there are no standard procedures for dealing with high-volume, high-velocity information in real-time. Rich, multidimensional, data-driven analysis is impossible without a consistent data structure and standard procedures for data integration and fusion. Furthermore, obtaining optimal data through testing is difficult and costly. Even if data is accessible, its low-quality renders it unsuitable for machine learning algorithms [34]. This makes feature selection for machine learning systems more difficult.

It is essential and challenging to develop a standard of qualification for sharing huge AM dataset required for ML model that will facilitate data sharing among AM community with any available ML frameworks in the market like Pytorch, Tensorflow, and Caffe for encouraging collaboration among them [35]. Defects identification in layer-wise manner using an optical or thermal camera is an expensive technique and can identify defects only on the surface of the AM part but the technique is not suitable for identifying defects within the part. Expensive and time-consuming X-ray methodologies like X-ray phase-contrast imaging (XPCI) and computed tomography (CT) can identify internal defects but clubbing them with ML models for real-time monitoring and defects detection in AM is a challenge [36].

6 Conclusion

AM is being used in many industrial sectors including aerospace and defense where geometry complexity, customization and near-net-shape fabrication are the key requirements. However, certain defects in additively manufactured components like porosity, balling, cracking, and residual stresses limit the realization of the full potential of 3D printed components as the defects directly affect the mechanical properties and may lead to its failure in near future. Therefore, researchers around the world have come up with advanced defect identification techniques in the AM sector. The widely used techniques have been reviewed in this study. Besides, the study also highlights some of the major challenges in implementing advanced machine learning-based defect identification techniques. For applying deep learning technique to AM, enormous data related to AM parts like angles, size, shape, position, and other relevant information is needed. Machine learning and deep learning-based techniques for identifying defects have both benefits and limitations but designing a ubiquitous deep learning or machine learning framework for identifying defects in AM is a challenging job and therefore researchers must study on this area in near future. Currently practiced defect identification techniques mainly detect the surface defects like scratches that are two dimensional. An inspection system with multiple cameras can be incorporated for probing and identifying defects in all three dimensions of additively manufactured components by improving their accuracy. Moreover, various advanced smart sensors and embedded systems can be interfaced with machine learning techniques to help evolve online and real-time defects identification techniques for AM.

References

1. Druzgalski CL, Ashby A, Guss G, King WE, Roehling TT, Matthews MJ (2020) Process optimization of complex geometries using feed forward control for laser powder bed fusion additive manufacturing. Addit Manuf 34:101169
2. Mercado Rivera FJ, Rojas Arciniegas AJ (2020) Additive manufacturing methods: techniques, materials, and closed-loop control applications. Int J Adv Manuf Technol. https://doi.org/10.1007/s00170-020-05663-6
3. Lee J-Y, An J, Chua CK (2017) Fundamentals and applications of 3D printing for novel materials. Appl Mater Today 7:120–133
4. Vafadar A, Guzzomi F, Rassau A, Hayward K (2021) Advances in metal additive manufacturing: a review of common processes, industrial applications, and current challenges. Appl Sci (Basel) 11:1213
5. Vaezi M, Chianrabutra S, Mellor B, Yang S (2013) Multiple material additive manufacturing—Part 1: a review: This review paper covers a decade of research on multiple material additive manufacturing technologies which can produce complex geometry parts with different materials. Virtual Phys Prototyp. 8:19–50
6. Pyka G, Kerckhofs G, Papantoniou I, Speirs M, Schrooten J, Wevers M (2013) Surface roughness and morphology customization of additive manufactured open porous Ti6Al4V structures. Materials (Basel). 6:4737–4757
7. Turner BN, Gold SA (2015) A review of melt extrusion additive manufacturing processes: II. Materials, dimensional accuracy, and surface roughness. Rapid Prototyp J 2:250–261
8. Leung CLA, Marussi S, Towrie M, Atwood RC, Withers PJ, Lee PD (2019) The effect of powder oxidation on defect formation in laser additive manufacturing. Acta Mater 166:294–305
9. Tian L, Fan Y, Li L, Mousseau N (2020) Identifying flow defects in amorphous alloys using machine learning outlier detection methods. Scr Mater 186:185–189
10. Bernhard R, Neef P, Wiche H, Hoff C, Hermsdorf J, Kaierle S, Wesling V (2020) Defect detection in additive manufacturing via a toolpath overlaid melt-pool-temperature tomography. J Laser Appl 32:022055
11. Deng Q, Fu W, Chen D, Cao P (2014) Measurement of the molten pool image during laser cladding process. In: Proceedings of the 2014 international conference on mechatronics, electronic, industrial and control engineering. pp 630–634. Atlantis Press, Paris, France
12. Grasso M, Colosimo BM (2017) Process defects and in situ monitoring methods in metal powder bed fusion: a review. Meas Sci Technol 28:044005
13. Edwards P, O'Conner A, Ramulu M (2013) Electron Beam additive manufacturing of titanium components: Properties and performance. J Manuf Sci Eng 135:061016
14. Telang VS, Pemmada R, Thomas V, Ramakrishna S, Tandon P, Nanda HS (2021) Harnessing additive manufacturing for magnesium-based metallic bio implants: Recent advances and future perspectives. Curr Opin Biomed Eng 17:100264
15. Günther J, Krewerth D, Lippmann T, Leuders S, Tröster T, Weidner A, Biermann H, Niendorf T (2017) Fatigue life of additively manufactured Ti–6Al–4V in the very high cycle fatigue regime. Int J Fatigue 94:236–245

16. Demir AG, Previtali B (2017) Investigation of remelting and preheating in SLM of 18Ni300 maraging steel as corrective and preventive measures for porosity reduction. Int J Adv Manuf Technol 93:2697–2709
17. Qiu Y-D, Wu J-M, Chen A-N, Chen P, Yang Y, Liu R-Z, Chen G, Chen S, Shi Y-S, Li C-H (2020) Balling phenomenon and cracks in alumina ceramics prepared by direct selective laser melting assisted with pressure treatment. Ceram Int 46:13854–13861
18. Malekipour E, El-Mounayri H (2018) Common defects and contributing parameters in powder bed fusion AM process and their classification for online monitoring and control: a review. Int J Adv Manuf Technol 95:527–550
19. Gao Y, Wu C, Peng K, Song X, Fu Y, Chen Q, Zhang M, Wang G, Liu J (2021) Towards superior fatigue crack growth resistance of TC4-DT alloy by in-situ rolled wire-arc additive manufacturing. J Mater Res Technol
20. Kruth JP, Froyen L, Van Vaerenbergh J, Mercelis P, Rombouts M, Lauwers B (2004) Selective laser melting of iron-based powder. J Mater Process Technol 149:616–622
21. Zhou, Z (2017) New progress of the study and application of advanced ultrasonic testing technology. Jixie Gongcheng Xuebao (Chin J Mech Eng) 53(1)
22. Millon C, Vanhoye A, Obaton A-F, Penot J-D (2018) Development of laser ultrasonics inspection for online monitoring of additive manufacturing. Weld World. 62:653–661
23. Waller JM, Saulsberry RL, Parker BH, Hodges KL, Burke ER, Taminger KM (2015) Summary of NDE of additive manufacturing efforts in NASA. Presented at the AIP Publishing LLC
24. Schwerdtfeger J, Singer RF, Körner C (2012) In situ flaw detection by IR-imaging during electron beam melting. Rapid Prototyp J. 18:259–263
25. Du W, Bai Q, Wang Y, Zhang B (2018) Eddy current detection of subsurface defects for additive/subtractive hybrid manufacturing. Int J Adv Manuf Technol 95:3185–3195
26. Lin J, Yao Y, Ma L, Wang Y (2018) Detection of a casting defect tracked by deep convolution neural network. Int J Adv Manuf Technol 97:573–581
27. Westphal E, Seitz H (2021) A machine learning method for defect detection and visualization in selective laser sintering based on convolutional neural networks. Addit Manuf 41:101965
28. Lu H, Liu S, Wei H, Tu J (2020) Multi-kernel fuzzy clustering based on auto-encoder for fMRI functional network. Expert Syst Appl 159:113513
29. Ren Z, Zhu Y, Yan K, Chen K, Kang W, Yue Y, Gao D (2020) A novel model with the ability of few-shot learning and quick updating for intelligent fault diagnosis. Mech Syst Signal Process 138:106608
30. Tang Y, Huang J, Zhang F, Gong W (2020) Deep residual networks with a fully connected reconstruction layer for single image super-resolution. Neurocomputing 405:186–199
31. Chen Y, Peng X, Kong L, Dong G, Remani A, Leach R (2021) Defect inspection technologies for additive manufacturing. Int J Extrem Manuf 3:022002
32. Lei J, Gao X, Feng Z, Qiu H, Song M (2018) Scale insensitive and focus driven mobile screen defect detection in industry. Neurocomputing 294:72–81
33. Koeppe A, Hernandez Padilla CA, Voshage M, Schleifenbaum JH, Markert B (2018) Efficient numerical modeling of 3D-printed lattice-cell structures using neural networks. Manuf Lett. 15:147–150
34. Everton SK, Hirsch M, Stravroulakis P, Leach RK, Clare AT (2016) Review of in-situ process monitoring and in-situ metrology for metal additive manufacturing. Mater Des 95:431–445
35. Goh GD, Sing SL, Yeong WY (2021) A review on machine learning in 3D printing: applications, potential, and challenges. Artif Intell Rev 54:63–94
36. Thompson A, Maskery I, Leach RK (2016) X-ray computed tomography for additive manufacturing: a review. Meas Sci Technol 27:072001

Design and Development of Heterogeneous Porous Scaffold—A Review

Anand Prakash Mall and Puneet Tandon

Abstract This paper presents a review of the strategies for the design and development of heterogeneous porous scaffolds. Once the lattice structure (scaffold) is designed with the help of computer-aided design (CAD) software, analysis is done to investigate its compressive response with finite element analysis (FEA). The stiffness of the scaffold is calculated and compared with that of a compact bone and validated experimentally with the help of 3D printing of the scaffold. Heterogeneity of both types, i.e. structural and material, helps mimic the structure of human bones as human bones are porous, heterogeneous and anisotropic. Topology optimization of lattice structure is performed to achieve a high surface area to volume ratio of the scaffold so that cell proliferation and vascularization can be increased. Porosity, pore interconnectivity and pore size are varied to achieve the desired properties in the scaffold to match the stiffness of the scaffold with that of humane bone. Metallic biomedical materials that are investigated in the reported works include stainless steel SS 316L (~210 GPa), Co–29Cr–6Mo alloy (~210 GPa), Ti–6Al–4V alloy (~110 GPa) and Mg alloy (~45 GPa). All the above materials have stiffness greater than that of the cancellous bone (0.02–2 GPa) and the cortical bone (3–30 GPa) which creates a stress shielding effect after implantation in the bone. To avoid the stress shielding problem, these materials are made porous to decrease the stiffness of the solid material. Porosity is not only used for reducing the stress shielding effect but is also used for cell proliferation and vascularization. The objective of this review is to reflect on the heterogeneous porous scaffold design process detailing various phases like material and structural heterogeneity. The current literature on the design of heterogeneous porous scaffold is mainly focused on discussing structural heterogeneity and not much focused on material heterogeneity and there is very little work that reports both structural and material heterogeneity. Thus, there is a clear lack of a comprehensive view. This paper acts as a bridge among the main research outcomes available in the literature related to the design optimization and modelling of the heterogeneous porous scaffolds.

A. P. Mall · P. Tandon (✉)
deLOGIC Lab, Discipline of Mechanical Engineering, PDPM Indian Institute of Information Technology, Design and Manufacturing, Jabalpur, MP 482005, India
e-mail: ptandon@iiitdmj.ac.in

V. K. Gupta et al. (eds.), *Recent Advances in Machines and Mechanisms*, Lecture Notes in Mechanical Engineering, https://doi.org/10.1007/978-981-19-3716-3_56

Keywords Structural heterogeneity · Material heterogeneity · Porous scaffold · Unit cell · Lattice structure · Topology optimization

1 Introduction

Biomedical materials are classified into two categories, e.g. implanted biomaterials and non-implanted materials. Implanted biomaterials are further classified as synthetic biomaterials and biological biomaterials. Metals that are used in biomedical engineering are broadly classified into four main categories, i.e. stainless steel, Co-based alloys, Ti-based alloys and miscellaneous, i.e. NiTi, Mg, Ta, etc. Ti-alloy, e.g. Ti–6Al–4V comes under the category of synthetic biomaterials. We are now in the third generation of biomaterials, where biodegradation and tissue regeneration are given importance, e.g. magnesium alloys and degradable polymers. In the first generation of biomaterials, importance was given to biologically inert materials which do not harm tissues, e.g. Co-alloys. In the second generation of biomaterials, importance was given to surface erosion and tissue bonding, e.g. titanium alloys and hydroxylapatite [1]. Tissues are mainly of two types, e.g. hard tissue and soft tissue. Bones come under the category of hard tissue while muscle, ligament, tendon and cartilages come under the category of soft tissue. The scaffold has a main role in tissue engineering, which is a branch of biomedical engineering. A scaffold is a three-dimensional (3D) lattice structure that is required for tissue repair when the size of the tissue defect is above a certain limit and it cannot be repaired on its own [2]. Tissue engineering has three important constituents, which are known as tissue engineering triad, i.e. scaffolds, cells and signals. Orthopaedic is a branch of medical science which deals with the study of bones. Human bones consist of two regions, e.g. the inner core is called cancellous bone and the outer peripheral region is called cortical bone. The compressive strength of the cortical bone is 130–290 MPa, whereas its tensile strength is 90–190 MPa. The compressive strength of cancellous bone is 02–38 MPa [3]. The segmental bone defect is a major problem in the orthopaedic. In large-size bone defect, the bones are separated by a large gap. The main cause of this type of defect is high energy impact and/or disease. The two separated bones can be repaired by a bone plate if the gap is not large. A scaffold is required when the gap between the two bones are not small. There are two types of scaffolds, e.g. natural scaffold and synthetic scaffold. The natural scaffold can be further classified as autograft and allograft. In the autograft, a graft of tissue is taken from one part of the body of a person and put at another body point of the same person. In allograft, a graft of tissue is taken from a donor and put on the recipient of the same species. In autograft, the harvesting process has been associated with preoperative and post-operative problems. Viral disease transmission and bacterial infection are the major risks in the allograft. These restrictions limit the use of autograft and allograft in orthopaedic. This is the reason for the use of synthetic scaffold nowadays.

1.1 Introduction to Scaffold

A scaffold is used to provide temporary initial structural support and it should degrade during organ tissue regeneration [4]. Any porous structure cannot be considered as a scaffold unless it is biodegradable. The rate of degradation of the scaffold should match the rate of regeneration of tissue. The scaffold has a conflicting requirement of porosity and mechanical strength. Mechanical strength of scaffold decreases when porosity increases. Three things need to be considered in the biomimetic design for load-bearing tissue scaffolds, i.e. biological requirement, mechanical requirement and anatomical requirement. The scaffold should have the following desired properties which are (i) support and deliver cells, (ii) induce, differentiate and channel tissue growth, (iii) target cell adhesion substrates, (iv) stimulate cellular responses, (v) biocompatible and biodegradable, (vi) large surface/volume ratio, (vii) mechanically strong and structurally stable, (viii) processable and malleable and (ix) sterilizable [5]. The traditional methods of scaffold preparations are (i) solvent casting and particulate leaching (SC/PL), (ii) electrospinning (ES), which includes (a) solution electrospinning and (b) melt electrospinning, (iii) emulsion freeze-drying, (iv) thermally induced phase separation (TIPS), (v) melt moulding and (vi) three-dimensional printing (3DP). Ti–alloy, e.g. Ti–6Al–4V scaffold can be fabricated by three additive manufacturing methods, namely selecting laser melting (SLM), electron beam melting (EBM) and laser engineering net shaping (LENS). EBM uses an electron beam as an energy source, while SLM uses a laser beam as an energy source. EBM can only process conductive metals, while SLM can process metals, polymers, ceramics and composites. As EBM has more diffuse energy, so it results in a larger heat affected zone and minimum feature size larger than that of SLM. Scaffolds, which are made by Ti–6Al–4V show very good biocompatibility, corrosion resistance and low density. Ti–6Al–4V alloy of Ti is very popular in bioimplant but it has one issue of short lifespan due to erosion of vanadium (V) and aluminium (Al) in the body fluid. Maximum acceptable corrosion rate for bioimplant is 0.13 mm/year [6]. Release of aluminium (Al) up to a certain limit in the body fluid is not found to have any carcinogenic effect on the body but release of vanadium (V) in the body fluid has bad effect on human health. So it is required to have an alternative element in place of vanadium (V) so that the properties of material will remain same or improve. Co-Cr alloy is used as a bioimplant for its good wear resistance property. So cobalt (Co) can be an alternative of vanadium (V). So Ti–6Al–2Co alloy can be an alternative of Ti–6Al–4V alloy. Singh et al. [6], Abhash et al. [7] and Jain et al. [8] have done good work on synthesis of Ti-alloy (Ti–6Al–2Co) by using space holder particle in powder metallurgy technique. It is observed that Ti-foam at room and low temp (200 °C) shows brittle behaviour and Ti-foam at higher temp (400–600 °C) shows ductile behaviour [8].

Heterogeneity: Heterogeneity in a material can be defined based on the properties that are different in all directions at any two points in the material, whereas anisotropy in the material can be defined as properties that are different in different directions at a point. Heterogeneity in scaffold or lattice structure can be of two types, e.g. structural

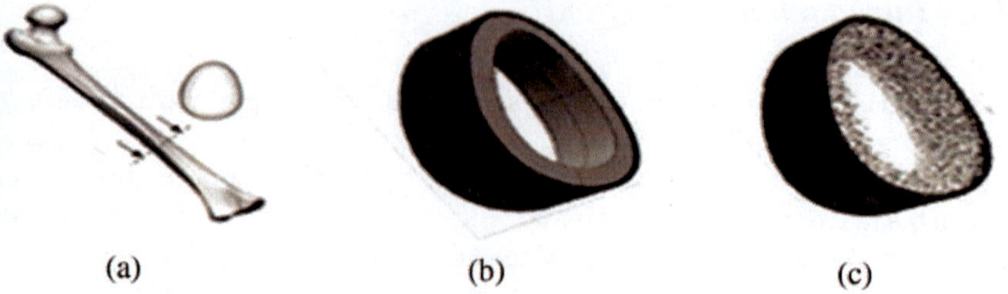

Fig.1 **a** Human femur bone, **b** cross section of the bone, **c** cancellous and cortical (Reproduced with permission from [10] Copyright Elsevier, 2016)

heterogeneity and material heterogeneity [9]. Human bones are heterogeneous, i.e. cancellous bone is the inner core of the bone which is more porous than the cortical bone which is the outer peripheral region of the bone (see Fig. 1).

Heterogeneous Porous Scaffold: Heterogeneous porous scaffold can best mimic the properties of human bone. Structural heterogeneity can be produced by using honeycomb geometry. Gupta and Tandon [11, 12] unfolded the development of a standalone convolution surface-based modelling approach for complex heterogeneous objects with the varied material distribution.

2 Design and Development

The major challenges in the design of the scaffolds include the selection of biocompatible material, need for metallic scaffold, controlled porosity, compressive strength, heterogeneous lattice structures, patient-specific implant designs and the complexity of manufacturing and cost of manufacturing. Design parameters that are used in the manufacturing of scaffold or lattice structure are (i) pore size, porosity and pore interconnectivity, (ii) ratio of surface area to volume and (iii) mechanical strength. The difference of stiffness of bone and implant gives rise to a new term known as the stress shielding effect. Implant and bone act as a composite bar. The implant has a higher stiffness due to which the load shared by the implant is higher than that of the bone which causes stress shielding. Stress shielding means the load shared by the bone after implantation is less than that of before implantation [13]. The cause of aseptic loosening is stress shielding. Aseptic loosening causes micro-motion of the implant which is not only painful to the patient but it can lead to failure of the implant also. If a force is applied on the composite bar which consists of implant and bone, and if we apply the equality of strain criteria in both the bars and Hooke's law, then the load which is shared by the implant and bone can be separately obtained by using the following equation [14]:

$$F_{\mathrm{i}} = \frac{A_{\mathrm{i}} E_{\mathrm{i}} F}{A_{\mathrm{i}} E_{\mathrm{i}} + A_{\mathrm{b}} E_{\mathrm{b}}} \quad F_{\mathrm{b}} = \frac{A_{\mathrm{b}} E_{\mathrm{b}} F}{A_{\mathrm{i}} E_{\mathrm{i}} + A_{\mathrm{b}} E_{\mathrm{b}}}$$

In the above equation, F is the force applied to the composite which consists of bone and implant. E is the stiffness of the material and the cross-sectional area of composite is denoted as A. Subscript i denotes the implant and b denotes the bone. Then,

$$\frac{F_\mathrm{i}}{F_\mathrm{normal}} = \frac{A_\mathrm{i} E_\mathrm{i}}{A_\mathrm{i} E_\mathrm{i} + A_\mathrm{b} E_\mathrm{b}} \frac{M_\mathrm{i}}{M_\mathrm{bending}} = \frac{I_\mathrm{i} E_\mathrm{i}}{I_\mathrm{i} E_\mathrm{i} + I_\mathrm{b} E_\mathrm{b}}$$

In the literature survey, it can be found that porous biomaterials are mostly homogeneous in terms of pore size, porosity, composition and mechanical property. However, we can find some literature on porous biomaterials which are heterogeneous in terms of pore size and composition. These parameters, e.g. porosity, pore size, composition and mechanical property can be graded or gradient to get the heterogeneous porous scaffold. Graded means step-wise change, while the gradient means continuous change [15, 16].

2.1 *Stiffness Measurements for Scaffold*

Scaffolds have reduced stiffness and density due to porous structure. Gibson and Ashby have described it in the form:

$$\boldsymbol{E} = \boldsymbol{E}_0 \left[\frac{\boldsymbol{\rho}}{\boldsymbol{\rho}_0} \right]^n$$

where E is the scaffold stiffness which is having a density ρ, whereas E_0 and ρ_0 are stiffness and density of fully dense material, respectively. Ti-based alloy (Ti–6Al–4V) has stiffness (E_0) and density (ρ_0) as 110 GPa and 4.43 g/cm^3, respectively. Co-based alloy (Co–29Cr–6Mo) has stiffness (E_0) and density (ρ_0) of 210 GPa and 8.44 g/cm^3, respectively. The exponent n varies from ~ 1.8 to 2.2 for aluminium and aluminium alloy scaffolds. The exponent n varies from ~ 2.0 to 2.3 for metals and metal alloys, e.g. Ti-based alloy (Ti–6Al–4V), Cu and Co-based alloy (Co–29Cr–6Mo). The value of exponent n has been assumed as 2. Systematic tapping of an optimum specimen size results in resonant frequency or vibration which is used in the measurement of dynamic stiffness in non-destructive testing according to the expression:

$$\boldsymbol{E} = \varsigma \boldsymbol{m} \boldsymbol{f}_\mathrm{r}^2$$

where E is the stiffness of the scaffold, ς: scaffold shape factor, m: scaffold mass and f_r frequency of resonance.

The structural stiffness E_s of the scaffolds is calculated using the linear elastic response of the load–displacement relation according to the following equation:

$$E_s = \frac{F_R . l_0}{A . \Delta l}$$

where F_R is the applied load to compress the scaffolds, l_o is the initial length of the scaffold, Δl is the shortening of the scaffold and A is the bottom surface of the scaffold.

Relative density of the lattice structure (scaffold): Relative density is defined as the ratio of density of the lattice structure (ρ) to density of the solid material (ρ_0) with which lattice structure is made. Relative density is very important parameter for the lattice structure because it is the main parameter for controlling the mechanical properties of the lattice structure like elastic modulus, yield strength and Poisson's ratio. Relative density differs with different morphology of a unit cell and with different cross sections of the strut like circular, square and triangular. Relative density calculated for a unit cell can be considered as the relative density of the lattice structure since a unit cell is the smallest representative unit of the lattice structure. Method of calculating relative density has been discussed for a particular morphology of a unit cell with struts of circular cross section (see Fig. 2). Unit cell shape parameters are length (a), breadth (b) and depth (d), whereas the radius and length of the strut of the unit cell are r and L, respectively.

$$\frac{\rho}{\rho_0} = \frac{\text{Density of the cellular structure}}{\text{Density of the solid that encloses it}}$$

Cuboid Unit Cell

$$\frac{\rho}{\rho_0} = \frac{(\pi . r^2) . \frac{\left(\sqrt{a^2+b^2+d^2}\right)}{2} . 8}{a . b . d}$$

Cubic Unit Cell

$$\frac{\rho}{\rho_0} = \left(\pi . 4 . \sqrt{3}\right) . \left(\frac{r}{l}\right)^2$$

Porosity and Pore Interconnectivity: Pores can be of two types, i.e. closed pores and open pores. Closed pores are not connected, while open pores are connected. In the manufacturing of scaffold, pore interconnectivity is a very important parameter that is responsible for cell proliferation and vascularization. Scaffolds are made porous not only for reducing the stress shielding effect since porosity reduces stiffness but also for biological requirements (see Fig. 3).

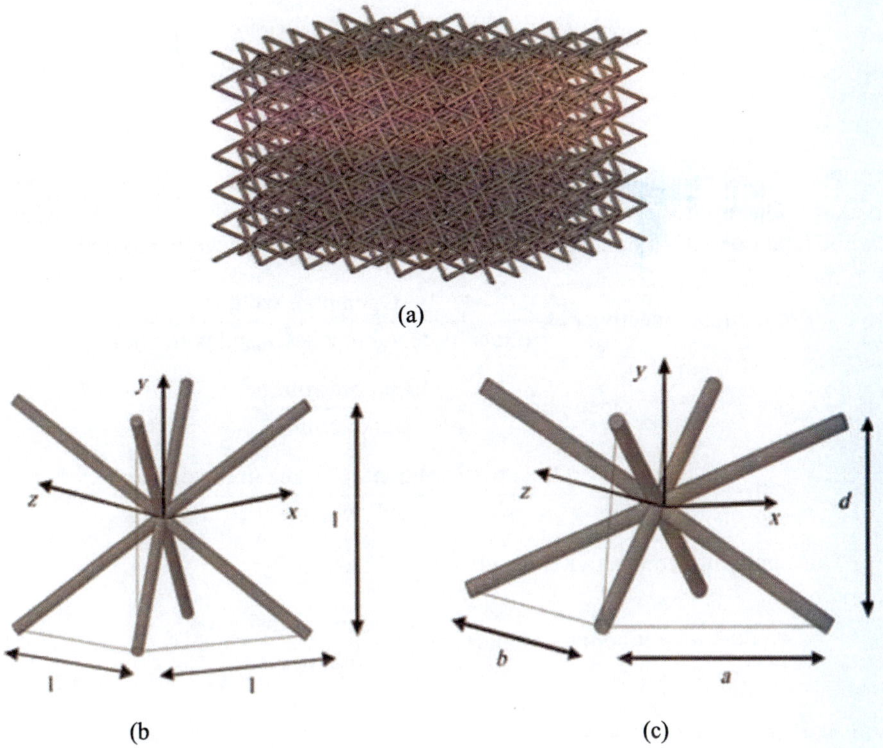

Fig.2 **a** Lattice structure, **b** cubic unit cell geometry (cubic with $a = \mathrm{b} = d = l$), **c** cuboid unit cell geometry (cuboid with $a \neq b \neq d$) (Reproduced with permission from [17] Copyright John Wiley and Sons, 2012)

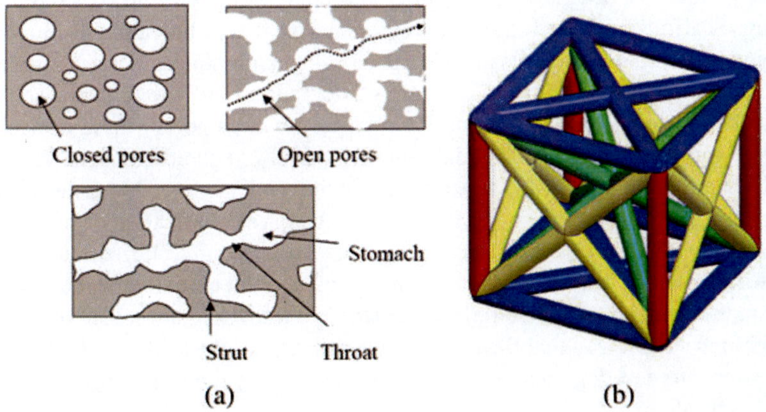

Fig.3 **a** Closed pores, open pores and pore interconnectivity [18], **b** three possible strut inclination angles, blue 0^0, green 35.3^0, yellow 45^0 and red 90^0 (Reproduced with permission from [19] Copyright Elsevier, 2016)

$$\text{Porosity} = \left(1 - \frac{V_{\text{strut}}}{V_{\text{unit cell}}}\right) = \left(1 - \frac{\rho_{\text{unit cell}}}{\rho_{\text{solid material}}}\right)$$

$$\text{Porosity} = (1 - \text{Relative density})$$

Pore interconnectivity is defined as the ratio of open pore volume to total pore volume. Open porosity is defined as the ratio of open pore volume to bulk volume, while total porosity is defined as the ratio of total pore volume to bulk volume.

$$\text{Pore interconnectivity} = \frac{\text{Open pore volume}}{(\text{Open pore volume} + \text{Closed pore volume})}$$

$$\text{Open porosity} = \frac{\text{Open pore volume}}{\text{Bulk volume}}$$

$$\text{Total porosity} = \frac{(\text{Open pore volume} + \text{Closed pore volume})}{\text{Bulk volume}}$$

Thus, pore interconnectivity can also be defined as:

$$\text{Pore interconnectivity} = \frac{\text{Open porosity}}{\text{Total porosity}}$$

3 Discussions

Figure 4 shows the method of designing and analysis of scaffolds for tissue engineering. Scaffold must be biodegradable otherwise it will be called an implant. A combination of scaffolds, cells and signals is called a prosthesis. This is also known as the tissue engineering triad. The rate of degradation of the scaffold should fit the rate of regeneration of the tissue. The scaffold is a temporary three-dimensional structure for initial support. The material of the scaffold should be such that it will cause no harm to the human body after degradation and the degraded material should either absorb in the body or excreted from the body. Vascularization of tissue means developing a capillary network that is capable of delivering nutrients to the cells. The surface area to volume ratio is an important parameter for the vascularization of tissue in the scaffold. The materials of the struts of a scaffold are selected by genetic algorithm in such a way that the curve of degradation of the scaffold should fit with the desired curve of degradation which is determined by the tissue regeneration rate in the scaffold.

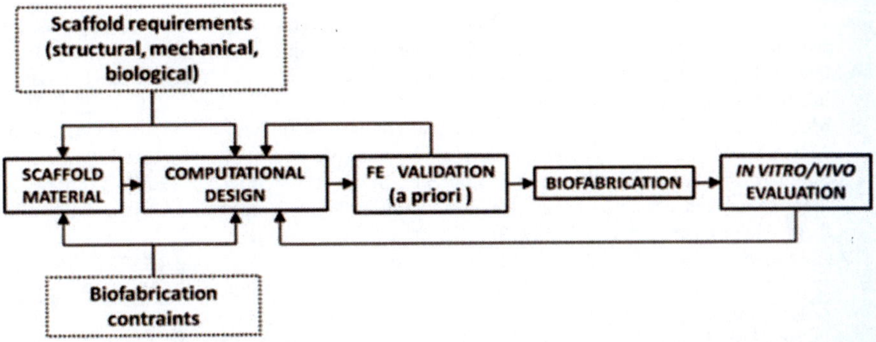

Fig.4 Current trends in the design of tissue engineering scaffolds (Reproduced with permission from [20] Copyright Elsevier, 2014)

4 Concluding Remarks

A scaffold can be made of different unit cells and the position of unit cells in a scaffold should be such that the scaffold stiffness value should come close to the stiffness of the bone in the case of a bone scaffold. The selection of scaffold material should be such that it should be biodegradable. A polymeric scaffold is a very good choice. The selection of strut material in scaffold should be such that it should degrade in such a way that the resultant stiffness of the degraded scaffold should be the same as required at that point of time during tissue regeneration in the scaffold. Manufacturing of scaffold by using different unit cells may cause the problem of stress concentration which results in low fatigue strength. In the literature, it can be found that graded or gradient porosity in a scaffold results in a reduction in fatigue life but it is good from the biological point of view since a particular pore size is responsible for particular tissue growth. There is an optimum pore size for the growth of each tissue. The heterogeneous porous scaffold can help in growing more than one type of tissue if required. Triply periodic minimal surfaces (TPMS) is a mathematical tool for generating smooth surface. TPMS-derived scaffolds are good in fatigue strength because of less stress concentration due to smooth surfaces. Degradation rate of a scaffold can be controlled by controlling its porosity because increase of porosity in the scaffold will increase its corrosion rate (degradation rate) due to increase of more number of micro pores in the surfaces which increases the scaffold effective surface area.

References

1. Chen Q, Thouas GA (2015) Metallic implant biomaterials. Mater Sci Eng R 87:1–57
2. Slaoui I, Stephenson MK, Rauf HA, Dow DE, Shady SS (2015) Stress analysis of bone scaffold designed for segmental bone defects. In: Proceedings of the ASME international mechanical engineering congress and exposition IMECE2015, November, Houston, Texas pp 13–19

3. Calori GM, Mazza E, Colombo M, Ripamonti C (2011) The use of bone-graft substitutes in large bone defects: any specific needs? Injury. Int J Care Injured 42:S56–S63
4. Ghomi ER, Khosravi F, Neisiany RE, Singh S, Ramakrishna S (2020) Future of additive manufacturing in healthcare. Curr Opin Biomed Eng 1–15
5. Muthuvijayan V (2021) Tissue engineering scaffolds: extracellular matrix. A NPTEL online course, IIT Madras. www.youtube.com/watch?v=x5qgWRW-xm8, Last accessed 29 May 2021
6. Singh P, Shrivastava V, Abhash A, Yadav BN, Singh IB, Mondal DP (2021) Compressive deformation and corrosion behaviour of moderate to highly porous Ti-4Al-4Co (wt%) alloy foam. Mater Chem Phys 257(123718):1–12
7. Abhash A, Singh P, Ch VAN, Sathaiah S, Kumar R, Gupta GK, Mondal DP (2020) Study of newly developed Ti-Al-Co alloys foams for bioimplant application. Mater Sci Eng A 774(138910):1–17
8. Jain H, Mondal DP, Gupta G, Kothari A, Kumar R, Pandey A, Shiva S, Agarwal P (2021) Microstructure and high temperature compressive deformation in lightweight open cell titanium foam. Manuf Lett 27:67–71
9. Liu Y, Zheng G, Letow N, Zhao YF (2020) A survey of modeling and optimization methods for multi-scale heterogeneous lattice structures. J Mech Des
10. Gomez S, Vlad MD, Lopez J, Fernandez E (2016) Design and properties of 3D scaffolds for bone tissue engineering. Acta Biomater 42:341–350
11. Gupta V, Tandon P (2015) Heterogeneous object modeling with material convolution surfaces. Comput Aid Des 62:236–247
12. Gupta V, Tandon P (2017) Heterogeneous composition adaptation with material convolution control Features. J Comput Inf Sci Eng 17(2):1–10
13. Singh SK, Tandon P (2017) Comparison of heterogeneous modeling based different patterns of hip prosthesis design. In: Proceedings of the world congress on engineering and computer science, vol 2(October). San Francisco, USA, pp 25–27
14. Singh SK, Tandon P (2018) Heterogeneous modeling based prosthesis design with porosity and material variation. J Mech Behav Biomed Mater 87:124–131
15. Dong G, Tang Y, Li D, Zhao YF (2020) Design and optimization of solid lattice hybrid structures fabricated by additive manufacturing. Addit Manuf 33:1–12
16. Tan C, Cheq Y, Duan Y, Weng F, Sui S, Ng FL, Du Z, Bi G (2021) Additive manufacturing of multi-scale heterostructured high-strength steels. Mater Res Lett 1–10
17. Ptochos E, Labeas G (2012) Shear modulus determination of cuboid metallic open-lattice cellular structures by analytical, numerical and homogenization methods. J Strain 48:415–429
18. Miao X, Sun D (2010) Graded/gradient porous biomaterials. Materials 3(1):26–47
19. Leary M, Mazur M, McMillan M, Chirent T, Sun Y, Qian M, Easton M, Brandt M (2016) Selective laser melting (SLM) of ALSI12MG lattice structures. Mater Des 98:344–357
20. Giannitellia SM, Accotob D, Trombettaa M, Rainer A (2014) Current trends in the design of scaffolds for computer-aided tissue engineering. Acta Biomater 10(2):580–594

Fabrication and Characterization of Aluminum Oxide-Based Polypropylene Filaments

Bikram Singh Solanki, Prakhar Khemka, Harpreet Singh, and Tanuja Sheorey

Abstract Polypropylene (PP) is a widely used engineering plastic material that has lightweight and excellent stiffness, but low load-carrying capacity. In order to enhance the load-carrying capacity of pure PP, different types of metallic and non-metallic particles can be added as reinforcement. In fact, mixing and uniform distribution of reinforcing particles in a matrix polymer is a major challenge. Therefore, effective control of fillers and matrix polymer mixing is essential to obtain a better PP composite. In present work, the PP granules are cryogenically grounded into PP powder and mixed with 4, 8, and 12 wt.% of alumina (Al_2O_3) micro-particles using ball milling process. The single-screw extruder was used to manufacture PP composite filaments of cylindrical cross-section. To examine the distribution of Al_2O_3 particles on PP, the SEM analysis was conducted which shows that 8 wt.% of Al_2O_3 got uniformly mixed with PP powders. In addition, tensile test results confirmed that filament having 8 wt.% of Al_2O_3 holds 34.3% improvement in ultimate tensile strength and 5.8% improvement in break strain.

Keywords Polypropylene · Al_2O_3 · Grinding · Extrusion · Tensile strength · Yield strength · SEM

B. S. Solanki (✉) · P. Khemka · H. Singh · T. Sheorey
Department of Mechanical Engineering, PDPM IIITDM, Jabalpur, Madhya Pradesh 482005, India
e-mail: bikram112.singh@gmail.com

P. Khemka
e-mail: 1713309@iiitdmj.ac.in

H. Singh
e-mail: hps.dme@iiitdmj.ac.in

T. Sheorey
e-mail: tanush@iiitdmj.ac.in

V. K. Gupta et al. (eds.), *Recent Advances in Machines and Mechanisms*, Lecture Notes in Mechanical Engineering, https://doi.org/10.1007/978-981-19-3716-3_57

1 Introduction

Polypropylene (PP) is a multipurpose-type thermoplastic polymer and having many essential properties, namely low water absorption, good chemical resistance, higher service temperature, and resistance against bending crack. Consequently, it is widely used in applications like households, packaging of goods, appliances, water and cold drink bottles, and automotive uses such as dome lights, kick panels, car battery cases, and washing machine parts [1]. Engineering applications especially polymer gear in power transmission require adequate load-carrying capacity and wear resistance to increase the gear life [2]. To enhance load-carrying capacity and reduced wear in pure PP, different types of metallic and non-metallic particles can be added as reinforcement. These fillers improve mechanical, thermal, and tribological properties of PP [3]. The fillers are either added in molten or in solid state of the matrix polymer. However, uniform mixing of these fillers in PP polymer is a challenge.

Extensive research is required to be conducted to uniformly distribute the reinforcing fillers, namely aluminum oxide (Al_2O_3), carbon fiber, graphite [4, 5] carbon nanotube [2, 6], and nano-magnesium oxide [7] in PP matrix. The twin-screw extrusion process is vastly used to homogeneously add reinforcing fillers with powder or granule of matrix polymer to get the composite filament. These filaments can be used in fused deposition modeling (FDM) [8], and their granule form can be used in injection molding machines [9]. In extrusion spheronization process, filament cross-section depends upon end shape and size of extruder nozzle. Preferably, nozzle is made of cylindrical cross-section in which end diameter varies from 0.5 to 7 mm [10]. During this process, critical solid volume concentration has vital role as it decides the ideal polymer filler mixture [11]. To date, mostly, twin-screw extruders have been used to mix matrix polymer and filler and extrude the composite filament, which ensure the uniform mixing of filler in the polymer was a vital challenge. In the present study, an effort has been made to uniformly mix the PP powder and Al_2O_3 micro-particle to get composite powder. Further, composite powder was used to manufacture the filament using single-screw extruder. In addition, the study aimed to analyze tensile and morphological properties of PP-Al_2O_3 composite filament. Some insights into extrusion spheronization of composite filaments are also discussed.

2 Experimental Methodology and Methods

2.1 Material

The polypropylene (Halen—P M108) with the solid density of 0.9 g/cc and MFI of 8 g/10 min (measured at 230 °C and 2.16 kg load) is acquired from Haldia petrochemical Ltd., India, to use as the matrix polymer [12]. The average diameter of the

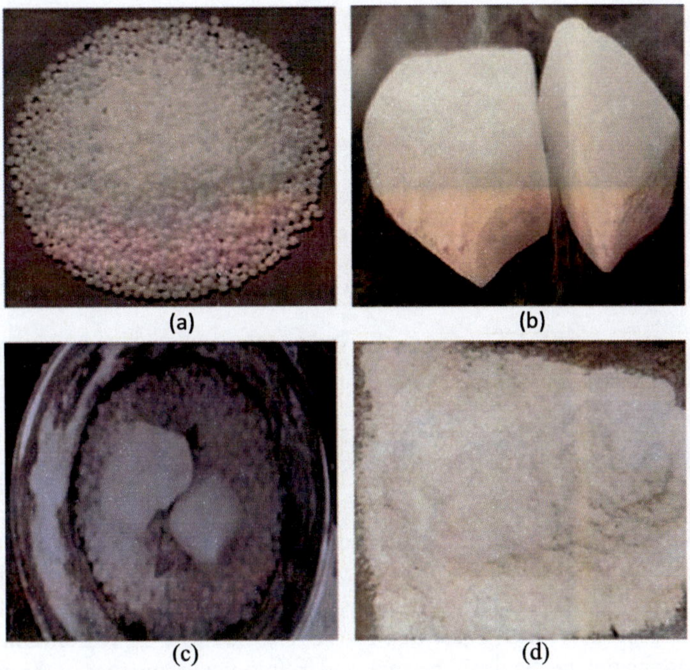

Fig. 1 **a** PP granule, **b** dry ice, **c** in-house grinding **d** PP powder

granules (Fig. 1a) was 3.5 mm. The Al_2O_3 was used as filler material with average particle size of 44 μm.

2.2 *Preparation of Powders*

The PP pellets were grounded with the help of dry ice (Fig. 1b using a home grinding machine (Fig. 1c), which converted PP pellets into a fine powder. In the first phase of grinding, the size of the pellets was reduced by 400%. These small size granules were further grounded with dry ice to get the finer powder. This powder was a mixture of micro-particles of different sizes. Next, three-step sieving was carried out with different sieves for maintaining the final size of the PP powder particle to 210 μm. Fine and uniform size particles help in even mixing of matrix and reinforcement, which may enhance the properties of the composite. Figure 1d shows grounded PP powder.

(a) (b)

Fig. 2 Shows rotary ball mill **a** machine, **b** steel balls

2.3 *Mixing and Characterization of Powders*

The mixing of powders plays a decisive role in the uniform distribution of the Al_2O_3 reinforcement particles on PP matrix powder that helps in enhancing the properties of a composite. In the present work, rotary ball milling (Fig. 2a) was used to mix the Al_2O_3 micro-particles and PP powder. The Al_2O_3 wt.% of 0–12 was chosen to make a balance among the mechanical, thermal, and tribological properties of the composite. These powders with different weight percentages were filled in a cylindrical rotary jar made of steel. Powder mixing was done at 150 RPM with the help of spherical steel balls having diameters between 5 and 10 mm as shown in Fig. 2b. The weight of "balls to powder" ratio is a crucial factor in the mixing and grinding of powders [13]. A 2:1 ratio was chosen to mix for three hours to achieve uniform mixing. The morphological characterization was conducted to study the distribution of Al_2O_3 in PP powder. Figure 3a, b, c, d shows SEM images of PP-Al_2O_3 composite powder with various combinations. Figure 3a clearly shows flakes in pure Al_2O_3 structure. Figure 3b shows non-uniform mixing of Al_2O_3. With increase in reinforcement particle concentration, Al_2O_3 particles can be seen well dispersed around PP particles as shown in Fig. 3c, while Fig. 3d shows agglomeration and entrapment of Al_2O_3 due to cold welding. It can be concluded here that 8wt.% of Al_2O_3 composite powder provides uniform mixing.

2.4 *Extrusion of Filament*

To manufacture virgin PP and PP-Al_2O_3 composite filament, an in-house developed extrusion spheronization machine was used. The powder was fed into machine hopper which then passes through lead screw and nozzle to get desired shape. The lead screw was surrounded by a barrel through which controlled heating was given to melt polymer using four-band heaters. The temperatures of band heaters were controlled

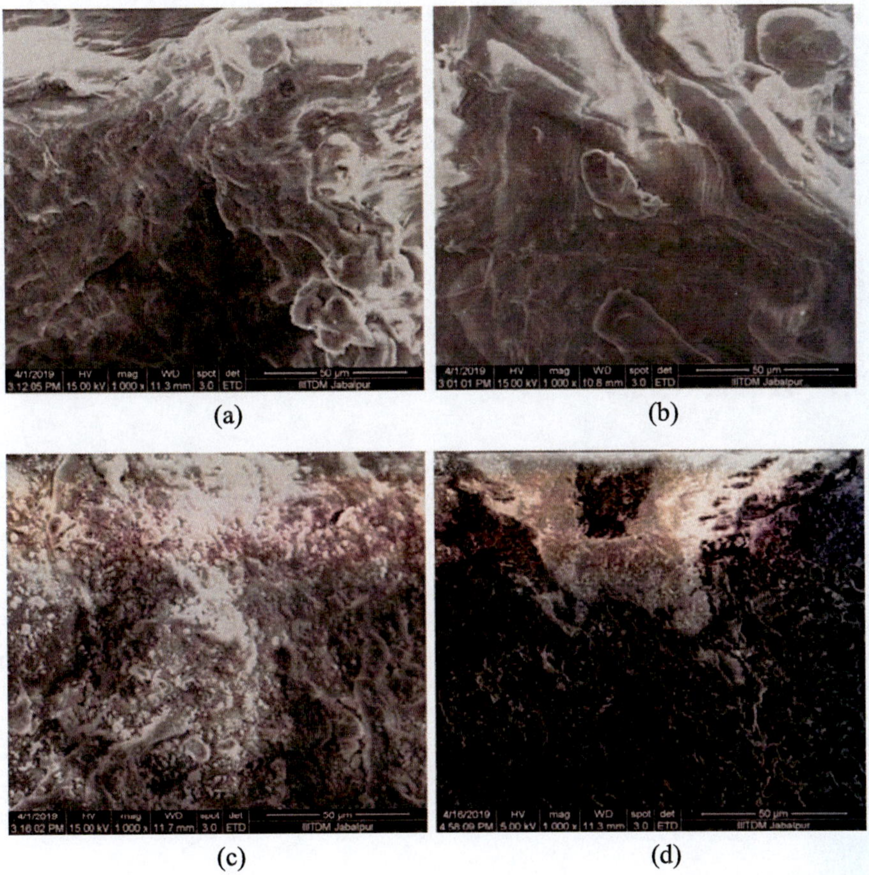

Fig. 3 SEM image of **a** Al_2O_3 powder, **b** 96 wt.% PP + 4 wt.% Al_2O_3, **c** 92 wt.% PP + 8 wt.% Al_2O_3, **d** 88 wt.% PP + 12 wt.% Al_2O_3

by PID controller so that the molten polymer when reaches near the nozzle it becomes semi-solid. The molten polymer moves ahead with the lead screw by 16 mm per revolution and passes through the nozzle to get filament of cylindrical cross-section. Figure 4a, b, c, d shows the extruded filaments with various combinations of PP and reinforcement.

3 Testing and Characterization Methods

To analyze the mechanical and morphological properties of PP and PP-Al_2O_3 composite, different testing are performed.

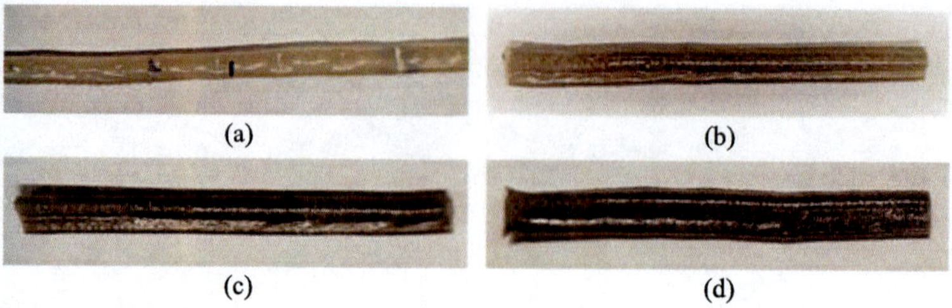

Fig. 4 Pictorial view of extruded filament **a** Virgin PP, **b** 96 wt.% PP + 4 wt.% Al_2O_3, **c** 92 wt.% PP + 8 wt.% Al_2O_3, **d** 88 wt.% PP + 12 wt.% Al_2O_3

3.1 Tensile Testing

The tensile test was performed to analyze the effect of Al_2O_3 content on mechanical properties, namely yield strength, ultimate strength, break strength, and respective strains of manufactured filaments. For the same, 3-mm-diameter filament of 90 mm length was taken as tensile test sample. The Tinus Olsen Universal testing machine was used for testing. After fixing the tensile test specimen in the holding jaws, diameter and length of the specimen and ram speed of 5 mm per minute were set in the machine software. With the help of these parameters, stress and strain developed in the specimen were determined.

3.2 SEM Analysis

To analyze the reinforcing Al_2O_3 particles distribution in PP matrix, SEM morphological test was conducted using FEI Quanta 200 machine. The fractured filament obtained from tensile testing was taken as a sample for this study. This filament has diameter of 3 mm and length of 5 m. The fractured surface was directly used in the SEM study without any further modification, and the images were captured at voltage of 2 kV and at magnification of 50 μm.

4 Results and Discussion

4.1 Tensile Testing

Figures 5a, b, c, d show stress–strain curve of different filaments. It was observed that as long as Al_2O_3 was added from 0 to 8 wt.% in PP matrix, the ultimate tensile stress of filament increased from 23.3 to 30.6 MPa (Fig. 6). However, addition of Al_2O_3 with 12 wt.% causes reduction in tensile strength to 24.4 MPa. Similar variation in yield stress was also observed and maximum yield stress of 30.5 MPa was recorded at wt.% of Al_2O_3 loading. This may be due to the fact that the addition of Al_2O_3 up to 8 wt.% gets uniformly distributed in PP matrix, which in turn resists the applied load. The addition of Al_2O_3 more than 8 wt.% may be promoting agglomeration of Al_2O_3 particles (as seen in case of 12 wt.%) which increases brittleness of the composite filament at specific locations and reduces tensile load-bearing capacity. Similarly, maximum yield strain of 51.1% was found in the filament having 8 wt.% of Al_2O_3. However, ultimate tensile strain continuously increased from 51.1 to 54.3% by adding 0 to 12 wt.% Al_2O_3 with PP (Fig. 7). This may be caused by instantaneous slipping of Al_2O_3 particle from PP lattice at yield point and thereafter again entrapment of same Al_2O_3 particles in PP lattice that enhanced load-bearing capacity. It was also observed that 8 wt.% of Al_2O_3 brings only 0.1 MPa difference in yield strength and ultimate strength and 2.9% difference in yield strain and ultimate strain which is considered to be better for tensile loading. It signified minimum strain hardening zone. 8 wt.% Al_2O_3 showed maximum working strength below which loading could be given without deformation. Hence, PP containing 8 wt.% of Al_2O_3 was found to be the best composite filament to fabricate polymer granules as it shows 34.3% improvement in ultimate tensile strength and 5.8% improvement in break strain.

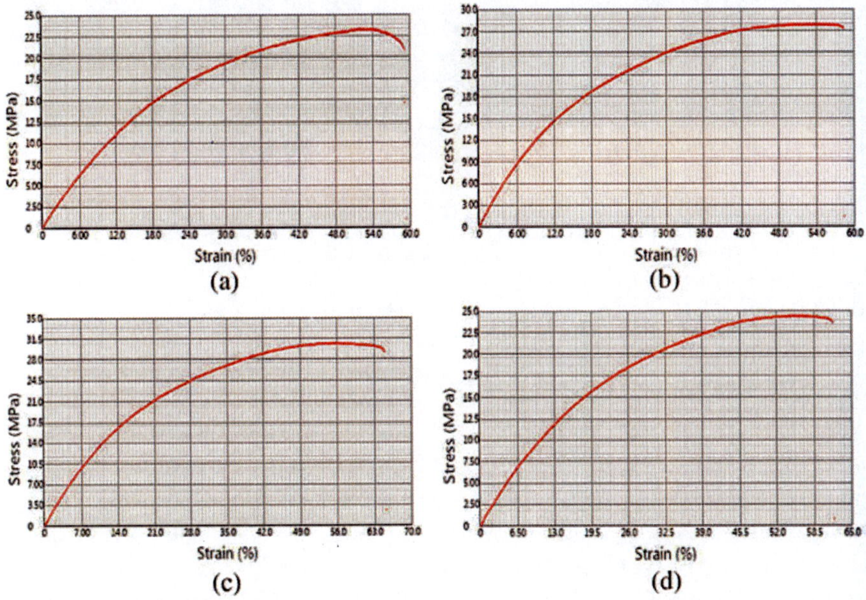

Fig. 5 Showing stress versus strain graph of **a** Virgin PP, **b** 96 wt.% PP + 4 wt.% Al_2O_3, **c** 92 wt.% PP + 8 wt.% Al_2O_3, **d** 88 wt.% PP + 12 wt. % Al_2O_3

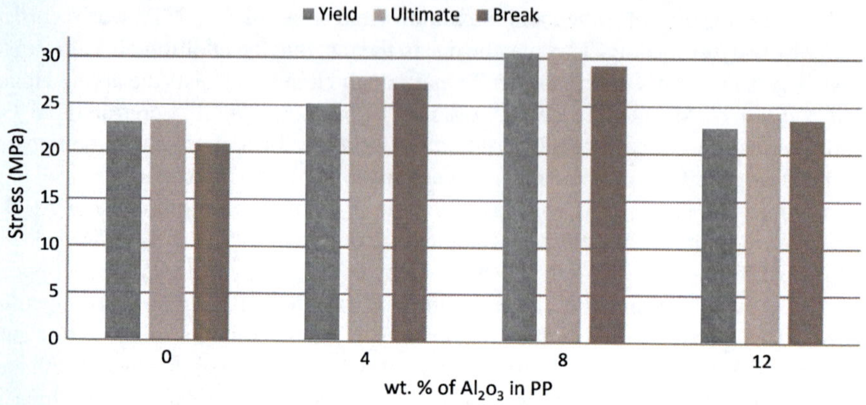

Fig. 6 Showing the strength with respect to various wt.% of Al_2O_3

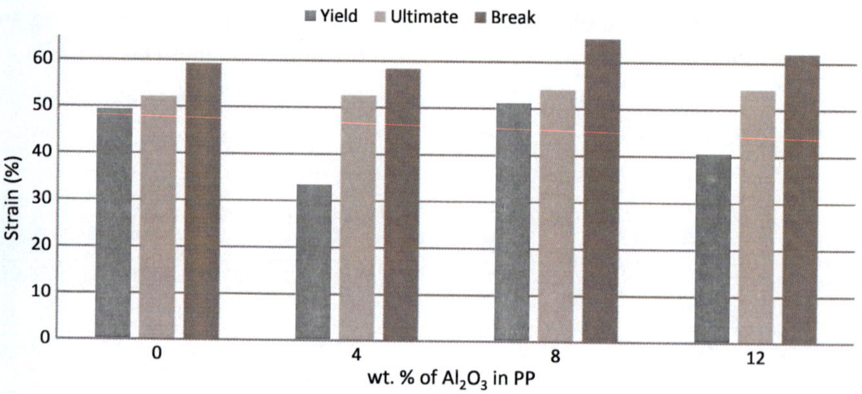

Fig. 7 Showing the strain with respect to various wt.% of Al_2O_3

4.2 SEM Characterization

Figure 8a, b, c shows SEM micrograph of fractured surface of PP-Al_2O_3 composite filament. From Fig. 8a, it can be observed that fractured surface of filament has some voids which were shown by red circle. These locations indicate void spaces created in place of Al_2O_3 particles in PP matrix during fracture. This may be due to weak bonding between 96 PP and 4 wt.% Al_2O_3 particles. However, fractured surface of filament containing 92 PP and 8 wt.% Al_2O_3 has flat surfaces without voids (Fig. 8b). It may be caused by uniform mixing and fine bonding between matrix and reinforcement particles. Furthermore, it was observed from Fig. 8c that filament containing 88 PP and 12 wt. % Al_2O_3 has many voids, a direct consequence of agglomeration of reinforcement particles that weakens the filament against tensile loading.

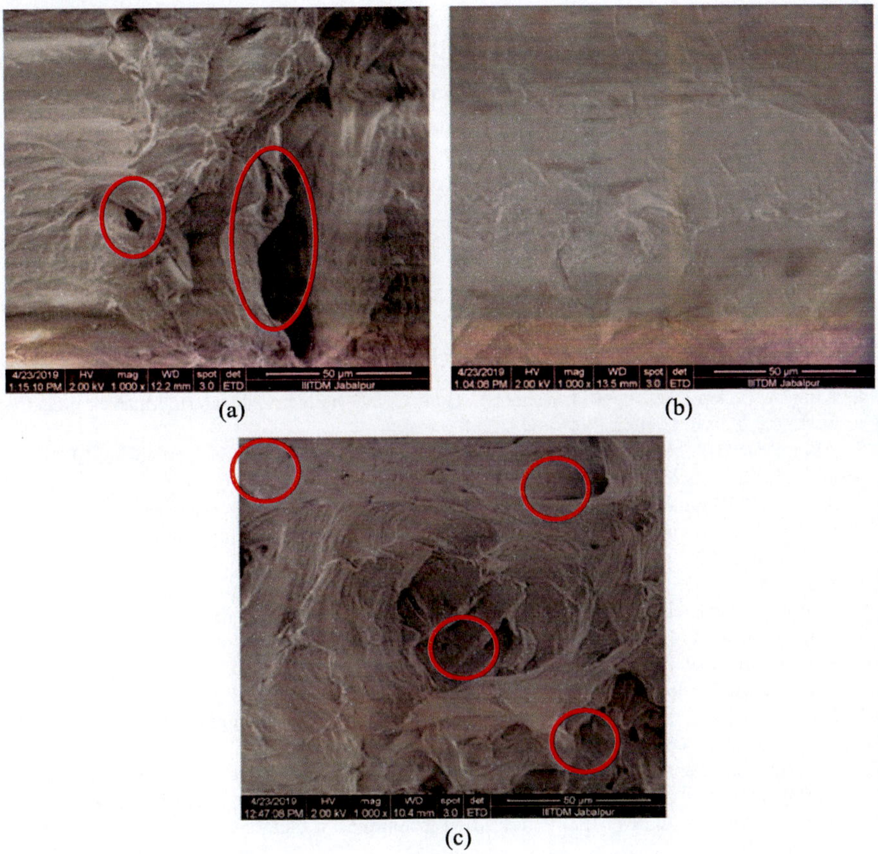

Fig. 8 SEM image of extruded filament of **a** 96 wt.% PP + 4 wt.% Al_2O_3, **b** 92 wt.% PP + 8 wt.% Al_2O_3, **c** 88 wt.% PP + 12 wt.% Al_2O_3

5 Conclusions

The present experimental work was undertaken with the objective of achieving uniform distribution of reinforcement particles in PP matrix for enhancement of load-carrying capacity and service life of the product. Fabrication and characterization of aluminum oxide-based polypropylene filaments have been conducted on in-house available facilities. The following conclusions may be drawn from the present work:

- PP powder with 8 wt.% of Al_2O_3 powder provides uniform mixing of reinforcement particles.
- Tensile strength of PP can be improved by adding reinforcement material, namely Al_2O_3 particles in appropriate proportion.

- Filament containing 92 of PP and 8 wt.% of Al_2O_3 was found to be the best composite filament to fabricate polymer granules as it shows 34.3% improvement in ultimate tensile strength and 5.8% improvement in break strain.

References

1. Baker AM, Mead J (1999) Modern plastics handbook. McGraw-Hill, Lowell, Massachusetts
2. Mertens AJ, Senthilvelan S (2018) Surface durability of injection-moulded carbon nanotube–polypropylene spur gears. Proc Inst Mech Eng Part L J Mater Des Appl 1–13
3. Joseph J (2012) Polymer composite, Wiley
4. King JA, Tucker KW, Vogt BD, Weber EH, Quan C (1999) Electrically and thermally conductive nylon 6,6. Polym Compos 20(5):643–654
5. Akinci A (2009) Mechanical and structural properties of polypropylene composites filled with graphite flakes. Arch Mater Sci Eng 35(2):91–94
6. Ashok GR, Palanikumar K, Ragunath BK, Davim JP (2013) Role of carbon nanotubes (CNTs) in improving wear properties of polypropylene (PP) in dry sliding condition. Mater Des 48(2013):52–57
7. Alateyah AI (2018) Thermal properties and morphology of polypropylene based on exfoliated graphite nanoplatelets/nanomagnesium oxide. Open Eng 8(1):432–439
8. Singh R, Ranjan N (2017) Experimental investigations for preparation of biocompatible feedstock filament of fused deposition modeling (FDM) using twin screw extrusion process. J Thermoplast Compos Mater 1–15
9. Sohn JS, Ryu Y, Yun C, Zhu K, Cha SW (2019) Extrusion compounding process for the development of eco-friendly SCG/PP composite pellets 11:1–12
10. Ganjyal GM, Hanna MA (2004) Effects of extruder die nozzle dimensions on expansion and micrographic characterization during extrusion of acetylated starch. Starch 56(2004):108–117
11. Punia S, Rajni B, ACR (2012) Pelletization techniques: a literature review. Int Res J Pharm 3(3):43–46
12. Halene P (2020) Technical data sheet M 108, Haldia petrochemical limited. http://www.haldiapetrochemicals.com/pdf/polypropylene/M108.pdf. Last accessed 05 Sep 2020
13. Suryanarayana C (2001) Mechanical alloying and milling. Prog Mater Sci 46(1–2):1–184

Ballistic Impact Behavior of 3D Hybrid Composite Laminates

Roopendra Kumar Pathak, Shivdayal Patel, and Vijay Kumar Gupta

Abstract This paper presents a numerical study of damages of hybrid basalt/Kevlar polypropylene composites under high velocity impact. The numerical results are carried out using the user-defined subroutine (VUMAT) to predict the residual velocity of the different composite laminates (K3H3, H3K3, H3-16 and K3-16) with different lay-ups of plies. The comparison has been done on 3D basalt/Kevlar woven fiber hybridization and Kevlar/polypropylene laminates impacted with 9 mm full metal jacket bullet. The energy absorbing capacity and ballistic limit velocity of the composite panels have been evaluated at different velocities. The high velocity impact with different circumstances is numerically simulated and reliability of the model is verified. The results are compared with the literature results presenting good agreements. As hybridization of basalt with Kevlar fibers, the ballistic limit impacting with 9 mm FMJ is not enhanced if compared to pure Kevlar laminates, but natural fibers over synthetic fibers have found better response due to less expensive and better mechanical properties in the field of research.

Keywords Basalt fiber composites · Kevlar fiber composites · High velocity impact · Finite element method

1 Introduction

Numerous studies have been done for finding the resistance against low, medium and high velocity impact of composites in the field of engineering. High specific stiffness and high specific strength are the key factors of composite structure in many applications. Ignatova et al. [1] found that the ballistic performance increased by coating of polyvinyl acetate on the surface of composite laminate. Gilson et al. [2] studied about kinetic energy of projectile after impact, displacement of laminas, failure phenomena of fibers were influenced by projectile geometry and material. Wei et al. [3] reported three-dimensional woven fiber ballistic performance was evaluated in the form of

R. K. Pathak · S. Patel (✉) · V. K. Gupta
Discipline of Mechanical Engineering, PDPM IIITDM Jabalpur, Jabalpur, Madhya Pradesh, India
e-mail: shivdayal@iiitdmj.ac.in

V. K. Gupta et al. (eds.), *Recent Advances in Machines and Mechanisms*, Lecture Notes in Mechanical Engineering, https://doi.org/10.1007/978-981-19-3716-3_58

residual velocity, damage distribution area and energy absorption mechanisms. Liu et al. [4] compared deformation and energy absorption of multi-phase of STF with graphene oxide, carbon nanotubes and found that graphene oxide additives gave better results. Patel et al. [5] determined the crashworthiness analysis for homogeneous and heterogeneous composites under axial and oblique impact. Pundhir et al. [6] determined residual velocity of UHMWPE/alumina composite gave lower residual velocity if projectile impacts on UHMWPE side plate. Chen and Yang [7] investigated that yarn crimp affected the ballistic performance of multi-layer laminate and energy absorption of layers increased of low crimp fabrics compared to high crimp fabrics. Patel and Guedes Soares [8, 9] successfully reported the uncertainty of material properties and initial velocity for composite laminate under low velocity impact. Patel and Guedes Soares [9] compared the probabilistic design optimization for body armors. Patel et al. [10] suggested that anti-symmetric cross-ply is more reliable than symmetric cross, symmetric angle and anti-symmetric angle plies. Soutis and Shi [11] investigated dynamic and static properties of carbon/basalt composite in intercalated. 3D fibers composite had higher nonlinearity than the 2D fiber composites [12]. Symmetric lay-ups gave better ballistic limit than non-symmetric lay-ups [13]. Chakrabarti and Ansari [14] studied the damage pattern in the composite panel and projectile, failure in thickness direction, deflection, front and back surface damaged area and effects of length, width and thickness of layers on ballistic limit velocity. Behavior of damage phenomena and erosion of impactor analyzed under high velocity impact for spherical projectile, FSP and FMJ projectile strike the Kevlar composite helmet with different impact velocities [15]. Rahman et al. [16] compared the FE models for 7.62 mm armor piercing projectile for ballistic limit velocity of different layered configuration. Ahmad and Bandaru [17] have been carried out the results in the field of simulation with cohesive elements exhibit better delamination progression in reduced integration and full integration hexahedra elements. Børvik et al. [18] studied the penetration and perforation for granular materials. Ballistic limit of E-glass/phenolic composite laminate if influenced by the thickness of laminate [19]. Stress-based failure criterion [20] and energy-based criteria were used to classify the damage initiation and propagation as intra and inter-laminar failure (delamination). Punch shear and crushing of fibers are the modified damage modes in composites [21]. Less work on ballistic was reported, and the recent study by Bandaru et al. has been done the characterization of Kevlar/basalt fiber composite.

In this study, woven 3D Basalt/Kevlar/polypropylene composite panels have been compared with different stacking sequences. The focus of this study is to observe the damage phenomena at different time intervals of 3D basalt/Kevlar woven fibers. In the present study, the strain-based Yen failure criteria are used to predict the failure initiation and propagation failure behavior of the hybrid composites. Presently, 3D fiber damage analysis of hybrid laminates with deformable impactor has been done. There are limited study presented on measurement of ballistic velocity on natural fibers. Basalt is an economically good replacement of synthetic fiber to improve the ballistic performance.

2 Numerical and Material Modeling

K3H3, H3K3, H3-16 and K3-16 were hybrid composite panels impacted with 9 mm FMJ projectile. Individual ply dimension is of 300 × 300 × 2 mm. Ply was meshed with SC8R (8-node quadrilateral) elements. There were 16 plies used in this model. Finite element mesh of a 9 mm FMJ bullet (mass = 8.2 g) made of copper cover with lead core [19]. Meshing for brass cover (shell thickness = 0.25 mm) and lead core (diameter of core = 8.5 mm) were used S4R (0.001 m) and C3D8R (0.001 m), respectively (Table 1).

2.1 Material Damage Constitutive Models for Composite Laminate and Projectile

The response of basalt/Kevlar woven hybrid panels analyzed by user-defined subroutine. Continuum damage mechanics approach was used to analyze the propagation of damage by reducing the material stiffness and quadratic interaction between strains. Fiber tension, fiber compression, punch shear, crushing, in-plane shear and delamination were the major modes of failures. Yen [21] used the damage initiation criteria for composite laminates. The copper jacket material is modeled by the Johnson–Cook (J-C) model and lead was assumed as elastic with using Steinberg–Guinan [15].

Table 1 Nomenclature used for different plies and stacking sequences

K3H3	Eight 3D Kevlar laminas and eight 3D hybrid laminas	K3D	3D woven Kevlar lamina
H3K3	Eight 3D hybrid laminas and eight 3D Kevlar laminas	H3D	3D woven hybrid lamina
H3-16	16 hybrid laminas	KPL	2D plane Kevlar
K3-16	16 Kevlar laminas	BPL	2D plane basalt
B3D	3D woven basalt lamina	HPL	2D hybrid lamina

Fig. 1 **a** Top view and **b** side view of numerical model

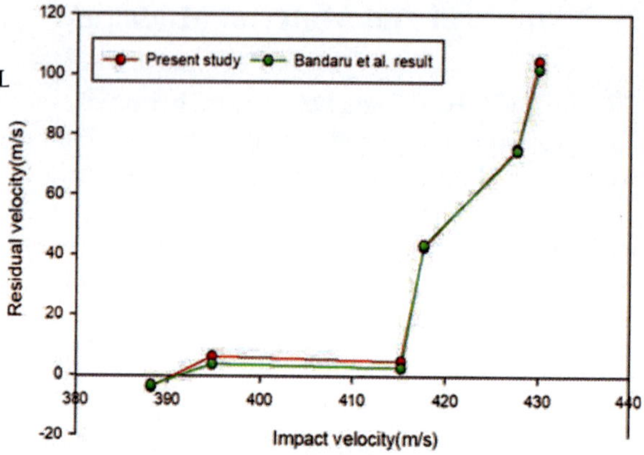

Fig. 2 Response of composite panel [B3D/BPL2/H3D/HPL2/K3D/KPL impacting with FMJ projectile of Impact velocity versus residual velocity

3 Numerical Results and Discussion

3.1 Validation

The impact velocity versus residual velocity plot is shown in Fig. 2. It can be seen that the results of present study are in close match with Bandaru et al. [13]. Figure 2 shows velocity versus time for [B3D/BPL2/H3D/HPL2/K3D/KPL]s stacking sequence composite laminate.

3.2 Change in Residual Velocity and Energy Absorption

Finite element analysis was carried out for the K3H3, H3K3, H3-16 and K3-16 laminates subjected to the ballistic impact of 9 mm FMJ projectile. The panel and projectile stored total internal energy. Another important observation was that H3K3 panel absorbed the 82.14% of the total internal energy at its ballistic limit velocity while K3-16, K3H3 and H3-16 absorbed 78.75%, 75%, 71.33%, respectively of the total internal energy at their ballistic limit velocity. During the ballistic impact, composite laminate absorbed kinetic energy also from the bullet and spread to all area resulting as failure of the laminate. It was observed among the four different composite panels with deformable FMJ projectile, K3-16 absorbed the maximum kinetic energy because of its high ballistic limit velocity. Impact velocity of the impactor was varied from 250 to 520 m/s and the numerical simulation also carried out in the same velocity range. Figure 3 shows velocity–time plots of FMJ impactor with strike velocity 450 m/s, 360 m/s, 350 m/s and 250 m/s impacting K3-16, K3H3, H3K3 and H3-16, respectively, the respective residual velocity approaches to zero. The ballistic limit velocity of K3-16 obtained 450 m/s which is more than 20%, 22.22% and 44.44% of K3H3, H3K3 and H3-16, respectively. Ballistic limit velocity of K3-16 among different laminate for FMJ projectile was found higher than other panels.

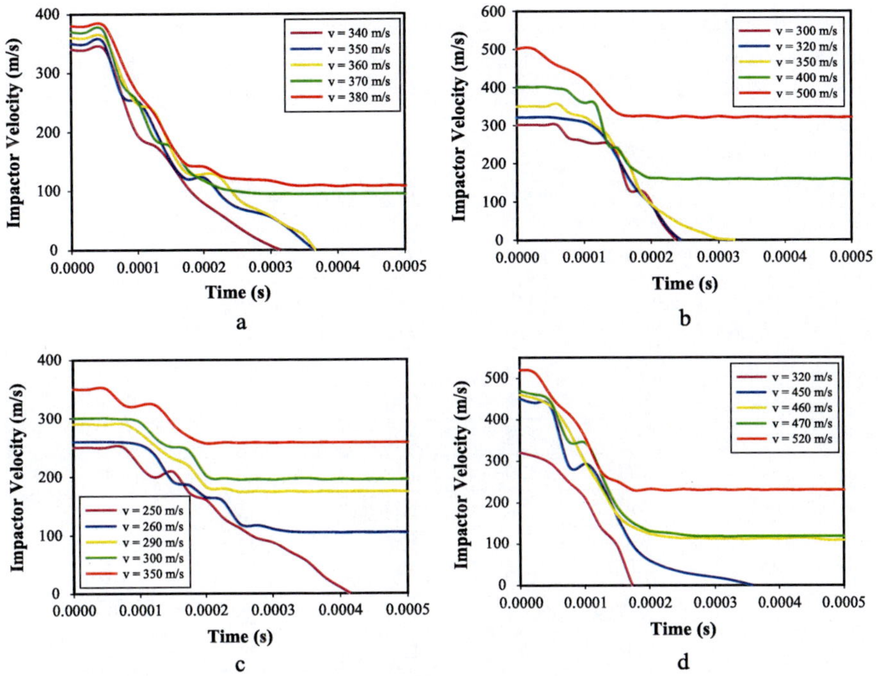

Fig. 3 Velocity versus time plots for **a** K3H3, **b** H3K3, **c** H3-16 and **d** K3-16

3.3 Failure Behavior of Laminate and Projectile

Four composite laminates with different arrangement of lamina have been impacted by FMJ impactor to study the impact resistance. The numerical model predicted perforation of laminate and residual velocity, when strike velocity reaches at the ballistic limit velocity, the effect of impact becomes very sensitive to impact speed, material defects and angle. Basalt laminate promotes natural fibers backing with Kevlar fibers and shows good agreement of damage tolerance compared to other laminates. Erosion of the materials started at the projectile front face and progressed the laminas. Flattened and mushroomed shape of impactor after impact occurred due to dissipation of kinetic energy. When projectile leaves the laminate contact, the laminate continuous to deform.

Numerical simulation results of ballistic limit velocity of K3H3, H3K3, H3-16 and K3-16 were shown in Fig. 4. The projectile impacts the front surface of the laminate and initiates Damage modes occurred in the form of matrix cracking, delamination, fiber failure, fiber crushing and internal energy (IE) of whole model and bullet are shown in Fig. 5 and Table 2 respectively. Extensive displacement of back side was observed due to delamination and stretching of fibers. Delamination was a dominating mode of back side deformation of fibers. In plane and out of plane compressive forces were found in the form of inter-laminar stresses. Projectile travel distance after hitting

Table 2 Ballistic impact performance for different four composite laminates

S. No.	Type of laminate	Impact velocity (m/s)	Residual velocity (m/s)	IE of whole model (J)	IE of bullet (J)	IE of panel (J)
1	K3H3	340	0	280	60	220
		350	0	285	64	221
		360	0	300	75	225
		370	95	315	82	233
		380	110	325	88	237
2	H3K3	300	0	215	40	175
		320	0	257	45	212
		350	2	280	50	230
		400	158	285	52	233
		500	320	322	60	262
3	H3-16	250	0	157	45	112
		260	104	128	22	106
		290	175	128	24	104
		300	194	118	20	98
		350	257	129	22	107
4	K3-16	320	0	270	72	198
		450	0	480	102	378
		460	100	490	104	386
		470	118	490	109	381
		520	230	520	117	403

was approximately more for B3K3 panel compared to others. Front basalt fiber layers absorb more energy due to more damages of the fibers. Characteristic length is the ratio of the fracture energy (G) of the element to the specific energy (g) in failed element [17, 21].

The thickness view of all panels subjected to ballistic limit velocity. The damage modes like matrix failure at front face, tensile failure in yarns and delamination can be seen. It is evident from figure that damage zone area along thickness direction maximum for K3-16 at middle layers. Larger volume of fibers interaction increases the contact time of projectile and laminate. It happens because of extra thickening of laminate due to elongation of fibers and delamination. One important observation is also found that the delamination increased with lowering the velocity. At the lower range of ballistic limit, delamination observed more than above the ballistic limit. A considerable amount of impact energy also consumed in the form of projectile deformation with changing the impact velocity (Table 3).

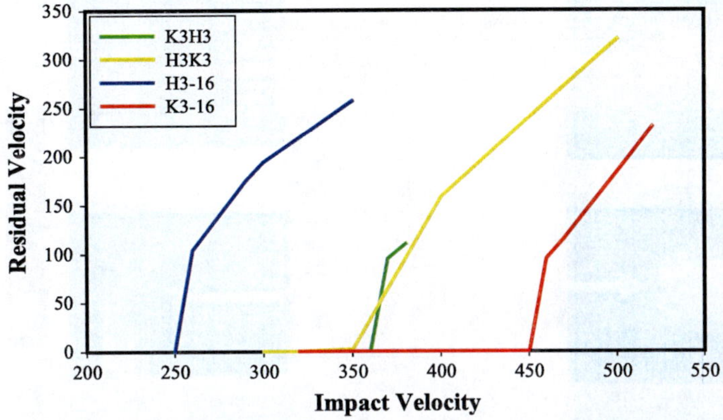

Fig. 4 Impact velocity versus residual velocity for different panel

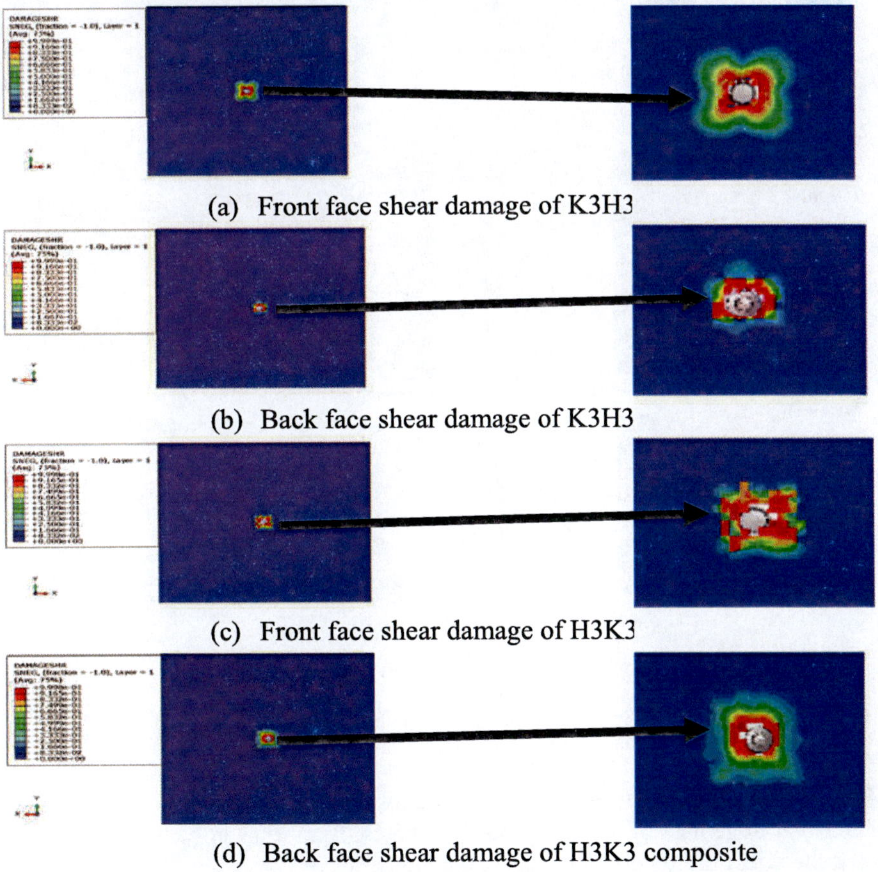

Fig. 5 Front and back face shear damage of K3H3 and H3K3 composite laminates

Table 3 Shear failure mechanisms of different laminates for FMJ impactor

Laminate	$t = 7.5e{-}5$ s	$t = 1.5e{-}4$ s	$t = 5e{-}4$ s
K3H3 (v = 360 m/s)	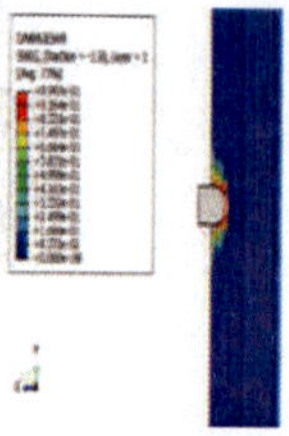	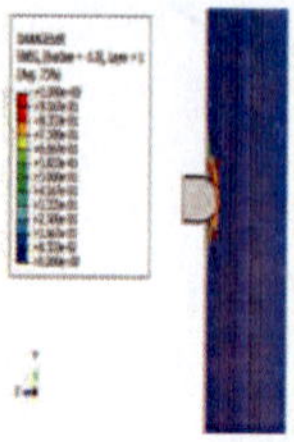	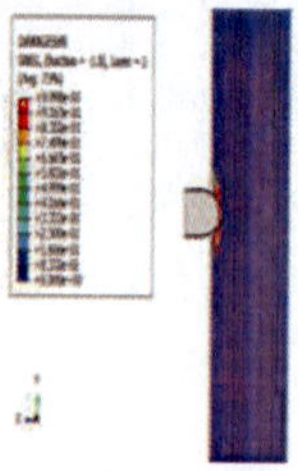
H3K3 (v = 350 m/s)	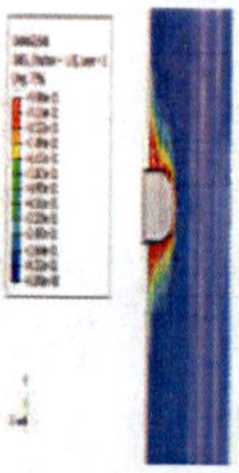	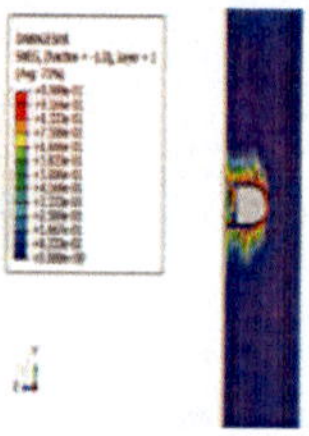	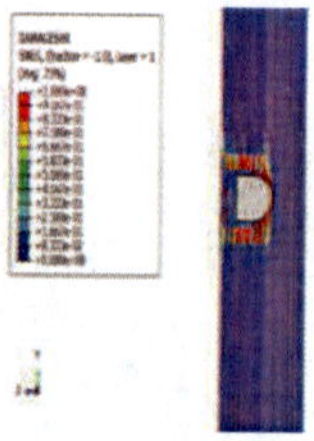
H3-16 (v = 250 m/s)	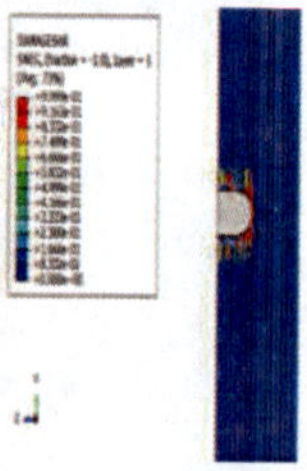	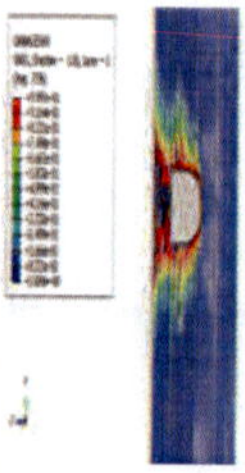	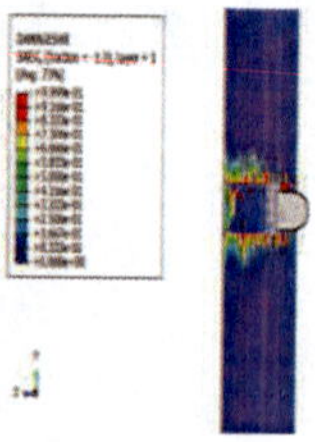
K3-16 (v = 450 m/s)	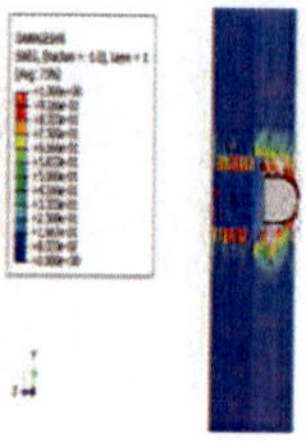	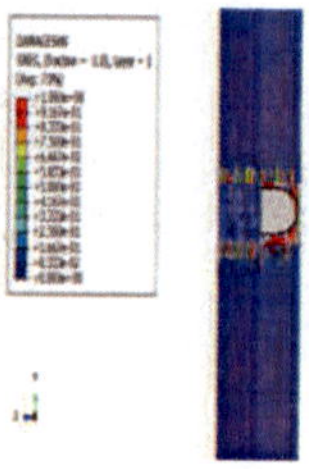	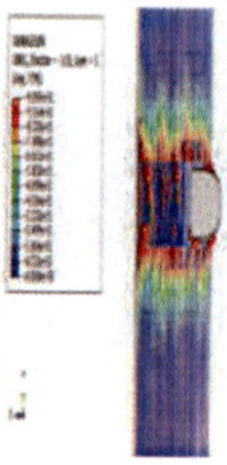

4 Conclusions

Ballistic impact behavior has been compared for different four (K3-16, K3H3, H3K3, H3-16) arrangement of the composite laminates. Basalt has good toughness property, the effect of higher toughness at the interfaces of layers gives more resistance and the efficient response to ballistic impact. The K3-16 finds suitable combination of dynamic and static properties but the highest fraction of expensive Kevlar reinforcement while basalt promotes the natural fibers with approximately same impact resistance. Numerical simulations extracted the results in the form of energy absorption, damage area and ballistic limit for K3H3, H3K3, H3-16 and K3-16 composite panels. It can be seen that ballistic limit increases with increasing the Kevlar layers. Strain rate effect of H3-16 composite panel easily penetrated due to less resistive forces at the interfaces while higher resistive forces observed at the interfaces of K3-16 composite panel. For hybrid layers with Kevlar layers combination (K3H3 and H3K3) with 9 mm FMJ impactor have not been much affected in the form of ballistic limit. Interaction of projectile with layers of composite panel can be understood. The behavior of damage pattern has been studied for hybrid Kevlar/basalt composites. It has been concluded that ballistic performance of Kevlar and basalt woven form in a single lamina (H3) is not better than pure Kevlar lamina (K3-16) while hybrid form of these laminas have been given a good agreement for replacing the synthetic fibers from natural fibers.

References

1. Ignatova AV, Kudryavtsev OA, Zhikharev MV (2020) Influence of surface polymer coating on ballistic impact response of multi-layered fabric composites: experimental and numerical study. Int J Impact Eng 144:103654. https://doi.org/10.1016/j.ijimpeng.2020.103654
2. Gilson L, Imad A, Rabet L, Coghe F (2020) On analysis of deformation and damage mechanisms of DYNEEMA composite under ballistic impact. Compos Struct 253:112791. https://doi.org/10.1016/j.compstruct.2020.112791
3. Wei Q, Yang D, Gu B, Sun B (2021) Numerical and experimental investigation on 3D angle interlock woven fabric under ballistic impact. Compos Struct 266:113778. https://doi.org/10.1016/j.compstruct.2021.113778
4. Liu L, Cai M, Liu X, Zhao Z, Chen W (2020) Ballistic impact performance of multi-phase STF-impregnated Kevlar fabrics in aero-engine containment. Thin-Walled Struct 157:107103. https://doi.org/10.1016/j.tws.2020.107103
5. Patel S, Vusa VR, Guedes SC (2019) Crashworthiness analysis of polymer composites under axial and oblique impact loading. Int J Mech Sci 156:221–234. https://doi.org/10.1016/j.ijmecsci.2019.03.038
6. Pundhir N, Goyal D, Singh P, Pathak H, Zafar S (2019) Numerical simulation of composite armor subjected to ballistic impact. Mater Today Proc 18:696–703. https://doi.org/10.1016/j.matpr.2019.06.470
7. Yang Y, Chen X (2019) Influence of fabric architecture on energy absorption efficiency of soft armour panel under ballistic impact. Compos Struct 224:111015. https://doi.org/10.1016/j.compstruct.2019.111015
8. Patel S, Guedes SC (2018) Reliability assessment of glass epoxy composite plates due to low velocity impact. Compos Struct 200:659–668. https://doi.org/10.1016/j.compstruct.2018.05.131
9. Patel S, Guedes SC (2017) System probability of failure and sensitivity analyses of composite plates under low velocity impact. Compos Struct 180:1022–1031. https://doi.org/10.1016/j.compstruct.2017.08.054
10. Patel S, Ahmad S, Mahajan P (2017) Safety assessment of composite armor under ballistic impact. Proc Eng 173:1901–1908. https://doi.org/10.1016/j.proeng.2016.12.249

11. Shi Y, Soutis C (2017) Modelling low velocity impact induced damage in composite laminates. Mech Adv Mater Mod Process 3. https://doi.org/10.1186/s40759-017-0029-x
12. Bandaru AK, Mittal VK, Ahmad S, Bhatnagar N (2017) Influence of hybridization on in-plane shear properties of 2D and 3D thermoplastic composites reinforced with Kevlar/basalt fabrics. Polym Test 61:396–403. https://doi.org/10.1016/j.polymertesting.2017.06.001
13. Bandaru AK, Ahmad S, Bhatnagar N (2017) Ballistic performance of hybrid thermoplastic composite armors reinforced with Kevlar and basalt fabrics. Compos Part A Appl Sci Manuf 97:151–165. https://doi.org/10.1016/j.compositesa.2016.12.007
14. Ansari MM, Chakrabarti A (2016) Impact behavior of FRP composite plate under low to hyper velocity impact. Compos Part B Eng 95:462–474. https://doi.org/10.1016/j.compositesb.2016.04.021
15. Olmedo A, Romualdo G, Feito N, Loya JA, Miguélez MH (2016) Behavior of a new combat helmet design against ballistic impact: experimental and numerical analysis. Pass 2016
16. Rahman NA, Abdullah S, Zamri WFH, Abdullah MF, Omar MZ, Sajuri Z (2016) Ballistic limit of high-strength steel and Al7075-T6 multi-layered plates under 7.62 mm armour piercing projectile impact. Lat Am J Solids Struct 13:1658–1676. https://doi.org/10.1590/1679-78252657
17. Bandaru AK, Ahmad S (2016) Modeling of progressive damage for composites under ballistic impact. Compos Part B Eng 93:75–87. https://doi.org/10.1016/j.compositesb.2016.02.053
18. Børvik T, Dey S, Olovsson L (2015) Penetration of granular materials by small-arms bullets. Int J Impact Eng 75:123–139. https://doi.org/10.1016/j.ijimpeng.2014.07.016
19. Reddy PRS, Reddy TS, Madhu V, Gogia AK, Rao KV (2015) Behavior of E-glass composite laminates under ballistic impact. Mater Des 84:79–86. https://doi.org/10.1016/j.matdes.2015.06.094
20. Shi Y, Swait T, Soutis C (2012) Modelling damage evolution in composite laminates subjected to low velocity impact. Compos Struct 94:2902–2913. https://doi.org/10.1016/j.compstruct.2012.03.039
21. Yen CF (2012) A ballistic material model for continuous-fiber reinforced composites. Int J Impact Eng 46:11–22. https://doi.org/10.1016/j.ijimpeng.2011.12.007

Numerical Analysis on Hexagonal Honeycomb Sandwich Structure Under Air-Blast Loading

Murlidhar Patel and Shivdayal Patel

Abstract Sandwich constructions with a honeycomb core have recently become popular for high strength and dynamic load. The purpose of this study is to look at the shock wave resistance performance of two distinct types of honeycomb core sandwich structures in terms of face plate deflection and energy absorption when subjected to a blast load. This study employed square and hexagonal honeycomb core structures to determine the minimal face deflection under blast conditions. The honeycomb sandwich panel is composed of steel that is very ductile. In the sandwich construction, both the front and rear plates are solid, and the core structure is of the shell type. To administer the air-blast loads of 1, 2, and 3 kg TNT, a 10-cm stand-off distance is used from the front face. The dynamic response of the sandwich constructions is determined using the ABAQUS/explicit finite element method (FEM). For both square honeycomb and hexagonal honeycomb core sandwich panels, the front and rear face plate deflections were measured. Under comparable blast loading circumstances, the front and rear plates of the hexagonal sandwich panel showed less deformation than the square honeycomb core sandwich panel.

Keywords Blast · Deflection · Hexagonal honeycomb · Sandwich

1 Introduction

The sandwich structure can be used as a protective cover against the high-intensity shock waves to provide the target with safety and reliability. The high-pressure energy dissipated from the blast causes the inelastic deformation and the shear deformation of the plate and core, respectively. In recent times, landmines and bomb blasts have been used by terrorists to harm human lives, public properties, and army vehicles. Hence, the safety and reliability analysis of the structural health condition shows that the sandwich panel design is necessary to achieve the target protection level of the structures. Some research work is also going on the sandwich structures to

M. Patel · S. Patel (✉)
Department of Mechanical Engineering, Design and Manufacturing, Indian Institute of Information Technology, Jabalpur 482005, India
e-mail: shivdayal@iiitdmj.ac.in

V. K. Gupta et al. (eds.), *Recent Advances in Machines and Mechanisms*, Lecture Notes in Mechanical Engineering, https://doi.org/10.1007/978-981-19-3716-3_59

improve the blast-resistance characteristics. Zhu and Lu [1] provided a brief review of the dynamic blast load and the failure behavior of sandwich structures. They concluded that, due to the blast wave, the sandwich structure was subjected to high elastic deformation, transverse, and shear tearing failure nearer to the supports. The researchers used ABAQUS and LSDYNA software for finite element (FE) modeling and finite element analysis (FEA) of the sandwich structures [1–4]. Dharmasena et al. [2] studied the blast load experiment on the metallic sandwich structure incorporated with a square honeycomb core. They analyzed the deformation of the face plates under the blasting of 1–3 kg TNT. They reported that the metallic sandwich structure has more blast resistance than a solid plate of the same mass. Patel and Patel [3] numerically analyzed the blast behavior of honeycomb steel sandwich panels. They noted that the use of different types of honeycomb cores with varying thickness affects the blast resistance properties of the sandwich structures. Rathbun et al. [4], Fleck and Deshpande [5] also presented the dynamic behavior of the metallic honeycomb sandwich structures under the air-blast and water-blast load. They observed that sandwich structures subjected to water blasts have very good blast resistance properties. Nahshon et al. [6], Murugesan and Jung [7], Turkmen and Mecitoglu [8] compared the dynamic response under explosive load with and without a stiffened composite plate. They also determined the Johnson–Cook as well as strain hardening parameters to analyze the plastic behavior of AISI-1045 sandwich panel. Dear et al. [9] and Kelly et al. [10] reported the dynamic response of the composite sandwich structures under the shock waves. The sandwich panel with different polymeric foam cores and the face sheet which is subjected to air as well as underwater blasts is studied. They used 100 kg of TNT and a stand-off distance of 15 m, and concluded that the styrene acrylonitrile foam as a core has a minimum deflection. Wang et al. [11] investigated the ballistic performance of different honeycombs in out-of-plane. They used triangular, hexagonal, square, reentrant, and circular honeycomb panels with identical mass and thickness. They noted that the square, reentrant, and triangular honeycombs absorb less kinetic energy as compared to hexagonal honeycomb.

Now, the whole blast phenomenon is classified into three phases. In the first phase, the incident of the blast wave on the front surface provides momentum. In the second phase, the blast wave moves downward, and at the same time, the core resists the motion of the front plate. During this phase, the core cruses and also minimizes energy. In the third phase, if the blast wave has high intensity, then the back plate will also bend due to the high impulse generated. Now, the basic aim of this paper is to mitigate the high intensity of the blast load with a sandwich panel made up of high-ductility stainless steel alloy. The front as well as back faces and cores (square/hexagonal) structure of the sandwich panel is modeled by the solid element and the shell element, respectively, using FEA to study the plastic behavior of the sandwich panel.

In this analysis, the blast source of three different masses of 1, 2, and 3 kg of TNT is placed at a stand-off distance of 10 cm from the center of the front face of the sandwich structure. This study was performed by using ConWep in-built blast load simulation in ABAQUS/explicit. The focus of this study is to minimize the deflection

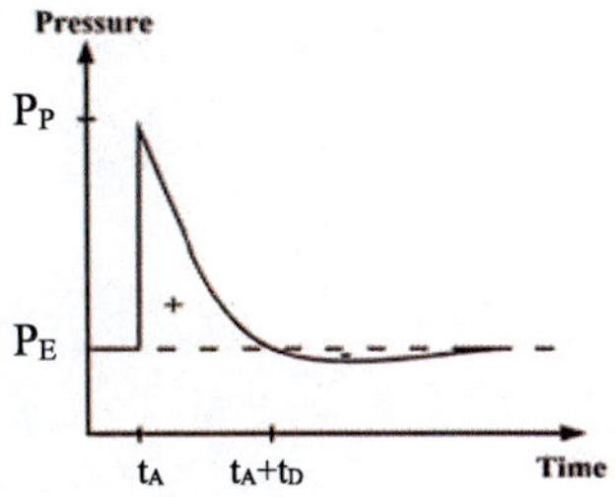

Fig. 1 Variation of pressure during blast

of both the front and bottom face plates by considering the different core structures (square and hexagonal) of the sandwich subjected to the blast load.

2 Numerical Modeling

2.1 Air-Blast Shock Wave

The high intensity of the shock wave generated by the bomb blast in the air compressed the atmospheric air molecules due to the high pressure and the high temperature in the surroundings, which may have caused damage to structures. This characteristic of the air blast can be understood by Fig. 1. In this figure, P_E is the environmental pressure, and P_P is the peak generated during the blast. Suddenly, increase in the pressure from P_E to P_P after the arrival time t_A, and then, this pressure decays to the environmental pressure for t_D duration. After that, this positive pressure further decays to a negative pressure. The modification of the Friedlander equation for the response of the pressure with respect to time is represented by

$$P(t) = (P_P - P_E)\left[1 - \frac{t - t_A}{t_D}\right]e^{-\frac{(t-t_A)}{\lambda}}$$

Here, λ is time decay constant.

2.2 Finite Element Modeling of Sandwich Structures

Dharmasena et al. [2] have investigated the air-blast resistance of a high-ductility steel square honeycomb sandwich panel. An air-blast load is applied at the center point of a sandwich structure having a 6% relative density. The dimensions of the sandwich panel are 610 mm × 610 mm with the thickness of the face plates and core height being 5 mm and 51 mm, respectively. The wall thickness is 0.76 mm. These dimensions were also shown in Fig. 2. The charges of 1–3 kg TNT explosives are

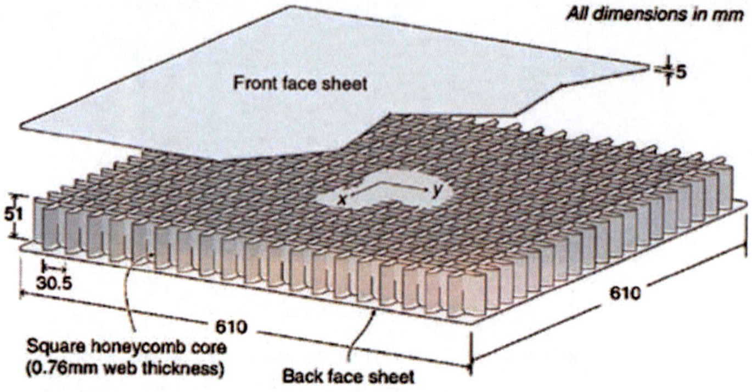

Fig. 2 Dimensions of the sandwich structure parts [2]

used for a 100 mm constant stand-off distance to perform the air-blast experiment. In the present study, to simulate the Dharmasena et al. [2] experimental results of front face plate deflections at 1, 2, and 3 kg TNT, a numerical analysis has been done using ABAQUS/explicit finite element calculations. All dimensions of the modeled square honeycomb sandwich structure are the same as those of Dharmasena et al. [2] fabricated sandwich panels. Because of the symmetry of the structure with respect to both the *X*–*Z* and *Z*–*Y* planes, only, a quarter part of the sandwich panel needs to be modeled to avoid complexity in numerical analysis. Through the *X*–*Z* and *Y*–*Z* planes, a fixed boundary condition is applied. The other two faces in the direction of the thickness of the sandwich panel are defined with the *X*-SYMM and *Y*-SYMM boundary conditions individually.

The numerical predicted results at 1, 2, and 3 kg TNT blast load are firstly validated to find the best performance of the honeycomb sandwich panels. After that, the FE modeling of the hexagonal honeycomb core used in sandwich structures was also modeled to investigate the effect of different metallic honeycomb cores on the blast resistance at 1, 2, and 3 kg TNT with a 100 mm stand-off distance. These FE models were prepared by using ABAQUS/CAE (version 6.14). Figure 3 shows the 1/4th FE models of square and hexagonal honeycomb core used sandwich structures. All FE-modeled sandwich panels (length = width = 305 mm) have a 51-mm honeycomb core height and a 30.5-mm cell wall spacing with a wall thickness of 0.76 mm. The thickness of each face sheet is 5 mm. The AL6XN highly ductile steel with the same composition as the material used by Dharmasena et al. [2] has been assigned to the metallic honeycomb core and face plates. The AL6XN steel properties with Johnson–Cook model parameters are listed in Table 1. The plastic deformation effect at elevated temperatures of the materials is modeled by using the Johnson–Cook model [7].

The node to surface tie constrains were used between the contact surfaces of the face plates and the core. The C3D8R element type is assigned to the solid plates,

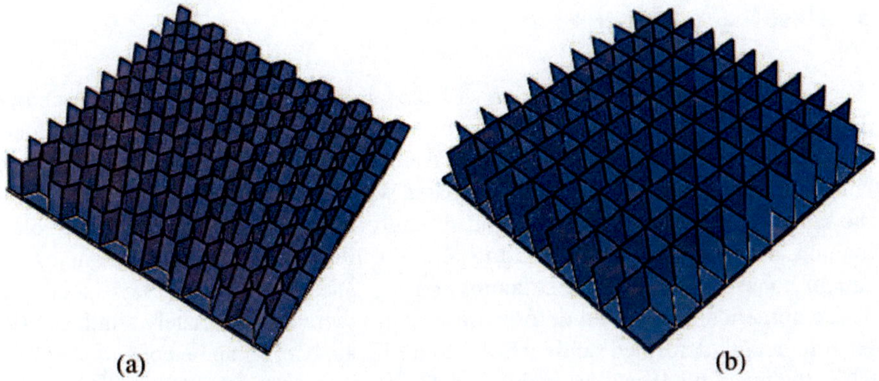

Fig. 3 1/4th FE models (without front face) of **a** hexagonal and **b** square honeycomb core used sandwich structure

Table 1 Assigned properties with the parameters of Johnson–Cook model [3, 12]

Properties	Symbol	Values
Yield stress	A	400 MPa
Strain hardening constant	B	1500 MPa
Strengthening coefficient of strain rate	C	0.045
Young's modulus	E	161 GPa
Poisson's ratio	υ	0.35
Density	ρ	7850 kg/m^3
Strain hardening coefficient	n	0.4
Thermal softening coefficient	m	1.2
Strain rate	$\dot{\varepsilon}_0$	0.001
Melting temperature	T_{m}	1800 K
Transition temperature	T_{tran}	293 K

and the S4R element type is assigned to the honeycomb core structures. Each honeycomb sandwich structure undergoes three air-blast loadings, and after each explosion, deflections of both the front and back plates of the sandwich structure are recorded for comparison of results. The sandwich structures with lower face deflection withstand the high explosive load intensity easily. The results obtained by simulation of the square honeycomb sandwich were compared with Dharmasena et al. [2] published experimental results of the square honeycomb panel.

3 Results and Discussion

After stand-off distance specification with the target surface and the explosive charge definition, the modeled sandwich structure undergoes the ConWep air-blast load. The face deflection experimental results available in the literature [2] for the square honeycomb sandwich structure are validated with the present simulation results for the same square honeycomb sandwich structure. The deformed shapes during blast loading are shown in Fig. 4. The Fig. 4a, b, c illustrated the deformations of the square honeycomb sandwich structure at 1 kg, 2 kg, and 3 kg TNT, respectively. These numerically modeled deformation shapes were approximately similar to the experimentally deformed sandwiches. From Fig. 4a, b, c, it can be noted that at 1 kg TNT, the buckling as well as inelastic deformation of core; at 2 kg TNT, the shear deformation of core with stretching of faces, and at 3 kg TNT, core crushing with large bent in faces are predominant, respectively. The simulation results are also approximately matched with the experimental results, but in the case of the 3 kg TNT blast, the simulation shows small variations in the deflections. The difference in deflections at 3 kg TNT air-blast load could be due to the larger separation of the front plate from the webs of the core. The quarter modal is analyzed in FE simulation to also save the computational time. Figure 5 represents the validation of the present simulation results of front face plate deflections with experimental deflections results for the square honeycomb sandwich structure at different air-blast loading conditions such as 1 kg, 2 kg, and 3 kg TNT, respectively. Figure 6 showed the crushing behavior of the square honeycomb core used sandwich structure under the blast loads of 1 kg TNT, 2 kg TNT, and 3 kg TNT. The hexagonal honeycomb core used sandwich structure (Fig. 7) is also subjected to the same air-blast load (1, 2, and 3 TNT). Figure 8 showed the crushing behavior of the hexagonal honeycomb core used sandwich structure under the blast loads of 1, 2, and 3 kg TNT. From Figs. 6 and 8, it can be observed that the hexagonal honeycomb core used in the sandwich structure shows less deformation than the square honeycomb core used in the sandwich structure for the same blast load.

Therefore, the hexagonal core used in the sandwich structure shows better protection against shock waves as compared to the square honeycomb core used in the sandwich structure. The hexagonal sandwich structure has less springback effect in comparison with the square sandwich structure on both the front and back faces.

The face deflections of the hexagonal sandwich structure at 3, 2, and 1 kg TNT are plotted in Fig. 9. The center point of the front face deflections are 110 mm, 85 mm, and 38 mm at 3 kg, 2 kg, and 1 kg TNT blast load, respectively. Similarly, the center point deflections of the back face of the hexagonal sandwich structure for the same blast loads are 75 mm, 55 mm, and 20 mm, respectively. The front face deflections of the hexagonal honeycomb core used in the sandwich structure for different air-blast loadings of 3 kg, 2 kg, and 1 kg TNT are 21.4%, 12.4%, and 15.5%, respectively, lower than the square honeycomb core used in the sandwich structure. In the case of back face plate deflections, the hexagonal honeycomb sandwich structure shows 28.6, 9.09, and 10% lower than the square honeycomb sandwich structure.

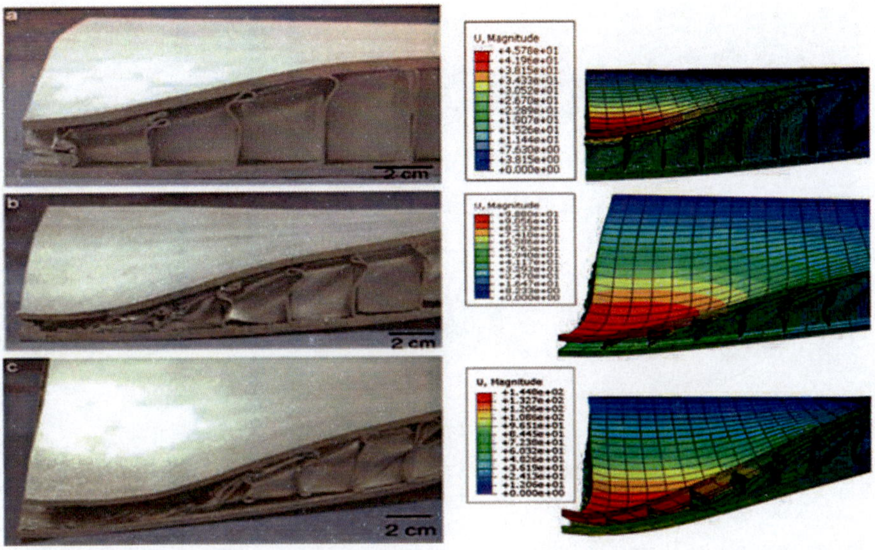

Fig. 4 Deformations of square honeycomb sandwich structure at **a** 1 kg, **b** 2 kg, **c** 3 kg TNT. Experimental deformation structures [2]. Numerical deformation structures

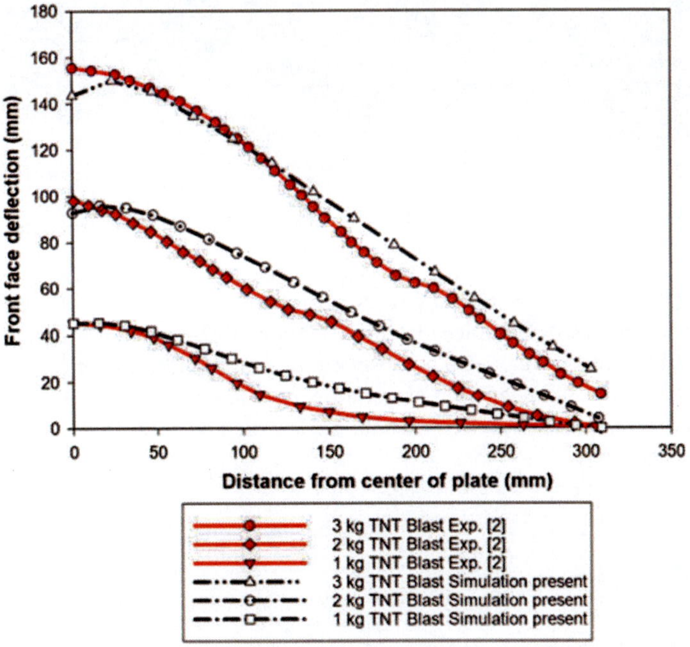

Fig. 5 Validation of Dharmasena et al. [2] experimental results with present simulation results for front face deflections of square honeycomb sandwich structure

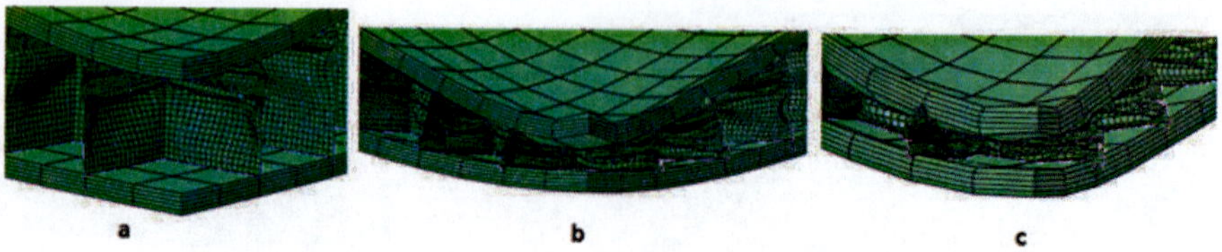

Fig. 6 Square honeycomb core crushing at **a** 1 kg, **b** 2 kg, **c** 3 kg TNT blast load

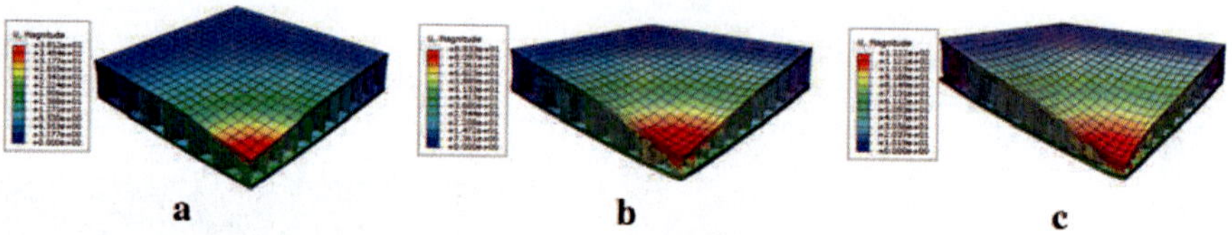

Fig. 7 Deformed shapes of hexagon honeycomb core used sandwich structure at **a** 1 kg, **b** 2 kg, **c** 3 kg TNT blast load

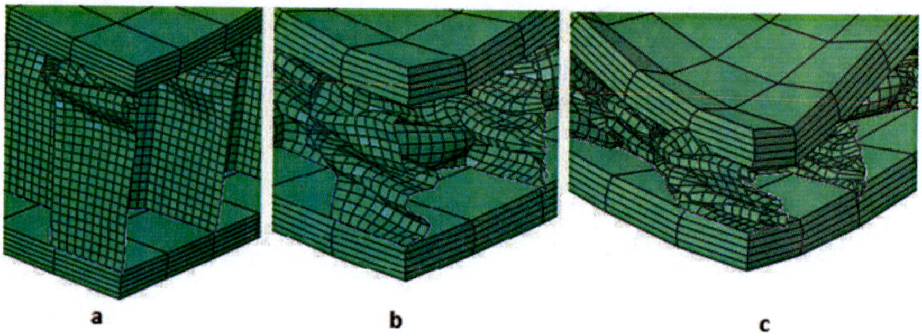

Fig. 8 Hexagonal honeycomb core crushing at **a** 1 kg, **b** 2 kg, **c** 3 kg TNT blast load

From Fig. 10, it can be observed that the honeycomb cores are subjected to higher stresses as compared to their face plates due to their crushing deformation phenomena during the blast load. It can also be observed that at all blast loading conditions, the hexagonal honeycomb sandwich structure shows lower peak values of stresses as compared to the square honeycomb sandwich structure because of their high blast resistance property.

4 Conclusions

The present numerical analysis deals with the comparison of both front and back face deflections of the quarter FE models of a square and a hexagonal honeycomb core used sandwich structures. Both the models were subjected to the ConWep air-blast load in ABAQUS/explicit. From the current analysis, it is concluded that the front and back face deflections of the hexagonal honeycomb sandwich structure are less

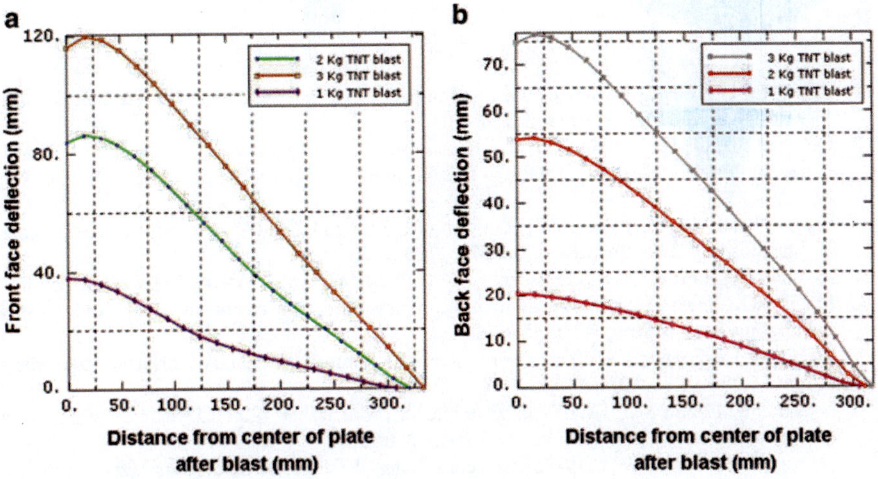

Fig. 9 Deflection versus distance graphs for **a** front face and **b** back face at 1, 2, and 3 kg TNT blast load

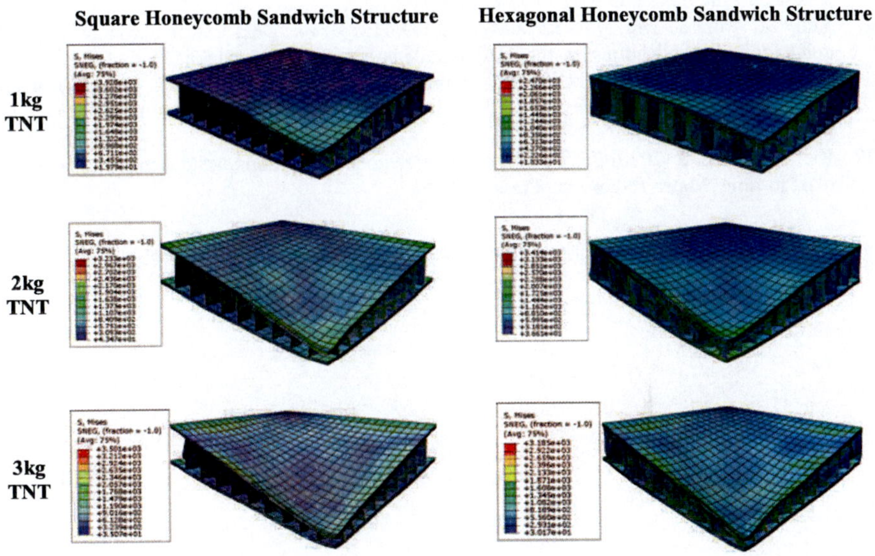

Fig. 10 Stress distribution in the modeled sandwich structures at 1, 2, and 3 kg TNT blast loads

in comparison with the square honeycomb sandwich structure for the three different air-blast loads of 3, 2, and 1 kg TNT. The hexagonal honeycomb sandwich structure is the better candidate for protection against the blast load, and it can be used for the naval industries and for making military vehicles.

References

1. Zhu F, Lu G (2007) A review of blast and impact of metallic and sandwich structures. EJSE Special Issue: Loading on Structures 92–101
2. Dharmasena KP, Wadley HNG, Xue Z, Hutchinso JW (2008) Mechanical response of metallic honeycomb sandwich panel structures to high-intensity dynamic loading. Int J Impact Eng 35:1063–1074
3. Patel S, Patel M (2022) The efficient design of hybrid and metallic sandwich structures under air blast loading. J Sandw Struct Mater 24:1706–1725
4. Rathbun HJ, Radford DD, Xue Z, Yang MYH, Deshpande V, Flek NA, Hutchinson Zok FW, Evans AG (2006) Performance of metallic honeycomb-core sandwich beams under shock loading. Int J solids Struct 43:1746–1763
5. Fleck NA, Deshpande VS (2004) The resistance of clamped sandwich beams to shock loading. J Appl Mech 71:386–401
6. Nahshon K, Pontin MG, Evans AG, Hutchinson JW, Zok FW (2007) Dynamic shear rupture of steel plates. J Mech Mat Struct 2(10):2049–2066
7. Murugesan M, Jung DW (2019) Johnson cook material and failure model parameters estimation of AISI-1045 medium carbon steel for metal forming applications. Materials 12:609
8. Turkmen HS, Mecitoglu Z (1999) Dynamic response of a stiffened laminated composite plate subjected to blast loading. J Sound Vib 221(3):371–389
9. Dear JP, Rolfe E, Kelly M, Arora H, Hooper PA (2017) Blast performance of composite sandwich structures. Proc Eng 173:471–478
10. Kelly M, Arora H, Worley A, Kaye M, Linz PD, Hooper PA, Dear JP (2016) Sandwich panel cores for blast applications: materials and graded density. Exp Mech 56:523–544
11. Wang Y, Yi Y, Chunyan W, Guan Z, Aminreza K, Wanzhong Z (2019) On the out-of-plane ballistic performances of hexagonal, reentrant, square, triangular and circular honeycomb panels. Int J Mech Sci 173:2–11
12. Kumar R, Patel S (2020) Failure analysis on octagonal honeycomb sandwich panel under air blast loading. Mater Today Proc Press

Experimental Data-Based Model of Fracture of Adhesive Joint for a Link of Mechanism Made from Bamboo

S. M. Gondane, P. N. Belkhode, M. P. Joshi, P. B. Maheshwary, and J. P. Modak

Abstract Development of bamboo product is playing the significant role in the market due to easy availability at lower cost and environmental friendly. Apart from the traditional uses, now, bamboo is used in the building construction and mechanism. Bamboo development and enhancement benefit the rural people to start their own setup generate the employability and sustainable livelihoods in rural communities. In view, above stated is to lay an emphasis on developing appropriate adhesives for joinery of structural members made for chemically appropriated treated bamboo. Anyone would say that the prerequisite for this would be decide proper chemicals which would enhance mechanical strength (elevation of tensile strength, compressive strength and shear strength). However, the literature indicated that this has been already achieve to great extent by proper treatment of raw bamboo with borax chemical. Paper details the formulation of mathematical model of the sic chain mechanism with bamboo linkage and bamboo joint with appropriate adhesives in the bamboo joints. In our life, we come across very many activities. These activities have some environmental system in which these activities take place. The environment or system can be specified in terms of its parameters some of which are always constant in their magnitudes, whereas some are variable. The activities are set in action by some parameters which are considered as causes. These causes interact with parameters of

S. M. Gondane
TGPCET, Nagpur, India

P. N. Belkhode
Laxminarayan Institute of Technlogy, Nagpur, India

M. P. Joshi
Priyadarshini College of Engineering, Nagpur, India

P. B. Maheshwary (✉)
Faculty of Science and Technology, RTMNU, Nagpur, India
e-mail: prashantmaheshwary51@gmail.com

J. P. Modak
JD College of Engineering and Management, Nagpur, India

V. K. Gupta et al. (eds.), *Recent Advances in Machines and Mechanisms*, Lecture Notes in Mechanical Engineering, https://doi.org/10.1007/978-981-19-3716-3_60

the system; as a result of this interaction, some effects are produced. To investigate the shear strength of the bamboo joints in terms of length, weight, cross-section and speed. The investigated result predicts the performance of four-bar mechanism to replace the existing material with bamboo.

Keywords Bamboo · Adhesive · Mechanism · Mathematical modeling · Joinery

1 Introduction

Appropriately treated bamboo will have immense impact resistance, strength, or resilience. Treated bamboo can be suitable structural member having very high impact resistance strength; it should be tried as material substitute for steel in the design and development of process machines of low capacity. The proposed power range is process machines up to 10 HP capacities. Further, in a view this property of bamboo, it is proposed that the frames of such machines be replaced by bamboo material.

The steel is replaced by bamboo because the values of tensile, compressive, bending, shearing are very closer to mechanical properties of steel; several samples of bamboo and steel were tested, and steel was tested and analyzed to examine their tensile properties. This test was carried out on sic chain mechanism in which bamboo linkages with the adhesive joint are used to investigate the performance at varying load with different speed. The mathematical model is formulated based on the observed reading to know the behavior of dependent variable shear strength related to various independent pie terms involved in the experimentation such as weight applied, operational speed, link length, adhesive applied, and cross-sectional area. The indices of each pie term indicate the influence of corresponding pie term on the output variable to determine the performance of bamboo.

2 Mathematical Model

2.1 *Identification of Variables*

Identification of experimental variable is the important step to investigate the performance of the experimental setup. Identified variables correlate to formulate the mathematical model. The identification of the variables is:

(1) Independent variables
Independent variables can be defined as nay experimental variable that influences the test and can be changed or altered independently of the other test variables.
(2) Dependent variables
The variables which change due to the change in the values of the independent variables are called the response or the dependent variables.

(3) Extraneous variables
An extraneous variable influences the process, but it cannot be changed or altered at our level examples such as pressure, effect of humidity, and human effects.
(4) Controlled variables
Controlled variables are basically fundamental or independent variables, but due to practical reasons, they are not altered, or it is not possible to alter them examples such as acceleration due to gravity.

2.2 Dimensional Analysis

The dimensional analysis was rightly to reduce the number of experimental variables then mainly applied to the fluid mechanics and heat transfer for almost all the experiments. Dimensional analysis technique is making experimentation shorter. Fundamental dimensions (Fixed dimensions or fixed quantity) mass (*M*), length (*L*), and time (*T*) are three fixed dimensions. If heat is involved, then temperature (θ) is also taken as fixed dimensions.

2.3 Dimension Equation

If in an equation containing physic quantity, each quantity is represented by its dimensional formula; the resulting equation is known as dimensional equation.

$$\text{Kinetic energy} = \tfrac{1}{2}\,\text{mv2}$$

Here, m is the mass of the body; v is velocity. Writing the formula for kinetic energy in the dimensional equation form, we have

$$[\text{M}] \times \left[\text{MLT}^{-1}\right]^2 = \left[\text{ML}^2\text{T}^{-2}\right]$$

Rayleigh's method.

Let *Y* be an independent variable which depends on $x1, x2, x3, x4$, etc. According to Rayleigh's method, *Y* is a function of $x1, x2,\ x3, x4\ldots$,etc., and mathematically, it can be written as

$$Y = f(x1,\ x2,\ x3,\ x4)$$

Equation can be written as follows

$$Y = k\,x1a,\ x2b,\ x3c,\ x4d$$

where k is constant and a, b, c, and d are arbitrarily powers. The values of a, b, c, and d are obtained by comparing the powers of the fundamental dimension on both side.

Buckingham pi theorem method. This theorem can be used for reducing number of variable affecting the process. In this method, m numbers of repeated variables are selected and dimension-less groups obtained by each one of the remaining variables one at a time.

2.4 *Significance of Dimension Equation*

Dimensional analysis is selected for three prominent reasons: (1). Dimensional analysis is used to check the consistency of the equation; (2). Dimensional analysis is used to form the relation between physical quantities involved in the experimentation, and (3). Dimensional analysis is used to change the units from one system to another. As it become very difficult to evolve logic-based design data for such a machine system. Hence, only approach left over was to generate experimental data-based design data. In this approach, all the independent quantities should be experimentally varied over widest possible range collect the response data and based on this collected data to formulate. "Experimental data Based Mathematical Models." Obviously, these mathematical models can new work as design data for such systems. The approach of experimental data-based modeling is deduced based on application of approach proposed in the book Theories of Engineering Experimentation.

3 Experimental Setup

3.1 *Proposed Mechanism*

It is evident from the schematics of the experimental setup that a mechanical hardware as described in Fig. 1 comprises of:

Fig. 1 Six chain mechanism with bamboo links

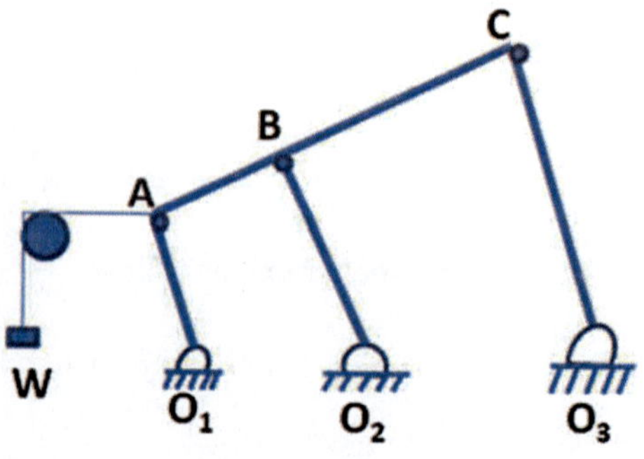

Electric motor and associated of crank-rocker six-link chain $O_1ABO_2CO_3$ loaded with a band brake. It is well known in machine dynamic that a points O_1, O_2, and O_3 time variation forces will be exerted by a six chain bar chain. These time variation forces act as impact loading on the frame of the bar chain. The various structural members of the frame are proposed to be varied (a) Material wise (of course, basic material is various species of bamboo appropriately chemically treated) (b) Dimension wise; (c) Type of cross-section wise. These structural members are proposed to be joined investigation in the domain of chemistry part of this research.

The experimental parametric variations are proposed to be in the context of (a) Different adhesives (b) Different geometry of the frame structure and (c) Different impact loading on the frame. Thus, this time investigation is most likely to generate huge experimental data. Based on which, the experimental finding would be formulated by mathematical modeling.

3.2 *Establishment of Dimensionless Group of π Terms*

The appropriateness of replacement of steel/plastic by properly chemically treated bamboo and compatible adhesives for bamboo structural members joinery so as to replace mainly steel as a material for the stationary frames of process machines in the horsepower range 0–10 hp. This is under the presumption that the frames are subjected to transient/at times random loading including severe unpredictable random vibrations in the frame structure. This is so because the end result of the investigation is to indicate appropriate resilient strength of properly chemically treated bamboo frame structure manufactured using such a chemically treated bamboo raw material and compatible adhesives.

The variables affecting the effectiveness of the phenomenon under consideration are interacting physical system. The dependent or the response variables are the ratio of shear stress to tensile yield strength (Table 1).

3.3 *Establishment of Dimensionless Group of π Terms*

Total seventeen independent terms were analyzed which is involved in this experimental analysis. As it is difficult to handle the total seventeen independent variables, therefore, these seventeen variables were grouped to the four numbers of independent variables which are dependent on one variable.

The dependent and independent variables are formulating in Tables 2 and 3.

Table 1 Independent and dependent variables

S. No.	Description	Variables	Symbol	Dimension
1	Link length O_1O_3	Independent	L1	$[M^0 L^1 T^0]$
2	Link length O_1A	Independent	L2	$[M^0 L^1 T^0]$
3	Link length AB	Independent	L3	$[M^0 L^1 T^0]$
4	Link length O_2B	Independent	L4	$[M^0 L^1 T^0]$
5	Link length BC	Independent	L5	$[M^0 L^1 T^0]$
6	Link length O_3C	Independent	L6	$[M^0 L^1 T^0]$
7	Adhesive layer depth	Independent	D	$[M^0 L^1 T^0]$
8	Adhesive layer thickness	Independent	T	$[M^0 L^1 T^0]$
9	Adhesive layer length	Independent	L	$[M^0 L^1 T^0]$
10	Elasticity adhesive	Independent	E_{adh}	$[M^1 L^1 T^{-2}]$
11	Elasticity bamboo	Independent	E_b	$[M^1 L^1 T^{-2}]$
12	Weight applied	Independent	W	$[M^1 L^1 T^{-2}]$
13	Mass of adhesive	Independent	M_{adh}	$[M^1 L^0 T^0]$
14	Oscillating angle	Independent	B	$[M^0 L^0 T^0]$
15	Instant force of breaking	Independent	F_{65}	$[M^1 L^1 T^{-2}]$
16	Shear stress	Independent	S	$[M^1 L^{-1} T^{-2}]$
17	Tensile yield strength	Independent	Ss	$[M^1 L^{-1} T^{-2}]$
18	Ratio of shear stress to tensile yield strength	Dependent	Pi_D	$[M^0 L^0 T^0]$

Table 2 Independent dimensionless π terms

Sr. No.	Independent dimensionless ratios	Nature of basic physical quantities
01	$\pi_1 = [(L_2/L_1)(L_3/L_1)(L_4/L_1)(L_5/L_1)(L_6/L_1)(D/t)(l/L_1)]$	Pie term related to linkages length
02	$\pi_2 = [E_{ADH}/E_B]$	Pie related to elasticity
03	$\pi_3 = [(\mathrm{T} * \mathrm{L})/(\mathrm{Mass}_{(adhesive)} * \mathrm{L}_1]$	Pie term related to overlapping area of adhesive
04	$\pi_4 = [\beta * (\mathrm{P}/180)]$	Pie term related to oscillating angle

Table 3 Dependent dimensionless π terms

Sr. No.	Dependent dimensionless ratios or π terms	Nature of basic physical quantities
01	$Z_1 = \mathbf{[S/Ss]}$	Ratio of shear stress to tensile yield strength

3.4 Formulation of Field Data-Based Model

Four independent π terms ($\pi_1, \pi_2, \pi_3, \pi_4$) and one dependent π terms (Z_1) have been identified for model formulation.

$$Z_1 = \text{function of } (\Pi_1,\ \Pi_2,\ \Pi_3,\ \Pi_4)$$

where

$$Z_1 = \Pi_{D1}, \text{First dependent } \pi \text{ term} = S/Ss$$

$$\begin{aligned}(Z) &= K * [(L_2/L_1)(L_3/L_1)(L_4/L_1)(L_5/L_1)(L_6/L_1)(D/t)(l/L_1)]^a \\ &, [E_{ADH}/E_B]^b,\ \left[(T * L)\ / \left(\text{Mass}_{(\text{adhesive})} * L_1\right]^c,\ \left[\beta * (\Pi/180)\right]^d\end{aligned} \tag{1}$$

$$(Z_1) = K_1 * [(\pi_1)^{a1} * (\pi_2)^{b1} * (\pi_3)^{c1} * (\pi_4)^{d1}] \tag{2}$$

To find the values of a1, b1, c1, and d1, Eq. 2 is presented as follows:

$$SZ_1 = nK_1 + a_1 * SA + b_1 * SB + c_1 * SC + d_1 * SD$$

$$\begin{aligned}SZ_1 * A = {} & K_1 * SA + a_1 * SA * A + b_1 * SB * A + c_1 * SC * A \\ & + d_1 * SD * A\end{aligned}$$

$$\begin{aligned}SZ_1 * B = {} & K_1 * SB + a_1 * SA * B + b_1 * SB * B + c_1 * SC * B \\ & + d_1 * SD * B\end{aligned}$$

$$\begin{aligned}SZ_1 * C = {} & K_1 * SC + a_1 * SA * C + b_1 * SB * C + c_1 * SC * C \\ & + d_1 * SD * C\end{aligned}$$

$$\begin{aligned}SZ_1 * D = {} & K_1 * SD + a_1 * SA * D + b_1 * SB * D + c_1 * SC * D \\ & + d_1 * SD * D\end{aligned}$$

The above equations are tabulated in the matrix form

$$X_1 = \text{inv}\ (W)\ x\ P_1$$

W is the matrix with constant K_1, a_1, b_1, c_1, and d_1.

P_1 is the matrix on L H S and.

X_1 is the matrix of values of K_1, a_1, b_1, c_1, and d_1.

Then, the matrix obtained is given by,

Matrix

$$z_1 x \begin{bmatrix} 1 \\ A \\ B \\ C \\ D \end{bmatrix} = \begin{bmatrix} n & A & B & C & D \\ A & A^2 & BA & CA & DA \\ B & AB & B^2 & CB & DB \\ C & AC & BC & C^2 & DC \\ D & AD & BD & CD & D^2 \end{bmatrix} x \begin{bmatrix} K_1 \\ a_1 \\ b_1 \\ c_1 \\ d_1 \end{bmatrix}$$

After solving the matrix, the unknown is find out which can be expressed as

$$(Z_1) = k_1 * [(\pi_1)^{a1} * (\pi_2)^{b1} * (\pi_3)^{c1} * (\pi_4)^{d1}] \quad (5)$$

To determine the values of a_1, b_1, c_1, and d_1 and to arrive at the regression hyper-plane, the above equations are presented as follows:

$$SZ_1 = nK_1 + a_1 * SA + b_1 * SB + c_1 * SC + d_1 * SD$$

$$SZ_1 * A = K_1 * SA + a_1 * SA * A + b_1 * SB * A + c_1 * SC * A + d_1 * SD * A$$

$$SZ_1 * B = K_1 * SB + a_1 * SA * B + b_1 * SB * B + c_1 * SC * B + d_1 * SD * B$$

$$SZ_1 * C = K_1 * SC + a_1 * SA * C + b_1 * SB * C + c_1 * SC * C + d_1 * SD * C$$

$$SZ_1 * D = K_1 * SD + a_1 * SA * D + b_1 * SB * D + c_1 * SC * D + d_1 * SD * D$$

Above equations is expressed as

$$Z_1 = W_1 x X_1$$

Here,

W is the matrix with constant K_1, a_1, b_1, c_1, and d_1.

P_1 is the matrix on L H S and.

X_1 is the matrix of values of K_1, a_1, b_1, c_1, and d_1.

Then, the matrix obtained is given by,
Matrix

$$z_1\,x \begin{bmatrix} 1 \\ A \\ B \\ C \\ D \end{bmatrix} = \begin{bmatrix} n & A & B & C & D \\ A & A^2 & BA & CA & DA \\ B & AB & B^2 & CB & DB \\ C & AC & BC & C^2 & DC \\ D & AD & BD & CD & D^2 \end{bmatrix} x \begin{bmatrix} K_1 \\ a_1 \\ b_1 \\ c_1 \\ d_1 \end{bmatrix}$$

In the above equations, *n* is the number of sets of readings; *A*, *B*, *C*, and *D* represent the independent π terms π_1, π_2, π_3, and π_4, while Z_1 matrix represents dependent π term.

Substituting the values of *A*, A^2, *BA*, *CA*...up to D^2 in the above matrix value of constant K_1 and indices a_1, b_1, c_1 and d_1 are evaluated.

$K_1 = 0.1764$, $a_1 = 2.9810$, $b_1 = -1.7182$, $c_1 = 0.8010$ and $d_1 = 2.9245$.

The exact form of model obtained is as under:

$$(Z_1) = 0.1764 * (\pi_1)^{2.9810} * (\pi_2)^{-1.7182} * (\pi_3)^{0.8010} * (\pi_4)^{2.9245}$$

4 Analysis of the Model for Dependent π Term Z_1

1. The absolute index of π_1 is the highest, viz., 2.9810. Thus, the term related to the specification of the mechanism of linkages involved the most influencing π term in this model. The value of this index is positive indicating that the ratio of shear stress (Z_1) is directly proportional to term related to the specification of stirrer linkage is involved, i.e., π_4 indicates that more the length of linkage length, the cross-section of the linkage, material selected for the linkage, and weight of the link. Suggesting that high process parameters such as the length of the links increase the shear stress increased.
2. The absolute index of π_3 is the lowest, viz., 0.8010. Thus, the term related is the effect influencing π term in this model. The value of this index is positive indicating that the ratio of shear stress (Z_1) is directly proportional to the term related to linkage [π_1]. The linkages increases as [π_1] increases on the effect of predetermining parameter. Suggestions regarding approval of link length will reduce the shear strength.
3. 3. The sequence of influence of another independent π terms present on this model is π_1, π_2, and π_4 having absolute indices as 2.9810, 0.8010, and 2.9245 in the order, respectively.

5 Conclusion

Accuracy of models is dependent on magnitude of curve fitting constants *K*. Ideally, if these are numerically 1, then the model very rightly simulates the man machine system. If it is too low, the causes are over estimated; if it is too high, the causes are under estimated. This would decide when to repeat the investigation again or to refine the approach in subsequent attempts. The magnitude of exponents of the causes, i.e., of quantities on right hand side of equations indicates the degree of influences of these causes on the specific response. The ratio of indices indicates the relative influences of causes are more influential than that of quality of tools used.

A six-link compound chain is fabricated from timber bamboo species appropriately joined by specially developed synthesized adhesive. The compound chain brings about rotary to oscillatory motion transformation in the stages that is partly with crank rocker converting rotary to oscillatory motion of 43° and in series is a double-lever four-bar mechanism converting 43°–52°. The output link is sustaining load torque organized by a dead weight *W*, pulley *P*, and a rope *R* around the pulley.

References

1. International training workshop on sustainable bamboo and processing techniques for small size bamboo enterprises GCT04 – 16, (2000) Hangzhou, Zhejiang of China
2. Hain K (1967) Applied kinematics. 2nd edn. Mc-Graw Hills
3. Schenck J (1967) Theories of engineering experimentation. 2nd edn. Mc-Graw Hills International
4. Gondane SM, Belkhode PN, Joshi MP, Modak JP (2020) Synthesis of novel adhesive for bamboo joining and comparison of its mechanical proper ties with adhesives available in the market. Int J Grid Distrib Comput 13(2):2203–2210
5. Krishna G, Rao AM, K (2006) Optimization of machining parameters for milling operations using a scatter search approach. Int J Adv Manuf Technol 3(4):219–224
6. Belkhode P (2019) Development of mathematical model and artificial neural network simulation to predict the performance of manual loading operation of underground mines. J Market Res 8(2):2309–2315
7. Hinduja S, Petty DJ, Tester M, Barrow G (1985) Calculation of optimum cutting conditions for turning operations. Proc Inst Mech Eng 199(B2):81–92
8. Kim JD, Kim DS (1995)Theoretical analysis of micro-cutting characteristics in ultra-precision machining. J Mater Process Technol 49:387–398
9. Belkhode P, Vidyasagar V (2014) Mathematical model for face drilling in underground mining operation. Int J Eng Res Sci Technol 3(2)
10. Belkhode P (2017) Mathematical modelling of liner piston maintenance activity using field data to minimize overhauling time and human energy consumption. J Inst Eng (India): Ser C 1–9
11. Liu W, Zheng Y (2019) Interfacial bonding enhancement on the epoxy adhesive joint between engineered bamboo and steel substrates with resin pre-coating surface treatment. wood science and technology Wood Sci Technol 53(1)
12. Ghadge RR, Prakash S, Ganorkar SA (2021) Experimental investigations on fatigue life enhancement of composite (e-glass/epoxy) single lap joint with graphene oxide modified adhesive, IOP Publishing Ltd, 8(2)

13. Belkhode PN, Bejalwar A (2018) Analysis of experimental setup of a small solar chimney power plant. Elsevier Proc Manuf 20:481–486
14. Belkhode PN, Bejalwar A (2019) Evaluation of the experimental data to determine the performance of a solar chimney power plant. MATPR J
15. Rajak DK, Pagar DD, Menezes PL, Linul E (2019) Fiber-reinforced polymer composites: manufacturing, properties, and applications. MDPI 11(1)
16. Chen Q, Dai C, Fang C, Chen M (2019) Mode I interlaminar fracture toughness behavior and mechanisms of bamboo. Mater Des 183
17. Maheshwary PB, Handa CC, Nemade KR (2017) A comprehensive study of effect of concentration, particle size and particle shape on thermal conductivity of titania/water based nanofluid. Appl Therm Eng 119(5):79–88
18. Maheshwary PB, Handa CC, Nemade KR, Chaudhary SR (2020) Role of nanoparticle shape in enhancing the thermal conductivity of nanofluids. Mater Today Proc 28:873–878
19. Maheshwary PB, Handa CC, Nemade KR (2018) Effect of shape on thermophysical and heat transfer properties of ZnO/R-134a nanorefrigerant. Mater Today Proc 5(1):1635–1639